# ADVANCES IN CHEMICAL PHYSICS

VOLUME 117

## EDITORIAL BOARD

# Advances in CHEMICAL PHYSICS

*Edited by*

**I. PRIGOGINE**

Center for Studies in Statistical Mechanics and Complex Systems
The University of Texas
Austin, Texas
and
International Solvay Institutes
Université Libre de Bruxelles
Brussels, Belgium

*and*

**STUART A. RICE**

Department of Chemistry
and
The James Franck Institute
The University of Chicago
Chicago, Illinois

**VOLUME 117**

AN INTERSCIENCE® PUBLICATION
**JOHN WILEY & SONS, INC.**
NEW YORK • CHICHESTER • WEINHEIM • BRISBANE • SINGAPORE • TORONTO

This book is printed on acid-free paper. ♾

An Interscience® Publication



Published simultaneously in Canada.



***Library of Congress Catalog Number: 58-9935***

ISBN 0-471-40542-6

Printed in the United States of America.

10 9 8 7 6 5 4 3 2 1

# CONTRIBUTORS TO VOLUME 117

W. T. COFFEY, School of Engineering, Department of Electronic and Electrical Engineering, Trinity College, Dublin 2, Ireland

J. L. DEJARDIN, Centre d'Etudes Fondamentales, Universite de Perpignan, 52 Avenue de Villeneueve, 66860 Perpignan Cedex, France

P. M. DEJARDIN, Department of Applied Mathematics and Theoretical Physics, The Queen's University if Belfast, Belfast BT7 1NN, Northern Ireland

C. A. DE LANGE, Laboratory for Physical Chemistry, University of Amsterdam, Nieuwe Achtergracht 127-129, 1018 WS Amsterdam, The Netherlands

JOHN T. FOURKAS, Eugene F. Merkert Chemistry Center, Boston College, Chestnut Hill, MA 02467

D. A. GARANIN, Max-Planck-Institut fur Physik komplexer Systeme, Nothnitzer Strasse 38, D-01187 Dresden, Germany

YU P. KALMYKOV, Centre d'Etudes Fondamentales, Universite de Perpignan, 52 Avenue de Villeneueve, 66860 Perpignan Cedex, France

D. J. MCCARTHY, School of Mathematics, Dublin Institute of Technology, Kevin Street, Dublin 8, Ireland

HIROKI NAKAMURA, Department of Theoretical Studies, Institute for Molecular Science, Myodaiji, Okazaki 444-8585, Japan

YOSHIAKI TERANISHI, Department of Theoretical Studies, Institute for Molecular Science, Myodaiji, Okazaki 444-8585, Japan

CHAOYUAN ZHU, Department of Theoretical Studies, Institute for Molecular Science, Myodaiji, Okazaki 444-8585, Japan

# INTRODUCTION

Few of us can any longer keep up with the flood of scientific literature, even in specialized subfields. Any attempt to do more and be broadly educated with respect to a large domain of science has the appearance of tilting at windmills. Yet the synthesis of ideas drawn from different subjects into new, powerful, general concepts is as valuable as ever, and the desire to remain educated persists in all scientists. This series, Advances in Chemical Physics, is devoted to helping the reader obtain general information about a wide variety of topics in chemical physics, a field that we interpret very broadly. Our intent is to have experts present comprenhensive analyses of subjects of interest and to encourage the expression of individual points of view. We hope that this approach to the presentation of an overview of a subject will both stimulate new research and serve as a personalized learning text for beginners in a field.

I. PRIGOGINE
STUART A. RICE

# CONTENTS

# ADVANCES IN CHEMICAL PHYSICS

VOLUME 117

# LASER PHOTOELECTRON SPECTROSCOPY: SPECTROSCOPY AND DYNAMICS OF EXCITED STATES IN SMALL AND MEDIUM-SIZED MOLECULES

C. A. DE LANGE

*Laboratory for Physical Chemistry, University of Amsterdam, 1018 WS Amsterdam, The Netherlands*

## CONTENTS



*Advances in Chemical Physics, Volume 117*, Edited by I. Prigogine and Stuart A. Rice.
ISBN 0-471-40542-6 

## I. INTRODUCTION

For obvious experimental reasons, the electronic ground state of a molecule is known in much greater detail than its infinity of excited states in virtually every case. This is not because molecular excited states are unimportant. In fact, when it comes to chemical reactivity the essential role of molecular excited states with their enormous variation in spectroscopic and dynamic properties can hardly be overestimated. It is therefore not surprising that since the advent of molecular spectroscopy the experimental observation and reliable characterization of excited states of molecules has become an important goal in chemical physics. Clearly, the complications associated with such a task increase with molecular size, and hence novel spectroscopic methods are usually first applied to small molecular species. Molecules belonging to this category can either be stable or short lived. The problems encountered when stable molecules are studied by spectroscopic means tend to be magnified appreciably when experimental attention is focused on short-lived species. Such short-lived molecules often play a central role in reactions in the earth's atmosphere under the influence of solar radiation, or in combustion processes. In interstellar space, under conditions where molecular collisions are rare, such species, which may be very unstable in the laboratory, can acquire appreciable lifetimes. Although the importance of studying many unstable molecules is undisputed, the experimental detection of most short-lived species is very challenging and often requires significant modifications of experimental methods and procedures that have proved their worth in the case of stable molecules.

A detailed understanding of the spectroscopy and dynamics of excited electronic states of small molecules is not achieved easily. A truly impressive body of experimental data has been collected over many years through the use of optical absorption techniques [1–3]. Of course, optical observation methods depend heavily on optical selection rules that can be viewed as either a curse or a blessing. Without selection rules, molecular spectroscopy would probably be close to impossible. However, the limitations set by the one-photon selection criteria that rule conventional optical absorption processes put severe restrictions on the excited states that can be accessed in practice. When we consider stable homonuclear diatomics, their electronic ground states usually possess *gerade* symmetry. Hence, it is not surprizing that the data collected with one-photon absorption spectroscopy for *gerade* excited states of these molecules are scarce at best, since approximately one-half of the number of molecular excited states cannot be reached from the ground state on the basis of one-photon electric-dipole selection rules. Clearly, more sophisticated experimental methods than conventional photoabsorption spectroscopy are required to study molecular excited states in a general fashion.

In this chapter, we shall explore the uses and advantages of resonance-enhanced multiphoton ionization (REMPI) spectroscopy in some detail. With the availability of lasers with high-pulse energies in the visible (vis) and ultraviolet (UV) portion of the spectrum, the experimentalist can now go beyond the realm of one-photon methods and utilize processes involving the absorption of more than one photon. The optical selection rules associated with multiphoton absorption are strict, but do depend on the number of absorbed photons. This provides for a degree of flexibility, which removes many of the limitations that plague one-photon absorption spectroscopy, in that a much wider range of molecular excited states can be accessed. Another feature to be stressed is the fact that with the absorption of one or more additional photons the ionization threshold of the molecule can be exceeded and that essentially every molecule can thus be photoionized. This ionization step is again subject to selection rules, but these do not form a serious constraint on the final state reached, since the outgoing electron can take away the required excess energy and angular momentum. A schematic diagram of a $(2 + 1)$ resonance-enhanced multiphoton ionization process is presented in Figure 1. Because the excited states are mapped efficiently and with few restrictions on to an ionic state that is often well known, the REMPI method is very general and highly suitable for the detailed observation of such states. The occurrence of the ionization process can be monitored either by relatively simple mass-resolved ion detection [4, 5], or by experimentally much more demanding kinetic energy resolved photoelectron spectroscopy (PES) [6–11]. With the latter

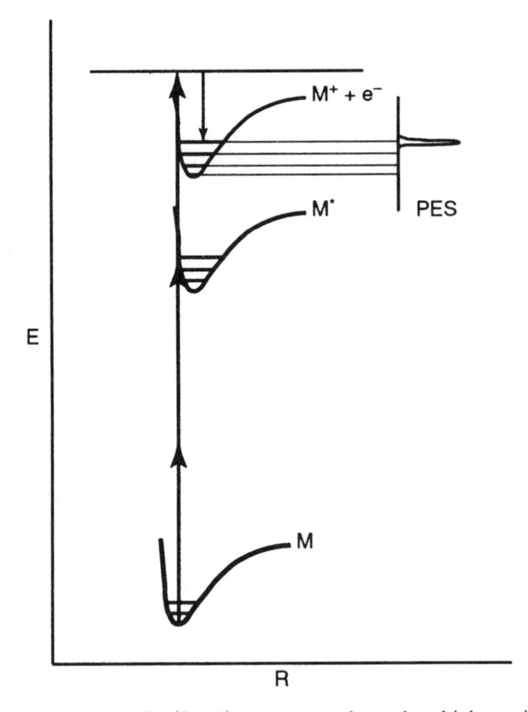


**Figure 1.** Schematic diagram of a $(2 + 1)$ resonance-enhanced multiphoton ionization process.

method, considerations of energy and momentum conservation allow the experimentalist to determine the internal energies of the ions formed in the two-step process. In this chapter, it will be argued and demonstrated that the determination of ion internal energies (electronic, vibrational, rotational) is a true asset of the method, and that the use of REMPI–PES, also often termed laser PES, offers very significant advantages for the study of molecular excited states that are not easily matched in its broad applicability by other techniques.

As an abundance of examples will show, laser PES in the frequency domain is nowadays a reliable source of detailed spectroscopic information about molecular excited states. In addition to the spectroscopy, a very interesting and relatively recent development should be emphasized as well. With the advent of ultrafast lasers it is now feasible to apply an adjustable time delay between laser pulses used for photoabsorption and photoionization (of the same or different colors), in order to obtain detailed real-time dynamical information about molecular excited states via a pump–probe procedure. Time-resolved laser PES already has shown appreciable promise [12–14] and is expected to be a very fruitful scientific activity in the years to come.

The method of laser PES is highly sensitive when appropriate experimental procedures are followed. In this paper, we shall rely heavily on the use of a "magnetic bottle" analyzer that was developed into a very flexible tool for high-resolution kinetic energy resolved electron detection [15–17]. In addition, this type of analyzer can be easily adapted to mass-resolved time-of-flight ion detection, and to zero kinetic energy electron detection (ZEKE) with essentially laser-limited resolution [18]. For more detailed information on ZEKE, the reader is referred to the literature [19–22]. With the available techniques the resolution in PE spectra is usually sufficient to resolve molecular electronic and vibrational levels, while in favorable cases separate rotational levels in the ion can be observed as well. When this is the case the essential role of angular momentum in the photoionization process can be studied experimentally, as will be evidenced in various examples. The two-step character of the photoionization process has one very attractive feature in that it essentially "converts" optical selectivity to chemical selectivity. When short-lived species are generated *in situ* in the spectrometer from appropriate parent compounds as is often the case, the optical excitation process can be tuned to a transition that is particular to the radical rather than to the parent compound. The electrons that are ejected in the second step then solely arise from the short-lived molecule, despite the fact that the parent compound is usually present in much larger concentrations.

The study of molecular excited states employing laser PES has to a large extent focused on molecular Rydberg states for a number of reasons. First, molecular Rydberg states can often be viewed as possessing an ionic core with an electron orbiting around this core at relatively large distance. In such a picture, molecular Rydberg states become quite atomic-like and their description is essentially very simple [23]. Second, on removal of the Rydberg electron an ionic state is formed with a potential energy curve that is very similar in shape and that has its minimum at the same internuclear distance as the Rydberg state itself. The photoionization process under these circumstances is usually dominated by Franck–Condon factors, which are very close to diagonal, leading to very simple one-line PE spectra. These simple PE spectra are then used as a diagnostic tool for the reliable assignment of Rydberg states.

The ionic core of a Rydberg state can correspond to the ground ionic state. Depending on the principal quantum number of the molecular orbital to which the Rydberg electron was excited, infinite series of excited levels exist, all located below the lowest ionic threshold and converging upon the lowest ionic state. However, when the Rydberg core corresponds to that of a higher lying ionic state, similar Rydberg series but now converging upon a higher ionic state will also exist. This will automatically lead to the existence of metastable neutral states, which are situated above the lowest ionic limit. Clearly, such superexcited states occur in infinite numbers in every molecule, and their role in photoionization processes turns out to be very significant in many cases. In fact, the properties of Rydberg states converging upon higher ionic limits are crucial when it comes to carrying out ZEKE experiments on higher ionic states. The success or failure of such a ZEKE experiment depends entirely on the importance of a variety of decay processes of such states. The significance of superexcited states and the part they play in laser PES can hardly be overestimated [24].

When considering multiphoton absorption in a molecule a variety of processes can occur. Of course, depending on the total photon energy deposited into the molecule one or more molecular ionization continua can be addressed and direct molecular ionization occurs. However, in a significant number of cases a competing process involving multiphoton absorption to neutral metastable superexcited states takes place. These superexcited states can undergo a variety of decay processes. These exit channels include autoionization in which an electron is lost into lower-lying ionization continua, or molecular (pre)dissociation into smaller fragments, either in their ground or excited states. Fragments formed in sufficiently highly excited states can be subsequently photoionized in the strong photon field required to induce the multiphoton process in the first place. The unique power of laser PES is that all these competitive decay channels can be observed and compared quantitatively at the same time [24]. Recent research on molecular hydrogen and its isotopomers forms a good illustration [25, 26].

The picture of a Rydberg state as consisting of a single ionic core and an outer electron orbiting around it may be too simple minded in some cases. There are various ways in which molecular orbitals can interact with each other, and when such interactions are important the simple one-electron picture is no longer upheld. The power of laser PES is that deviations from the simple one-electron picture show up directly in the PE spectrum. An analysis of the more complicated PE spectra that are obtained in these cases produces a wealth of detailed information about orbital interactions that may arise either from direct electronic mixing between orbitals, for example, via the electron repulsion operator, or that result from interactions mediated by the nuclei, for example, in vibronic coupling. Similar to the situation with conventional PES where molecular orbitals became experimentally accessible in a direct way [27–34], laser PES has now developed to the point where orbital interactions between orbitals with different cores can be made experimentally "visible" for the first time. Many examples exist to date in systems as diverse as N [35] and S atoms [36, 37], the NH radical [38, 39], the SH radical [40, 41], $N_2O$ [42], $NH_3$ [43, 44], methanethiol [45], and dimethyl sulfide [46]. The role of orbital interactions in laser PES will prove to be one of the key elements of the research described in this contribution.

The energies of members of Rydberg series are given by the Rydberg formula, which underscores the atomic-like character of Rydberg states:

$$\nu = \mathrm{IE}_i - \mathrm{R}/(n - \delta)^2$$

where $IE_i$ stands for the ionization energy of ionic state $i$, R is for the Rydberg constant, $n$ is for the principal quantum number, and $\delta$ is for the quantum defect. When one-photon optical absorption methods or REMPI are employed for the detection of Rydberg states, assignments must rely on the determination of quantum defects, the use of isotope effects, and where possible the observation of rotational band contours. However, the position of the origin of a Rydberg series often remains unclear. Moreover, in the presence of orbital interactions the observed quantum defects are distinctly perturbed, and assignments of Rydberg series become unreliable. Such interactions between different states are expected to occur for low-lying Rydberg states where Rydberg-valence interactions can be prominent, and for high-lying Rydberg states where Rydberg–Rydberg interactions come into play. In the intermediate region, the assignment problems encountered with one-photon absorption methods or REMPI are usually least severe. The problems indicated above can be alleviated to a large extent by the application of laser PES. In addition to the assignment tools listed above, it can now be established whether a particular state is pure Rydberg in character, or whether orbital interactions are dominant. These unique features of REMPI–PES allow the precise determination of the origin of a Rydberg series and make it the method of choice for reliable assignments of individual Rydberg states [43, 44, 47–50].

In addition to the conservation laws for linear momentum and energy, which are normally employed in laser PES, conservation of angular momentum is an issue when at the final state level rotationally resolved PES can be performed. In conventional PES, the experimental resolution that can be achieved is usually insufficient for rotational resolution in the ionic state, and few examples of rotationally resolved PES are known. With laser PES the situation is more favorable. Often, short-lived molecular species are produced with a significant degree of rotational excitation in a photofragmentation process. On photoionization much of this rotational excitation is conserved, and ionic rotational levels with high values of $J^+$ are accessed. For sufficiently high $J^+$ the spacing between ionic rotational levels exceeds the experimental resolution and rotationally resolved PES can be performed. Especially for molecules with relatively large rotational constants such as the diatomic hydrides such studies are very informative, because they produce a plethora of information on the dynamics of the photoionization process [17, 51–56]. With the advent of ZEKE the observation of rotationally resolved PES of small molecules has become a very feasible proposition [18, 57, 58].

Although laser PES has been employed effectively to study atoms [35–37], the main emphasis of the present contribution will be on gas-phase molecules ranging from molecular hydrogen and its isotopomers to dimethyl sulfide. What the species discussed will have in common is that they often occur as atmospheric molecules with some relevance for atmospheric chemistry. Moreover, every molecule treated will be seen to display one or more of the physical phenomena and dynamical processes described above. The broad range of spectroscopic and dynamical issues that can be addressed with laser PES is a convincing demonstration of the scientific value of the method [24]. It is therefore safe to predict that the next few years will see many more applications of this technique, particularly in the realm of time-resolved laser PES.

## II. EXPERIMENTAL

Our laser PES experiments in Amsterdam are carried out with two time-of-flight "magnetic bottle" electron analyzers, one equipped with an effusive beam sample inlet system, the other interfaced with a pulsed molecular nozzle beam. These spectrometers are based on the original design of Kruit and Read [15] in which the multiphoton ionization process takes place in the focused beam of a pulsed laser with a high photon flux. The focus is situated in the high-field portion of an inhomogeneous magnetic field, which decreases from 1 T in the focus to $10^{-3}$ T in the 50-cm long flight tube. The trajectories of the ejected electrons with a velocity component in the direction of the flight tube are made parallel in such a way that electrons ejected with the same kinetic energy arrive at virtually the same time at the microchannel plates that are used as detectors at the end of the flight tube. Under appropriate experimental conditions, the collection efficiency approaches 50% and for this reason the term "$2\pi$ analyzer" is often used. The parallelization process takes place in the first few millimeters of the electron flight path, a small distance compared to the total length of the flight tube. Two pole faces to which insulated grids are attached for accelerating and decelerating voltages supply the magnetic field in the focal region. Furthermore, electrons can be retarded in the flight tube, after their trajectories have been made parallel. Employing two pairs of mutually perpendicular Helmholtz coils compensates the earth's magnetic field in the ionization region and the flight tube. Using judiciously chosen retarding voltages a typical electron kinetic energy resolution of $\sim 10$ meV can be achieved for appreciable periods of time even under experimental conditions involving the presence of reactive short-lived species that may react with critical surfaces of the spectrometer [16, 17].

In the "magnetic bottle" electron analyzer zero kinetic energy electron detection following pulsed-field ionization (ZEKE–PFI) is also a feasible option. This can be achieved by applying appropriate voltages to the various grids of the spectrometer [18]. Since our spectrometer is not specifically designed for ZEKE–PFI experiments, the resolution that can be achieved ($\sim 5\,\mathrm{cm}^{-1}$) is not as good as in a dedicated ZEKE instrument ($\sim 0.1\,\mathrm{cm}^{-1}$) where it is truly laser limited. However, the fact that PES both with more conventional time-of-flight and sophisticated ZEKE methods can be carried out in the same instrument has proved to be an invaluable asset when it comes to comparing intensities measured with both methods. In a sense, the ZEKE–PFI facility allows us to "zoom in" on spectral features that on the basis of the time-of-flight PE spectrum merit improved resolution.

Although the spectrometer was primarily designed for electron detection, we have added a capability for mass-resolved time-of-flight ion detection. The sensitivity in this mode appears to be mass dependent and is somewhat lower than with electron detection. Ion detection has proved to be a valuable tool in monitoring the various ions present under conditions where photofragmentation plays a significant role. With the high laser pulse energies that are commonly used, the occurrence of photodissociation is the rule rather than the exception [17].

The laser system usually employed consists of a XeCl excimer laser (Lumonics HyperEx-460) operating at a 30-Hz repetition rate and providing typically 170 mJ in a 10-ns pulse. The excimer laser can pump two dye lasers (Lumonics HyperDye-500) with a bandwidth of $\sim 0.08\,\text{cm}^{-1}$ operating on various dyes. The dye laser output can be frequency doubled using a Lumonics Hypertrak-1000 unit equipped with a beta barium borate (BBO) or potassium dihydrogen phosphate (KDP) crystal. The output, which is polarized parallel to the time-of-flight axis of the spectrometer and which again has a width of $\sim$ 10 ns, is focused into the "magnetic bottle" spectrometer using a quartz lens with a focal length of 25 mm. In order to avoid space-charge effects, the laser power is kept as low as possible. The polarization of the laser light can be changed from linear to circular by means of a home-made Fresnel rhomb. For calibration of both the wavelengths and the PE kinetic energies, well-known resonances of noble gas atoms are usually employed [17].

## III. MOLECULAR HYDROGEN AND DEUTERIUM

### A. General Aspects

In order to understand the relative abundance of excited hydrogen atoms in interstellar space the process of dissociative recombination (DR) could well be important. In DR, free electrons recombine with molecular ions followed by a break up of the neutral excited molecule formed into fragments. Some of these fragments can be in excited states. For molecular hydrogen, DR at low electron collision energies ($< 3$ eV) can be viewed as consisting of three subsequent stages: (1) electron capture into a doubly excited repulsive state of the neutral molecule; (2) competition between autoionization and dissociation along this curve, the so-called survival process; and (3) half-collision/dissociation processes at large internuclear distances leading to two fragments. The first two steps have to do with the electron capture process. The third step leads to the final state distributions. Despite the perceived simplicity of molecular hydrogen this third step in particular is very complicated. As described by Chupka [59] in a discussion of multiphoton ionization of $H_2$, this process is affected by a large series of interactions between doubly excited repulsive states and singly excited bound molecular Rydberg states at large internuclear separations. In this process, many outgoing channels are addressed, leading to a distribution that is difficult to predict. For a complete description, the interaction strengths at every avoided crossing should be known. The only hope for an adequate description then lies in a full multichannel treatment [60] based on such couplings.

Experimentally it is not easy to study the process of dissociative recombination and to measure cross sections. Electron energies have to be controlled very accurately, and only recently have storage ring experiments developed to the point where this is feasible. Applications to DR of molecular hydrogen have now been performed [60–66], and (absolute) cross-sections [61–64] and product state information [60, 65, 66] have been obtained. However, since the product state resolution inherent in these storage ring experiments is limited, a complete description of the dissociation process still remains difficult.

The molecular dissociation process is formally independent from the way in which dissociative states are formed. In order to get some insight into the factors that play a role in the dissociation process we can access relevant excited states in a multiphoton absorption scheme from below and study the subsequent dissociation dynamics. Such a multiphoton study requires an intermediate state as a stepping stone. Multiphoton absorption studies of molecular hydrogen via the $B^1\Sigma_u^+$ [67–71], $C^1\Pi_u$ [69, 72–75], $EF^1\Sigma_g^+$ [76–84], $B'^1\Sigma_u^+$ [83, 85], $B''^1\Sigma_u^+$ [83], and $D^1\Pi_u$ [83, 85] have been utilized for that purpose. This work has amply demonstrated the importance of doubly excited repulsive states on the photoionization and photodissociation dynamics. In our experiments, to be described below, we shall study the photoionization and dissociation dynamics of molecular hydrogen and deuterium via excitation of the $B^1\Sigma_u^+(1s\sigma_g)(2p\sigma_u)(v', N')$ state from the $X^1\Sigma_g^+(1s\sigma_g)^2$ ground state. This transition is allowed with an odd number of photons and we shall either employ a one-color (3 + 1), or a two-color (1 + 1′) scheme. From the $B^1\Sigma_u^+$ state a variety of one-photon processes can be envisaged. First, the molecule can ionize to various rovibrational levels of the ground ionic state $X^2\Sigma_g^+(1s\sigma_g)$. Second, doubly excited states with configuration$(2p\sigma_u)(n\ell\lambda)$ may be excited. These states may (1) autoionize into the ionic ground state, or (2) lead to molecular dissociation resulting in a ground state ($1s$) atom and an excited ($n'\ell$) atom where $n'$ may differ from $n$. Finally, a possibility that until now has received little attention is that singly excited Rydberg states of configuration $(1s\sigma_g)(n''\ell\lambda)$ and their dissociation continua may be accessed.

In our experiments, both molecular ionization and dissociation into a ground-state fragment and an excited atom are observed at the same time. Laser PES offers a distinct advantage in that the final states reached in the dissociation process can be monitored reliably through one-photon ionization of excited hydrogen atoms. With our resolution of $\sim$ 10-meV electrons ejected from excited atoms with principal quantum numbers from $n = 2$ to $n = 8$ can be resolved, while signals arising from atoms with higher $n$ merge. The same technique allows us to determine the internal energies of the molecular ions formed in the direct ionization process with rotational resolution. In the following, we shall describe $(3 + 1)$ REMPI–PES and $(1 + 1')$ REMPI experiments via $B^1\Sigma_u^+$. In general, our work on the hydrogens provides a convincing example of the crucial role played by superexcited states in multiphoton absorption processes. More specifically, our experiments will emphasize the role of dissociative continua of singly excited Rydberg states belonging to series converging upon the $X^2\Sigma_g^+$ ionic ground state in the photoionization and photodissociation dynamics of the $B^1\Sigma_u^+$ state [25]. A schematic potential energy diagram with the relevant states of molecular hydrogen is presented in Figure 2.

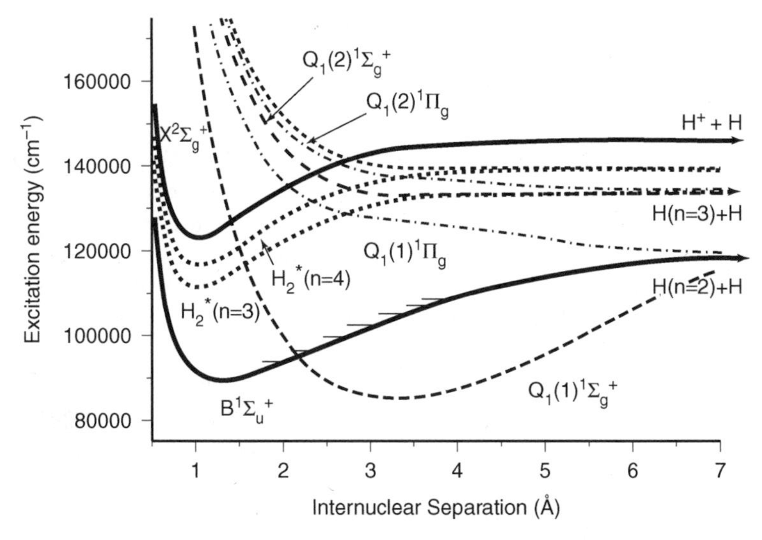


**Figure 2.** Schematic potential energy diagram of molecular hydrogen. The $v' = 3$–22 levels of the $B^1\Sigma_u^+$ state, reached from the ground state with three photons, are dissociated or ionized using a one-photon excitation via the doubly excited repulsive $Q_1$- states [91] of $^1\Sigma_g^+$ (dashed lines) or $^1\Pi_g$ symmetry (dashed–dots) or via dissociation continua of bound Rydberg states [$H_2^*(n = 3)$ and higher, dotted lines]. A quasidiabatic representation is presented, ignoring the interactions between the Rydberg states and the doubly excited states. The $v' = 3$, 5, 8, 11, 14, 17 and 20 are indicated in the $B$ state. (Reproduced with permission from [25], copyright 1998 American Institute of Physics.)

## B. (3 + 1) REMPI–PES of Molecular Hydrogen

Dihydrogen (99.9995%) (Air Liquide) was introduced effusively into the "magnetic bottle" spectrometer. In a three-photon absorption process the $B^1\Sigma_u^+(1s\sigma_g)(2p\sigma_u)$ ($v' = 3$–$22$, $N' = 0$–$3$) levels were accessed from the ground-state. Excitation spectra of the $B^1\Sigma_u^+$ state were obtained by scanning the laser wavelength and by monitoring either the $H^+$ and $H_2^+$ ion channels, or an energy-selected part of the electron current. The measured positions of these resonances were found to be in good agreement with data from the literature [86–88]. Subsequently, photoionization took place by absorption of a fourth photon. In this one-color experiment, the spectral region at the four-photon level between 125,000 and 150,000 cm$^{-1}$ (15.5–18.6 eV) was probed in a step-wise manner [25].

Photoelectron spectra were recorded for ionization via several rotational branches. Spectra were obtained for ionization after excitation of the $v' = 3$–$22$ vibrational levels, but only those measured via the $R(1)$ transition using $v' = 3$–$13$ are shown in Figure 3. Inspection of these spectra clearly shows that the absorption of an additional photon from $B^1\Sigma_u^+(v')$ levels leads to a number of competing processes. The PE spectra obtained for ionization via $B^1\Sigma_u^+$ vibrational levels up to $v' = 8$, for instance, are dominated by peaks arising from a molecular photoionization process yielding $H_2^+$ in the various accessible vibrational ($v^+$) and rotational ($N^+$) levels of its electronic ground ionic state $X^2\Sigma_g^+$. Apart from molecular ionization, a competing dissociation process leading to excited H($n = 2$) atoms, which are subsequently ionized by one-photon absorption, is also visible in these spectra. The width of the H($n = 2$) peak is due to the large kinetic energy of the H fragments, $\approx 0.9$ eV per H atom.

For ionization via the vibrational levels up to $v' = 8$ dissociation plays a minor role, a situation, which changes dramatically when ionization is performed via higher vibrational levels. Figure 3 shows that above the $n = 3$ threshold ($v' > 8$) dissociation in fact dominates over direct ionization of molecular hydrogen, in agreement with previous results [68].

In the following sections, we shall first discuss the molecular photoionization process, which can be observed for ionization via the vibrational levels up till $v' = 9$. Subsequently, the dissociation process will be considered.

### 1. *Molecular Photoionization*

For ionization via the $v' = 3$–$8$ levels of the $B^1\Sigma_u^+$ state, the PE spectra exhibit well-resolved rotational structure in the transitions deriving from molecular photoionization to the various accessible vibrational levels in the ionic state. Comparable structure, though far less resolved, has been observed by Pratt and co-workers [9] for ionization via the $R(3)$ and $P(3)$ transitions to the $v' = 7$ level. Rotational selection rules predict that in the one-photon ionization process

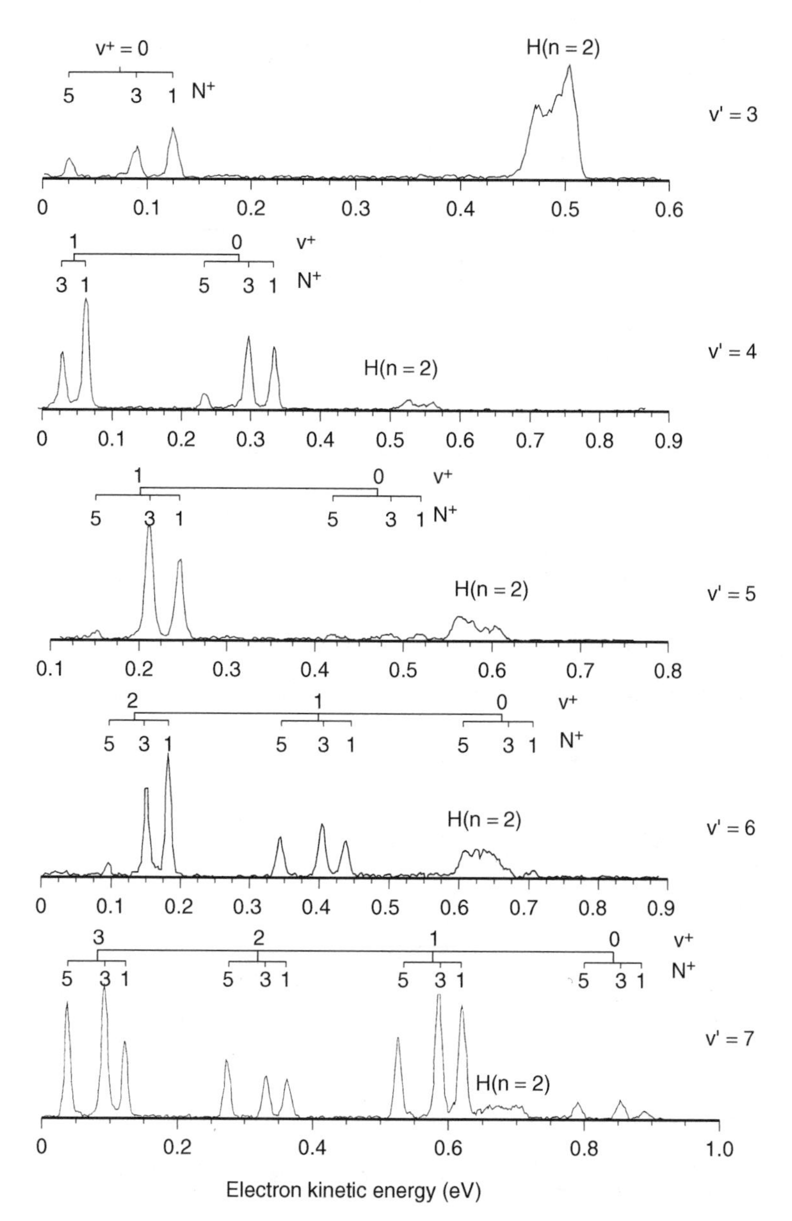


**Figure 3.** Laser PES of molecular hydrogen obtained following three-photon excitation of the $B^1\Sigma_u^+(v' = 3\text{–}13)$ levels via the $R(1)$ rotational transitions. (Reproduced with permission from [25], copyright 1998 American Institute of Physics.)

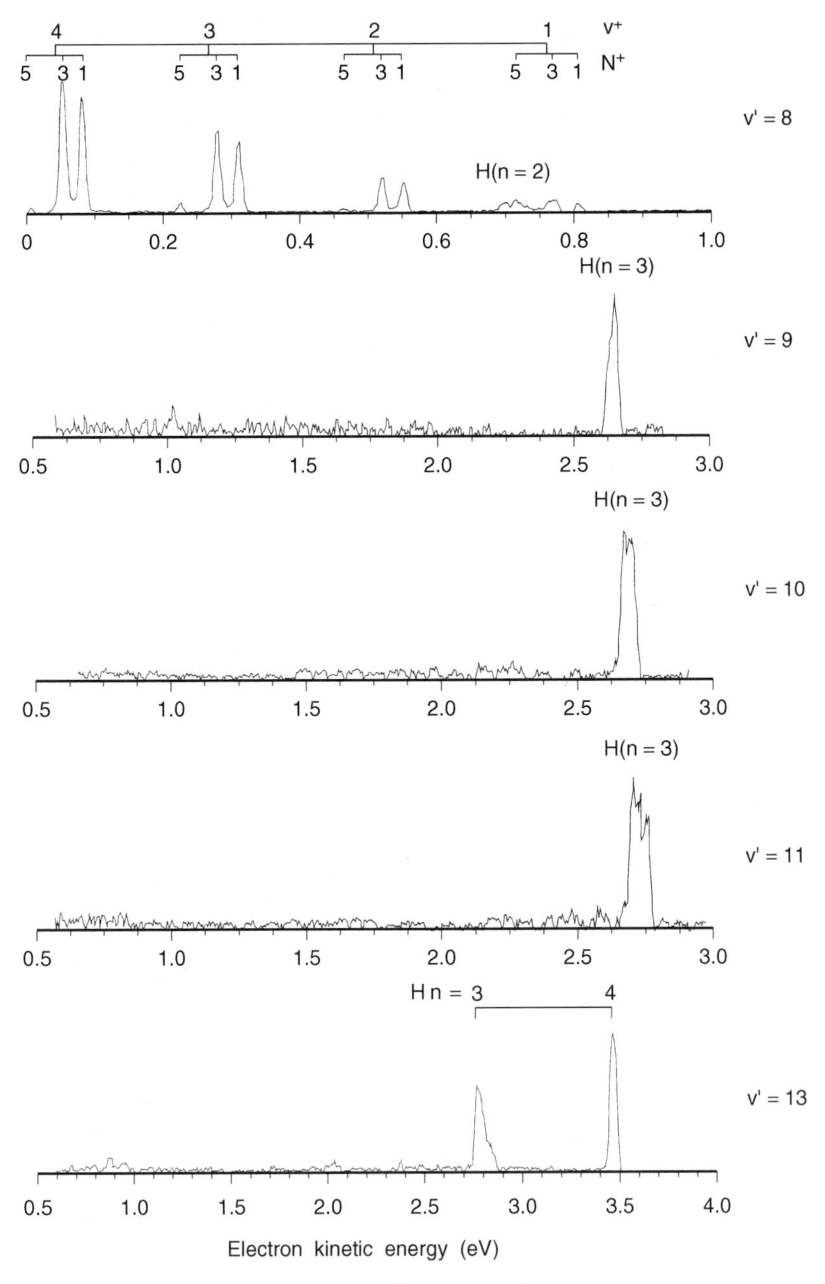


**Figure 3.** *(Continued)*

from $B^1\Sigma_u^+$ ($v'$) levels only $\Delta N + \ell = odd$ transitions should be allowed, where $\Delta N = N^+ - N'$ and $\ell$ a partial wave component of the photoelectron [89]. In an atomic-like picture, ionization of the $2p\sigma_u$ electron is expected to lead to $s(\ell = 0)$ and $d(\ell = 2)$ partial waves. Accordingly, only $\Delta N = odd$ transitions should be present in the PE spectra. In the $B^1\Sigma_u^+$ state, *ortho*-hydrogen can only exist in the $N' = 0, 2, 4\ldots$ and *para*-hydrogen only in the $N' = 1, 3, 5\ldots$ rotational levels. In the ionic $X^2\Sigma_g^+$ state, *ortho*-hydrogen has only $N^+ = 1, 3, 5\ldots$ and *para*-hydrogen only $N^+ = 0, 2, 4\ldots$ available to it. Therefore, in our experiments the PE spectra show either formation of odd or even $N^+$ levels, due to ionization of *ortho*- or *para*-hydrogen, respectively.

Conservation of total angular momentum requires that $s$ partial waves are accompanied by $\Delta N = \pm 1$, while $d$ partial waves may lead to changes of $\pm 1$ and 3. Superficially, one might be tempted to derive the relative importance of the $s$ and $d$ partial waves from the intensities of the $\Delta N = \pm 1$ and 3 transitions. The results of an *ab initio* study by Lynch et al. [90] of the rovibrational branching ratios resulting from $(3 + 1)$ REMPI via the $v' = 7$ level of the $B^1\Sigma_u^+$ state have shown, however, that such an approach is not valid. Here, it was found that $\Delta N = \pm 3$ transitions are largely suppressed as a result of dynamic interference between the $d\sigma$ and $d\pi$ channels, even though the $d$ partial wave is in fact stronger than the $s$ partial wave. As a result, the PE spectra measured for ionization via the $R(3)$ and $P(3)$ transitions to the $v' = 7$ level only exhibit $\Delta N = \pm 1$ peaks.

The PE spectra obtained in the present study for ionization via the $v' = 3$–$8$ levels of the $B^1\Sigma_u^+$ state demonstrate that the ionization dynamics as observed and calculated previously for ionization via the $P(3)$ and $R(3)$ transitions to the $v' = 7$ level are by no means exemplary for the other transitions. For example, Figure 3 shows that for ionization via the $R(1)$ transition to the $v' = 3$–$6$ and $v' = 8$ levels $\Delta N = \pm 3$ transitions are much weaker than the $\Delta N = \pm 1$ transitions, but are of similar intensity for $v' = 7$. Analogous behavior occurs for ionizing transitions via the $P(1)$ transition to the various vibrational levels in the excited state (not shown): ionization via the $v' = 7$ level leads to dominant intensity in the $\Delta N = \pm 3$ transitions, while these peaks are significantly less intense for ionization via other vibrational levels.

Surprisingly, the intensities of $\Delta N = \pm 3$ transitions depend not only on the initial vibrational level in the excited state, but also on the rotational level, as well as the rotational branch employed to populate this rotational level. Consider, for example, the spectra measured for ionization via the $v' = 7$ level shown in Figure 4. Ionization via the $P(1)$, $R(1)$, and, to a lesser extent, $R(0)$ transition results here in significant intensities of the $\Delta N = \pm 3$ transitions, while ionization via the $R(2)$, $R(3)$ [67], and $P(3)$ [67] transitions leads to dominant $\Delta N = \pm 1$ transitions. These observations are the more peculiar when it is realized that the $R(1)$ transition, for which intense $\Delta N = \pm 3$ peaks are observed, populates

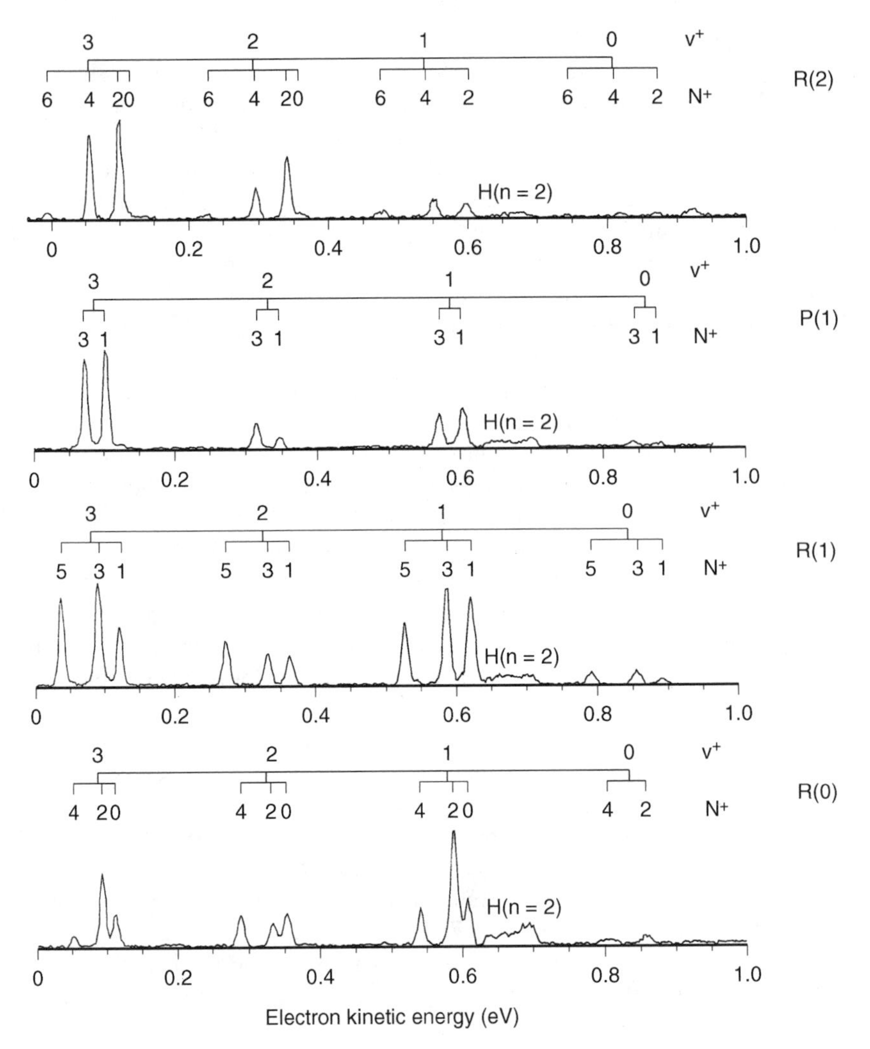


**Figure 4.** The MPI–PE spectra of molecular hydrogen obtained following three-photon excitation of the $B^1\Sigma_u^+(v' = 7)$ level via various rotational transitions. (Reproduced with permission from [25], copyright 1998 American Institute of Physics.)

the same rotational level in the excited state as the $P(3)$ transition, where $\Delta N = \pm 3$ transitions are completely absent. Although the two rotational branches lead to a different alignment of the $N' = 2$ level, one does not expect such large differences in ionization dynamics solely on the basis of a different initial alignment.

A final striking observation is that the vibrational branching ratios upon ionization may also depend strongly on the rotational transition used to excite a particular vibrational level in the excited state. This was most apparent for ionization via the $v' = 4$ level shown in Figure 5. A previous study [71] found in this case that ionization via the $R(0)$ and $P(1)$ transitions leads to dominant population of the $v^+ = 1$ level, while ionization via the $R(1)$ transition results in equal intensities of the $v^+ = 0$ and $v^+ = 1$ peaks. In the present study, these observations are supported and extended. Ionization via the $P(2)$ and $R(3)$ transitions leads to a small $v^+ = 0{:}v^+ = 1$ vibrational branching ratio, ionization via the $R(2)$ transition to a large ratio. Moreover, we observe that a large $v^+ = 0{:}v^+ = 1$ branching ratio is accompanied by a significant reduction of excited H($n = 2$) fragments. Interestingly, when the PE spectra are considered in order of the final energies reached in these experiments at the four-photon level, that is, $P(2)$, $P(1)$, $R(0)$, $R(1)$, $R(2)$, $R(3)$, it is clear that the $v^+ = 0{:}v^+ = 1$ branching ratio and the amount of dissociation do not vary randomly, but show a maximum and minimum, respectively, $\sim$ 127,100–127,150 $\text{cm}^{-1}$.

All of the above observations indicate that the ionization dynamics depend sensitively on the *four-photon* level reached in the experiments. *Ab initio* calculations have shown that dynamic interference between the $d\sigma$ and $d\pi$ ionization channels plays an important role in describing ionization from the $B^1\Sigma_u^+(v' = 7)$ level [90]. The delicate balance between these two channels will certainly be electron kinetic energy dependent, but it is hard to imagine that this energy dependence would be so large that a mere change of a few 100 $\text{cm}^{-1}$ in kinetic energy, as is occurring, for example, when ionizing via the $R(1)$ or the $P(3)$ transition to the $v' = 7$ level, would make so much difference. Another possible explanation might be found in the influence of dissociative Rydberg states converging upon the $^2\Sigma_u^+(2p\sigma_u)$ ionic state. For excitation of the levels investigated here the only doubly excited states that can be accessed at the four-photon level are the $^1\Sigma_g^+(2p\sigma_u)^2$ and $^1\Pi_g(2p\sigma_u)(2p\pi_u)$ Rydberg states. Calculations of the Franck–Condon factors between the vibrational wave functions of $B^1\Sigma_u^+(v')$ levels and these dissociative $^1\Sigma_g^+$ and $^1\Pi_g$ states show that these factors do not vary enough to account for the observed rotational branch dependence of one particular $B^1\Sigma_u^+(v')$ level. These doubly excited states might, however, be involved in an explanation of the $v'$ dependence of the PE spectra, which involve a much larger range of energy.

The fact that features observed in the PE spectra change so drastically over a relatively small energy interval strongly suggests that bound states at the four-photon level are of influence. At this energy level, vibrationally excited Rydberg states with an $X^2\Sigma_g^+(1s\sigma_g)$ ionic core are present, which converge upon $v^+$ levels of the ground ionic state that lie higher in energy than the employed four-photon energy. From the point of view of vibrational overlap these levels are well accessible, since the potential energy surface of the $B^1\Sigma_u^+$ state is significantly

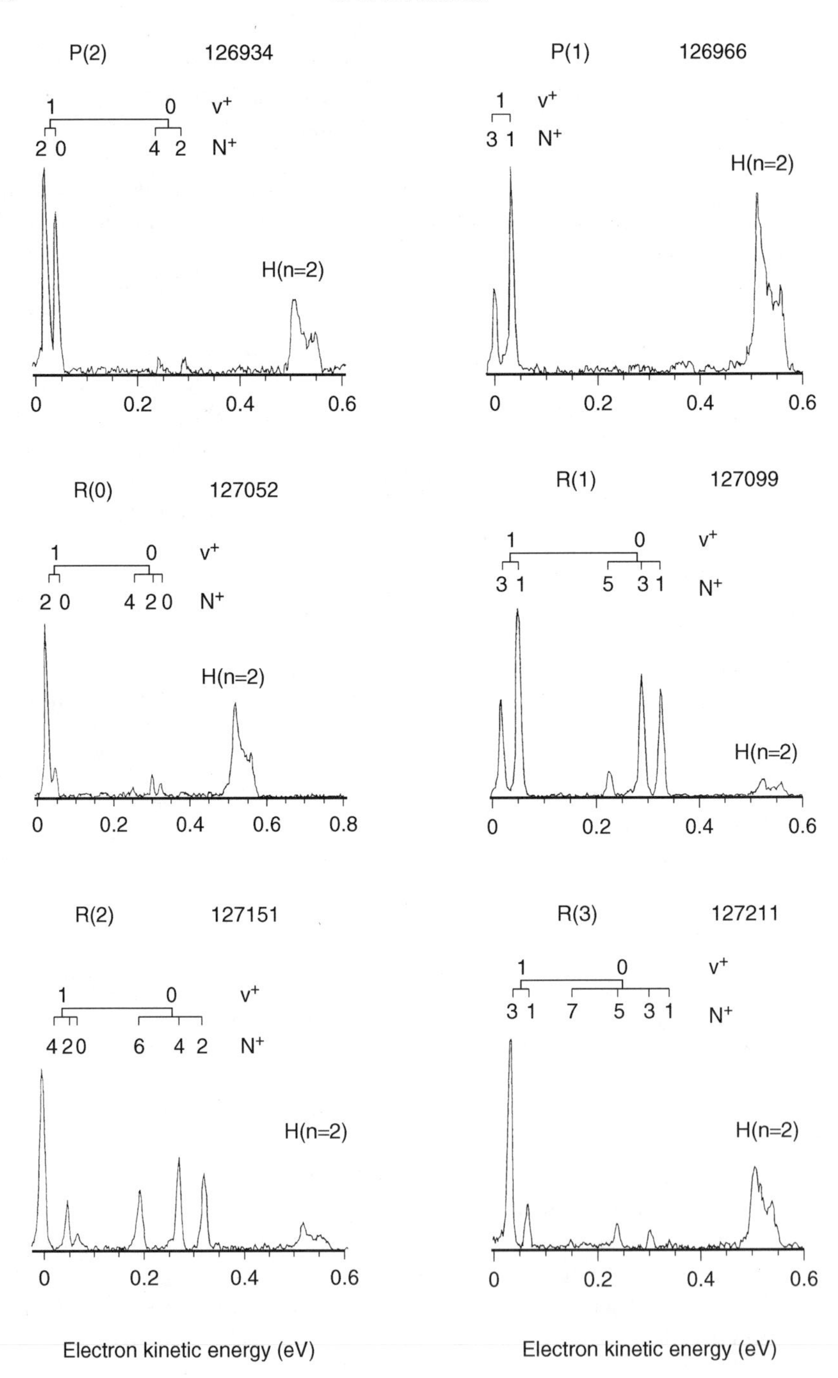
P(2)
126934
1
0
v+
2 0
4 2
N+
H(n=2)
0
0.2
0.4
0.6
P(1)
126966
1
v+
3 1
N+
H(n=2)
0
0.2
0.4
0.6
R(0)
127052
1
0
v+
2 0
4 2 0
N+
H(n=2)
0
0.2
0.4
0.6
0.8
R(1)
127099
1
0
v+
3 1
5
3 1
N+
H(n=2)
0
0.2
0.4
0.6
R(2)
127151
1
0
v+
4 2 0
6
4 2
N+
H(n=2)
0
0.2
0.4
0.6
R(3)
127211
1
0
v+
3 1
7
5
3 1
N+
H(n=2)
0
0.2
0.4
0.6
Electron kinetic energy (eV)
Electron kinetic energy (eV)

different from that of the bound Rydberg states. Upon excitation of these states, the PE spectra will not merely display the photoionization dynamics of the $B^1\Sigma_u^+$ state, but also the decay dynamics of the bound Rydberg state. A complete unraveling of the influence of bound Rydberg states on the molecular photoionization spectra clearly demands theoretically the application of high-quality *ab initio* calculations. In addition, experimental studies, in which $B^1\Sigma_u^+(v')$ levels are excited by one-photon absorption and subsequently ionized in a tuneable one-photon absorption step, are very valuable as will be shown below.

### 2. *Photodissociation*

Figure 3 shows that dissociation occurring at the four-photon level plays an important role in the one-photon excitation dynamics from the $B^1\Sigma_u^+$ state. As can be seen in this figure, in one-photon absorption from the $v' = 3$ to $v' = 13$ levels excited hydrogen atoms—on energy grounds accompanied by ground-state hydrogen atoms—are seen to be formed in all accessible Rydberg states. One notable exception concerns the $n = 2$ fragments, which disappear as soon as the $n = 3$ dissociation channel becomes accessible. It can be shown that under our experimental conditions the ionization rates are such that all excited fragments formed are actually ionized. This implies that quantitative excited-atom contributions can be derived from the peak areas observed in our PE spectra [25].

Our PE spectra reveal that most energetically allowed dissociation limits are observed with the clear exception of the $H(n = 1) + H(n = 2)$ dissociation limit as soon as the production of $H(n = 3)$ is possible. A quantitative analysis of these distributions is very complex. A logical starting point forms a calculation of the relative excitation probabilities of the different doubly excited repulsive states. These states have Rydberg character and converge to the first excited state in $H_2^+$, the $^2\Sigma_u^+$ state (see Fig. 2). The correlation of these states with the various dissociation limits is nontrivial. The presence of the bound Rydberg series results in a large number of (avoided) crossings. Part of the double-well structures found in Born–Oppenheimer calculations of the $^1\Sigma_g^+$ Rydberg states is caused by the doubly excited repulsive character (see also [60]). Due to these interactions dissociation flux will be distributed over more than one dissociation limit even if one doubly excited state is excited in the Franck–Condon region.

We have attempted to model the excited atom distributions in two separate calculations: (1) on the basis of cross-sections for excitation of the doubly excited repulsive states, described in an uncoupled diabatic representation,

←

**Figure 5.** The MPI–PE spectra of molecular hydrogen obtained following three-photon excitation of the $B^1\Sigma_u^+(v' = 4)$ level via various rotational transitions. The final energy ($cm^{-1}$) reached at the four-photon level is given for each rotational transition. (Reproduced with permission from [25], copyright 1998 American Institute of Physics.)

and (2) on the basis of cross-sections to the dissociation continuum of singly excited Rydberg states. Both calculations describe in zero order a one-electron transition, $(n\ell\lambda_u \leftarrow 1s\sigma_g)$ for the excitation to the doubly excited states and $(n\ell\lambda_g \leftarrow 2p\sigma_u)$ for the singly excited Rydberg states.

1. We have performed approximate calculations on the relative transition probabilities for excitation from the $B^1\Sigma_u^+(v')$ levels to the one-photon accessible ${}^1\Sigma_g^+(2p\sigma_u)(np\sigma_u)$ and ${}^1\Pi_g(2p\sigma_u)(np\pi_u)$ repulsive states with dissociation limits up to $n < 7$. We calculated integrals $\int \chi_k(R)D(R)\chi_{v'}(R)dR$, with $\chi_k(R)$ an—energy normalized—nuclear wave function in the doubly excited state under consideration, $\chi_{v'}(R)$ a vibrational wave function of the $B^1\Sigma_u^+(v')$ level, and $D(R)$ the one-photon electronic transition moment. The potential energy curves for the $Q_1$ ${}^1\Sigma_g^+$ states with $\mathrm{H}(n=1)+\mathrm{H}(n=2)$ and $\mathrm{H}(n=1)+\mathrm{H}(n=3)$ dissociation limits were taken from Guberman [91]. The higher ${}^1\Sigma_g^+$ states were simply obtained by raising the $R$-dependent quantum defect of the ${}^1\Sigma_g^+$ state with the $\mathrm{H}(n=1)+\mathrm{H}(n=3)$ dissociation limit with one for each successive state. The potential energy curves of the ${}^1\Pi_g$ states were calculated using the $R$-dependent quantum defect of the ${}^1\Pi_g$ state with the $\mathrm{H}(n=1)+\mathrm{H}(n=2)$ dissociation limit. Calculations on the relevant electronic transition moment have not been published. In the one-electron approximation, this transition moment is given by the $\langle 1s\sigma_g|z|2p\sigma_u\rangle$ matrix element. These matrix elements are also the dominant contribution to the $B^1\Sigma_u^+(1s\sigma_g)(2p\sigma_u) \leftarrow\leftarrow\leftarrow X^1\Sigma_g^+(1s\sigma_g)^2$ and $C^1\Pi_u(1s\sigma_g)(2p\pi_u) \leftarrow\leftarrow\leftarrow X^1\Sigma_g^+(1s\sigma_g)^2$ transitions. The $X$–$B$ and $X$–$C$ transition moments may approximate the *magnitude* of the transition moments from the $B$ state to the doubly excited states. The approximation becomes highly questionable at larger internuclear distances where the electronic character of the involved states changes. These arguments and the calculated internuclear distance dependence of the calculated $X$–$B$ and $X$–$C$ electronic transition moments [92], led us to use transition moments, independent of internuclear distance, of 1.6 and 1.0 a.u. for the ${}^1\Sigma_g^+(2p\sigma_u)^2 \leftarrow B^1\Sigma_u^+(1s\sigma_g)(2p\sigma_u)$ and ${}^1\Pi_g(2p\sigma_u)(2p\pi_u) \leftarrow B^1\Sigma_u^+(1s\sigma_g)(2p\sigma_u)$ transitions. For transitions to the higher doubly excited states, the electronic transition moments can be scaled by $(n^*/n^{**})^{3/2}$, where $n^*$ is the effective quantum number of the ${}^1\Sigma_g^+(2p\sigma_u)^2$ or ${}^1\Pi_g(2p\sigma_u)(2p\pi_u)$ state and $n^{**}$ is the effective quantum number of the higher doubly excited state. Comparison of the distributions simulated with the above model and those obtained experimentally show poor agreement. In particular, the simulations predict that the $\mathrm{H}(n=2)$ fragments should form an important exit channel [25]. Simulations based upon $R$-dependent $X$–$B$ and $X$–$C$ transition moments do not improve the qualitative picture at all. Recent DR results for $H_2^+$ show

that at total energies above the H($n = 3$) energy, still a significant fraction of H($n = 2$) fragments is found [60]. The theoretical treatment in that paper implies that the lowest $Q_1\,{}^1\Sigma_g^+$ state produces a significant fraction of H($n = 2$) fragments at energies above the H($n = 3$) limit [60]. The absence of H($n = 2$) fragments in the present experiments has to be due to another mechanism than direct excitation of the doubly excited repulsive states.

2. Above it was concluded that simulations based solely on excitation of doubly excited Rydberg states could not explain our experimental results. Moreover, the rapidly changing dynamics observed in the excitation from the lower $B$ state vibrational levels were indicative for the influence of vibrationally excited levels of singly excited Rydberg states on molecular photoionization. This inspired us to calculate the influence of the *vibrational continua* of singly excited Rydberg states with ${}^1\Sigma_g^+(1s\sigma_g)(ns\sigma_g)$, ${}^1\Sigma_g^+(1s\sigma_g)(nd\sigma_g)$, and ${}^1\Pi_g(1s\sigma_g)(nd\pi_g)$ symmetry on the final fragment-state distribution. As mentioned earlier, these states can be accessed by one-photon absorption from the $B^1\Sigma_u^+$ state. We note that the potential energy curves of the $B^1\Sigma_u^+$ state differ considerably from the bound Rydberg states. Hence, the overlap between the vibrational wave functions of $B^1\Sigma_u^+(v')$ levels and the vibrational continua is not necessarily much smaller than the vibrational overlap for transitions to doubly excited repulsive Rydberg states. Also, the $2p\sigma_u \rightarrow (ns\sigma_g, nd\sigma_g, nd\pi_g)$ electronic transition moments are not expected to differ by orders of magnitude from the $1s\sigma_g \rightarrow (np\sigma_u, np\pi_u)$ electronic transition moments involved in the transition to doubly excited states. Excitation of these states has been ignored in previous studies on the $B^1\Sigma_u^+$ state, even though it was concluded to be important in one-photon excitation studies from the $EF^1\Sigma_g^+$ state [84].

To put these arguments on a somewhat more quantitative basis, we have calculated the transition moments to the vibrational continua of $n = 2$ and higher Rydberg states. As we observe hydrogen fragments, it is assumed that an $n = n'$ Rydberg state correlates with an excited H($n = n'$) fragment. Model bound Rydberg states have been made, assuming an R-independent integral quantum defect. Constant electronic transition moments were also assumed. We have used the electronic transition moments calculated for the $GK^1\Sigma_g^+(1s\sigma_g)(3d\sigma_g) \leftarrow B$ and $H\overline{H}^1\Sigma_g^+(1s\sigma_g)(3s\sigma_g) \leftarrow B$ transitions by Wolniewicz and Dressler [93] taken at the equilibrium separation of the $B$ state. Lynch et al. [90] calculated for ionization of the $B^1\Sigma_u^+$ state that the electronic transition moments to the $\varepsilon s\sigma$ and $\varepsilon d\sigma$ continua are in the ratio of $\sim$ 1:7. Since this corresponds to excitation of $n = \infty$ states, we assume that this ratio is equally applicable to the presently considered states. Support for this assumption is found in the calculations of Wolniewicz and Dressler [93], where a ratio of 1:8 was found for the electronic transition moments of the $GK \leftarrow B$ and $H\overline{H} \leftarrow B$ transitions at the equilibrium distance of the $B$ state.

For transitions to Rydberg states with higher principal quantum numbers the electronic transition moments are scaled in the same way as described for the doubly excited states. Using the Wigner–Eckart theorem [94] it is derived that the electronic transition moment to $nd\pi_g$ states is 1.6 times smaller than that to $nd\sigma_g$ states. The probability for formation of H($n$) is calculated by multiplying the Franck–Condon factor for excitation of the Rydberg state with principal quantum number $n$ with the squared transition moments to $ns\sigma_g$, $nd\sigma_g$, and $nd\pi_g$ respectively, and by adding these contributions incoherently.

The simulation of the excited-atom distributions following from this mechanism is depicted in Figure 6. The most striking differences with the previous model is that the contribution of H($n = 2$) fragments to the distributions is now absent, in agreement with our experimental observations. These fragments can within the present model only be produced by excitation of the vibrational continuum of the $EF^1\Sigma_g^+$ state, but our calculations indicate that the Franck–Condon factor for this pathway is negligibly small compared to excitation of the vibrational continua of $n = 3$ and higher Rydberg states. This calculation also suggests that the closing down of the molecular ionization channel for excitation via $B^1\Sigma_u^+(v' > 8)$ levels, is due to the large cross-section for excitation of the vibrational continua of the $n \geq 3$ states in comparison with that for excitation of the $n = 2$ states. Although the dominant features observed in photoionization and photodissociation are thus nicely reproduced, we notice that with this model

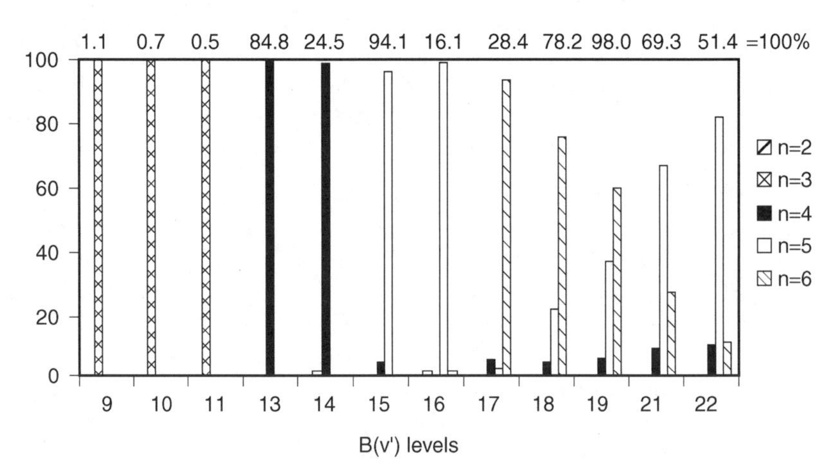


**Figure 6.** Excited-atom distributions H($n$) calculated for excitation of molecular hydrogen via the $B^1\Sigma_u^+(v')$ levels, and assuming that dissociation uniquely occurs by excitation of the vibrational continua of bound singly excited Rydberg states with $^1\Sigma_g^+(1s\sigma_g)(ns\sigma_g)$, $^1\Sigma_g^+(1s\sigma_g)(nd\sigma_g)$, and $^1\Pi_g(1s\sigma_g)(nd\pi_g)$ symmetry. For dissociation via each vibrational level the sum of the cross sections for the pathways leading to H($n = 2$) to H($n = 6$) is set to 100% in the stick diagrams. The absolute value of this sum is given at the top of the diagram for each vibrational level. (Reproduced with permission from [25], copyright 1998 American Institute of Physics.)

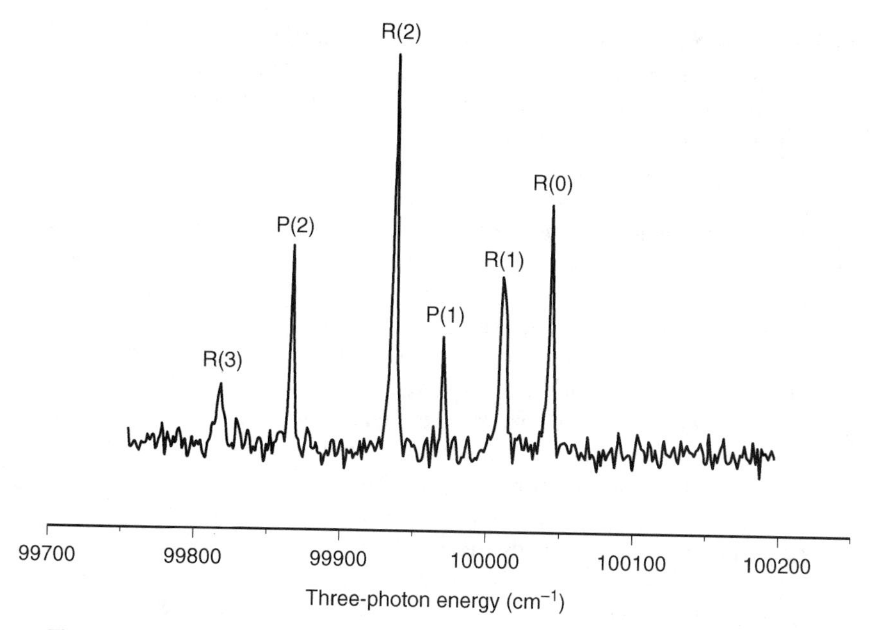


**Figure 7.** Three-photon excitation spectrum of molecular deuterium between the three-photon energies of 99,750 and 100,200 cm$^{-1}$ following the $B^1\Sigma_u^+(1s\sigma_g)(2p\sigma_u)(v' = 11) \leftarrow\leftarrow\leftarrow X^1\Sigma_g^+(1s\sigma_g)^2$ transition monitored with electron detection. (Reproduced with permission from [26], copyright 2000 Elsevier Science B.V.)

cross-sections for H($n = 3$) production via $v' = 9$–11 are smaller than those obtained for H($n = 2$) and H($n = 3$) via the same transitions in the previous model [25]. Quantitatively the agreement with the observed distributions is still poor. We note that a complete treatment has to combine the excitation of doubly excited repulsive curves, the excitation of the dissociation continua of bound Rydberg states and the interactions between these states at large internuclear separations. We believe that these conclusions hold even if correct $R$-dependent electronic transition moments would be employed.

### C. (3 + 1) REMPI–PES of Molecular Deuterium

In a subsequent experiment, molecular deuterium (Hoekloos, $D_2$ 2.8) was introduced into the "magnetic bottle" spectrometer effusively. This time, REMPI–PES was performed in a one-color $(3 + 1)$ experiment via the rovibrational levels ($v' = 10$–13, $N' = 0$–3) of the $B^1\Sigma_u^+(1s\sigma_g)(2p\sigma_u)$ state, covering the four-photon energy region below and above the D($n = 1$) + D($n = 3$) threshold. Because of the smaller vibrational and rotational spacings between the energy levels of molecular deuterium this can be achieved in smaller steps than previously possible in the $(3 + 1)$ REMPI–PES study of molecular hydrogen.

Moreover, the possible influence of the different nuclear spin statistics can be investigated [26]. In order to guide the discussion, the relevant electronic energy levels of molecular deuterium are similar (but not identical) to those of molecular hydrogen depicted in Figure 2.

Three-photon excitation spectra of the $B^1\Sigma_u^+(1s\sigma_g)(2p\sigma_u) \leftarrow\leftarrow\leftarrow X^1\Sigma_g^+(1s\sigma_g)^2$ transition were obtained by scanning the laser wavelength and by monitoring either an energy-selected part of the electron current, or the $D^+$ or $D_2^+$ ion fluxes in the respective mass channels. As an illustration, the wavelength spectrum of molecular deuterium between the three-photon energies of 99,750 and 100,200 $cm^{-1}$ obtained with electron detection is presented in Figure 7. The positions of the resonances in these excitation spectra agreed to $\sim 0.5\,cm^{-1}$ with previous data from vacuum ultraviolet (VUV) one-photon absorption spectra [87]. Assignments were taken from this previous study.

PE spectra were obtained following absorption of a fourth photon, with electrons arising from ionization of $D_2$ into the $X^2\Sigma_g^+(v^+, N^+)$ continua, either directly or through autoionization, and from subsequent one-photon ionization of D atoms in excited states. Rovibrational levels associated with the $X^1\Sigma_g^+(v'' = 0, N'')$ electronic ground-state of molecular deuterium and the relevant vibrational and rotational constants are very accurately known experimentally [3]. The measured molecular transitions observed in our PE spectra could easily be assigned using the results of *ab initio* calculations of the vibrational and rotational levels of molecular deuterium in its $X^2\Sigma_g^+$ ground ionic state [95]. These calculations employed the adiabatic potentials of Hunter and Pritchard [96] as their starting point, and used a Numerov method to obtain the vibrational and rotational levels with a numerical accuracy of better than 0.01 $cm^{-1}$. The agreement between the results of these computations in the adiabatic approximation and experimental data was estimated to be better than 0.1 or 0.2 $cm^{-1}$. The assignment procedure generally allowed a reliable determination of the rovibrational levels in $X^1\Sigma_g^+$, $B^1\Sigma_u^+$, and $X^2\Sigma_g^+$ states pertinent to the observed four-photon transition.

In addition to the molecular ionization channel, the PE spectra show transitions arising from one-photon ionization of D atoms formed in excited states ($n = 2, 3$). Since in the photodissociation process excited D($n = 2$) atoms were formed with considerable excess kinetic energy (typically $\sim 0.9$ eV per D atom), this was reflected in the Doppler widths of the corresponding atomic peaks. Apart from the small influence of reduced mass and of relativistic effects, which differ between H and D, the excited-state energies of D atoms are identical to those of H atoms. With the current resolution achievable in our PE spectra, these shifts for D atoms could not be observed.

A summary of PE spectra obtained via $B^1\Sigma_u^+(v' = 10\text{–}13, N' = 0\text{–}3)$ and involving different rotational branches in the transition from $B(v', N') \leftarrow\leftarrow\leftarrow X(v'' = 0, N'')$ is presented in Figure 8. These spectra are organized in order of

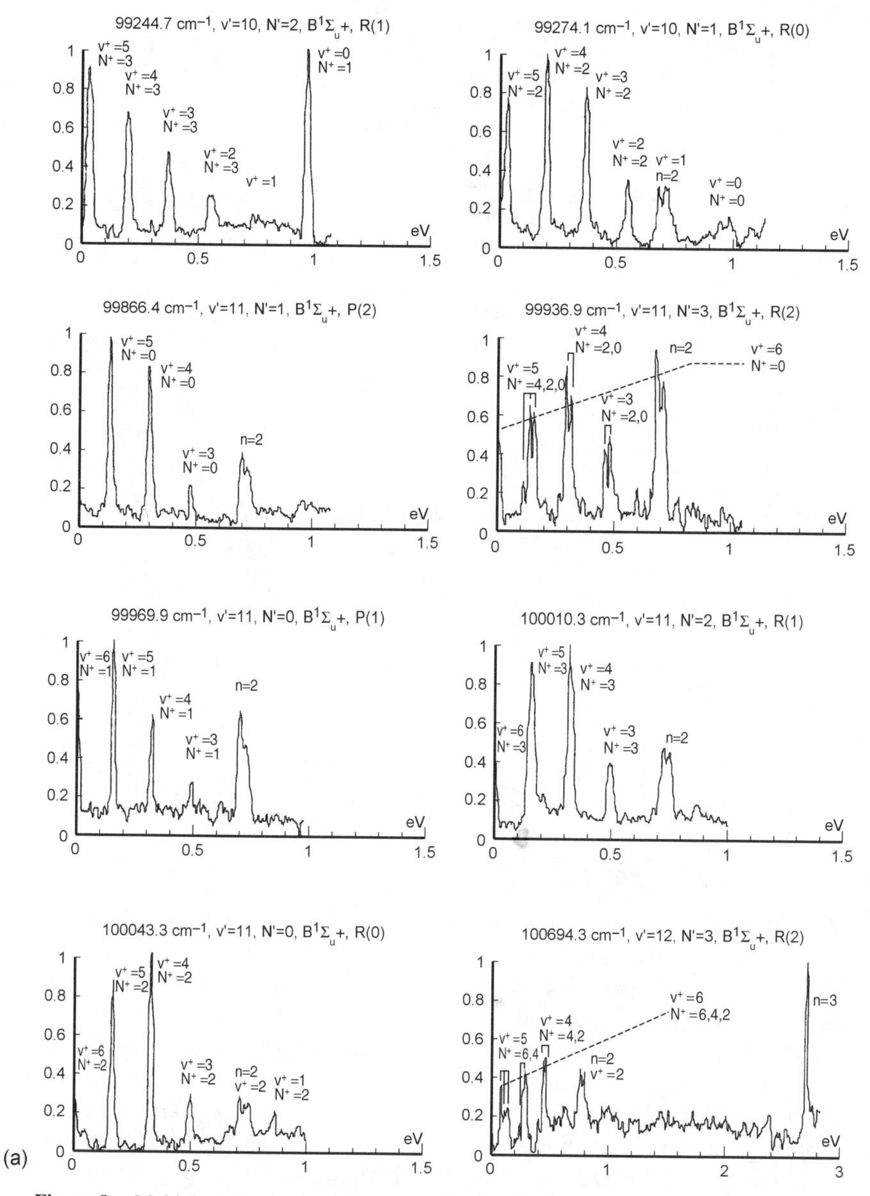


**Figure 8.** Multiphoton ionization PE spectra of molecular deuterium obtained following three-photon excitation of the $B^1\Sigma_u^+(1s\sigma_g)(2p\sigma_u)(v' = 10\text{–}13)$ levels via various rotational transitions. The three-photon energy ($\text{cm}^{-1}$) is given for each rotational transition. (Reproduced with permission from [26], copyright 2000 Elsevier Science B.V.)

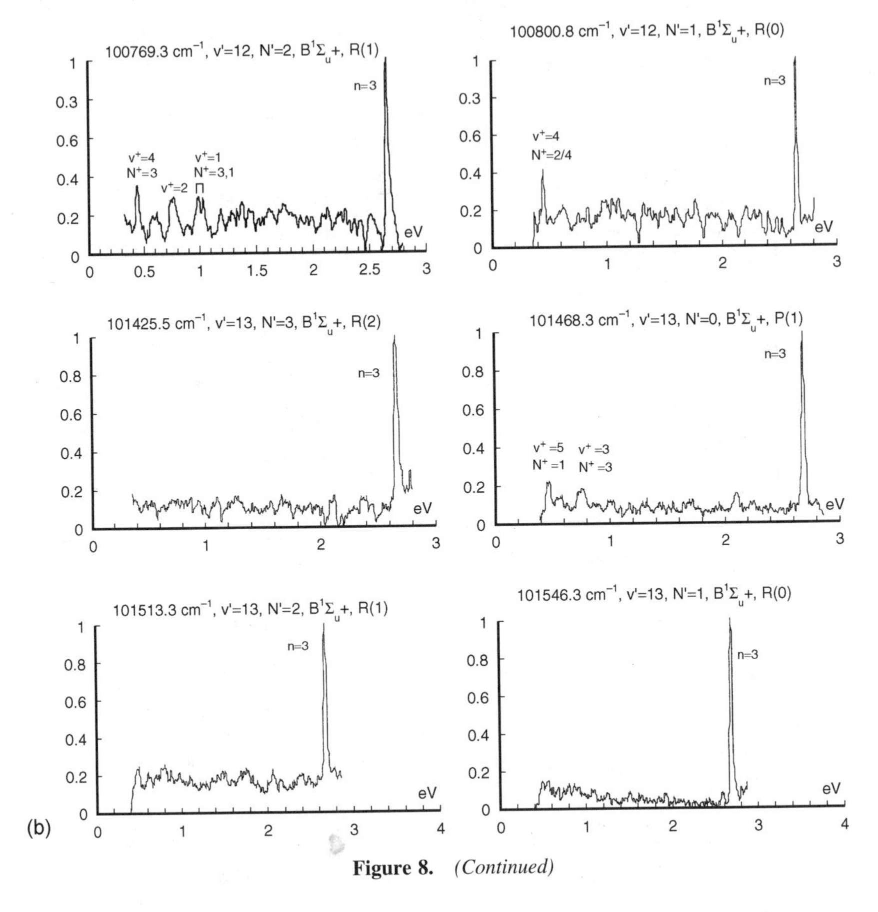


**Figure 8.** *(Continued)*

increasing three-photon energies. In addition, electron peaks arising from excited D atoms are also apparent in the spectra. Assignments of the PE spectral features are presented in the spectra. From these PE spectra, a number of distinct trends can be deduced.

The spectra clearly show that the absorption of an additional photon from $B^1\Sigma_u^+(v', N')$ leads to competing photoionization and photodissociation processes. The PE spectra obtained via $B^1\Sigma_u^+(v' = 10, 11)$ are dominated by molecular photoionization signals, with atomic peaks derived from one-photon ionization of D($n = 2$) atoms. For PE spectra obtained via $B^1\Sigma_u^+(v' = 12, 13)$ there is a dramatic change in that both the molecular photoionization channel and the production of D($n = 2$) atoms are no longer observed. Instead, the production of D($n = 3$) atoms and their subsequent one-photon ionization is now the dominant process. Since the energy onset for the formation of D($n = 1$) + D($n = 3$) is located at 134,266.7 cm$^{-1}$, the sum of the dissociation energy of $D_2$ in its electronic, vibrational, and rotational ground-state $^1\Sigma_g^+(v'' = 0,$

$N''=0$) [3] and the D($n=3$) $\leftarrow$ D($n=1$) excitation energy, it is apparent that as soon as this energy threshold is surpassed by the absorption of the fourth photon, the production of D($n=3$) through photodissociation is essentially the only exit channel.

When the PE spectra that show significant molecular photoionization are inspected in more detail, it is apparent that for quite small changes in the four-photon energy very large changes in vibrational and rotational branching ratios are observed. Moreover, the relative importance of the competing molecular photoionization and photodissociation processes is also seen to vary strongly with relatively small changes in the four-photon energy. This behavior is again indicative of the role played by superexcited states at the four-photon level.

In a few of the PE spectra obtained at four-photon energies above the D($n=1$) + D($n=3$) threshold (at 100,694.3, 100,769.3, 100,800.8, and 101,468.3 $\text{cm}^{-1}$) spectral features are observed that arise from molecular photoionization. This situation, seems at odds with the statement made above that above this threshold the molecular photoionization channel closes completely. Closer inspection shows that these molecular signals can be invariably ascribed to the close presence of levels associated with the $C^1\Pi_u$ state of $D_2$ [26].

### 1. *Molecular Photoionization*

The molecular ground and excited states involved in our experiments, as well as the lowest ionic state, have all $\Sigma$ symmetry. These states can therefore be conveniently treated in a Hund's case (b) coupling scheme [1] in which we only focus on the end-over-end rotation $N''$, $N'$, and $N^+$. Rotational structure in the PE spectra of molecular deuterium ionized in a $(3+1)$ multiphoton absorption process is subject to rotational selection rules [89]. These selection rules predict that for one-photon ionization from the $B^1\Sigma_u^+(v')$ levels, only $\Delta N + \ell = odd$ transitions are allowed, where $\Delta N = N^+ - N'$ and $\ell$ is a partial wave component of the photoelectron. In an atomic-like picture removal of the $2p\sigma_u$ electron is expected to lead to $s(\ell=0)$ and $d(\ell=2)$ partial waves. In the PE spectra only $\Delta N = odd$ transitions would then be expected. In the $B^1\Sigma_u^+$ state of molecular deuterium even values of $N'$ can only exist in conjunction with nuclear spin wave functions which are antisymmetrical under permutation of the spins. Conversely, odd values of $N'$ can only exist in combination with nuclear spin functions which are symmetrical under permutation. The ratio of symmetrical to antisymmetrical nuclear spin functions is 2:1. In the ionic $^2\Sigma_g^+$ state of molecular deuterium even $N^+$ levels exist in combination with symmetrical, odd $N^+$ levels with antisymmetrical nuclear spin functions. In our experiments therefore, either even or odd $N^+$ levels are accessed in the ionic state. In addition to the parity selection rules, conservation of angular momentum is an issue. Since $\Delta N = \ell+1, \ell, \ldots, -\ell-1$, with the photoelectron leaving as an $s$ partial wave, the change in overall rotation is limited to $\Delta N = \pm 1$, while for an electron leaving as a $d$ wave $\Delta N = \pm 1, \pm 3$ holds. All our observations agree with these predictions.

In our molecular photoionization spectra, vibrational and rotational branching ratios are very dependent on small changes in the four-photon energy. Franck–Condon factors obtained with simple calculations were computed for the direct ionization process from various $B^1\Sigma_u^+(v', N')$ levels to the $X^2\Sigma_g^+(v^+)$ continua and compared to our experimental vibrational branching ratios. Although in these calculations we have neglected the changes of the electronic transition moment with internuclear distance, it was apparent that a direct photoionization process, which only allowed for gradual changes in vibrational branching ratios as a function of final-state energy could not account for our observations. The situation is very similar to that encountered previously for molecular hydrogen [25]. Clearly, another mechanism, capable of explaining the fluctuations in branching ratios as well as the variations in the competition with the photodissociation channel is required. The explanation is again sought in the role of bound superexcited states located above the lowest $X^2\Sigma_g^+$ ionic threshold, but below the $D(n = 1) + D(n = 3)$ energy threshold, of type ${}^1\Sigma_g^+(1s\sigma_g)(ns\sigma_g)$, ${}^1\Sigma_g^+(1s\sigma_g)(nd\sigma_g)$, and ${}^1\Pi_g(1s\sigma_g)(nd\pi_g)$. The singly excited states at large internuclear distances can again couple with dissociative doubly excited ${}^1\Sigma_g^+(2p\sigma_u)(np\sigma_u)$ and ${}^1\Pi_g(2p\sigma_u)(np\pi_u)$ Rydberg states belonging to series converging upon the ${}^2\Sigma_u^+(2p\sigma_u)$ excited ionic state. Depending on whether in the excitation process such a superexcited state is accessed at the four-photon level or not, the balance between direct photoionization on the one hand, and photodissociation on the other, can be influenced to a very significant extent.

## 2. *Photodissociation*

Once the $D(n = 1) + D(n = 3)$ threshold is exceeded in our experiments, the molecular ionization channel and the dissociation channel that involves excited $D(n = 2)$ atoms close. Instead, the PE spectra show convincing evidence for a single exit channel in which $D(n = 3)$ is formed and subsequently ionized with one photon. These observations again are very similar in nature to what has been observed previously in our extensive $(3 + 1)$ REMPI–PES study on molecular hydrogen [25]. Excitation of the vibrational continua of the ${}^1\Sigma_g^+(1s\sigma_g)(ns\sigma_g)$, ${}^1\Sigma_g^+(1s\sigma_g)(nd\sigma_g)$, and ${}^1\Pi_g(1s\sigma_g)(nd\pi_g)$ singly excited Rydberg states is again thought to play a key role in the photodissociation of molecular deuterium into the $D(n = 1) + D(n = 3)$ channel. Clearly, in $D_2$ the nuclear spin statistics are different from the $H_2$ case. Also, the dissimilar vibrational and rotational constants allow for an experiment in which the energy region below and above the $D(n = 1) + D(n = 3)$ threshold can be probed in much smaller steps than with $H_2$. However, our current study shows convincingly that these differences have no bearing on the fundamental outcome of the experiments [26].

### D. $(1 + 1')$ REMPI of Molecular Hydrogen

A disadvantage of our $(3 + 1)$ REMPI–PES experiments discussed above is that the energy region above and below the $n = 3$ limit was scanned in a stepwise manner, since the rovibrational levels of the $B$ state were used as stepping stones for absorption of the fourth photon. We now present a $(1 + 1')$ REMPI study of the $H(n = 1) + H(n = 3)$ dissociation process in molecular hydrogen, in which various rovibrational levels $(v' = 19, 20; N' = 0, 1)$ of the $B^1\Sigma_u^+(1s\sigma_g)(2p\sigma_u)$ state were accessed with a single extreme ultraviolet (XUV) photon, and where the photon energy of the second photon was scanned across the $n = 3$ limit continuously. By monitoring the $H^+$ signal from laser-induced ionization of $n = 3$ fragments, information about the $H(n = 1) + H(n = 3)$ dissociation process at the $n = 3$ threshold was obtained [97]. The present results will again be discussed in terms of the electronic states that contribute to the $H(n = 3)$ yield in this energy region.

For the experimental setup and the performance of the XUV laser source the reader is referred to previous publications [98, 99]. The XUV frequency was fixed on transitions from the $X^1\Sigma_g^+$ electronic ground-state to high vibrational levels of the $B^1\Sigma_u^+$ state. The photons for the second step, whose energy was scanned continuously across the $H(n = 1) + H(n = 3)$ dissociation limit, were obtained from a second Nd:YAG pumped PDL. The experiments were performed via rotational levels associated with the $v' = 19$ and 20 states, since in that case the photon energy for the second step ranges from 23,000 to 23,800 $\text{cm}^{-1}$, which is too small for one-photon ionization of $n = 2$ fragments. The intensity was kept low enough to avoid multiphoton ionization processes that give rise to parasitic signals in the detection of $n = 3$ fragments. The frequency-doubled output of a Nd:YAG laser was used to photoionize the $n = 3$ fragments. A delay of 20 ns was used for these 532-nm pulses with respect to the two other laser beams, in order to avoid the influence of the 532-nm beam on molecular ionization processes. This frequency was again large enough to ionize $H(n = 3)$ with a single photon, but too small to ionize $H(n = 2)$. In this way, the yield of $n = 3$ fragments as a function of the wavelength could be determined simply by detecting $H^+$ ions.

In our $(1 + 1')$ REMPI experiment the frequency of the second photon had to fulfill certain requirements as discussed above. These requirements limit the frequencies that can be used for the XUV photon and restrict the vibrational levels of the $B$ state that can be employed as stepping stones in the REMPI process. The vibrational levels used were $v' = 19$ and $v' = 20$, which are relatively high in the potential energy well of the $B$ state. The photoexcitation process therefore addressed molecular ionization continua and superexcited states at larger internuclear distances than in the case of excitation via lower vibrational levels of the $B$ state.

Examples of two-color excitation spectra are depicted in Figures 9 and 10, showing spectra for one-photon absorption from the $B(v' = 19, N' = 2)$ and the $B(v' = 20, N' = 2)$ levels, respectively. The spectra obtained by monitoring the $H^+$ channel show a sharp onset for dissociation into $n = 3$ fragments. The $H_2^+$ spectra show intense structure below the $n = 3$ dissociation limit, which we have not assigned. At higher excitation energies, less intense structure was observed in both $H_2^+$ spectra (not shown in the figures). The structure in the $H_2^+$ spectra arises from excitation of bound states of the neutral molecule, located above the first ionization energy. These so-called superexcited states [24, 100] can decay via autoionization and predissociation.

The measurements in Figures 9 and 10 are concerned with a $(1 + 1')$ REMPI process for *ortho*-hydrogen starting from its $X^1\Sigma_g^+(v'' = 0, N'' = 1)$ ground-state. The onset for dissociation into $H(n = 3)$ fragments determined from Figure 9 is 23,779.91(6) $cm^{-1}$. The onset determined from Figure 10 is

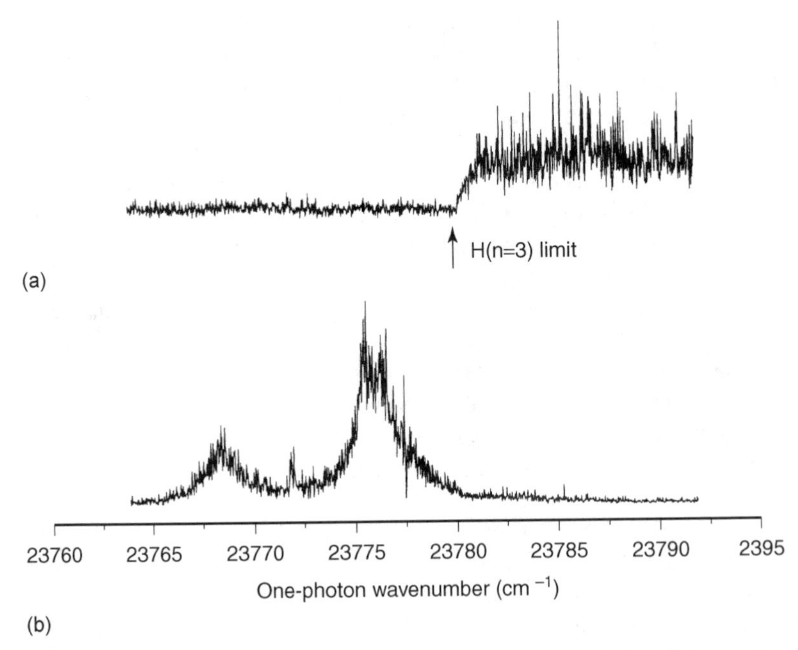


**Figure 9.** Two-color excitation spectra of molecular hydrogen obtained by fixing one laser on the $B^1\Sigma_u^+(v' = 19, N' = 2) \leftarrow X^1\Sigma_g^+(v'' = 0, N'' = 1)$ transition, while the second laser is scanned across the $H(n = 1) + H(n = 3)$ dissociation limit. The frequency of the second laser is given along the horizontal axis. In (*a*) and (*b*), the $H^+$ signal from ionization of $n = 3$ fragments and the $H_2^+$ signal are shown, respectively. (Reproduced with permission from [97], copyright 1999 Elsevier Science B.V.)

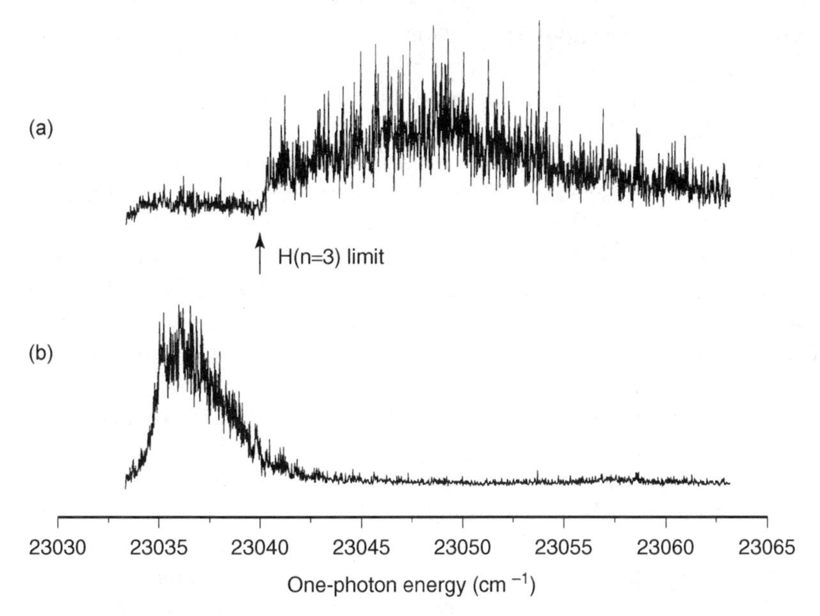


**Figure 10.** Two-color excitation spectra of molecular hydrogen obtained by fixing one laser on the $B^1\Sigma_u^+(v'=20, N'=2) \leftarrow X^1\Sigma_g^+(v''=0, N''=1)$ transition, while the second laser is scanned across the $H(n=1)+H(n=3)$ dissociation limit. The frequency of the second laser is given along the horizontal axis. In (*a*) and (*b*), the $H^+$ signal from ionization of $n=3$ fragments and the $H_2^+$ signal are shown, respectively. (Reproduced with permission from [97], copyright 1999 Elsevier Science B.V.)

23,040.02(6) cm$^{-1}$. From the rotational constants of the $X^1\Sigma_g^+(v''=0)$ state [101] and the $B^1\Sigma_u^+(v'=19,20; N'=2) \leftarrow X^1\Sigma_g^+(v''=0, N''=1)$ transition frequencies [102, 103], it can be calculated that the formation of $n=3$ fragments must occur at 133,610.28(6) or 133,610.34(6) cm$^{-1}$, above the $X^1\Sigma_g^+(v''=0, N''=0)$ ground-state, based on the onsets measured in Figures 9 and 10. The onset determined for one-photon absorption from the $B(v'=20, N'=1)$ level, excited from the ground-state of *para*-hydrogen, is 133,610.13(10) cm$^{-1}$. By using results from [101–103], it was calculated that the excitation energy required to reach the $H(n=1)+H(n=3)$ dissociation limit from the $X^1\Sigma_g^+(v''=0, N''=0)$ level ranges from 133,610.27 to 133,610.41 cm$^{-1}$ for the various fine structure levels of the $n=3$ fragments. From our results, it can be concluded that $n=3$ fragments were observed as soon as the $H(n=1)+H(n=3)$ dissociation threshold was surpassed.

The initial distribution of the $H(3\ell)$ atoms over the various angular-momentum substates is *a priori* unknown in our experiments. During the applied waiting period, the various H($3\ell$) atoms were also subject to radiative decay and possibly collisional deactivation processes with different transition probabilities. From the radiative lifetimes of the 3*s*, 3*p*, and 3*d* states [104], it can be estimated that after the applied delay time of 20 ns, 88% of the 3*s*, 2.5% of the 3*p* and 28% of the 3*d* atoms will remain. The initial populations of the various angular momentum substates of H($n = 3$) atoms, resulting from excitation between the H($n = 1$) + H($n = 3$) and the H($n = 1$) + H($n = 4$) dissociation limits by one-photon absorption from the $X^1\Sigma_g^+$ electronic ground-state, have been determined experimentally by Kouchi et al. [105] and Terazawa et al. [106] by analysis of the observed time-dependent Balmer-$\alpha$ intensity. In these experiments, states of *ungerade* symmetry are excited, while in the present experiments *gerade* states were accessed. These experimental results are therefore not directly applicable, but we can conclude that the observed $H^+$ signal mainly results from ionization of 3*s* and 3*d* atoms, which have survived radiative decay during the delay period. The experimental errors are, however, too large to distinguish between these two dissociation limits.

The onset of H(1*s*) + H($3\ell$) dissociation channels in the present experiment is influenced by the atomic interaction potentials ($V_a$) as calculated by Stephens and Dalgarno [107], and also by a centrifugal term ($V_c$) for continua with $J > 0$. The centrifugal barrier in the effective radial potential arising from the centrifugal term can be calculated as the maximum of the sum of $V_a$ and $V_c$.

All 3*s* and 3*d* configurations are dominated by attractive van der Waals interaction, although the 3*d* configurations contain a small quadrupole term as well. In contrast, the strongest contributions to the 3*p* potentials are dipole interaction terms. The interaction is repulsive for the $3p\sigma$, excluding this configuration from near-threshold dissociation. The $3p\pi$ and $3d\pi$ configurations are restricted to continuum states with $J \geq 1$; excitation of the $3d\delta$ configuration from the intermediate $B^1\Sigma_u^+$ state is impossible due to dipole selection rules. Therefore, $3s\sigma$ and $3d\sigma$ configurations may contribute to dissociation continua with angular momentum $J = 0$, excited via a $J' = 1$ intermediate state.

In continua with nonzero angular momentum of nuclear motion the centrifugal term $V_c = [J(J+1) - \Lambda^2]/2\mu R^2$ ($V$ and $R$ in atomic units) gives rise to a barrier in the effective radial potential. For $J = 1$ continuum states, the lowest possible value accessed via $J' = 2$ intermediates, the resulting centrifugal barrier heights are 0.10 cm$^{-1}$ ($3s\sigma$), 0.11 cm$^{-1}$ ($3d\sigma$), 0.0006 cm$^{-1}$ ($3p\pi$), and 0.04 cm$^{-1}$ ($3d\pi$). Continua with $J > 1$ show centrifugal barriers of more than 0.5 cm$^{-1}$, except for the $3p\pi$ configuration with 0.08 cm$^{-1}$ for $J = 2$. However, even at energies below the barrier but above the dissociation energy, tunneling can still lead to dissociation. The dissociation probability $p$ increases smoothly with the energy above threshold ($\varepsilon$), quantified by Wigner's threshold law $p \sim \varepsilon^{J+1/2}$. Therefore, the presence of centrifugal barriers may shift the onset of dissociation toward higher energies by no $> 0.05$ cm$^{-1}$, as observed from linear fits to signals arising from the intermediate states with $J' = 2$ (see Figs. 9 and 10).

In Figure 11, the $H^+$ and $H_2^+$ spectra are depicted for one-photon absorption from the $B(v' = 19, N' = 0)$ level. These spectra show that at the $H(n = 1) + H(n = 3)$ dissociation threshold, the autoionization channel closes, to be replaced by dissociation into $n = 1$ and $n = 3$ fragments. This observation is in agreement with the results from one-color $(3 + 1)$ REMPI–PES experiments in which dissociation dominates over ionization of molecular hydrogen above the $H(n = 3)$ dissociation limit [25, 68, 70]. However, at a one-photon energy of $23{,}858\,\mathrm{cm}^{-1}$, which corresponds with an energy of $133{,}630\,\mathrm{cm}^{-1}$ above the $X^1\Sigma_g^+(v'' = 0, N'' = 0)$ ground-state, some structure was still observed in the $H_2^+$ channel.

Following one-photon absorption from the $B$ state, excitation of doubly excited dissociative states with ${}^1\Sigma_g^+(2p\sigma_u)(np\sigma_u)$ and ${}^1\Pi_g(2p\sigma_u)(np\pi_u)$

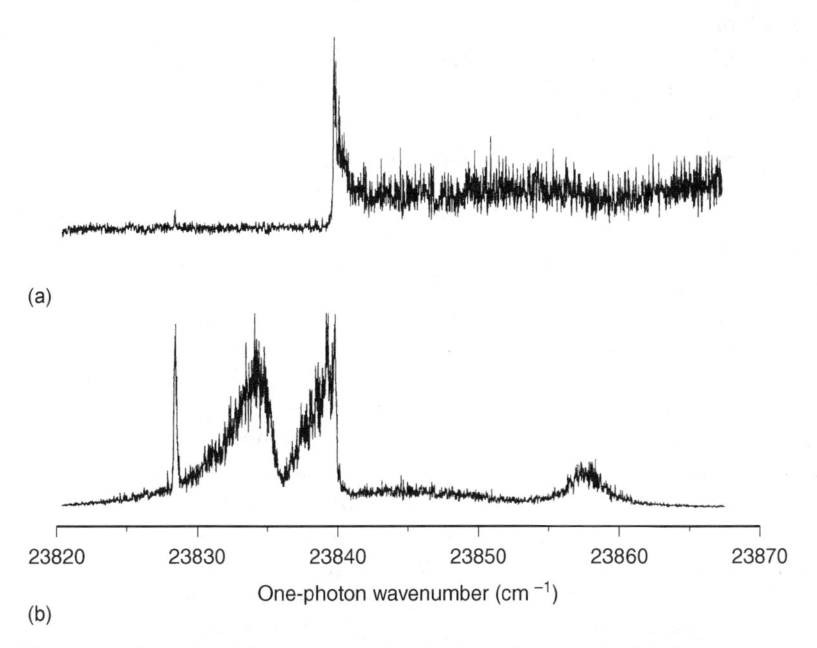


**Figure 11.** Two-color excitation spectra of molecular hydrogen obtained by fixing one laser on the $B^1\Sigma_u^+(v' = 19, N' = 0) \leftarrow X^1\Sigma_g^+(v'' = 0, N'' = 1)$ transition, while the second laser is scanned across the $H(n = 1) + H(n = 3)$ dissociation limit. The frequency of the second laser is given along the horizontal axis. In (*a*) and (*b*) the $H^+$ signal from ionization of $n = 3$ fragments and the $H_2^+$ signal are shown, respectively. This example clearly shows the closing of the autoionization channel on a broad resonance of $H_2$ at the $H(n = 3)$ limit. (Reproduced with permission from [97], copyright 1999 Elsevier Science B.V.)

configurations should be considered. Excitation of the $Q_1(2)(2p\sigma_u)(3p\sigma_u)$ or the $Q_1(2)(2p\sigma_u)(3p\pi_u)$ state may lead to dissociation into H($n = 3$) fragments. The $(2p\sigma_u)(3p\pi_u)$ potential is repulsive at all internuclear separations $R$. As a consequence, for a transition from any bound state into the dissociative continuum the Franck–Condon factor vanishes as the excitation energy approaches the dissociation threshold. In contrast, the $(2p\sigma_u)(3p\sigma_u)$ potential is slightly bound with respect to the $n = 3$ dissociation limit for $R > 6$ a.u. [91], giving rise to Franck–Condon overlap with the $B\ v' = 19$ and 20 states with an outer classical turning point of $\sim 8$ a.u.

The observation of $n = 3$ fragments in this energy region can be explained, alternatively, by the excitation of vibrational continua of singly excited Rydberg states converging upon the ground ionic state $X^2\Sigma_g^+$. By one-photon absorption from the $B$ state, Rydberg states with ${}^1\Sigma_g^+(1s\sigma_g)(ns\sigma_g)$, ${}^1\Sigma_g^+(1s\sigma_g)(nd\sigma_g)$, and ${}^1\Pi_g(1s\sigma_g)(nd\pi_g)$ configurations can be accessed. Excitation of vibrational continua of those Rydberg states that possess a H($n = 1$) + H($n = 3$) dissociation limit results in the observation of $n = 3$ fragments. In a diabatic picture, the dissociation of the molecule following excitation of the doubly excited $Q_1(1)\ {}^1\Sigma_g^+(2p\sigma_u)^2$ or $Q_1(1)\ {}^1\Pi_g(2p\sigma_u)(2p\pi_u)$ states results in $n = 2$ excited fragments. Interaction with Rydberg states that exhibit (diabatic) potential crossings with the doubly excited states, however, may redistribute the flux in the exit channels. This has been observed in DR experiments where exclusive population of the $Q_1(1)\ {}^1\Sigma_g^+(2p\sigma_u)^2$ state leads to considerable population in all energetically allowed dissociation channels [60]. Since $(3 + 1)$ REMPI–PES experiments have shown unambiguously that $n = 2$ dissociation products completely disappear above the $n = 3$ dissociation limit [25, 70], the dissociation into $n = 3$ fragments via the dissociation mechanism that involves doubly excited states with a $n = 2$ dissociation limit can be excluded.

We therefore conclude that the formation of $n = 3$ fragments at the H($n = 1$) + H($n = 3$) dissociation threshold and at energies just above this limit either originates from the long-range part of a higher doubly excited state, or from direct excitation of vibrational continua of singly excited Rydberg states. The latter interpretation of our experimental results is supported by experiments in which the Balmer-$\alpha$ radiation of $n = 3$ fragments was monitored as a function of the excitation wavelength [108, 109]. This radiation was observed as soon as the H($n = 3$) dissociation threshold is reached by one-photon absorption from the $X^1\Sigma_g^+$ ground-state. In this way, *ungerade* states are excited at small internuclear distances, so that doubly excited states with a $n = 3$ dissociation limit cannot be accessed at these excitation energies. The spectra obtained in this experiment, in combination with other experimental results, lead us to conclude that a very significant contribution to the production of $n = 3$ fragments at the H($n = 1$) + H($n = 3$) dissociation threshold originates from the direct excitation of vibrational continua of singly excited Rydberg states.

## IV. SHORT-LIVED DIATOMIC RADICALS

The great sensitivity and selectivity with which laser PES can be performed makes the technique perfectly suitable for the study of excited electronic states of small short-lived radicals. In this section, several examples of radical species of atmospheric and interstellar significance will be discussed. Excited states, which were hitherto unobserved can be detected and studied in significant detail. For these small molecules in particular, the interplay between careful experimentation and sophisticated quantum chemical calculations is extremely fruitful. The assignment and characterization of molecular excited states leads to a detailed understanding of the electronic structure of the species involved. In this context, the feature to demonstrate experimentally the occurrence of orbital interactions is a valuable asset. Often these excited states of small molecules can be viewed as a mini-laboratory where interesting physical phenomena are demonstrated in a way, is not obscured or complicated by additional effects often dominate the picture in the case of larger molecules. Laser PES is capable of studying such physical processes and the way in which they often compete. Also, for diatomic hydrides their relatively large rotational constants often allow rotationally resolved PES. Under these circumstances, detailed angular momentum considerations can provide much fundamental insight in the dynamics of the photoionization process. Finally, since short-lived radicals are usually generated in a photofragmentation process of a suitable parent compound, laser PES can make contributions to the understanding of photodissociation processes.

### A. The OH Radical

The hydroxyl radical OH plays an important role in a wide variety of chemical and physical processes occurring in environments, which range from liquid solution to interstellar space [110–115]. It is therefore important to determine the spectroscopic, physical, and dynamical properties of as many of its excited electronic states as possible. Many studies involving the OH radical have used laser-induced fluorescence (LIF) for its detection. Single-photon excitation of the 0–0 band of the $A^2\Sigma^+ \leftarrow X^2\Pi$ system at $\sim 308\,\text{nm}$ has proved to be convenient for this purpose.

Only a limited number of five excited states is known for the OH radical [116]. The ground configuration of OH is $(1\sigma)^2(2\sigma)^2(3\sigma)^2(1\pi)^3$, which leads to the $X^2\Pi$ ground-state. The ground-state of the radical is spin–orbit split, with the $\Omega = \frac{3}{2}$ component located $123\,\text{cm}^{-1}$ below the $\Omega = \frac{1}{2}$ level. The $A$, $B$, and $C$ states are all of ${}^2\Sigma^+$ symmetry, and are valence states of which the $A$ state is the best characterized. The only known Rydberg state until the present laser PES study was performed, was the $D^2\Sigma^-$ state. Here we shall focus on the $D$ state, and on the detection and characterization of a novel state of $^2\Sigma^-$ symmetry, the $3^2\Sigma^-$ state. In a $(2+1)$ REMPI–PES study, we shall map these $^2\Sigma^-$ excited states onto the lowest ionic state of $OH^+$ with a configuration $(1\sigma)^2(2\sigma)^2(3\sigma)^2(1\pi)^2$ leading to the $X^3\Sigma^-$ ground-state.

*Ab initio* calculations by van Dishoeck and Dalgarno [116] showed that the $3s\sigma$ state is the lowest $^2\Sigma^-$ state that is repulsive in nature. The $\Sigma^-$ states can be formed by adding $\sigma$ Rydberg orbitals to the $X^3\Sigma^-$ ion core. Rydberg orbitals of diatomic hydrides evolve rapidly into their united or separated atom limits over a range of internuclear distances spanned by low vibrational levels. The angular momentum composition of the Rydberg orbitals therefore varies rapidly with $R$ and strong mixing is not unusual. The $D^2\Sigma^-$ state is the lowest stable Rydberg state with configuration $(1\sigma)^2(2\sigma)^2(3\sigma)^2(1\pi)^2(3p\sigma)^1$ at the equilibrium internuclear distance. This state was first observed by Douglas in one-photon absorption [117] and lies 81,797.95 $\text{cm}^{-1}$ above the ground-state. *Ab initio* calculations indicate that the next lowest $^2\Sigma^-$ Rydberg state, designated as $3^2\Sigma^-$, is located $\sim 0.8\,\text{eV}$ above the $D$ state and possesses a $4s\sigma$ Rydberg orbital. For a state that conforms to Hund's case (b) the two-photon selection rules are $\Delta N = 0, \pm 1, \pm 2$. Transitions not obeying the $\Delta N$ selection rules only have appreciable probability as long as $N$ is not a good quantum number.

Hydroxyl radicals in their $X^2\Pi$ ground-state were produced via photodissociation of hydrogen peroxide ($H_2O_2$), or formic acid (HCOOH) [118]. The excess energy available in the fragmentation process will be distributed over the degrees of freedom. The internal state distribution upon photolysis is not clear *a priori.* When hydrogen peroxide is used it has been shown [119] that with a photolysis wavelength of 248-nm OH is formed in its electronic and vibrational ground-state, but with significant rotational excitation. When formic acid is photolyzed at 222 nm, OH is predominantly formed in its electronic and vibrational ground-state, and with a much slighter degree of rotational excitation [120, 121]. Since all the experiments are carried out with only one laser, the dissociation wavelength is not constant. Therefore both the internal state distribution and total yield of OH may vary as the wavelength is scanned. Hydrogen peroxide is known to possess an absorption continuum extending from 280 to $< 200$ nm [122, 123], with the cross-section showing a smooth increase toward shorter wavelength. The OH production is not expected to be strongly influenced under these conditions. Formic acid is known to dissociate via its first $^1A''$ excited state in the 260–200 nm region [124]. This absorption band exhibits fairly well resolved vibrational structure superimposed on a diffuse background, which may lead to wavelength dependent OH production.

In Figure 12, the excitation spectrum of OH in the two-photon energy range 82,500–80,750 $\text{cm}^{-1}$ is presented. The radical was produced via photodissociation of $H_2O_2$, and the spectrum was obtained by measuring the total electron current. The assignment of the various rotational branches is given above the

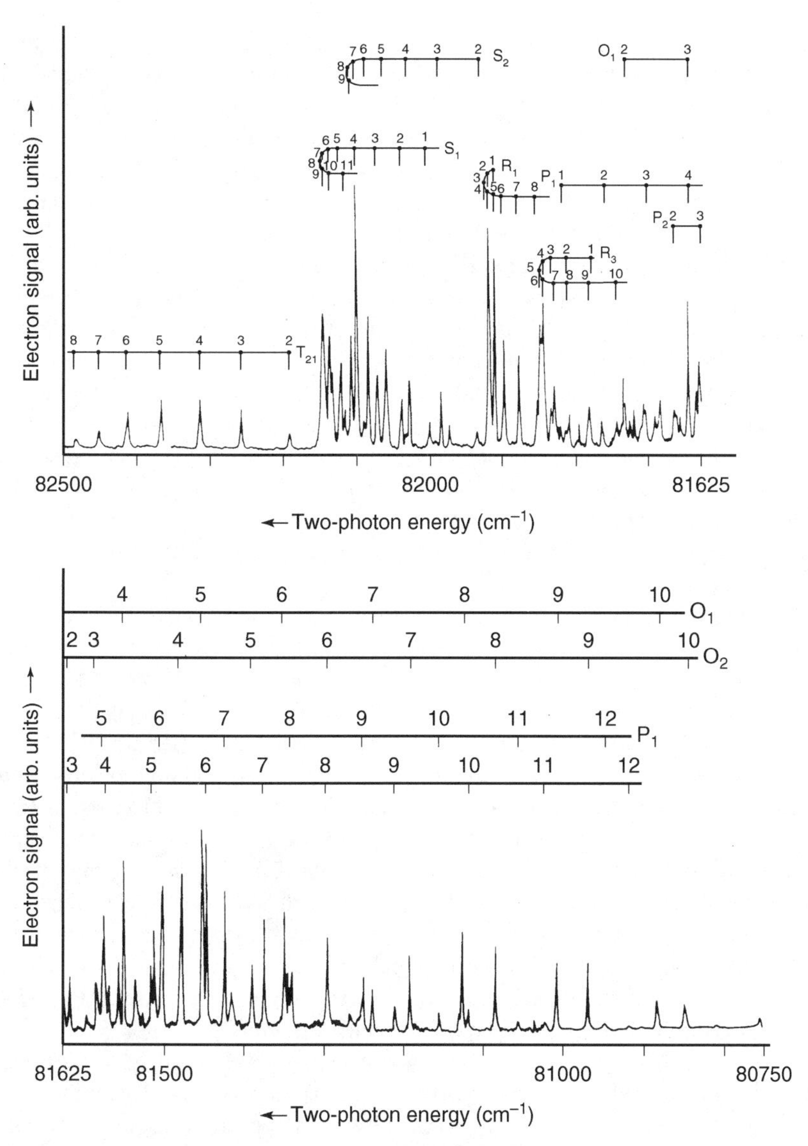


**Figure 12.** The $(2 + 1)$ REMPI spectrum of the $D^2\Sigma^-(v' = 0)$ Rydberg state of OH produced from photodissociation of hydrogen peroxide. The spectrum was obtained by monitoring the total electron count. The $Q_1$, $Q_2$, and $N_{12}$ branches have not been indicated. (Reproduced with permission from [118], copyright 1991 American Institute of Physics).

spectrum. This excitation spectrum represents OH in its $D^2\Sigma^-(v'=0)$ Rydberg state. Similar spectra were obtained for the transition from the electronic ground-state to $D^2\Sigma^-(v'=1)$ in the two-photon energy range 85,000–82,750 cm$^{-1}$, and to $D^2\Sigma^-(v'=2)$ in the two-photon energy range 87,250–85,500 cm$^{-1}$. In Figure 13, the OH excitation spectrum in the two-photon energy range 88,000–86,500 cm$^{-1}$ of a hitherto unknown state, assigned as $3^2\Sigma^-(v'=0)$ is presented. The spectrum is again obtained by employing $H_2O_2$ as the precursor and by monitoring the total electron current. The rotational structure was again analyzed and assignments are indicated above the spectrum. In Figure 14 the assigned $(2+1)$ REMPI spectrum of the $3^2\Sigma^-(v'=0)$ state of the OD radical, produced from photodissociation of HCOOD, is presented in the two-photon energy range 87,920–87,200 cm$^{-1}$. This spectrum was obtained by measuring the $OD^+$ ion count. When the excitation spectra are simulated on the basis of reasonable assumptions, the spectra obtained from the photodissociation of $H_2O_2$ suggest an OH rotational temperature in agreement with that obtained before [119]. For OH produced from formic acid, the simulated rotational temperatures seem consistently lower than those obtained previously [120, 121].

By fixing the laser wavelength at a particular transition in the excitation spectrum and by analyzing the electrons according to their kinetic energies, PE spectra can be obtained. The PE spectra for the $D^2\Sigma^-$ and $3^2\Sigma^-$ states measured for transitions involving low rotational quantum numbers show the $\Delta v = 0$ single line behavior expected for ionization of unperturbed Rydberg states possessing a $^3\Sigma^-$ core. No electron signals corresponding to ionizations into higher ionic states were observed. The OH radicals produced from fragmentation of $H_2O_2$ appear to be translationally and rotationally hot. The appreciable translational energy leads to significant Doppler broadening of the observed PE bands. The rotational excitation with which the OH radicals are formed has the interesting consequence that for sufficiently high values of the rotational quantum number the rotational spacing in the ion will exceed the observed translational broadening, thus allowing rotationally resolved PES [51].

Figure 15($a$) and ($c$) show PE spectra for the $X^3\Sigma^-(v^+=0, N^+) \leftarrow D^2\Sigma^-(v'=0, N'=9)$ and $X^3\Sigma^-(v^+=0, N^+) \leftarrow 3^2\Sigma^-(v'=0, N'=9)$ transitions for the $O_{11}(11)$ rotational branch. For ionization of these Rydberg levels, the ion rotational distribution is governed by the propensity rule $\Delta N + \ell = odd$, where $\Delta N = N^+ - N'$ is the change of rotational quantum number and $\ell$ a partial wave component of the photoelectron [72, 125–127]. For ionization of the $D^2\Sigma^-$ state, we observe strong $\Delta N = even$ signals in the PE spectrum, in contrast to the $\Delta N = odd$ signals expected for ionization of a $3p\sigma$ Rydberg electron in an atomic-like picture, that is, $3p\sigma \rightarrow ks, kd$. Figure 15($c$) shows a rotationally resolved REMPI–PES spectrum for the $3^2\Sigma^-$ state, which reveals a qualitatively different and much broader distribution of ionic rotational levels.

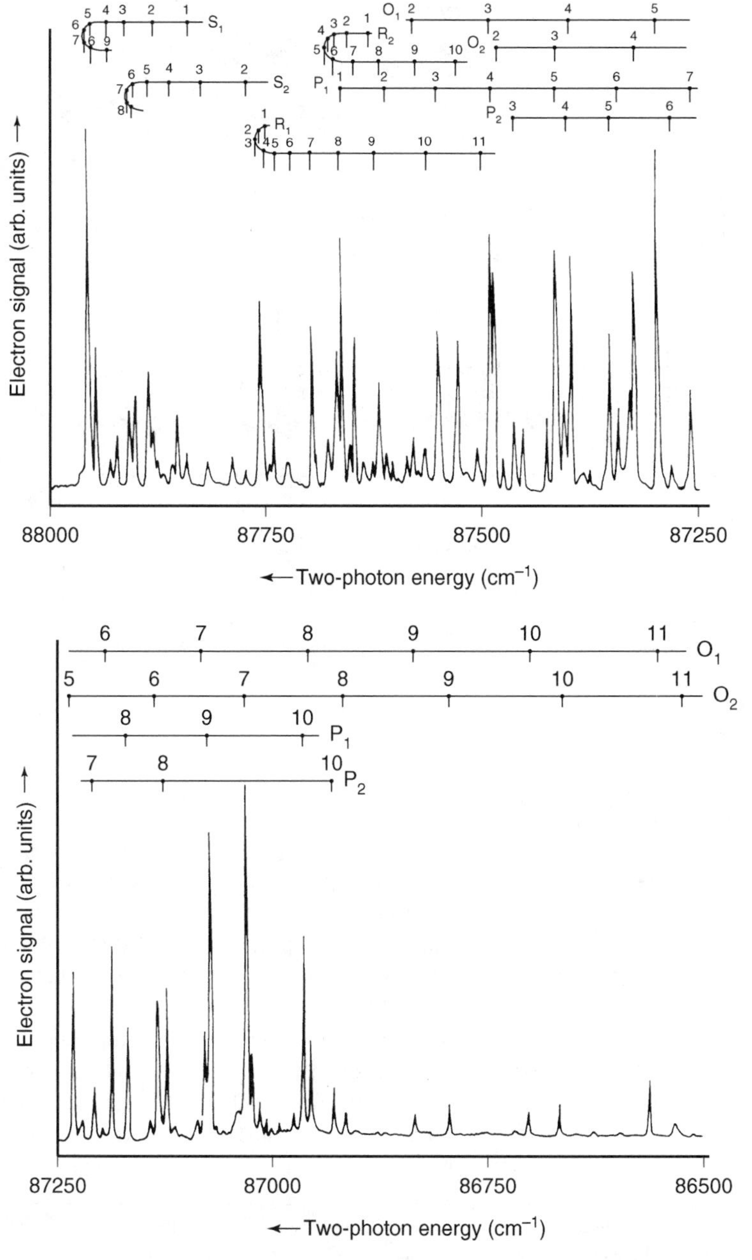


**Figure 13.** The $(2+1)$ REMPI spectrum of the $3^2\Sigma^-(v'=0)$ Rydberg state of OH produced from photodissociation of hydrogen peroxide. The spectrum was obtained by monitoring the total electron count. The $Q_1$, $Q_2$, and $N_{12}$ branches have not been indicated. (Reproduced with permission from [118], copyright 1991 American Institute of Physics.)

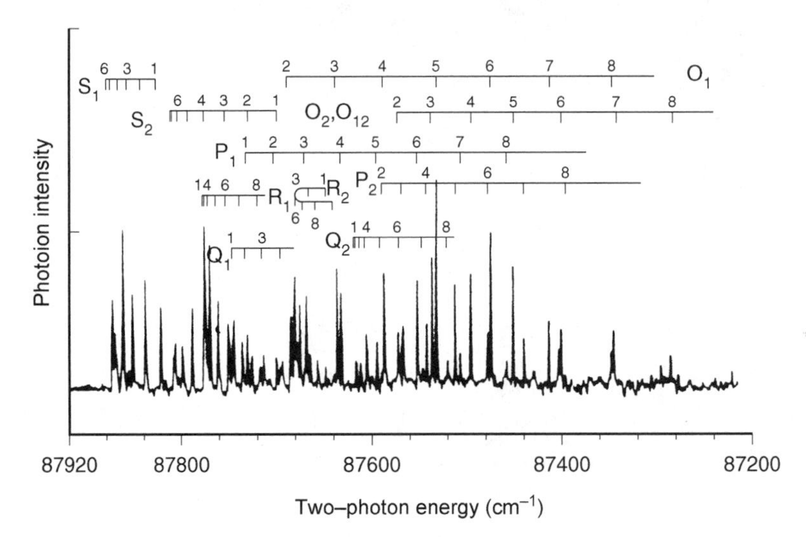


**Figure 14.** The $(2+1)$ REMPI spectrum of the $3^2\Sigma^-(v'=0) \leftarrow X^2\Pi(v''=0)$ transition of OD produced from photodissociation of HCOOD. The spectrum was obtained by monitoring the $OD^+$ ion count. (Reproduced with permission from [118], copyright 1991 American Institute of Physics.)

In contrast to transitions observed for the $D^2\Sigma^-$ state, both $\Delta N = even$ and $\Delta N = odd$ transitions are prominent.

Figure 15(*b*) and (*d*) show results of *ab initio* calculations of the rotationally resolved PE spectra to be compared with experiment. In these calculations, the resonant intermediate state and electronic continuum final states are treated at the Hartree–Fock level using multiplet-specific potentials. The effect of vibrational motion and the alignment induced by a two-photon absorption step are also included in these studies [127–129]. We have also studied the effect of alignment induced in the hydroxyl $X^2\Pi$ fragment via photofragmentation [130, 131] on these rotational distributions and found that this had no influence on the present spectra. Analysis of the results for the $D^2\Sigma^-$ state reveals that this rotational distribution arises from a Cooper minimum in the $3p\sigma \rightarrow k\pi(\ell=2)$

**Figure 15.** Experimental and calculated rotationally resolved PE spectra for $(2+1)$ REMPI of OH. (*a*) Observed spectra for the $D^2\Sigma^-$ state, $v^+=0 \leftarrow v'=0$, $O_{11}(11)$ rotational branch; (*b*) calculated $D^2\Sigma^-(3p\sigma)$ PE spectra, assuming a Gaussian line shape with a WHM of 30 meV; (*c*) observed spectra for the $3^2\Sigma^-$ state, $v^+=0 \leftarrow v'=0$, $O_{11}(11)$ rotational branch; (*d*) calculated $3^2\Sigma^-(4s\sigma)$ PE spectra, assuming a Gaussian line shape with a WHM of 35 meV. The labeling of peaks in the calculated spectra indicates the change of rotational quantum number $\Delta N = N^+ - N'$. (Reproduced with permission from [51], copyright 1991 American Institute of Physics.)

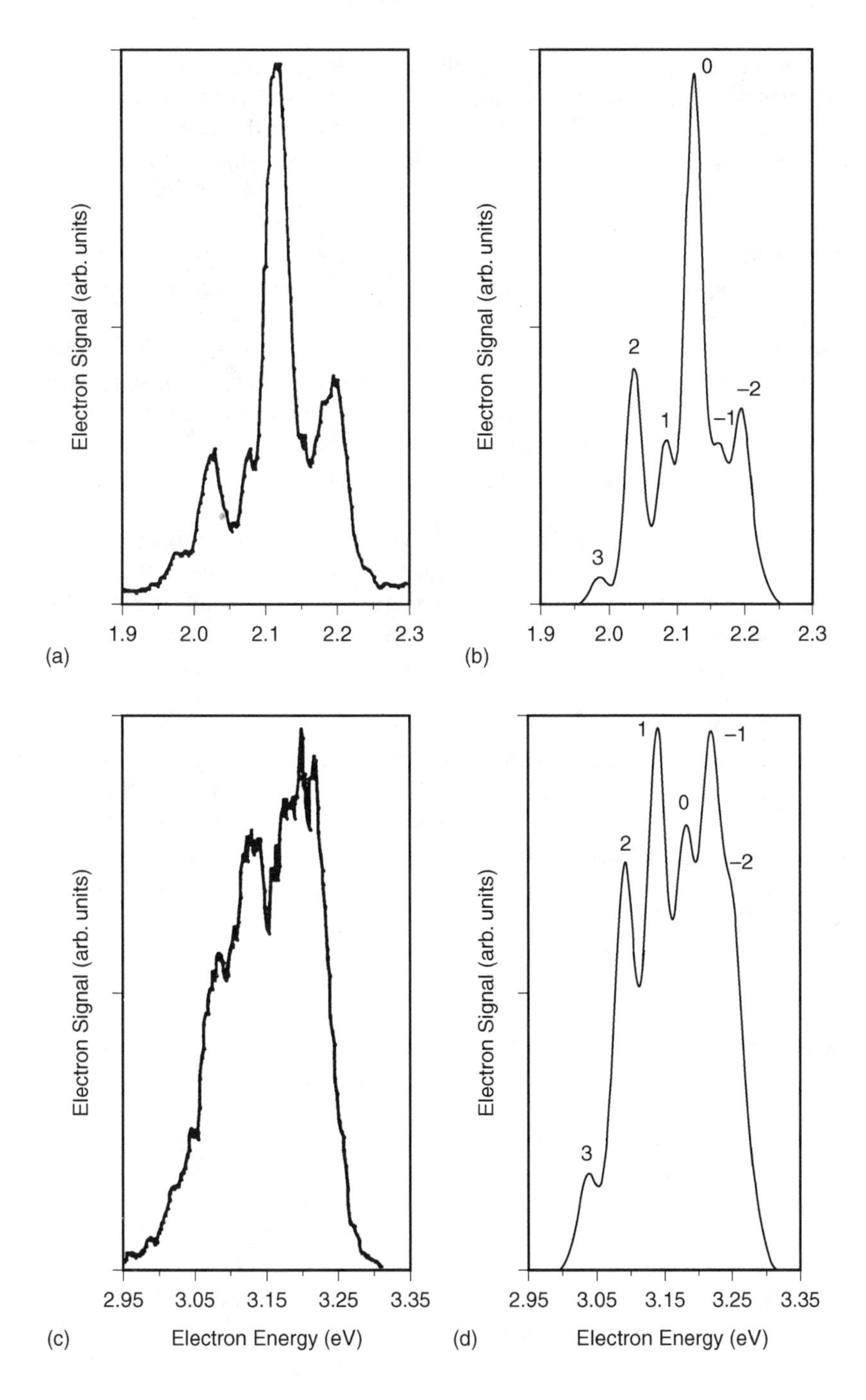

Electron Signal (arb. units)
1.9
2.0
2.1
2.2
2.3
(a)
0
2
1
−1
−2
3
(b)
2.95
3.05
3.15
3.25
3.35
Electron Energy (eV)
(c)
(d)

ionization channel [128, 129] near the PE kinetic energies accessed in our experiments. As the PE energy increases from threshold, severe cancelations occur in the $3p\sigma \rightarrow k\pi(\ell = 2)$ channel, at internuclear distances less than $\sim 2$ $a_0$. Since the $\ell = 2$ continuum wave becomes depleted upon formation of the Cooper minimum, the odd PE waves from the $3p\sigma \rightarrow k\sigma$ channel dominate. The strong $\Delta N = 0$ peak is associated mainly with the $\ell = 1$ ($p$ wave) component of the PE orbital. At $R_e = 2.043a_0$ the "$3p\sigma$" Rydberg wave function is 63.3% $p$ and 34.9% $s$ character, and the $3s\sigma \rightarrow kp\sigma$ component contributes substantially to the total transition moment. The continuum $\ell = 3$ ($f$ wave) is also important, and contributes principally to the weaker $\Delta N = \pm 2$ peaks of Figure 15($a$). The agreement between experimental, Figure 15($a$), and theoretical, Figure 15($b$), PE spectra is encouraging. Note that the $p$ and $s$ contributions to the "$3p\sigma$" Rydberg wave function are associated with the same core, hence leading to a single transition (apart from rotational structure) in the PE spectra.

Experimental and theoretical PE spectra for photoionization of the $3^2\Sigma^-(v' = 0)$ state are shown in Figure 15($c$) and ($d$). In contrast to the $D^2\Sigma^-$ state, the PE spectra reveal a broader spectral distribution for rotational transitions $\Delta N = 0, \pm 1, \pm 2, \pm 3$. The appearance of this spectrum arises from the greater degree of $\ell$ mixing in the higher Rydberg orbital (54.3% $s$ and 42.7% $p$ at $R = 2.043a_0$) and much weaker Cooper minima for photoionization from this state [129].

Cooper minima in the REMPI–PES spectra of diatomic molecules were first predicted for the $D^2\Sigma^-$ Rydberg state of OH [59, 128, 129]. The close connection between Cooper minima and rotational state selection in REMPI–PES was identified using the example of NO [132–134]. The present work provides the first experimental evidence of these features in molecular REMPI–PES.

### B. The SH Radical

The SH radical is a molecule of environmental importance and plays a role in the UV photochemistry of various sulfur-containing compounds, which are released into the earth's atmosphere from natural and anthropogenic sources. In view of its similarity to the abundant OH radical, the mercapto radical may well have astrophysical relevance [135]. Several sulfur-containing molecules have already been observed in interstellar space, but despite serious attempts to detect SH [136, 137], the molecule until now has remained elusive under extraterrestrial conditions.

When $H_2S$ is photolyzed following excitation to its first dissociative absorption band in the wavelength range between 270 and 180 nm [138], one of the fragments formed is the SH radical. The SH radical is generated in its $^2\Pi$ electronic ground-state resulting from a $(1\sigma)^2(2\sigma)^2(3\sigma)^2(1\pi)^4(4\sigma)^2(5\sigma)^2(2\pi)^3$ electron configuration. This ground-state is spin–orbit split and possesses an inverted splitting with $^2\Pi_{3/2}$ below $^2\Pi_{1/2}$ due to a spin-orbit constant of $A_0'' = -376.835\,\text{cm}^{-1}$ [139]. At near-UV photodissociation wavelengths, SH fragments are formed in their ground vibrational level with a $\sim$ 3:2 ratio over the $^2\Pi_{3/2}$ and $^2\Pi_{1/2}$ components, and with a rotational distribution that can be described by a Boltzmann temperature of $\sim$ 300 K. Removal of an outer electron that is predominantly localized on the sulfur atom gives rise to a $\ldots(2\pi)^2$ electron configuration that leads to three close-lying ionic states, an $X^3\Sigma^-$ ground-state, and $a^1\Delta$ and $b^1\Sigma^+$ excited states [140–143].

The electronic absorption spectrum of SH has been investigated from the near-UV into the vacuum UV region, and 10 electronic transitions, all arising from the ground-state, have been identified [3, 144, 145]. With $(2 + 1)$ laser PES 14 Rydberg states of SH (SD) were investigated that had not been observed before, and their spectroscopic properties and/or photoionization dynamics were studied. All these states have in common that they are Rydberg states based on an excited ionic core, either $a^1\Delta$ or $b^1\Sigma^+$. Some of these states are located below the lowest ionization limit, but several are situated above the $X^3\Sigma^-$ ground ionic threshold. A summary of these newly observed states is given in Figure 16 [40, 41].

Next, we shall focus on a number of excited and superexcited states of the SH radical that demonstrate specific physical phenomena. We shall commence with the $[b^1\Sigma^+]4p\sigma\ ^2\Sigma^+(v' = 0)$ and $[a^1\Delta]3d\delta\ ^2\Sigma^+(v' = 1)$ states, both located below the lowest ionic threshold, which serve as illustrative examples of how laser PES can make electronic interactions between molecular orbitals of the same overall symmetry and with different cores experimentally "visible". Then, the $[a^1\Delta]3d\pi\ ^2\Phi(v' = 0)$ state, also situated below the lowest $X^3\Sigma^-$ ionic threshold, will be discussed. This state shows rotationally resolved PE spectra with large asymmetries apparent in the ionic distributions [54]. With the help of *ab initio* calculations the observed features can be understood in detail. In a subsequent ZEKE–PFI study, involving for the first time an excited ionic state of a radical species, the same state has been studied under conditions of much better resolution [140]. Subsequently, we shall focus on a number of superexcited states located above the $X^3\Sigma^-$ or $a^1\Delta$ ionic thresholds. As representative examples of superexcited states of SH with $a^1\Delta$ or $b^1\Sigma^+$ cores, the $[a^1\Delta]5p\pi\ ^2\Phi(v' = 0)$ state [24, 41, 53] and the $[b^1\Sigma^+]3d$, $[b^1\Sigma^+]5p$, $[b^1\Sigma^+]4d$, and $[b^1\Sigma^+]4f$ complexes will be discussed [24, 40, 41, 53]. All these states appear to be remarkably long lived and, in contrast to the examples of $H_2$ discussed above, their two-photon excitation spectra above the lowest ionic limit can be observed directly.

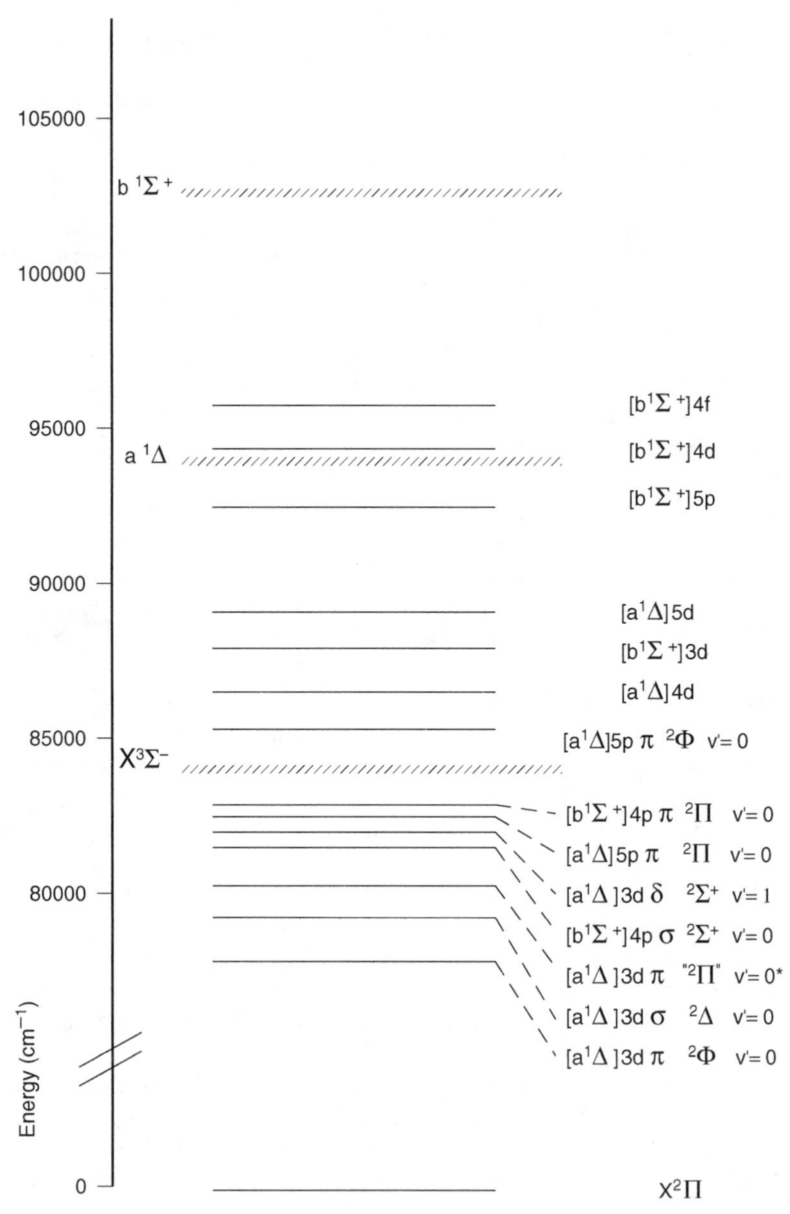


**Figure 16.** Energy level diagram showing 14 newly observed Rydberg states of SH and their relevant ionization limits. Most of these states have also been observed for SD. The state marked with an asterisk was only seen for SD. (Reproduced with permission from [40], copyright 1996 American Institute of Physics.)

### 1. *Electronic Interaction between the SH* $[b^1\Sigma^+]4p\sigma\ ^2\Sigma^+(v'=0)$ *and* $[a^1\Delta]3d\delta\ ^2\Sigma^+(v'=1)$ *States*

In recent years, there has been an increasing interest in molecular Rydberg states, motivated to a considerable extent because of the essential role played by such states as long-lived stepping stones in ZEKE–PFI studies. Laser PES is the method of choice for studying molecular Rydberg states. Since Rydberg states and the ionic state upon which the series of Rydberg states converge are very similar in shape and position, the PE spectrum is usually dominated by virtually diagonal Franck–Condon factors leading to a one-line spectrum. Such very simple PE spectra can therefore be used as a diagnostic tool for the reliable assignment of Rydberg states. On the other hand, when deviations from the expected one-line spectra are observed, such deviations usually arise from orbital interactions between two or more states with different cores in a rather straightforward fashion. It is exactly this feature of laser PES, which has developed into a unique method for the direct experimental observation of electronic and vibronic coupling between different molecular states.

In a simple two-state interaction model molecular orbitals of the same overall symmetry $\Psi_1$ and $\Psi_2$ can couple to form new states $\Phi_1$ and $\Phi_2$ when the energy separation between the levels is small and when a term in the Hamiltonian, which has not been considered explicitly so far (e.g., electron repulsion, vibronic coupling), possesses significant coupling matrix elements between the wave functions.

$$\Phi_1 = \Psi_1 + \Psi_2 < \Psi_1|H^{\text{interaction}}|\Psi_2 > /\Delta E$$

$$\Phi_2 = \Psi_2 + \Psi_1 < \Psi_2|H^{\text{interaction}}|\Psi_1 > /\Delta E$$

The PE spectra expected before the orbital coupling is switched on depend on the (square of the) electric dipole matrix elements between the excited states and the manifold of ionic states $\langle\Psi_1|\mu|\Psi^+_{\text{ion}};\ \Psi_{\text{photoelectron}}\rangle$ and $\langle\Psi_2|\mu|\Psi^+_{\text{ion}};\ \Psi_{\text{photoelectron}}\rangle$, and for pure Rydberg states consist of single lines. When configuration interaction becomes operative, the matrix elements $\langle\Phi_1|\mu|\Psi^+_{\text{ion}};\ \Psi_{\text{photoelectron}}\rangle$ and $\langle\Phi_2|\mu|\Psi^+_{\text{ion}};\ \Psi_{\text{photoelectron}}\rangle$ must now be considered for the two states. The ensuing PE spectra then generally will no longer consist of single transitions, but will show satellite lines. The intensities of these satellite lines reflect the (square of the) mixing coefficients. Here, we shall focus on a representative experimental example of the observation of orbital interaction arising from electronic coupling between two states in the SH radical. Later in this contribution electronic coupling in the NH radical, and vibronic coupling in somewhat larger molecules such as methanethiol, $HSCH_3$, dimethyl sulfide, $S(CH_3)_2$, and $N_2O$ will be discussed.

In the two-photon energy range between 81,200 and 82,750 $\mathrm{cm}^{-1}$, two partly overlapping rotationally resolved excitation spectra of the SH radical were found. When only photoelectrons with a kinetic energy of $\sim 2.5$ eV were monitored, a $(2+1)$ process in which $SH^+$ is formed in its second excited $b^1\Sigma^+$ ionic state was observed. A rotational analysis of this excitation spectrum and quantum defect considerations indicate that the corresponding state can be assigned as $[b^1\Sigma^+]4p\sigma\ ^2\Sigma^+$. The PE spectrum of this state is presented in Figure 17 and shows a number of unusual features. The leading transition represents the expected peak corresponding to the core-preserving transition to the vibrationless level of the $b^1\Sigma^+$ state. However, weaker peaks whose kinetic energies are consistent with ionization into the $v^+ = 1$ level of the same ionic state, as well as transitions associated with the formation of $SH^+$ in its lower lying $a^1\Delta(v^+ = 0, 1)$ ionic state are observed as well. The PE spectrum provides the clear signature of a mixed configuration, and labels the state predominantly as a $[b^1\Sigma^+]4p\sigma\ ^2\Sigma^+(v' = 0)$ configuration, but mixed with a state with an $a^1\Delta$ ionic core in its $v' = 1$ level. Considering the energy region, and assuming an electronic interaction, a perturber which also possesses $^2\Sigma^+$ symmetry is obviously called for. A likely candidate on energy grounds would be the $[a^1\Delta]3d\delta\ ^2\Sigma^+(v' = 1)$ state, but it would be worthwhile if the inferred presence of this state could be confirmed by direct detection [24, 40, 41].

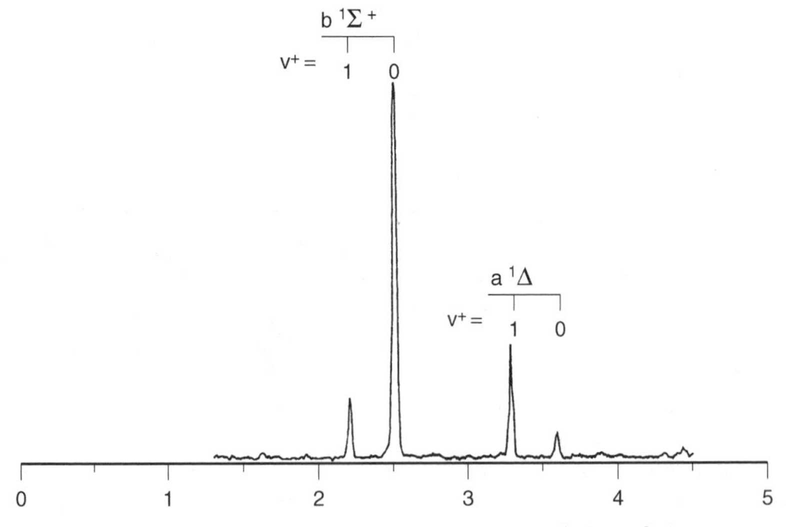


**Figure 17.** PE spectrum obtained for $(2+1)$ ionization via the $[b^1\Sigma^+]4p\sigma\ ^2\Sigma^+(v' = 0)$ state of SH, taken at a one-photon energy of 40,908 $\mathrm{cm}^{-1}$. (Reproduced with permission from [40], copyright 1996 American Institute of Physics.)

By monitoring photoelectrons with a kinetic energy of $\sim 3.3\,\mathrm{eV}$ in the two-photon energy region between 81,100–81,750 $\mathrm{cm}^{-1}$ for the SD radical a rotationally resolved excitation spectrum was observed. For SH a weaker and somewhat more fragmentary structure was seen $\sim 800\,\mathrm{cm}^{-1}$ to higher energy. The PE spectrum of this SH state is presented in Figure 18 and again shows the signature of a state with mixed electronic character. Moreover, this spectrum readily confirms the presence of the $[a^1\Delta]3d\delta\ ^2\Sigma^+(v' = 1)$ state inferred above. Not only is a strong peak present that represents the core-preserving transition to the $a^1\Delta(v^+ = 1)$ ionic state, but also transitions to the $[b^1\Sigma^+](v^+ = 0, 1)$ states are apparent. The direct observation of the interacting states through their corresponding PE spectra shows the unique capacity of laser PES in experimentally elucidating the effects of configuration interaction [24, 40, 41]. It is tempting to try to extract the wave function composition of the two interacting states in a more quantitative manner also. However, the spectral intensities associated with the branching into the different ionic states depend on the unknown relative cross-sections for ionization to the two ionic cores.

### 2. *The SH* $[a^1\Delta]3d\pi\ ^2\Phi(v' = 0)$ *State*

A rotationally resolved REMPI–PES study of the SH radical in its $[a^1\Delta]3d\pi\ ^2\Phi(v' = 0)$ state is an illustrative example of the wealth of information

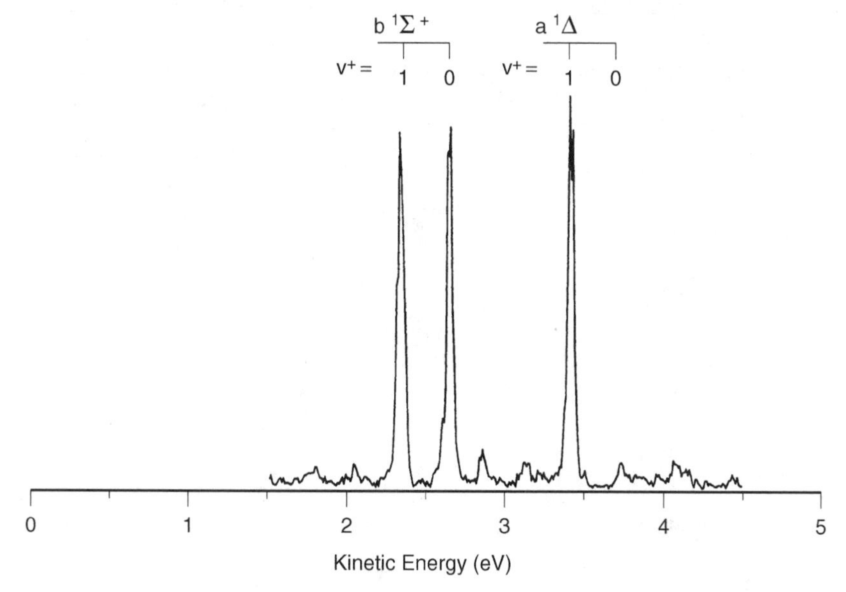


**Figure 18.** PE spectrum obtained for $(2 + 1)$ ionization via the $[a^1\Delta]3d\sigma\ ^2\Sigma^+(v' = 1)$ state of SH, taken at a one-photon energy of 41,255 $\mathrm{cm}^{-1}$. (Reproduced with permission from [40], copyright 1996 American Institute of Physics.)

that can be obtained from the interplay between such experiments and sophisticated *ab initio* quantum-chemical calculations [17, 51, 52, 56, 146]. Figure 19 shows the two-photon excitation spectrum of the SH radical in the two-photon energy range of 77,500–78,500 $cm^{-1}$ obtained by monitoring photoelectrons with a kinetic energy of $\sim$ 2.9 eV. These electrons derive from a core-preserving photoionization process in which a $(2+1)$ photon transition from the $X^2\Pi(v''=0)$ ground-state to the $a^1\Delta(v^+=0)$ excited ionic state occurs. At the two-photon level, resonance enhancement via the $[a^1\Delta]3d\pi\ ^2\Phi(v'=0)$ Rydberg state takes place. Spectra obtained by time-of-flight ion detection in which $SH^+$ ions are monitored are essentially identical to those measured with electron detection, thus supporting the assignment. The excitation spectrum shows clear substructure with energy separations corresponding to the well-known spin–orbit splitting of the ground-state. In addition, abundant rotational structure is observed. Combs above the experimental spectra indicate the assignments of the various rotational branches.

Figure 20 shows the measured (left column) and calculated (right column) PE spectra obtained in a $(2+1)$ photoionization process via the $S_{11}(\frac{3}{2})$ to $S_{11}(\frac{11}{2})$, $S_{11}(\frac{15}{2})$, and $S_{11}(\frac{19}{2})$ rotational branches of the $[a^1\Delta]3d\pi\ ^2\Phi \longleftarrow\!\!\!\longleftarrow X^2\Pi$ transition of the SH radical leading to the $a^1\Delta$ excited ionic state of $SH^+$. Our experiments have established that ionization takes place to the $a^1\Delta(v^+=0)$ excited state. No transitions to higher vibrational levels of the $a^1\Delta$ state, nor to the $X^3\Sigma^-$ ground ionic state were observed. From the spectra, it is apparent that the amount of resolved rotational structure increases dramatically as the PE spectra are measured via rotational lines corresponding to higher rotational quantum numbers in the intermediate $^2\Phi$ state. The calculated spectra were convoluted with a Gaussian

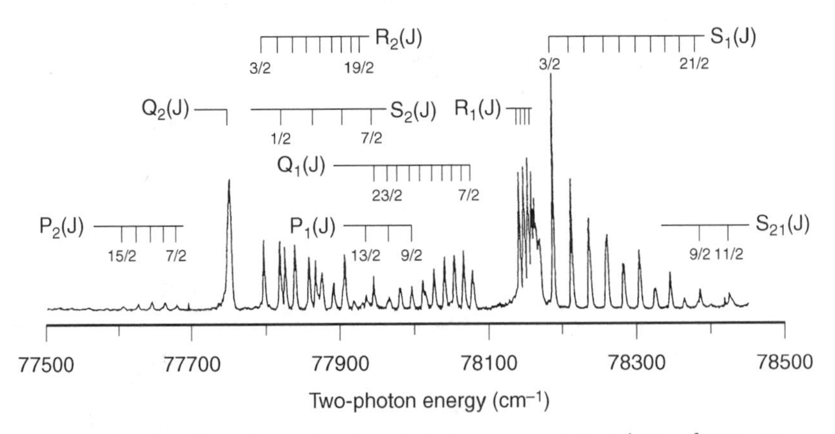


**Figure 19.** Experimental two-photon excitation spectrum of the $[a^1\Delta]3d\pi\ ^2\Phi(v'=0) \longleftarrow\!\!\!\longleftarrow X^2\Pi(v''=0)$ transition in SH. Individual line assignments are given by the combs above the spectrum. (Reproduced with permission from [54], copyright 1997 American Institute of Physics.)

detector function with a width at half-maximum (WHM) of 12 meV. Possible effects on the rotational ion distributions of alignment in the ground-state of the SH fragments resulting from the $H_2S$ photodissociation step were neglected in our calculations. However, the effects of alignment following the two-photon absorption to the excited $^2\Phi$ state were included. Since the $e/f$ parities of the rotational levels in the $^2\Phi$ intermediate state and the $a^1\Delta$ ionic state cannot be resolved experimentally, the contributions to the cross-sections from both the $e$ and $f$ parities of each rotational level were summed [54].

Agreement between the measured and calculated spectra in Figure 20 is excellent. The spectra only show small changes in total angular momentum, $|\Delta N| \leq 2$, for every rotational branch. From angular momentum conservation, this observation suggests that the PE continua are dominated by the $s$ ($\ell = 0$) and $p$ ($\ell = 1$) partial waves. This differs from expectations on the basis of atomic-like propensity rules, since $p$ and $f$ partial waves would be expected for photoionization of the $3d\pi$ orbital with 95% $d$ character.

To understand the underlying dynamics of these rotationally resolved PE spectra, it is useful to examine the angular momentum composition of the PE matrix element. It appears that in all three ($k\sigma, k\pi$, and $k\delta$) continuum channels available on one-photon ionization from the $3d\pi$ orbital, the $f(\ell = 3)$ partial wave clearly dominates. The $p$ wave is relatively weak in both the $k\sigma$ and $k\pi$ channels near threshold, which may indicate the presence of a Cooper minimum in the discrete region. Also, a Cooper minimum is calculated in the $d(\ell = 2)$ wave of the $k\delta$ channel around an electron kinetic energy of 1.5 eV. However, the effects of this Cooper minimum on the rotational ion distributions may be weak since its amplitude is relatively small compared to that of the $f$ wave. That the expected angular momentum changes up to $|\Delta N| = 4$ are not seen in the spectra, despite the significant $f$ wave participation, is due to interference between the $f$ waves in the different PE continua. To test this assumption, we arbitrarily varied the phase factor in one of the three ($\sigma, \pi$, and $\delta$) continuum channels, and the resulting spectra showed strong $\Delta N = \pm 3$ and $\Delta N = \pm 4$ transitions. It is interesting to point out that this interference effect in the PE matrix element persists over a broad energy range. With hindsight, the results can be interpreted on a more approximate level. The underlying physical picture that could explain the main features of our PE spectra of the $[a^1\Delta]3d\pi\ ^2\Phi$ state of SH is a simple sudden approximation model in which the $a^1\Delta$ core acts as a spectator as the $3d\pi$ Rydberg electron is ionized by the absorption of a single photon. In such a model, the angular momentum of the photon is transferred only to the Rydberg electron [54].

The most striking features in the PE spectra of the $[a^1\Delta]3d\pi\ ^2\Phi$ state are the strong asymmetries between the $\Delta N = \pm 1$ and $\Delta N = \pm 2$ peaks. Even though slight discrepancies in intensities are expected in many molecular systems, such a large difference in intensities for losing or gaining the same amount of angular

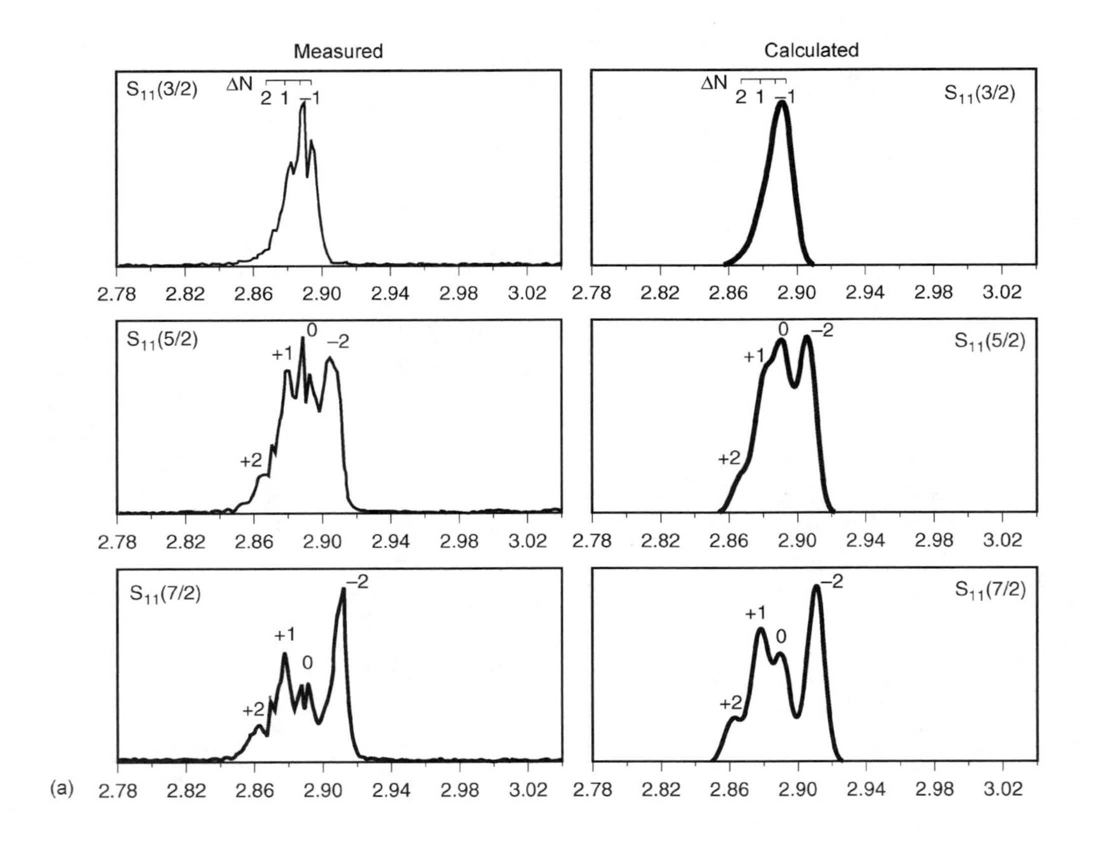
Measured
Calculated
$S_{11}(3/2)$
ΔN
2 1 –1
$S_{11}(5/2)$
+1
0
–2
+2
$S_{11}(7/2)$
(a)
2.78
2.82
2.86
2.90
2.94
2.98
3.02

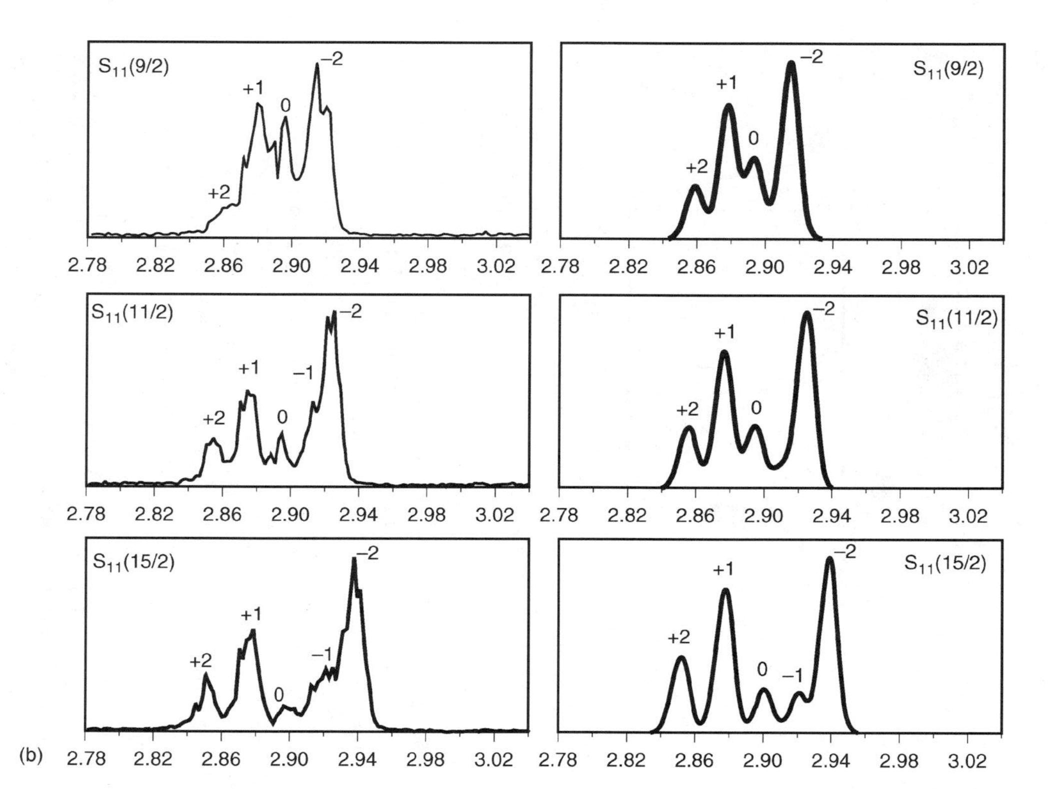


**Figure 20.** The measured (left column) and calculated (right column) PE spectra obtained in a $(2+1)$ photoionization process via the $S_{11}(\frac{3}{2})$ to $S_{11}(\frac{11}{2})$, $S_{11}(\frac{15}{2})$, and $S_{11}(\frac{19}{2})$ rotational lines of the $[a^1\Delta]3d\pi$ $^2\Phi(v'=0) \leftarrow\leftarrow X^2\Pi(v''=0)$ two-photon transition of the SH radical leading to the $a^1\Delta$ excited ionic state of $SH^+$. The labeling denotes the changes in rotational angular momentum $\Delta N$. (Reproduced with permission from [54], copyright 1997 American Institute of Physics.)

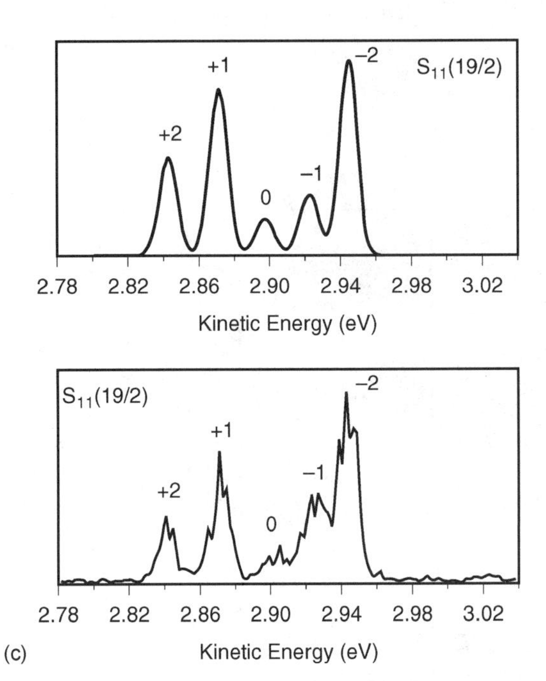


**Figure 20.** *(Continued)*

momentum is highly unusual. This phenomenon is especially evident at smaller rotational quantum numbers of the intermediate $^2\Phi$ state. Since the resonant $^2\Phi$ state has a large component of electronic angular momentum $(\Lambda' = 3)$ along the internuclear axis, it could play an important role in these asymmetries. To reach the classical limit for large $\Lambda'$ requires a correspondingly larger rotational angular momentum. To test this idea, we have also calculated the PE spectra for rotational levels up to $N' = 50$. Indeed, the spectra revealed very symmetrical patterns at such high $N'$s. Therefore, the current study represents a situation that is rather different from the one described by the classical picture of rotational motion.

Next, we shall be concerned with a two-color ZEKE–PFI experiment via the $[a^1\Delta]3d\pi\ ^2\Phi$ state of SH to the $a^1\Delta$ excited ionic limit [140]. This work represents the first ZEKE–PFI experiment involving an excited ionic state of a short-lived radical. A series of ZEKE–PFI spectra, shown in Figure 21, was obtained by two-photon excitation of various rotational levels of the $[a^1\Delta]3d\pi\ ^2\Phi(v' = 0)$ state via the $S_1(\frac{3}{2})$ to $S_1(\frac{13}{2})$ rotational transitions, and probing the excited $a^1\Delta$ rotational ionization thresholds by subsequent one-photon pulsed-field ionization. For the SD radical, a considerably lower signal-to-noise ratio was experienced and ZEKE–PFI could only be observed via the

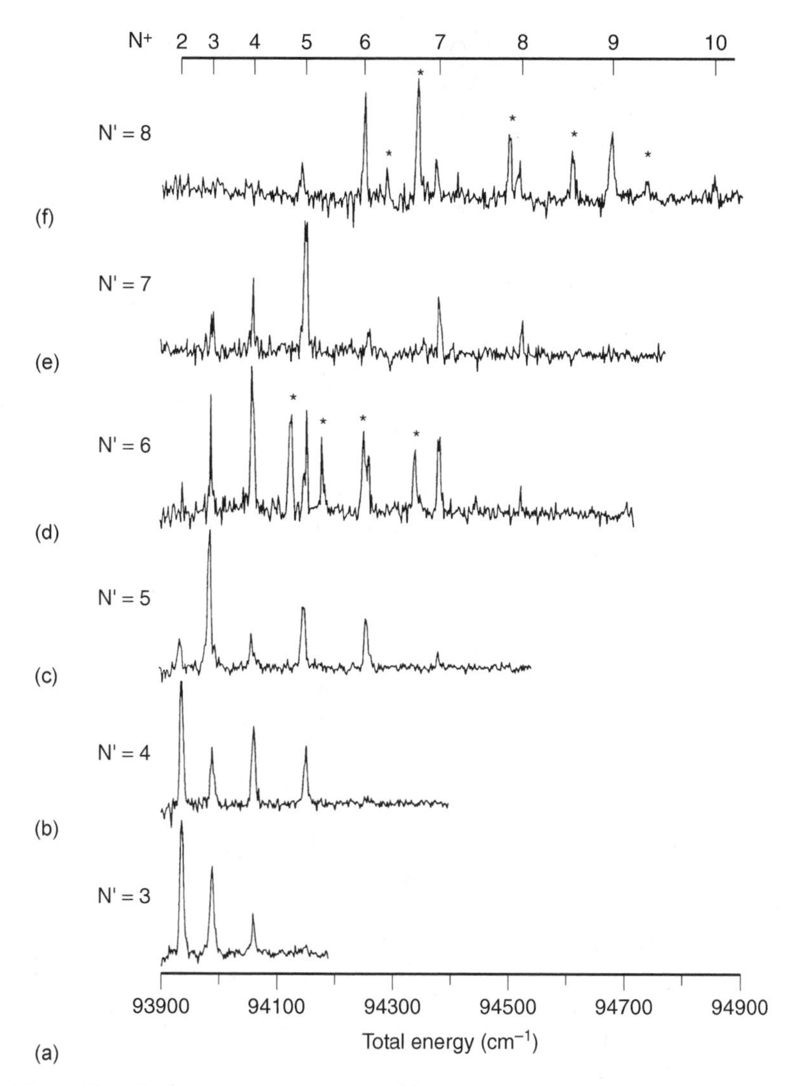


**Figure 21.** The ZEKE–PFI spectra of the $a^1\Delta$ excited ionic state of $SH^+$ obtained after two-photon excitation of the $[a^1\Delta]3d\pi\ ^2\Phi(v' = 0)$ state via the (*a*) $S_1(\frac{3}{2})$, (*b*) $S_1(\frac{5}{2})$, (*c*) $S_1(\frac{7}{2})$, (*d*) $S_1(\frac{9}{2}) + {}^tS_{21}(\frac{5}{2})$, (*e*) $S_1(\frac{11}{2})$, and (*f*) $S_1(\frac{13}{2}) + {}^tS_{21}(\frac{5}{2})$ rotational transitions. Individual line assignments are indicated by the comb above the spectrum. (Reproduced with permission from [140], copyright 1996 American Institute of Physics.)

$S_1(\frac{3}{2})$ rotational transition. The ZEKE–PFI spectra obtained by pumping the intermediate rotational levels $N' = 3$ and 4 via the $Q_1$ and $R_2$ rotational branches are barely different from those obtained by pumping via the corresponding $S_1$ rotational branches shown in Figure 21(*a*) and (*b*). Similarly, the ZEKE–PFI spectrum obtained by pumping the $N' = 6$ rotational level via the $Q_1(\frac{13}{2})$ rotational transition hardly differs from that obtained via $S_1(\frac{9}{2})$, Figure 21(*d*). It can therefore be concluded that alignment effects play a minor role in the present experiments.

The energy scales in the spectra of Figure 21 have been corrected for Stark shifts arising from the applied pulsed electric field, the "stray" electric fields present in the "magnetic bottle" spectrometer, and the motional Stark field induced in molecules moving in the 1-T magnetic field. Energy scales are given with respect to the lowest rotational level of the $X^2\Pi_{3/2}$ ground-state of the neutral molecule. Field-free ionization energies of the transition $a^1\Delta_2(v^+ = 0, J^+ = 2) \leftarrow X^2\Pi_{3/2}(v'' = 0, J'' = \frac{3}{2})$ for $SH^+$ and $SD^+$ are $93{,}925 \pm 3\,\text{cm}^{-1}$ ($11.6453 \pm 0.0004$ eV) and $93{,}944 \pm 3\,\text{cm}^{-1}$ ($11.6476 \pm 0.0004$ eV), respectively. Rotational analysis of the ZEKE–PFI spectra for $SH^+$ and $SD^+$ allows for an accurate determination of the $B_0$ rotational constant and the $D_0$ centrifugal distortion constant [140].

When we compare the ZEKE–PFI spectra with those obtained in the previous REMPI–PES study via the same intermediate $[a^1\Delta]3d\pi\ ^2\Phi(v' = 0)$ state of SH, the correspondence is striking. First, the only ionization process that is observed is the core-preserving ionization to $a^1\Delta(v^+ = 0)$, indicating that the intermediate state is a "pure" Rydberg state with a single core. Second, the expected transitions $\Delta N = \pm 3, \pm 4$ are virtually absent, as was the case with the REMPI–PES study. Finally, the ZEKE–PFI spectra show dramatic differences between the intensities of the various $\Delta N = N^+ - N'$ transitions. However, these differences compare very well with the asymmetries observed with REMPI–PES. The last observation is especially interesting and even unusual, because there are countless examples in the literature where REMPI–PES and ZEKE–PFI intensities show major differences. This is in fact hardly surprising, since the mechanisms, which lie at the root of REMPI–PES (direct ionization) and ZEKE–PFI (employing high-*n* Rydberg states) are very different. In the present ZEKE–PFI experiment, high-*n* Rydberg states belonging to series converging upon the $a^1\Delta$ excited ionic state of SH play a crucial role. These high-lying Rydberg states are clearly superexcited states, which may undergo decay processes such as (pre)dissociation, or autoionization into the underlying $X^3\Sigma^-$ ionic continua. The fact that the final ionic state distributions are so similar for both independent experimental methods is a strong indication that decay processes of the Rydberg states used as "reservoir" states in ZEKE–PFI are of minor importance. Clearly, autoionization into the $X^3\Sigma^-$ ionic continua is relatively slow. We shall come back to this point later.

### 3. *The SH* $[a^1\Delta]5p\pi\ ^2\Phi(v' = 0)$ *State*

In the two-photon energy range between 84,800 and 85,600 cm$^{-1}$ an SH excitation spectrum, shown in Figure 22, was obtained by only monitoring photoelectrons with a kinetic energy of $\sim 4.2$ eV. As confirmed by PES these electrons derive from a core-preserving photoionization process in which a $(2+1)$ transition from the $X^2\Pi(v'' = 0)$ ground-state to the $a^1\Delta(v^+ = 0)$ excited ionic state occurs. Despite the fact that the resonance-enhancing state at the two-photon energy is located above the lowest ionization threshold, the excitation spectrum shows extensive rotational structure. This highly unusual observation indicates that the absorption of a third photon after the two-photon excitation step is a more probable process than autoionization into the underlying continua of the $X^3\Sigma^-$ ionic ground-state. The spectroscopic parameters of this state were determined from an analysis of the rotational structure. The results of a detailed simulation of the excitation spectrum and of PES lead to the unambiguous conclusion that this Rydberg state must be assigned as $[a^1\Delta]5p\pi\ ^2\Phi(v' = 0)$ [53].

In general, autoionization processes prevent the observation of rotationally resolved excitation spectra as these processes lead to short lifetimes of the states, and hence to excessive broadening of the spectra. Only few examples are known in which autoionization, in particular spin–orbit autoionization, proceeds on a sufficiently slow timescale to allow for rotational resolution to be observed. Generally, the spectral line widths and line shapes are then clearly determined by the rate of autoionization. However, in this work a careful consideration of

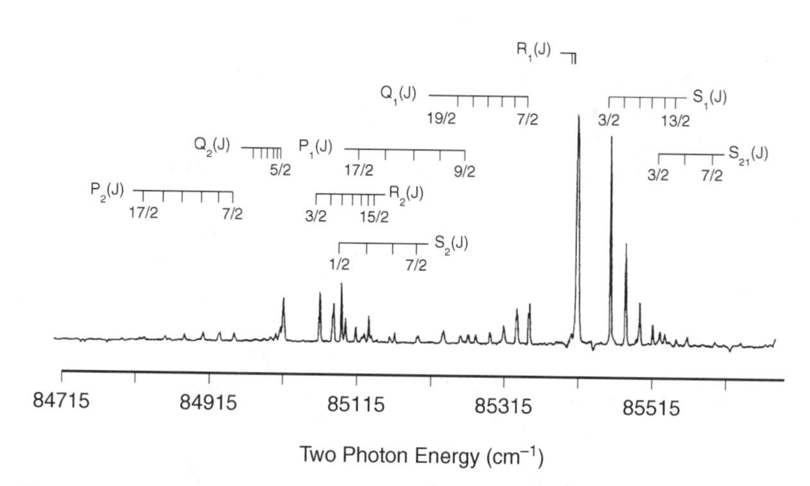


**Figure 22.** Two-photon excitation spectrum of the $[a^1\Delta]5p\pi\ ^2\Phi(v' = 0) \leftarrow\leftarrow X^2\Pi(v'' = 0)$ transition of SH obtained by collecting photoelectrons deriving from a $(2+1)$ ionization process to the $a^1\Delta(v^+ = 0)$ state. Individual line assignments are indicated by the combs above the spectrum. (Reproduced with permission from [53], copyright 1995 Elsevier Science B.V.)

the line widths as a function of laser power suggests that the autoionization rate of the $[a^1\Delta]5p\pi\ ^2\Phi(v' = 0)$ state is not large enough to have a measurable influence on the line widths.

When the laser wavelength was fixed at rotational levels of the $^2\Phi$ state associated with relatively high rotational angular momentum, rotationally resolved PE spectra were obtained. These PE spectra showed strong asymmetries between $\Delta N = +1$ and $-1$ peaks, and between $\Delta N = +2$ and $-2$ peaks, while transitions associated with larger changes in $\Delta N$ were not observed [54]. The rotationally resolved PE spectra show strong similarities with those of the SH $[a^1\Delta]3d\pi\ ^2\Phi(v' = 0)$ state [52, 54, 140], which is located below the ionic ground-state. In an atomic-like picture, the selection rules on removal of a $5p$-type Rydberg electron would predict a dominant $d$ wave for the outgoing electron on photoionization, and hence rotational ion distributions with transitions up to $\Delta N = \pm 3$. However, quantum-chemical calculations on this state of SH show a dominant $f$ wave contribution. The absence of large angular momentum transfers is again ascribed to interferences between the various outgoing continuum channels. The strong asymmetries observed in the rotational ion distributions again arise from the fact that combination of a large electronic angular momentum of the superexcited state $(\Lambda' = 3)$ and the relatively low end-over-end rotational angular momentum lead to a situation that is far removed from the classical limit [53, 54].

### *4. The SH $[b^1\Sigma^+]3d$, $[b^1\Sigma^+]5p$, $[b^1\Sigma^+]4d$, and $[b^1\Sigma^+]4f$ Complexes*

In the two-photon energy range between 87,650 and 96,000 cm$^{-1}$, several new excitation spectra of the SH radical were observed. These spectra were detected either by mass-resolved ion detection of the $SH^+$ fragment, or by monitoring photoelectrons resulting from autoionization into the continua of the $X^3\Sigma^-$ ionic ground-state after two-photon resonant excitation. This electron signal was superimposed on a direct ionization background signal due to $H_2S$, SH, and atomic sulfur, but was large enough to be distinguished from it. From their positions, these spectra are assigned in order of increasing energy to the $[b^1\Sigma^+]3d$, $[b^1\Sigma^+]5p$, $[b^1\Sigma^+]4d$, and $[b^1\Sigma^+]4f$ complexes and shown in Figure 23. The excitation spectra associated with the $[b^1\Sigma^+]3d$ and $[b^1\Sigma^+]5p$ complexes are located above the $X^3\Sigma^-$ ionic ground-state, but below the $a^1\Delta$ excited ionic state. Those associated with $[b^1\Sigma^+]4d$ and $[b^1\Sigma^+]4f$ lie above the $a^1\Delta$ and below the $b^1\Sigma^+$ excited ionic states [40].

→

**Figure 23.** Two-photon excitation spectra of (*a*) the $[b^1\Sigma^+]3d$ Rydberg complex of SH; (*b*) the $[b^1\Sigma^+]5p$ Rydberg complex of SH; (*c*) the $[b^1\Sigma^+]4d$ Rydberg complex of SH; and (*d*) the $[b^1\Sigma^+]4f$ Rydberg complex of SH. Spectrum (*a*) was obtained using mass-resolved $SH^+$ ion detection, (b–d) with electron detection. Arrows indicate the subbands $F_1$ and $F_2$ arising from the ground-state spin–orbit splitting. (Reproduced with permission from [40], copyright 1996 American Institute of Physics.)

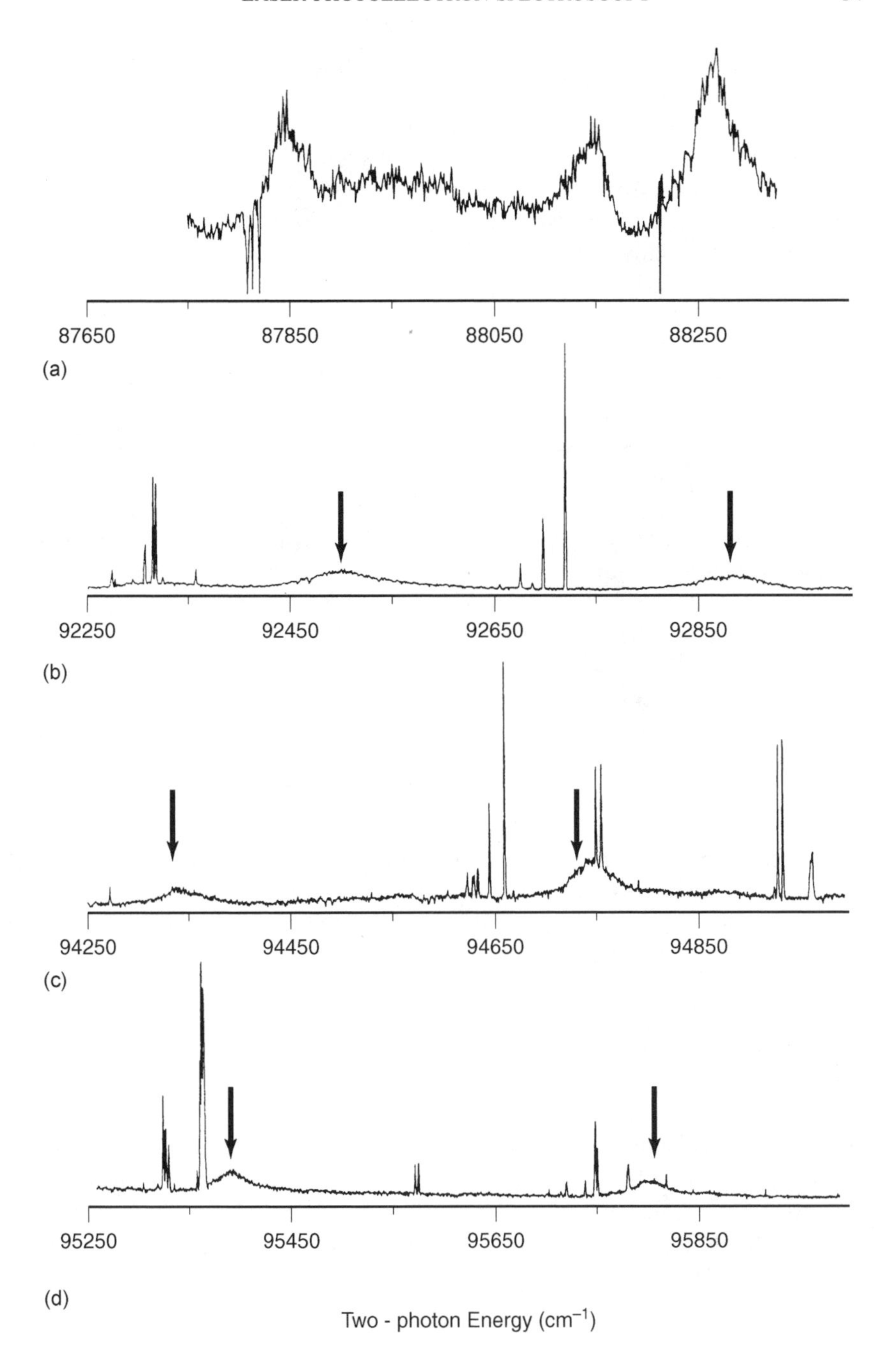
87650
87850
88050
88250
(a)
92250
92450
92650
92850
(b)
94250
94450
94650
94850
(c)
95250
95450
95650
95850
(d)
Two - photon Energy (cm⁻¹)

In remarkable contrast to the excitation spectra that possess an $a^1\Delta$ core, the excitation spectra arising from levels with a $b^1\Sigma^+$ core in the present energy range show no rotational structure whatsoever. The only features that can be distinguished, and are indicated by arrows in Figure 23, are two subbands $F_1$ and $F_2$, which arise from the ground-state spin–orbit splitting. Clearly, the rate of autoionization into the underlying ionic continua is fast enough that the observation of rotational structure in the excitation spectra is prevented [40].

## 5. *Summarizing Remarks*

The new Rydberg states observed in the SH radical show a plethora of interesting physical phenomena. The experimental observation of electronic interactions in the PE spectra of the SH $[b^1\Sigma^+]4p\sigma\ ^2\Sigma^+(v'=0)$ and $[a^1\Delta]3d\delta\ ^2\Sigma^+(v'=1)$ states is but one striking example.

The Rydberg states of SH that have been observed above the lowest $X^3\Sigma^-$ ionic threshold fall into two categories, which show a distinctly different dynamical behavior. Surprisingly, the ones with an excited $a^1\Delta$ ionic core (of which the $[a^1\Delta]5p\pi\ ^2\Phi$ state is but one example) show rotationally resolved excitation spectra. The results for such states indicate that autoionization into the underlying $X^3\Sigma^-$ continua cannot compete with the process in which a third photon is absorbed in a core-preserving ionization process. The conclusion that autoionization is relatively slow for states with an $a^1\Delta$ core is supported by the fact that the REMPI–PES and ZEKE–PFI spectra of the $[a^1\Delta]3d\pi\ ^2\Phi$ state below the $X^3\Sigma^-$ threshold show very similar intensities for the transitions to rotational levels of the $a^1\Delta$ excited ionic state. For states that possess a $b^1\Sigma^+$ core, regardless of whether they are located below or above the $a^1\Delta$ threshold, only severely broadened excitation spectra are measured. This observation demonstrates that autoionization into the underlying continua on the one hand is slow enough to make the excitation spectra observable, but on the other hand fast enough to prevent the observation of rotational structure.

The most realistic explanation for the overall behavior of superexcited Rydberg states above the lowest ionic limit in SH is found in the involvement of spin–orbit interaction. This interaction in itself is not all that large in SH, but it is the only mechanism available to provide the necessary coupling between the low-lying SH ionic cores. The $X^3\Sigma^-$, $a^1\Delta$, and $b^1\Sigma^+$ ionic states all derive from the same $\ldots(2\pi)^2$ configuration. The $X^3\Sigma^-$ and $b^1\Sigma^+$ ionic states can interact via off-diagonal matrix elements of the spin–orbit operator, whereas such matrix elements are zero for the interaction between $X^3\Sigma^-$ and $a^1\Delta$ ionic states. Consequently, Rydberg states built upon a $b^1\Sigma^+$ core can be viewed as having some $X^3\Sigma^-$ ionic character, leading to a coupling to the autoionization continua of the $X^3\Sigma^-$ ground ionic state, and providing a decay mechanism for the superexcited states with the $b^1\Sigma^+$ core. In contrast, Rydberg states built upon an $a^1\Delta$ ionic core are expected to autoionize very slowly at best [40, 41].

### C. The NH Radical

The neutral NH or imidogen radical is an important representative of the diatomic hydrides, of interest in combustion chemistry [147], and found in the sun, comets, and planetary atmospheres [148, 149]. The NH radical possesses a lowest energy electron configuration $(1\sigma)^2(2\sigma)^2(3\sigma)^2(1\pi)^2$, leading to an $X^3\Sigma^-$ electronic ground-state, and to excited $a^1\Delta$ and $b^1\Sigma^+$ states at higher energy. The metastable $a^1\Delta$ state is located 12,688.4 $cm^{-1}$ above the rotationless ($v'' = 0$, $N'' = 0$) level of the $X^3\Sigma^-$ ground-state [150]. Reasonably characterized higher excited valence states of NH include $A^3\Pi$ [3, 151], $c^1\Pi$ [3, 150, 152], and $d^1\Sigma^+$ [3, 152, 153]. The production method of the NH (ND) radical often employed involves photodissociation of hydrazoic acid, $HN_3$ ($DN_3$), prepared by the dropwise addition of $H_3PO_3$ ($D_3PO_3$) to $NaN_3$. In this process, the imidogen radical is exclusively generated in its lowest singlet excited $a^1\Delta$ state, with some vibrational excitation [154]. The Rydberg states originating from excitation of this $a^1\Delta$ state thus all belong to the singlet manifold. Some of them have been studied with REMPI [155–157]. The three singlet Rydberg states observed with REMPI so far, $f^1\Pi$, $g^1\Delta$, and $h^1\Sigma$, have all been assigned as members of the "3$p$" complex, with $a \ldots (3\sigma)^2(1\pi)^1(3p\lambda)^1$ electron configuration [155, 156]. These assignments are based on the fact that transitions from the $a^1\Delta$ originating state are two-photon allowed, on PE spectra that show Franck–Condon diagonal transitions to the $X^2\Pi$ ionic state [55], on the observed quantum defects that lie in the range 0.6–0.8, and on the rotational analyses that in all cases support $a \ldots (3\sigma)^2(1\pi)^1(3p\lambda)^1$ configuration, leading to $\ldots (3\sigma)^2(1\pi)^1(3p\sigma)^1\ {}^{1,3}\Pi$, and $\ldots (3\sigma)^2(1\pi)^1(3p\pi)^1\ {}^{1,3}\Delta$, ${}^{1,3}\Sigma^+$, and ${}^{1,3}\Sigma^-$ states. In an independent experiment NH (ND) was produced in its $X^3\Sigma^-$ ground-state from the reaction between fluorine atoms and $NH_3$ ($ND_3$). Three new excited states for NH and five for ND belonging to the triplet manifold were observed in the energy region 85,000–91,000 $cm^{-1}$ [158].

Next, we shall describe a REMPI–PES study of the $(1\sigma)^2(2\sigma)^2(3\sigma)^0(1\pi)^4 d^1\Sigma^+(v' = 0)$ valence state of the NH radical where configuration interaction is apparent [39]. Then, we shall discuss the laser PES of two NH singlet excited Rydberg states at higher energies than employed before. These new states must belong to the "3d" complex, which gives rise to the $\ldots (3\sigma)^2(1\pi)^1(3d\sigma)^1\ {}^{1,3}\Pi$, $\ldots (3\sigma)^2(1\pi)^1(3d\pi)^1\ {}^{1,3}\Delta$, ${}^{1,3}\Sigma^+$, and ${}^{1,3}\Sigma^-$, and $\ldots (3\sigma)^2(1\pi)^1(3d\delta)^1\ {}^{1,3}\Phi$, ${}^{1,3}\Pi$ electronic states [38]. In an atomic-like picture two-photon excitations to these states should not be favored. Also, it should be realized here that states arising from the $4s\sigma \leftarrow 1\pi$ orbital promotion occur in much the same energy range. Finally, we shall discuss rotationally resolved PE spectra of the "3p" $f^1\Pi(3p\sigma)$, $g^1\Delta(3p\pi)$, and $h^1\Sigma^+(3p\pi)$ states [56].

### *1. The NH* $d^1\Sigma^+(1\sigma)^2(2\sigma)^2(3\sigma)^0(1\pi)^4$ *State*

The $d^1\Sigma^+(v' = 0)$ excited valence state, which possesses an electronic configuration $(1\sigma)^2(2\sigma)^2(3\sigma)^0(1\pi)^4$, ionizes into the ground ionic state with main configuration $(1\sigma)^2(2\sigma)^2(3\sigma)^2(1\pi)^1$ via a formally forbidden one-photon transition in a $(2 + 1)$ REMPI process. The associated PE spectrum is shown in Figure 24 and deviates strongly from the simple one-line behavior expected for a Rydberg state. These observations arise from the occurrence of configuration interaction, both in the $d^1\Sigma^+$ state and in the $X^2\Pi$ ground ionic state [39].

By employing the Wuppertal Bonn self-consistent field (SCF) plus multi-reference single and double excitation configuration interaction (MRD–CI) package of programmes [159], both the compositions of the excited and ionic states were computed over a large range of internuclear distances. The wave functions obtained for the $d^1\Sigma^+$ state indicate that a good description requires significant contributions ( > 10%) of several configurations, the most important of which are $\ldots(3\sigma)^0(1\pi)^4$, $\ldots(3\sigma)^2(1\pi)^2$, and $\ldots(3\sigma)^2(1\pi)^1(2\pi)^1$. For the $X^2\Pi$ ground ionic state the dominant configurations are $\ldots(3\sigma)^2(1\pi)^1$ and $\ldots(3\sigma)^0(1\pi)^3$. With these theoretical results, various transitions from a "minor" configuration of the $d$ state to the "major" configuration of the ionic state, and of the "major" configuration of the $d$ state to a "minor" configuration of the ionic state now become formally one-photon allowed. The calculation of

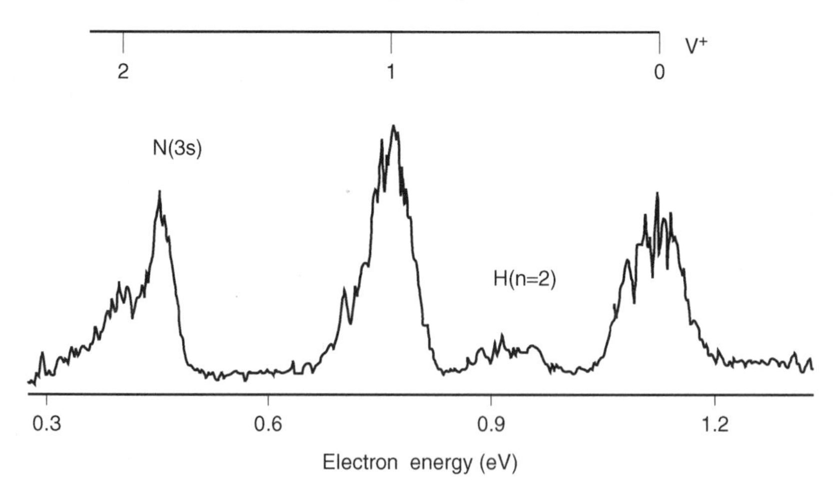


**Figure 24.** PE spectrum following the two-photon excitation of the $d^1\Sigma^+(v' = 0)$ state of the NH radical via the bandhead of the $R$ branch. Various vibrational levels ($v^+ = 0$–$2$) of the $X^2\Pi$ ionic state are reached and one-photon ionization of excited H and N atoms is also observed. (Reproduced with permission from [39], copyright 1994 American Institute of Physics.)

intensities of transitions in the PE spectrum was seriously hampered by the fact that the electric dipole transition operator strongly depends on internuclear distance. When approximate calculations were carried out in which this operator is assumed constant, the Franck–Condon factors between the main configuration of the *d* state to the main configuration of the ionic state were diagonal. When the remaining three possible pathways arising from the above configuration interaction contributions in both states were included, it is encouraging to see that, despite the approximate character of the calculations, large deviations from $\Delta v = 0$ are predicted. The same lack of $\Delta v = 0$ propensity was observed in the PE spectra of the $X^2\Pi \leftarrow d^1\Sigma^+$ transition measured via different rotational branches. These PE spectra also show the signature of the one-photon ionization of excited H and N atoms produced from photofragmentation of NH, as illustrated in Figure 24 [39].

### 2. *The NH* $i^1\Pi \ldots (3\sigma)^2(1\pi)^1(3p\sigma/3d\sigma)^1$ *State*

Figure 25 displays the excitation spectrum of the $i^1\Pi(v' = 0) \leftarrow\leftarrow a^1\Delta(v'' = 0)$ transition for the NH radical obtained with electron detection in the two-photon

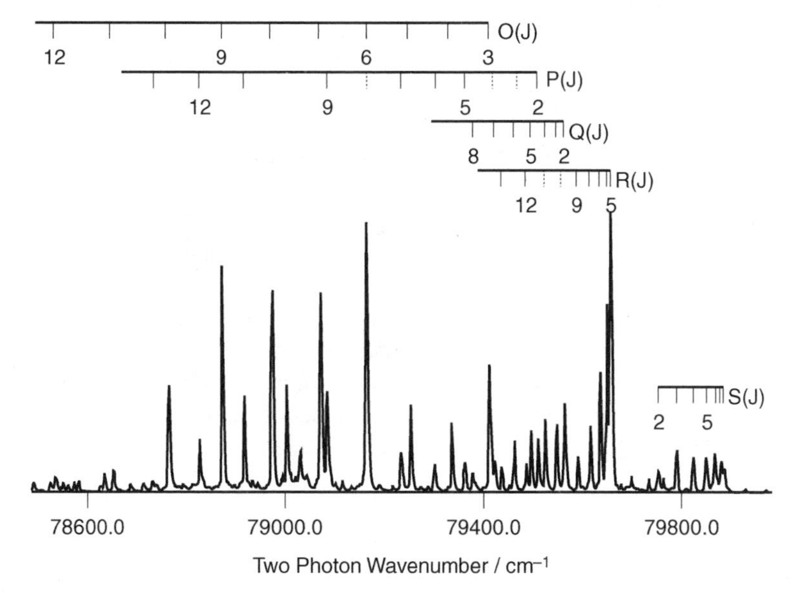


**Figure 25.** Two-photon resonant excitation spectrum of the NH $i^1\Pi(v' = 0) \leftarrow\leftarrow a^1\Delta(v'' = 0)$ transition obtained by monitoring photoelectrons with a kinetic energy of $\sim 2.85$ eV as a function of laser wavelength. This energy filtering ensures collection of only those photoelectrons formed in association with $NH^+(^2\Pi)$ ions in their ground vibrational state, thereby discriminating against electrons arising from $(2+1)$ REMPI via the overlapping $j^1\Delta(v' = 0) \leftarrow\leftarrow a^1\Delta(v'' = 1)$ hot band. Individual line assignments are given by the combs above the spectrum. (Reproduced with permission from [38], copyright 1992 American Institute of Physics.)

energy range 78,600–80,000 $cm^{-1}$. Individual line assignments are given by combs above the spectrum. Even for higher values of $J'$ the $\Lambda$ doubling in the $i^1\Pi$ state was not resolved. This is in contrast to what was observed for the same transition for the ND radical in the two-photon energy range 79,100–79,800 $cm^{-1}$. In the latter excitation spectrum (not shown), a sizeable $\Lambda$ doubling, which amounts to $\sim 13.5\,cm^{-1}$ for the highest observed $J'$ levels, was discernible. The $\Lambda$ doubling in the $a^1\Delta$ state cannot be resolved in the present experiments [38].

PE spectra from the NH $i^1\Pi(v' = 0)$ state to the ionic continua showed a transition consistent with the formation of the lowest $NH^+$ $^2\Pi$ ionic state. Moreover, the $\Delta v = v' - v^+ = 0$ transition dominated over the progressively weaker transitions to $^2\Pi(v^+ = 1, 2)$ levels. This can be taken as evidence that the $i^1\Pi$ state possesses a $\ldots(3\sigma)^2(1\pi)^1$ ionic core. *Ab initio* calculations for the lower lying $f^1\Pi$ state have indicated that the $^1\Pi$ Rydberg states may show significant configuration interaction [160]. For the higher lying $i^1\Pi$ state, the dominant configuration is thought to be $\ldots(3\sigma)^2(1\pi)^1(3p\sigma)^1$ or $\ldots(3\sigma)^2(1\pi)^1(3d\sigma)^1$ in the Franck–Condon region, with increased $3p$ participation as the system evolves to larger internuclear distances. Such an idea is broadly consistent with the observed quantum defect of $\sim 0.42$ (assuming $n = 3$), and would also make the two-photon transition to this "$3d$" state somewhat allowed.

The observation of substantial $\Lambda$ doubling in the $i^1\Pi$ state of ND, but not of NH, merits some comment. This requires the $i^1\Pi$ origin of ND (but not of NH) to be in accidental resonance with a state of $\Sigma$ symmetry. Also, since the observed $\Lambda$ doubling increases smoothly with $J'$, a local perturbation on the ND level structure can be ruled out. We therefore anticipate that the perturbing $\Sigma$ state has a rotational constant fairly similar to that of $i^1\Pi$. Because there is no obvious electronic state in this energy region that could cause this perturbation, an almost resonant vibrationally excited level of a lower lying $\Sigma$ state would seem the most probable candidate. Hopefully, future REMPI–PES experiments starting from the $a^1\Delta$ state will provide a firm identification of the configuration and parity of the perturbing state.

### 3. *The NH $j^1\Delta$ state, mixture of $\ldots(3\sigma)^2(1\pi)^1(3d\pi)^1$ and $\ldots(3\sigma)^1(1\pi)^2(3s\sigma)^1$ Configurations*

Figure 26 displays the excitation spectrum of the $j^1\Delta(v' = 0) \leftarrow\leftarrow a^1\Delta(v'' = 0)$ transition for the NH radical obtained with electron detection in the two-photon energy range 82,000–83,000 $cm^{-1}$. Individual line assignments are given by combs above the spectrum. In the same spectrum transitions corresponding to the $(1, 1)$ hot band are also discernible. In addition, weak signals arising from a perturbing state are observed and assigned as a $Q$ branch, labeled in the figure as $Q'$. Both the (0,0) and (1,1) transitions also take the form of strong $Q$ branches,

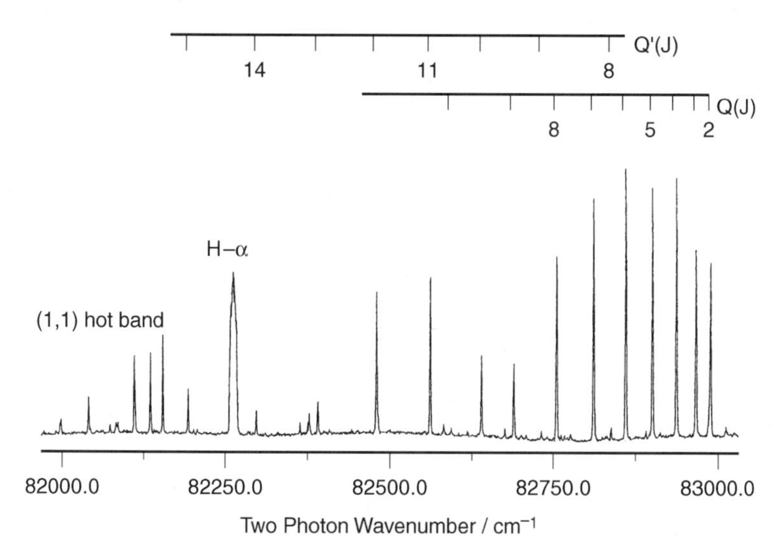


**Figure 26.** Two-photon resonant excitation spectrum of the NH $j^1\Delta(v' = 0) \longleftarrow\!\!\leftarrow a^1\Delta(v'' = 0)$ transition obtained by monitoring the $m/z = 15$ ion mass channel as a function of laser wavelength. Combs indicating the $Q$ branches associated with the $j^1\Delta$ state (unprimed) and a perturbing state (primed) are indicated. The $Q$ branch associated with the (1,1) hot band is also shown. The broad band at 82,259 $cm^{-1}$ is due to ionization of translationally hot H photofragments, resonance enhanced at the two-photon energy by the $n = 2$ state. (Reproduced with permission from [38], copyright 1992 American Institute of Physics.)

associated with the $T_0^0(A)$ component of the two-photon transition tensor. Their peak positions show clear evidence of local perturbations affecting both the lowest $J'$ levels and the $J' = 9$ level. The term value plot for the $j^1\Delta$ origin and the perturbing state presented in Figure 27 is a benchmark example of an "avoided crossing" in the rotational term values. The dashed curves in Figure 27 were calculated assuming for simplicity a perturbing state of $^1\Delta$ symmetry. However, because the perturbing state shows few rotational levels, its symmetry cannot be assigned unambiguously, and may be $^1\Pi$, $^1\Delta$, or $^1\Phi$ [38].

Analogous $Q$ branch features attributable to the (1,0), (1,1), and (2,1) hot bands of the $j^1\Delta(v') \longleftarrow\!\!\leftarrow a^1\Delta(v'')$ transition in NH were also observed. One noteworthy feature concerning the latter transition is the increased contribution made by the $T_0^2(A)$ component of the two-photon transition tensor. Whereas the origin band in this NH transition is dominated by its $Q$ branch, the (2,1) hot band shows more extensive $O, P, R$, and $S$ branch structure. The spectra involving the $v' = 1$ level show clear evidence of one or more perturbations affecting the $J' =$ 5–8 levels, while analysis of the $Q$ branch of the (2,1) hot band suggests that the $v' = 2$ level of $j^1\Delta$ is perturbed for $J' \geq 12$. These vibrational assignments

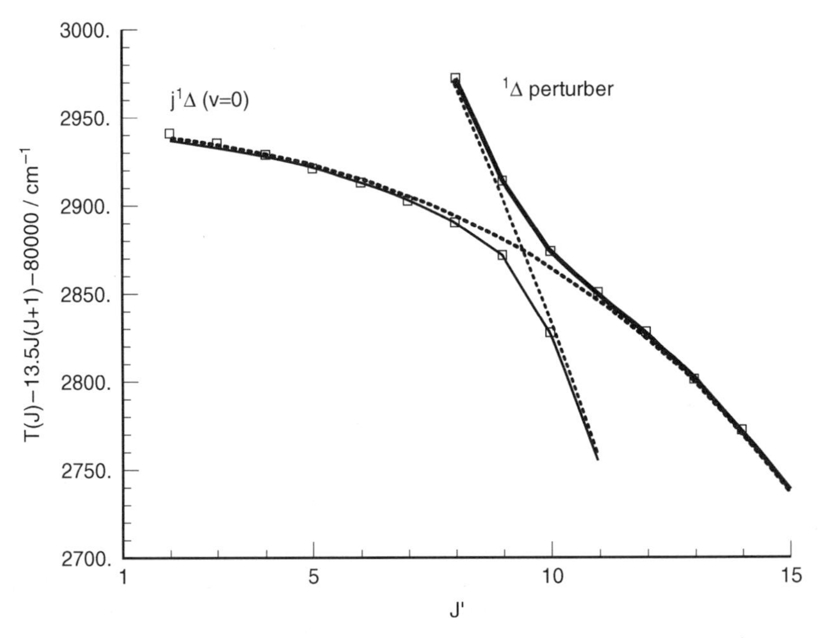


**Figure 27.** Reduced term value plot showing the observed energy levels of the NH $j^1\Delta(v' = 0)$ state, and of the perturbing state assumed to have $^1\Delta$ symmetry. The solid curves are calculated [38], while the dashed curves show the energy level pattern predicted for the two zero-order $^1\Delta$ states in the absence of the perturbation. (Reproduced with permission from [38], copyright 1992 American Institute of Physics.)

together imply a positive vibrational anharmonicity in the $j^1\Delta$ state, which in itself suggests that the state is perturbed.

PE spectra obtained by tuning the laser wavelength to the $Q(2)$ lines of the (0,0), (1,0), and (2,1) bands, respectively, are presented in Figure 28 and are very informative. These spectra provide confirmation for the $v'$ numbering of the resonant intermediate state, and, in addition, give important clues as to the mixed electronic character of the $j^1\Delta$ state. Inspection of the spectra of Figure 28 shows that, in each case, the $(2 + 1)$ REMPI process yields $NH^+$ in a range of $v^+$ levels of the $X^2\Pi$ ionic ground-state. The deviation from Franck–Condon behavior is evident. Moreover, we see some weak probability for forming $NH^+$ in its $A^2\Sigma^-$ excited ionic state, and, most importantly, strong ionizing transitions into the higher $B^2\Delta$ ionic state, specifically to levels that have the same vibrational quantum number as that of the resonance enhancing $j^1\Delta$ state. Confirmation that the most intense features are indeed associated with vibrational levels of the $B^2\Delta$ (and *not* the $A^2\Sigma^-$) state of the ion is provided by analysis of REMPI–PES spectra via higher $J'$ levels of the intermediate $j^1\Delta$ state. These

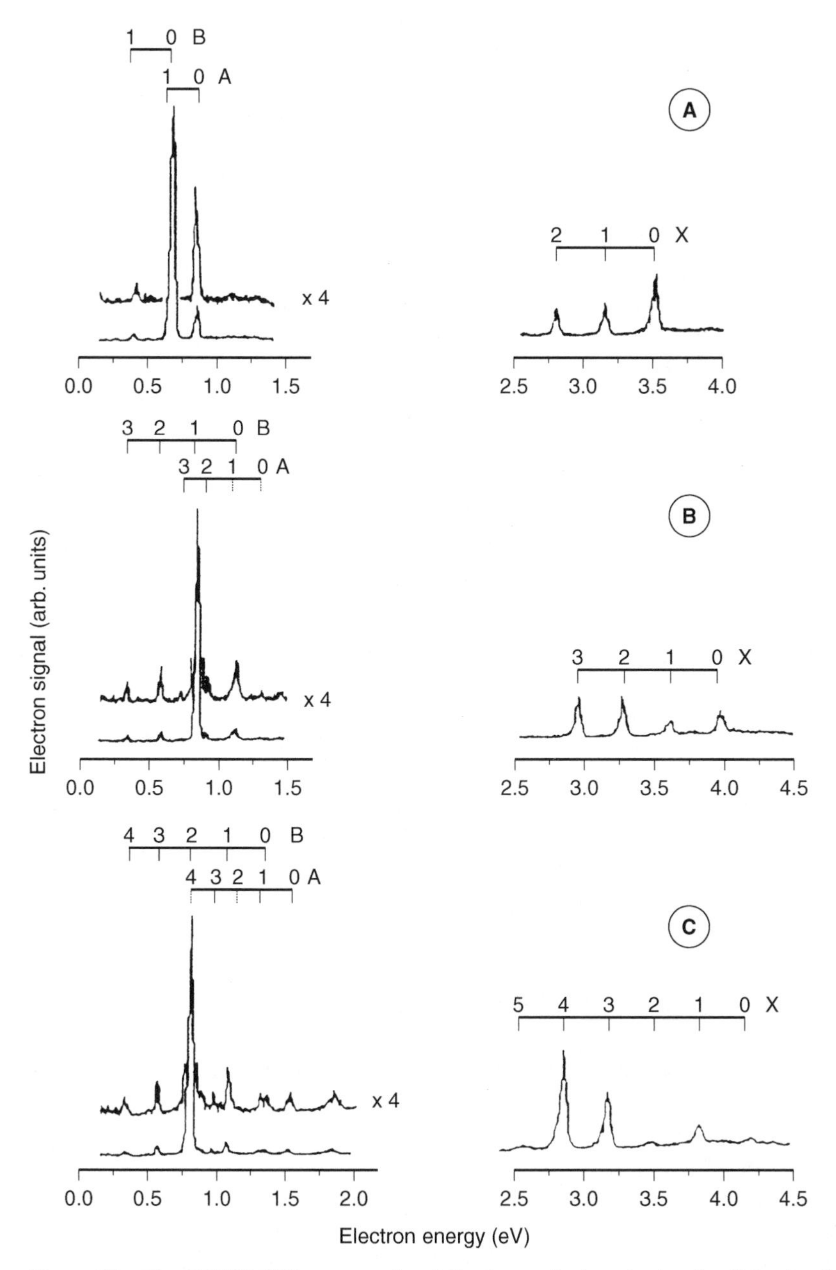


**Figure 28.** The REMPI–PES spectra taken following excitation via the $J' = 2$ levels of, respectively, the (*a*) $v' = 0$, (*b*) $v' = 1$, and (*c*) $v' = 2$ vibrational levels of the NH $j^1\Delta$ state. Combs indicating the various vibronic levels of the accompanying $NH^+$ ions are indicated above each spectrum. In spectra (*b*) and (*c*), small signals arising from one-photon ionization of $H(n = 2)$ are also observed. (Reproduced with permission from [38], copyright 1992 American Institute of Physics.)

exhibit resolvable rotational structure [17], the peak separations of which lead to a rotational constant that is only consistent with that of the $B^2\Delta$ state [3, 161]. From these results, it can be concluded that the $j^1\Delta$ state is based on two dominant configurations, $\ldots(3\sigma)^2(1\pi)^1(3d\pi)^1$ with a ground-state core, and $\ldots(3\sigma)^1(1\pi)^2(3s\sigma)^1$ with an excited ion core. The observed quantum defect of $\sim 0.1$ (assuming $n = 3$) is in line with this idea. The conclusion that the $j^1\Delta$ state shows significant configuration interaction involving different cores is further supported by the fact that the rotational constants associated with the observed $v' \leq 2$ vibrational levels of the $j^1\Delta$ state are not in line with and significantly smaller than those obtained for all other NH Rydberg states, and the positive anharmonicity referred to above which suggests an "unusual" shape for the $j^1\Delta$ potential energy curve. Moreover, results obtained for the ND radical explicitly confirm this picture [38].

### 4. *Rotationally Resolved Photoelectron Spectroscopy of the NH $f^1\Pi\ldots(3\sigma)^2(1\pi)^1(3p\sigma)^1$, $g^1\Delta\ldots(3\sigma)^2(1\pi)^1(3p\pi)^1$, and $h^1\Sigma^+\ldots(3\sigma)^2(1\pi)^1(3p\pi)^1$ States*

The $f^1\Pi\ldots(3\sigma)^2(1\pi)^1(3p\sigma)^1$, $g^1\Delta\ldots(3\sigma)^2(1\pi)^1(3p\pi)^1$, and $h^1\Sigma^+\ldots(3\sigma)^2(1\pi)^1(3p\pi)^1$ states of the NH radical have been observed first with $(2+1)$ REMPI spectroscopy [55, 155, 156]. From the excitation spectra it can be concluded that neither in the $f$ and $g$ states, nor in the originating $a^1\Delta$ state, $\Lambda$ doubling was resolved experimentally. Here, we shall illustrate how rotationally resolved laser PES can characterize these states in more detail, and leads to a better understanding of the dynamics of the photoionization process. The intermediate $f$, $g$, and $h$ states are accessed by two-photon absorption from the originating $a^1\Delta$ state, and subsequently ionized via absorption of one additional photon. The PE spectra of the three states have in common that ionization is observed solely into the $X^2\Pi$ ground ionic state, and that $\Delta v = 0$ propensity dominates. These are clear indications that the $f$, $g$, and $h$ states are "pure" Rydberg states belonging to series converging upon the lowest ionic state.

In Figure 29, experimental and calculated rotationally resolved PE spectra of the $X^2\Pi(v^+ = 0, 1, N^+) \leftarrow f^1\Pi(v' = 0, N' = 13)$ transition via the $R(12)$ rotational branch at a two-photon energy of 73,840.2 cm$^{-1}$ in the $(2+1)$ excitation spectrum of NH [56] are shown. The transition to $v^+ = 1$ in the ionic state has low intensity. With regard to the rotational structure, the $\Delta N = 0$ feature dominates, and $\Delta N = \pm 2$ transitions show a larger intensity than $\Delta N = \pm 1$. In the calculated spectra the rotational distributions are convoluted with a Gaussian line width of 30-meV WHM. Apart from the slight "skewing" of the spectra that is not reproduced by the calculations, the agreement between experimental and calculated spectra is encouraging. In order to gain more insight in the underlying photoionization dynamics, we shall examine the angular momentum composition of the photoionization matrix element. The channels

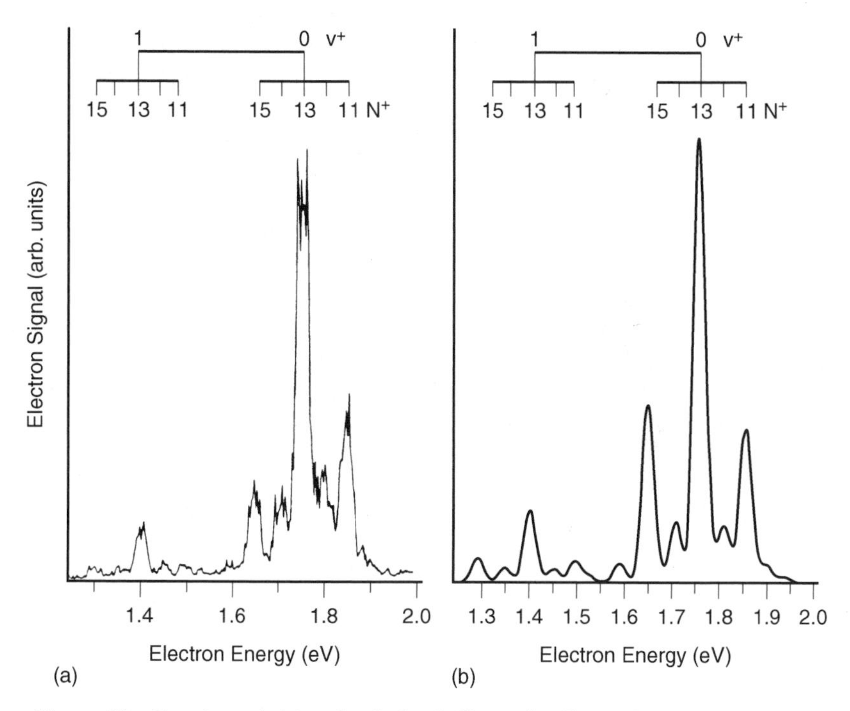


**Figure 29.** Experimental (*a*) and calculated (*b*) rotationally resolved PE spectra of the $X^2\Pi(v^+ = 0, 1, N^+) \leftarrow f^1\Pi(v' = 0, N' = 13)$ transition via the $R(12)$ rotational branch at a two-photon energy of 73,840.2 cm$^{-1}$ in the $(2 + 1)$ excitation spectrum of NH. (Reproduced with permission from [56], copyright 1992 American Institute of Physics.)

relevant for photoionization of $f^1\Pi(3p\sigma)$ are $3p\sigma \rightarrow k\sigma(^1\Pi)$, $3p\sigma \rightarrow k\pi(^1\Sigma^+)$, $3p\sigma \rightarrow k\pi(^1\Sigma^-)$, and $3p\sigma \rightarrow k\pi(^1\Delta)$. Detailed calculations show that distinct Cooper minima occur in the $\ell = 2$ outgoing electron waves in the $3p\sigma \rightarrow k\pi(^1\Sigma^-)$ channel at an electron kinetic energy of $\sim 0.5$ eV, and in the $3p\sigma \rightarrow k\pi(^1\Sigma^+)$ channel at an electron kinetic energy of $\sim 1.8$ eV. A less pronounced Cooper minimum occurs for $\ell = 2$ in the $3p\sigma \rightarrow k\pi(^1\Delta)$ channel at $\sim 0.5$ eV [56]. In the vicinities of these Cooper minima odd partial *p* and *f* waves are seen to dominate. In a simple atomic picture, in contrast, ionization from a *p*-type orbital would give rise to even partial waves arising from $p \rightarrow s$ and $p \rightarrow d$ transitions. Since the Cooper minima occur very close to the electron kinetic energies at which the experiments were performed, the suppression of the $\ell = 2$ waves in this region allows the $\ell = 1$ and $\ell = 3$ waves to dominate. The strong $\Delta N = 0$ peak is associated mainly with the *p* (76%) and *f* (24%) wave components of the photoionization matrix element. At $R_e = 2.06a_0$, the $3p\sigma$ orbital possesses 24.8% *s* and 67.2% *p* character, and thus the $3s\sigma \rightarrow kp\sigma$ transition will contribute significantly to the photoionization matrix element. The continuum $f$ wave, which arises from strong $\ell$ mixing in the electronic continuum, is also important and entirely molecular in origin. It contributes mainly to the $\Delta N = \pm 2$ transitions. Since the spin–orbit splitting between $X^2\Pi_{3/2}$ and $X^2\Pi_{1/2}$ ionic states could not be resolved, *even* and *odd* partial wave contributions to the ionic rotational distributions, although calculated separately, were summed in the computed spectra of Figure 29.

In Figure 30 experimental and calculated rotationally resolved PE spectra of the $X^2\Pi(v^+ = 0, 1, N^+) \leftarrow g^1\Delta(v' = 0, N' = 16)$ transition via the $Q(16)$ rotational branch at a two-photon energy of 75,098 $\mathrm{cm}^{-1}$ in the $(2+1)$ excitation spectrum of NH [56] are shown. As with the $f^1\Pi$ state, the transition to $v^+ = 1$ in the ionic state has low intensity. Compared to the $f^1\Pi$ state, the rotational ion distributions here look rather different. In the calculated spectra, the rotational distributions are convoluted with a Gaussian line width of 30-meV WHM, and the agreement is again pleasing.

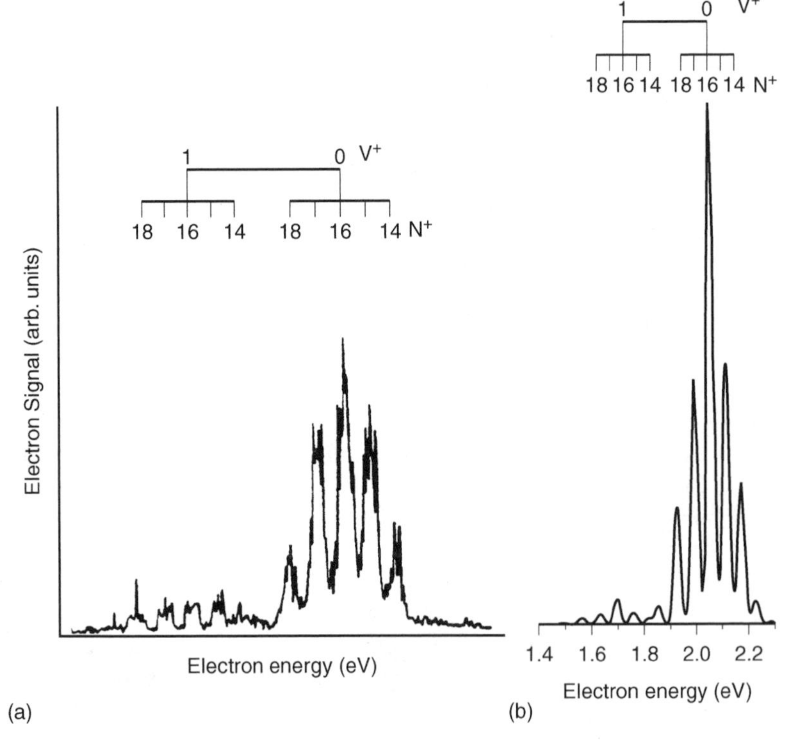


**Figure 30.** Experimental (*a*) and calculated (*b*) rotationally resolved PE spectra of the $X^2\Pi(v^+ = 0, 1, N^+) \leftarrow g^1\Delta(v' = 0, N' = 16)$ transition via the $Q(16)$ rotational branch at a two-photon energy of 75,098 $\mathrm{cm}^{-1}$ in the $(2+1)$ excitation spectrum of NH. (Reproduced with permission from [56], copyright 1992 American Institute of Physics.)

On one-photon ionization of the $g^1\Delta(3p\pi)(v' = 0)$ state of the NH radical, the relevant outgoing channels for the photoelectron are $3p\pi \rightarrow k\sigma(^1\Pi)$, $3p\pi \rightarrow k\pi(^1\Delta)$, $3p\pi \rightarrow k\delta(^1\Pi)$, and $3p\pi \rightarrow k\delta(^1\Phi)$. Cooper minima are evident in the $\ell = 2$ wave of the $k\pi(^1\Delta)$, $k\delta(^1\Pi)$, and $k\delta(^1\Phi)$ channels. Strong $\ell$ mixing in the electronic continuum is reflected in the significant $p$ and $f$ wave components of the photoionization matrix element. A crucial difference with the situation in the $f^1\Pi$ state is that the kinetic energies at which the Cooper minima occur are somewhat outside the energy range addressed in our experiments. The suppression of the $\ell = 2$ partial wave, and the concurrent relative dominance of $\ell = 1$ and $\ell = 3$ waves plays therefore no role in the case of the $g^1\Delta$ state. The $\Delta N = even$ and $\Delta N = odd$ contributions are now of comparable importance as is evident from the experimental and calculated ion rotational distributions [56]. Results for the $v' = 1$ vibrational state associated with the $g^1\Delta$ show very similar behavior with regard to the ion rotational distributions [56].

In Figure 31, experimental and calculated rotationally resolved PE spectra of the $X^2\Pi(v^+ = 0, N^+) \leftarrow h^1\Sigma^+(v' = 0, N' = 11)$ transition via the $R(10)$ rotational branch at a two-photon energy of 77,131.1 cm$^{-1}$ in the $(2 + 1)$ excitation spectrum of NH [56] are shown. In the calculated spectra, the rotational

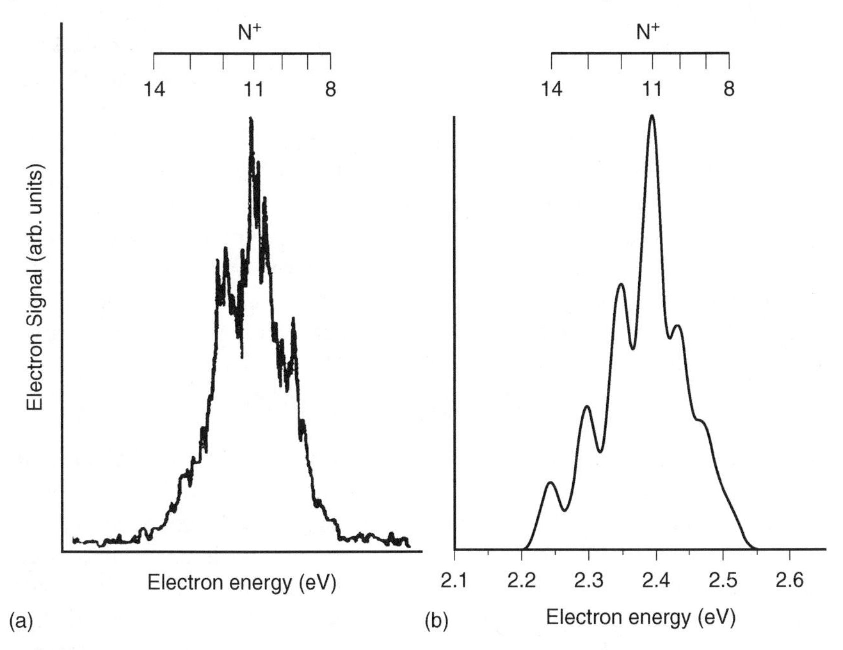


**Figure 31.** Experimental (*a*) and calculated (*b*) rotationally resolved PE spectra of the $X^2\Pi(v^+ = 0, N^+) \leftarrow h^1\Sigma^+(v' = 0, N' = 11)$ transition via the $R(10)$ rotational branch at a two-photon energy of 77,131.1 cm$^{-1}$ in the $(2 + 1)$ excitation spectrum of NH. (Reproduced with permission from [56], copyright 1992 American Institute of Physics.)

distributions are convoluted with a Gaussian line width of 35-meV WHM. The $h^1\Sigma^+$ and $g^1\Delta$ states differ only in the overall term symbols, but have Rydberg electrons of the same spatial symmetry. Strong similarities in the ionic rotational branching ratios are therefore expected. For the $h^1\Sigma^+(3p\pi)$ state the $3p\pi \rightarrow k\pi(^1\Sigma^+)$ and $3p\pi \rightarrow k\delta(^1\Pi)$ photoionization channels closely resemble the corresponding channels of the $g^1\Delta$ state, and show Cooper minima at virtually the same electron kinetic energies. As was the case for the $g^1\Delta$ state, the kinetic energies addressed in the case of the $h^1\Sigma^+$ state are again somewhat outside the range where the Cooper minima reside. The good agreement between the experimental and calculated ion rotational branching ratios can therefore hardly come as a surprise [56].

### 5. *Summarizing Remarks*

The ability of laser PES to study configuration interaction in molecular excited states was demonstrated for the $d^1\Sigma^+$ valence excited state of the imidogen radical. The formally forbidden one-photon ionization to the $X^2\Pi$ ground ionic state becomes possible through configuration interaction, both in the excited state and the ionic state. The higher Rydberg states $i^1\Pi$ and $j^1\Delta$ states, belonging to the "3*d*" complex, also show effects of orbital mixing. For the $i^1\Pi$ state of ND (but not of NH) a significant $\Lambda$ doubling is observed that is ascribed to an interaction with an almost resonant vibrationally excited level of a lower lying $\Sigma$ state. The $j^1\Delta$ state provides a benchmark example of an "avoided crossing" in the rotational term values caused by interaction with another state. This perturbation shows up in both the excitation and the PE spectra, and gives insight into the mixed-orbital composition of the $j^1\Delta$ state. The lower lying $f^1\Pi$, $g^1\Delta$, and $h^1\Sigma^+$ Rydberg states, which belong to the "3*p*" complex, show rotationally resolved PE spectra. A combined experimental and theoretical study gives detailed information on the photoionization dynamics, and highlights the role of Cooper minima. The situation is strongly reminiscent of a laser PES study on the $D^2\Sigma^-$ Rydberg state of the OH radical [31].

## D. The ClO Radical

The physical and chemical processes in the earth's atmosphere that are at the root of phenomena such as ozone depletion and global warming are becoming increasingly understood. In the stratosphere, short-wavelength solar radiation causes fragmentation of parent molecules leading to short-lived reactive species that may be formed in electronically, vibrationally, and rotationally excited states, and that can initiate chain reactions leading to seriously depleted ozone concentrations. The ClO radical is a case in point, because under Antarctic conditions this radical and higher chlorine oxides resulting from the breakdown of chlorofluorocarbons (CFCs) under the influence of solar radiation are now known to be crucial in the catalytic ozone breakdown cycles in that part of the globe [162]. The Nobel Prize in Chemistry, awarded to Crutzen, Molina, and Rowland in 1995, recognized the scientific and societal importance of the elucidation of such atmospheric processes.

From the point of view of the chemical physicist, the ClO radical in its excited states is not all that well characterized. In the laboratory, the ClO radical can be produced in a number of ways. The earliest method was introduced by Pannetier and Gaydon [163] who seeded hydrogen–oxygen flames with molecular chlorine. Later Porter and Wright [164] observed ClO production after flash photolysis of $Cl_2/O_2$ mixtures. Another method of generating ClO is the reaction between Cl atoms, often produced in an electrodeless microwave discharge, and ozone [165–167]. Also, the reactions between O atoms and $Cl_2O$ [168], and between H and Cl atoms and $ClO_2$ [169, 170], have been used to produce ClO. In the work discussed in the present contribution, we generate ClO through photolysis of $ClO_2$ via its dissociative $\tilde{A}^2A_2$ excited state [171, 172]. We note that the main photodissociation channel, $ClO_2(\tilde{A}^2A_2) \rightarrow ClO(X^2\Pi, v'') + O(^3P)$, is enhanced when bands associated with the antisymmetric stretching mode are excited in the $\tilde{A}^2A_2$ state of $ClO_2$ [171, 173, 174]. The competing, but minor, dissociation channel giving $Cl(^2P) + O_2(^3\Sigma_g^-, ^1\Delta_g, ^1\Sigma_g^+)$ has also been investigated [175–177]. In all these production methods, the ClO radical is formed with different degrees of electronic, vibrational, and rotational excitation.

The ground-state of the ClO radical is characterized by the outer valence electron configuration $\ldots(7\sigma)^2(2\pi)^4(3\pi)^3$. This configuration leads to an inverted $^2\Pi_\Omega$ ground-state, with $^2\Pi_{3/2}$ below $^2\Pi_{1/2}$ separated by a spin–orbit splitting of $-321.77\,\text{cm}^{-1}$. The ClO in its $X^2\Pi_{3/2}$ state has been well-characterized with various experimental physical methods [178–183]. The lowest ionic states obtained by ejecting an electron out of the $3\pi$ antibonding orbital give a $\ldots(7\sigma)^2(2\pi)^4(3\pi)^2$ configuration leading to three relatively close-lying $^3\Sigma^-$, $^1\Delta$, and $^1\Sigma^+$ ionic states.

The lowest valence excited state, labeled $A$, with valence shell configuration $\ldots(7\sigma)^2(2\pi)^3(3\pi)^4$ has $^2\Pi_\Omega$ symmetry, and is inverted with a spin–orbit splitting of $-519.5\,\text{cm}^{-1}$ [184]. Promotion of a bonding $2\pi$ electron to an antibonding $3\pi$ orbital enlarges the bond length and lowers the vibrational frequency. The $A$ state has been studied extensively by absorption and emission spectroscopy [184–188], and is known to be predissociative [185]. The existence of a second valence excited state, $B^2\Sigma^+$, presumably with electron configuration $\ldots(7\sigma)^1(2\pi)^4(3\pi)^4$ [185], has not been confirmed directly. It is estimated to be located $\sim 31{,}000\,\text{cm}^{-1}$ above the ground-state, and is quite likely predissociative as well. The location of the $B$ state is inferred from the $\Lambda$ doubling observed in the electronic ground-state of the ClO radical [189].

Higher lying excited states of the ClO radical possess Rydberg character and six states labeled C to H have been characterized so far. These states are all formed by promoting an antibonding $3\pi$ electron into a predominantly nonbonding

Rydberg orbital. The lowest lying Rydberg state is the $C^2\Sigma^-$ state, with configuration $\ldots(7\sigma)^2(2\pi)^4(3\pi)^2(8\sigma)^1$. The Rydberg electron possesses predominantly $4s\sigma$ character at small internuclear distance, and significant $3d\sigma$ character at larger internuclear distances [190]. This state has been investigated previously by VUV absorption spectroscopy [182], laser induced fluorescence (LIF) [191, 192], and (2 + 1) REMPI spectroscopy [193]. Accurate spectroscopic constants are known from [181, 182, 184]. The $D(3\pi \rightarrow 9\sigma)$, $E(3\pi \rightarrow 10\sigma)$, and $F(3\pi \rightarrow 11\sigma)$ Rydberg states are all believed to have $^2\Sigma^-$ symmetry [190, 194]. The Rydberg electron in the $D$ state has dominant $4p\sigma$ character at small internuclear distance, and strong $4s\sigma$ character at larger $R$ [190]. The Rydberg electron in the $E$ state evolves rapidly from an $s, p, d$ admixture for small $R$ to dominant $4p\sigma$ character at larger $R$ [190]. For the $F$ state, there is a gradual change with internuclear distance from dominant $s\sigma$ and $p\sigma$ character to stronger $d\sigma$ character [190]. These Rydberg states have been studied with single-photon VUV absorption spectroscopy [182], and with (3 + 1) REMPI spectroscopy [195]. The $G$ and $H$ states are even less well known [193, 196].

Depending on the way in which ClO $X^2\Pi$ is generated, the distribution over the ground-state $^2\Pi_{3/2}$ and $^2\Pi_{1/2}$ spin–orbit components, and vibrational and rotational levels can vary substantially. It is important to obtain experimental information about this distribution. Not only does this provide significant clues to the mechanisms that produce ClO, in addition it will affect the chemical reactivity of the ClO radical under atmospheric conditions. The ClO is observed to be formed vibrationally cold after reaction of Cl with $O_3$ [195], or vibrationally excited up to $v' = 3, 4$ using LIF detection [191, 192]. When ClO is formed following photodissociation of $ClO_2$, the amount of ClO generated and its state distribution is known to vary with the photolysis wavelength [191, 197, 198]. The one-photon absorption spectrum of the precursor compound shows significant (vibrational) structure, and intensity changes due to variations in the photolysis efficiency can not be distinguished reliably from those due to intrinsic changes in the excitation spectrum of the ClO radical formed. When employing the latter production mechanism it may therefore be important to use a separate photolysis laser whose wavelength is optimised for ClO generation. This can be achieved by tuning the photolysis laser to appropriate vibrational features of the $\tilde{A}^2A_2(\nu_1, \nu_2, \nu_3) \leftarrow \tilde{X}^2B_1$ dissociative transition of $ClO_2$ [171, 197].

Next, we shall discuss a one-color REMPI–PES study of the $E$, $F$, and $G$ Rydberg states of ClO. From this work, the origins of the transitions to these states can be assigned unambiguously. The $E$ and $F$ states will be shown to possess an $X^3\Sigma^-$ ionic core, while the $G$ state has an excited $a^1\Delta$ core [194]. Subsequently, we shall describe a REMPI–PES study of the $C$ Rydberg state of ClO in which a separate fixed laser is used to control the ClO production during the experiment [199]. This state is also found to possess an $X^3\Sigma^-$ ionic core, and a Rydberg electron of $k\sigma$ symmetry.

### *1. The ClO $E^2\Sigma^-$, $F^2\Sigma^-$, and G Rydberg States*

Excitation spectra are commonly recorded either by employing mass-resolved ion detection or by monitoring the total electron count. The latter method in the present case leads to overlapping spectra from the $^{35}$ClO and $^{37}$ClO isotopomers (ratio 3:1) and so ion detection, preferably on $^{35}ClO^+$, was employed when advantageous. In Figure 32 the three-photon excitation spectrum in the one-photon energy region between 21,650 and 23,050 cm$^{-1}$, obtained with $^{35}ClO^+$ detection, is depicted. Photoelectron spectra (not shown) were obtained for the various features in this excitation spectrum and showed predominantly simple one-line character. Clearly, ionization takes place to the lowest $X^3\Sigma^-$ ionic state with $\Delta v = 0$ propensity, indicating an $X^3\Sigma^-$ ionic core for the $E^2\Sigma^-$ state. On the basis of the electron kinetic energies the peaks in the excitation spectrum could be consistently assigned as three-photon transitions from vibrational levels of the $X^2\Pi_{3/2,1/2}$ ground-state to vibrational levels of the $E$ state of ClO [195, 196]. The origin of the $E$ Rydberg state, observed at $\sim$ 67,467 cm$^{-1}$, agrees with the value proposed previously [196]. An improved value of the energy of

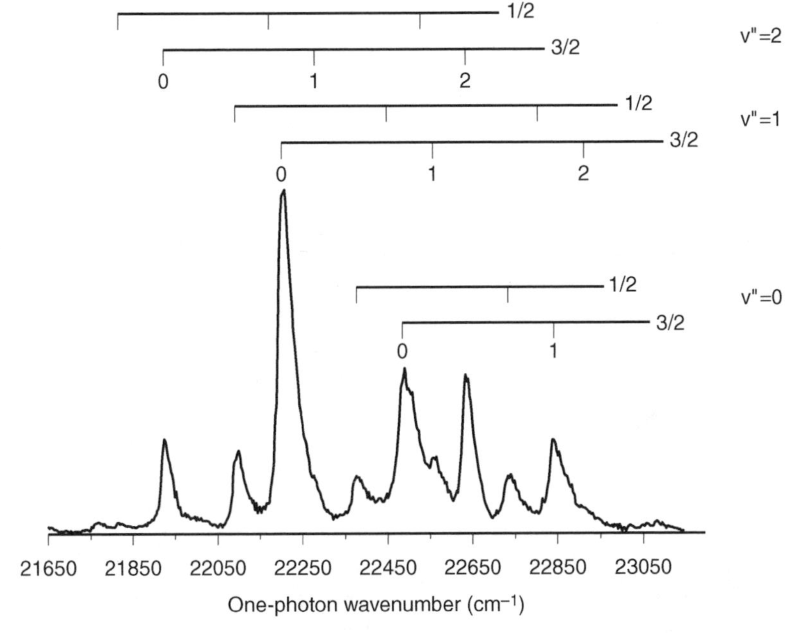


**Figure 32.** Three-photon excitation spectrum of the $E^2\Sigma^-$ Rydberg state of ClO in the one-photon energy region 21,650–23,050 cm$^{-1}$, employing mass-resolved ion detection monitoring the $m/z = 51$ ion mass channel. The various combs indicate the vibrational numbering of the excited states accessed via various vibrational levels of the $X^2\Pi_{3/2,1/2}$ ground-state (indicated on the right-hand side of the figure). (Reproduced with permission from [194], copyright 1996 Elsevier Science B.V.)

the lowest ionic state is found to be $10.887 \pm 0.005$ eV, which compares well with the previous value of $10.87 \pm 0.01$ from a He(I) PES study [166]. The simple one-line character of the PE spectra is suggestive of a "pure" Rydberg state. Previous *ab initio* calculations indicate a rather complex composition for the Rydberg orbital, with an *s*, *p*, *d* admixture for small $R$ evolving to dominant $4p\sigma$ character at larger $R$ [190]. However, this mixing is not reflected in the PE spectra since on ionization from every orbital contribution the same ionic state is reached.

Figure 33 shows excitation spectra of ClO in the one-photon energy region of 23,800–24,500 $\text{cm}^{-1}$, employing electron detection of low [$E_k < 0.6$ eV, Fig. 33(*a*)], and of high [$E_k > 0.6$ eV, Fig. 33(*b*)] kinetic energy electrons. Due to a dip in the dye gain curve of the Excalite 416 dye $\sim 23{,}900\ \text{cm}^{-1}$, a second scan with Excalite 416 employing mass-resolved ion detection was carried out and is shown as an insert in Figure 33(*a*). The high kinetic energy electrons are associated with ionization from higher vibrational levels of the $F^2\Sigma^-$ Rydberg state. The low kinetic energy electrons, on the other hand, not only derive from ionization from high vibrational levels of the $E^2\Sigma^-$ Rydberg state, but also from the $G$ Rydberg state. Excitation spectra in the one-photon energy region 22,800–23,800 $\text{cm}^{-1}$ were also obtained [194]. They show low vibrational levels associated with the $F^2\Sigma^-$ state, and intermediate vibrational levels associated with the $E^2\Sigma^-$ state. From our entire series of excitation spectra, it is clear that the ClO radical following photofragmentation of $ClO_2$ is produced with vibrational excitation, at least to $v'' = 4$ [194].

In Figure 34, the PE spectra associated with the various vibrational levels of the $F^2\Sigma^-$ Rydberg state are depicted. From these spectra, the origin of the $F$ state, previously suggested at $\sim 70{,}179\ \text{cm}^{-1}$ [195], is confirmed. Apart from photoelectrons deriving from ionization of the $F$ state into the $X^3\Sigma^-$ continua, photoelectrons are also observed from ionization of higher vibrational levels of the $E$ state into the $X^3\Sigma^-$ continua, and from vibrational levels of the $G$ state into the $a^1\Delta$ continua (see below). These $E$ and $G$ vibrational levels are in accidental coincidence with those of the $F$ state, as can be seen from the various excitation spectra. For the $F$ state ionization the $\Delta v = 0$ propensity is not as strictly obeyed as seen before for the $E$ state. This behavior implies that, in contrast to the $E$ state, the $F$ Rydberg state has a slightly different potential energy curve from that of the $X^3\Sigma^-$ ionic state. As with the $E$ state, the calculated orbital mixing [190] cannot be observed with PES, since on ionization from every orbital contribution the same ionic state is reached.

The symmetry of the $G$ state, which has been observed before in a one-photon VUV study [196], is unknown. The suggested symmetry of $^2\Pi$ or $^2\Delta$ cannot be determined in our experiments, but its previously proposed origin [196] at $\sim 73{,}997\ \text{cm}^{-1}$ is confirmed. Our PE spectrum (not shown) from the $G(v' = 0)$ level shows a single line, indicating $\Delta v = 0$ propensity, but to the

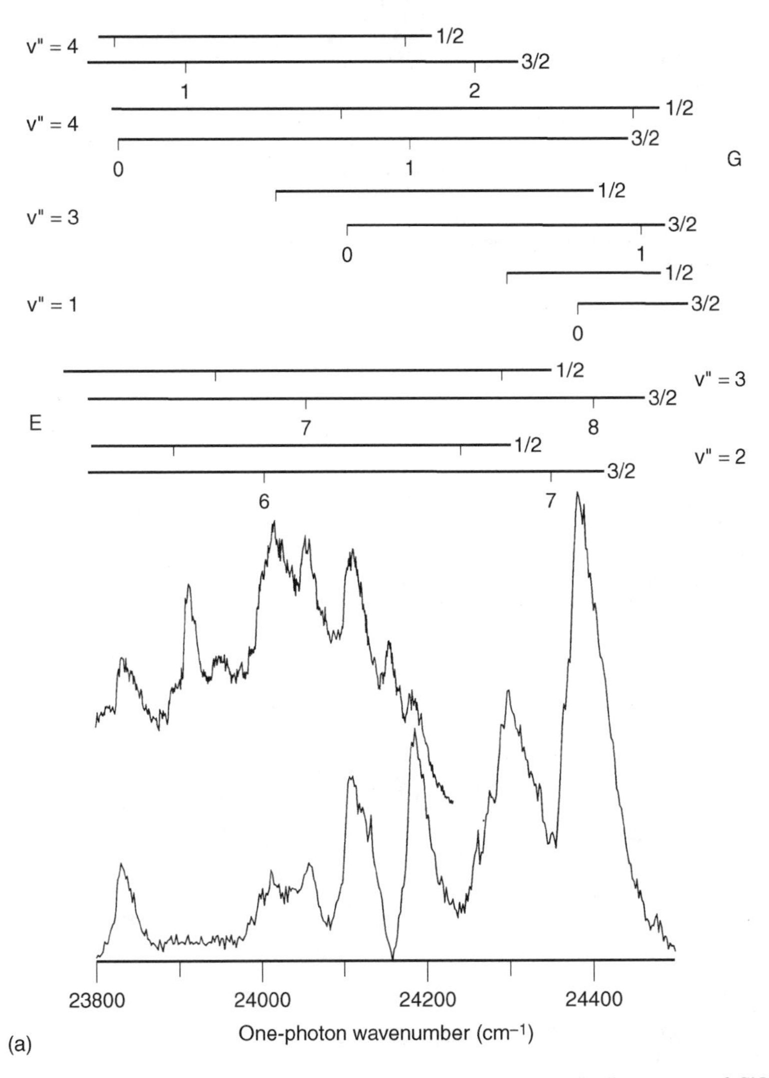


**Figure 33.** Three-photon excitation spectrum of the $E$, $F$, and $G$ Rydberg states of ClO in the one-photon energy region 23,800–24,500 cm$^{-1}$, employing electron detection of ($a$) low ($E_k < 0.6$ eV); and ($b$) of high ($E_k > 0.6$ eV) kinetic energy electrons. The insert in ($a$) is a scan obtained with mass-resolved $m/z = 51$ ion detection. The various combs indicate the vibrational numbering of the excited states accessed via various vibrational levels of the $X^2\Pi_{3/2,1/2}$ ground-state (indicated on the left- and right-hand side of the figure). (Reproduced with permission from [194], copyright 1996 Elsevier Science B.V.)

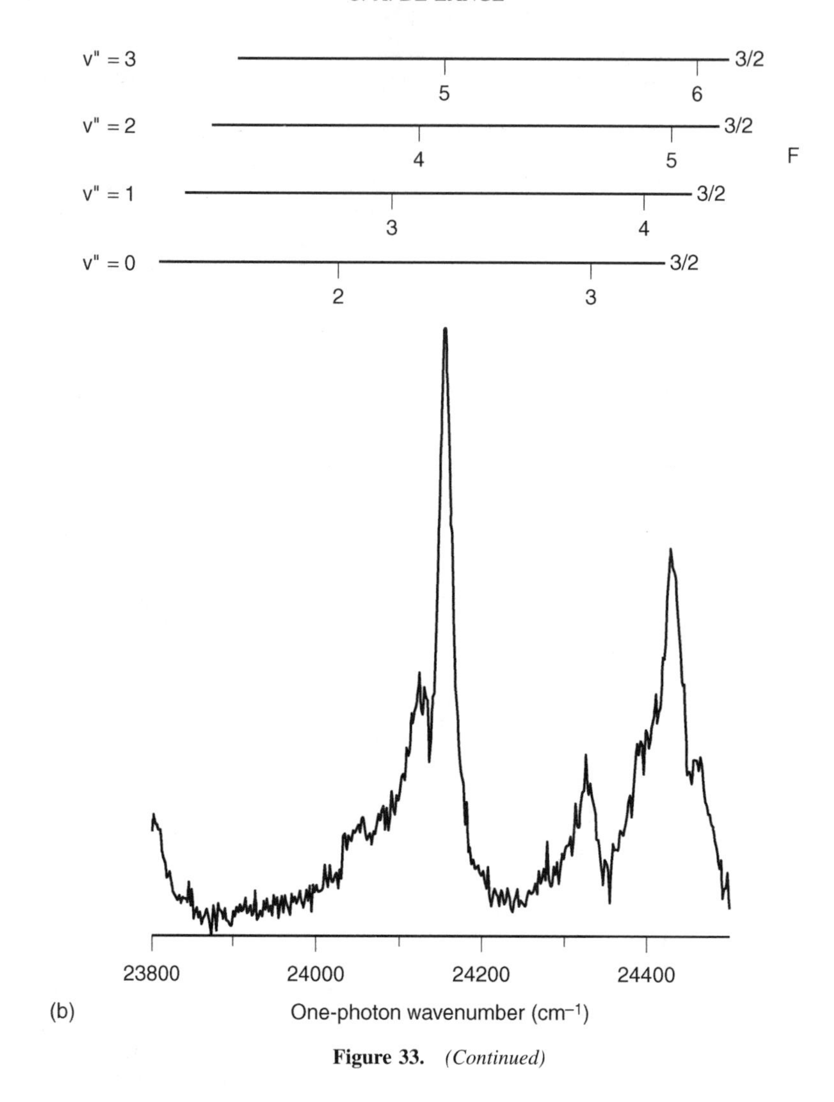


**Figure 33.** *(Continued)*

excited $a^1\Delta$ ionic state. In all the PE spectra to higher vibrational levels of the $G$ state, $\Delta v = 0$ transitions prevail, suggesting a potential energy curve of the $G$ state that is very similar in shape and with a minimum at an internuclear distance very comparable to that of the $a^1\Delta$ ionic state. From this work an improved value of the ionization energy of the $a^1\Delta(v^+ = 0) \leftarrow X^2\Pi_{3/2}(v'' = 0)$ transition is determined, and found to be $11.750 \pm 0.005$ eV. This value compares well with the previous one of $11.74 \pm 0.01$ from He(I) PES [200].

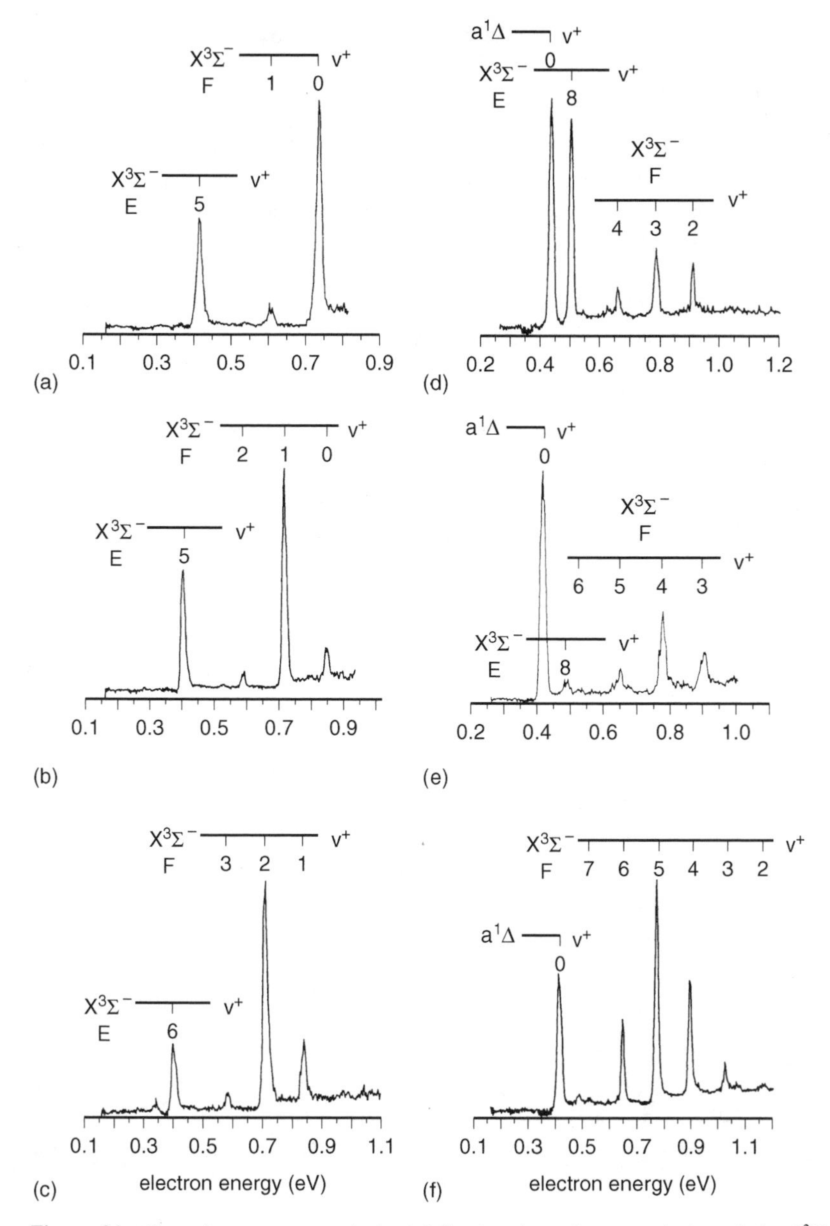


**Figure 34.** Photoelectron spectra obtained following three-photon excitation of the $F^2\Sigma^-$ Rydberg state of ClO via the *(a)* $v' = 0$; *(b)* $v' = 1$; *(c)* $v' = 2$; *(d)* $v' = 3$; *(e)* $v' = 4$; and *(f)* $v' = 5$ vibrational levels. (Reproduced with permission from [194], copyright 1996 Elsevier Science B.V.)

### 2. *The ClO $C^2\Sigma^-$ Rydberg State*

In Figure 35, we present an excitation spectrum of the ClO radical in the two-photon energy range 59,000 and 62,500 $\mathrm{cm}^{-1}$. This spectrum was obtained in a $(2+1)$ REMPI process employing $^{35}\mathrm{ClO}^+$ ion detection. However, the crucial difference with the spectra shown for the $E$, $F$, and $G$ states in the Section IV.D.1 is that here a separate photolysis laser with a wavelength of 359.2 nm ($\sim 27{,}840\,\mathrm{cm}^{-1}$) was added which pumps the $\tilde{A}^2A_2(10,0,0) \leftarrow \tilde{X}^2B_1(0,0,0)$ one-photon transition of $ClO_2$. This procedure ensures that the production of ClO is approximately constant, leading to an excitation spectrum that is more representative for the spectroscopy of the ClO radical. Several features arising from the $C^2\Sigma^-(v') \leftarrow\leftarrow X^2\Pi_\Omega(v'')$ transition are observed and extensive and not completely resolved rotational structure is present. In order to aid with the analysis, simulations of the excitation spectrum were attempted using rotational constants $B_e$, rotation–vibration coupling constants $\alpha$ and $\gamma$, and vibrational constants $\omega_e$ and $\omega_e x_e$ for the $X^2\Pi$ [181, 182, 184] and $C^2\Sigma^-$ [182] states, as well as the ground-state spin–orbit splitting $A$ [182]. An estimated temperature

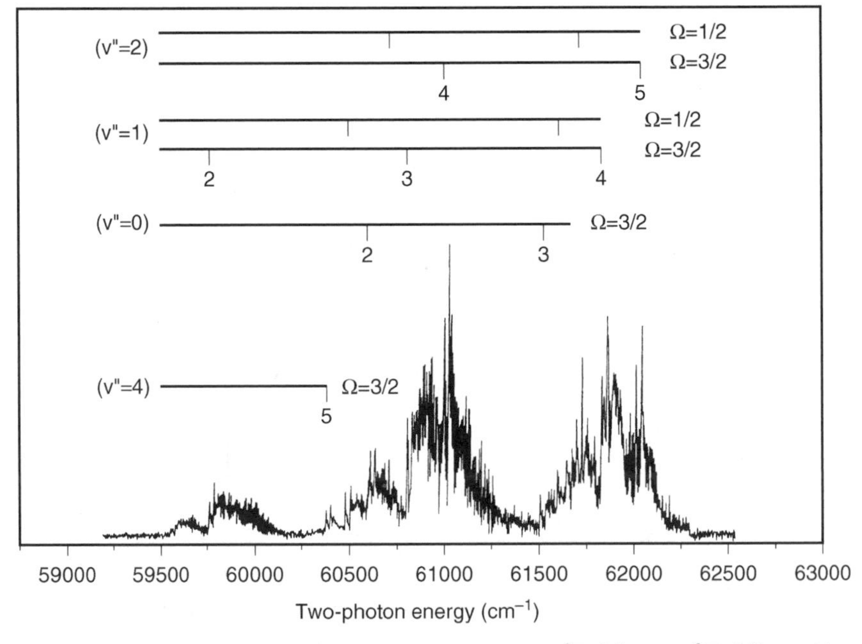


**Figure 35.** Experimental excitation spectrum of the ClO $C^2\Sigma^-(v') \leftarrow\leftarrow X^2\Pi_\Omega(v'')$ transition using a separate laser at 359.2 nm for the photolysis of $ClO_2$ and a second tunable laser to scan the ClO transition. This $(2+1)$ REMPI spectrum was obtained using $^{35}$ClO ion detection. Assignments are given above the spectrum. (Reproduced with permission from [199], copyright 2000 The Royal Society of Chemistry.)

of 500 K was employed. The agreement with the experimental spectrum is quite reasonable. On the basis of our simulations we assign various transitions of the type $C^2\Sigma^-(v' = 2\text{–}5) \leftarrow\leftarrow X^2\Pi_{1/2,3/2}(v'' = 0\text{–}4)$, as indicated in Figure 35 [199].

At our photolysis wavelength of 359.2 nm the excess kinetic energy available is $\sim 9050\,\text{cm}^{-1}$, more than sufficient for the observed vibrational excitation. From our spectra it is not easy to decide whether the ClO radical is formed in higher vibrational levels than $v'' = 4$. The observable transitions expected from these levels either occur outside the spectral range accessed during the present experiments, or are predicted to be weak. The expected low intensities arise from the small Franck–Condon factors that play a role in the two-photon excitation processes $C^2\Sigma^-(v') \leftarrow\leftarrow X^2\Pi_{1/2,3/2}(v'' > 4)$, which appear in our spectral range [201]. Our results, therefore, do not allow any conclusions about ClO $X^2\Pi_{1/2,3/2}(v'' > 4)$ formation.

In a recent study, $ClO_2$ was photolyzed in the spectral region between 335 and 370 nm [198] in a single-laser experiment, and $ClO^+$ was monitored with REMPI. The ClO was found to be generated in its spin–orbit split electronic ground-state in vibrational levels $v'' = 3\text{–}6$. With a laser wavelength of 360.2 nm or 27,762 cm$^{-1}$ one-photon excitation of $ClO_2$ to its $\tilde{A}^2A_2(10,0,0)$ state takes place with significant ClO $X^2\Pi_\Omega(v'' = 5)$ production. Previous experiments that involved the wavelength-dependent formation of vibrationally excited ClO from the photolysis of $ClO_2$ include a study in the spectral region 356–370 nm [202]. Employing a laser at 351 nm to excite the $\tilde{A}^2A_2(11,0,0)$ dissociative band of $ClO_2$ the formation of ClO $X^2\Pi_\Omega$ in vibrational levels $v'' = 0\text{–}4$ was observed [197]. Our experiments essentially confirm such results.

Fixing the probe laser wavelength on specific features in the excitation spectrum of the $C^2\Sigma^-$ state and analyzing the ejected electrons according to their kinetic energies, PE spectra from the $C^2\Sigma^-$ state to the accessible ionic states were observed. In Figure 36, these spectra are depicted, obtained at two-photon energies of (a) 60,488 cm$^{-1}$, which corresponds to the $O$ branch band head in the $C^2\Sigma^-(v' = 3) \leftarrow\leftarrow X^2\Pi_{1/2}(v'' = 1)$ transition; (b) 60,808 cm$^{-1}$, which corresponds to the $O$ branch band head in the $C^2\Sigma^-(v' = 3) \leftarrow\leftarrow X^2\Pi_{3/2}(v'' = 1)$ transition; (c) 61,832 cm$^{-1}$, which corresponds to the $O$ branch band head in the $C^2\Sigma^-(v' = 4) \leftarrow\leftarrow X^2\Pi_{3/2}(v'' = 1)$ transition; and (d) 62,012 cm$^{-1}$, which corresponds to the $O$ branch band head in the $C^2\Sigma^-(v' = 5) \leftarrow\leftarrow X^2\Pi_{3/2}(v'' = 2)$ transition [199].

The PE spectra only show evidence for one-photon transitions from the $C^2\Sigma^-$ state to the $X^3\Sigma^-$ ionic state. This is not surprizing, since the $C^2\Sigma^-$ state can be viewed as the lowest Rydberg state belonging to a series converging upon the lowest ionic state. Often for a Rydberg state the shape and position of the potential energy curve are very similar to that of the ionic state upon which the series converges, leading to Franck–Condon diagonal spectra with

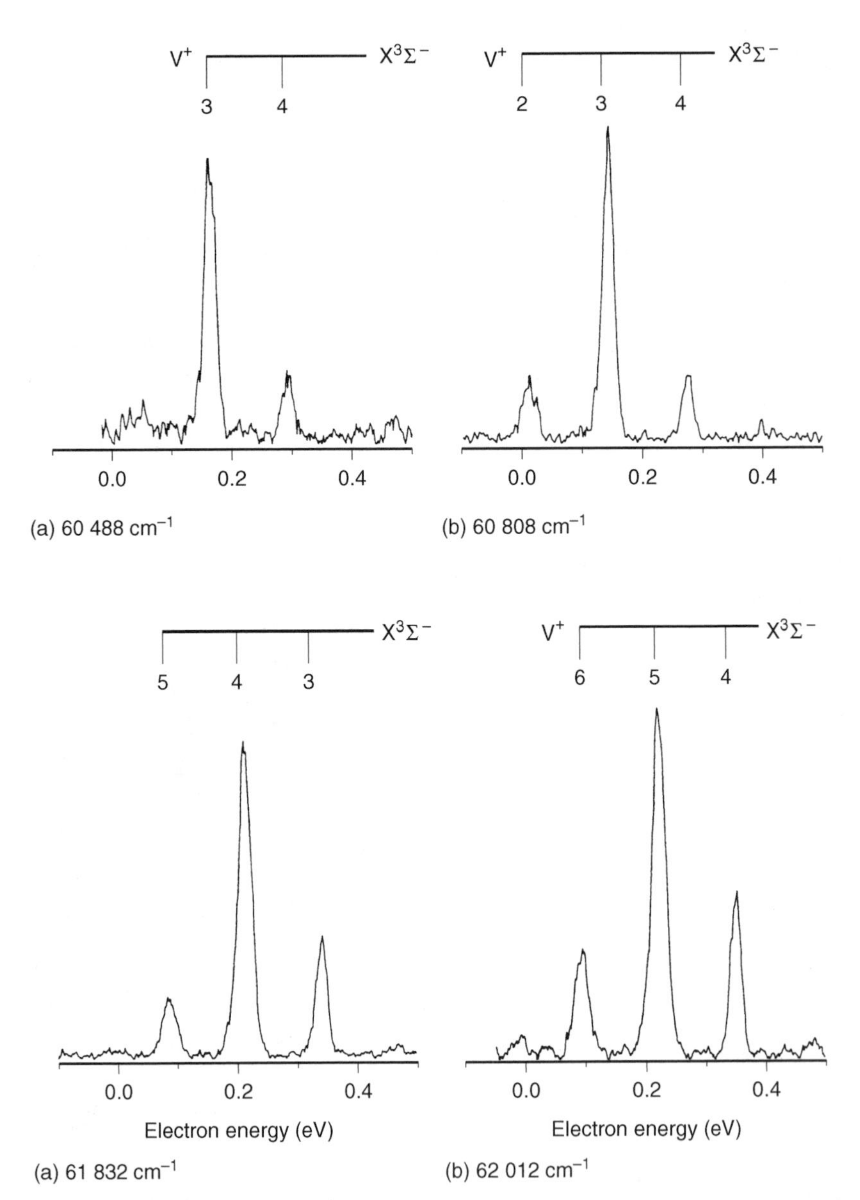


**Figure 36.** PE spectra for ClO obtained at two-photon energies of (*a*) 60,488 cm$^{-1}$, which corresponds to the *O* branch band head in the $C^2\Sigma^-(v'=3) \leftarrow\leftarrow X^2\Pi_{1/2}(v''=1)$ transition; (*b*) 60,808 cm$^{-1}$ which corresponds to the *O* branch band head in the $C^2\Sigma^-(v'=3) \leftarrow\leftarrow X^2\Pi_{3/2}(v''=1)$ transition; (*c*) 61,832 cm$^{-1}$, which corresponds to the *O* branch band head in the $C^2\Sigma^-(v'=4) \leftarrow\leftarrow X^2\Pi_{3/2}(v''=1)$ transition; and (*d*) 62,012 cm$^{-1}$, which corresponds to the *O* branch band head in the $C^2\Sigma^-(v'=5) \leftarrow\leftarrow X^2\Pi_{3/2}(v''=2)$ transition. (Reproduced with permission from [199], copyright 2000 The Royal Society of Chemistry.)

$\Delta v = v^+ - v' = 0$. Indeed, all PE spectra show a maximum intensity at the diagonal transition. However, the off-diagonal Franck–Condon factors are not negligible and indicate a slight difference in bond lengths between the $C^2\Sigma^-$ state and the lowest $X^3\Sigma^-$ ionic state. Moreover, the PE spectra confirm unambiguously that the $C^2\Sigma^-$ state can be described predominantly as a ${}^3\Sigma^-$ core with an external Rydberg electron, which is removed on photoionization [199].

The PE spectra arising from $C^2\Sigma^-(v' = 3)$ are similar in appearance, with the dominant transition to $X^3\Sigma^-(v^+ = 3)$ as expected for photoionization from a Rydberg state. The $C^2\Sigma^-(v' = 3)$ state in (a) is reached from ClO $X^2\Pi_{1/2}$, and in (b) from $X^2\Pi_{3/2}$. For one-photon ionization the difference in the positions of the dominant peaks in the PE spectra shown in Figure 36(*a*) and (*b*) should therefore correspond to one-half of the spin–orbit splitting in the neutral ground-state. Indeed, a shift of $19 \pm 1$ meV is observed in going from Figure 36(*a*) to (*b*). The PE spectra arising from $C^2\Sigma^-(v' = 4)$ and $C^2\Sigma^-(v' = 5)$ show dominant transitions to $v^+ = 4$ and $v^+ = 5$, respectively, but for higher $v'$ off-diagonal Franck–Condon factors become progressively important. In our PE spectra, we observed vibrational spacings in the $X^3\Sigma^-$ lowest ionic state which correspond well with an earlier experimental value of $1040\,\text{cm}^{-1}$ obtained by conventional PES [166]. The slight differences that exist for the $^{35}$ClO and $^{37}$ClO isotopomers are not resolved.

Finally, our experimental Franck–Condon factors measured for the $X^3\Sigma^-(v^+) \leftarrow C^2\Sigma^-(v')$ one-photon ionization transitions can be compared with the results of calculations based on *ab initio* and Morse potential energy curves [199]. The *ab initio* calculated Franck–Condon factors are more diagonal than those computed from the Morse curves, but the overall correspondence is reasonable. The agreement between our experimental Franck–Condon factors and those computed from our Morse type potentials is optimal for a bond length of $1.475 \pm 0.005$ Å for $ClO^+$ in its $X^3\Sigma^-$ state, a value in agreement with the previous experimental estimate of $1.48 \pm 0.01$ Å [166].

## V. LINEAR THREE-ATOMIC MOLECULES WITH 16 VALENCE ELECTRONS

### A. General Aspects

In this section, we shall be concerned with three-atomic molecules that possess 16 valence electrons and have a linear electronic ground-state, namely, $CO_2$, OCS, $CS_2$, and $N_2O$. These molecules have all been studied extensively with laser PES and our knowledge about their excited states has been amplified considerably [48–50, 203]. Compared to diatomics the multiphoton excitation spectra of these species tend to exhibit much more extensive vibronic structure, and the associated assignment problems are usually far from trivial. However, many of the physical processes and mechanisms that were illustrated in some detail for the diatomic molecules, will return in some shape or another for these three-atomic species, albeit often in a more disguised form. With laser PES many of these features have been brought to light in detail. Since the valence electronic structures of the linear three-atomics show strong similarities, it is not surprising that results obtained with REMPI–PES have many features in common. Rather than concentrating on every molecule in turn, we shall therefore summarize the important phenomena and processes which occur in every molecule by taking one of the species as a representative example for the entire series.

Discovering and assigning new molecular Rydberg series, or establishing unambiguous assignments for Rydberg series that have been detected before with optical absorption spectroscopy is one of the strong points of laser PES. In particular, the accurate determination of origins and the precise numbering of the various members of the series has helped to create a reliable body of data on excited molecular Rydberg states. The three-atomic molecules to be discussed here form no exception in this regard. In $CO_2$ the $[\tilde{X}^2\Pi_{g3/2,1/2}]nf$ series has been studied extensively for $n = 4$–$13$ with $(3+1)$ REMPI–PES in the three-photon energy range 103,500–111,000 $cm^{-1}$ [203]. In OCS $(2+1)$ and $(3+1)$ REMPI–PES in the energy range 70,500–85,000 $cm^{-1}$ was carried out. Five Rydberg origins originating from the $4p\lambda \leftarrow 3\pi$ orbital promotion were studied and assigned unambiguously in the 70,500–73,000 $cm^{-1}$ energy range, and a further 21 origins arising from $s$, $p$, $d$ and (possibly) $f$ Rydberg complexes were observed at higher energies [49]. In $CS_2$ $(2+1)$ and $(3+1)$ REMPI–PES in the energy range 56,000–81,000 $cm^{-1}$ was carried out. Rydberg states arising from the $[\tilde{X}^2\Pi_{g3/2,1/2}]4p\sigma_u$ $(^{1,3}\Pi_u)$ and $[\tilde{X}^2\Pi_{g3/2,1/2}]4p\pi_u$ $(^{1,3}\Delta_u, ^1\Sigma_u^+)$ were studied and partly reassigned, while at higher energies higher members of the $[\tilde{X}^2\Pi_g]np$ complexes, as well as members of the $[\tilde{X}^2\Pi_g]ns$ and $[\tilde{X}^2\Pi_g]nf$ complexes for $n \leq 10$ were observed [48]. In $N_2O$ a $(3+1)$ REMPI–PES study was performed in the spectral range from 80,000 $cm^{-1}$ up to the lowest ionization limit of 103,963 $cm^{-1}$. Rydberg states of the type $[\tilde{X}^2\Pi]ns\sigma$ $(^{1,3}\Pi)$, $[\tilde{X}^2\Pi]np\sigma$ $(^{1,3}\Pi)$, $[\tilde{X}^2\Pi]np\pi$ $(^{1,3}\Sigma^+)$, and $[\tilde{A}^2\Sigma^+]ns\sigma$ $(^1\Sigma^+)$, were observed and characterized [42]. In Section V.B the $[\tilde{X}^2\Pi_{g3/2,1/2}]nf$ $(n = 4$–$10)$ complexes of $CS_2$ will be treated as an example.

It has been emphasized before that laser PES possesses a unique capability in that orbital interactions can be made experimentally "visible" when the various configurations contributing to the molecular orbital derive from different ionic cores. Examples of configuration interaction due to electronic coupling have been discussed and illustrated for the NH and SH radicals in Section IV. Vibronic coupling between different states is another mechanism that can lead to orbital mixing in excited states observable in the corresponding PE spectra. In molecules larger than diatomics the mechanism of vibronic coupling, rather than being an exception, appears to be omnipresent. In Section IV.C, we shall treat illustrative examples taken from laser PES of $CS_2$ [48] and $N_2O$ [42].

The role of superexcited states was a central topic in the case of molecular hydrogen and it isotopomers treated in Section III, and of the SH radical discussed in Section IV. Optical excitation of superexcited states and subsequent decay processes such as autoionization into the underlying ionic continua and dissociation into (excited) fragments in many cases seriously interfere with the simple direct ionization process. These superexcited states can either be directly visible employing laser PES (as shown in the SH case), or their effect can be observed in an indirect fashion (as illustrated in the case of molecular hydrogen). As can be expected, for larger molecules the importance of superexcited states will only increase. In Section V.D, the crucial role played by superexcited states belonging to Rydberg series of $CO_2$ leading up to the $\tilde{A}^2\Pi_u$, $\tilde{B}^2\Sigma_u^+$, and $\tilde{C}^2\Sigma_g^+$ ionic limits and situated in the $\tilde{X}^2\Pi_{g3/2,1/2}$ ionization continua will be discussed [203].

In laser PES high laser pulse energies are routinely employed. Multiphoton absorption does not usually terminate at the desired excited state level, but absorption of more photons to dissociative states can take place. Alternatively, competing dissociative processes at the level of the excited state itself, or at lower photon levels, can also cause extensive fragmentation. It is therefore important to monitor the fragments originating from this multitude of possible decay processes. Often a study of the fragments generated can give insight into the photodissociation processes. Again, the linear three-atomics of this section form no exception, and extensive fragmentation can be observed in the entire series. For $CO_2$, the production of C and O atoms generated in a variety of excited states has been established unambiguously. Moreover, the experiments suggest strongly that CO is formed in lower excited Rydberg states from which one-photon ionization into the $CO^+$ $X^2\Sigma^+(v^+)$ ionic continua takes place [203]. For $CS_2$, strong signals due to ground-state C ($^3P$) and ground and excited state S ($^3P$, $^1D$) atoms, as well as transitions arising from excited CS ($a^3\Pi$) fragments were observed [48]. For $N_2O$ at $\sim$ 203 nm, an excitation takes place to a bent dissociative state, resulting in fragmentation into excited O ($^1D$) atoms and highly rotationally excited ground-state $N_2$ ($X^1\Sigma_g^+$) for which rotationally resolved PES is observed [204]. This process is of some relevance to atmospheric chemistry. In Section V.E, we shall treat a similar dissociation process in OCS as an example [49, 50].

Laser PES has significant advantages over the technically simpler experiment of multiphoton ionization with mass-resolved ion detection. A convincing demonstration of what can be achieved with kinetic energy resolved ion detection will be illustrated in Section V.F, for $CS_2$ [48, 50].

### B. The $[\tilde{X}^2\Pi_{g3/2,1/2}]nf$ ($n = 4$–$10$) Rydberg Complexes of $CS_2$

Carbon disulfide, a molecule that possesses an inversion center, has a ground-state valence electron configuration $\ldots(5\sigma_g)^2(4\sigma_u)^2(6\sigma_g)^2(5\sigma_u)^2\ (2\pi_u)^4(2\pi_g)^4$

$\tilde{X}^1\Sigma_g^+$. The highest occupied $2\pi_g$ orbital is largely nonbonding with the electron density mainly on the terminal sulfur atoms. The molecule has three fundamental vibrational modes: $\nu_1$, the symmetric stretch (which tranforms as $\sigma_g^+$), $\nu_2$, the degenerate bend ($\pi_u$), and $\nu_3$, the asymmetric stretch ($\sigma_u^+$), the ground-state frequencies of which are 658, 396, and 1535 cm$^{-1}$, respectively [2]. The linear ionic ground-state, formed by removing a $2\pi_g$ electron, is a spin–orbit split state with $\tilde{X}^2\Pi_{g3/2} \sim 440\,\text{cm}^{-1}$ below $\tilde{X}^2\Pi_{g1/2}$, with a lowest ionization threshold of $81{,}286 \pm 5\,\text{cm}^{-1}$, and with vibrational normal mode frequencies of 620, 332, and 1195 cm$^{-1}$ [205–208].

In the region 290–410 nm, one-photon absorption techniques have revealed a wealth of resolved rovibronic structure, much of which has now been assigned to bent valence states arising from the $3\pi_u \leftarrow 2\pi_g$ electronic excitation [209–213]. In the range 190–210 nm, both indications for a bent valence state [214] and of Rydberg states converging upon the two spin–orbit components of the ground-state ion have been identified [215]. The energy range up to the first ionization limit has been studied subsequently in one-photon absorption [215–219] and with electron impact spectroscopy [220–221], but multiphoton studies have remained relatively scarce and fragmentary [222–226].

For centrosymmetric molecules such as $CS_2$, multiphoton absorption is valuable, because the number of photons used helps in the assignment. Starting from the $\tilde{X}^1\Sigma_g^+$ ground-state with an odd number of photons Rydberg states with a $^2\Pi_g$ ion core and an *ungerade* (e.g., *p* or *f*) Rydberg electron can be accessed, while with an even number of photons Rydberg states with a $^2\Pi_g$ ion core, but with a *gerade* (e.g., *s* or *d*) Rydberg electron can be reached. In a $(3+1)$ REMPI–PES experiment we have been able to observe the $[\tilde{X}^2\Pi_{g3/2,1/2}]nf\lambda_u$ ($n = 4$–10) Rydberg complexes of $CS_2$ in the energy range 68,400–81,000 cm$^{-1}$. The excitation spectrum, obtained by monitoring the total PE yield as a function of laser wavelength, is presented in Figure 37. Since the total spectrum exceeds the tuning range of a single laser dye, the spectrum has been constructed by splicing together spectra recorded using a number of different dyes. No undue emphasis should therefore be placed on relative intensities. Assignments are given by the combs above the spectrum. Transitions associated with carbon and sulfur atoms are also identified. Above 74,000 cm$^{-1}$ the $(3+1)$ spectrum shows a dense progression of sharp bands that can be arranged into features belonging to two well-defined series, split by $\sim 440\,\text{cm}^{-1}$, converging upon the two spin–orbit components of the ground-state ion. The accompanying REMPI–PES spectra show single line character, and indicate these transitions to be origins. The assigments are supported by the experimental quantum defects, and by the similarities noted with the $[\tilde{X}^2\Pi_{g3/2,1/2}]nf\lambda_u \leftarrow\leftarrow\leftarrow \tilde{X}^1\Sigma_g^+$ transitions observed in the related centrosymmetric molecule $CO_2$ [203].

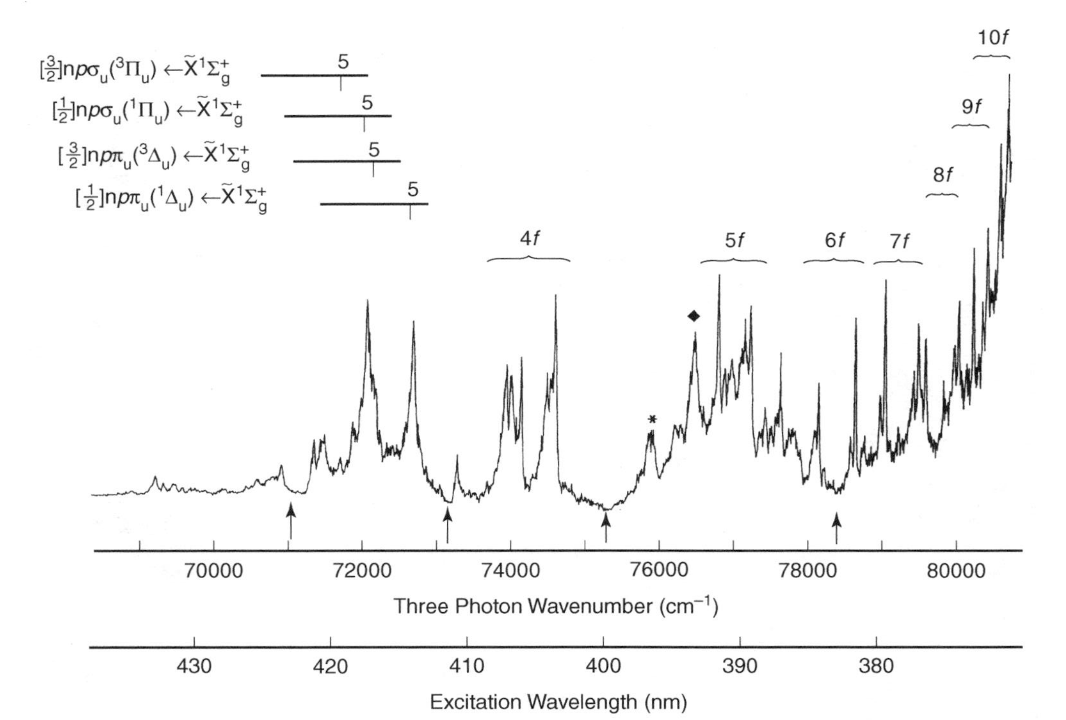


**Figure 37.** The $(3+1)$ REMPI excitation spectrum of a near room temperature sample of $CS_2$ in the energy range 68,400–81,000 $cm^{-1}$. The spectrum was obtained by monitoring the total PE yield as a function of laser wavelength. Assignments are indicated by combs above the spectrum. The peaks identified by the symbols * and ◆ correspond to carbon and sulfur atom transitions, respectively. (Reproduced with permission from [48], copyright 1996 American Institute of Physics.)

### C. Vibronic Coupling in $N_2O$ and $CS_2$

The $N_2O$ molecule is without an inversion center, and has a ground-state valence electron configuration $\ldots(4\sigma)^2(5\sigma)^2(6\sigma)^2(1\pi)^4(7\sigma)^2(2\pi)^4\ \tilde{X}^1\Sigma^+$. The molecule possesses three fundamental vibrational modes: $\nu_1$, the symmetric stretch (which tranforms as $\sigma^+$), $\nu_2$, the degenerate bend ($\pi$), and $\nu_3$, the asymmetric stretch ($\sigma^+$), the ground-state frequencies of which are 1284.91, 588.77, and 2223.76 cm$^{-1}$, respectively [227]. The linear ionic ground-state, formed by removing a $2\pi$ electron, is a spin–orbit split state with $\tilde{X}^2\Pi_{3/2} \sim 134\,\text{cm}^{-1}$ below $\tilde{X}^2\Pi_{1/2}$, with a lowest ionization threshold of $103{,}963 \pm 5\,\text{cm}^{-1}$ [228]. For $\tilde{X}^2\Pi_{3/2}$ the vibrational normal mode frequencies are 1126.51 cm$^{-1}$ for $\nu_1$, and 1737.6 cm$^{-1}$ for $\nu_3$, with very similar values for the $\tilde{X}^2\Pi_{1/2}$ state [227]. The degenerate $\nu_2$ mode shows extensive Renner–Teller splitting [229].

The study of excited states of $N_2O$ below its lowest ionic $\tilde{X}^2\Pi$ limit has been carried out with one-photon absorption [230–232], electron impact [233–234], and REMPI spectroscopy [42, 235, 236]. Extensive Rydberg structure has been observed, especially above 80,000 cm$^{-1}$. For the present discussion, we shall focus on our $(3+1)$ REMPI–PES results obtained for $N_2O$ in the energy range 100,800–104,200 cm$^{-1}$ [42]. In this range, $[\tilde{X}^2\Pi_{3/2,1/2}]ns\sigma\ (n = 7\text{–}11) \longleftarrow\!\!\longleftarrow\!\!\longleftarrow \tilde{X}^1\Sigma^+$ and $[\tilde{X}^2\Pi_{3/2,1/2}]np\sigma\ (n = 8\text{–}9) \longleftarrow\!\!\longleftarrow\!\!\longleftarrow \tilde{X}^1\Sigma^+$ transitions were observed, as shown in Figure 38. This spectrum was obtained by monitoring the $N_2O^+$ mass channel. As can be seen from the proposed assignments, indicated by

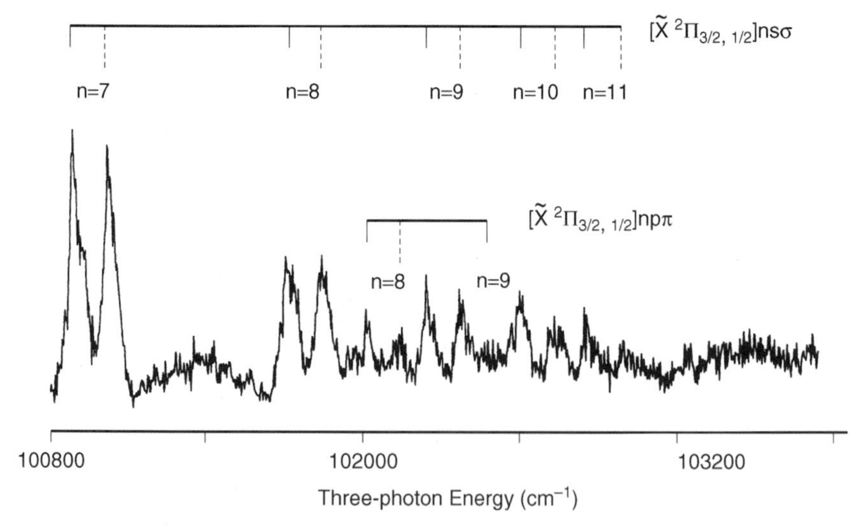


**Figure 38.** The $(3+1)$ REMPI spectrum from 100,800 up to 104,200 cm$^{-1}$ for $N_2O$ obtained by monitoring the $N_2O^+$ mass channel. Assignments are indicated as combs above the spectrum. (Reproduced with permission from [42], copyright 1998 American Institute of Physics.)

combs above the spectra, the experimental bands can be arranged into well-defined series, split by $\sim 134\,\text{cm}^{-1}$, and converging upon the two spin–orbit components of the ground-state ion. At lower energies lower members of these Rydberg series were observed as well [42].

By employing PES, the various origins indicated in Figure 38 could be assigned unambiguously. Since we are approaching the ionic limit, the number of close-lying excited states becomes rather large, and this situation sets the scene for the occurrence of extensive vibronic coupling. In Figure 39, the PE spectrum, obtained in a $(3+1)$ ionization process via the $[\tilde{X}^2\Pi_{3/2}]11s\sigma$ state at $102{,}872\,\text{cm}^{-1}$, is shown. In addition to the expected $\Delta v = 0$ transition at 4.105 eV, which in this case has relatively low intensity, there is a plethora of slower electron kinetic energy peaks. Closer inspection reveals that the energy splittings in the PE spectrum correspond rather accurately to separations between the Rydberg origin under consideration with lower lying features in the excitation spectra assigned as different Rydberg origins. Apparently, there is a degree of near-resonant vibronic mixing between the Rydberg origin under consideration and the various manifolds of vibrational levels based on other Rydberg origins that lie lower in energy. Since all Rydberg states that belong to series converging upon the same ionic threshold possess almost identical potential energy curves, diagonal Franck–Condon factors and consequently near-preservation of vibrational energy are expected. Hence, vibronic coupling would explain the close similarities in energy separations observed in both the excitation and PE spectra. A schematic picture of this vibronic coupling mechanism is presented in Figure 40. This observation offers the somewhat unusual possibility of exploiting vibronic coupling effects by observing large parts of a Rydberg series through the measurement of a single PE spectrum [42].

We now turn to $CS_2$ for another, albeit more speculative example of vibronic coupling. In Figure 41, the $(3+1)$ REMPI spectrum of a jet-cooled sample of $CS_2$ is shown in the energy range 61,850–64,500 $\text{cm}^{-1}$, obtained while monitoring only those ions with $m/z = 76$. The spectrum is a composite, constructed by splicing together spectra recorded using two different dyes. No effort was made to ensure correct normalization of the relative intensities appearing within the tuning range of either dye, nor between the one dye tuning curve and the other. In the excitation spectrum, four members of the $[^2\Pi_g]4p\lambda_u \longleftarrow\!\!\longleftarrow\!\!\longleftarrow \tilde{X}^1\Sigma_g^+$ transition are indicated by combs above the spectrum.

In order to arrive at the assignments given in the excitation spectrum of Figure 41, PE spectra are invaluable. The PE spectra, presented in Figure 42, were taken at four positions marked in the excitation spectrum, and will be discussed in turn. The first REMPI–PE spectrum, obtained following excitation at 483.1 nm ($3\nu = 62{,}100\,\text{cm}^{-1}$), shows one dominant peak centered at a kinetic energy of 0.188 eV. Given the value of $81{,}286 \pm 5\,\text{cm}^{-1}$ for the lowest ionization energy, the PE peak corresponds to a transition to the ground vibrational level of

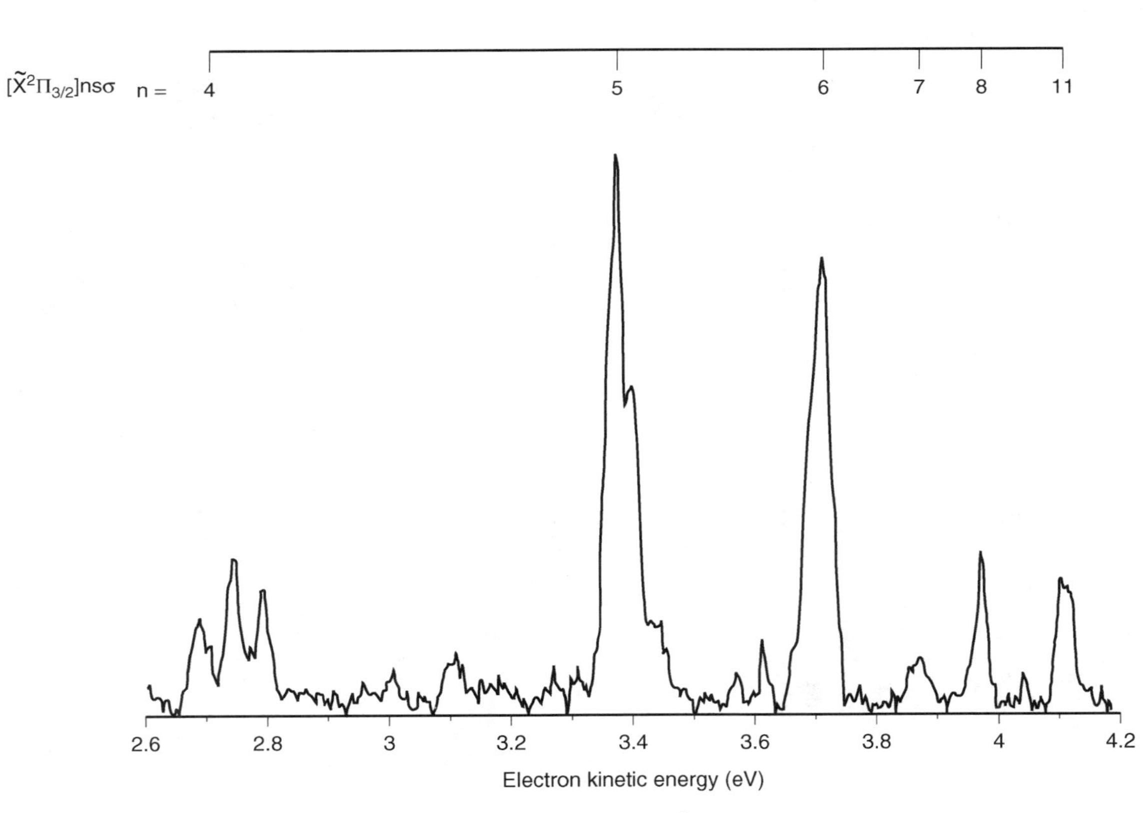


**Figure 39.** PE spectrum for $N_2O$ from $(3 + 1)$ REMPI via the $[\tilde{X}^2\Pi_{3/2}]11s\sigma$ state located at a three-photon energy of 102,872 $cm^{-1}$. (Reproduced with permission from [42], copyright 1998 American Institute of Physics.)

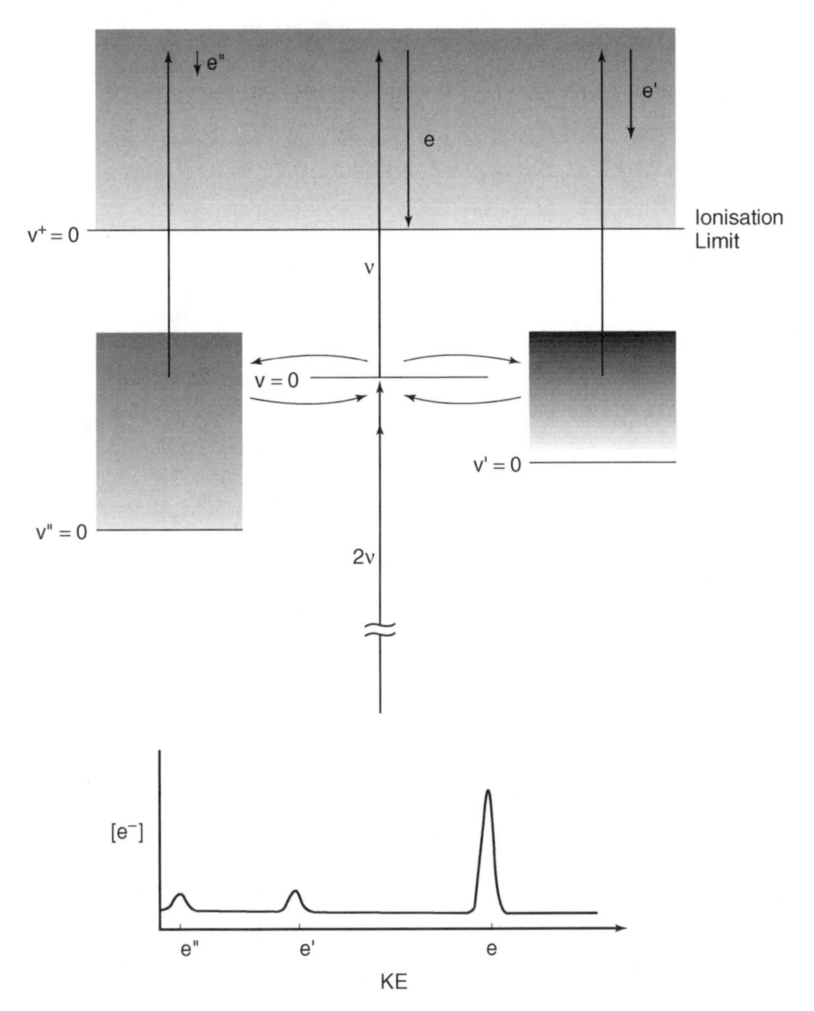


**Figure 40.** Schematic illustration showing vibronic coupling between a two-photon resonant origin ($v = 0$) level of one Rydberg state and accidentally near-resonant excited vibrational levels of two lower lying Rydberg states whose origins are labeled $v' = 0$ and $v'' = 0$. The vibrational quantum numbers involved will be preserved on ionization, and the vibronic coupling will show up in the kinetic energy distribution of the accompanying photoelectrons. (Reproduced with permission from [46], copyright 1995 The Royal Society of Chemistry.)

the $^2\Pi_{3/2}$ ionic ground-state. This $\Delta v = 0$ behavior clearly indicates an electronic origin, built on a (predominantly) $^2\Pi_{3/2}$ core. The assignment to the $[^2\Pi_{g3/2}]4p\sigma_u\ (3\Pi_u) \leftarrow\leftarrow\leftarrow \tilde{X}^1\Sigma_g^+$ transition is entirely consistent with that of previous analyses [217, 225]. The second spectrum, obtained following excitation at 477.9 nm ($3\nu = 62,780\,\mathrm{cm}^{-1}$), shows a major peak at a kinetic energy of

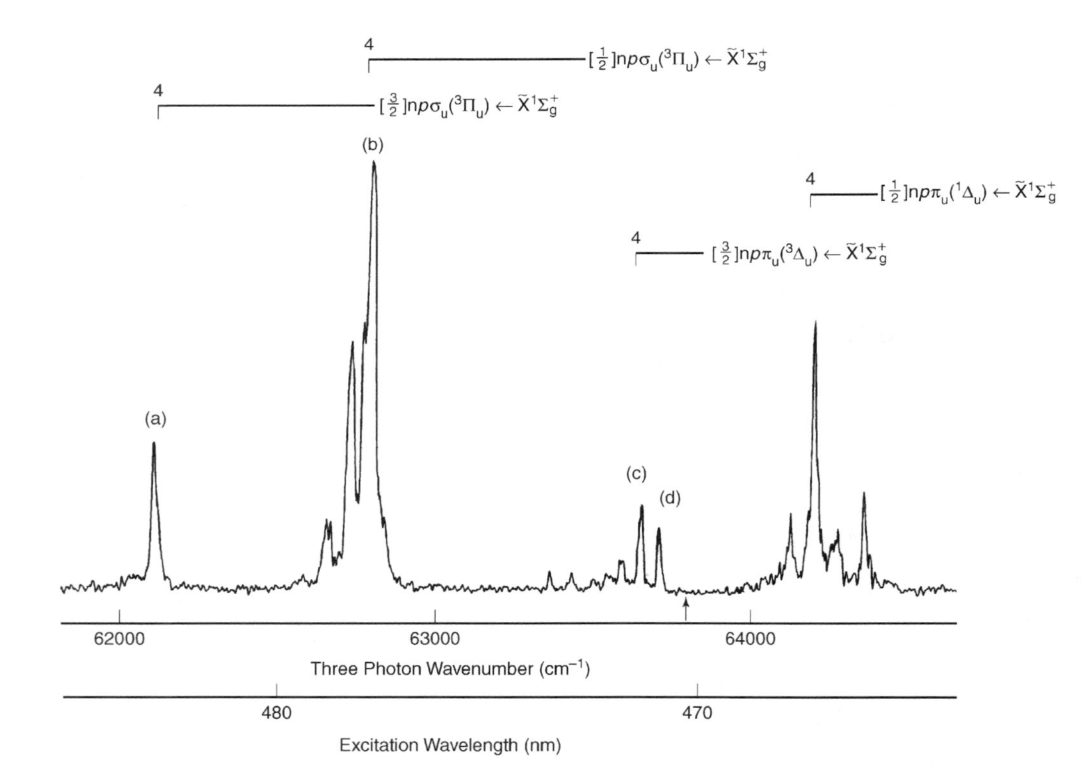


**Figure 41.** The $(3+1)$ REMPI spectrum of a jet-cooled sample of $CS_2$ is shown in the energy range 61,850–64,500 $cm^{-1}$, obtained while monitoring only those ions with $m/z = 76$. The spectrum is a composite, constructed by splicing together spectra recorded using two different dyes. The vertical arrow indicates where the two scans have been joined. Assignments are indicated by the combs above the spectrum. The REMPI–PES spectra were measured at wavelengths corresponding to the peaks marked (*a*)–(*d*). (Reproduced with permission from [50], copyright 1996 Elsevier Science B.V.)

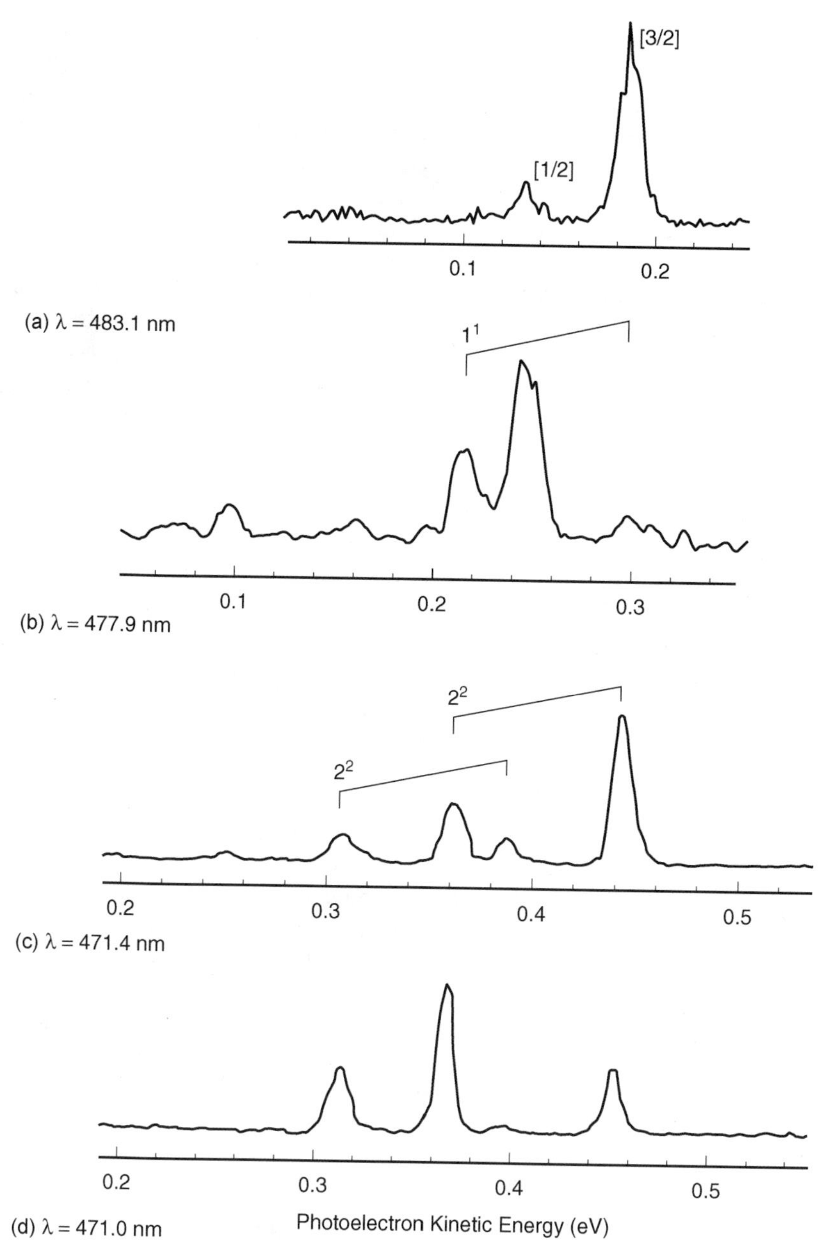


**Figure 42.** The (3 + 1) REMPI–PE spectra of $CS_2$ following excitation at wavelengths of (*a*) 483.1 nm ($3\nu = 62{,}100\,\mathrm{cm}^{-1}$); (*b*) 477.9 nm ($3\nu = 62{,}780\,\mathrm{cm}^{-1}$); (*c*) 471.4 nm ($3\nu = 63{,}644\,\mathrm{cm}^{-1}$); and (*d*) 471.0 nm ($3\nu = 63{,}700\,\mathrm{cm}^{-1}$). The excitation wavelengths correspond to those of the peaks marked (*a*)–(*d*) in Figure 41. The kinetic energy scales have been offset so that the various vibronic states of the ion align vertically. (Reproduced with permission from [50], copyright 1996 Elsevier Science B.V.)

0.246 eV, suggesting ionization to the vibrationless level of the $^2\Pi_{g1/2}$ ionic ground-state. This suggest that the resonance at 62,780 cm$^{-1}$ again corresponds to an origin, this time with a $^2\Pi_{g1/2}$ core. The assignment to the $[^2\Pi_{g1/2}]4p\sigma_u\,(^1\Pi_u) \longleftarrow\longleftarrow\longleftarrow \tilde{X}^1\Sigma_g^+$ transition is again consistent with that of previous work [48, 225, 226]. The additional feature at $\sim$ 0.214 eV in the same PE spectrum can either be assigned to formation of ions in their $^2\Pi_{g3/2}$ ionic ground-state with some 650 cm$^{-1}$ of internal energy, or to formation of ions in their $^2\Pi_{g1/2}$ spin–orbit excited state with some 210 cm$^{-1}$ of internal energy. In view of the normal mode vibrational energies available in the ion, the former explanation seems the more plausible. This suggests some resonance enhancement at 62,780 cm$^{-1}$ by a lower lying origin possessing one quantum of vibrational excitation, notably the symmetric stretch. The level responsible is most likely the $\nu_1 = 1$ level of the $[^2\Pi_{g3/2}]4p\sigma_u$ $(^3\Pi_u)$ state at 62,720 cm$^{-1}$ [48, 50].

The assignments of the two remaining PE spectra of Figure 42 are less obvious. The spectrum of Figure 42(*c*) was taken at 471.4 nm ($3\nu =$ 63,644 cm$^{-1}$), that of Figure 42(*d*) at 471.0 nm ($3\nu = $63,700 cm$^{-1}$). In the spectrum taken at 471.4 nm, we observe four peaks, the largest of which (0.443 eV) is consistent with ionization to the vibrationless ionic ground-state $^2\Pi_{g3/2}$. The peak at 0.388 eV indicates that ionization to the $^2\Pi_{g1/2}$ spin–orbit excited state occurs with much lower probability, indicating that the ionic core is predominantly $^2\Pi_{g3/2}$. The two additional peaks at 0.360 and 0.306 eV can be associated with formation of ions with both $^2\Pi_{g3/2}$ and $^2\Pi_{g1/2}$ cores, each with $\sim$ 650 cm$^{-1}$ of internal energy. The PE spectrum taken at 471.0 nm in Figure 42(*d*) exhibits the same fast and slow peaks, but their intensities are reversed compared to those in Figure 42(*c*). This has been explained by the suggestion that these observations result from vibronic mixing between an electronic origin located at 63,644 cm$^{-1}$, assigned as $[^2\Pi_{g3/2}]4p\pi_u$ $(^3\Delta_u)$ and a vibrationally excited level of an electronic state located at $\sim$ 63,050 cm$^{-1}$ [48, 50]. In view of the fact that the level ordering arising from a $(2\pi_g)^3(2\pi_u)^1$ configuration is $^1\Sigma_u^+$, $^1\Delta_u$, $^1\Sigma_u^-$, $^3\Sigma_u^-$, $^3\Delta_u$, and $^3\Sigma_u^+$ in order of decreasing energy [237], the hitherto unobserved $[^2\Pi_g]4p\pi_u$ $(^3\Sigma_u^+)$ state, involving a spin–orbit "mixed" core, is thought to be responsible. Considering that in order for the states to mix the *vibronic* symmetries ought to be identical, the $[^2\Pi_g]4p\pi_u$ $(^3\Sigma_u^+)$ state must possess two quanta of the degenerate bending mode $(\nu_2^2)$. This puts the $^3\Sigma_u^+$ origin, somewhat speculatively, at $\sim$ 63,050 cm$^{-1}$.

### D. The Role of Superexcited States in $CO_2$

Carbon dioxide possesses an inversion center and has a ground-state valence electron configuration $\ldots(3\sigma_g)^2(2\sigma_u)^2(4\sigma_g)^2(3\sigma_u)^2(1\pi_u)^4(1\pi_g)^4\ \tilde{X}^1\Sigma_g^+$. The molecule has three fundamental vibrational modes: $\nu_1$, the symmetric stretch (which tranforms as $\sigma_g^+$), $\nu_2$, the degenerate bend $(\pi_u)$, and $\nu_3$, the asymmetric stretch $(\sigma_u^+)$, the ground-state frequencies of which are 1388.17, 667.40, and 2349.16 cm$^{-1}$, respectively [2]. The linear ionic ground-state, formed by removing a $2\pi_g$ electron, is a spin–orbit split state with $\tilde{X}^2\Pi_{g3/2} \sim 160\,\mathrm{cm}^{-1}$ below $\tilde{X}^2\Pi_{g1/2}$, with a lowest ionization threshold of $81{,}286 \pm 5\,\mathrm{cm}^{-1}$, and with vibrational normal mode frequencies of 1244.27, 511.35, and $1423.08\,\mathrm{cm}^{-1}$ [238, 239]. Higher ionic states with $\tilde{A}^2\Pi_u$, $\tilde{B}^2\Sigma_u^+$, and $\tilde{C}\,^2\Sigma_u^+$ symmetry and situated within 6 eV of each other originate from the removal of an electron out of the $1\pi_u$, $3\sigma_u$, and $4\sigma_g$ orbital, respectively.

The lower excited valence states arise from the $2\pi_u \leftarrow 1\pi_g$ excitation ($^{1,3}\Sigma_u^-$, $^{1,3}\Sigma_u^+$, $^{1,3}\Delta_u$), and the $5\sigma_g \leftarrow 1\pi_g$ excitation ($^{1,3}\Pi_g$). Several experimental [216, 240–243] and theoretical [244–248] studies have investigated these states, in particular with respect to bent geometries expected in some cases. The study of higher lying states of $CO_2$ is considerably hindered by (pre)dissociative processes. A comprehensive one-photon study of excited states in the energy region from $87{,}600\,\mathrm{cm}^{-1}$ up to the lowest ionization limit $\sim 111{,}201\,\mathrm{cm}^{-1}$ has been performed, and *np* and *nf* Rydberg series have been assigned. The *np* Rydberg series seem to be heavily perturbed by dissociative valence states, while the *nf* series is relatively unaffected in comparison [249].

In Figure 43, we present $(3+1)$ excitation spectra of $CO_2$ in the one-photon energy region $34{,}650$–$34{,}900\,\mathrm{cm}^{-1}$, obtained with both electron and ion detection. The spectra show the $[^2\Pi_{g3/2,1/2}]4f \longleftarrow\longleftarrow\longleftarrow {}^1\Sigma_g^+$ transition. The sharp peaks on the high-energy side derive from $(2+1)$ REMPI of atomic carbon [203]. In Figure 44, PE spectra are depicted taken at one-photon energies of 34,771, 34,781, 34,791, 34,804, and $34{,}811\,\mathrm{cm}^{-1}$, which correspond to different positions in the three-photon resonance line of the $[^2\Pi_{g1/2}]4f$ transition. In all five PE spectra, two main progressions labeled as $^2\Pi_{g1/2}(v_1^+, 0, 0)$ and $^2\Pi_{g1/2}(v_1^+, 0, 2)$ can be discerned, as indicated above the spectra. The large changes in the PE spectra occurring for small changes in the excitation wavelength are striking and are reminiscent of the situation found in hydrogen and its isotopomers [25, 26]. The explanation can be found in the role of superexcited states at the four-photon level, which exert a strong influence on the PE spectra of Figure 44. These superexcited states are expected to be neutral states above the lowest $^2\Pi_{g3/2,1/2}$ ionic threshold and belonging to Rydberg series converging upon the excited $\tilde{A}^2\Pi_u$, $\tilde{B}^2\Sigma_u^+$, and $\tilde{C}^2\Sigma_u^+$ ionic states. It is important to realize that excitation to the dense manifold of these superexcited states can be considered as a one-photon excitation from the $[^2\Pi_{g3/2,1/2}]4f$ states. As a consequence, we do not necessarily excite the same superexcited states as in one-photon excitation studies from the $^1\Sigma_g^+$ ground-state of $CO_2$ [203].

### E. Fragmentation Processes in OCS

The OCS molecule is without an inversion center, and has a ground-state valence electron configuration $\ldots(6\sigma)^2(7\sigma)^2(8\sigma)^2(2\pi)^4(9\sigma)^2(3\pi)^4\ \tilde{X}^1\Sigma^+$. The molecule possesses three fundamental vibrational modes: $\nu_1$, the symmetric stretch (which tranforms as $\sigma^+$), $\nu_2$, the degenerate bend ($\pi$), and $\nu_3$, the asymmetric

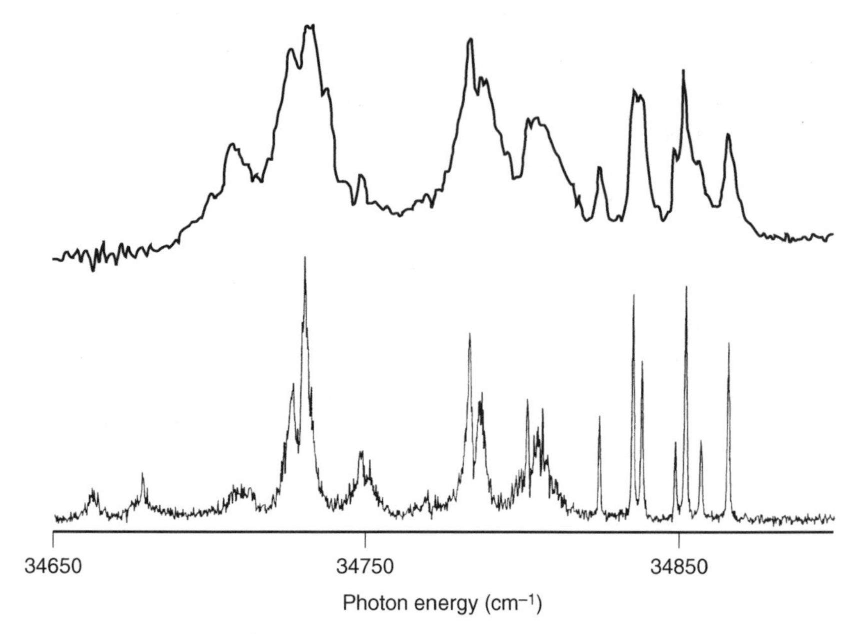


**Figure 43.** The $(3+1)$ REMPI spectra of $CO_2$ in the one-photon energy region 34,650–34,900 $cm^{-1}$, obtained with both electron (upper trace) and ion (lower trace) detection. The spectra show the $[^2\Pi_{g3/2,1/2}]4f \leftarrow\leftarrow\leftarrow {}^1\Sigma_g^+$ vibrationless origin transitions. The sharp peaks on the high-energy side derive from $(2+1)$ REMPI of atomic carbon. (Reproduced with permission from [203], copyright 1994 American Institute of Physics.)

stretch ($\sigma^+$), the ground-state frequencies of which are 2062, 520, and 859 $cm^{-1}$, respectively [2]. The linear ionic ground-state, formed by removing a $3\pi$ electron, is a spin–orbit split state with $\tilde{X}^2\Pi_{g3/2} \sim 368\,cm^{-1}$ below $\tilde{X}^2\Pi_{g1/2}$, with a lowest ionization threshold of $90,121 \pm 12\,cm^{-1}$ [250]. The normal modes $\nu_1$, $\nu_2$, and $\nu_3$ in the lowest ionic state are 2080, 510, and 698 $cm^{-1}$, respectively [49, 251].

The lower valence states arise from the $4\pi \leftarrow 3\pi$ and from the $10\sigma \leftarrow 3\pi$ transitions and have been studied with conventional one-photon absorption techniques [2, 216, 232, 252–254]. Broad continua centered $\sim$ 222, 167, and 152 nm are assigned in terms of $^1\Delta \leftarrow X^1\Sigma^+$, $^1\Pi \leftarrow \tilde{X}^1\Sigma^+$, and $^1\Sigma^+ \leftarrow \tilde{X}^1\Sigma^+$, respectively. The $^1\Delta$ state has been shown to possess a bent geometry [255]. At shorter wavelengths Rydberg states of OCS dominate. The energy region 70,000–75,000 $cm^{-1}$ has been studied extensively by one-photon absorption spectroscopy [2, 216, 232, 252, 253, 256], electron loss spectroscopy [254], and with REMPI [257–258]. More recently, we have performed REMPI–PES studies in the energy range 70,500–86,000 $cm^{-1}$ [49].

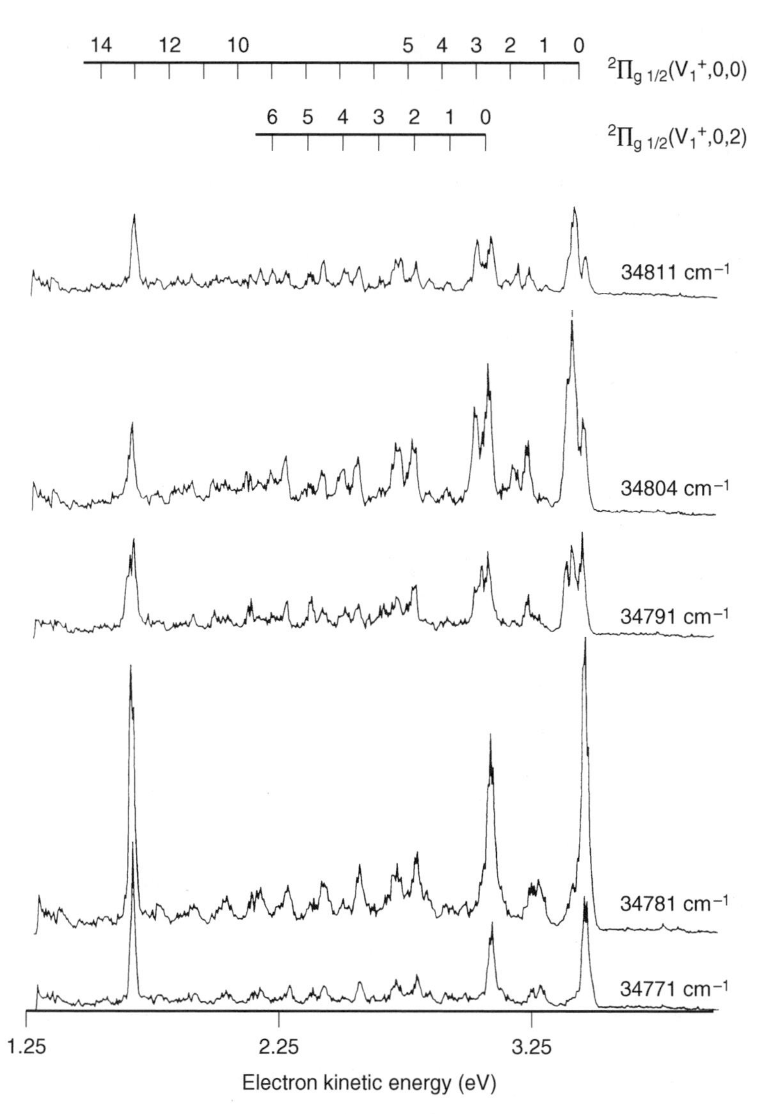


**Figure 44.** PE spectra of $CO_2$ obtained at one-photon energies of 34,771, 34,781, 34,791, 34,804, and 34,811 $cm^{-1}$ (lower to upper trace), which correspond to different positions in the three-photon resonance line of the $[^2\Pi_{g1/2}]4f$ origin transition in Figure 43. The two main progressions are labeled by the combs above the spectra as $^2\Pi_{g1/2}(v_1^+, 0, 0)$ and $^2\Pi_{g1/2}(v_1^+, 0, 2)$. (Reproduced with permission from [203], copyright 1994 American Institute of Physics.)

When OCS is excited with a single photon of $\sim 230\,\text{nm}$ to the long wavelength limit of the bent $^1\Delta$ state, appreciable fragmentation takes place. One of the sensitive and selective methods to monitor the fragments formed is with mass-resolved REMPI. In Figure 45, the $m/z = 28$ ($CO^+$) and $m/z = 32$ ($S^+$) mass channels are shown when OCS is subjected to photons of $\sim 230\,\text{nm}$. In both channels, strong resonances appear that can be ascribed to $(2+1)$ resonance-enhanced processes starting from CO ($X^1\Sigma^+$) ground-state molecules and S ($^1D$) excited atoms [259]. The REMPI spectrum of CO corresponds to the $Q$ branch of the $B^1\Sigma^+(v'=0) \leftarrow\leftarrow X^1\Sigma^+(v''=0)$ transition [260]. Clearly, the dissociation via the bent excited $^1\Delta$ state leads to ground-state CO produced with an appreciable bimodal excited rotational state distribution with rotational

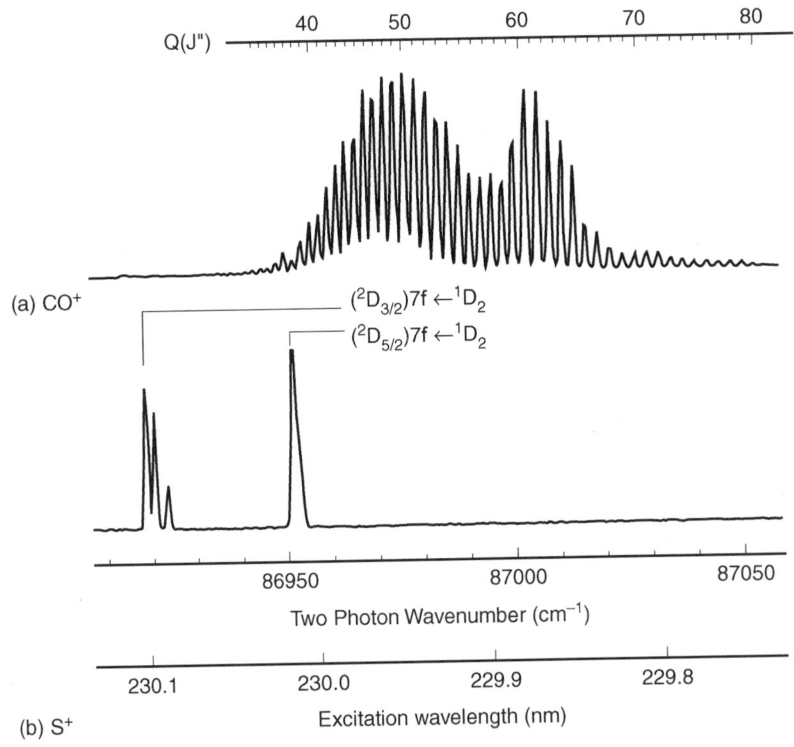


**Figure 45.** The $(2+1)$ REMPI spectra of OCS in the two-photon energy region 86,900–87,050 cm$^{-1}$ showing (*a*) the $Q$ branch of the CO $B^1\Sigma^+(v'=0) \leftarrow\leftarrow X^1\Sigma^+(v''=0)$ transition; and (*b*) various resonances of the S ($^1D$) fragments generated in the one-photon dissociation of OCS at the same excitation wavelength. The combs above the spectra highlight the bimodal rotational distribution associated with the CO fragment. (Reproduced with permission from [50], copyright 1996 Elsevier Science B.V.)

quantum numbers between $\sim 35$ and 70 [50]. This behavior is consistent with previous observations [260, 261].

### F. Some Advantages of REMPI–PES as Exemplified by $CS_2$

In order to illustrate the benefits of REMPI–PES, we shall discuss an example taken from the spectroscopy of $CS_2$. In Figure 46, a portion of the $(2+1)$ excitation spectrum is presented in the two-photon energy range 56,000–57,500 $cm^{-1}$. The upper spectrum, obtained in a jet-cooled sample of $CS_2$ with mass-resolved ion detection at $m/z = 76$ ($CS_2^+$), appears to be remarkably complicated. Closer inspection shows that this spectrum is a superposition of resonance enhancements at the one-photon level to bent valence states, and at the two-photon level to origins of linear Rydberg states. The vibronic assignments given above this spectrum are taken from [209]. Franck–Condon considerations dictate that only the latter states can ionize to $v^+ = 0$ ground-state ions, and thus give rise to photoelectrons with relatively high kinetic energies. The lower spectrum in Figure 46 was obtained in an effusive beam of $CS_2$ by monitoring only those photoelectrons with kinetic energies greater than $\sim 0.35$ eV. These kinetic energies are chosen in such a way that only electrons ejected in the ionization process of linear Rydberg states are measured. The two peaks observed in this spectrum correspond to the $[\tilde{X}^2\Pi_{g3/2}]4s\sigma_g \longleftarrow\!\!\longleftarrow \tilde{X}^1\Sigma_g^+$ and $[\tilde{X}^2\Pi_{g1/2}]4s\sigma_g \longleftarrow\!\!\longleftarrow \tilde{X}^1\Sigma_g^+$ two-photon transitions to two Rydberg origins [48, 50].

## VI. AMMONIA

Ammonia has a pyramidal ($C_{3v}$) structure in its electronic ground-state, with the corresponding electronic configuration usually written as $(1a_1')^2 (2a_1')^2(1e')^2(1a_2'')^2\ \tilde{X}^1A_1'$. Excited states arising as a result of electron promotion from the highest occupied $1a_2''$ nitrogen lone-pair orbital all have planar ($D_{3h}$) equilibrium geometries. This, and the modest size of the potential energy barrier to inversion in the ground-state, justifies a description (as above) of the ground-state configuration in terms of $D_{3h}$ symmetry labels.

The excited electronic states of ammonia have been the subject of extensive experimental and theoretical study. Much of the early spectroscopic work involved conventional UV and, particularly, VUV gas-phase absorption measurements of the vertical electronic spectrum [262–265]. Definitive assignments proved difficult because of the congested nature of the spectra, which in part is due to the long vibronic progressions associated with excitation of the $v_2'$ out-of-plane mode. This is an inevitable consequence of the Franck–Condon principle and of the pyramidal $\leftarrow$ planar geometry change that accompanies the excitation from the $1a_2''$ nitrogen lone pair orbital. Moreover, many of these excited states show spectral line broadening due to predissociation. Our current

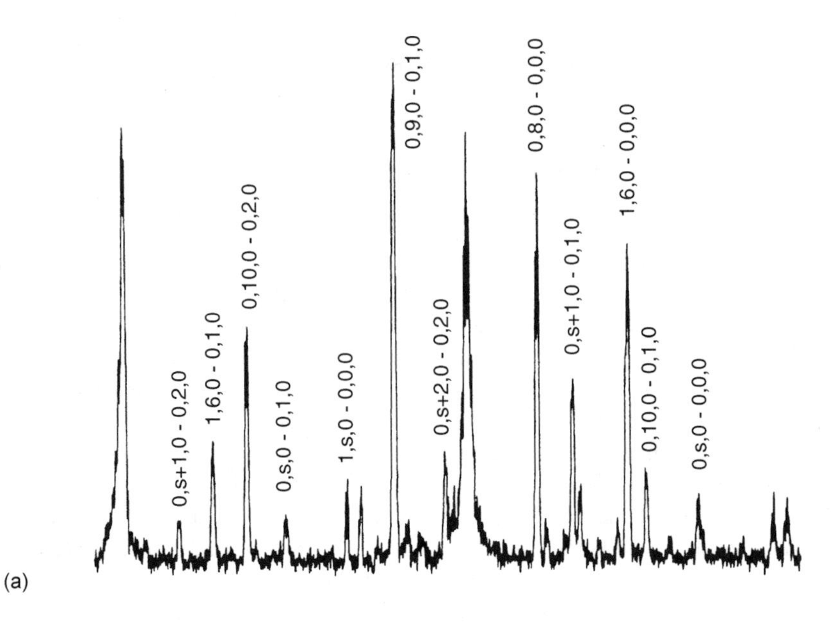

0,s+1,0 - 0,2,0
1,6,0 - 0,1,0
0,10,0 - 0,2,0
0,s,0 - 0,1,0
1,s,0 - 0,0,0
0,9,0 - 0,1,0
0,s+2,0 - 0,2,0
0,8,0 - 0,0,0
0,s+1,0 - 0,1,0
1,6,0 - 0,0,0
0,10,0 - 0,1,0
0,s,0 - 0,0,0
(a)


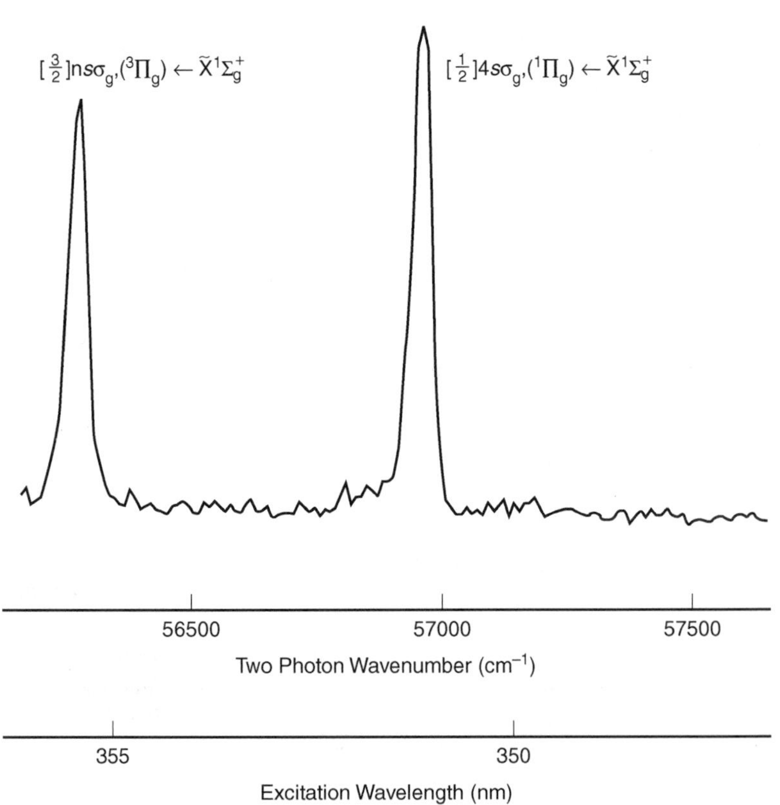

$[\frac{3}{2}]ns\sigma_g,(^3\Pi_g) \leftarrow \tilde{X}^1\Sigma_g^+$
$[\frac{1}{2}]4s\sigma_g,(^1\Pi_g) \leftarrow \tilde{X}^1\Sigma_g^+$
56500
57000
57500
Two Photon Wavenumber ($cm^{-1}$)
355
350
(b)
Excitation Wavelength (nm)

level of understanding is based to a considerable extent on the introduction of jet-cooling and REMPI techniques to the problem by Colson and co-workers [266–269]. In this work, the energy range between $\sim$ 62,500 and 80,000 cm$^{-1}$ was covered with $(3+1)$ REMPI, comparisons with previous absorption spectra were made, and a reasonably complete global description of the excited states of ammonia in this energy region was obtained. Much of the later work has been concerned with extending and refining parts of their spectroscopic analysis, and/or with the study of the predissociation mechanisms affecting some of the lower lying Rydberg states.

The small size and high symmetry of ammonia allow a description of the molecular orbitals in the framework of a united atom (neon) scheme situated in a field of $D_{3h}$ symmetry. All the documented electronic states can be incorporated in such a scheme, with electron promotion to a Rydberg orbital with $n \geq 3$. In Figure 47, we present the lowest energy part of the states predicted by this model. In order not to complicate the figure unduly, only states with $n \leq 5$ and $\ell \leq 3$ are shown. Rydberg states with higher values of $n$ converge upon the lowest ionic limit indicated at 82,159 cm$^{-1}$ [270, 271]. States with higher $\ell$ cannot be accessed in our simple model via a two-photon excitation (as we shall consider below) from the electronic ground-state that possesses $s/p$ character.

The spectroscopy and photochemistry of the first excited $\tilde{A}^1A_2''$ $(3sa_1' \leftarrow 1a_2'')$ state of ammonia have been studied in great detail with a variety of physical techniques. The predissociative character of the $\tilde{A}$ state is apparent in many studies. For an overview of what is currently known about the $\tilde{A}$ state we refer to the literature [272, 273]. The $3pe' \leftarrow 1a_2''$ and $3a_2'' \leftarrow 1a_2''$ electronic promotions give rise to the $\tilde{B}^1E'' \leftarrow \tilde{X}^1A_1'$ and $\tilde{C}'^1A_1' \leftarrow \tilde{X}^1A_1'$ band systems. Members of the $2_0^n$ vibronic progression associated with the transition to the $\tilde{B}$ state have been observed via VUV absorption spectroscopy [262, 274–276], but much of our current knowledge derives from two-photon REMPI studies both under molecular beam conditions [277] and, with sub-Doppler resolution, in the bulk gas phase [278]. The transition from the ground to the $\tilde{C}'^1A_1'$ state possesses negligible oscillator strength in one-photon absorption, but shows a well-resolved progression in $\nu_2'$, interspersed among the various $\tilde{B} \leftarrow \tilde{X}$ features, in both two-and three-photon REMPI [266, 267, 279, 280]. The first REMPI–PES study on jet-cooled ammonia helped identify the symmetric stretching mode $\nu_1'$

**Figure 46.** Portions of $(2+1)$ excitation spectra of $CS_2$ in the two-photon energy range 56,000–57,500 cm$^{-1}$. In (*a*), a jet-cooled sample is employed using mass-resolved ion detection at $m/z = 76$ ($CS_2^+$); in (*b*) an effusive beam is used and only photoelectrons with kinetic energies greater than $\sim$ 0.35 eV are monitored. The transitions observed in (*a*) arise from a superposition of one- and two-photon resonances. The two peaks observed in (*b*) correspond to the $[\tilde{X}^2\Pi_{g3/2}]4s\sigma_g \leftarrow\leftarrow \tilde{X}^1\Sigma_g^+$ and $[\tilde{X}^2\Pi_{g1/2}]4s\sigma_g \leftarrow\leftarrow \tilde{X}^1\Sigma_g^+$ two-photon transitions to two Rydberg origins (see text). (Reproduced with permission from [50], copyright 1996 Elsevier Science B.V.)

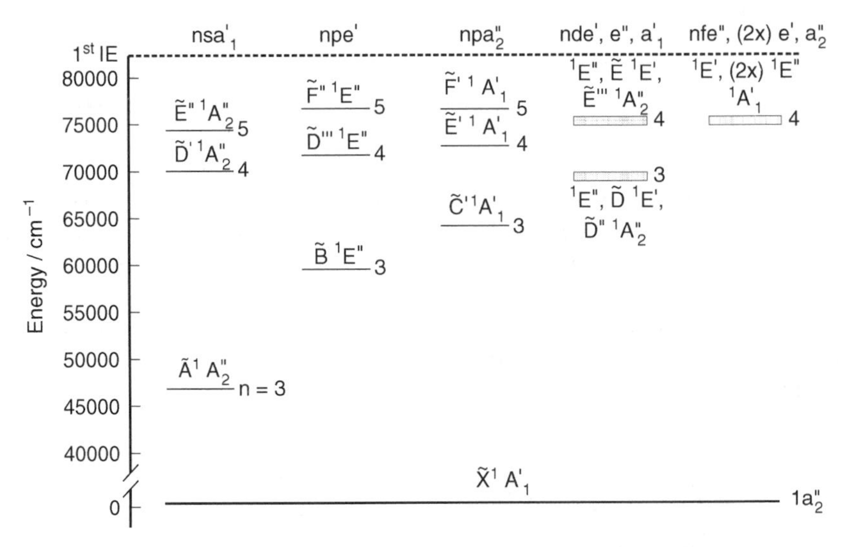


**Figure 47.** Diagram showing observed and/or predicted members (with $n \leq 5$) of the first few Rydberg series of ammonia. Labels and symmetries are indicated, along with the appropriate values of the principal quantum number, $n$. (Reproduced with permission from [43], copyright 1998 American Institute of Physics.)

in the $\tilde{C}'$ state [281]. Selected rovibrational levels of both the $\tilde{B}$ and $\tilde{C}'$ states can predissociate via coupling to the lower lying $\tilde{A}$ state. A picosecond pump–probe REMPI–PES study has allowed real-time observation of these decay processes [282].

Progressing to higher energies, the spectrum of ammonia becomes ever more complex, reflecting the increased electronic state density and the consequent increase in Franck–Condon accessible vibronic levels as we approach the first ionization limit. The availability of rotationally "cold" spectra, obtained using samples cooled in a molecular beam, has been essential in unraveling a significant part of the congested level structure. Li and Vidal [275, 276] identified a $2_0^n$ vibronic progression in the VUV excitation spectrum by monitoring either the fluorescence from the fraction of $NH_2$ photofragments that are formed in their excited $\tilde{A}^2A_1$ state, or parent ions that result from absorption of a visible photon by the VUV prepared state (i.e., $1 + 1'$ two-color REMPI). Rotational analysis showed that this progression must be associated with the $\tilde{D}^1E' \leftarrow \tilde{X}^1A_1'$ $(3de'' \leftarrow 1a_2'')$ transition. Parallel analyses of one-photon VUV absorption spectra and/or $(3 + 1)$ REMPI spectroscopy [268, 269] indicate that another member of the $3d \leftarrow 1a_2''$ complex (the $\tilde{D}''^1A_2'' \leftarrow \tilde{X}^1A_1'$ transition, associated with the orbital promotion $3da_1' \leftarrow 1a_2''$), has its electronic origin in the same energy region, as do the $\tilde{D}'^1A_2'' \leftarrow \tilde{X}^1A_1'$ $(4sa_1' \leftarrow 1a_2'')$ and $\tilde{D}'''^1E'' \leftarrow \tilde{X}^1A_1'$

($4pe' \leftarrow 1a_2''$) transitions. Energetic considerations dictate that the "missing" $^1E''$ ($3de' \leftarrow 1a_2''$) excited state must also lie in this energy range, but it remains to be identified. The situation at still higher energies becomes even more complex. Virtually all recent progress has come from $(2+1)$ REMPI spectroscopy, much of it in conjunction with REMPI–PES studies [43, 44, 283].

At higher excitation energies the level congestion increases, predissociation and the associated loss of rotational resolution tend to become more prominent, and even with sample cooling provided by molecular beam methods the assignment problems become increasingly insurmountable. Even the use of isotope effects and quantum defect considerations reach their limits of applicability at progressively higher energies, because origins and higher vibrational members of the various progressions become hard to distinguish. In addition, orbital interactions may play a role as well. Under such circumstances, the unique power of laser PES can be employed successfully in alleviating the analysis problems by experimentally determining the precise internal energy of the final ionic state reached in a resonance-enhanced multiphoton ionization process that uses the intermediate state under consideration as a stepping stone. Next, we shall focus on new states of ammonia ($NH_3$) detected at these higher energies, as well as on their properties. For similar results obtained for deuterated ammonia ($ND_3$) the reader is referred to the literature [43, 44].

In Figure 48 the $(2+1)$ REMPI spectrum of a jet-cooled sample of $NH_3$ in the energy range 68,700–82,160 cm$^{-1}$, obtained by monitoring ions with $m/z = 17$, is depicted [43]. In this spectrum, members of vibrational progressions involving $\nu_1'$ and $\nu_2'$, the upper state symmetric stretching and out-of-plane bending quantum numbers, respectively, are indicated. Because of the high sensitivity, for the $\tilde{B}^1E''$ ($3pe' \leftarrow 1a_2''$) and $\tilde{E}'^1A_1'$ ($4pa_2'' \leftarrow 1a_2''$) Rydberg states the $2^n$ progressions are extended to higher vibrational quantum numbers than observed before [277, 278, 283]. For the $\tilde{C}'^1A_1'$ ($3pa_2'' \leftarrow 1a_2''$) Rydberg state both the $2^n$ and $1^12^n$ vibronic progressions are extended beyond what was known previously [281].

We shall now focus our attention on the REMPI spectrum of ammonia above the $\tilde{E}'$ state origin at $\sim$ 72,995 cm$^{-1}$. In this region, several clear progressions have been identified amidst an abundance of novel features. Again, REMPI–PES will be used for assignment purposes, leading to considerable insight into the vibrational and electronic characteristics of the Rydberg states involved. In Figure 49, representative PE spectra are depicted. They will be discussed in turn.

In Figure 49(*a*), a REMPI–PE spectrum obtained for a room temperature sample of $NH_3$ is shown, obtained at an excitation wavelength of 260.77 nm ($2\nu = 76{,}674$ cm$^{-1}$). The kinetic energy of the dominant peak in the spectrum at 4.08 eV is entirely consistent with a three-photon ionization to the zero-point vibrational level of the lowest ionic state. The observed $\Delta v = 0$ propensity, together with REMPI–PES results for other members of a progression

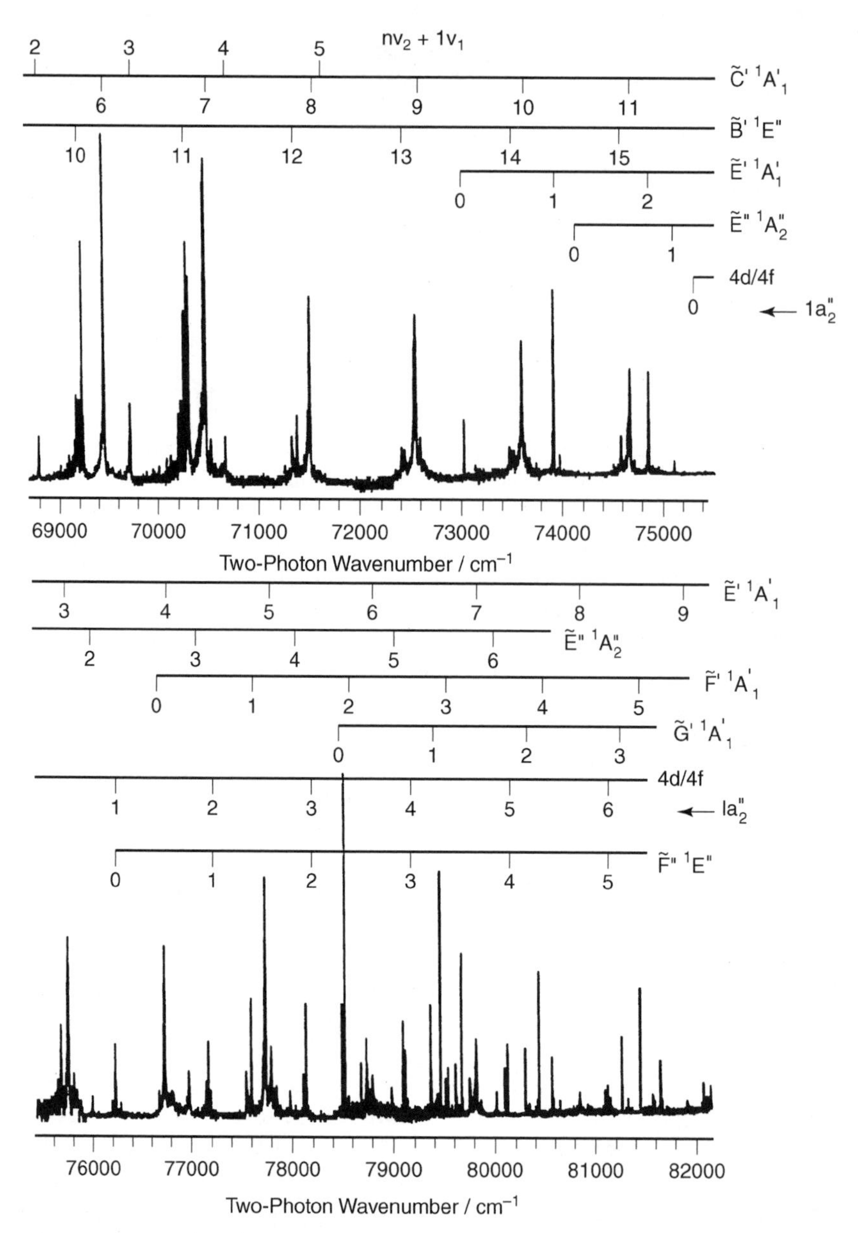


**Figure 48.** The $(2+1)$ REMPI spectrum of a jet-cooled sample of $NH_3$ in the energy range 68,700–82,160 $cm^{-1}$, obtained by monitoring ions with $m/z = 17$. The spectrum is a composite obtained by splicing together spectra using more than one dye. Not too much value should therefore be attached to relative intensities. The numbers above the spectra indicate $v_2'$, the upper state out-of-plane bending quantum number of a $2^n$ vibronic level. The numbers above the $\tilde{C}'$ state comb indicate the $v_2'$ number of a $1^1 2^n$ vibronic progression. (Reproduced with permission from [43], copyright 1998 American Institute of Physics.)

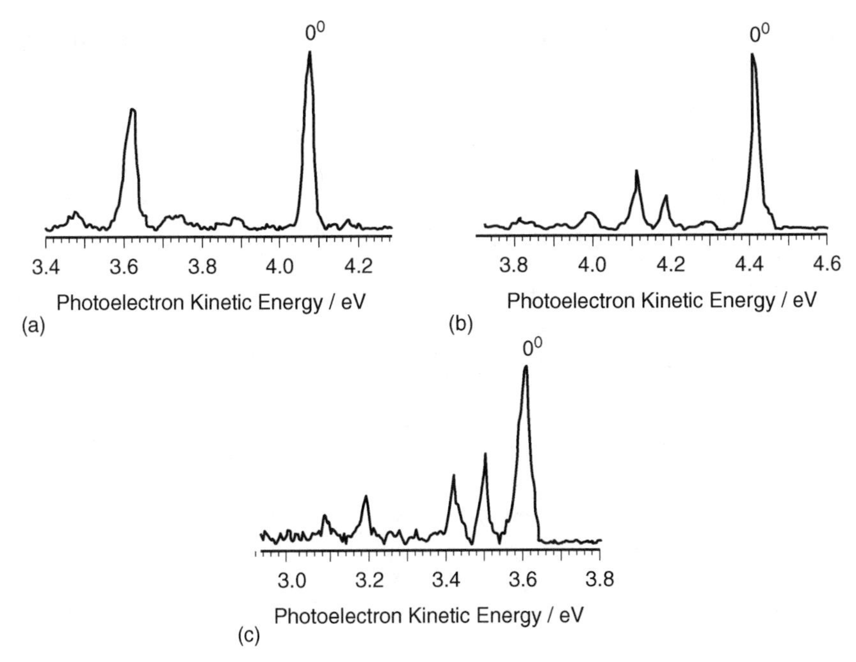


**Figure 49.** The REMPI–PE spectra of a room temperature sample of $NH_3$ obtained following excitation (*a*) at 260.77 nm ($2\nu = 76{,}674\,cm^{-1}$); (*b*) at 254.72 nm ($2\nu = 78{,}495\,cm^{-1}$); and (*c*) at 269.75 nm ($2\nu = 74{,}119\,cm^{-1}$). On the basis of the measured PE kinetic energies we deduce that these two-photon resonances involve the $0_0^0$ bands of, respectively, transitions to the $\tilde{F}'^1A_1'$, $\tilde{G}'^1A_1'$, and $\tilde{E}''^1A_2''$ state of $NH_3$. (Reproduced with permission from [43], copyright 1998 American Institute of Physics.)

converging upon the same ion core, indicates that the transition at $76{,}674\,cm^{-1}$ in the excitation spectrum corresponds to the origin of a series. The other PE peak at 3.62 eV indicates formation of ions with an internal energy of $\sim 0.46$ eV. This corresponds to four quanta of $\nu_2^+$, the parent ion bending vibration, not unexpected in view of the fact that the excitation spectrum in Figure 48 shows an overlapping $\tilde{E}' \leftarrow \tilde{X}\ 2_0^4$ transition at the two-photon energy. In fact, the $\tilde{C}' \leftarrow \tilde{X}\ 2_0^{13}$ transition also overlaps in the two-photon excitation spectrum, leading to a PE peak at $\sim 2.45$ eV (not shown). By assuming that the origin of the series lies at $76{,}674\,cm^{-1}$, a quantum defect of $\delta = 0.52$ (assuming $n = 5$) is derived. This quantum defect is very similar to those obtained for the $\tilde{C}'^1A_1'$ ($\delta = 0.56$) and $\tilde{E}'^1A_1'$ ($\delta = 0.54$) Rydberg states. The transition is therefore assigned as the next ($n = 5$) member of the $npa_2'' \leftarrow 1a_2''$ Rydberg series of ammonia, which we label as $\tilde{F}'^1A_1'$. The rotational structure associated with the $\tilde{F}'^1A_1'$ progression is not resolved. However, rotational contours can be observed, and band contour simulations based on rotational constants can be performed. It appears that parameters very similar to those for the $\tilde{C}'^1A'_1$ and $\tilde{E}'^1A'_1$ states give reasonable agreement [43].

In Figure 49(*b*), a REMPI–PE spectrum obtained for a room temperature sample of $NH_3$ is shown, obtained at an excitation wavelength of 254.72 nm ($2\nu = 78{,}495\,cm^{-1}$). A strong peak at a kinetic energy of 4.41 eV is consistent with three-photon ionization to the lowest vibrational level of the ionic ground-state, suggesting that the feature in the excitation spectrum at $78{,}495\,cm^{-1}$ corresponds to a two-photon transition to a vibrationless level. Weaker peaks in the PE spectrum are all explicable in terms of $(2+1)$ REMPI via near resonant levels of the $\tilde{C}', \tilde{D}'''^1E'', \tilde{E}'$, and $\tilde{F}'$ states, whose room temperature band contours contribute to the two-photon absorption at this wavelength. Additional PE results for higher members of the observed progression in $\nu'_2$ in the excitation spectrum confirm the vibrational numbering. A quantum defect of $\delta = 0.53$ (assuming $n = 6$) is obtained, and band contour simulations could be carried out with parameters very similar to those of the $\tilde{C}'^1A'_1, \tilde{E}'^1A'_1$, and $\tilde{F}'^1A'_1$ Rydberg states. We therefore designate this state as the next ($n = 6$) member of the $npa''_2 \leftarrow 1a''_2$ Rydberg series of ammonia, which we label as $\tilde{G}'^1A'_1$ [43].

In Figure 49(*c*), a REMPI–PE spectrum obtained for a room temperature sample of $NH_3$ is shown, obtained at an excitation wavelength of 269.75 nm ($2\nu = 74{,}119\,cm^{-1}$). The dominant peak at 3.59 eV is consistent with three-photon ionization to the $\nu^+ = 0$ level of the lowest ionic state, and again confirms the presence of an origin in the excitation spectrum at this wavelength. Weaker features in the PE spectrum can again be assigned in terms of $\Delta v = 0$ ionization following transitions to near resonant levels associated with the $\tilde{C}', \tilde{D}'''$, and $\tilde{E}'$ states. The band contours of these levels contribute to the two-photon absorption at the given wavelength. The PE spectra obtained for the observed members of a $\nu'_2$ progression in the excitation spectrum establish the vibrational numbering beyond doubt. From the determination of the origin quantum defects of $\delta = 1.41$ (assuming $n = 5$), corresponding to the $5sa' \leftarrow 1a''_2$ transition, or $\delta = 0.41$ (assuming $n = 4$), corresponding to the $4da' \leftarrow 1a''_2$ excitation, can be derived. In view of the fact that an *nd* series would be expected to have a quantum defect of $\delta \sim 0$, the former assignment is preferred. We therefore designate this state as the next ($n = 5$) member of the $nsa' \leftarrow 1a''_2$ Rydberg series of ammonia, which we label as $\tilde{E}''^1A''_2$ [43].

The plethora of resonances observed for $NH_3$ in Figure 48 (and for $ND_3$, see [43]) can be studied and assigned along lines very similar to those discussed above. This work has led to the first observation of many new members of vibrational progressions associated with Rydberg states that had been identified before, as well as to the characterization of many new Rydberg states hitherto unobserved. In addition, in many places features in the laser PE spectra indicative of orbital interactions due to vibronic coupling were identified [43, 44]. The novel states observed for $NH_3$ in the energy region between

TABLE I
Electronic Origins and Quantum Defects, δ, of New Rydberg States of Ammonia ($NH_3$) Observed in [43]

| State | Origin ($cm^{-1}$) | δ |
|---|---|---|
| $\tilde{E}''^1A''_2$ ($5sa' \leftarrow 1a''_2$) | 74,119(2) | 1.41 |
| ($4d/4f \leftarrow 1a''_2$) | 75,340(10) | 0.01 |
| $\tilde{F}''^1E''$ ($5pe' \leftarrow 1a''_2$) | 76,220(50) | ~0.70 |
| $\tilde{F}'^1A'_1$ ($5pa''_2 \leftarrow 1a''_2$) | 76,674(1) | 0.56 |
| $\tilde{G}'^1A'_1$ ($6pa''_2 \leftarrow 1a''_2$) | 78,494(1) | 0.53 |
| ($5d/5f \leftarrow 1a''_2$) | 77,860(10) | 0.05 |
| $\tilde{H}''^1E''$ ($7pe' \leftarrow 1a''_2$) | 79,295(5) | 0.81 |
| $\tilde{H}'^1A'_1$ ($7pa''_2 \leftarrow 1a''_2$) | 79,560(5) | 0.50 |
| $\tilde{I}''^1E''$ ($8pe' \leftarrow 1a''_2$) | 80,020(10) | 0.82 |
| $\tilde{I}'^1A'_1$ ($8pa''_2 \leftarrow 1a''_2$) | 80,225(5) | 0.47 |

$\sim$ 73,000 $cm^{-1}$ and the lowest ionic threshold at 82,159 $cm^{-1}$ are summarized in Table I [43].

## VII. LARGER SULFUR CONTAINING MOLECULES

In this section, we shall be concerned with a series of somewhat larger molecules than discussed in this contribution so far, namely, the sulfur containing compounds thiirane, methanethiol, and dimethyl sulfide. Thiirane (ethylene sulfide, $H_2\overline{CSCH_2}$) is a three-membered heterocylic species with $C_{2v}$ symmetry. Methanethiol (methyl mercaptan, $CH_3SH$) possesses $C_s$ symmetry. Dimethyl sulfide (DMS, $H_3CSCH_3$) has $C_{2v}$ symmetry as well. The electronic spectroscopy of these compounds is in many respects similar to that of $H_2S$. The molecules of our series have in common that the highest occupied molecular orbital is largely comprised of a sulfur 3$p$ atomic orbital. Electronically excited states are usually obtained by excitation from this sulfur lone-pair orbital into higher $n\ell$ Rydberg states, giving rise to a variety of Rydberg series converging upon the lowest ionic limit. Two of the species possess $C_{2v}$ symmetry, and when they are subjected to multiphoton ionization methods the spectroscopy will be expected to be different for absorption of an even or an odd number of photons. For methanethiol, which possesses lower $C_s$ symmetry, (2 + 1) and (3 + 1) REMPI spectra are expected to be more similar. Next, we shall discuss each molecule in turn. We shall primarily focus on REMPI and REMPI–PES studies of excited Rydberg states. Spectral assignments will be based on quantum defect considerations, on relative intensities of the Rydberg series in two- and three-photon excitation spectra, and, in appropriate cases, on the observed changes in the two-photon band intensities upon switching the polarization of the excitation laser from linear to circular. Of course, kinetic energy resolved electron detection after absorption of an additional photon from the Rydberg state under study will again play a key role. Moreover, prominent examples of vibronic coupling, which in these larger molecules can be quite important, will be treated in some detail.

The research described in this section illustrates once more how, in addition to the familiar assignment tools such as quantum defects considerations, REMPI and, in particular, REMPI–PES can be employed successfully in unraveling and assigning the highly congested electronic spectra of medium-sized molecules such as thiirane, methanethiol, and dimethyl sulfide. The use of the polarization dependence of the transitions involving two and three photons will be seen to provide additional information invaluable in establishing reliable state assignments. Moreover, REMPI–PES provides a unique tool when it comes to understanding the often prominent effects of vibronic coupling.

### A. Thiirane

Ground-state thiirane has $C_{2v}$ symmetry. Adopting the convention that $z$ corresponds to the $C_2$ axis and that the sulfur atom lies in the $yz$ plane, the lowest electronic configuration may be written as $\ldots(2b_1)^2(7a_1)^2(1a_2)^2(8a_1)^2(4b_2)^2(3b_1)^2\ \tilde{X}^1A_1$. The highest occupied $3b_1$ MO is predominantly the sulfur $3p_x$ atomic orbital, directed perpendicular to the plane containing the ring of heavy atoms. Promoting one $3b_1$ electron to higher $n\ell$ Rydberg orbitals can be expected to give rise to one '$s$' ($\ell = 0$) series involving states of $^1B_1$ electronic symmetry and $n \geq 4$, three 'p' ($\ell = 1$) series (of $^1B_1$, $^1A_1$, and $^1A_2$ symmetries and $n \geq 4$), five '$d$' ($\ell = 2$) series (two of $^1B_1$ symmetry, and one each of $^1A_1$, $^1A_2$, and $^1B_2$ symmetry, all with $n \geq 3$), seven '$f$' ($\ell = 3$) series (two each transforming as $^1A_1$, $^1A_2$, and $^1B_1$, the other with $^1B_2$ symmetry, all with $n \geq 4$), and so on. The quotation marks are included as a reminder that configuration mixing may render $\ell$ an approximate quantum number only, while the state count remains unaffected. Quantum defects associated with "pure" $\ell$ series are expected to be $\sim 2.0$ ($s$ series), $\sim 1.6$ ($p$ series), and $\sim 0.0$ (for $d$,$f$, and higher $\ell$ series). We shall focus on the energy range from $\sim 50{,}000\,\text{cm}^{-1}$ up to the lowest ionic limit at $73{,}090 \pm 60\text{cm}^{-1}$, obtained by extrapolation of Rydberg series observed in one-photon absorption spectroscopy [284]. This ionization energy is in good accord with that of conventional photoelectron spectroscopy [285, 286]. Rydberg series converging upon higher ionic limits are not expected to appear below $\sim 68{,}000\,\text{cm}^{-1}$ [47].

The one-photon absorption spectra of thiirane exhibit a region of continuous absorption in the near-UV, followed at shorter wavelengths by numerous sharper vibronic features associated with Rydberg series converging upon the lowest ionic limit. An unambiguous assignment of the Rydberg structure has been complicated by spectral congestion and the lack of any rotational structure, presumably compounded by predissociation broadening [284, 287–289]. An *ab initio* theoretical study should also be mentioned [290].

We have performed $(2+1)$ and $(3+1)$ REMPI and REMPI–PES studies in the energy range from $\sim 50{,}000\,\text{cm}^{-1}$ up to the lowest ionic limit at $73{,}090 \pm 60\,\text{cm}^{-1}$. The two- and three-photon excitation spectra show striking differences. We shall concentrate on the two-photon excitation spectrum presented in Figure 50. Combs above the spectra indicate members belonging to Rydberg series designated as II to VI. An additional Rydberg series labeled I is only visible in the three-photon excitation spectrum (not shown). Rydberg origins were assigned unambiguously by REMPI–PES in a way that should by now be familiar.

The lowest energy feature occurs as a two-photon resonance at 51,986 cm$^{-1}$, with $\delta \approx 1.72$ assuming $n = 4$. We assign this feature to the two-photon allowed $4pb_1 \longleftarrow\!\!\longleftarrow 3pb_1$ transition (series III), but the 'purity' of the $\ell = 1$ description of this Rydberg state is debatable considering its strong showing in one-photon absorption spectroscopy [284, 287–289]. The *ab initio* study supports our assignment [290]. A few higher members of this series appear very weakly in the three-photon excitation spectrum, which may reflect the short lifetime of these states.

The next weak feature in the two-photon excitation spectrum occurs as a broad diffuse band centered ~ 54,800 cm$^{-1}$, not observed in the one-photon

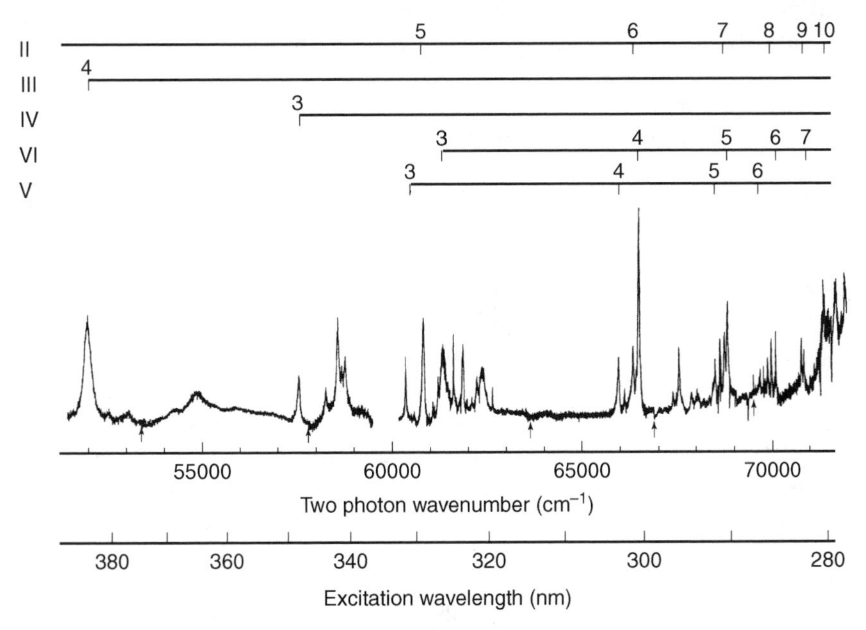


**Figure 50.** The (2 + 1) REMPI spectrum of a jet-cooled sample of thiirane in the wavelength region 50,000–72,000 cm$^{-1}$ using linearly polarized light and obtained by monitoring ions with $m/z = 60$. The displayed spectrum is a composite measured with various different dyes. No serious attempts were made at normalizing intensities. Spectral transition strengths should therefore be viewed with some caution. Arrows below the spectrum indicate where we switched from one dye to the next. Combs above the spectrum show our assignments. (Reproduced with permission from [47], copyright 1994 The Royal Society of Chemistry.)

absorption spectrum. We speculate that this band, with $\delta \approx 1.55$ assuming $n = 4$, corresponds to the transition arising from the promotion of a $3pb_1$ electron to the $4pb_2$ orbital that is formally both one- and two-photon forbidden.

The feature at 57,481 cm$^{-1}$ is also seen in one- [284, 287–289] and three-photon [47] absorption. This fact alone indicates that this state should possess hybrid $\ell$ character. Higher members of the series do not appear in the two-photon spectrum. The entire series is assigned as the two-photon forbidden $n$'$d$'$a_1 \longleftarrow\!\!\longleftarrow 3pb_1$ transition, with $\delta \approx 0.35$ assuming $n = 3$ (series IV).

Series V, which dominates the three-photon excitation spectra, is also prominent in two-photon absorption as shown by the transition observed at 60,358 cm$^{-1}$. When circularly polarized light is used, the intensity of the members of series V decreases. The situation is indicative of a two-photon transition linking two states of the same symmetry. This is the only way in which the scalar $[T_0^0(A)]$ part of the two-photon transition tensor (the component which is forbidden in circularly polarized excitation) can lead to a nonzero amplitude for the transition [291, 292]. We therefore designate series V, with $\delta \approx 0.06$, as arising from an $n$'$d$'$b_1 \longleftarrow\!\!\longleftarrow 3pb_1$ transition, which in an atomic picture would be formally two-photon forbidden. A degree of $p/d$ mixing is invoked to explain our observations.

As is clear from Figure 50, close to the transitions associated with series V other transitions assigned as members of series II and VI are found. The REMPI–PES studies in this region indicate that the transitions at 60,805 cm$^{-1}$ (series II) and 61,316 cm$^{-1}$ (series VI) and the vibronic structure built upon these origins show extensive vibronic coupling [47], which will not be expanded on here. Weighing the available evidence our assignments are $n$'$p$'$a_1 \longleftarrow\!\!\longleftarrow 3pb_1$ for series II ($\delta \approx 2.0$), and $n$'$d$'$b_1 \longleftarrow\!\!\longleftarrow 3pb_1$ for series VI ($\delta \approx 0.05$) [47]. The first member of series II falls outside the range studied here, and was characterized in separate $(1 + 1)$ REMPI and REMPI–PES experiments [47]. Again hybrid $\ell$ character is required to explain the occurrence of these transitions in two-photon excitation. The fact that the assignments of both series V and VI involve $n$'$d$'$b_1$ excited states, notwithstanding the fact that there exists only one $d$ orbital resulting from an $ndb_1 \leftarrow 3pb_1$ promotion, is a further illustration of the high degree of $\ell$ mixing.

Finally, series I observed in three-, but not in two- and (surprisingly) one-photon excitation, is designated as arising from the $nsa_1 \longleftarrow\!\!\longleftarrow 3pb_1$ transition, with $\delta \approx 1.81$. The first $n = 4$ member of this two-photon forbidden series, which is outside the present range, has been identified previously in one-photon absorption around $\sim 47{,}137$ cm$^{-1}$ [284].

## B. Methanethiol

Ground-state methanethiol has $C_s$ symmetry. The lowest electronic configuration is normally written as $\ldots(2a'')^2(9a')^2(10a')^2(3a'')^2\ \tilde{X}^1A_1$. The highest occupied $3a''$ MO is predominantly the sulfur $3p_x$ atomic orbital, directed perpendicular to the plane of the molecule. Promoting one $3a''$ electron to higher $n\ell$ Rydberg orbitals can be expected to give rise to one '$s$' ($\ell = 0$) series involving states of $^1A''$ electronic symmetry and $n \geq 4$, three '$p$' ($\ell = 1$) series (two of $^1A''$ symmetry, the other $^1A'$, all with $n \geq 4$), five '$d$' ($\ell = 2$) series (three of $^1A''$ symmetry, two of $^1A'$, all with $n \geq 3$), seven '$f$' ($\ell = 3$) series (four of $^1A''$, three of $^1A'$, all with $n \geq 4$), and so on. The quotation marks are included, as a reminder that configuration mixing may render $\ell$ an approximate quantum number only, while the total number of states remains valid. Quantum defects associated with "pure" $\ell$ series are expected to be $\sim 2.0$ ($s$ series), $\sim 1.6$ ($p$ series), and $\sim 0.0$ (for $d, f$, and higher $\ell$ series). We shall focus on the energy range from $\sim$ 55,000–75,000 cm$^{-1}$. The lowest ionic limit was obtained in our REMPI–PES work at 76,260 $\pm$ 40 cm$^{-1}$ [45]. This value is in reasonable agreement with previous ones obtained by extrapolation of Rydberg series observed in one-photon absorption spectroscopy [293, 294], by conventional PES [33], and by photoionization mass spectrometry [295].

The one-photon absorption spectrum of methanethiol exhibits two overlapping regions of continuous absorption in the near-UV, followed at shorter wavelengths by many sharper vibronic features associated with Rydberg series converging upon the lowest ionic limit. Unambiguous assignment of the Rydberg structure has been complicated by spectral congestion and by the lack of any rotational structure, presumably exacerbated by predissociation broadening. Previous analyses have therefore relied heavily on quantum defects obtained in one-photon absorption studies [287, 293, 294]. Several theoretical studies concerning the electronic spectrum of this molecule should also be mentioned [296–298].

We have performed (2 + 1) and (3 + 1) REMPI and REMPI–PES studies in the energy range from 55,000–75,000 cm$^{-1}$. In contrast to what was observed for thiirane with $C_{2v}$ symmetry, the two- and three-photon excitation spectra for the less symmetric species methanethiol are fairly similar. Both excitation spectra are presented in Figure 51. Members belonging to Rydberg series designated as I–V are indicated by combs above the spectra. Rydberg origins and associated vibronic structure were assigned unambiguously by REMPI–PES [45], following procedures as discussed several times before.

The progression labeled I in Figure 51 shows a first member at 56,902 cm$^{-1}$ with a quantum defect $\delta \approx 1.6$ assuming $n = 4$. The earlier suggestion from one-photon experiments that this series derives from an excitation from $3a''$ to Rydberg states with predominantly $\ell = 1$ character is supported [287, 293, 294]. The fact that these states are observed in both two- and three-photon absorption indicates that an atomic picture based on sulfur atomic orbitals is inadequate, as illustrated before for thiirane [47]. The identification of Rydberg series III, IV, and V is not particularly controversial. All three possess near-integer quantum defects, suggesting that they should be viewed as excitations from $3a''$ to

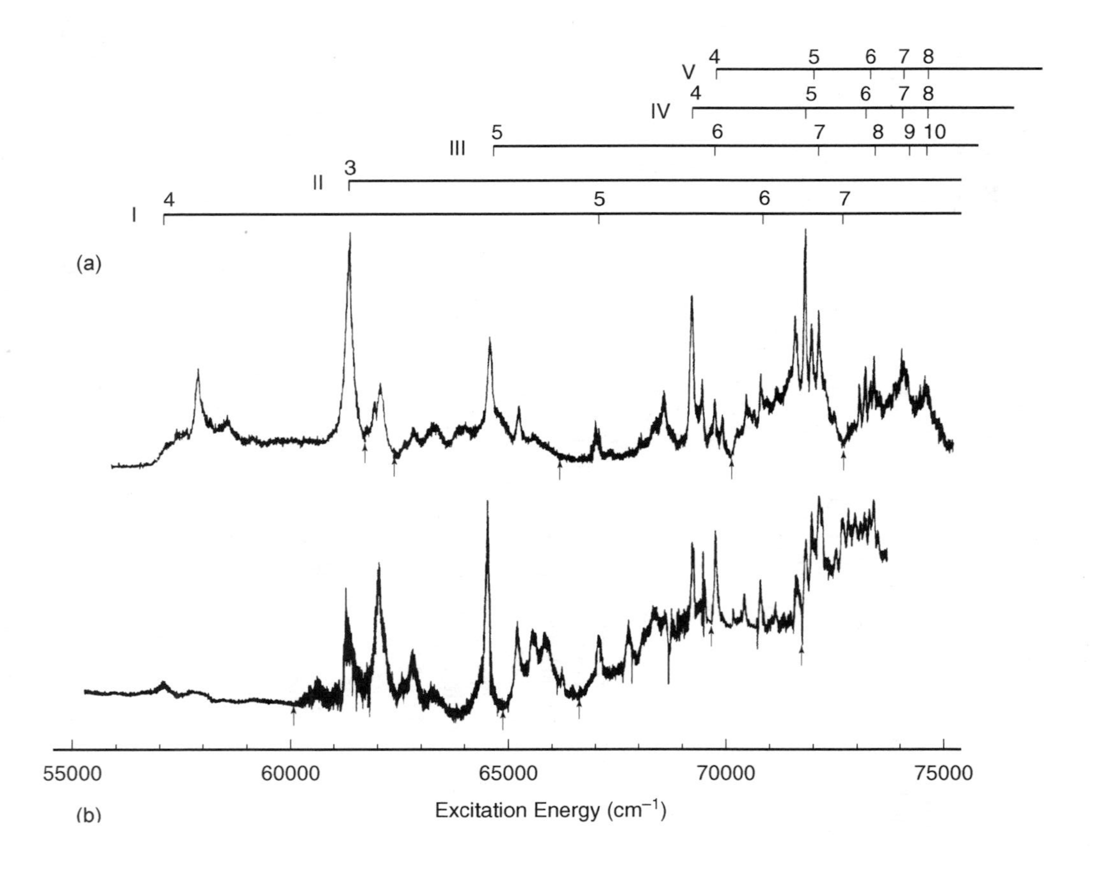

V 4 5 6 7 8
IV 4 5 6 7 8
III 5 6 7 8 9 10
II 3
I 4 5 6 7
(a)
(b)
55000
60000
65000
70000
75000
Excitation Energy (cm−1)

Rydberg orbitals with mainly $s, d$, or $f$ character. Several members of series III and IV have been observed in one-photon absorption also [294]. We assign series III as $n$‘$s$’ $\leftarrow 3a''$ and series IV as $n$‘$d$’ $\leftarrow 3a''$ excitations. Most strikingly, the progression we observe as series V does not seem to have a counterpart in the one-photon work. The latter series is best described as an $n$‘$f$’ $\leftarrow 3a''$ Rydberg series. The most controversial spectral feature is the one at 61,140 $cm^{-1}$ and its associated vibronic structure, labeled as II in Figure 51. The interpretation of this spectral structure has caused difficulties in the previous one-photon studies [287, 293, 294]. We have measured REMPI–PE spectra for all the transitions in this region. Our results are consistent with a picture in which terms corresponding to one dominant vibrational progression are built on a Rydberg origin at 61,140 $cm^{-1}$. The observed vibronic activity probably arises from Rydberg-valence mixing. Since higher members of what in Figure 51 is termed as series II appear to be absent, it is tempting to speculate that the 61,140 $cm^{-1}$ feature could be the first member of the progression labeled as series IV [45].

The laser PE spectra of methanethiol taken at several features observed in the excitation spectra show strong evidence for orbital mixing caused by vibronic coupling. This behavior in polyatomic molecules, especially when the molecular size increases, by now appears to be the rule rather than the exception [24]. An illustrative example can be viewed when the PE spectrum obtained by fixing the laser wavelength at the transition in the excitation spectrum at a three-photon energy of 66,991 $cm^{-1}$ is studied. The main peak observed in the PE spectrum (not shown) at 1.619 eV is fully consistent with an interpretation in which a $(3+1)$ process takes place via an origin of a Rydberg series located at 66,991 $cm^{-1}$. Three “slower” peaks are measured, separated from the main peak by amounts that correspond rather accurately to energy splittings between specific features in the excitation spectrum assigned on the basis of separate PE spectra as lower lying Rydberg origins at 64,413 (series III), 61,140 (series II), and 56,902 $cm^{-1}$ (series I) [45]. Once more the results for methanethiol underline how through vibronic coupling, observed directly in laser PE spectra, Rydberg origins can be obtained via a novel, “single-shot” route. For the coupling mechanism at play the reader is again referred to Figure 40.

**Figure 51.** (*a*) $(3+1)$ and (*b*) $(2+1)$ REMPI spectra of a jet-cooled sample of methanethiol over the energy range 55,000–75,000 $cm^{-1}$, obtained by monitoring ions with $m/z = 48$. The displayed spectra are composites obtained by splicing together spectra recorded using a number of different dyes. Vertical arrows below the spectra indicate where the various scans have been joined. Normalization of relative intensities was not attempted. The combs above the spectra indicate the assignments in terms of $n$‘$\ell$’$\leftarrow 3a''$ Rydberg excitations. (Reproduced with permission from [45], copyright 1995 The Royal Society of Chemistry.)

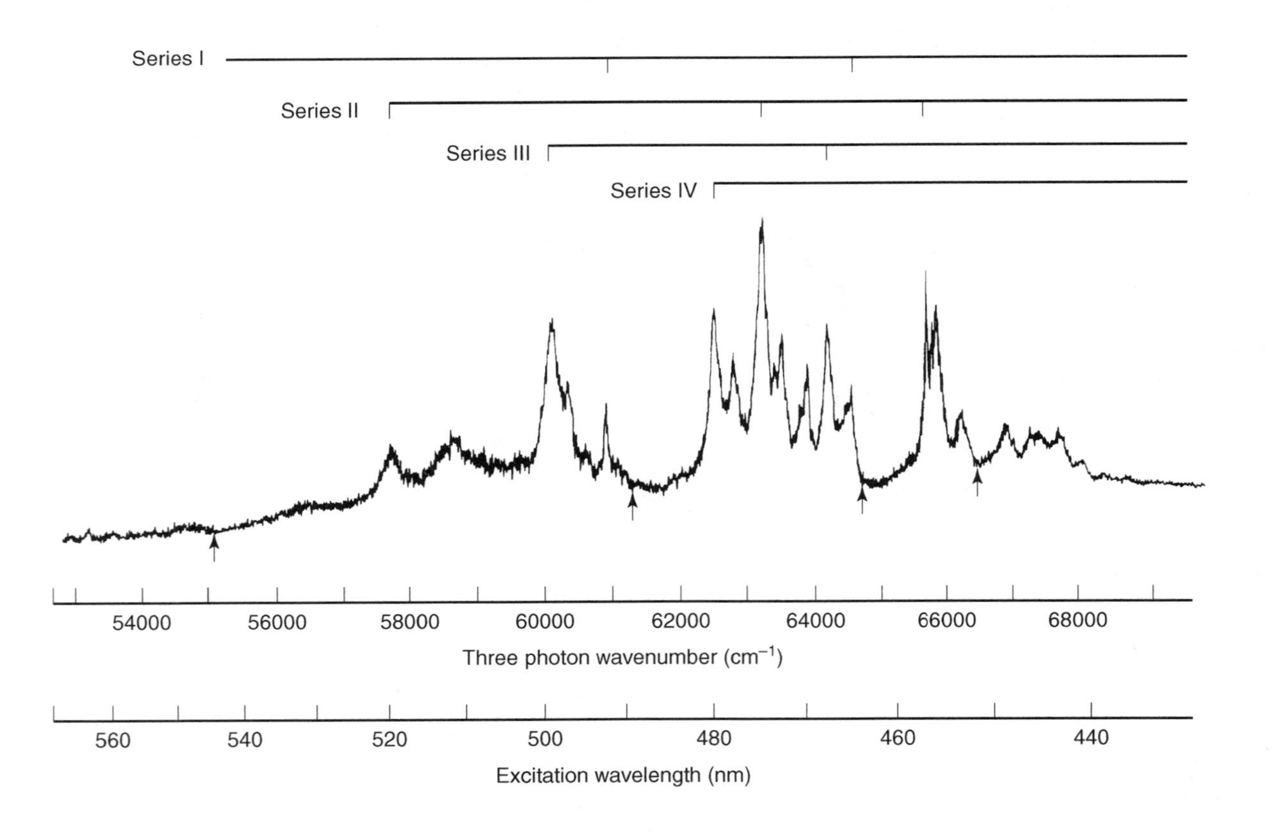
Series I
Series II
Series III
Series IV
54000
56000
58000
60000
62000
64000
66000
68000
Three photon wavenumber (cm$^{-1}$)
560
540
520
500
480
460
440
Excitation wavelength (nm)

### C. Dimethyl sulfide

Ground-state dimethyl sulfide has $C_{2v}$ symmetry. Adopting the convention that $z$ corresponds to the $C_2$ axis and that the sulfur atom lies in the $yz$ plane, the lowest electronic configuration may be written as $\ldots(1a_2)^2(8a_1)^2(4b_2)^2(3b_1)^2\ \tilde{X}^1A_1$. The highest occupied $3b_1$ MO is predominantly the sulfur $3p_x$ atomic orbital, directed perpendicular to the plane containing the C–S–C framework. Promoting one $3b_1$ electron to higher $n\ell$ Rydberg orbitals can be expected to give rise to one 's' ($\ell = 0$) series involving states of $^1B_1$ electronic symmetry and $n \geq 4$, three 'p' ($\ell = 1$) series (of $^1B_1$, $^1A_1$, and $^1A_2$ symmetries and $n \geq 4$), five 'd' ($\ell = 2$) series (two of $^1B_1$ symmetry, and one each of $^1A_1$, $^1A_2$ and $^1B_2$ symmetry, all with $n \geq 3$), seven '$f$' ($\ell = 3$) series (two each transforming as $^1A_1$, $^1A_2$, and $^1B_1$, the other with $^1B_2$ symmetry, all with $n \geq 4$), and so on. The quotation marks are again included, as a reminder that configuration mixing may render $\ell$ an approximate quantum number only, while the state count remains unaffected. Quantum defects associated with "pure" $\ell$ series are expected to be $\sim 2.0$ ($s$ series), $\sim 1.6$ ($p$ series), and $\sim 0.0$ (for $d$,$f$, and higher $\ell$ series). We shall focus on the energy range from $\sim$ 50,000–69,000 cm$^{-1}$. The lowest ionic limit was obtained in our REMPI–PES work at 70,250 $\pm$ 40 cm$^{-1}$ [46]. This value agrees very well with previous ones obtained by extrapolation of Rydberg series observed in one-photon absorption spectroscopy [299–301], and by conventional PES [33].

The electronic structure of dimethyl sulfide has been studied experimentally with one-photon absorption methods in the UV and VUV regions [287, 293, 299–301]. The one-photon spectra of dimethyl sulfide show a region of weak diffuse absorption with a long wavelength onset at $\sim$ 230 nm, followed at shorter wavelengths by a broad intense feature at $\sim$ 203 nm. In the VUV region, a number of sharp vibronic features can be arranged into Rydberg series converging upon the lowest ionic limit. As with thiirane and methanethiol, unambiguous assignments of the Rydberg structure have been complicated by spectral congestion and by the lack of any rotational structure. Previous analyses have therefore again relied heavily on quantum defects obtained in the one-photon absorption studies.

We have performed $(2 + 1)$ and $(3 + 1)$ REMPI and REMPI–PES studies in the energy range from 50,000–69,000 cm$^{-1}$. As was the case with thiirane with

←

**Figure 52.** The $(3 + 1)$ REMPI spectrum of a jet-cooled sample of dimethyl sulfide in the energy range 50,000–69,000 cm$^{-1}$, obtained by monitoring ions with $m/z = 62$. The displayed spectra are composites obtained by splicing together spectra recorded using a number of different dyes. Vertical arrows below the spectra indicate where the various scans have been joined. Normalization of relative intensities was not attempted. The combs above the spectra indicate assignments in terms of $n$'$\ell$'$\leftarrow 3b_1$ Rydberg excitations. (Reproduced with permission from [46], copyright 1995 The Royal Society of Chemistry.)

$C_{2v}$ symmetry [47], the two- and three-photon excitation spectra for dimethyl sulfide show striking differences. We shall concentrate on the three-photon excitation spectrum presented in Figure 52. Members belonging to Rydberg series designated as I–IV are indicated by combs above the spectra. In the two-photon excitation spectrum (not shown separately) series I dominates, and an additional Rydberg series labeled V showing a long vibrational progression is visible. As usual, Rydberg origins were assigned unambiguously by REMPI–PES [46].

The first strong feature in the two-photon excitation spectrum at 51,184 cm$^{-1}$ has also been clearly observed in the one-photon VUV absorption spectra [299–301]. The fact that this transition is not observed in the three-photon excitation spectrum of Figure 52 is ascribed to the fact that at the requisite photon energies ionization would involve a less probable five-photon $(3+2)$ process. With a quantum defect of $\delta \approx 1.6$ this origin transition can be viewed as the first $(n = 4)$ member of the $n$'$p$' $\leftarrow 3b_1$ Rydberg series. The appearance of this feature in both one- and two-photon spectra again casts doubt on the validity of the "atomic" description. Two higher members of this series were also assigned. The lowest energy feature in the three-photon spectrum in Figure 52 is a broad band occurring $\sim$ 57,705 cm$^{-1}$. A corresponding peak has also been observed in a one-photon study [299]. PES shows this feature to be an origin $(n = 5)$, while the quantum defect is near integer. The lowest member of this series $(n = 4)$ has been observed in one-photon absorption spectroscopy at $\sim$ 228 nm [289, 299, 300]. Higher energy members of this series, labeled II, are observed in Figure 52 and have been seen in previous one-photon absorption studies [299–301]. The assignment is taken as $n$'$s$' $\leftarrow 3b_1$. The congested spectral structure in the energy region $>$ 60,000 cm$^{-1}$ is much more difficult to assign. With REMPI–PES is was established that the features at 60,064, 64,187 (both labeled series III), and 62,150 cm$^{-1}$ (labeled IV) represent origins. Their definitive assignments remains undecided, since characterization in terms of $n$'$d$' $\leftarrow 3b_1$ and $n$'$p$' $\leftarrow 3b_1$ transitions runs into problems. The well-developed long vibrational progression associated with series V, only observed in the three-photon excitation spectrum, with near-zero quantum defects for the members of the series, is assigned as $n$'$f$' $\leftarrow 3b_1$ [46].

A final example of orbital mixing caused by vibronic coupling observed with laser PES is provided by dimethyl sulfide. In Figure 53, the wavelength-resolved $(3+1)$ and $(2+1)$ excitation spectra between 49,000–65,000 cm$^{-1}$ are plotted on the same energy scale as the PE spectrum obtained by fixing the laser wavelength at a two-photon energy of 63,300 cm$^{-1}$. The dominant peak at 3.06 eV is consistent with a $(2+1)$ process via a Rydberg origin situated at 63,300 cm$^{-1}$ in the $(2+1)$ excitation spectrum. As in previous cases, we notice the presence of several relatively broad slower electron kinetic energy features. Closer inspection again reveals that the energy splittings in the PE spectrum

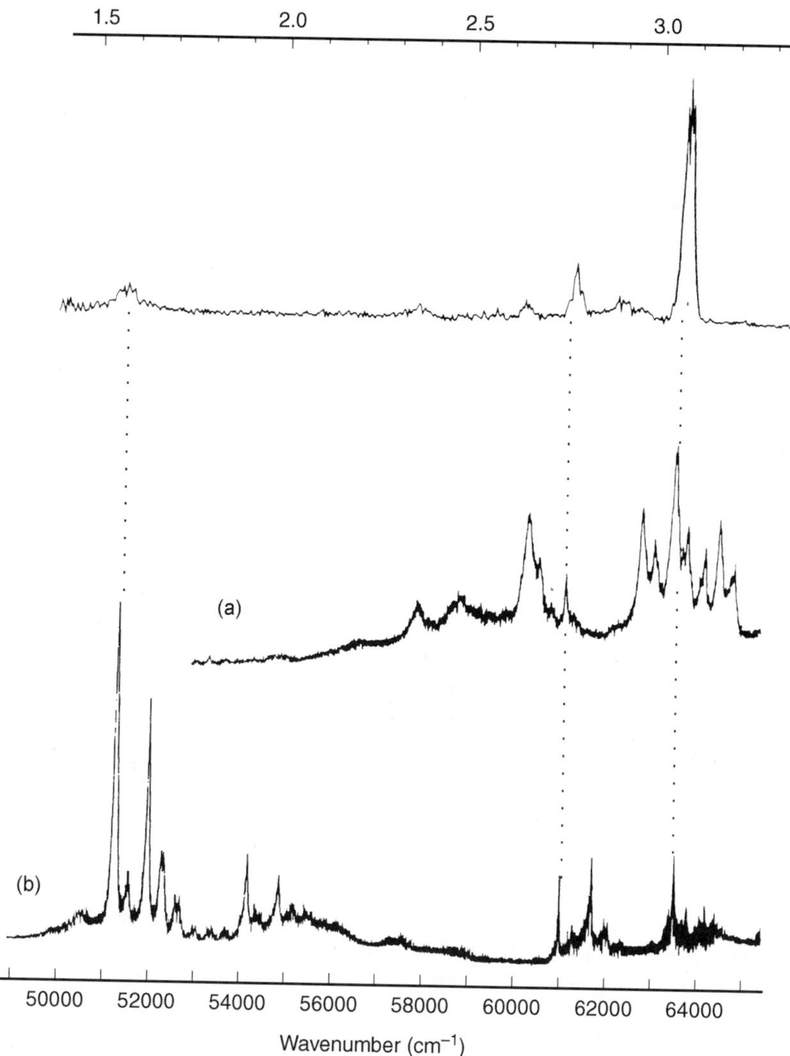


**Figure 53.** PE spectrum of dimethyl sulfide for $(2+1)$ ionization following excitation of a Rydberg origin level at a two-photon energy of $63{,}300\,\text{cm}^{-1}$; (*a*) the relevant portion of the three-photon excitation spectrum obtained with ion detection, plotted on the same energy scale; (*b*) the relevant portion of the two-photon excitation spectrum obtained with ion detection, plotted on the same energy scale. (Reproduced with permission from [46], copyright 1995 The Royal Society of Chemistry.)

correspond rather accurately to separations between the Rydberg origin under consideration with lower lying features in the excitation spectra assigned as different Rydberg origins on the basis of PES. Most notably, the origins of the $n = 4$ and 5 members of series I observed in the $(2 + 1)$ excitation spectrum appear in the PE spectrum. The coupling mechanism pictured in Figure 40, which involves near-resonant vibronic mixing between the Rydberg origin under consideration and the various manifolds of vibrational levels based on other Rydberg origins that lie lower in energy is again at the root of our observations. Clearly, in dimethyl sulfide the correlation between the slow photoelectron peaks with the Rydberg origins observed in the $(2 + 1)$ excitation spectrum is much more outspoken than with those detected in $(3 + 1)$ REMPI. Since Rydberg origins belonging to different irreducible representations of the $C_{2v}$ group may be reached in the two- and three-photon excitation processes, the vibronic levels available for vibronic mixing may well be associated with different normal modes. This could explain why the slow electron features correlate with one excitation spectrum rather than the other [46].

## VIII. CONCLUSIONS

The unique power of laser PES in elucidating the detailed spectroscopy and dynamics of excited states in small and medium-sized molecules was demonstrated. Examples were chosen from a wide selection of stable and short-lived molecules that all have relevance for atmospheric or interstellar chemistry. In the future, the application of laser PES to chemical and physical phenomena on an ultrafast time scale can be envisaged. In this context, the experimental and theoretical methods developed in the frequency domain and described in this contribution will prove to be crucial.

### Acknowledgments

The author wishes to acknowledge countless fruitful collaborations and enlightening discussions with all the senior colleagues who appear as co-authors on a large number of joint papers. Special thanks are due to the graduate students of my research group, past and present, Esther de Beer, Bianca Milan, Connie Scheper, Anouk Rijs, Nicolas Wales, and Marc Smits. Their tireless efforts contributed greatly to the research described in this contribution.

### References

1. G. Herzberg, *Molecular Spectra and Molecular Structure, Vol. I. Spectra of Diatomic Molecules*, Van Nostrand, Princeton, 1950.
2. G. Herzberg, *Molecular Spectra and Molecular Structure, Vol. III. Electronic Spectra of Polyatomic Molecules*, Van Nostrand, Princeton, 1967.
3. K. P. Huber and G. Herzberg, *Molecular Spectra and Molecular Structure, Vol. IV. Constants of Diatomic Molecules*, Van Nostrand-Reinhold, New York, 1979.

4. G. S. Voronov and N. B. Delone, *Sov. Phys. JETP* **23**, 54 (1966).
5. S. H. Lin, Y. Fujimura, H. J. Neusser, and E. W. Schlag, *Multiphoton Spectroscopy of Molecules*, Academic, New York, 1984.
6. J. P. Reilly, *Israel J. Chem.* **24**, 266 (1984).
7. K. Kimura, *Adv. Chem. Phys.* **60**, 161 (1985).
8. K. Kimura, *Int. Rev. Phys. Chem.* **6**, 195 (1987).
9. P. M. Dehmer, J. L. Dehmer, and S. T. Pratt, *Comm. At. Mol. Phys.* **19**, 205 (1987).
10. S. T. Pratt, P. M. Dehmer, and J. L. Dehmer, *Advances in Multiphoton Processes and Spectroscopy, Vol. 4*, S. H. Lin, Ed., World Scientific Press, Singapore, 1988.
11. R. N. Compton and J. C. Miller, in *Laser Applications in Physical Chemistry*, D. K. Evans, Ed., Marcel Dekker, New York, 1989, p. 121.
12. W. Domcke and G. Stock, *Adv. Chem. Phys.* **100**, 1 (1997).
13. V. Blanchet and A. Stolow, *J. Chem. Phys.* **109**, 4371 (1998).
14. V. Blanchet, M. Z. Zgierski, T. Seideman, and A. Stolow, *Nature* (*London*) **401**, 52 (1999).
15. P. Kruit and F. H. Read, *J. Phys. E* **16**, 313 (1983).
16. B. G. Koenders, D. M. Wieringa, K. E. Drabe, and C. A. de Lange, *Chem. Phys.* **118**, 113 (1987).
17. C. A. de Lange, in *High Resolution Laser Photoionization and Photoelectron Studies*, I. Powis, T. Baer and C. Y. Ng, Eds., Wiley, New York, 1995, p. 195.
18. J. B. Milan, W. J. Buma, and C. A. de Lange, *J. Chem. Phys.* **104**, 521 (1996).
19. K. Müller-Dethlefs, M. Sander, and E. W. Schlag, *Chem. Phys. Lett.* **112**, 291 (1984).
20. K. Müller-Dethlefs and E. W. Schlag, *Annu. Rev. Phys. Chem.* **42**, 109 (1991).
21. K. Müller-Dethlefs, E. W. Schlag, E. R. Grant, K. Wang, and B. V. McKoy, *Adv. Chem. Phys.* **90**, 1 (1995).
22. E. W. Schlag, *ZEKE Spectroscopy: the Linnett Lectures*, Cambridge University Press, Cambridge, UK, 1998.
23. T. F. Gallagher, *Rydberg Atoms (Cambridge Monographs on Atomic, Molecular, and Chemical Physics, No. 3)*, Cambridge University Press, Cambridge, UK, 1994.
24. C. A. de Lange, *J. Chem. Soc., Faraday Trans.* **94**, 3409 (1998).
25. C. R. Scheper, W. J. Buma, C. A. de Lange, and W. J. van der Zande, *J. Chem. Phys.* **109**, 8319 (1998).
26. C. R. Scheper, M. F. Somers, and C. A. de Lange, *J. El. Spec. Relat. Phenom.* **108**, 123 (2000).
27. D. W. Turner, A. D. Baker, C. Baker, and C. R. Brundle, *Molecular Photoelectron Spectroscopy*, Wiley, London, 1970.
28. A. D. Baker and D. Betteridge, *Photoelectron Spectroscopy*, Pergamon, Oxford, 1972.
29. J. H. D. Eland, *Photoelectron Spectroscopy*, Butterworths, London, 1974.
30. J. W. Rabalais, *Principles of Ultraviolet Photoelectron Spectroscopy*, Wiley, New York, 1977.
31. R. E. Ballard, *Photoelectron Spectroscopy and Molecular Orbital Theory*, Adam Hilger, Bristol, 1978.
32. J. Berkowitz, *Photoabsorption, Photoionization and Photoelectron Spectroscopy*, Academic, New York, 1979.
33. K. Kimura, S. Katsumata, Y. Achiba, T. Yamazaki, and S. Iwata, *Handbook of He(I) Photoelectron Spectroscopy of Fundamental Organic Molecules*, Halsted, New York, 1981.
34. P. K. Gosh, *Chemical Analysis, Introduction to Photoelectron Spectroscopy*, P. J. Elving and J. D. Winefordner, Eds., vol. 67, Wiley, New York, 1983.

35. E. de Beer, C. A. de Lange, and N. P. C. Westwood, *Phys. Rev. A* **46**, 5653 (1992).
36. S. Woutersen, J. B. Milan, W. J. Buma, and C. A. de Lange, *Phys. Rev. A* **54**, 5126 (1996).
37. S. Woutersen, J. B. Milan, W. J. Buma, and C. A. de Lange, *J. Chem. Phys.* **106**, 6831 (1997).
38. S. G. Clement, M. N. R. Ashfold, C. M. Western, E. de Beer, C. A. de Lange, and N. P. C. Westwood, *J. Chem. Phys.* **96**, 4963 (1992).
39. N. P. L. Wales, E. de Beer, N. P. C. Westwood, W. J. Buma, C. A. de Lange, and M. C. van Hemert, *J. Chem. Phys.* **100**, 7984 (1994).
40. J. B. Milan, W. J. Buma, and C. A. de Lange, *J. Chem. Phys.* **105**, 6688 (1996).
41. C. A. de Lange, in *The Role of Rydberg States in Spectroscopy and Reactivity*, C. Sandorfy, Ed., Kluwer Academic Publishers, Dordrecht, The Netherlands 1999, p. 457.
42. C. R. Scheper, J. Kuijt, W. J. Buma, and C. A. de Lange, *J. Chem. Phys.* **109**, 7844 (1998).
43. S. R. Langford, A. J. Orr-Ewing, R. A. Morgan, C. M. Western, M. N. R. Ashfold, A. Rijkenberg, C. R. Scheper, W. J. Buma, and C. A. de Lange, *J. Chem. Phys.* **108**, 6667 (1998).
44. M. N. R. Ashfold, S. R. Langford, R. A. Morgan, A. J. Orr-Ewing, C. M. Western, C. R. Scheper, and C. A. de Lange, *Eur. Phys. J. D* **4**, 189 (1998).
45. R. A. Morgan, P. Puyuelo, J. D. Howe, M. N. R. Ashfold, W. J. Buma, N. P. L. Wales, and C. A. de Lange, *J. Chem. Soc., Faraday Trans.* **91**, 2715 (1995).
46. R. A. Morgan, A. J. Orr-Ewing, M. N. R. Ashfold, W. J. Buma, N. P. L. Wales, and C. A. de Lange, *J. Chem. Soc., Faraday Trans.* **91**, 3339 (1995).
47. R. A. Morgan, P. Puyuelo, J. D. Howe, M. N. R. Ashfold, W. J. Buma, J. B. Milan, and C. A. de Lange, *J. Chem. Soc., Faraday Trans.* **90**, 3591 (1994).
48. R. A. Morgan, M. A. Baldwin, A. J. Orr-Ewing, M. N. R. Ashfold, W. J. Buma, J. B. Milan, and C. A. de Lange, *J. Chem. Phys.* **104**, 6117 (1996).
49. R. A. Morgan, A. J. Orr-Ewing, D. Ascenzi, M. N. R. Ashfold, W. J. Buma, C. R. Scheper, and C. A. de Lange, *J. Chem. Phys.* **105**, 2141 (1996).
50. R. A. Morgan, M. A. Baldwin, D. Ascenzi, A. J. Orr-Ewing, M. N. R. Ashfold, W. J. Buma, J. B. Milan, C. R. Scheper, and C. A. de Lange, *Int. J. Mass Spectrom. Ion Processes* **159**, 1 (1996).
51. E. de Beer, C. A. de Lange, J. A. Stephens, K. Wang, and V. McKoy, *J. Chem. Phys.* **95**, 714 (1991).
52. J. B. Milan, W. J. Buma, C. A. de Lange, K. Wang, and V. McKoy, *J. Chem. Phys.* **103**, 3262 (1995).
53. J. B. Milan, W. J. Buma, C. A. de Lange, C. M. Western, and M. N. R. Ashfold, *Chem. Phys. Lett.* **239**, 326 (1995).
54. J. B. Milan, W. J. Buma, C. A. de Lange, K. Wang, and V. McKoy, *J. Chem. Phys.* **107**, 2782 (1997).
55. E. de Beer, M. Born, C. A. de Lange, and N. P. C. Westwood, *Chem. Phys. Lett.* **186**, 40 (1991).
56. K. Wang, J. A. Stephens, V. McKoy, E. de Beer, C. A. de Lange, and N. P. C. Westwood, *J. Chem. Phys.* **97**, 211 (1992).
57. N. P. L. Wales, W. J. Buma, C. A. de Lange, H. Lefebvre-Brion, K. Wang, and V. McKoy, *J. Chem. Phys.* **104**, 4911 (1996).
58. N. P. L. Wales, W. J. Buma, C. A. de Lange, and H. Lefebvre-Brion, *J. Chem. Phys.* **105**, 5702 (1996).

59. W. A Chupka, *J. Chem. Phys.* **87**, 1488 (1987).
60. D. Zajfman, Z. Amitay, M. Lange, U. Hechtfischer, L. Knoll, D. Schwalm, R. Wester, A. Wolf, and X. Urbain, *Phys. Rev. Lett.* **79**, 1829 (1997).
61. T. Tanabe, I. Katayama, N. Inoue, K. Chida, Y. Arakaki, T. Watanabe, M. Yoshizawa, S. Ohtani, and K. Noda, *Phys. Rev. Lett.* **70**, 422 (1993).
62. P. Forck, M. Grieser, D. Habs, A. Lampert, R. Repnow, D. Schwalm, A. Wolf, and D. Zajfman, *Phys. Rev. Lett.* **70**, 426 (1993).
63. M. Larsson, H. Danared, J. R. Mowat, P. Sigray, G. Sundström, L. Broström, A. Filevich, A. Källberg, S. Mannervik, K. G. Rensfelt, and S. Datz, *Phys. Rev. Lett.* **70**, 430 (1993).
64. M. Larsson, M. Carlson, H. Danared, L. Broström, S. Mannervik, and G. Sundström, *J. Phys. B* **27**, 1397 (1994).
65. D. Zajfman, Z. Amitay, C. Broudem, P. Forck, B. Seidel, M. Grieser, D. Habs, D. Schwalm, and A. Wolf, *Phys. Rev. Lett.* **75**, 814 (1995).
66. W. J. van der Zande, J. Semaniak, V. Zengin, G. Sundström, S. Rosén, C. Strömholm, S. Datz, H. Danared, and M. Larsson, *Phys. Rev. A* **54**, 5010 (1996).
67. S. T. Pratt, P. M. Dehmer, and J. L. Dehmer, *J. Chem. Phys.* **78**, 4315 (1983).
68. J. H. M. Bonnie, J. W. J. Verschuur, H. J. Hopman, and H. B. van Linden van den Heuvell, *Chem. Phys. Lett.* **130**, 43 (1986).
69. E. Y. Xu, T. Tsuboi, R. Kachru, and H. Helm, *Phys. Rev. A* **36**, 5645 (1987).
70. J. W. J. Verschuur, L. D. Noordam, J. H. M. Bonnie, and H. B. van Linden van den Heuvell, *Chem. Phys. Lett.* **146**, 283 (1988).
71. J. W. J. Verschuur and H. B. van Linden van den Heuvell, *Chem. Phys.* **129**, 1 (1989).
72. S. T. Pratt, P. M. Dehmer, and J. L. Dehmer, *Chem. Phys. Lett.* **105**, 28 (1984).
73. S. T. Pratt, P. M. Dehmer, and J. L. Dehmer, *J. Chem. Phys.* **85**, 3379 (1986).
74. M. A. O'Halloran, S. T. Pratt, P. M. Dehmer, and J. L. Dehmer, *J. Chem. Phys.* **87**, 3288 (1987).
75. S. T. Pratt, P. M. Dehmer, and J. L. Dehmer, *J. Chem Phys.* **87**, 4423 (1987).
76. C. W. Zucker and E. E. Eyler, *J. Chem. Phys.* **85**, 7180 (1986).
77. D. Normand, C. Cornaggia, and J. Morellec, *J. Phys. B* **19**, 2881 (1986).
78. C. Cornaggia, D. Normand, J. Morellec, G. Mainfray, and C. Manus, *Phys. Rev. A* **34**, 207 (1986).
79. W. L. Glab and J. P. Hessler, *Phys. Rev. A* **35**, 2102 (1987).
80. J. D. Buck, D. C. Robie, A. P. Hickman, D. J. Bamford, and W. K. Bischel, *Phys. Rev. A* **39**, 3932 (1989).
81. E. Y. Xu, H. Helm, and R. Kachru, *Phys. Rev. A* **39**, 3979 (1989).
82. E. Xu, A. P. Hickman, R. Kachru, T. Tsuboi, and H. Helm, *Phys. Rev. A* **40**, 7031 (1989).
83. M. A. Buntine, D. P. Baldwin, and D. W. Chandler, *J. Chem. Phys.* **96**, 5843 (1992).
84. E. F. McCormack, S. T. Pratt, P. M. Dehmer, and J. L. Dehmer, *J. Chem. Phys.* **98**, 8370 (1993).
85. S. T. Pratt, P. M. Dehmer, and J. L. Dehmer, *J. Chem. Phys.* **86**, 1727 (1987).
86. P. G. Wilkinson, *Can. J. Phys.* **46**, 1225 (1968).
87. I. Dabrowski and G. Herzberg, *Can. J. Phys.* **52**, 1110 (1974).
88. T. Namioka, *J. Chem. Phys.* **40**, 3154 (1964).
89. J. Xie and R. N. Zare, *J. Chem. Phys.* **93**, 3033 (1990).

90. D. L. Lynch, S. N. Dixit, and V. McKoy, *Chem. Phys. Lett.* **123**, 315 (1986).
91. S. L. Guberman, *J. Chem. Phys.* **78**, 1404 (1983).
92. L. Wolniewicz, *J. Chem. Phys.* **51**, 5002 (1969).
93. L. Wolniewicz and K. Dressler, *J. Mol. Spectrosc.* **96**, 195 (1982).
94. R. N. Zare, *Angular Momentum*, Wiley, New York, 1988.
95. G. Hunter, A. W. Yau, and H. O. Pritchard, *Atomic Data and Nuclear Data Tables*, vol. **14**, No. 1, July 1974.
96. G. Hunter and H. O. Pritchard, *J. Chem. Phys.* **46**, 2153 (1967).
97. C. R. Scheper, C. A. de Lange, A. de Lange, E. Reinhold, and W. Ubachs, *Chem. Phys. Lett.* **312**, 131 (1999).
98. W. Ubachs, K. S. E. Eikema, and W. Hogervorst, *Appl. Phys. B* **57**, 411 (1993).
99. P. C. Hinnen, *XUY-Laser Spectroscopy of $H_2$ and the Mystery of the Diffuse Interstellar Bands*. Ph.D. Thesis, Vrije Universiteit, Amsterdam, The Netherlands, 1997.
100. N. Kouchi, M. Ukai, and Y. Hatano, *J. Phys. B* **30**, 2319 (1997).
101. S. L. Bragg, J. W. Brault, and W. H. Smith, *Astrophys. J.* **263**, 999 (1982).
102. C. E. Moore, *Atomic Energy Levels*, National Bureau Standards (US) circ. 467, vol. I (U.S. GPO, Washington, DC, 1958).
103. L. Wolniewicz, *J. Chem. Phys.* **103**, 1792 (1995).
104. H. A. Bethe and E. E. Salpeter, *Quantum Mechanics of One and Two Electron Atoms*, Springer, Berlin, 1957.
105. N. Kouchi, N. Terazawa, Y. Chikahiro, M. Ukai, K. Kameta, Y. Hatano, and K. Tanaka, *Chem. Phys. Lett.* **190**, 319 (1992).
106. N. Terazawa, N. Kouchi, M. Ukai, K. Kameta, Y. Hatano, and K. Ito, *J. Chem. Phys.* **100**, 7036 (1994).
107. T. L. Stephens and A. Dalgarno, *Mol. Phys.* **28**, 1049 (1974).
108. P. Borrell, P. M. Guyon, and M. Glass-Maujean, *J. Chem. Phys.* **66**, 818 (1977).
109. M. Ukai, private communication, 1999.
110. A. Vallance Jones, *Space Sci. Rev.* **15**, 335 (1973).
111. E. J. Llewellyn and B. H. Long, *Can. J. Phys.* **56**, 581 (1978).
112. D. J. W. Kendall and T. A. Clark, *J. Quant. Spectrosc. Radiat. Transfer* **21**, 511 (1979).
113. J. A. Coxon and S. C. Foster, *Can. J. Phys.* **60**, 41 (1982).
114. W. J. Wilson and A. H. Battett, *Science* **161**, 778 (1968).
115. B. J. Robinson and R. X. McGee, *Annu. Rev. Astron. Astrophys.* **5**, 183 (1967).
116. E. F. van Dishoeck and A. Dalgarno, *J. Chem. Phys.* **79**, 873 (1983).
117. A. E. Douglas, *Can. J. Phys.* **52**, 318 (1974).
118. E. de Beer, M. P. Koopmans, C. A. de Lange, Yumin Wang, and W. A. Chupka, *J. Chem. Phys.* **94**, 7634 (1991).
119. G. Ondrey, N. van Veen, and R. Bersohn, *J. Chem. Phys.* **78**, 3732 (1983).
120. T. Ebata, A. Fujii, T. Amano, and M. Ito, *J. Phys. Chem.* **91**, 6095 (1987).
121. T. Ebata, A. Fujii, T. Amano, and M. Ito, *J. Phys. Chem.* **92**, 2394 (1988).
122. R. B. Holt, C. K. Lane, and O. Oldenberg, *J. Chem. Phys.* **16**, 638 (1948).
123. M. Schurgers and K. H. Welge, *Z. Naturforsch. A* **23**, 1508 (1968).
124. T. L. Ng and S. Bell, *J. Mol. Spectrosc.* **50**, 166 (1974).

125. K. S. Visnawathan, E. Sekreta, E. R. Davidson, and J. P. Reilly, *J. Phys. Chem.* **90**, 5078 (1986) and references cited therein.

126. S. N. Dixit and V. McKoy, *Chem. Phys. Lett.* **128**, 49 (1986).

127. K. Wang and V. McKoy, *J. Chem. Phys.* **95**, 8718 (1991).

128. J. A. Stephens and V. McKoy, *Phys. Rev. Lett.* **62**, 889 (1989).

129. J. A. Stephens and V. McKoy, *J. Chem. Phys.* **93**, 7863 (1990).

130. J. August, M. Brouard, M. P. Docker, A. Hodgson, C. J. Milne, and J. P. Simons, *Phys. Chem.* **92**, 264 (1988).

131. S. Klee, K.-H. Gericke, and F. J. Comes, *Phys. Chem.* **92**, 429 (1988).

132. H. Rudolph and V. McKoy, *J. Chem. Phys.* **91**, 7995 (1989).

133. H. Rudolph and V. McKoy, *J. Chem. Phys.* **93**, 7054 (1990).

134. S. Fredin, D. Gauyacq, M. Horani, C. Jungen, G. Lefebvre, and F. Masnou-Seeuws, *Mol. Phys.* **60**, 825 (1987).

135. K. Sinha, *Proc. Astr. Soc. Aust.* **9**, 32 (1991).

136. M. L. Meeks, M. A. Gordon, and M. M. Litvak, *Science* **163**, 173 (1969).

137. C. E. Heiles and M. B. E. Turner, *Astrophys. Lett.* **8**, 89 (1971).

138. L. C. Lee, X. Wang, and M. Suto, *J. Chem. Phys.* **86**, 4353 (1987).

139. S. H. Ashworth and J. M. Brown, *J. Mol. Spectrosc.* **153**, 41 (1992).

140. J. B. Milan, W. J. Buma, and C. A. de Lange, *J. Chem. Phys.* **104**, 521 (1996).

141. S. J. Dunlavey, J. M. Dyke, N. K. Fayad, N. Jonathan, and A. Morris, *Mol. Phys.* **38**, 3 (1979).

142. S. J. Dunlavey, J. M. Dyke, N. K. Fayad, N. Jonathan, and A. Morris, *Mol. Phys.* **44**, 265 (1981).

143. C. W. Hsu, D. P. Baldwin, C. L. Liao, and C. Y. Ng, *J. Chem. Phys.* **100**, 8047 (1994).

144. B. Morrow, *Can. J. Phys.* **44**, 2447 (1966).

145. M. N. R. Ashfold, B. Tutcher, and C. M. Western, *Mol. Phys.* **66**, 981 (1989).

146. K. Wang, J. A. Stephens, and V. McKoy, *J. Chem. Phys.* **97**, 9874 (1993).

147. W. R. Anderson, L. J Decker, and A. J. Kotlar, *Combustion Flame* **48**, 1079 (1982).

148. M. M. Litvak and E. N. Rodriguez Kuiper, *Astrophys. J.* **253**, 622 (1982).

149. D. E. Jennings, *J. Quant. Spectrosc. Radiat. Transfer* **40**, 221 (1988).

150. R. S. Ram and P. F. Bernath, *J. Opt. Soc. Am. B* **3**, 1170 (1986).

151. C. R. Brazier, R. S. Ram, and P. F. Bernath, *J. Mol. Spectrosc.* **120**, 381 (1986).

152. W. R. M. Graham and H. Lew, *Can. J. Phys.* **56**, 85 (1978).

153. M. N. R. Ashfold, S. G. Clement, J. D. Howe, and C. M. Western, *J. Chem. Soc., Faraday Trans.* **87**, 2515 (1991).

154. K.-H. Gericke, T. Haas, M. Lock, R. Theinl, and F. J. Comes, *J. Phys. Chem.* **95**, 6104 (1991).

155. R. D. Johnson, III and J. W. Hudgens, *J. Chem. Phys.* **92**, 6420 (1990).

156. S. G. Clement, M. N. R. Ashfold, and C. M. Western, *J. Chem. Soc., Faraday Trans.* **88**, 3121 (1992).

157. J.-J. Chu, P. Marcus, and P. J. Dagdigian, *J. Chem. Phys.* **93**, 257 (1990).

158. S. G. Clement, M. N. R. Ashfold, C. M. Western, R. D. Johnson, III, and J. W. Hudgens, *J. Chem. Phys.* **97**, 7064 (1992).

159. R. J. Buenker, in *Studies in Physical and Theoretical Chemistry*, R. Carbo, Ed., vol. 21 Elsevier, Amsterdam, The Netherlands, 1982, p. 17.

160. K. Wang, J. A. Stephens, and V. McKoy, *J. Chem. Phys.* **93**, 7874 (1990).
161. R. Colin and A. E. Douglas, *Can. J. Phys.* **46**, 61 (1968).
162. J. C. Farman, B. G. Gardiner, and J. D. Shanklin, *Nature (London)* **315**, 207 (1985).
163. G. Pannetier and A. G. Gaydon, *Nature (London)* **161**, 242 (1948).
164. G. Porter and J. Wright, *J. Phys. Chem.* **14**, 23 (1953).
165. M. J. Molina and F. S. Rowland, *Nature (London)* **249**, 810 (1974).
166. D. K. Bulgin, J. M. Dyke, N. Jonathan, and A. Morris, *Mol. Phys.* **32**, 1487 (1976).
167. J. Zhang and Y. T. Lee, *J. Phys. Chem. A* **101**, 6485 (1997).
168. C. G. Freeman and L. F. Phillips, *J. Chem. Phys.* **72**, 3025 (1968).
169. M. A. A. Clyne and R. T. Watson, *J. Chem. Soc., Faraday Trans. I* **70**, 2250 (1974).
170. P. P. Bemand, M. A. A. Clyne, and R. T. Watson, *J. Chem. Soc., Faraday Trans. I* **69**, 1356 (1973).
171. E. Bishenden and D. J. Donaldson, *J. Chem. Phys.* **101**, 9565 (1994).
172. C. J. Kreher, R. T. Carter, and J. R. Huber, *Chem. Phys. Lett.* **286**, 389 (1998).
173. E. Bishenden and D. J. Donaldson, *J. Chem. Phys.* **99**, 3129 (1993).
174. R. Flesch, B. Wassermann, B. Rothmund, and E. Rühl, *J. Phys. Chem.* **98**, 6263 (1994).
175. E. Bishenden, J. Haddock, and D. J. Donaldson, *J. Phys. Chem.* **95**, 2113 (1991).
176. H. F. Davis and Y. T. Lee, *J. Phys. Chem.* **96**, 5681 (1992).
177. T. Baumert, J. L. Herek, and A. H. Zewail, *J. Chem. Phys.* **99**, 4430 (1993).
178. A. Loewenschluss, J. C. Miller, and L. Andrews, *J. Mol. Spectrosc.* **80**, 351 (1980).
179. F. K. Chi and L. Andrews, *J. Phys. Chem.* **77**, 3062 (1973).
180. R. T. Menzies, J. S. Margolis, E. D. Hinkley, and R. A. Toth, *Appl. Opt.* **16**, 523 (1977).
181. A. G. Maki, F. J. Lovas, and W. B. Olson, *J. Mol. Spectrosc.* **92**, 410 (1982).
182. J. A. Coxon, *Can. J. Phys.* **57**, 1538 (1979).
183. R. K. Kakar, E. A. Cohen, and M. Geller, *J. Mol. Spectrosc.* **70**, 243 (1978).
184. J. A. Coxon, W. E. Jones, and E. G. Skolnik, *Can. J. Phys.* **54**, 1043 (1976).
185. R. A. Durie and D. A. Ramsey, *Can. J. Phys.* **36**, 35 (1958).
186. J. A. Coxon and D. A. Ramsey, *Can. J. Phys.* **54**, 1034 (1976).
187. J. A. Coxon, *J. Photochem.* **5**, 337 (1976).
188. S. A. Barton, J. A. Coxon, and V. K. Roychowdhury, *Can. J. Phys.* **62**, 473 (1984).
189. T. Amano, S. Saito, E. Hirota, Y. Morino, D. R. Johnson, and F. X. Powell, *J. Mol. Spectrosc.* **30**, 275 (1969).
190. K. Wang and V. McKoy, *J. Phys. Chem.* **99**, 1727 (1995).
191. S. Baumgärtel and K. H. Gericke, *Chem. Phys. Lett.* **227**, 461 (1994).
192. Y. Matsumi, S. M. Shamsuddin, and M. Kawasaki, *J. Chem. Phys.* **101**, 8262 (1994).
193. M. J. Cooper, T. Diez-Rojo, L. J. Rogers, C. M. Western, M. N. R. Ashfold, and J. W. Hudgens, *Chem. Phys. Lett.* **272**, 232 (1997).
194. N. P. L. Wales, W. J. Buma, and C. A. de Lange, *Chem. Phys. Lett.* **259**, 213 (1996).
195. M. T. Duignan and J. W. Hudgens, *J. Chem. Phys.* **82**, 4426 (1985).
196. N. Basco and R. D. Morse, *J. Mol. Spectrosc.* **45**, 35 (1973).
197. R. F. Delmdahl, S. Baumgärtel, and K. H. Gericke, *J. Chem. Phys.* **104**, 2883 (1996).
198. E. Rühl, A. Jefferson, and V. Vaida, *J. Phys. Chem.* **94**, 2990 (1990).

199. D. H. A. ter Steege, M. Smits, C.A. de Lange, N. P. C. Westwood, J. B. Peel, and L. Visscher, *Faraday Discuss.* **115**, 259 (2000).
200. D. K. Bulgin, J. M. Dyke, N. Jonathan, and A. Morris, *J. Chem. Soc., Faraday Trans. II* **75**, 456 (1978).
201. J. B. Nee and K. J. Hsu, *J. Photochem. Photobiol. A: Chem.* **55**, 269 (1991).
202. R. J. Delmdahl, S. Welcker, and K.-H. Gericke, *Ber. Bunsen-Ges. Phys. Chem.* **102**, 244 (1998).
203. M. R. Dobber, W. J. Buma, and C. A. de Lange, *J. Chem. Phys.* **101**, 9303 (1994).
204. A. M. Rijs, C. A. de Lange, and M. H. M. Janssen, unpublished results.
205. I. Reineck, B. Wanneberg, H. Veerhuizen, C. Nohre, R. Maripuu, K. E. Norell, L. Mattson, L. Karlsson, and K. Siegbahn, *J. El. Spec.* **34**, 235 (1984).
206. M. Endoh, M. Tsuji, and Y. Nishimura, *Chem. Phys. Lett.* **109**, 35 (1984).
207. L. S. Wang, J. E. Reutt, Y. T. Lee, and D. A. Shirley, *J. Electron Spectrosc. Relat. Phenom.* **47**, 167 (1988).
208. I. Fischer, A. Lochschmidt, A. Strobel, G. Niedner-Schatteburg, K. Müller-Dethlefs, and V. E. Bondybey, *Chem. Phys. Lett.* **202**, 542 (1993).
209. B. Kleman, *Can. J. Phys.* **41**, 357 (1963).
210. Ch. Jungen, D. N. Malm, and A. J. Merer, *Chem. Phys. Lett.* **16**, 302 (1972).
211. Ch. Jungen, D. N. Malm, and A. J. Merer, *Can. J. Phys.* **51**, 1471 (1973).
212. A. J. Merer, S. A. Morris, and Ch. Jungen, *J. Mol. Spectrosc.* **127**, 425 (1988).
213. H. Bitto, A. Ruzicic, and J. R. Huber, *Chem. Phys.* **189**, 713 (1994).
214. A. E. Douglas and I. Zanon, *Can. J. Phys.* **42**, 627 (1964).
215. W. C. Price and D. M. Simpson, *Proc. R. Soc., London, Ser. A* **165**, 272 (1938).
216. J. W. Rabalais, J. M. McDonald, V. Scherr, and S. P. McGlynn, *Chem. Rev.* **71**, 73 (1971).
217. F. R. Greening and G. W. King, *J. Mol. Spectrosc.* **59**, 312 (1976).
218. R. McDiarmid and J. P. Doering, *J. Chem. Phys.* **91**, 2010 (1989).
219. K. O. Lantz, V. Vaida, and D. J. Donaldson, *Chem. Phys. Lett.* **184**, 152 (1991).
220. D. G. Wilden and J. Comer, *Chem. Phys.* **53**, 77 (1980).
221. M.-J. Hubin-Franskin, J. Delwiche, A. Poulin, B. Leclerc, P. Roy, and D. Roy, *J. Chem. Phys.* **78**, 1200 (1983).
222. S. Couris, E. Patsilinakou, M. Lotz, E. R. Grant, C. Fotakis, C. Cossart-Magos, and M. Horani, *J. Chem. Phys.* **100**, 3514 (1994).
223. L. Li, X.T. Wang, and X. B. Xie, *Chem. Phys.* **164**, 305 (1992).
224. L. Li, X. T. Wang, and X. B. Xie, *Chem. Phys. Lett.* **202**, 115 (1993).
225. J. Baker, M. Konstantaki, and S. Couris, *J. Chem. Phys.* **103**, 2436 (1995).
226. J. Baker and S. Couris, *J. Chem. Phys.* **103**, 4847 (1995).
227. J. H. Callomon and F. Creutzberg, *Philos. Trans. R. Soc., London, Ser. A* **277**, 157 (1974).
228. R. T. Wiedmann, E. R. Grant, R. G. Tonkyn, and M. G. White, *J. Chem. Phys.* **95**, 746 (1991).
229. J. F. M. Aarts and J. H. Callomon, *Chem. Phys. Lett.* **91**, 419 (1982).
230. A. B. F. Duncan, *J. Chem. Phys.* **4**, 638 (1936).
231. M. Zelikoff, K. Watanabe, and E. C. Y. Inn, *J. Chem. Phys.* **21**, 1643 (1953).
232. Y. Tanaka, A. S. Jurza, and F. J. Leblanc, *J. Chem. Phys.* **32**, 1205 (1960).
233. E. N. Lassettre, A. Skerbele, M. A. Dillon, and K. J. Ross, *J. Chem. Phys.* **48**, 5066 (1968).

234. R. H. Hueber, R. J. Celotta, S. R. Mielczarek, and C. E. Kuyatt, *J. Chem. Phys.* **63**, 4490 (1975).
235. E. Patsilinakou, R. T. Wiedmann, C. Fotakis, and E. R. Grant, *J. Chem. Phys.* **91**, 3916 (1989).
236. M. G. Szarka and S. C. Wallace, *J. Chem. Phys.* **95**, 2336 (1991).
237. H. Lefebvre-Brion and R. W. Field, *Perturbations in the Spectra of Diatomic Molecules*, Academic, New York, 1986.
238. K. Kawaguchi, C. Yamada, and E. Hirota, *J. Chem. Phys.* **82**, 1174 (1985).
239. J. M. Frye and T. J. Sears, *Mol. Phys.* **62**, 919 (1987).
240. R. N. Dixon, *Proc. R. Soc., London, Ser. A* **275**, 431 (1963).
241. C. Cossart-Magos and S. Leach, *J. Chem. Soc., Faraday Trans.* **2**, 78, 1477 (1982).
242. C. Cossart-Magos, F. Launay, and J. E. Parkin, *Mol. Phys.* **75**, 835 (1992)
243. X.-F. Yang, J.-L. Lemaire, F. Rostas, and J. Rostas, *Chem. Phys.* **164**, 115 (1992).
244. W. B. England, B. J. Rosenberg, P. J. Fortune, and A. C. Wahl, *J. Chem. Phys.* **65**, 684 (1976).
245. P. J. Knowles, P. Rosmus, and H.-J. Werner, *Chem. Phys. Lett.* **146**, 230 (1988).
246. H.-J. Werner, A. Spielfiedel, N. Feautrier, G. Chambaud, and P. Rosmus, *Chem. Phys. Lett.* **175**, 203 (1990).
247. A. Spielfiedel, N. Feautrier, G. Chambaud, P. Rosmus, and H.-J. Werner, *Chem. Phys. Lett.* **183**, 16 (1991).
248. A. Spielfiedel, N. Feautrier, C. Cossart-Magos, G. Chambaud, P. Rosmus, H.-J. Werner, and P. Botschwina, *J. Chem. Phys.* **97**, 8382 (1992).
249. C. Cossart-Magos, M. Jungen, and F. Launay, *Mol. Phys.* **61**, 1077 (1987).
250. Y. Ono, E. A. Osuch, and C. Y. Ng, *J. Chem. Phys.* **74**, 1645 (1981).
251. B. Kovac, *J. Chem. Phys.* **78**, 1684 (1983).
252. W. C. Price and D. M. Simpson, *Proc. R. Soc., London, Ser. A* **169**, 50 (1939).
253. F. M. Matsunaga and K. Watanabe, *J. Chem. Phys.* **46**, 4457 (1967).
254. B. Leclerc, A. Poulin, D. Roy, M.-J. Hubin-Franskin, and J. Delwiche, *J. Chem. Phys.* **75**, 5329 (1981).
255. J. A. Joens, *J. Chem. Phys.* **89**, 5366 (1985).
256. I. Kopp, *Can. J. Phys.* **45**, 4011 (1967).
257. B. Yang, M. H. Eslami, and S. L. Anderson, *J. Chem. Phys.* **89**, 5527 (1988).
258. R. Weinkauf and U. Boesl, *J. Chem. Phys.* **98**, 4459 (1993).
259. S. T. Pratt, *Phys. Rev. A* **38**, 1270 (1988).
260. Y. Sato, Y. Matsumi, M. Kawasaki, K. Tsukiyama, and R. Bersohn, *J. Phys. Chem.* **99**, 16307 (1989).
261. N. Sivakumar, I. Burak, W.-Y. Cheung, P. L. Houston, and J. W. Hepburn, *J. Phys. Chem.* **89**, 3609 (1985).
262. A. E. Douglas and J. M. Hollas, *Can. J. Phys.* **39**, 479 (1961).
263. A. E. Douglas, *Discuss. Faraday Soc.* **35**, 158 (1963).
264. A. D. Walsh and P. A. Warsop, *Trans. Faraday Soc.* **57**, 345 (1961).
265. M. N. R. Ashfold, C. L. Bennett, and R. J. Strickland, *Comm. At. Mol. Phys.* **19**, 181 (1987).
266. G. C. Nieman and S. D. Colson, *J. Chem. Phys.* **68**, 5656 (1978).
267. G. C. Nieman and S. D. Colson, *J. Chem. Phys.* **71**, 571 (1979).
268. J. H. Glownia, S. J. Riley, S. D. Colson, and G. C. Nieman, *J. Chem. Phys.* **72**, 5998 (1980).

269. J. H. Glownia, S. J. Riley, S. D. Colson, and G. C. Nieman, *J. Chem. Phys.* **73**, 4296 (1980).
270. W. Habenicht, G. Reiser, and K. Müller-Dethlefs, *J. Chem. Phys.* **95**, 4809 (1991).
271. W. Habenicht, G. Reiser, and K. Müller-Dethlefs, *J. Chem. Phys.* **98**, 8462 (1993).
272. M. N. R. Ashfold, R. N. Dixon, D. H. Mordaunt, and S. H. S. Wilson, in *Advances in Photochemistry*, D. C. Neckers, D. H. Volman and G. von Brünau, Eds., vol. 21, Wiley, New York, 1996, p. 217.
273. M. N. R. Ashfold, R. N. Dixon, M. Kono, D. H. Mordaunt, and C. L. Reed, *Philos. Trans. R. Soc., London, Ser. A* **355**, 1659 (1997).
274. V. Vaida, W. Hess, and J. L. Roebber, *J. Phys. Chem.* **88**, 3397 (1984).
275. X. Li and C. R. Vidal, *J. Chem. Phys.* **101**, 5523 (1994).
276. X. Li and C. R. Vidal, *J. Chem. Phys.* **102**, 9167 (1995).
277. M. N. R. Ashfold, R. N. Dixon, R. J. Stickland, and C. M. Western, *Chem. Phys. Lett.* **138**, 201 (1987).
278. M. N. R. Ashfold, R. N. Dixon, N. Little, R. J. Stickland, and C. M. Western, *J. Chem. Phys.* **89**, 1754 (1988).
279. M. N. R. Ashfold, R. N. Dixon, and R. J. Stickland, *Chem. Phys.* **88**, 463 (1984).
280. M. N. R. Ashfold, R. N. Dixon, K. N. Rosser, R. J. Stickland, and C. M. Western, *Chem. Phys.* **101**, 467 (1986).
281. P. J. Miller, S. D. Colson, and W. A. Chupka, *Chem. Phys. Lett.* **145**, 183 (1988).
282. M. R. Dobber, W. J. Buma, and C. A. de Lange, *J. Phys. Chem.* **99**, 1671 (1995).
283. M. N. R. Ashfold, C. M. Western, J. W. Hudgens, and R. D. Johnson, III, *Chem. Phys. Lett.* **260**, 27 (1996).
284. I. Tokoue, A. Hiraya, and K. Shobotake, *J. Chem. Phys.* **91**, 2808 (1989).
285. L. Karlsson, L. Mattson, R. Jadrny, T. Bergmark, and K. Siegbahn, *Phys. Scr.* **13**, 229 (1976).
286. D. H. Aue, H. M. Webb, W. R. Davidson, M. Vidal, M. T. Bowers, H. Goodwhite, L. E. Vertal, J. E. Douglas, P. A. Kollman, and G. L. Kenyon, *J. Am. Chem. Soc.* **102**, 5151 (1980).
287. L. B. Clark and W. T. Simpson, *J. Chem. Phys.* **43**, 3666 (1965).
288. N. Basco and R. D. Morse, *Chem. Phys. Lett.* **20**, 404 (1973).
289. D. D. Altenloh and B. R. Russell, *Chem. Phys. Lett.* **77**, 217 (1981).
290. M. Carnell and S. D. Peyerimhoff, *Chem. Phys. Lett.* **212**, 654 (1993).
291. W. M. McClain and R. A. Harris, in *Excited States*, E. C. Lim, Ed., Academic, New York, 1977, p. 1.
292. R. N. Dixon, J. M. Bayley, and M. N. R. Ashfold, *Chem. Phys.* **84**, 21 (1984).
293. W. C. Price, J. P. Teegan, and A. D. Walsh, *Proc. R. Soc., London, Ser. A* **201**, 600 (1950).
294. I. Tokue, A. Hiraya, and K. Shobatake, *Chem. Phys.* **116**, 449 (1987).
295. R. E. Kutina, A. K. Edwards, G. L. Goodman, and J. Berkowitz, *J. Chem. Phys.* **77**, 5508 (1982).
296. A. Rauk and S. Collins, *J. Mol. Spectrosc.* **105**, 438 (1984).
297. B. Mouflih, C. Larrieu, and M. Chaillet, *New J. Chem.* **12**, 65 (1988).
298. B. Mouflih, C. Larrieu, and M. Chaillet, *Chem. Phys.* **119**, 221 (1988).
299. J. D. Scott, G. C. Causley, and B. R. Russell, *J. Chem. Phys.* **59**, 6577 (1973).
300. R. McDiarmid, *J. Chem. Phys.* **61**, 274 (1974).
301. I. Tokue, A. Hiraya, and K. Shobatake, *Chem. Phys.* **130**, 401 (1989).

# NONADIABATIC TRANSITIONS DUE TO CURVE CROSSINGS: COMPLETE SOLUTIONS OF THE LANDAU–ZENER–STUECKELBERG PROBLEMS AND THEIR APPLICATIONS

CHAOYUAN ZHU, YOSHIAKI TERANISHI,*
and HIROKI NAKAMURA

*Department of Theoretical Studies, Institute for Molecular Science, Myodaiji, Okazaki 444-8585, Japan*

## CONTENTS



*Present address: The Institute of Physical and Chemical Research(RIKEN), Wako, Saitama 351-0198, Japan.

*Advances in Chemical Physics, Volume 117*, Edited by I. Prigogine and Stuart A. Rice.
ISBN 0-471-40542-6 

## I. INTRODUCTION

As will be explained in more detail in Section II, nonadiabatic transition due to potential curve crossing presents a very important fundamental mechanism of state and/or phase changes in various fields of natural sciences. Without nonadiabatic transition, this world would have been dead, because no basic chemical and biological processes, such as electron and proton transfer, could have occurred. Nonadiabatic transition is certainly an origin of mutability of this world. Because of the importance of nonadiabatic transition in all branches of natural sciences, a lot of theoretical studies have been carried out to formulate this phenomenon. The pioneering work was done independently by Landau [1], Zener [2], and Stueckelberg [3] for the two-state curve crossing problem in the same year of 1932. Interestingly, the basic theory for the two-state noncurve crossing problem was also first formulated by Rosen and Zener (RZ) [4] in the same year. These pioneering works initiated the subsequent big flow of theoretical studies of nonadiabatic transitions covering a lot of quantum mechanical, semiclassical, and numerical investigations. Readers should refer to recent books and review articles [5–11].

In spite of these numerous efforts in the last 60 years some essential problems have remained unsolved. First of all, no good formulas were available to evaluate nonadiabatic transition probabilities at energies near or lower than the crossing point that represents the physically and chemically important region. In that energy region, the nonadiabatic transition probability takes the value of the order 0.5 in many cases. The celebrated Landau–Zener formula cannot work in this energy region and behaves correctly only in high energies. Second, no theory could work well for the entire region of coupling strength and actually no good theory was available at all in the region of strong diabatic coupling. Third, in the conventional theory nonunique diabatization procedure and the inconvenient complex contour integral, which is quite annoying for nonspecialist users, were required. Furthermore, various phases, which play important roles in many cases, were not necessarily provided properly.

Starting with the most basic linear potential model, in which the two diabatic potentials are linear functions of the spatial coordinate and the diabatic coupling is a constant, Zhu and Nakamura [12–23] have solved all the above mentioned problems for the first time in 60 years since the pioneering works was done by Landau, Zener, and Stuckelberg [1–3]. They have not only obtained the quantum mechanically exact analytical solutions for the two-state linear potential model [12, 13], but also carried out the semiclassical analysis carefully to derive a useful set of compact semiclassical analytical solutions for the linear potential model [14–16]. Based on these solutions, they further extended the theory to present a complete set of compact analytical solutions that are applicable to general two-state curved potential systems. The theory can cover both the Landau–Zener (LZ) case [17, 18], in which the two diabatic potentials cross with the same sign of slopes, and the nonadiabatic tunneling (NT) case [19, 20], in which the two diabatic potentials cross with the opposite slope signs and create a potential barrier. In the LZ case, the whole energy range is divided into two regions: (1) higher and (2) lower than the crossing energy, for each of which compact analytical formulas are obtained not only for the nonadiabatic transition probability but also for the various phases. In the NT case, three energy regions are considered (1) higher than the bottom of the upper adiabatic potential, (2) lower than the top of the lower adiabatic potential, and (3) the region in between. All necessary quantities are expressed in terms of real-phase integrals along adiabatic potentials on the real axis. The two basic parameters designated as $a^2$ and $b^2$, which were originally defined in terms of the diabatic coupling and slopes of the diabatic potentials, can be determined in the final version of the theory from shapes of the adiabatic potentials on the real axis in the vicinity of the avoided crossing. The theory is not only very simple but also very accurate and should be very useful for practical applications. The parameter $a^2$ represents the effective coupling strength and turns out to be a very important parameter to judge the overall significance of the nonadiabatic transition. The parameter $b^2$ represents the energy effectively; $b^2 = 0$ corresponds to the crossing point. Very interestingly, the theory does not require any information on the nonadiabatic coupling and everything can be estimated only from the adiabatic potentials on the real axis. Furthermore, the following semiclassical idea has been confirmed to work well [17, 19, 21–23]: the whole molecular process, whatever it is, involving nonadiabatic transitions, can be decomposed into a series of basic phenomena that include nonadiabatic transition at avoided crossing, wave propagation along the adiabatic potential, reflection at the turning point, potential barrier penetration, and so on. The present theory presents, of course, the transition probability amplitude including phases at the avoided crossing. We know how to deal with the other elements mentioned above and the corresponding matrices can be easily incorporated. In the case of inelastic scattering, we can formulate the scattering matrix by incorporating the nonadiabatic transition matrix that we call the $I$ matrix. In the case of elastic scattering with resonance due to nonadiabatic transition, the elastic scattering phase shift can be formulated. In the case of the bound state problem involving nonadiabatic transition, a secular matrix equation can be obtained. That is to say, whenever nonadiabatic transitions are involved in whatever the problem, the present theory can be nicely incorporated into the framework and should work well. These achievements are summarized in recent review articles [21–23].

The next natural question is whether this new theory can be applied to general multichannel and even multidimensional problems, since this is very important in view of the applications to various practical problems in chemical dynamics beyond simple diatomic systems. As can be easily conjectured from the above explanation, in the case of multichannel problems each avoided crossing is treated independently by the two-state theory and the whole problem can be formulated by combining all of them. Namely, by incorporating the two-state $I$ matrices, any problem such as elastic or inelastic scattering with or without resonances and multichannel bound state problems can be formulated. Zhu and Nakamura [24, 25] actually demonstrated that their theory works very well in multichannel problems, nicely reproducing even heavily overlapping resonances in the NT case. This means that not only the nonadiabatic transition probability but also the various phases can be accurately estimated by the theory. The theory works surprisingly well even when two avoided crossings lie very close together and the nonadiabatic couplings overlap significantly on the real axis. This is because the new two-state theory takes into account the analytical structure of the problem correctly. The next challenging application is, of course, treatment of multidimensional problems. There are two possible ways of treating multidimensional problems: (1) reduction of the problem into a one-dimensional (1D) multichannel system by expanding the total wave function in terms of appropriate internal state wave functions, and (2) direct treatment of the problem by using classical trajectories in an appropriate way. The first one is more accurate naturally, because we do not rely on classical trajectories. But, from the view point of applications to a variety of chemical dynamical processes such as electronically nonadiabatic chemical reactions in higher dimensions, the second method with inclusion of important quantum mechanical effects such as interference and tunneling would be much more promising. As an example of the first method, Zhu et al. [26] analyzed the electronically *adiabatic* three-dimensional (3D) chemical reactions, $O(^3P) + HCl \rightarrow OH + Cl$ and $Cl + HBr \rightarrow HCl + Br$, fully analytically with use of the Zhu–Nakamura theory within the framework of the hyperspherical coordinate approach. They treated $> 100$ channels and almost 1000 avoided crossings. The parameter $a^2$ can be usefully utilized to pick up important avoided crossings; actually $\sim 100$ avoided crossings are found to be important and are treated analytically by the theory. The cumulative reaction probabilities are nicely reproduced by the theory. For the more important second method, theoretical studies have just started. The nonadiabatic transition matrix $I$ can be incorporated into various semiclassical propagation schemes such as trajectory surface hopping (TSH), the semiclassical method based on initial value representation (IVR), and cellular frozen Gaussian wave packet propagation (CFGW) method [27–30]. In the case of TSH, only the probabilities are required, but in the other methods, phases also play important roles. Since the new theory of nonadiabatic transition is very simple and accurate, it is expected that the combination of the second or the third semiclassical propagation method with the new theory could present a nice practically useful methodology to attack large chemical dynamical systems.

Based on the exponential potential model of Nikitin [6], which can cover the LZ and RZ formulas in certain limits, further studies of various exponential potential models have been carried out in an attempt to hopefully formulate a unified theory [31–34].

So far, we have been discussing the time-independent framework of nonadiabatic transitions. In view of the recent remarkable progress of laser technology, however, a time-dependent version of nonadiabatic transitions has attracted much attention and is gaining importance more and more than before. Since it is possible to create new curve crossings by shifting energy levels up and down with use of the time-dependent external fields and also to vary the diabatic coupling strength there by changing the field strengh, it is probably easily understood that various versions of time-dependent nonadiabatic transitions become very important in this sense. The importance is more emphasized because of the possibility of controlling molecular processes by manipulating external fields. Time-dependent theory of nonadiabatic transitions is, however, simpler than the time-independent one. This is simply because time is unidirectional and the time-dependent Schrödinger equation is just a first-order differential equation. Since the Zener's treatment of the time-dependent linear potential model [2], various types of time-dependent nonadiabatic transition models have been proposed and solved such as Demkov–Osherov multilevel problem, band-crossing model, Demkov–Kunike model, exponential model, and bowtie model [35–43]. The new time-independent theory developed by Zhu and Nakamura can also be transferred to the time-dependent version. Actually, the exact solution of the time-independent linear potential model can provide the exact solution for the time-dependent quadratic potential model, and thus the whole set of the Zhu–Nakamura theory can be easily transferred to the time-dependent framework [44]. Corresponding to the fact that the theory can deal with the energy region near and lower than the crossing point, the time-dependent version can treat the cases of tangential touching and diabatic avoided crossing of two diabatic potentials. These time-dependent theories can be nicely used to control molecular processes based on the dressed or Floquet state representation [45]. Actually, Teranishi and Nakamura [45,47] proposed a new idea of controlling nonadiabatic processes by sweeping the frequency and/or the intensity of the external field. The control conditions can be formulated analytically with the help of the theory of time-dependent nonadiabatic transitions.

In this chapter, we will explain in more detail what has been outlined above. In Section II, the significance of nonadiabatic transitions in natural sciences are emphasized more. In Section III, the complete solutions of the two-state Landau–Zener–Stueckelberg (LZS) problems by Zhu and Nakamura are summarized together with a brief description of history. The final set of practically useful formulas are presented. Applications of the theory to multichannel and multidimensional problems are discussed in Section IV. Other types of nonadiabatic transitions are explained in Section V. Time-dependent theories of nonadiabatic transitions are presented in Section VI, and a new way of controlling molecular processes by time-dependent external fields is discussed in Section VII. Future perspectives are mentioned in Section VIII.

## II. PHYSICAL SIGNIFICANCE OF LEVEL CROSSING

As was briefly mentioned in the introduction, it is needless to say that nonadiabatic transitions due to energy level or curve crossings play very crucial roles in various branches of natural sciences, representing an origin of mutability of this universe. Especially, the nonadiabatic transitions in the NT type curve crossing, in which two diabatic potentials cross with opposite signs of slope and create a potential barrier in the lower adiabatic potential curve, must play a significant role in various fields, inducing changes of state, conformation, and phase.

The best known examples are, of course, collision and spectroscopic processes in atomic and molecular physics, which occur most effectively through potential curve crossings. Needless to say, all kinds of chemical dynamics proceed effectively only through potential energy surface crossings. Even organic chemical reactions are tried to be classified in terms of curve crossing schemes [48]. Electron and proton transfer, which play significant roles in chemical and biological systems, are nothing but a curve crossing problem in which the ordinate is not the ordinary potential energy but free energy [49, 50]. In solid-state physics, there are also many examples such as desorption of molecules from a solid surface [51], ion neutralization in collision with a solid surface [52], quenching of color center [53, 54], self-trapped localized state of exciton [55], and many other nonradiative energy relaxation processes in solids [56]. In many of these processes in condensed medium, the abscissa is some sort of renormalized coordination coordinate. Nuclear collision and reaction mechanisms are also often clarified in terms of potential curve crossings based on the picture of a nuclear molecular orbital [57]. Curve crossings represent, on the other hand, one of the causes to create quantum chaotic behavior [58].

If we assume the adiabatic parameter (which is usually a certain nuclear coordinate in the above examples) to be a classical time-dependent variable, then the nonadiabatic transition processes are transformed into time-dependent problems. On the other hand, when we explicitly apply time-dependent external fields to various systems, there arise a variety of intrinsically time-dependent nonadiabatic transition problems. Molecular processes in laser fields are one of the most typical examples. Within the picture of the dressed or Floquet state [45], dynamic processes due to artificially created curve crossings can be induced and thus it becomes possible to control those dynamic processes by manipulating the external fields. Controlling molecular processes by lasers has actually become an attractive and important subject nowadays [59]. There are many other examples of time-dependent nonadiabatic transitions induced by external fields such as quantum mechanical effects in current driven Josephson junctions in a magnetic field and Zener transitions in flux-driven normal metallic rings or current biased tunnel junctions in a magnetic field [60–66], and quantum spin tunneling in a magnetic field [67]. Even electron spin resonance (ESR) and nuclear magnetic resonance (NMR) can be considered as time-dependent nonadiabatic transitions. The neutrino conversion among various kinds of neutrinos is a subject attracting much attention recently in elementary particle physics [68]. This is again an example of time-dependent nonadiabatic transition in which the ordinate is the neutrino mass squared specifying different kinds of neutrinos and the abscissa is a time-dependent electron density in the matter through which the neutrinos fly.

As explained above, nonadiabatic transitions appear in all branches of the natural sciences and actually play crucial roles there. The concept of nonadiabatic transition is very deep and multidisciplinary. The concept can also be applied to social sciences, for instance, to economics [69]. We can find and encounter many examples of nonadiabatic transitions even in daily life.

## III. COMPLETE SOLUTIONS OF THE TWO-STATE LANDAU–ZENER–STUECKELBERG PROBLEMS

### A. Brief Historical Survey

The simplest two-state curve crossing model was discussed independently, for the first time, by Landau [1], Zener [2], and Stueckelberg [3]. Landau dealt with the time-independent problem by using the perturbation theory and the complex integral method through which the exponent of the nonadiabatic transition probability was expressed. This complex integral method is now called the Landau method [70]. The simplest linear potential model in the time domain was solved exactly by Zener. When we apply various approximations such as linear diabatic potentials, constant diabatic coupling, and the straight-line trajectory with constant velocity for the relative motion to Landau's expression of nonadiabatic transition probability, then we can get the same probability expression as that of Zener, which is now known as the Landau–Zener formula. Stueckelberg analyzed the general two-state time-independent problem by using the WKB (Wentzel–Kramers–Brillouin) type semiclassical phase-integral method. He obtained the overall transition probability for the whole collision process, which includes the phase interference effect due to the two possible paths. This phase is now called the Stueckelberg phase. The LZS theory has been reviewed and further extended by many authors since their pioneering works in 1932. In spite of such numerous efforts, however, the semiclassical theory has not been complete and has many problems, as mentioned in the introduction [5–11].

### B. Complete Solutions

Let us start with the basic coupled equations for the two-state curve crossing problems in the diabatic representation:

$$\left(\frac{\hbar^2}{2\mu}\right)\frac{d^2\psi_1(R)}{dR^2} + [E - V_{11}(R)]\psi_1(R) = V_{12}(R)\psi_2(R) \tag{3.1}$$

and

$$\left(\frac{\hbar^2}{2\mu}\right)\frac{d^2\psi_2(R)}{dR^2} + [E - V_{22}(R)]\psi_2(R) = V_{21}(R)\psi_1(R) \tag{3.2}$$

where $\mu$ is the reduced mass of the system and $R$ is a spatial coordinate defined in the region $0 < R < +\infty$. This region is not essential, because the most important quantity is the nonadiabatic transition matrix at the avoided crossing, which can be used even if the potentials extend to $-\infty$. The coupling terms are symmetric, $V_{12}(R) = V_{21}(R)$. In this chapter, we present our semiclassical theory in adiabatic representation. For this purpose, what we need is just the following two adiabatic potentials,

$$E_2(R) = \tfrac{1}{2}[V_{22}(R) + V_{11}(R)] + \tfrac{1}{2}\sqrt{[V_{22}(R) - V_{11}(R)]^2 + 4V_{12}^2(R)} \tag{3.3}$$

and

$$E_1(R) = \tfrac{1}{2}[V_{22}(R) + V_{11}(R)] - \tfrac{1}{2}\sqrt{[V_{22}(R) - V_{11}(R)]^2 + 4V_{12}^2(R)} \tag{3.4}$$

As explained before, there are two types of avoided crossings shown in Figure 1: one is the LZ type and the other is the NT type.

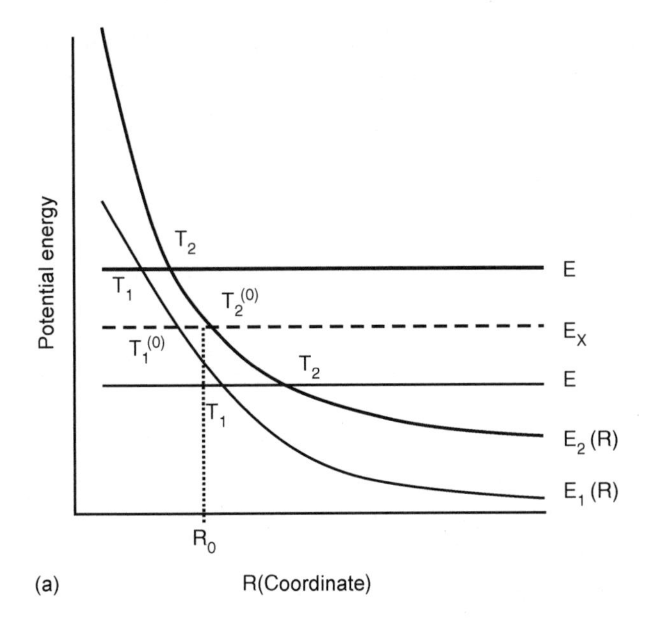


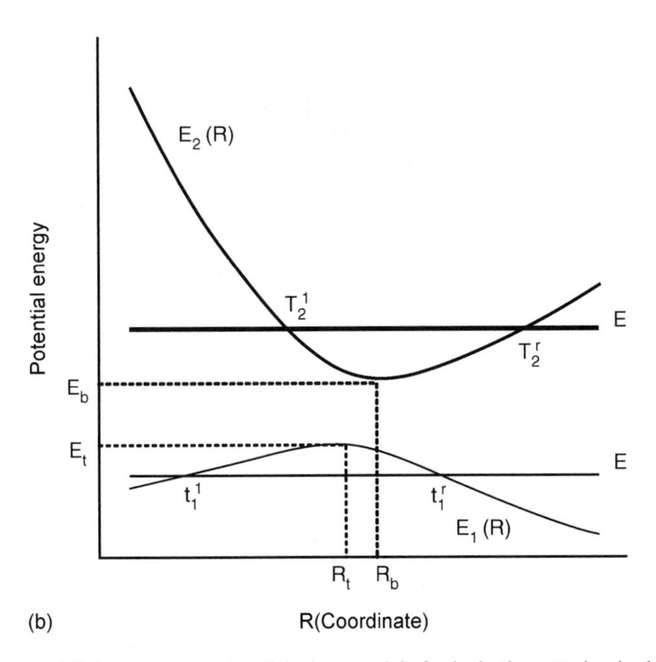


**Figure 1.** (*a*) Schematic two-state adiabatic potentials for inelastic scattering in the LZ case. (*b*) Schematic two-state adiabatic potentials for transmission and reflection in the NT case.

In the LZ case we choose $R_0$, which corresponds to the minimum separation between the two adiabatic potentials as a reference point. The WKB type of wave functions for Eqs. (3.1) and (3.2) can be written in the form,

$$\psi_1(R) = \frac{A_1}{\sqrt{K_1(R)}} e^{i\int_{T_1}^{R} K_1(R)dR - i(\pi/4)} + \frac{B_1}{\sqrt{K_1(R)}} e^{-i\int_{T_1}^{R} K_1(R)dR + i(\pi/4)} \tag{3.5}$$

and

$$\psi_2(R) = \frac{A_2}{\sqrt{K_2(R)}} e^{i\int_{T_2}^{R} K_2(R)dR - i(\pi/4)} + \frac{B_2}{\sqrt{K_2(R)}} e^{-i\int_{T_2}^{R} K_2(R)dR + i(\pi/4)} \tag{3.6}$$

at $R \gg R_0$ [see Fig. 1($a$)], where $T_i (i = 1, 2)$ is the turning point on the adiabatic potential $E_i(R)$ and

$$K_i(R) = \frac{\sqrt{2\mu}}{\hbar}\sqrt{E - E_i(R)} \qquad i = 1, 2 \tag{3.7}$$

The reduced $S$ matrix, $S^R$, can be defined by

$$\begin{pmatrix} A_1 \\ A_2 \end{pmatrix} = \begin{pmatrix} S_{11}^R & S_{12}^R \\ S_{21}^R & S_{22}^R \end{pmatrix} \begin{pmatrix} B_1 \\ B_2 \end{pmatrix} \equiv S^R \begin{pmatrix} B_1 \\ B_2 \end{pmatrix} \tag{3.8}$$

which includes all the necessary information about nonadiabatic transitions. A simple manipulation can prove that $S^R$ satisfies [71, 72]

$$S_{11}^R = (S_{22}^R)^* \qquad \text{and} \qquad S_{12}^R = (S_{21}^R)^* = \text{pure imaginary} \tag{3.9}$$

The scattering matrix $S$ in the semiclassical theory is written as [73]

$$S_{mn} = S_{mn}^R \exp[i(\eta_m + \eta_n)] \qquad n, m = 1, 2 \tag{3.10}$$

where $\eta_n$ represents the WKB phase shift for elastic scattering in the channel $n$, and is well defined as

$$\eta_1 = \lim_{R\to\infty}\left[\int_{T_1}^{R} K_1(R)dR - K_1(R)R + \frac{\pi}{4}\right] \tag{3.11}$$

and

$$\eta_2 = \lim_{R \to \infty} \left[ \int_{T_2}^{R} K_2(R) dR - K_2(R) R + \frac{\pi}{4} \right] \tag{3.12}$$

Under the two-state linear curve crossing model, Zhu et al. [13] for the first time, derived the exact quantum mechanical solution, which is expressed in terms of one parameter, $U_1$, called the Stokes constant in mathematics [74],

$$S^R = \begin{bmatrix} (1 + U_1 U_2) e^{-2i\sigma} & -U_2 \\ -U_2 & (1 - U_1^* U_2) e^{2i\sigma} \end{bmatrix} \tag{3.13}$$

where

$$U_2 = \frac{U_1 - U_1^*}{1 + |U_1|^2} \tag{3.14}$$

with

$$U_1 = U_1(a^2, b^2; \sigma, \delta) \tag{3.15}$$

The overall nonadiabatic transition probability is given by

$$P_{12} = |S_{12}^R|^2 = \frac{4(\mathrm{Im}\, U_1)^2}{(|U_1|^2 + 1)^2} = 4p(1 - p) \sin^2 \psi \tag{3.16}$$

with

$$\psi = \arg(U_1) \tag{3.17}$$

and

$$p = \frac{1}{1 + |U_1|^2} \tag{3.18}$$

The exact quantum solution for $U_1$ was obtained as a convergent infinite series as a function of the two parameters $a^2$ and $b^2$ defined below, but its form is not very convenient for practical applications to the general two-state curve crossing problems (an explicit expression of $U_1$ is not given here [13]). In the case of the two-state linear model, they can be expressed only in terms of $a^2$ and $b^2$. In order to generalize the theory so as to be applicable to general curved potentials, we have carefully analyzed the expression of $U_1$ and divided the roles of the parameters into two parts: a portion directly responsible for nonadiabatic transition and a portion responsible for phases. In the case of general curved potentials, the first part can still be expressed in terms of $a^2$ and $b^2$, and the second part was rewritten in terms of phase integrals $\sigma$ and $\delta$, whose definitions are given below. By generalizing the expressions of $\sigma$ and $\delta$ as phase integrals along curved adiabatic potentials, the theory became applicable to general cases.

The two dimensionless parameters $a^2$ and $b^2$ are originally defined as follows in terms of diabatic potentials:

$$a^2 = \left(\frac{\hbar^2}{2m}\right)\frac{F(F_1 - F_2)}{8V_X^3} \tag{3.19}$$

and

$$b^2 = (E - E_X)\frac{F_1 - F_2}{2FV_X} \tag{3.20}$$

with $F = \sqrt{|F_1F_2|}$, where $F_i$ $(i = 1, 2)$, $V_X$, and $E_X$, all of which are defined at the crossing point, represent the slope of $V_i(R)$, diabatic coupling, and energy, respectively. In our semiclassical theory, however, these parameters can be reexpressed in terms of adiabatic potentials [17],

$$a^2 = \sqrt{d^2 - 1}\,\frac{\hbar^2}{\mu(T_2^0 - T_1^0)^2[E_2(R_0) - E_1(R_0)]} \tag{3.21}$$

and

$$b^2 = \sqrt{d^2 - 1}\,\frac{E - [E_2(R_0) + E_1(R_0)]/2}{[E_2(R_0) - E_1(R_0)]/2} \tag{3.22}$$

where

$$d^2 = \frac{[E_2(T_1^0) - E_1(T_1^0)][E_2(T_2^0) - E_1(T_2^0)]}{[E_2(R_0) - E_1(R_0)]^2} \tag{3.23}$$

This means that the theory does not require diabatization. Here, $a^2$ represents effective nonadiabatic coupling strength; $a^2 \to 0(\infty)$ corresponds to the adiabatic (diabatic) limit. These two limiting cases are dynamically not interesting, because everything proceeds either adiabatically or diabatically. The most effective range for nonadiabatic transition lies in $0.01 \lesssim a^2 \lesssim 100$. The parameter $b^2$ in Eq. (3.22) represents effective collsion energy, and when $b^2 > 0(b^2 < 0)$ corresponds to the energy higher (lower) than the crossing point: thus $b^2 < 0$ is a nonadiabatic transition accompanied by tunneling. Definitions of $T_1^{(0)}$ and $T_2^{(0)}$ are given by [see Fig. 1($a$)]

$$E_X = [E_2(R_0) + E_1(R_0)]/2 = E_2(T_2^{(0)}) = E_1(T_1^{(0)}) \tag{3.24}$$

where $E_X$ represents the crossing energy as mentioned above.

In the NT case, the WKB type of wave function on the lower adiabatic potential $E_1(R)$ [see Fig. 1($b$)] can be written as

$$\psi_1(R)^{R \gtrsim t_1^r} \frac{A_1}{\sqrt{K_1(R)}} e^{i\int_{t_1^r}^R K_1(R)dR - i(\pi/4)} + \frac{B_1}{\sqrt{K_1(R)}} e^{-i\int_{t_1^r}^R K_1(R)dR + i(\pi/4)} \tag{3.25}$$

and

$$\psi_2(R)^{R \lesssim t_1^l} - \frac{B_2}{\sqrt{K_2(R)}} e^{i\int_{t_1^l}^R K_2(R)dR + i(\pi/4)} + \frac{A_2}{\sqrt{K_2(R)}} e^{-i\int_{t_1^l}^R K_2(R)dR - i(\pi/4)} \tag{3.26}$$

where $K_i(R)$ $(i = 1, 2)$ is defined in Eq. (3.7). Two reference points in Eqs. (3.25) and (3.26) are chosen as turning points $t_1^r$ and $t_1^l$ on the lower adiabatic potential $E_1(R)$ at $E \leq E_t$, where $E_t$ represents the top of the lower adiabatic potential. If $E \geq E_t$, the reference points are fixed at $R = R_t$. Now, let us define the reduced scattering matrix as

$$\begin{pmatrix} A_1 \\ A_2 \end{pmatrix} = \begin{pmatrix} S_{11}^R & S_{12}^R \\ S_{21}^R & S_{22}^R \end{pmatrix} \begin{pmatrix} B_1 \\ B_2 \end{pmatrix} \equiv S^R \begin{pmatrix} B_1 \\ B_2 \end{pmatrix} \tag{3.27}$$

Again, the simple manipulation proves [75]:

$$|S_{11}^R| = |S_{22}^R| \qquad \text{and} \qquad S_{12}^R = S_{21}^R \tag{3.28}$$

under the two-state linear curve crossing model. Note that Eq. (3.9) for the LZ case differs from Eq. (3.28) for the NT case. It should be noted that the $S$ matrix here represents the transmission amplitude on the lower adiabatic potential.

Exact quantum solution was found again to be expressed in terms of one Stokes constant $U_1$ [13]:

$$S^R = \frac{1}{1 + U_1 U_2} \begin{pmatrix} e^{i\Delta_{11}} & U_2 e^{i\Delta_{12}} \\ U_2 e^{i\Delta_{12}} & e^{i\Delta_{22}} \end{pmatrix} \tag{3.29}$$

where

$$U_2 = \frac{U_1 - U_1^*}{|U_1|^2 - 1} \tag{3.30}$$

with

$$U_1 = U_1(a^2, b^2; \sigma, \delta) \tag{3.31}$$

The additional phases $\Delta_{11}$, $\Delta_{22}$, and $\Delta_{12}$ are defined later. As in the LZ case, $U_1$ is expressed only in terms of $a^2$ and $b^2$ in the linear potential model, but was generalized by introducing $\sigma$ and $\delta$. The overall transmission probability is given by

$$P_{12} = |S_{12}^R|^2 = \frac{4(\operatorname{Im} U_1)^2}{(|U_1|^2 - 1)^2 + 4(\operatorname{Im} U_1)^2} = \frac{4\cos^2\psi}{4\cos^2\psi + p^2/(1-p)} \quad \text{for} \quad b^2 \geq 1 \tag{3.32}$$

with

$$p = 1 - |U_1|^2 \qquad \text{for} \qquad b^2 \geq 1 \tag{3.33}$$

and

$$\psi = \arg(U_1) - \pi/2 \tag{3.34}$$

The scattering matrix $S$ in the semiclassical theory is written as [73]

$$S_{mn} = S_{mn}^R \exp[i(\eta_m + \eta_n)] \qquad n, m = 1, 2 \tag{3.35}$$

where $\eta_n$ represents the WKB phase shift for elastic scattering in the channel $n$, and is given by

$$\eta_1 = \lim_{R \gg t_1^r} \left[ \int_{t_1^r}^{R} K_1(R)dR - K_1(R)R + \frac{\pi}{4} \right] \tag{3.36}$$

and

$$\eta_2 = \lim_{R \ll t_1^l} \left[ -\int_{t_1^l}^{R} K_1(R)dR + K_1(R)R + \frac{\pi}{4} \right] \tag{3.37}$$

Note that $t_1^l$ and $t_1^r$ must be replaced by $R_t$ if $E \geq E_t$. The original definitions of the two basic dimensionless parameters $a^2$ and $b^2$ are the same as before in diabatic representation; but in our semiclassical theory these are again re-expressed in terms of the adiabatic potentials as [19]

$$a^2 = \frac{(1-\gamma^2)\hbar^2}{\mu(R_b - R_t)^2(E_b - E_t)} \tag{3.38}$$

and

$$b^2 = \frac{E - (E_b + E_t)/2}{(E_b - E_t)/2} \tag{3.39}$$

where

$$\gamma = \frac{E_b - E_t}{E_2(\frac{R_b+R_t}{2}) - E_1(\frac{R_b+R_t}{2})} \tag{3.40}$$

Definitions of $R_b$, $R_t$ and $E_b$, $E_t$ are shown in Fig. 1(*b*). The physical meaning of $a^2$ and $b^2$ are the same as in the LZ case mentioned before. In the two-state linear curve crossing model, the exact quantum solution of the Stokes constant $U_1$ is again obtained in an infinite series and is not very convenient for general cases; but it still provides a unique basis for generalization and for checking the validity of the semiclassical solution. The LZS type of nonadiabatic transition corresponds to the Stokes phenomenon in the four-transition-point asymptotic expansion problem in mathematics [13, 74]. Although we have solved this problem exactly in the case of the linear potential model, the final expression is not simple and not very convenient as mentioned above. In the semiclassical approximation, we have treated the four transition points as two pairs of two transition points, since the two-transition-point problem is exactly solved in terms of the Weber function. The semiclassical theory thus derived by Zhu and Nakamura [19] works very well even when the two pairs come close together, but naturally does not work well when the two-pair sturcture disappears and the four points almost coalesce. This extreme situation occurs at $a^2 \gg 1$ at $b^2 \sim 0$. In order to cover even this kind of situation uniformly, we have introduced certain empirical corrections. In Section III.B.1, we present our compact semiclassical solutions directly applicable to the two-state general curved potentials.

### 1. *Landau–Zener Case*

*a.* $E \geq E_X (b^2 \geq 0)$. The Stokes constant $U_1$ in Eq. (3.15) can be written in the form

$$U_1 = \sqrt{\frac{1}{p} - 1}\, e^{i\psi} \tag{3.41}$$

where $p$ is the nonadiabatic transition probability for one passage of the crossing point and is given as

$$p = \exp\left[ -\frac{\pi}{4a} \left( \frac{2}{b^2 + \sqrt{b^4 + 0.4a^2 + 0.7}} \right)^{1/2} \right] \tag{3.42}$$

The phase $\psi$ is given by

$$\psi = \sigma + \phi_s = \sigma - \frac{\delta}{\pi} + \frac{\delta}{\pi} \ln\left(\frac{\delta}{\pi}\right) - \arg\Gamma\left(i\frac{\delta}{\pi}\right) - \frac{\pi}{4} \tag{3.43}$$

where the parameters $a^2$ and $b^2$ are defined in Eqs. (3.21) and (3.22), and two other parameters $\sigma$ and $\delta$ originally defined as the real and the imaginary part of the complex phase integral can now be expressed by the simple real quantities as follows [18]:

$$\sigma + i\delta = \left[ \int_{T_-}^{R_0} K_-(R)dR - \int_{T_+}^{R_0} K_+(R)dR \right] + \sigma_0 + i\delta_0 \tag{3.44}$$

with

$$\sigma_0 + i\delta_0 \equiv \int_{R_0}^{R_*} [K_-(R) - K_+(R)]dR \simeq \frac{1}{\sqrt{a^2}} \frac{\sqrt{2}\pi}{4} \frac{1}{F_+^2 + F_-^2} [F_-^c + iF_+^c] \tag{3.45}$$

where

$$F_\pm = \sqrt{\sqrt{(b^2 + \gamma_1)^2 + \gamma_2} \pm (b^2 + \gamma_1)} + \sqrt{\sqrt{(b^2 - \gamma_1)^2 + \gamma_2} \pm (b^2 - \gamma_1)} \tag{3.46}$$

$$F_+^c = F_+[b^2 \longrightarrow (b^2 - b_c^2)] \qquad \text{with} \qquad b_c^2 = \frac{0.16 b_x}{\sqrt{b^4 + 1}} \tag{3.47}$$

$$F_-^c = F_-(\gamma_2 \longrightarrow \gamma_2') \qquad \text{with} \qquad \gamma_2' = \sqrt{d^2} \frac{0.45}{1 + 1.5 e^{2.2 b_x |b_x|^{0.57}}} \tag{3.48}$$

$$b_x = b^2 - 0.9553 \tag{3.49}$$

and

$$\gamma_1 = 0.9\sqrt{d^2 - 1} \qquad \text{and} \qquad \gamma_2 = \frac{7}{16}\sqrt{d^2} \tag{3.50}$$

where $d^2$ is defined by Eq. (3.23). In deriving Eq. (3.45), $R_0$ was found to be better as the reference point [18] than $T_j^{(0)}$ $(j = 1, 2)$ [see Fig. 1($a$)], which was used previously [17]. In order to cover the very strong coupling region well, $a^2 \lesssim 0.05$ at $|b^2| < 1$, some empirical corrections are introduced as can be seen in Eqs. (3.47)–(3.50).

*b.* $E \leq E_X (b^2 \leq 0)$. The reduced $S$ matrix is given in the same way as before in terms of Stokes constant $U_1$, but $U_1$ itself is now given by

$$\text{Re}\, U_1 = \cos(\sigma) \left\{ \sqrt{B(\sigma/\pi)} e^{\delta} - g_1 \sin^2(\sigma) \frac{e^{-\delta}}{\sqrt{B(\sigma/\pi)}} \right\} \tag{3.51}$$

and

$$\text{Im}\, U_1 = \sin(\sigma) \left\{ B(\sigma/\pi) e^{2\delta} - g_1^2 \sin^2(\sigma) \cos^2(\sigma) \frac{e^{-2\delta}}{B(\sigma/\pi)} + 2 g_1 \cos^2(\sigma) - g_2 \right\}^{1/2} \tag{3.52}$$

The nonadiabatic transition probability for one passage of the crossing point and the phase $\psi$ are given by

$$p = [1 + B(\sigma/\pi) e^{2\delta} - g_2 \sin^2(\sigma)]^{-1} \tag{3.53}$$

and

$$\psi = \arg(U_1) \tag{3.54}$$

where

$$g_1 = 1.8(a^2)^{0.23} e^{-\delta} \tag{3.55}$$

$$g_2 = \frac{3\sigma}{\pi\delta} \ln(1.2 + a^2) - 1/a^2 \tag{3.56}$$

and

$$B(X) = \frac{2\pi X^{2X} e^{-2X}}{X\Gamma^2(X)} \tag{3.57}$$

where $\Gamma(X)$ is the gamma function. The two parameters $\sigma$ and $\delta$ are given by Eq. (3.44).

*c. I Matrix (for Both $b^2 \geq 0$ and $b^2 \leq 0$).* The reduced scattering matrix in Eq. (3.8) represents a full scattering process that actually includes both incoming and outgoing segments of the propagation. In order to apply the two-state theory to multichannel curve crossing problems, we can semiclassically extract the incoming part of propagation that is called the $I$ matrix given by

$$I = \begin{pmatrix} \sqrt{1-p}\, e^{-i(\sigma-\psi)} & -\sqrt{p}\, e^{i\sigma} \\ \sqrt{p}\, e^{-i\sigma} & \sqrt{1-p}\, e^{i(\sigma-\psi)} \end{pmatrix} \tag{3.58}$$

in the adiabatic representation. In the case of the two-state linear curve crossing model, we can directly compare the present Zhu–Nakamura formula for $p$ in Eqs. (3.42) and (3.53) with the LZ formula in Figure 2. The Zhu–Nakamura formulas are much better than the LZ formula. In terms of the above $I$ matrix, the reduced scattering matrix $S^R$ can be rewritten as

$$S^R = I^t I \quad (t = \text{transposed}). \tag{3.59}$$

The nonadiabatic transition matrix $I_X$ at the avoided crossing is given by

$$I_X = \begin{pmatrix} \sqrt{1-p}\, e^{i\phi_{s\to(\psi-\sigma)}} & -\sqrt{p}\, e^{i\sigma_0} \\ \sqrt{p}\, e^{-i\sigma_0} & \sqrt{1-p}\, e^{-i\phi_{s\to(\psi-\sigma)}} \end{pmatrix} \tag{3.60}$$

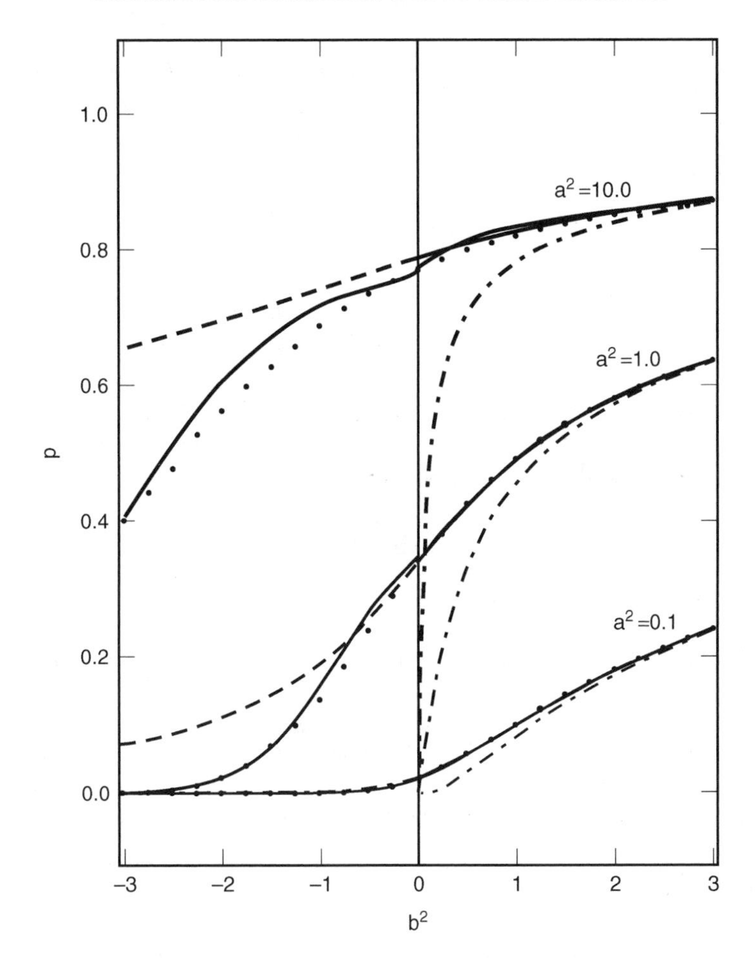


**Figure 2.** Nonadiabatic transition probability $p$ for one passage of crossing point in the LZ type of two-state linear curve crossing. Dots(exact numerical result); solid line [the present semiclassical formulas Eq. (3.42) for $b^2 \geq 0$ and Eq. (3.53) for $b^2 \leq 0$]; dash dotted line [the LZ formula $p = \exp(-\pi/(4\sqrt{a^2b^2}))$]; dash line [extension of Eq. (3.42)].

In terms of this matrix, the total scattering matrix $S$ is expressed as

$$S = P_{\infty X} O_X P_{XTX} I_X P_{X\infty} \tag{3.61}$$

where $P_{A\ldots B}$ is a diagonal matrix, representing the adiabatic wave propagation from B to A, and is defined as

$$(P_{\infty X})_{nm} = (P_{X\infty})_{nm} = \delta_{nm} \exp\left\{ i\int_{R_0}^{\infty} [K_n(R) - K_n(\infty)]dR - iK_n(\infty)R_0) \right\} \tag{3.62}$$

and

$$(P_{XTX})_{nm} = \delta_{nm} \exp\left[ 2i\int_{T_n}^{R_0} K_n(R)dR + i\frac{\pi}{2} \right] \tag{3.63}$$

where $T_n$ is the turning point on the $n$th adiabatic potential. Since the $I_X$ matrix describes the transition locally at the avoided crossing, this would be useful in applications to multidimensional problems.

### 2. *Nonadiabatic Tunneling Case*

This case was treated in [19, 20].

*a.* $E \geq E_b(b^2 \geq 1)$. The Stokes constant $U_1$ in Eq. (3.31) is given by

$$U_1 = i\sqrt{1-p}\,e^{i\psi} \tag{3.64}$$

where the nonadiabatic transition probability for one passage of the crossing point takes the form

$$p = \exp\left[ -\frac{\pi}{4a}\left( \frac{2}{b^2 + \sqrt{b^4 - 0.72 + 0.62a^{1.43}}} \right)^{1/2} \right] \tag{3.65}$$

and the phase $\psi$ is given by

$$\psi = \sigma - \phi_s = \sigma + \frac{\delta}{\pi} - \frac{\delta}{\pi}\ln\left(\frac{\delta}{\pi}\right) + \arg\Gamma\left(i\frac{\delta}{\pi}\right) + \frac{\pi}{4} \tag{3.66}$$

in which $\sigma$ and $\delta$ are estimated from

$$\sigma = \int_{T_2^l}^{T_2^r} K_2(R)dR \tag{3.67}$$

and

$$\delta = \frac{\pi}{8ab}\frac{1}{2}\frac{\sqrt{6 + 10\sqrt{1 - (1/b^4)}}}{1 + \sqrt{1 - (1/b^4)}} \tag{3.68}$$

Note that $a^2$ and $b^2$ in above equations are now defined in Eqs. (3.38) and (3.39). The additional phases appearing in Eq. (3.29) take the form

$$\Delta_{11} = 2\int_{T_2^l}^{R_b} K_2(R)dR - 2\sigma_0$$

$$\Delta_{22} = 2\int_{R_b}^{T_2^r} K_2(R)dR + 2\sigma_0 \tag{3.69}$$

and

$$\Delta_{12} = \sigma \tag{3.70}$$

with

$$\sigma_0 = \left(\frac{R_b - R_t}{2}\right)\left\{K_1(R_t) + K_2(R_b) + \frac{1}{3}\frac{[K_1(R_t) - K_2(R_b)]^2}{K_1(R_t) + K_2(R_b)}\right\} \tag{3.71}$$

Note that the $I$ matrix for the NT case can be defined only for $b^2 \geq 1$ (i.e., $E \geq E_b$), and is given by

$$I = \begin{pmatrix} \sqrt{1-p}\, e^{i\phi_s} & \sqrt{p}\, e^{i\sigma} \\ -\sqrt{p}\, e^{-i\sigma} & \sqrt{1-p}\, e^{-i\phi_s} \end{pmatrix} \tag{3.72}$$

Here, it should be noted that the sign in off-diagonal elements of Eqs. (3.58) and (3.72) differ.

The $I_X$ matrix is given in the same way as before [see Eq. (3.60)] as

$$I_X = \begin{pmatrix} \sqrt{1-p}\, e^{i\phi_s} & \sqrt{p}\, e^{i\sigma_0} \\ -\sqrt{p}\, e^{-i\sigma_0} & \sqrt{1-p}\, e^{-i\phi_s} \end{pmatrix} \tag{3.73}$$

The overall transmission probability is given by

$$P_{12} = |S^R_{12}|^2 = \frac{4\cos^2\psi}{4\cos^2\psi + p^2/(1-p)} \tag{3.74}$$

This indicates that an interesting phenomenon of complete reflection ($P = 0$) occurs at energies that satisfy $\psi = (n + 1/2)\pi$ $(n = 0, 1, 2, \ldots)$ and gives a possibility of molecular switching [76–78].

*b.* $E_b \geq E \geq E_t$ ($|b^2| \leq 1$). In this case, the energy $E$ is in between the two adiabatic potentials [see Fig. 1(*b*)]. The Stokes constant $U_1$ is given by

$$U_1 = i\left[\sqrt{1 + W^2}e^{i\phi} - 1\right]\Big/W \tag{3.75}$$

where

$$\phi = \sigma + \arg\Gamma\left(\frac{1}{2} + i\frac{\delta}{\pi}\right) - \frac{\delta}{\pi}\ln\left(\frac{\delta}{\pi}\right) + \frac{\delta}{\pi} - g_3 \tag{3.76}$$

with

$$g_3 = 0.34\,\frac{a^{0.7}(a^{0.7} + 0.35)}{a^{2.1} + 0.73}\,(0.42 + b^2)\left(2 + \frac{100b^2}{100 + a^2}\right)^{0.25} \tag{3.77}$$

The quantity $W$ in Eq. (3.75) is defined by

$$W = \frac{1 + g_5}{a^{2/3}}\int_0^\infty \cos\left[\frac{t^3}{3} - \frac{b^2}{a^{2/3}}t - \frac{g_4}{2a^{2/3}}\,\frac{t}{0.61\sqrt{2 + b^2 + a^{1/3}t}}\right]dt \tag{3.78}$$

where

$$g_4 = \frac{\sqrt{a^2} - 3b^2}{\sqrt{a^2} + 3}\sqrt{1.23 + b^2} \tag{3.79}$$

and

$$g_5 = 0.38(1 + b^2)^{1.2 - 0.4b^2}/a^2 \tag{3.80}$$

The two parameters $\sigma$ and $\delta$ can be written in terms of $a^2$ and $b^2$ as

$$\sigma = -\frac{1}{\sqrt{a^2}}\left[0.057(1+b^2)^{0.25} + \frac{1}{3}\right](1-b^2)\sqrt{5+3b^2} \tag{3.81}$$

and

$$\delta = \frac{1}{\sqrt{a^2}}\left[0.057(1-b^2)^{0.25} + \frac{1}{3}\right](1+b^2)\sqrt{5-3b^2} \tag{3.82}$$

The additional phases $\Delta_{ij}$ in Eq. (3.29) have the form

$$\Delta_{11} = \sigma - 2\sigma_0$$

$$\Delta_{22} = \sigma + 2\sigma_0 \tag{3.83}$$

and

$$\Delta_{12} = \sigma \tag{3.84}$$

with

$$\sigma_0 = -\tfrac{1}{3}(R_t - R_b)K_1(R_t)(1+b^2) \tag{3.85}$$

The overall transmission probability takes the form,

$$P_{12} = \frac{W^2}{1+W^2} \tag{3.86}$$

*c.* $E \leq E_t$ $(b^2 \leq -1)$. The Stokes constant $U_1$ is given by

$$\mathrm{Re}\, U_1 = \sin(2\sigma_c)\left\{\frac{0.5\sqrt{a^2}}{1+\sqrt{a^2}}\sqrt{B(\sigma_c/\pi)}e^{-\delta} + \frac{e^{\delta}}{\sqrt{B(\sigma_c/\pi)}}\right\} \tag{3.87}$$

and

$$\mathrm{Im}\, U_1 = \cos(2\sigma_c)\sqrt{\frac{(\mathrm{Re}\, U_1)^2}{\sin^2(2\sigma_c)} + \frac{1}{\cos^2(2\sigma_c)}} - \frac{1}{2\sin(\sigma_c)}\left|\frac{\mathrm{Re}\, U_1}{\cos(\sigma_c)}\right| \tag{3.88}$$

where $B(X)$ is defined in Eq. (3.57), and

$$\sigma_c = \sigma(1 - g_6) \tag{3.89}$$

$$g_6 = 0.32 \times 10^{-2/a^2} e^{-\delta} \tag{3.90}$$

with

$$\sigma = \frac{\pi}{8a|b|} \frac{1}{2} \frac{\sqrt{6 + 10\sqrt{1 - (1/b^4)}}}{1 + \sqrt{1 - (1/b^4)}} \tag{3.91}$$

and

$$\delta = \int_{t_1^l}^{t_1^r} |K_1(R)| dR \tag{3.92}$$

The additional phases in this case are quite simple as

$$\Delta_{11} = \Delta_{22} = \Delta_{12} = -2\sigma \tag{3.93}$$

For one passage of transition probability we can define $p$ in Eq. (3.65) at $b^2 \geq 1$, but this loses the meaning completely at $b^2 < 1$ because of the tunneling process that can never be separated from the nonadiabatic transition. It should be emphasized, however, that $p$ in the LZ case can be mathematically extended to $b^2 < 0$. Equation (3.59) with the $I$ matrix given by Eq. (3.58) still holds at $b^2 < 0$, although $p$ is not physically the same as in the case of $b^2 \geq 0$. In the NT case, $p$ in Eq. (3.65) cannot be mathematically extended to the region $b^2 \leq 1$. However, it is interesting to note that the quantity

$$Q = \frac{1}{1 + |U_1|^2} \tag{3.94}$$

and $p$ in Eq. (3.65) approach each other at $b^2 = 1$ as the effective coupling constant $a^2$ gets larger (see Fig. 3). This indicates that the quantity $Q$ in Eq. (3.94) might be considered as a mathematical extension of $p$ in Eq. (3.65) for the NT case. Figure 3 also demonstrates good agreement between the exact quantum results and the semiclassical calculations. Discussions here actually address an important fact that nonadiabatic transition cannot be easily separated from

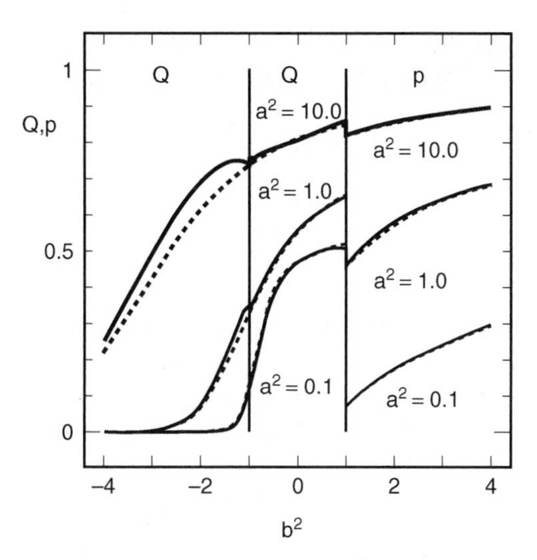


**Figure 3.** Nonadiabatic transition probability $p$ for $b^2 > 1$ and the quantity $Q$ for $b^2 \leq 1$, which might be interpreted as the nonadiabatic transition probability in the NT type of two-state linear curve crossing. Dashed line: exact numerical result, solid line: the present semiclassical result for $p$ in Eq. (3.65), and for $Q$ in Eq. (3.94).

tunneling whenever tunneling is involved. How to approximately separate these two transitions is still a big problem in the multidimensional case. The present semiclassical theory based on the two-state model can be considered as a nice illustration for studying this problem.

The overall transmission (nonadiabatic tunneling) probability is explicitly expressed as

$$P_{12} = |S_{12}^R|^2 = \frac{B(\sigma_c/\pi)e^{-2\delta}}{[1 + (0.5\sqrt{a^2}/(1 + \sqrt{a^2}))B(\sigma_c/\pi)e^{-2\delta}]^2 + B(\sigma_c/\pi)e^{-2\delta}} \tag{3.95}$$

It should be noted that this naturally contains both effects of quantum tunneling ($e^{-2\delta}$ is the Gamov factor) and nonadiabatic transition. This transmission probability is always smaller than the corresponding potential barrier penetration probability because of the nonadiabatic coupling effect. When the diabatic coupling is infinitely strong, that is, $a^2 \to 0$, Eq. (3.95) naturally goes to the ordinary tunneling probability $[= e^{-2\delta}/(1 + e^{-2\delta})]$.

# IV. HOW TO DEAL WITH MULTICHANNEL AND MULTIDIMENSIONAL PROBLEMS

## A. Multichannel Processes

A diatomic system is a typical example for multichannel curve crossing problems, in which the adiabatic potential curves, obtained with the internuclear coordinate $R$ fixed, may show many avoided crossings. The present two-state semiclassical theory can present not only physically meaningful interpretation about those avoided crossings, but also a nice computational tool to calculate physical quantities such as scattering matrix, resonance width, and bound state energies. When avoided crossings are well separated from each other, nonadiabatic transitions well localized at each avoided crossing can be treated as in a pure two-state problem. Even when some of the avoided crossings come close together and their nonadiabatic couplings overlap with each other, the present semiclassical theory can still work surprisingly well. This is because the present semiclassical theory can take multistate coupling effects into consideration by using adiabatic potentials and also the underlying analytical structure of the problem is most properly taken into account. Actually, the important two parameters $a^2$ and $b^2$, which are defined in terms of the two adiabatic potential curves at one avoided crossing, include all the interaction information coming from the other neighboring avoided crossings; namely, when two avoided crossings come close together, the corresponding $a^2$ and $b^2$ change from the corresponding values when they are far apart. This is simply because the adiabatic potentials are obtained by diagonalizing the whole multichannel electronic Hamiltonian matrix. In the diabatic representation the interactions among other states are completely neglected, so that the theory cannot work better than in the adiabatic representation [25].

In a certain special multichannel curve crossing model, Demkov and Osherov [35] proved that the overall state-to-state transition probabilities can be exactly expressed in terms of an appropriate multiplication of the LZ probabilities at crossings, and any phase is not necessary. Later, many people thought that in general phases may not be as important. This is not correct, of course. For example, oscillation and resonance structure of overall nonadiabatic transition probabilities depend on various phases strongly. The $I$ matrix propagation method developed by Nakamura [79, 80] made an important step to properly take phases into consideration. The present version of the $I$ matrix propagation scheme [24, 25] enables us to deal with multichannel curve crossing problems more conveniently by absorbing all adiabatic phases between avoided crossings into the redefined $I$ matrix. Generally speaking, the better the two-state theory is, the more accurate results we can obtain for multichannel curve crossing problems. We will first present a general framework for multichannel processes in which the transitions considered can be either LZ or NT, and then give a couple of examples to demonstrate the accuracy of the present theory. Actually, this general framework can be used for any type of transition other than LZ and NT as far as the corresponding $I$ matrix is known. The RZ type noncurve crossing and the exponential potential models are such examples.

### 1. *General Framework*

The multichannel WKB wave function can be defined almost anywhere for both incoming and outgoing branches, and the internal reduced scattering matrix can be defined at a certain finite distance $R = R_0$, where all channels are energetically open. This is a connection matrix that connects the coefficients associated with incoming WKB solutions to the coefficients associated with outgoing WKB solutions. Then, we further propagate the solutions to the asymptotic region where the final $S$ matrix is defined. As is well known, the exact quantum mechanical close-coupling calculations have to be carried out far into the asymptotic region to obtain converged solutions. In the semiclassical propagation method, however, we can terminate the propagation at the position just beyond the outmost avoided crossing.

Let us assume that we consider a general multichannel system that contains totally $n + m$ states, in which $n$ represents the number of asymptotically open channels and $m$ is the number of closed channels (see Fig. 4). Of course, $n$ and $m$ vary as the collision energy changes.

*a. Case of No Closed Channel* ($m = 0$). In this case, all channels ($n$) are open, and the avoided crossings are assumed to be distributed in the order,

$$R_N < R_{N-1} < \cdots < R_1 \tag{4.1}$$

where $R_N$ represents the innermost avoided crossing and $R_1$ is the outermost avoided crossing. Each avoided crossing at $R_i$ ($i = 1, 2, \ldots, N$) is identified by the channel indexes $\alpha$ and $\beta$ ($\alpha < \beta = 1, 2, \ldots, n$), and the $n \times n$ $I_{R_i}$ matrix is given by

$$(I_{R_i})_{\alpha\alpha} = I_{11} \qquad (I_{R_i})_{\alpha\beta} = I_{12}$$

$$(I_{R_i})_{\beta\alpha} = I_{21} \qquad \text{and} \qquad (I_{R_i})_{\beta\beta} = I_{22} \tag{4.2}$$

with the other elements

$$(I_{R_i})_{\nu\gamma} = \delta_{\nu\gamma} \qquad (\nu, \gamma) \neq (\alpha\alpha), (\beta\beta), (\alpha\beta), (\beta\alpha) \tag{4.3}$$

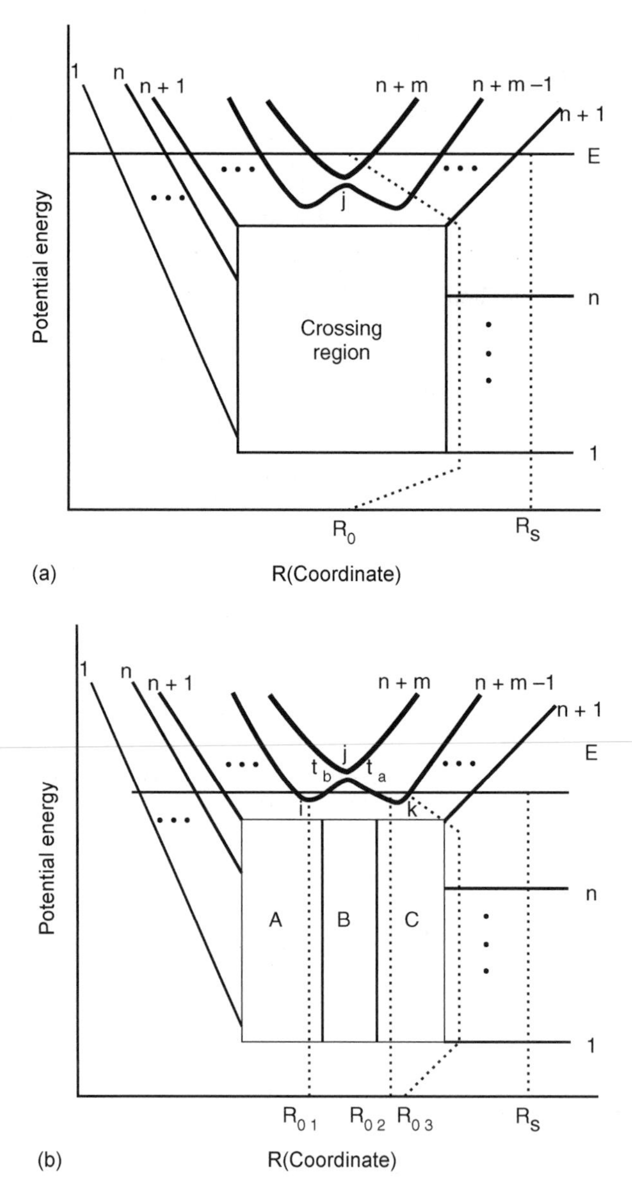


**Figure 4.** Schematic $(n+m)$-state potential curves with $n$ open and $m$ closed channels. The turning points $T_\alpha$ ($\alpha = 1, 2, \ldots, n+m$) and $t_\beta$ ($\beta = n+1, n+2, \ldots, n+m$) mentioned in the text (omitted in this figure) are, respectively, the leftmost and rightmost turning points on the $\alpha$th and $\beta$th adiabatic potentials. $1, \ldots, n, n+1, \ldots, n+m-1, n+m$ correspond to the adiabatic potentials $E_1(R) \ldots E_n(R)$, $E_{n+1}(R) \ldots E_{n+m-1}(R)$, $E_{n+m}(R)$. (*a*) In the case of energy higher than the bottom of $E_{n+m}(R)$. (*b*) In the case of energy lower than the bottom of $E_{n+m}(R)$.

where $I_{11}$, $I_{12}$, $I_{21}$, and $I_{22}$ are the matrix elements of the two-by-two $I$ matrix defined in Eq. (3.58) for the LZ case and Eq. (3.72) for the NT case. The final reduced scattering matrix can be expressed as

$$S^R = (I_{R_N} I_{R_{N-1}} \cdots I_{R_1})^t (I_{R_N} I_{R_{N-1}} \cdots I_{R_1}) \tag{4.4}$$

When there are closed channels, that is, $m \neq 0$, this matrix gives the internal reduced scattering matrix $\chi$. We will always denote it as $\chi$ in order to distinguish it from the final reduced scattering matrix $S^R$.

*b. Case of $m \neq 0$ With Energies Higher Than the Bottom of the Highest Adiabatic Potential.* In this case $m$ channels are closed, as is shown in Figure 4(*a*). Although in Figure 4(*a*), we have chosen the NT type avoided crossing at $R = R_j$ for the highest one, it can be the LZ type or any other type as far as we know its $I$ matrix. We first find the $\chi$ matrix at $R = R_0$, where all $n + m$ channels are open. Actually, $\chi$ can be obtained exactly in the same way as in Eq. (4.4), but now it becomes $(n + m) \times (n + m)$. It should be noted that this $\chi$ matrix may now also include contributions from the other avoided crossings, if any, lying in the region $R > R_0$. This means that $R_i$ does not have to be a definite single value, but can designate different positions depending on channels.

Let us write the $\chi$ matrix as

$$\chi = \left[ \begin{array}{c|c} \chi_{\mathrm{oo}}(n \times n) & \chi_{\mathrm{oc}}(n \times m) \\ \hline \chi_{\mathrm{co}}(m \times n) & \chi_{\mathrm{cc}}(m \times m) \end{array} \right] \tag{4.5}$$

where o(c) means open (closed). Then, the final $n \times n$ reduced $S^R$ matrix can be found as

$$S^R = \chi_{\mathrm{oo}} - \chi_{\mathrm{oc}} D^{-1} \chi_{\mathrm{co}} \tag{4.6}$$

with

$$D_{\alpha\beta} = \delta_{\alpha\beta} e^{-i2\theta_\alpha} + [\chi_{\mathrm{cc}}]_{\alpha\beta} \qquad (\alpha, \beta = 1, 2, \ldots, m) \tag{4.7}$$

where the additional adiabatic phase $\theta_\alpha$ in Eq. (4.7) represents the WKB phase integral on the $\alpha$-th adiabatic potential,

$$\theta_\alpha = \int_{T_{n+\alpha}}^{t_{n+\alpha}} K_{n+\alpha}(R) dR \qquad \alpha = 1, 2, \ldots, m \tag{4.8}$$

which is an ordinary-phase integral on a single adiabatic potential well. The derivation of the $S^R$ matrix is essentially the same as that of the multichannel quantum defect theory [8, 81], and actually has been used in the heavy particle scattering theory [80, 82]. But now we have compact analytical expressions for all elements.

Equation (4.6) indicates that resonance information is totally included in the $D$ matrix given by Eq. (4.7), and actually the complex zeros of $\det(D)$ provide positions and widths of resonances. It can be easily checked that the $D$ matrix goes to the correct adiabatic limit when all avoided crossings turn to be adiabatic, $a^2 \longrightarrow 0$.

*c. Case of $m \neq 0$ With Energies Lower Than the Bottom of the Highest Adiabatic Potential.* Now we turn to the situation shown in Figure 4(*b*) in which the highest avoided crossing $j$ is assumed to be the NT type and energy $E$ is lower than the bottom of the highest adiabatic potential. The NT and LZ cases require different approaches. In the case of LZ, we can still use the $I$ matrix even if energies are lower than the corresponding avoided crossing, and thus the whole procedure is the same as in section IV.A.1.b. But in the case of NT, the $I$ matrix no longer exists at energies lower than the corresponding avoided crossing, and we must use the transfer matrix $N^R$ [renamed from $S^R$ in Eq. (3.29) for the reason that it represents the local transmission phenomenon]. This $N^R$ matrix represents nonadiabatic tunneling through the avoided crossing $j$ in Figure 4(*b*), since the highest state $E_{n+m}(R)$ is closed everywhere. Now the propagation scheme becomes a little bit complicated because of this nonadiabatic tunneling process. We have to divide avoided crossings into three regions: A, B, and C in Figure 4(*b*);

$$\text{Avoided crossings in A: } R_\alpha < R_i$$

$$\text{Avoided crossings in B: } R_i < R_\alpha < R_k$$

and

$$\text{Avoided crossings in C: } R_k < R_\alpha$$

where $\alpha$ is a running index that covers all avoided crossings in the corresponding region. Thus, we can define the $I$ matrices for these three regions as

$$I_A = I_{R_N} I_{R_{N-1}} \cdots I_{R_i} \tag{4.9}$$

$$I_B = I_{R_{i-1}} \cdots I_{R_{j-1}} I_{R_{j+1}} \cdots I_{R_{k+1}} \tag{4.10}$$

and

$$I_C = I_{R_k} I_{R_{k-1}} \cdots I_{R_1} \tag{4.11}$$

where $R_N < R_{N-1} \cdots < R_i < R_{i-1} < R_{j-1} < R_{j+1} \cdots < R_{k+1} < R_k \cdots < R_1$, and each $I_{R_\alpha}$ $(\alpha \neq j)$ can be calculated in the same way as shown in Eqs. (4.2) and (4.3).

Let us define the internal $(n+m-1) \times (n+m-1)$ $\chi$ matrix at $R \lesssim R_{01}$ as

$$\chi^{[1]} = (I_A I_B)^t (I_A I_B) \tag{4.12}$$

where $R_{01} > R_i$ $(R_{02} < R_k)$ is a certain position in the left (right) well of $E_{n+m-1}(R)$ [see Fig. 4(*b*)]. Here, we have combined $I_B$ with $I_A$, since the tunneling through $j$ can commute with nonadiabatic transitions in region B. Next, we consider the WKB wave function $\psi_{n+m-1}(R)$ in the region of tunneling through the top barrier (crossing $j$) from $R_{01}$ to $R_{02}$, and then we have the internal $(n+m-1) \times (n+m-1)$ $\chi$ matrix, $\chi^{[2]}$:

$$\chi^{[2]}_{\alpha\beta} = \chi^{[1]}_{\alpha\beta} - \frac{\chi^{[1]}_{\alpha(n+m-1)} N^R_{22} \chi^{[1]}_{(n+m-1)\beta}}{D_b} \tag{4.13}$$

$$\chi^{[2]}_{\alpha(n+m-1)} = \frac{iN^R_{21} e^{-i\theta_b}}{D_b} \chi^{[1]}_{\alpha(n+m-1)} \tag{4.14}$$

$$\chi^{[2]}_{(n+m-1)\beta} = \chi^{[2]}_{\beta(n+m-1)} \tag{4.15}$$

and

$$\chi^{[2]}_{(n+m-1)(n+m-1)} = N^R_{11} - \frac{\chi^{[1]}_{(n+m-1)(n+m-1)} (N^R_{12})^2}{D_b} \tag{4.16}$$

where $\alpha, \beta = 1, 2, \ldots, (n+m-2)$, and $D_b$ is given by

$$D_b = e^{-i2\theta_b} + \chi^{[1]}_{(n+m-1)(n+m-1)} N^R_{22} \tag{4.17}$$

with

$$\theta_b = \int_{T_{n+m-1}}^{t_b} K_{n+m-1}(R)dR \tag{4.18}$$

which is the phase integral along the left well of $E_{n+m-1}(R)$ [see Fig. 4(*b*)]. $N_{11}^R$, $N_{22}^R$, and $N_{12}^R$ are evaluated from Eq. (3.29). The third step is to propagate the WKB wave functions to $R \sim R_{03}$ where all avoided crossings in region C contribute to give

$$\chi^{[3]} = I_C^t \chi^{[2]} I_C \equiv \left[ \begin{array}{c|c} \chi_{oo}(n \times n) & \chi_{oc}(n \times (m-1)) \\ \hline \chi_{co}((m-1) \times n) & \chi_{cc}[(m-1) \times (m-1)] \end{array} \right] \tag{4.19}$$

The final step is to propagate the WKB wave functions from $R_{03}$ to $R_s$ where the asymptotic region is reached, and we finally obtain the $S^R$ matrix $(n \times n)$ as

$$S^R = \chi_{oo} - \chi_{oc} D^{-1} \chi_{co} \tag{4.20}$$

with

$$D_{\alpha\beta} = \delta_{\alpha\beta} e^{-i2\theta_\alpha} + [\chi_{cc}]_{\alpha\beta} \qquad (\alpha, \beta = 1, 2, \ldots, m-1) \tag{4.21}$$

where the phase integrals along adiabatic potentials are defined by

$$\theta_\alpha = \int_{T_{n+\alpha}}^{t_{n+\alpha}} K_{n+\alpha}(R)dR \qquad (\alpha = 1, 2, \ldots, m-1) \tag{4.22}$$

Note that

$$\theta_{m-1} \equiv \theta_a = \int_{t_a}^{t_{n+(m-1)}} K_{n+(m-1)}(R)dR \tag{4.23}$$

which is the phase integral on the right well along $E_{n+m-1}(R)$ in Figure 4(*b*). Note that the turning points $t_a$ and $t_b$ in Figure 4(*b*) must be replaced by $R_{t_j}$ when the energy is located in the gap of the top avoided crossing $j$. Now, resonances come from the following two parts: one is from the complex zeros of $D_b = 0$ in Eq. (4.16) and the other is from the complex zeros of $\det D = 0$ in Eq. (4.20). The great advantage of the present semiclassical method is that the resonance part can be completely separated from the other transition processes and thus can be easily analyzed.

When the energy goes down further, lower than the bottom of the adiabatic potential $E_{n+m-1}(R)$ in Figure 4($b$), the present semiclassical theory cannot take the highest $j$-th crossing contribution into consideration. However, when energy is very low, the nonadiabatic tunneling at $j$ almost coincides with single potential barrier tunneling. In that case, we can neglect the adiabatic potential $E_{n+m}(R)$, or we can treat the effect of the highest avoided crossing perturbatively. Then, the present $I$-matrix propagation method can still be formulated in a similar way as above. In this way, the present semiclassical theory can deal with multichannel curve crossing problems without any restriction for energy and the number of channels, as long as all avoided crossings are relatively separated from each other.

### 2. *Numerical Applications*

We choose a model system of two Morse potentials crossed with two repulsive exponential potentials to demonstrate the accuracy of our semiclassical theory [25]. The model potentials in diabatic representation are given by

$$V_1(R) = 0.037e^{-1.3(R-3.25)} - 0.034$$

$$V_2(R) = 0.037e^{-1.3(R-3.25)} - 0.012$$

$$V_3(R) = 0.4057[1 - e^{-0.344(R-3)}]^2 - 0.03$$

and

$$V_4(R) = 0.4057[1 - e^{-0.344(R-3)}]^2 \tag{4.24}$$

Coupling terms are given as

$$V_{13}(R) = V_{14}(R) = V_{23}(R) = V_{24}(R) = \frac{2V_0}{1 + e^{R-3}}$$

and

$$V_{12}(R) = V_{34}(R) = 0 \tag{4.25}$$

All quantities are in atomic units, and the reduced mass of the system is chosen to be that of an oxygen molecule ($m = 29,377.3$). This model system was taken from some states of $O_2$ and the coupling represents the spin–orbit interaction among vibrational states of the oxygen molecule [83]. We have chosen $V_0 = 0.002$ at which any perturbation theory does not work at all, as it is realized as the strong coupling regime [83]. Figure 5 shows an adiabatic potential diagram with $V_0 = 0.002$ from which we estimate all effective coupling constants $a^2$ for four nonzero coupling terms from Eq. (3.38) as

$$V_0 = 0.002 \Rightarrow (a_1^2, a_2^2, a_3^2, a_4^2) = (7.88, 4.5, 3.61, 1.8) \tag{4.26}$$

Note that although the diabatic coupling constant $V_0$ is the same, their corresponding effective coupling parameters $a^2$ are very different from crossing to crossing. Actually, that $0.1 \gtrsim a^2 \gtrsim 10$ corresponds to the most significant region for nonadiabatic transition, and in this region any attempt to use perturbation method will fail. Figure 6 show an excellent agreement between exact quantum results [Fig. 6(*a*)] and the present semiclassical results [Fig. 6(*b*)] for a wide range of energy. Even very detailed resonance structures are very well reproduced by the present semiclassical theory. More examples can be seen in [25].

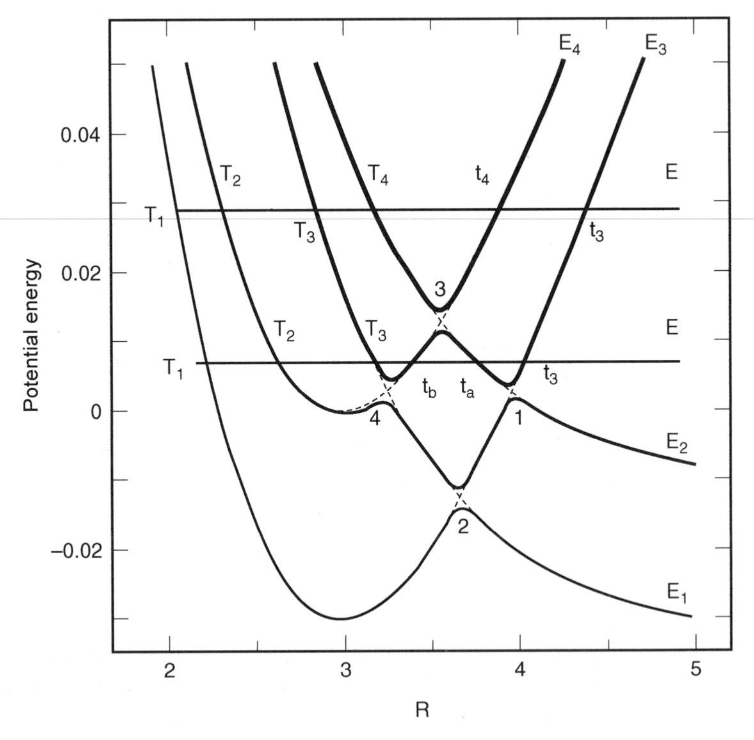


**Figure 5.** Four-state potential diagram of Eqs. (4.24) and (4.25) with $V_0 = 0.002$. Solid lines for adiabatic potentials $E_i(R)$ and dashed lines for diabatic potentials $V_i(R)(i = 1, 2, 3, 4)$.

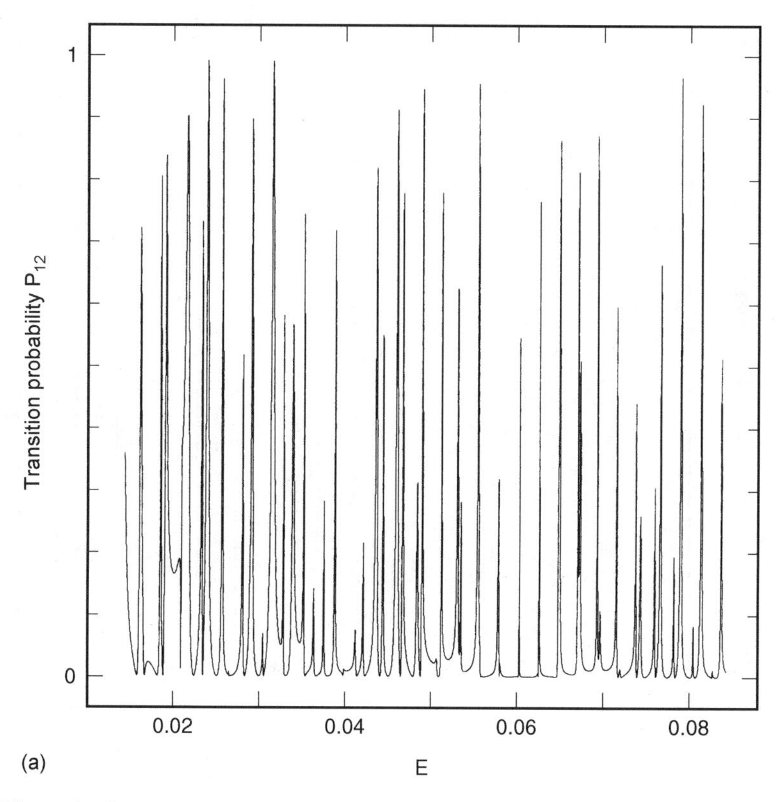


**Figure 6.** Overall transition probability $P_{12}$ against energy for the potential system given in Figure 5. (*a*) Quantum mechanically exact numerical solution of coupled equations. (*b*) Present semiclassical theory in the adiabatic representation. The starting energy and the energy step used in both calculations are exactly the same.

## B. Multidimensional Problems

Since most of chemical dynamical processes in reality proceed in multidimensional configuration space, a very natural question arises as to whether the simple 1D two-state theory can deal with such problems. Unfortunately, it is almost impossible to develop any intrinsically multidimensional analytical theory; and thus it is better to figure out some useful methods with the help of accurate 1D theories. In this sense, there are two ways: one is to reduce any multidimensional problem to a 1D multichannel problem by expanding the total wave function in terms of appropriate internal states, as is usually done in the quantum mechanical numerical solutions of the Schrödinger equation. The

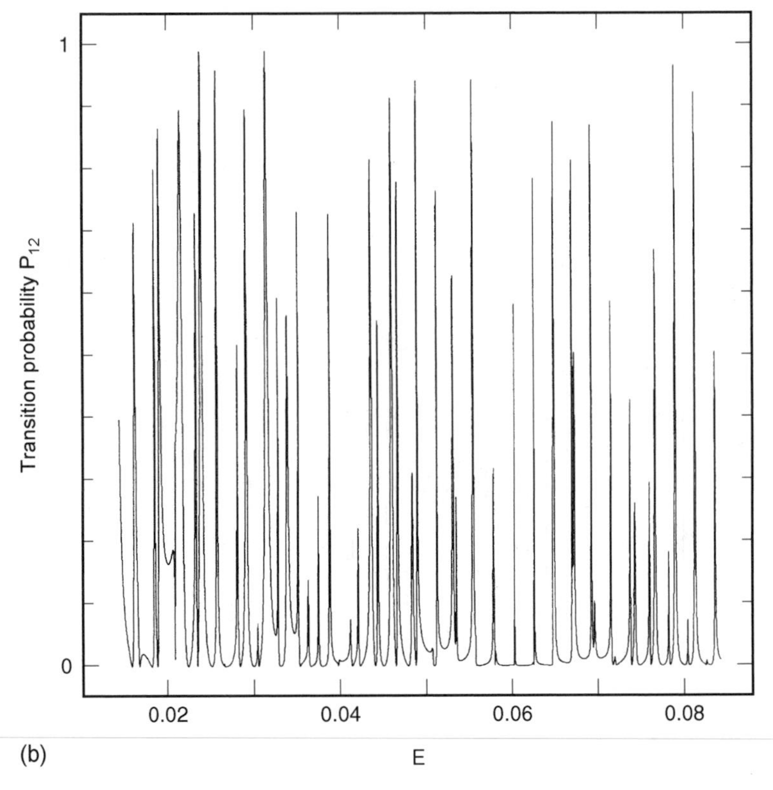


**Figure 6.** *(Continued)*

second, which is more approximate but may be practically more useful, is to use classical trajectories that define curvilinear 1D space and try to incorporate quantum mechanical effects as much as possible. The ordinary multidimensional semiclassical mechanics [84–90] belongs to this category. The present semiclassical theory of nonadiabatic transition can be incorporated in either way.

As for the first example, here we consider reactive scattering that presents a very interesting example, since different rearrangement channels are associated with different coordinates. The most convenient method within the framework of the collision theory is to use the hyperspherical coordinate approach in which all arrangement channels can be treated on equal footing. The hyperspherical coordinate approach implemented for tri- and tetraatomic reaction systems has shown remarkable progress recently [86–90]. Especially, in the 3D heavy–light–heavy (HLH) reactions on a single potential energy surface, a newly introduced coordinate system called "hyperspherical elliptic coordinate system" has made it possible to define the vibrationally adiabatic ridge lines and clarified the reaction mechanisms nicely in terms of the concept of vibrationally nonadiabatic transitions [89, 90]. Zhu et al. [26] treated, for the first time, such 3D HLH reactions analytically with the use of the present semiclassical theory. Needless to say, the conventional understanding of nonadiabatic transition is a transition between two electronically adiabatic states. This traditional adiabatic separation is, of course, based on the big mass disparity between the electron and the nucleus. This idea may be extended to reactive scattering on a single electronically adiabatic potential energy surface. Especially in the case of 3D HLH systems, the light atom has some analogy to an electron, and the adiabatic separation becomes the separation between hyperradius and hyperangles.

The Schrödinger equation for a triatomic system in hyperspherical coordinates can be written as (in atomic units)

$$\left[\frac{1}{2\mu}\frac{1}{\rho^5}\frac{\partial}{\partial\rho}\rho^5\frac{\partial}{\partial\rho}+\frac{\Lambda^2(\omega)}{2\mu\rho^2}+V(\rho,\omega_H)-E\right]\Psi(\rho)=0 \qquad (4.27)$$

where $\Lambda^2(\omega)$ is the so-called grand angular momentum operator, $\rho$ is the hyperradius, $\omega$ represents the five hyperangles that are divided into three Euler angles $\omega_E$ [not appearing in the interaction potential $V(\rho, \omega_H)$], and two geometric angles $\omega_H$ of a triatomic system. The reduced mass $\mu$ is defined by

$$\mu=\sqrt{\frac{m_A m_B m_C}{m_A+m_B+m_C}} \qquad (4.28)$$

where $m_A$, $m_B$, and $m_C$ are the masses of atoms A, B, and C, respectively.

The following expansion is used for solving Eq. (4.27):

$$\Psi(\rho)=\rho^{-5/2}\sum_{\nu}F_{\nu}(\rho)\Phi_{\nu}(\omega:\rho) \qquad (4.29)$$

where the adiabatic channel functions $\Phi_\nu(\omega:\rho)$ satisfy the hyperspherical adiabatic eigenvalue problem,

$$[H_{ad}(\omega,\rho)-\mu\rho^2\tilde{U}_{\nu}(\rho)]\Phi_{\nu}(\omega:\rho)=0 \qquad (4.30)$$

with

$$H_{ad}(\omega,\rho)=\tfrac{1}{2}\Lambda^2(\omega)+\mu\rho^2V(\rho,\omega_H) \qquad (4.31)$$

in which $\tilde{U}_\nu(\rho)$ is the eigenvalue to be determined at each fixed $\rho$, since $H_{ad}(\omega, \rho)$ depends on $\rho$ parametrically. The scattering wave function $F_\nu(\rho)$ in Eq. (4.29) turns out to satisfy

$$\left\{\frac{d}{d\rho^2} + 2\mu[E - U_\nu(\rho)]\right\}F_\nu(\rho) + \sum_\mu W_{\nu\mu}(\rho)F_\mu(\rho) = 0 \tag{4.32}$$

with

$$U_\nu(\rho) = \tilde{U}_\nu(\rho) + \frac{15}{8\mu\rho^2} \tag{4.33}$$

where $W_{\nu\mu}(\rho)$ is the nonadiabatic coupling term (not given here explicitly; see [89]). Now the problem becomes an ordinary multichannel scattering problem in the same way as in Section IV.A. We need only $U_\nu(\rho)$ in Eq. (4.33) to formulate the analytical solution of the reduced scattering matrix within the framework of the present semiclassical theory. Namely, we do not need any information about the nonadiabatic coupling $W_{\nu\mu}(\rho)$. The scattering wave function $F_\nu(\rho)$ in Eq. (4.32) can be written in the WKB form as

$$F_\nu(\rho) = \frac{A_\nu}{\sqrt{K_\nu(\rho)}} e^{i\int_{T_\nu}^{\rho} K_\nu(\rho)d\rho - i(\pi/4)} + \frac{B_\nu}{\sqrt{K_\nu(\rho)}} e^{-i\int_{T_\nu}^{\rho} K_\nu(\rho)d\rho + i(\pi/4)} \tag{4.34}$$

for $\rho \longrightarrow \infty$, where $T_\nu$ is the rightmost turning point on the adiabatic potential $U_\nu(\rho)$ and

$$K_\nu(\rho) = \sqrt{2\mu[E - U_\nu(\rho)]} \tag{4.35}$$

The reduced scattering matrix $S^R$ is defined as

$$\begin{pmatrix} A_1 \\ A_2 \\ \vdots \\ A_n \end{pmatrix} = S^R \begin{pmatrix} B_1 \\ B_2 \\ \vdots \\ B_n \end{pmatrix} \tag{4.36}$$

where $n$ represents the number of open channels at a given total collision energy $E$.

The $I$-matrix propagation method is directly implemented to obtain the reduced scattering matrix,

$$S^R = (I_1 I_2 \cdots I_N)^t (I_1 I_2 \cdots I_N) \tag{4.37}$$

where $N$ is the number of avoided crossings that can be as many as a thousand among the massive number of adiabatic potential curves $U_\nu(\rho)$. Those avoided crossings represent rovibrationally nonadiabatic transitions that represent reactive as well as nonreactive transitions. For 3D HLH systems, vibrationally adiabatic ridge lines can be extracted and the most important avoided crossings that represent reactive transitions are found to be located along or near these ridge lines. The lowest (ground vibrational, $\nu = 0$) ridge line defines the boundary of reaction zone. The avoided crossings outside this ridge line represent only nonreactive inelastic transitions. Those avoided crossings that are distributed far inside the ridge line represent a mixture of reactive and nonreactive transitions.

The effective coupling parameter $a^2$ defined in Eq. (3.21) for the LZ type and Eq. (3.38) for the NT type provides a very nice quantitative index of nonadiabatic coupling strength at each avoided crossing. Most of thousands of avoided crossings correspond to $a^2 > 1000$, which represent very sharp avoided crossings, and do not play meaningful roles in dynamics. Only $\sim$ 100 avoided crossings with $0.001 \lesssim a^2 \lesssim 1000$ among $100 \sim 200$ adiabatic potential curves contribute significantly to the reaction.

In the hyperspherical coordinate approach, all arrangement channels are treated equally and represented as adiabatic potential curves as a function of the hyperradius $\rho$. Therefore, important avoided crossings exist not only between adjacent adiabatic potentials, but can appear among nonadjacent adiabatic potentials. Besides, adiabatic potential curves belonging to physically separated arrangement channels avoid crossings very sharply outside the reaction zone and are better connected diabatically without any transitions. In order to extract these avoided crossings among nonadjacent adiabatic potentials, we have developed a certain diabatic decoupling method [26, 91]. We follow $U_\nu(\rho)$ inward from the asymptotic $\rho$, where each channel can be assigned. If avoided crossings between adjacent adiabatic potentials on the way in have $a^2 > a_0^2$, then we switch $U_\nu(\rho)$ to $U_{\nu-1}(\rho)$, or $U_{\nu+1}(\rho)$. By repeating this diabatic switching procedure even inside the reaction zone, we can finally obtain a diabatic potential manifold and pick up important avoided crossings between originally nonadjacent potential curves. The dependence of this decoupling procedure on the critical value $a_0^2$ is not as strong and $a_0^2 \cong 100$ was chosen. All the important avoided crossings are treated analytically to evaluate the scattering matrix. In [26], we studied the two examples of the 3D HLH reactions: $O(^3P) + HCl \rightarrow OH + Cl$ and $Br + HCl \rightarrow HBr + Cl$. Figure 7 shows a magnification of the reaction zone of adiabatic potential curve diagram of OHCl. A comparison between the exact quantum calculation and the present semiclassical result is shown in Figure 8 for cumulative reaction probability.

We can see that the agreement is quite good, considering that the semiclassical one is a completely analytical treatment. Similar agreement has

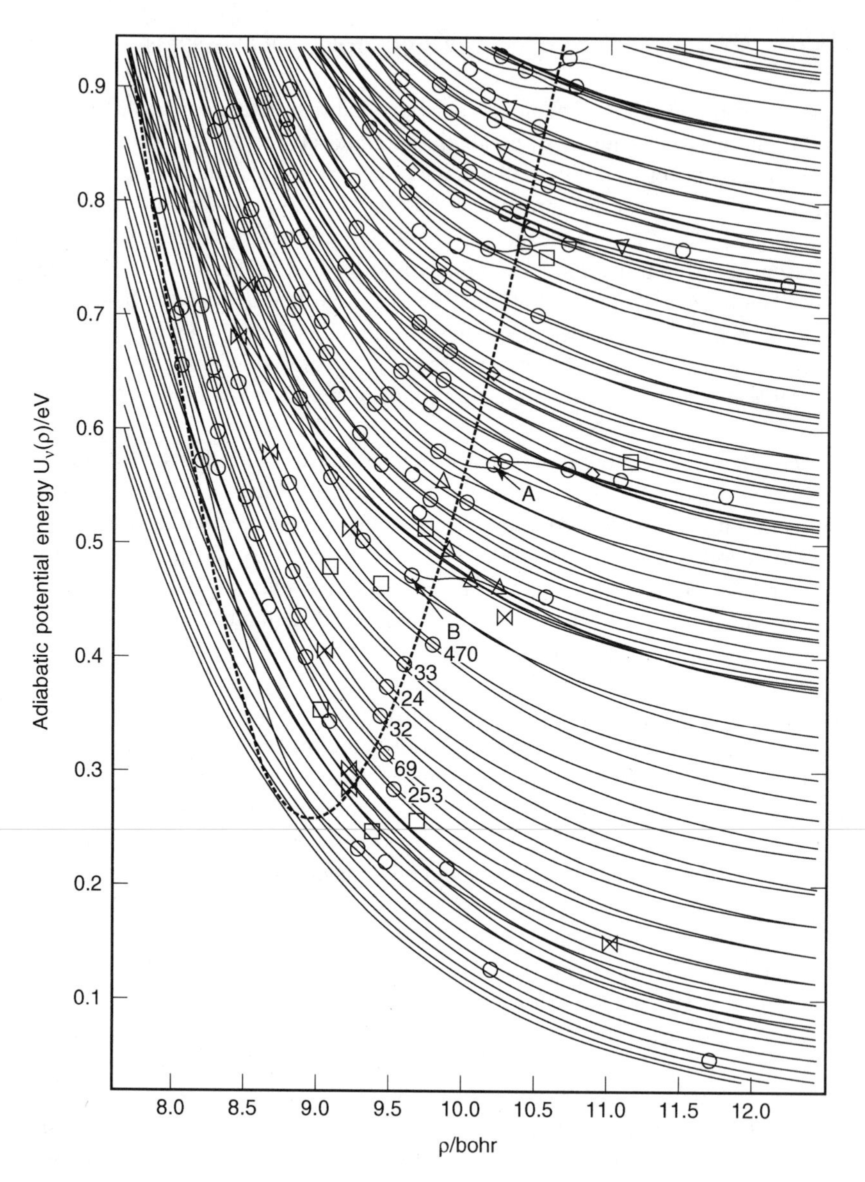


**Figure 7.** Magnification of the reaction zone of the adiabatic potential curves of OHCl. The dashed line represents the $\nu = 0$ ridge line. Circles represent important avoided crossings among the adjacent adiabatic potentials and some of the circles assigned with the values of $a^2$ are the most important ones. The circle signed by letter A (B) indicates the avoided crossing responsible for the peak of certain vibrationally specified cumulative reaction probabilities (see [26]). The symbols □, △, ▽, ⋈, and ◇ represent the avoided crossings among the nonadjacent adiabatic potentials. The symbols □, △, and ▽ are from the diabatic potential manifolds with $\nu = 0, 1, 2$, respectively, for HCl + O arrangement (see [26]). The symbols ⋈ and ◇ are from the diabatic potential manifolds, respectively, for OH + Cl arrangement (see [26]).

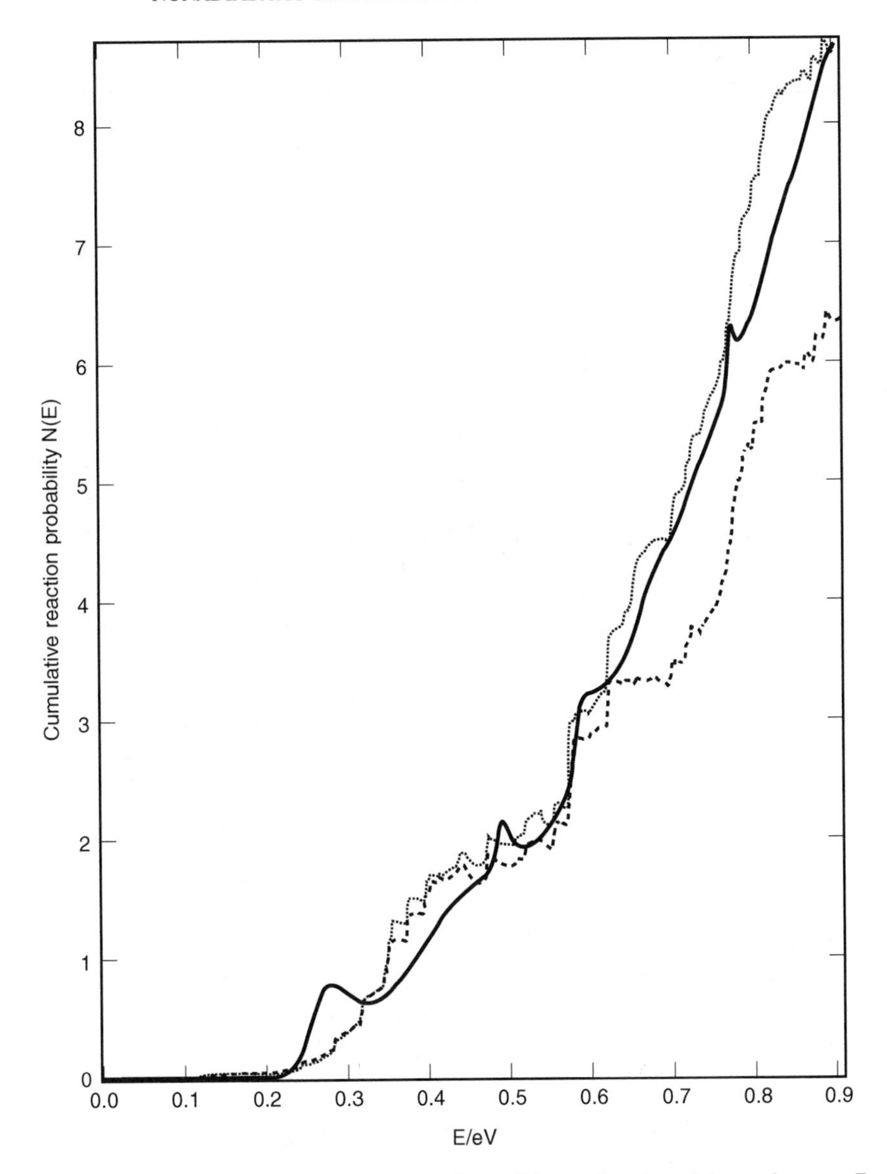


**Figure 8.** Total cumulative reaction probability $N(E)$ as a function of the total energy $E$ measured from the ground state of the reactant in the case of OHCl system: solid line (exact numerical result); dash line (semiclassical calculation with the avoided crossings among adjacent adiabatic potentials); dotted line (semiclassical calculation with the avoided crossings among both adjacent and nonadjacent adiabatic potentials included).

been obtained for BrHCl. The state-to-state reaction probabilities are not quantitatively well reproduced, however. This is because nonreactive inelastic transitions are not necessarily well represented by avoided crossings.

The second more approximate way of treating multidimensional dynamics is to incorporate the present semiclassical theory into a certain semiclassical propagation scheme based on classical trajectories. The best example will be an electronically nonadiabatic chemical reaction. Since a classical trajectory defines nothing but a curvilinear 1D space, the nonadiabatic transition $I$ matrix, actually $I_X$ matrix given by Eq. (3.60), can be incorporated whenever the trajectory comes to a potential energy surface crossing region. We can actually utilize various types of semiclassical propagation schemes such as TSH [27], the semiclassical IVR [28], and CFGWP [29]. The TSH is actually an ordinary quasiclassical trajectory (QCT) method without any incorporation of phases, but the electronically nonadiabatic transition probability or the surface hopping probability can be very much improved by the present semiclassical theory. Numerical solutions of time-dependent Schrödinger equations required in the case of fewest switches trajectory surface hopping (TFSH) [30] are not necessary at all. A more sophisticated framework is the semiclassical propagation based on IVR, in which all phases associated with each trajectory are taken into account to evaluate the transition amplitude. The IVR was figured out to avoid the divergence of the Van Vleck determinant at caustics [28]. With use of the present semiclassical theory even the dynamical phases due to nonadiabatic transition are accurately incorporated. For systems bigger than triatomic ones, the CFGW method will probably be useful [29]. In this method, the initial wave function is represented as a superposition of small wave packets and all of them are propagated along the classical trajectory of its center with the Gaussian shape frozen. This method was successfully applied to a 15-dimensional (15D) eigenvalue problem [29]. In any case, whatever the semiclassical propagation method is, the present semiclassical theory, especially the $I_X$ matrix, may be nicely incorporated into the framework in order to treat the electronically nonadiabatic transitions. Not only the phases but also the transition probability even at energies lower than the crossing point can be accurately and easily evaluated and incorporated.

## V. OTHER MODELS

From the previous sections, we have seen how the solutions of the two-state linear curve crossing model is generalized to deal with the two-state general curve crossing problems, as well as multichannel and multidimensional problems. These generalizations are made possible thanks to the fact that the compact solutions have been obtained for the linear potential model in a generalizable form and also the positions of nonadiabatic transitions can be easily identified as avoided crossings. As is well known, there is another kind of nonadiabatic transition generally termed as the noncurve crossing case. The RZ model is the most typical one in this category [4]. It is a very challenging subject to formulate a sort of unified theory that can cover both the curve crossing and noncurve crossing case in a unified way. The exponential potential model first studied by Nikitin [6] presents an interesting case in this sense, because the nonadiabatic transition probability contains two parameters and leads to either the LZ formula or the RZ formula in certain limits. Compared to the curve crossing case discussed in the previous sections, however, the theory of noncurve crossing case is actually far from complete.

### A. Exponential Potential Model

For the noncrossing model, it is convenient to start with the following exponential model. Diabatic potentials in Eqs. (3.1) and (3.2) now read as

$$V_{11}(x) = V_1 - \beta_1 V_0 e^{-\alpha x}$$

$$V_{22}(x) = V_2 - \beta_2 V_0 e^{-\alpha x}$$

and

$$V_{12}(x) = V_{21}(x) = V_0 e^{-\alpha x} \qquad (V_0 > 0) \qquad -\infty < x < \infty \tag{5.1}$$

where $V_2 > V_1$ is assumed without losing generality. Equation (5.1) actually includes both crossing (if $\beta_2 > \beta_1$) and noncrossing (if $\beta_2 < \beta_1$) cases. More generally, the three exponents in $V_{11}$, $V_{22}$, and $V_{12}$ can all be different, of course. Analytical solutions obtained so far are, however, restricted to the above case. The exponential model here is more general than the linear crossing model previously discussed in the sense that this can cover both crossing and noncrossing cases and the avoided crossing point can significantly differ from the original crossing point of the diabatic potentials. This means that the localizability of nonadiabatic transition is much better in the two-state linear crossing case.

Adiabatic potentials of Eq. (5.1) can be easily written as

$$E_2(x) = \tfrac{1}{2}[V_2 + V_1 - (\beta_2 + \beta_1)V_0 e^{-\alpha x}] + \tfrac{1}{2}\sqrt{[V_2 - V_1 - (\beta_2 - \beta_1)V_0 e^{-\alpha x}]^2 + 4V_0^2 e^{-2\alpha x}}$$

and

$$
\begin{aligned}
E_1(x) = {} & \tfrac{1}{2}[V_2 + V_1 - (\beta_2 + \beta_1)V_0 e^{-\alpha x}] \\
& - \tfrac{1}{2}\sqrt{[V_2 - V_1 - (\beta_2 - \beta_1)V_0 e^{-\alpha x}]^2 + 4V_0^2 e^{-2\alpha x}}
\end{aligned} \tag{5.2}
$$

There are an infinite number of complex zeros of $E_2(x) = E_1(x)$, all of which have the same real part, in contrast with the two complex zeros in the linear curve crossing case. As we know, the configuration structure of complex zeros completely determine the connection problem associated with the Stokes phenomenon and thus the physical phenomenon. The two-state exponential model has not been solved exactly quantum mechanically except for some special cases. By applying the Bessel integral transformation, Osherov et al. [32] transformed the coupled equations of Eqs. (3.1) and (3.2) with the potentials Eq. (5.1) in coordinate space into momentum space to obtain

$$
\left[\frac{d^2}{dp^2} + \frac{Q(p)}{p^2 S(p)}\right] f(p) = 0 \tag{5.3}
$$

where both $Q(p)$ and $S(p)$ are the fourth-order polynomials, whose explicit expressions can be found in [32]. The four transition points existing in $Q(p)$ are symmetrically distributed with respect to the imaginary axis in the complex $p$ plane, and the four poles (besides two at $p = 0$) in $S(p)$ are also distributed symmetrically with respect to the imaginary axis. The exact solution of the basic equation (5.3) is unfortunately unknown; but in the high-energy approximation, certain semiclassical solutions have been obtained with the use of the complex phase integral method [32]. There are generally two approaches to solve the two-state exponential model with Eq. (5.1): one is the method used in [32] and the other is that some special cases of the exponential model of Eq. (5.1) are first solved exactly, and then the exact solutions are tried to be generalized to general cases with use of the comparison-equation method (see, e.g., [5]). In fact, the two approaches mentioned above give almost the same final semiclassical solutions [31, 32, 34, 92–94].

Although the localizability of nonadiabatic transition on the real axis is worse than the linear curve crossing case, the semiclassical idea of nonadiabatic transition and adiabatic wave propagation still holds well and the $I_X$ matrix is given by

$$
I_X = \begin{pmatrix} \sqrt{1-p}\,e^{i\phi} & -\sqrt{p}\,e^{i\psi} \\ \sqrt{p}\,e^{-i\psi} & \sqrt{1-p}\,e^{-i\phi} \end{pmatrix} \tag{5.4}
$$

where $p$ is as usual the nonadiabatic transition probability for one passage of the transition region, and $\phi$ and $\psi$ are dynamical phases created by the nonadiabatic transition [32, 34]. The nonadiabatic transition or this $I_X$ matrix should be assigned at the real part of the complex crossing point $X^*$. The probability $p$ and the dynamical phases $\phi$ and $\psi$ are explicitly given by [6, 32, 34]

$$p = e^{-\pi\delta_2} \frac{\sinh(\pi\delta_1)}{\sinh(\pi\delta_2)} \tag{5.5}$$

$$\phi = \gamma(\delta_2) - \gamma(\delta) \tag{5.6}$$

and

$$\psi = \gamma(\delta_1) - \gamma(\delta) - 2\left[\sqrt{\delta\delta_2} + \frac{\delta_1}{2}\ln\frac{\sqrt{\delta} - \sqrt{\delta_2}}{\sqrt{\delta} + \sqrt{\delta_2}}\right] \tag{5.7}$$

with

$$\gamma(X) = X\ln(X) - X - \arg\Gamma(iX) \tag{5.8}$$

and

$$\delta = \delta_1 + \delta_2 \tag{5.9}$$

The parameters $\delta_1$ and $\delta_2$ are given by

$$\delta_1 = \frac{1}{\pi}\mathrm{Im}\int_{\mathrm{Re}X^*}^{X^*} [K_2(x) - K_1(x)]dx \tag{5.10}$$

and

$$\delta_2 = \frac{1}{2\pi i}\oint_{\infty} [K_2(x) - K_1(x)]dx - \delta_1 \tag{5.11}$$

where $K_j(x)$ $(j = 1, 2)$ are defined by Eq. (3.7). Equation (5.5) leads to the LZ probability $p = e^{-\pi\delta_2}$ in the limit $\delta \longrightarrow \infty$ or $\delta_2 \longrightarrow 0$, and covers the RZ probability $p = (1 + e^{2\pi\delta_2})^{-1}$ in the limit $\delta \longrightarrow 2\delta_2$.

Three exactly solvable cases follow with the potentials given by Eq. (5.1):

1. Rosen–Zener–Demkov model with $\beta_1 = \beta_2 = 0$ (see [94])
2. Attractive potential model with $\beta_1 = (1/\beta_2) > 0$ (see [31])
3. Repulsive potential model with $\beta_1 = (1/\beta_2) < 0$ (see [92])

By directly transforming the coupled equations (3.1) and (3.2) into a single fourth-order differential equation, we can solve equations exactly in terms of the Meijer's G functions [95] under the three conditions mentioned above. From the exact solution of wave functions, we can extract the reduced scattering matrix. In Section V.B, the above three exponential models will be briefly discussed.

## B. Rosen–Zener–Demkov Model ($\beta_1 = \beta_2 = 0$)

Two adiabatic potentials are shown in Figure 9($a$). Channel 3 is divergent, but it does not make any trouble to define the reduced scattering matrix $S^R$ that is actually given by [94]

$$S^R = \begin{pmatrix} -N_1 e^{i2\theta_1} & -\sqrt{N_1 N_2 - N_3} e^{i(\theta_1+\theta_2)} & -\sqrt{N_2 - N_1 N_3} e^{i(\theta_1+\theta_3)} \\ -\sqrt{N_1 N_2 - N_3} e^{i(\theta_1+\theta_2)} & -N_2 e^{i2\theta_2} & \sqrt{N_1 - N_2 N_3} e^{i(\theta_2+\theta_3)} \\ -\sqrt{N_2 - N_1 N_3} e^{i(\theta_1+\theta_3)} & \sqrt{N_1 - N_2 N_3} e^{i(\theta_2+\theta_3)} & N_3 e^{i2\theta_3} \end{pmatrix} \tag{5.12}$$

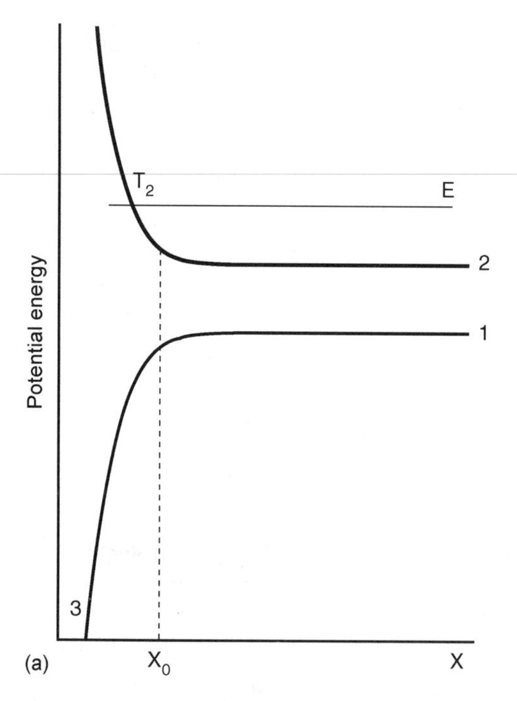


**Figure 9.** Adiabatic potentials for two special cases of the two-state (three-channel) exponential model. ($a$) Rosen–Zener–Demkov case with $\beta_1 = \beta_2 = 0$ in Eq. (5.1). ($b$) The repulsive case with $\beta_1 = 1/\beta_2 < 0$ in Eq. (5.1).

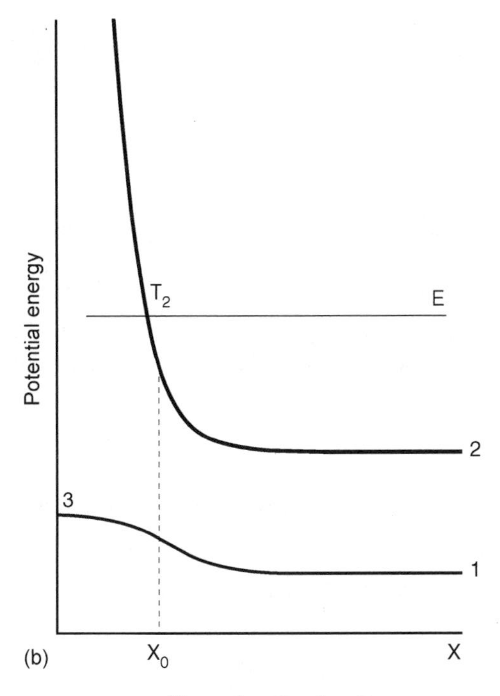


**Figure 9.** *(Continued)*

where the amplitudes are defined by

$$N_1 = e^{-\pi q_1} \frac{\cosh \pi(q_1 + q_2)/2}{\cosh \pi(q_1 - q_2)/2} \tag{5.13}$$

$$N_2 = e^{-\pi q_2} \frac{\cosh \pi(q_1 + q_2)/2}{\cosh \pi(q_1 - q_2)/2} \tag{5.14}$$

and

$$N_3 = e^{-\pi q_1 - \pi q_2} \tag{5.15}$$

with the phases given by

$$\theta_1 = 4q_1 \ln(2) + \arg\Gamma\left[\frac{1}{2} + i\frac{q_1 - q_2}{2}\right] - \arg\Gamma\left[\frac{1}{2} - i\frac{q_1 + q_2}{2}\right] + \arg\Gamma[iq_1] \tag{5.16}$$

$$\theta_2 = 4q_2 \ln(2) - \arg\Gamma\left[\frac{1}{2} + i\frac{q_1 - q_2}{2}\right] - \arg\Gamma\left[\frac{1}{2} - i\frac{q_1 + q_2}{2}\right] + \arg\Gamma[iq_2] \tag{5.17}$$

and

$$\theta_3 = \frac{\pi}{4} \tag{5.18}$$

It can be easily seen that the symmetrical $S^R$ matrix in Eq. (5.12) satisfies unitarity when $N_1$, $N_2$, and $N_3$ satisfy the following constraint:

$$N_1 + N_2 = 1 + N_3 \tag{5.19}$$

The parameters $q_1$ and $q_2$ in the above equations are given by

$$q_i = \frac{\sqrt{2\mu}}{\hbar\alpha}\sqrt{E - V_i}, \qquad (i = 1, 2) \tag{5.20}$$

In order to generalize this model so that we can deal with general two-state potentials in the noncrossing case, we have to close channel 3 in Figure 9($a$) by putting a repulsive wall far from the interaction region ($x \ll x_0$). In this way, we can have the following overall nonadiabatic transition probability:

$$P_{12} = \frac{\sinh(\pi q_1)\sinh(\pi q_2)\cos^2(\sigma)}{\cosh^2(\pi(q_1 - q_2)/2)[\cosh^2(\pi(q_1 + q_2)/2) - \cos^2(\sigma)]} \tag{5.21}$$

This formula is valid even in the threshold region where $q_2 \longrightarrow 0$. If we apply the high-energy approximation, Eq. (5.21) turns to be

$$P_{12} \cong \frac{\cos^2(\sigma)}{\cosh^2(\pi(q_1 - q_2)/2)} \equiv \frac{\cos^2(\sigma)}{\cosh^2\delta} \tag{5.22}$$

This is nothing but the famous RZ formula in which $\sigma$ and $\delta$ can be generally evaluated by the complex phase integral,

$$\sigma + i\delta = \int_{T_1}^{R^*} K_1(R)dR - \int_{T_2}^{R^*} K_2(R)dR \tag{5.23}$$

where $R^*$ is the complex crossing between two general noncrossing adiabatic potentials. The $I_X$ matrix for the RZ noncrossing problem can also be defined as

$$I_X = \begin{pmatrix} \sqrt{1-p}e^{-i\phi} & -\sqrt{p}e^{i\psi} \\ \sqrt{p}e^{-i\psi} & \sqrt{1-p}e^{i\phi} \end{pmatrix} \tag{5.24}$$

where

$$\phi = \gamma(\delta_{RZ}) - \gamma(2\delta_{RZ}) \tag{5.25}$$

$$\psi = \phi - \delta_{RZ}\left[2\sqrt{2} + \ln\frac{\sqrt{2}-1}{\sqrt{2}+1}\right] \tag{5.26}$$

$$\delta_{RZ} = \frac{\sqrt{\mu}}{\sqrt{2}\hbar\alpha}(\sqrt{E-V_1} - \sqrt{E-V_2}) \tag{5.27}$$

and one passage probability is given by

$$p = \frac{1}{1+e^{2\delta}} \tag{5.28}$$

### C. Special Cases of Exponential Potential Model ($\beta_1 = (1/\beta_2)$)

First we discuss the repulsive case. In this case, channel 3 is not divergent [see Fig. 9(*b*)] and the reduced scattering matrix is given by [93]

$$S = \begin{pmatrix} N_1 e^{i2\theta_1} & \sqrt{N_1N_2 - N_3}e^{i(\theta_1+\theta_2)} & \sqrt{N_2 - N_1N_3}e^{i(\theta_1+\theta_3)} \\ \sqrt{N_1N_2 - N_3}e^{i(\theta_1+\theta_2)} & N_2 e^{i2\theta_2} & -\sqrt{N_1 - N_2N_3}e^{i(\theta_2+\theta_3)} \\ \sqrt{N_2 - N_1N_3}e^{i(\theta_1+\theta_3)} & -\sqrt{N_1 - N_2N_3}e^{i(\theta_2+\theta_3)} & -N_3 e^{i2\theta_3} \end{pmatrix} \tag{5.29}$$

where the amplitude parts are defined by

$$N_1 = \frac{\sinh\pi(q_1 - q_3)\sinh\pi(q_1 + q_2)}{\sinh\pi(q_1 + q_3)\sinh\pi(q_1 - q_2)}$$

$$N_2 = \frac{\sinh\pi(q_3 - q_2)\sinh\pi(q_1 + q_2)}{\sinh\pi(q_3 + q_2)\sinh\pi(q_1 - q_2)}$$

and

$$N_3 = \frac{\sinh \pi(q_1 - q_3) \sinh \pi(q_3 - q_2)}{\sinh \pi(q_1 + q_3) \sinh \pi(q_3 + q_2)} \tag{5.30}$$

and the phases are given by

$$\begin{aligned}
\theta_1 &= -q_1 \ln(\gamma e^{-\alpha x_0}) + \arg \Gamma[2iq_1] + \arg \Gamma[i(q_1 - q_2)] + \arg \Gamma[i(q_1 + q_2)] \\
&\quad - \arg \Gamma[i(q_1 - q_3)] - \arg \Gamma[i(q_1 + q_3)] \\
\theta_2 &= -q_2 \ln(\gamma e^{-\alpha x_0}) + \arg \Gamma[2iq_2] - \arg \Gamma[i(q_1 - q_2)] + \arg \Gamma[i(q_1 + q_2)] \\
&\quad + \arg \Gamma[i(q_3 - q_2)] - \arg \Gamma[i(q_3 + q_2)]
\end{aligned}$$

and

$$\begin{aligned}
\theta_3 = q_3 \ln(\gamma e^{-\alpha x_0}) &+ \arg \Gamma[2iq_3] + \arg \Gamma[i(q_1 - q_3)] - \arg \Gamma[i(q_1 + q_3)] \\
&- \arg \Gamma[i(q_3 - q_2)] - \arg \Gamma[i(q_3 + q_2)]
\end{aligned} \tag{5.31}$$

Again, we can easily prove that $S^R$ in Eq. (5.29) satisfies unitarity and $N_1$, $N_2$, and $N_3$ satisfies the following relation:

$$N_1 + N_2 = 1 + N_3 \tag{5.32}$$

In the above equations, $q_1$ and $q_2$ are the same as in Eq. (5.20) with $q_3$ given by

$$q_3 = \frac{\sqrt{2\mu}}{\hbar\alpha} \sqrt{E - V_3} \tag{5.33}$$

with

$$V_3 = \frac{\beta_1 V_2 + V_1/\beta_1}{\beta_1 + 1/\beta_1} \tag{5.34}$$

The real part $x_0$ of the complex zero and $\gamma$ in Eq. (5.31) are defined by

$$x_0 = -\frac{1}{\alpha} \ln \left[ \frac{V_2 - V_1}{V_0(|\beta_1 + 1/\beta_1|)} \right] \tag{5.35}$$

and

$$\gamma = \frac{2\mu V_0}{\hbar^2 \alpha^2} (|\beta_1 + 1/\beta_1|) \tag{5.36}$$

It should be emphasized that the present repulsive model includes both the crossing case ($0 > \beta_1 > -1$) and the noncrossing case ($\beta_1 < -1$). In a similar way as carried out in Section V.B, we can generalize the present model to a general two-state problem with a repulsive wall. The overall transition probability is found as [93]

$$P_{12} = \frac{4(1 - N_1)(1 - N_2)}{(1 + N_3)^2} \frac{\sin^2(\theta + \theta_3)}{1 - (4N_3/(1 + N_3)) \sin^2(\theta + \theta_3)} \tag{5.37}$$

This formula is valid even in the threshold region ($q_2 \longrightarrow 0$). Under the high-energy approximation Eq. (5.37) turns to be

$$P_{12} \cong 4p(1 - p) \sin^2(\sigma + \phi) \tag{5.38}$$

where the one passage nonadiabatic transition probability $p$ is given by

$$p = \frac{\sinh(d^2 - 1)\delta}{\sinh(d^2\delta)} e^{-\delta} \tag{5.39}$$

and the phase $\phi$ by

$$\phi = \phi_s\left(\frac{\delta}{\pi}\right) - \phi_s\left[(d^2 - 1)\frac{\delta}{\pi}\right] \tag{5.40}$$

where

$$\phi_s(X) = X \ln(X) - X - \arg\Gamma(iX) - \frac{\pi}{4} \tag{5.41}$$

The two parameters $\delta$ and $\sigma$ are given by the complex phase integral as shown in Eq. (5.23). The probability expression given in Eq. (5.39) actually coincides with Eq. (5.5) and also with Nikitin's formula [6]. Here, however, the important parameter $d$ in the above equations, introduced by the comparison-equation method, represents a type of nonadiabatic transition in general two-state problems and is given by

$$d = \sqrt{1 + \frac{4V_{12}^2(x_0)}{[V_1(x_0) - V_{22}(x_0)]^2}} \tag{5.42}$$

where $x_0$ is the real part of the complex crossing point between the adiabatic potentials. In fact, it is easy to check that if we apply the two-state linear curve crossing model to Eq. (5.42) in which the diabatic crossing point coincides with the real part of the complex crossing, we have

$$d = \infty \tag{5.43}$$

under which the formula goes back to the LZ case, and that if we apply the two-state RZ model to Eq. (5.42), we have

$$d = \sqrt{2} \tag{5.44}$$

which leads the formula to the RZ case. In general, we have $1 < d < \infty$ so that type of nonadiabatic transition can change continuously depending on the value of $d$ in Eq. (5.42). In this way, a certain unified semiclassical theory for the general two-state problems can be formulated, but the theory is established only in the diabatic representation as is seen from Eq. (5.42). Some numerical calculations are presented in [93] in which the unified formula works very well especially for the case of $d < \sqrt{2}$, which is neither the LZ nor RZ case.

In the attractive case ($\beta_1 = \beta_2^{-1} > 0$), we have four- ($E > V_1$), three- ($V_1 > E > V_3$), two- ($V_3 > E > V_2$), and one- ($E < V_2$) channel problems, where $V_3$ is given in Eq. (5.34) (see Fig. 10) [31]. Here we only give the exact expressions of nonadiabatic transition probabilities $p_{ij}(= |(I_X)_{ij}|^2)$ in the four-channel case

$$p_{11} = \left[\frac{\sinh\pi(q_3 - q_1)\sinh\pi(q_2 + q_1)}{\sinh\pi(q_3 + q_1)\sinh\pi(q_2 - q_1)}\right]^2 e^{-4\pi q_1} \tag{5.45}$$

$$p_{12} = \frac{e^{-2\pi q_2 - 2\pi q_1}}{\sinh^2\pi(q_2 - q_1)} \frac{\sinh\pi(q_3 - q_1)\sinh(2\pi q_2)\sinh(2\pi q_1)\sinh\pi(q_2 - q_3)}{\sinh\pi(q_3 + q_2)\sinh\pi(q_3 + q_1)} \tag{5.46}$$

$$p_{13} = 2\frac{\sinh\pi(q_3 - q_1)\sinh\pi(q_2 + q_1)\sinh(2\pi q_1)}{\sinh\pi(q_3 + q_1)\sinh\pi(q_2 - q_1)} e^{-2\pi q_2 - 2\pi q_1} \tag{5.47}$$

$$p_{14} = \frac{e^{-2\pi q_3 - 2\pi q_1}}{\sinh^2\pi(q_1 + q_3)} \frac{\sinh(2\pi q_3)\sinh\pi(q_2 + q_1)\sinh(2\pi q_1)\sinh\pi(q_2 - q_3)}{\sinh\pi(q_3 + q_2)\sinh\pi(q_2 - q_1)} \tag{5.48}$$

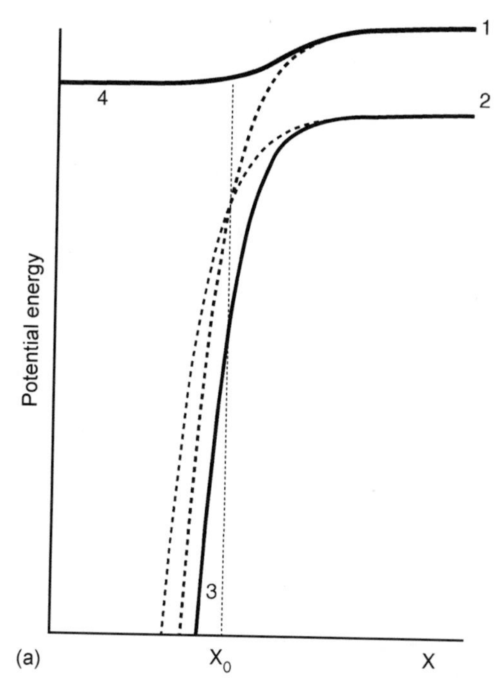


**Figure 10.** Two diabatic potentials (dotted line) and corresponding two adiabatic potentials (solid line) for special attractive cases of the two-state (four-channel) exponential model. (*a*) Crossing case, (*b*) Noncrossing case.

$$p_{22} = \left[\frac{\sinh \pi(q_2 - q_3) \sinh \pi(q_2 + q_1)}{\sinh \pi(q_2 + q_3) \sinh \pi(q_2 - q_1)}\right]^2 e^{-4\pi q_2} \tag{5.49}$$

$$p_{23} = 2 \frac{\sinh(2\pi q_2) \sinh \pi(q_2 + q_1) \sinh \pi(q_2 - q_3)}{\sinh \pi(q_2 + q_3) \sinh \pi(q_2 - q_1)} e^{2\pi q_3 - 2\pi q_2 - 2\pi q_1} \tag{5.50}$$

$$p_{24} = \frac{e^{2\pi q_3 - 2\pi q_2}}{\sinh^2 \pi(q_2 + q_3)} \frac{\sinh(2\pi q_2) \sinh(2\pi q_3) \sinh \pi(q_2 + q_1) \sinh \pi(q_3 - q_1)}{\sinh \pi(q_3 + q_1) \sinh \pi(q_2 - q_1)} \tag{5.51}$$

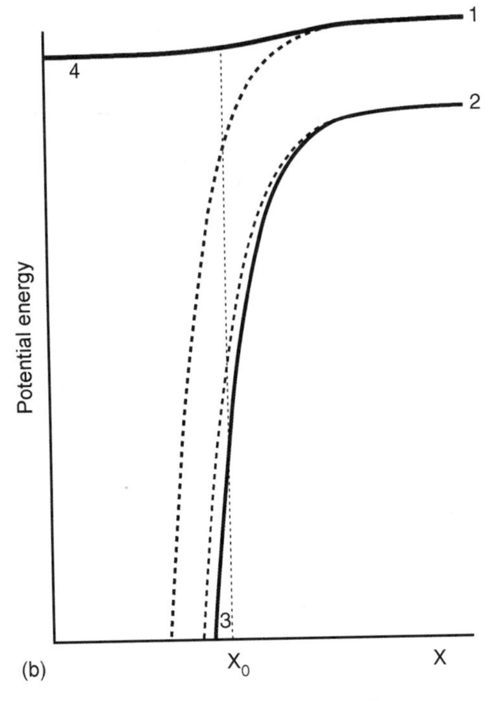


**Figure 10.** *(Continued)*

$$p_{33} = e^{4\pi(q_3 - q_2 - q_1)} \tag{5.52}$$

$$p_{34} = 2\frac{\sinh(2\pi q_3)\sinh\pi(q_3 - q_1)\sinh\pi(q_2 - q_3)}{\sinh\pi(q_2 + q_3)\sinh\pi(q_1 + q_3)} e^{2\pi q_3 - 2\pi q_1 - 2\pi q_2} \tag{5.53}$$

$$p_{44} = \left(\frac{\sinh\pi(q_3 - q_1)\sinh\pi(q_3 - q_2)}{\sinh\pi(q_3 + q_1)\sinh\pi(q_3 + q_2)}\right)^2 e^{4\pi q_3} \tag{5.54}$$

Semiclassically, $p_{13}$ and $p_{24}$ represent the nonadiabatic transition probability for one passage of the transition region, and is nicely given by the following formula except at the threshold:

$$p = e^{-\pi(q_2 - q_3)} \frac{\sinh\pi(q_3 - q_1)}{\sinh\pi(q_2 - q_1)} \xrightarrow{E \to +\infty} \frac{V_1}{V_1 + V_2} \tag{5.55}$$

Other probabilities are zeros except for

$$p_{14} = p_{23} = 1 - p \tag{5.56}$$

The $I_X$ matrix is given by Eq. (5.24) with $p$ given by Eq. (5.55). This $I_X$ matrix provides the nonadiabatic transition amplitude in the transition region and can be used even in multidimensional problems.

### D. Remarks

There is an important difference between Eq. (5.37) and (5.38) [the same is true between Eq. (5.21) and (5.22)]. The amplitude part of the overall transition probability in Eq. (5.37) has two independent quantities $N_1$ and $N_2$, but in Eq. (5.38) it has only one $p \cong N_1$. This means that Eq. (5.38) is not guaranteed to be valid in the threshold region ($q_2 \to 0$). Physically speaking, the overall transition process is always a mixture of nonadiabatic transition at avoided crossing and the threshold effect. It would be very good, if we could generalize both $N_1$ and $N_2$ for a general two-state problem; however, this generalization does not seem possible so far, since the threshold effect is not a localized transition at all. This fact can be seen more clearly from the special case of the exponential model. If we set $\beta_1 = \beta_2 = 1$ in Eq. (5.1), the formula Eq. (5.38) will turn to the RZ formula Eq. (5.22), which is obtained by applying $\beta_1 = \beta_2 = 0$ in Eq. (5.1). This is possible simply because we neglect the threshold effect completely. This is in good contrast with the fact that Eq. (5.37) with $\beta_1 = \beta_2 = 1$ will not become Eq. (5.21) with $\beta_1 = \beta_2 = 0$ due to the threshold effects. Fortunately, however, the threshold effects appear conspicuously only in a very small region, as was demonstrated in [31].

Another problem is the judgment of transition type when two adiabatic potentials are given. The parameter $d$ in Eq. (5.42) can do this, if a certain diabatization can be reasonably carried out. It is more desirable if we could do this only with the information of two adiabatic potentials. This might be possible by introducing another kind of complex phase integral, but this has not been successful and is actually one of the basic motivations in trying to formulate a unified theory within the adiabatic representation. As discussed earlier in this section, the exponential model may be a good candidate for this purpose, since a certain unified semiclassical theory has been already established in the diabatic representation [93]. Some tests were also carried out by somehow extending the exponential potential model at least at relatively high energies [34]. In any case, this is a very challenging subject.

Finally, we would like to draw attention to Eq. (5.3) in which the four transition points in $Q(p)$ correspond to definitely localized nonadiabatic transitions, and poles in $S(p)$ may contain information about the nonadiabatic transition type as well as the threshold effect. How to investigate the Stokes phenomenon of Eq. (5.3) more deeply without the high-energy approximation is still an unsolved problem. Moreover, Eq. (5.3) should also be a useful starting point for dealing with the case that energies are lower than the nonadiabatic transition points, which so far has never been attacked except for the linear curve crossing model.

## VI. TIME-DEPENDENT LEVEL CROSSINGS

### A. Complete Solutions of the Quadratic Model

As well as the time-independent nonadiabatic transitions, time-dependent ones play important roles in various fields of science. Examples are atomic and molecular dynamic processes in an external field [96], tunneling junctions of the Josephson element in an external electric field, metallic ring current in a magnetic field, and Zener transitions [60–66]. In these cases the external fields depend on time, and the processes are quantum mechanical time-dependent problems. Time-dependent and time-independent nonadiabatic transitions seem very different, but they are closely related to each other as easily conjectured from the fact that the latter can be reduced to the former by the classical approximation where the spatial coordinate is given by a function of time. The famous LZ formula, for example, is an exact solution of the linear potential model in the time domain, while it is a high-energy approximation in the time-independent linear potential model.

In this section, the complete solutions for time-independent curve crossing problems discussed in Sections II–IV are applied to time-dependent problems. It is shown that the exact solutions of the time-independent linear potential model can provide the exact solutions to the time-dependent quadratic potential model [44]. The compact semiclassical solutions developed for the general two-state time-independent potential curve crossing problems can be transferred to general time-dependent curve crossing problems. Time-dependent diabatically avoided crossing problems can also be accurately dealt with, since this case corresponds to the energy region lower than the crossing point in the time-independent curve crossing problem. Note that the NT type in the time-independent case does not show up in the time-dependent problems, since there is no bifurcation into transmission and reflection in the time-dependent case (time is unidirectional).

A general quadratic potential model of the time-dependent curve crossing problem in the diabatic representation is given by

$$i\hbar\frac{d}{dt}\begin{bmatrix} c_1(t) \\ c_2(t) \end{bmatrix} = \begin{bmatrix} \hat{\varepsilon}_1(t) & V_0 \\ V_0 & \hat{\varepsilon}_2(t) \end{bmatrix}\begin{bmatrix} c_1(t) \\ c_2(t) \end{bmatrix} \tag{6.1}$$

where $V_0$ is the constant diabatic coupling, $\hat{\varepsilon}_1(t)$ and $\hat{\varepsilon}_2(t)$ are quadratic diabatic energy levels given by

$$\hat{\varepsilon}_1(t) = \alpha_1 t^2 + \beta_1 \tag{6.2}$$

and

$$\hat{\varepsilon}_2(t) = \alpha_2(t - \gamma_2)^2 + \beta_2 \tag{6.3}$$

Equation (6.1) can be reduced to the following coupled equations:

$$i\hbar \frac{d}{dt} \begin{bmatrix} c_1(\tau) \\ c_2(\tau) \end{bmatrix} = \begin{bmatrix} 0 & \frac{1}{2} e^{i \int_{\tau_0}^{\tau} \Delta\varepsilon(t) d\tau} \\ \frac{1}{2} e^{-i \int_{\tau_0}^{\tau} \Delta\varepsilon(t) d\tau} & 0 \end{bmatrix} \begin{bmatrix} c_1(\tau) \\ c_2(\tau) \end{bmatrix} \tag{6.4}$$

where

$$\tau = 2V_0 \left( t + \frac{\alpha_2 \gamma_2}{\alpha_1 - \alpha_2} \right) \Big/ \hbar \tag{6.5}$$

$$\Delta\varepsilon(\tau) \equiv \frac{[\hat{\varepsilon}_1(\tau) - \hat{\varepsilon}_2(\tau)]}{V_0} = \alpha\tau^2 - \beta \tag{6.6}$$

$$\alpha \equiv \frac{(\alpha_1 - \alpha_2)\hbar^2}{8V_0^3} \tag{6.7}$$

and

$$\beta \equiv \left( \beta_2 - \beta_1 + \frac{\alpha_1 \alpha_2 \gamma_2^2}{\alpha_1 - \alpha_2} \right) \Big/ 2V_0 \tag{6.8}$$

Note that this problem can be described in terms of only two parameters, $\alpha$ and $\beta$, and that the two diabatic curves cross (do not cross), if $\beta > 0$ ($\beta < 0$). The parameter $\alpha$ is taken to be positive, that is, $\alpha_1 > \alpha_2$.

Now, we introduce the evolution matrix $F(z, z_0)$ defined by the following equation:

$$\begin{bmatrix} c_1(z) \\ c_2(z) \end{bmatrix} = F(z, z_0) \begin{bmatrix} c_1(z_0) \\ c_2(z_0) \end{bmatrix} \tag{6.9}$$

where $z$ represents a complex variable with $\tau = \mathrm{Re}(z)$. From the coupled equation (6.4), we can derive the following properties of $F(z, z_0)$:

$$\det F(z, z_0) = 1 \tag{6.10}$$

and

$$F(z, z_0) = \begin{pmatrix} 0 & -1 \\ 1 & 0 \end{pmatrix} F^*(z, z_0) \begin{pmatrix} 0 & 1 \\ -1 & 0 \end{pmatrix} \tag{6.11}$$

The transition matrix defined by

$$\begin{bmatrix} c_1(\infty) \\ c_2(\infty) \end{bmatrix} = T^R \begin{bmatrix} c_1(-\infty) \\ c_2(-\infty) \end{bmatrix} \tag{6.12}$$

is obviously equal to $F(\infty, -\infty)$. In addition to unitarity, the transition matrix $T^R$ satisfies the following symmetries, which are derived from the properties of the evolution matrix $F(z, z_0)$:

$$T_{11}^R = (T_{22}^R)^* \tag{6.13}$$

and

$$T_{12}^R = T_{21}^R = \text{pure imaginary} \tag{6.14}$$

In order to derive the explicit expression of $T^R$, let us consider the following single differential equation derived from the coupled equations (6.4):

$$\frac{d^2u(\tau)}{d\tau^2} + I(\tau)u(\tau) = 0 \tag{6.15}$$

where

$$u(\tau) = c_1(\tau) \exp\left[-\frac{i}{2}\int_{\tau_0}^{\tau} \Delta\varepsilon(\tau)d\tau\right] \tag{6.16}$$

and

$$I(\tau) = \frac{i}{2}\Delta\dot{\varepsilon}(\tau) + \frac{1}{4}[\Delta\varepsilon(\tau)]^2 + \frac{1}{4} \tag{6.17}$$

$$= \frac{1}{4} - i\alpha\tau + \frac{1}{4}(\alpha\tau^2 - \beta)^2 \tag{6.18}$$

A direct comparison of Eq. (6.15) with the corresponding differential equation in the time-independent linear potential model, see Eq. (2.3) of [13], can derive the following correspondence:

$$a^2 \Longleftrightarrow \alpha \tag{6.19}$$

and

$$b^2 \Longleftrightarrow \beta \tag{6.20}$$

It can also be confirmed that the transition matrix $T^R$ corresponds to the reduced scattering matrix $S^R$ of the time-independent case.

With this correspondence, the present time-dependent quadratic potential problem can be solved in exactly the same way as in the time-independent linear potential problem. Thus the transition matrix $T^R$ exactly corresponds to the reduced scattering matrix $S^R$ in the time-independent problem. It should be noted that since $a^2 > 0$, $\alpha_1$ and $\alpha_2$ must be chosen so that $\alpha > 0$, and that $b^2 > 0$ (crossing) corresponds to $\beta > 0$.

Following the solution method in the time-independent case, we briefly outline the procedure to obtain the exact solution of $T^R$ from Eqs. (6.4). First, we introduce the general WKB solutions, with which the asymptotic solutions are given by

$$u_1(z) = AI^{-1/4}(z)\exp\left[i\int_{z_0}^{z} I^{1/2}(z)dz\right] + BI^{-1/4}(z)\exp\left[-i\int_{z_0}^{z} I^{1/2}(z)dz\right] \tag{6.21}$$

and

$$u_2(z) = CI^{-1/4}(z)\exp\left[i\int_{-z_0}^{z} I^{1/2}(z)dz\right] + DI^{-1/4}(z)\exp\left[-i\int_{-z_0}^{z} I^{1/2}(z)dz\right] \tag{6.22}$$

where $A$, $B$, $C$, and $D$ are certain constants.

If we evaluate the phase integrals in Eqs. (6.21) and (6.22), we can rewrite them as

$$u_1(z) \xrightarrow{z\to\infty} AI^{-1/4}(z)\exp\left[iQ^+(z)\right] + BI^{-1/4}(z)\exp\left[-iQ^+(z)\right] \tag{6.23}$$

and

$$u_2(z) \xrightarrow{z\to-\infty} CI^{-1/4}(z)\exp\left[iQ^-(z)\right] + DI^{-1/4}(z)\exp\left[-iQ^-(z)\right] \quad (6.24)$$

where

$$i\int_{z_0}^{z} I^{\frac{1}{2}}(z)dz \xrightarrow{z\to\infty} iQ^+(z) \equiv iP(z) + i\ln(z) + i\delta_+(z_0) \quad (6.25)$$

$$-i\int_{-z_0}^{z} I^{\frac{1}{2}}(z)dz \xrightarrow{z\to-\infty} iQ^-(z) \equiv iP(z) + i\ln(z) + i\delta_-(z_0) \quad (6.26)$$

$$P(z) = \frac{1}{2}\int_0^z \Delta\varepsilon dz \quad (6.27)$$

and $\delta_\pm(z_0)$ are constants dependent on the reference point $z_0$. Let us now introduce the standard WKB solutions [22] in which the lower limit of the phase integral is not specified,

$$(\cdot, z) = z^{-1}\exp[iP(z) + i\ln z] \quad (6.28)$$

and

$$(z, \cdot) = z^{-1}\exp[-iP(z) + i\ln z] \quad (6.29)$$

In terms of these solutions, $u_1(z)$ and $u_2(z)$ can be expressed as

$$u_1(z) \xrightarrow{z\to\infty} A'(\cdot, z) + B'(z, \cdot) \quad (6.30)$$

and

$$u_2(z) \xrightarrow{z\to-\infty} C'(\cdot, z) + D'(z, \cdot) \quad (6.31)$$

From the analysis of the Stokes phenomenon, the following relation among the primed coefficients is obtained [13]:

$$\begin{aligned}\begin{bmatrix} C' \\ D' \end{bmatrix} &= \begin{bmatrix} 1 + U_2U_3 & U_1 + U_3 + U_1U_2U_3 \\ U_2 & 1 + U_3U_2 \end{bmatrix}\begin{bmatrix} A' \\ B' \end{bmatrix} \\ &\equiv \begin{bmatrix} L'_{11} & L'_{12} \\ L'_{21} & L'_{22} \end{bmatrix}\begin{bmatrix} A' \\ B' \end{bmatrix}\end{aligned} \quad (6.32)$$

where $U_1 \sim U_3$ are the Stokes constants that are functions of $\alpha$ and $\beta$. From Eqs. (6.23) and (6.24) with use of Eqs. (6.30)–(6.32), we can obtain the connection matrix $L$ as

$$\begin{bmatrix} C \\ D \end{bmatrix} \equiv \begin{bmatrix} L_{11} & L_{12} \\ L_{21} & L_{22} \end{bmatrix} \begin{bmatrix} A \\ B \end{bmatrix} = \begin{bmatrix} L'_{11}e^{i\delta_- + i\delta_+} & L'_{12}e^{-i\delta_- - i\delta_+} \\ L'_{21}e^{i\delta_- + i\delta_+} & L'_{22}e^{i\delta_- - i\delta_+} \end{bmatrix} \begin{bmatrix} A \\ B \end{bmatrix} \tag{6.33}$$

Thus, finally the transition matrix is given by

$$T^R = \begin{bmatrix} 1 + U_1U_2 & -(U_1 + U_3 + U_1U_2U_3)\frac{1}{2\alpha} \\ -2\alpha U_2 & 1 + U_2U_3 \end{bmatrix} \tag{6.34}$$

The following relations among the Stokes constants are obtained from the symmetries of $T^R$:

$$2\alpha U_2 = (U_1 + U_3 + U_1U_2U_3)\frac{1}{2\alpha} = \text{pure imaginary} \tag{6.35}$$

and

$$U_3 = -U_1^* \tag{6.36}$$

These relations enable us to express $T^R$ in terms of only one Stokes constant $U_1$,

$$T^R = \begin{bmatrix} 1 + U_1U_2 & -2\alpha U_2 \\ -2\alpha U_2 & 1 - U_1U_2 \end{bmatrix} \tag{6.37}$$

with

$$U_2 = \frac{U_1 - U_1^*}{4\alpha^2 - U_1U_1^*} \tag{6.38}$$

The total transition probability $P_{12}$ is expressed as,

$$P_{12} = |T_{12}^R|^2 = \frac{16\alpha^2(\Im U_1)^2}{(|U_1|^2 + 4\alpha^2)^2} \tag{6.39}$$

The Stokes constant $U_1$ is exactly expressed in the form of convergent infinite series as shown in [13]. It also has a compact and accurate semiclassical expression, as presented in Section III.

## B. Generalizations and Applications

The compact semiclassical formulas derived in the case of time-independent linear potential model can be transferred to the present time-dependent quadratic potential problem with the correspondences Eqs. (6.19) and (6.20).

The overall nonadiabatic transition probability $P_{12}$ is given by

$$P_{12} = 4p(1-p)\sin^2\psi \tag{6.40}$$

When the two diabatic potential curves cross as in Figure 11($a$), the Stokes constant $U_1$, the nonadiabatic transition probability $p$ for one passage of the

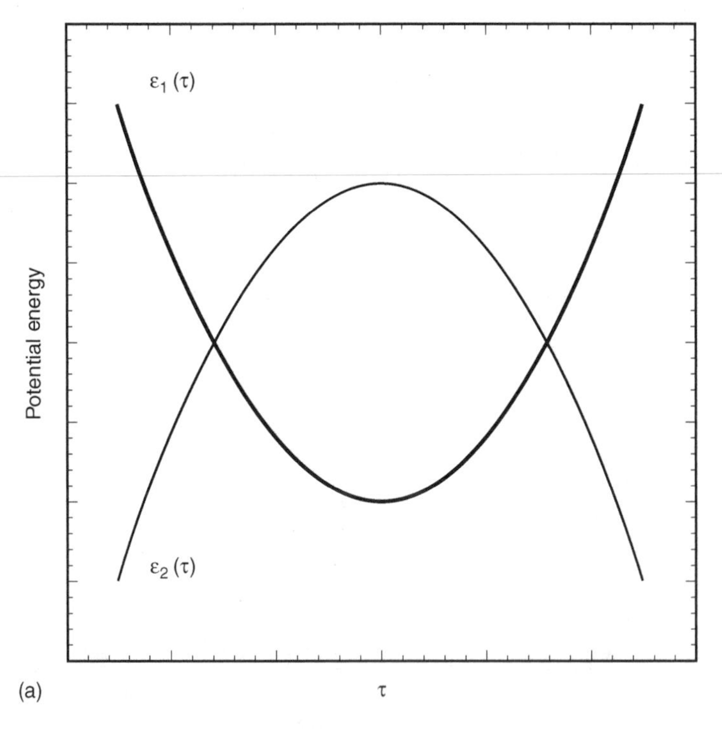


**Figure 11.** Schematic time-dependent quadratic potentials $\varepsilon_1(\tau)$ and $\varepsilon_2(\tau)$. ($a$) Crossing case. ($b$) Noncrossing case.

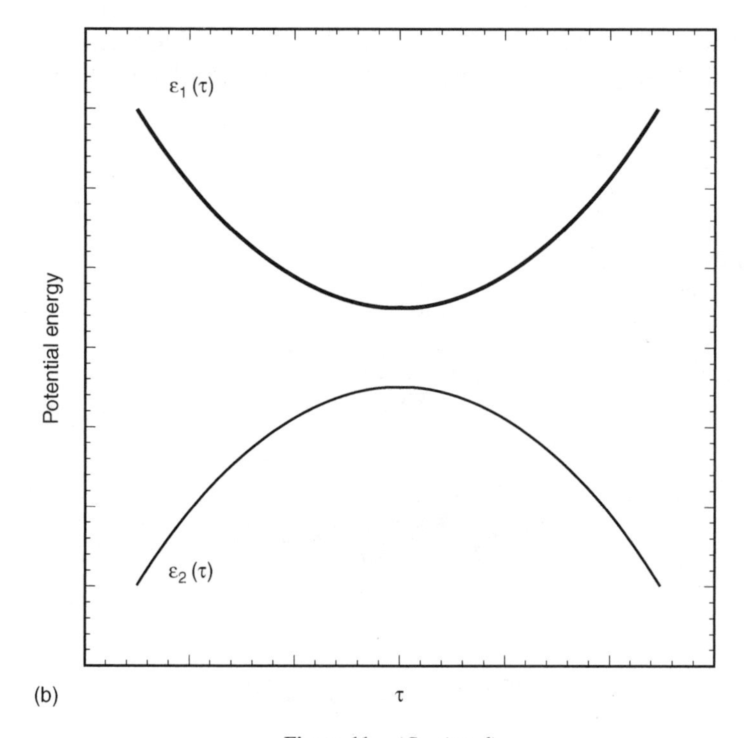


**Figure 11.** *(Continued)*

crossing, and the phase $\psi$, are the same as those given by Eqs. (3.41)–(3.43) with $\sigma$ and $\delta$ defined by

$$\sigma + i\delta = \frac{1}{\hbar}\int_{t_0}^{t^*} \Delta E(t)dt \tag{6.41}$$

with

$$\Delta E(t) \equiv E_+(t) - E_-(t) \tag{6.42}$$

where $E_\pm(t)$ are the adiabatic potentials obtained from the diabatic potentials as usual by

$$E_\pm(t) = \frac{1}{2}\{[\varepsilon_1(t) + \varepsilon_2(t)] \pm ([\varepsilon_1(t) - \varepsilon_2(t)]^2 + 4V_0^2)^{1/2}\} \tag{6.43}$$

The lower integral limit $t_0$ in Eq. (6.41) is the center of time between the two avoided crossings, and the upper limit $t^*$ is a complex solution of

$$\Delta E(t^*) = 0 \tag{6.44}$$

closest to the real axis.

In the case of general (nonquadratic) curve crossing problem, the parameters $\alpha$ and $\beta$ are given by

$$\alpha = \frac{\sqrt{d^2 - 1}\hbar^2}{2V_0^2(t_t^2 - t_b^2)} \tag{6.45}$$

$$\beta = -\sqrt{d^2 - 1}\frac{t_b^2 + t_t^2}{(t_t^2 - t_b^2)} \tag{6.46}$$

with

$$d^2 = \frac{[E_+(t_b) - E_-(t_b)][E_+(t_t) - E_-(t_t)]}{[E_+(t_x) - E_-(t_x)]^2} \tag{6.47}$$

and

$$V_0 = \tfrac{1}{2}[E_+(t_x) - E - (t_x)], \tag{6.48}$$

where $t_x$ is the real position at which $E_+(t) - E_-(t)$ becomes minimum, and $t_b(t_t)$ is the bottom(top) of the potential $E_+(\tau)(E_-(\tau))$, that is, the point where $(dE_+(t)/dt)$ $(E_-(t)/dt)$ becomes zero. If it is annoying to evaluate Eq. (6.41) directly, $\sigma$ and $\delta$ can be evaluated from $\alpha$, $\beta$, and $E_\pm(t)$ on the real axis by

$$\sigma + i\delta = \frac{1}{\hbar}\left[\int_0^{t_0} E_-(t)dt - \int_0^{t_0} E_-(t)dt + \Delta\right] \tag{6.49}$$

with

$$\Delta = \frac{\sqrt{2}\pi}{4\sqrt{\alpha}}\frac{1}{F_+^2 + F_-^2}[F_-^c + iF_+^c] \tag{6.50}$$

where $F_{\pm}^{c}$ are defined by Eqs. (3.46)–(3.50) with $a^2$ and $b^2$ replaced by $\alpha$ and $\beta$. The parameter $d^2$ is now given by Eq. (6.47). When the two crossings lie close together, $\alpha$ and $\beta$ are better estimated by the following expressions:

$$\alpha = \frac{\hbar^2 \sqrt{(\Delta E_M - \Delta E_m)(\Delta E_M + \Delta E_m)}}{(\Delta E_m)^3 t_m^2} \tag{6.51}$$

and

$$\beta = \frac{\sqrt{(\Delta E_M - \Delta E_m)(\Delta E_M + \Delta E_m)}}{\Delta E_m} \tag{6.52}$$

with

$$\Delta E_m = \tfrac{1}{2}\left[\Delta E(t_m^{(+)}) + \Delta E(t_m^{(-)}\right] \tag{6.53}$$

$$\Delta E_M = \Delta E(t_M) \tag{6.54}$$

and

$$t_m = \tfrac{1}{2}\left(t_m^{(+)} + t_m^{(-)}\right) \tag{6.55}$$

where $t_m^{(\pm)}$ are the positions at which $\Delta E(t) = E_2(t) - E_1(t)$ becomes minimum, and $t_M$ is the position at which $\Delta E(t)$ becomes maximum.

When, on the other hand, the two diabatic potentials do not cross as in Figure 11(*b*), we have to use the formulas that are valid at $E < E_X$, that is, Eqs. (3.51)–(3.57) with Eq. (6.41). Again, if it is difficult to evaluate Eq. (6.41), $\sigma$ and $\delta$ can be evaluated by Eq. (6.50) with $\beta < 0$.

The parameters $\alpha$ and $\beta$ are obtained by fitting the position of the minimum of $(\Delta E)^2$ by a quadratic polynomial as

$$\alpha = \hbar^2 \frac{XY}{(XZ - Y^2)^{3/2}} \tag{6.56}$$

and

$$\beta = -\frac{Y}{XZ - Y^2} \tag{6.57}$$

where

$$X = \frac{1}{24}\left\{\frac{d^4}{dt^4}[\Delta E(t)]^2\right\}_{t=0} \tag{6.58}$$

$$Y = \left\{\frac{d^2}{dt^2}[\Delta E(t)]^2\right\}_{t=0} \tag{6.59}$$

and

$$Z = [\Delta E(t)]_{t=0}^2 \tag{6.60}$$

In the numerical examples shown below [44], $\sigma$ and $\delta$ are evaluated with the use of Eqs. (3.14)–(3.17) of [44], which correspond to Eqs. (3.2)–(3.8) in [17]. Equation (6.50) is, however, finally recommended, since this is better as mentioned before [18]. The numerical examples shown here are not affected, since the values of $\alpha$ in the examples sit in the range where Eqs. (3.14)–(3.17) of [44] can work well. We would like to take this opportunity to show that Eq. (3.15) of [44] should contain $\sqrt{\beta/\alpha}$ and the term corresponding to the first term of Eq. (3.4) of [17].

Let us now consider, as an example, the following model:

$$\varepsilon_1(t) = -(A\cosh\omega t + B) \tag{6.61}$$

and

$$\varepsilon_2(t) = (A\cosh\omega t + B) \tag{6.62}$$

The coupling $V_0$ is taken to be constant. The adiabatic potentials are given by

$$E_{\pm} = \pm[(A\cosh\omega t + B)^2 + V_0^2]^{1/2} \tag{6.63}$$

There are an infinite number of complex crossing points given by

$$\omega t^* = \ln\left[\frac{-B\pm iV_0}{A} \pm \sqrt{\left(\frac{-B\pm iV_0}{A}\right)^2 - 1}\right] \pm 2\pi i \tag{6.64}$$

but here the crossing point nearest to the real axis is taken. This means that the other complex crossing points do not contribute very much to the transition probability. Numerical comparisons between the exact results and the semiclassical results are shown in Figure 12 for $A/V_0 = 1.0$ and $A/V_0 = 5.0$. The region $B/V_0 > -A/V_0$ corresponds to the non-crossing regime. This clearly demonstrates the usefulness of the present analytical formulas that cover nicely both crossing and noncrossing regimes even for nonquadratic problems.

As an additional example of general curved potentials with two crossing points, we take laser assisted surface ion neutralization (LASIN), which is a neutralization process induced by laser in the collision between an ion and a metal surface. We consider an electron transfer from a single valence bond state $\varepsilon_1$ of the surface to a state $\varepsilon_2$ of the atom. According to [52], this process can be described by the following coupled equations:

$$i\hbar \begin{bmatrix} \dot{c}_1(t) \\ \dot{c}_2(t) \end{bmatrix} = \begin{bmatrix} E_2(t) & g(t) \\ g(t) & E_1(t) + \hbar\eta \end{bmatrix} \begin{bmatrix} c_1(t) \\ c_2(t) \end{bmatrix} \tag{6.65}$$

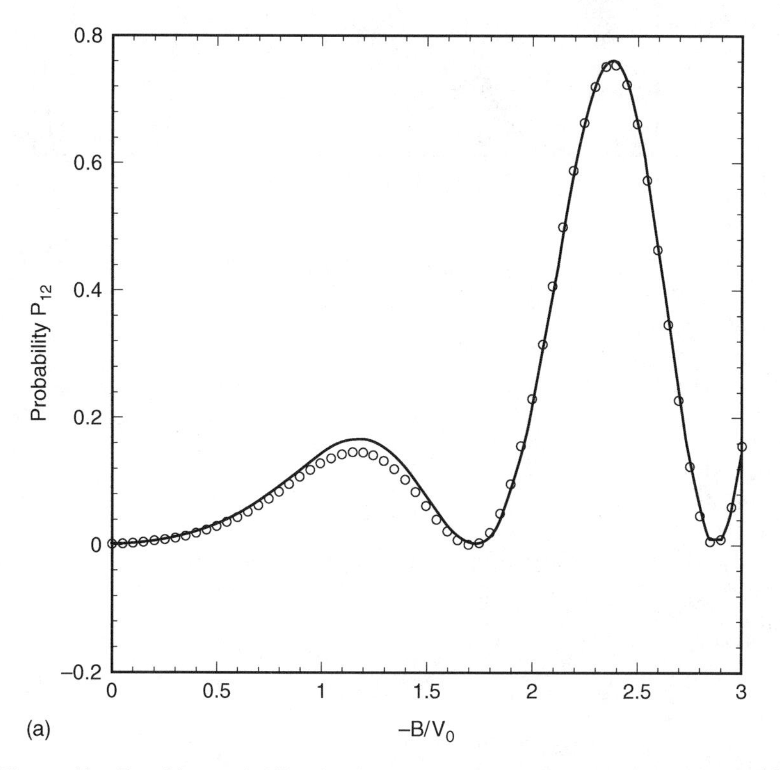


**Figure 12.** Transition probability in the case of hyperbolic cosine potential [$\Delta\varepsilon(\tau) = 2A/V_0 \cosh\tau + 2B/V_0$] against $B/V_0$. Solid line: exact numerical result, open circle: present semiclassical theory. (*a*) $A/V_0 = 1.0$ (*b*) $A/V_0 = 5.0$.

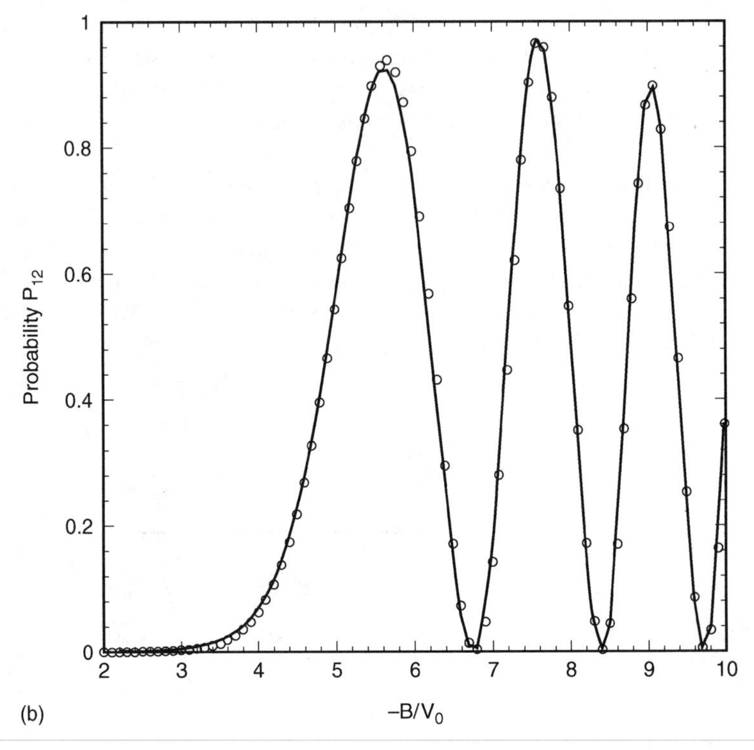


**Figure 12.** *(Continued)*

with

$$2E_1(t) = \varepsilon_1 + \varepsilon_2 - [4V^2(t) + \omega^2]^{1/2} \tag{6.66}$$

$$2E_2(t) = \varepsilon_1 + \varepsilon_2 + [4V^2(t) + \omega^2]^{1/2} \tag{6.67}$$

and

$$g(t) = \frac{1}{2} W_0 \omega f(\lambda t)/[4V_0^2(t)]f^2(\lambda t) + \omega^2)^{1/2} \tag{6.68}$$

where

$$V(t) = V_0 f(\lambda t) \tag{6.69}$$

$$f(\lambda t) = \text{sech}(\lambda t) \tag{6.70}$$

and $\omega$ is the energy defect defined by $\omega = \varepsilon_2 - \varepsilon_1 > 0$. The states $E_1(t)$ and $E_2(t)$ are the adiabatic states originating from $\varepsilon_1$ and $\varepsilon_2$ with the electronic hopping interaction potential $V(t)$ diagonalized. The state $E_1(t)$ is further dressed by one photon absorption to $E_1(t) + \hbar\eta$, where $\hbar\eta$ represents the photon energy. The quantity $g(t)$, given by Eq. (6.68), represents the interaction with a laser field after the first diagonalization procedure is carried out. The rotating wave approximation is also used.

The parameters $\sigma$ and $\delta$ are determined by the following complex integral:

$$\sigma + i\delta = \frac{1}{\hbar}\int_0^{t^*} dt[4g^2 + (E_2 - E_1 - \eta)^2]^{1/2} \tag{6.71}$$

By introducing the following dimensionless parameters:

$$X = \lambda t \tag{6.72}$$

$$r = \hbar\eta/2V_0 \tag{6.73}$$

$$q = \hbar\omega/2V_0 \tag{6.74}$$

$$v = 2V_0/\lambda \tag{6.75}$$

and

$$E = W_0/2V_0 \tag{6.76}$$

we can rewrite Eq. (6.71) as

$$\sigma + i\delta = v\int_0^{X^*} dX\left[\frac{E^2q^2f^2(X)}{f^2(X) + q^2} + \{[f^2(X) + q^2]^{1/2} - r\}^2\right]^{1/2} \tag{6.77}$$

where $X^*$ is the complex zero of the integrand, and is given by

$$X^* = \ln\left[\pm\frac{1}{\sqrt{x - q^2}} \pm \sqrt{\frac{1}{x - q^2} - 1}\right] \pm 2n\pi i \qquad (n = 0, 1, \ldots) \tag{6.78}$$

Here $x$ satisfies the following equation:

$$
\begin{aligned}
x^4 &+ (2E^2q^2 - 2r^2)x^3 + (E^4q^4 + 2E^2q^2r^2 + r^4 - 2E^2q^4)x^2 \\
&- (E^2q^6 + 2E^2q^4r^2)x + E^4q^8 = 0
\end{aligned}
\tag{6.79}
$$

It can be shown that the substitution of the real solutions of Eq. (6.79) gives $X^*$ with a large imaginary part, and that we need a complex solution of Eq. (6.79) to find $X^*$ closest to the real axis.

In the case of a weak laser field discussed in [52], the parameter $\alpha$ is extremely large ($\simeq 10^4$), and the ordinary simple perturbation theory with respect to $g$ works well. In order to demonstrate the effectiveness of the present semiclassical theory, we have chosen a much stronger laser field($\alpha \simeq 0.7$) for which the perturbation theory does not work at all. Figure 13 shows the results for $E = 0.25$, $q = 0.2923$, and $v = 23.585$. The present theory can nicely cover the diabatically avoided crossing region ($r > 1.042$). Note that when the parameter

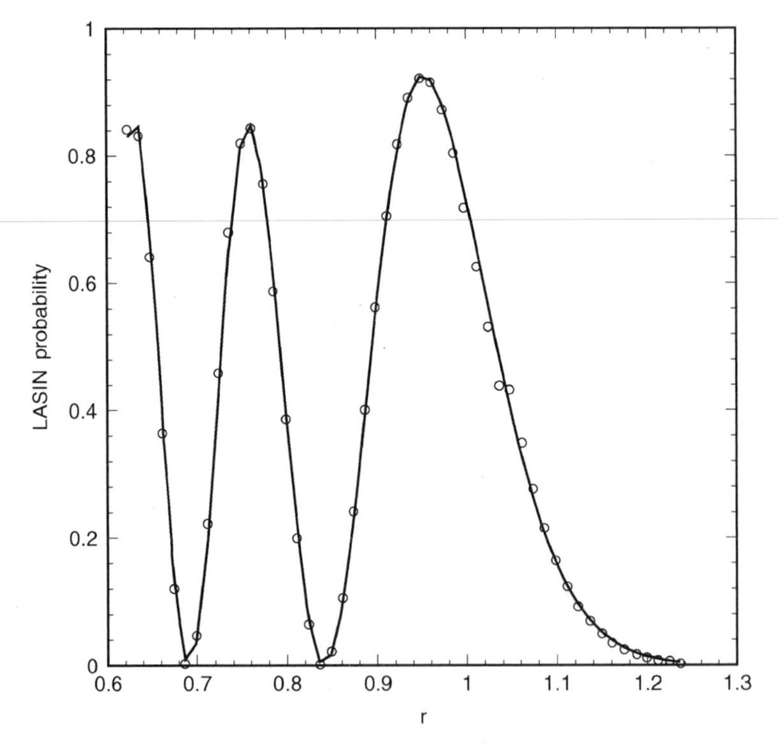


**Figure 13.** Transition probability as a function of the dimensionless laser photon energy $r$ in the case of LASIN. The parameters chosen are $v = 23.585$, $q = 0.29225$, and $E = 0.25$. The region $r > 1.04183$ corresponds to the diabatically avoided crossing.

$\alpha$ is extremely large ($\alpha > 10^3$) for which the perturbation theory works well, the present semiclassical theory produces an erroneous sharp peak near the point where the two diabatic potentials touch each other. This is because of the empirical correction, but the theory still gives an overall correct behavior if we neglect this sharp peak.

Multilevel problems and/or problems with more than two avoided crossings can be treated by dividing the total process into nonadiabatic transitions and adiabatic propagation, and total transition probability is given by the multiplication of transition matrices and adiabatic propagation matrices. This idea itself is not new at all. By utilizing our new formulas shown above, however, it is possible to treat diabatically avoided crossing and two closely lying avoided crossings as one unit. For isolated avoided crossing, the transition matrix $I_X$ is given by

$$I_X \equiv \begin{bmatrix} \sqrt{1-p}e^{i\phi_s \to (\psi - \sigma)} & -\sqrt{p}e^{i\sigma_0} \\ \sqrt{p}e^{-i\sigma_0} & \sqrt{1-p}e^{-i\phi_s \to (\psi - \sigma)} \end{bmatrix} \tag{6.80}$$

where $\psi$, $\sigma$ and $p$ are given, respectively, by Eqs. (3.43), (3.44) and (3.42) or by Eqs. (3.54), (344) and (3.53), and $\sigma_0$ is given by

$$\sigma_0 = \mathrm{Re}\left[\frac{1}{\hbar}\int_{\Re t^*}^{t^*} \Delta E(t)dt\right] \tag{6.81}$$

For an diabatically avoided crossing or a pair of two closely lying avoided crossings, the transition matrix is $T_{12}^R$ given by Eq. (6.37). The adiabatic propagation matrix $X$ is given by

$$X \equiv \begin{bmatrix} e^{-i\int_{t_A}^{t_B} E_+(t)dt} & 0 \\ 0 & e^{-i\int_{t_A}^{t_B} E_-(t)dt} \end{bmatrix} \tag{6.82}$$

As an example of a two state multicrossing problem, we take here the periodic crossing model in which the diabatic levels,

$$\varepsilon_1 = -(A \sin \omega t + B) \tag{6.83}$$

and

$$\varepsilon_2 = (A \sin \omega t + B) \tag{6.84}$$

are coupled by a constant coupling $V_0$. This model is frequently discussed for the systems with a periodic external field (Zener tunneling in a periodic external field is one of the examples). The adiabatic potentials of this model is given by

$$E_{\pm} = \pm[(A\sin\omega t + B)^2 + V_0]^{1/2} \tag{6.85}$$

and the complex crossing points closest to the real axis are expressed as

$$t^* = \frac{i}{\omega}\ln\left[\frac{V_0 - iB}{A} + \sqrt{\left(\frac{-iB \pm iV_0}{A}\right)^2 + 1}\right] \tag{6.86}$$

Finally, the total transition matrix is expressed as

$$T = I_X X I_X X I_X X \cdots \tag{6.87}$$

The matrix $X$ represents the adiabatic propagation between crossing points. When the diabatic potentials avoid crossing, we have

$$T = T^R X T^R X \cdots \tag{6.88}$$

The results are shown in Figure 14 for $A/V_0 = 15.0, B/V_0 = 0.0$ [Fig. 14($a$)] and $B/V_0 = 18.0$ [Fig. 14($b$)]. The region $B/V_0 > A/V_0$ corresponds to the noncrossing regime, and the famous LZ formula does not work at all there. Our theory, on the other hand, works well in all regions of $A/V_0$ and $B/V_0$. Even the time-evolution of the whole process can be well reproduced in the adiabatic representation.

The matrix multiplication method explained in Section IV.A can be easily extended to multilevel crossing problems. The matrices $I_X$, $T^R$, and $X$ become $N \times N$ matrices, where $N$ is the number of levels. From the analysis of the time-independent multi-channel curve crossing problems [24, 25], we can safely expect that this kind of two-by-two approximation works well, because our basic two-state theory is very accurate. The matrix $I_X$ (or $T^R$) contains a $2 \times 2$ submatrix given by Eq. (6.80) [or Eq. (3.60)], representing a transition at the relevant crossing; otherwise this matrix is diagonal. For the evaluation of the basic parameters $\sigma$ and $\delta$, the following three methods are possible. The first and the most accurate one is to diagonalize the whole $N \times N$ potential matrix to obtain fully adiabatic potentials and then evaluate the complex contour integral given by Eq. (6.41) for each relevant avoided crossing. This is quite difficult, unfortunately, because practically it is very hard to find complex crossing points accurately, especially when the number of states $N$ exceeds three. The simplest, yet still good method is the two-by-two diabatic approach, in which only the

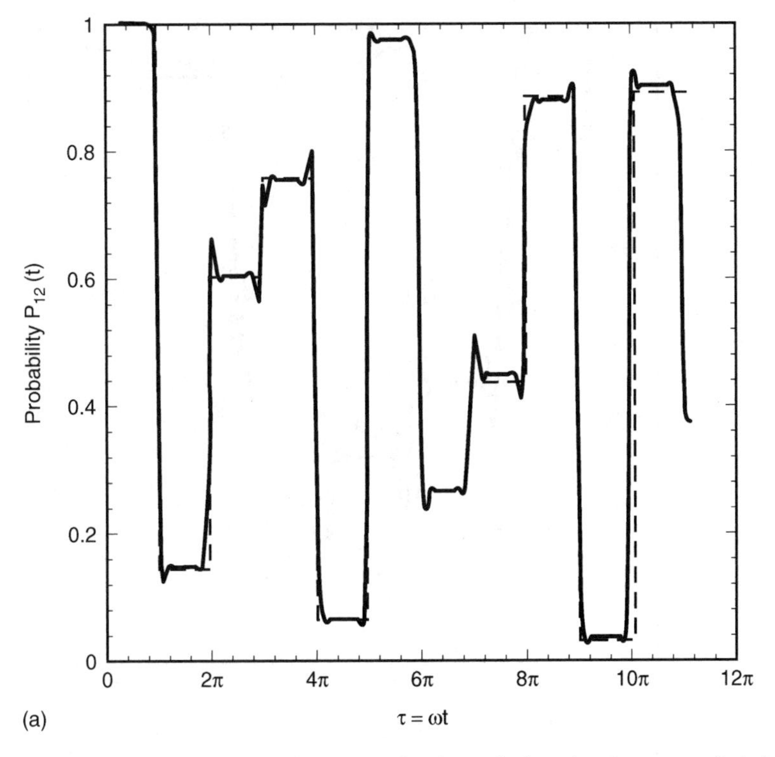


**Figure 14.** Transition probability as a function of time in the case of $\Delta\varepsilon(\tau) = 2A/V_0 \sin(\tau) + 2B/V_0$ with $A/V_0 = 15.0$. Solid line: exact, dashed line: present semiclassical theory. The region $B/V_0 > A/V_0$ corresponds to the noncrossing regime. (*a*) $B/V_0 = 0.0$, (*b*) $B/V_0 = 18.0$.

relevant two states are considered at each crossing. In the two-state case, it is, of course, not difficult to evaluate the integral in Eq. (6.41). If the couplings at other crossings are strong and affect the relevant crossing, however, this method naturally breaks down. The third method is to evaluate the parameters σ and δ from fully adiabatic potentials on the real axis only [Eqs. (6.49)–(6.50)]. This method can avoid the annoying complex calculus and is quite convenient, being more accurate than the two-by-two diabatic approach.

Now we take the following three-level problem:

$$H = \begin{bmatrix} -a_1 t^2 + b_1 & V_{12} & V_{13} \\ V_{12} & a_2 t^2 & 0 \\ V_{13} & 0 & a_3 t^2 + b_3 \end{bmatrix} \tag{6.89}$$

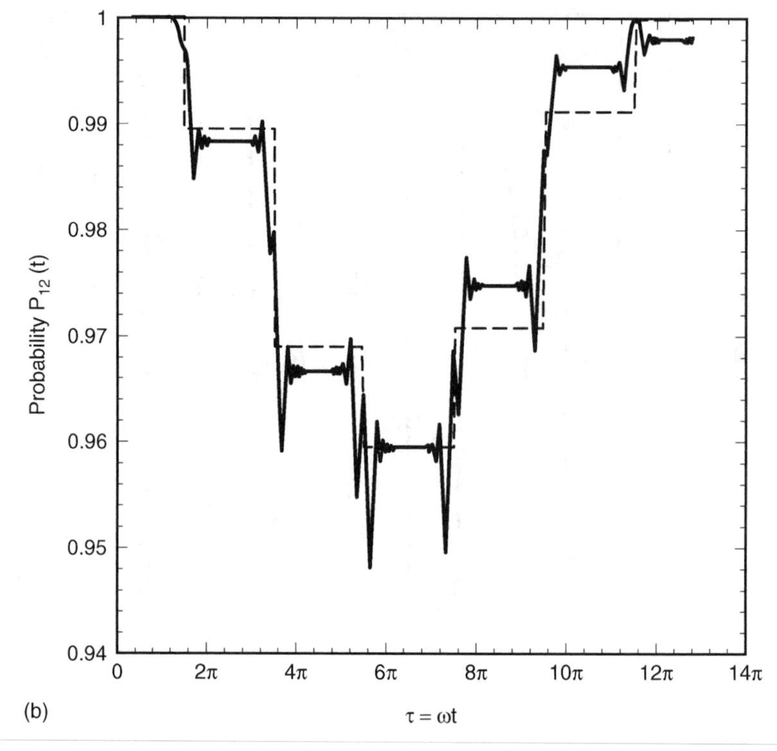


**Figure 14.** *(Continued)*

where $V_{ij}$ are constant couplings and $a_1, a_2, a_3, b_1, b_3 > 0$. The dimensionless parameters defined as $\alpha_j \equiv \hbar^2 a_j / 8V_{12}^3$, $\beta_j \equiv b_j / 2V_{12}$ are taken to be $\alpha_1 = 0.05$, $\alpha_2 = 0.061$, $\alpha_3 = 0.111$, $\beta_1 = 1.66$, and $\beta_3 = 1.0$. Figure 15 depict the diabatic and adiabatic potentials for (*a*) $V_{13}/V_{12} = 1.0$ and $V_{13}/V_{12} = 0.2$. Figure 16(*a*)–(*c*) shows the results of the three methods mentioned in Section III.C in comparison with the exact results as a function of $V_{13}/V_{12}$. The method based on the complex crossing points in the full adiabatic representation works very well even in a strong coupling region [see Fig. 16(*a*)]. The two-by-two diabatic method, on the other hand, dose not work well in a strong coupling region (see Fig. 16(*b*)]. The third method, which employs the parameters given by Eqs. (6.45)–(6.50) evaluated from the full adiabatic potentials on the real axis, gives much better results than the two-by-two diabatic method, even though the required computational effort is not much at all. This method works well until the full adiabatic potentials become flat [as in Fig. 15(*a*)] and the

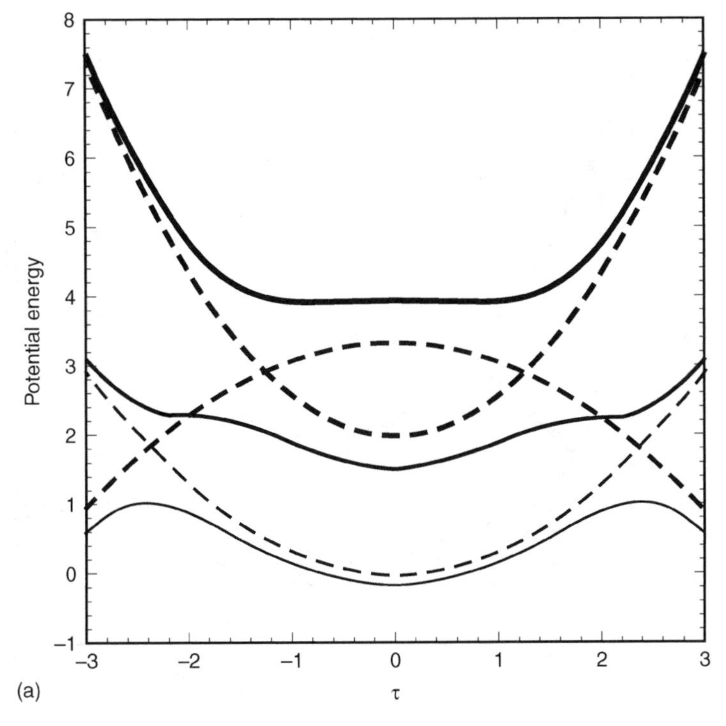


**Figure 15.** Diabatic (dashed line) and full adiabatic (solid line) potentials in the case of three-level problems of Eq. (6.89) with (*a*) $V_{13}/V_{12} = 1.0$ and (*b*) $V_{13}/V_{12} = 0.2$. The ordinate and abscissa are the scaled energy, $\alpha_j\tau^2 + \beta_j$ and the scaled time, $\tau = 2V_{12}t/\hbar$, respectively [see the text below Eq. (6.89)].

parameters cannot be estimated. Since the search of complex crossing points becomes extremely difficult when the number of states exceeds three, the third method based on Eqs. (6.45)–(6.50) is recommended.

## C. Other Models

Theoretical studies of time-dependent curve crossing problems naturally have a long history [6, 40, 41]. After the famous LZ and RZ models, various types of multilevel as well as two-state problems have been studied. Here, some important models are summarized.

Nikitin's exponential model [6] is a sort of mixture of LZ and RZ models, since it describes the nonadiabatic transition due to the time variations in both

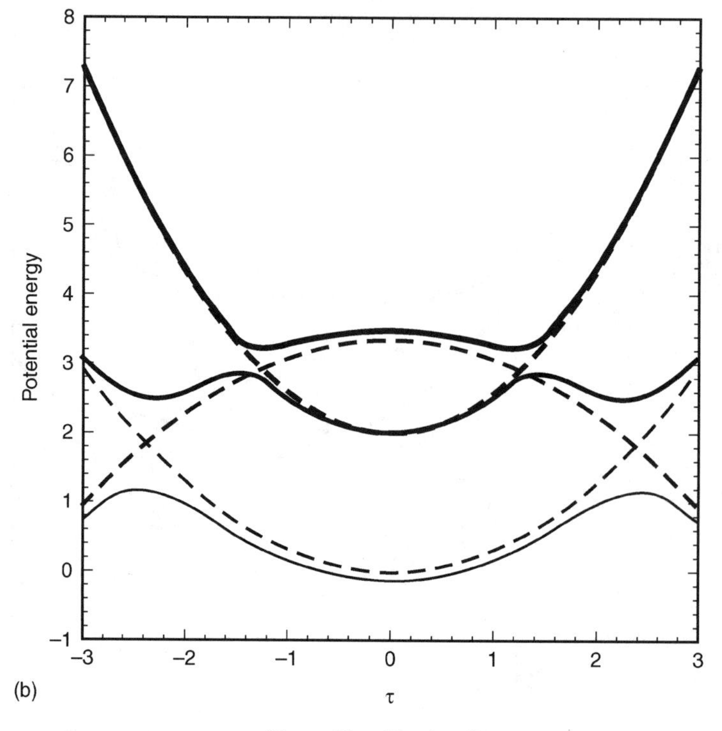


**Figure 15.** *(Continued)*

diabatic energy and diabatic coupling. The Hamiltonian of this model in the diabatic representation is given by [cf. Eq. (133)]

$$H_{\exp} = \begin{bmatrix} U_1 + V_1 e^{-\beta t} & V e^{-\beta t} \\ V e^{-\beta t} & U_2 + V_2 e^{-\beta t} \end{bmatrix} \tag{6.90}$$

The Shrödinger equation for this model can be solved in terms of confluent hypergeometric function, and the nonadiabatic transition probability is given by

$$p_{\exp} = \exp(-\pi\delta_2)\sinh(\pi\delta_1)/\sinh[\pi(\delta_1 + \delta_2)] \tag{6.91}$$

where

$$\delta_1 = \frac{U_1 - U_2}{2\hbar\beta}\left[1 + \frac{(V_1 - V_2)/2V}{\sqrt{1 + \left(\frac{V_1 - V_2}{2V}\right)^2}}\right] \tag{6.92}$$

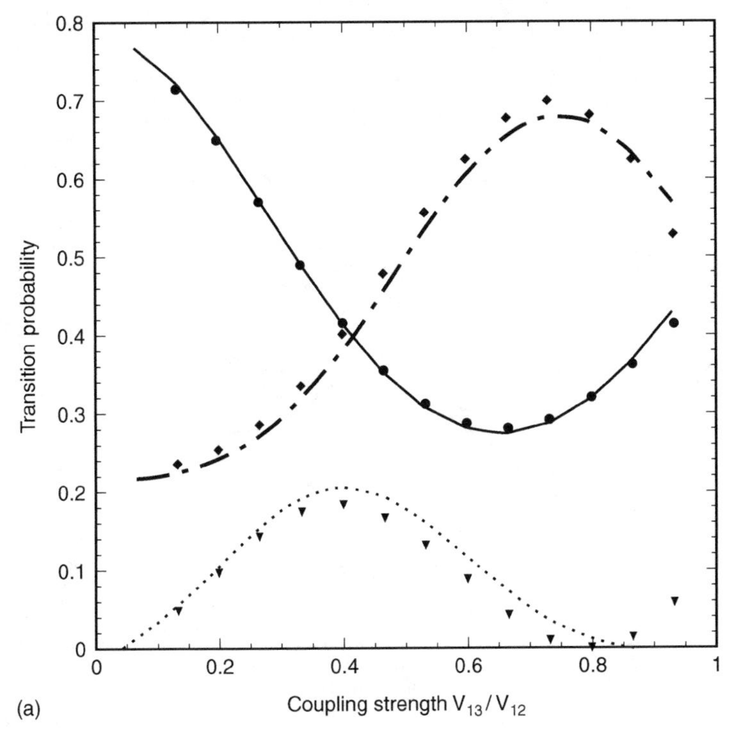


**Figure 16.** Transition probabilities against the coupling strength $V_{13}/V_{12}$ in the case of the three-level problem given in Eq. (6.89). Solid line: $P_{1\to1}$ (exact), dash–dot line: $P_{1\to2}$ (exact), dotted line: $P_{1\to1}$ (exact), solid circle: $P_{1\to1}$ (semiclassical), solid rhomb: $P_{1\to2}$ (semiclassical), solid triangle: $P_{1\to3}$ (semiclassical). (*a*) The semiclassical theory is based on the complex crossing points in the full adiabatic representation. (*b*) The semiclassical theory is based on the two-by-two diabatic approach. (*c*) The semiclassical theory is based on the parameter defined by Eqs. (6.45)–(6.50).

and

$$\delta_2 = \frac{U_1 - U_2}{2\hbar\beta}\left[1 - \frac{(V_1 - V_2)/2V}{\sqrt{1 + \left(\frac{V_1 - V_2}{2V}\right)^2}}\right] \tag{6.93}$$

Another important two-state model may be the one by Demkov–Kunike, in which two hyperbolic tangent diabatic levels are coupled by a hyperbolic secant potential [36].

The Demkov–Osherov model [35] is a generalization of the LZ model for multilevel problem. In this model, a band of flat parallel levels with an arbitrary

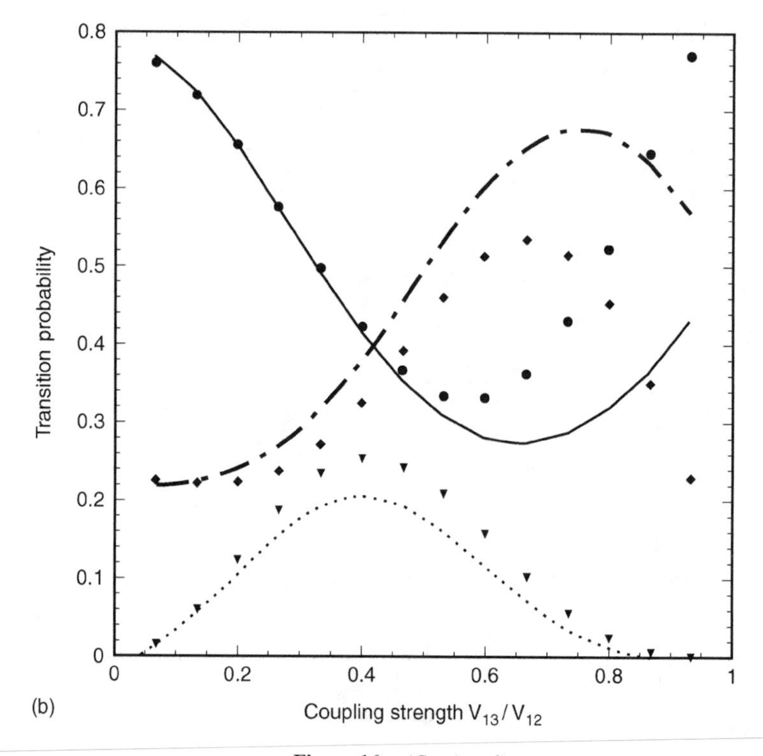


**Figure 16.** *(Continued)*

spacing cross another slant level. In this model, the various transition probabilities are simply expressed as a product of $p_{\mathrm{LZ}}$ or $1 - p_{\mathrm{LZ}}$, where $p_{\mathrm{LZ}}$ is the LZ probability. This is exact because of the peculiarity of the model. The next step of generalization was proposed by Demkov and Ostrovsky [37]. In their model, two bands of parallel levels cross each other, namely, the diabatic levels are given by

$$H_{an,an} = \beta_a t + \omega_{an} \tag{6.94}$$

where the index $a (= 1, 2)$ labels the band, and $n$ labels the states in each band, and the diabatic coupling is nonzero (constant) only between the levels belonging to different bands.

The bowtie model is another type of multilevel system, in which many linear potentials cross at one point. After the three-level case of this model was solved

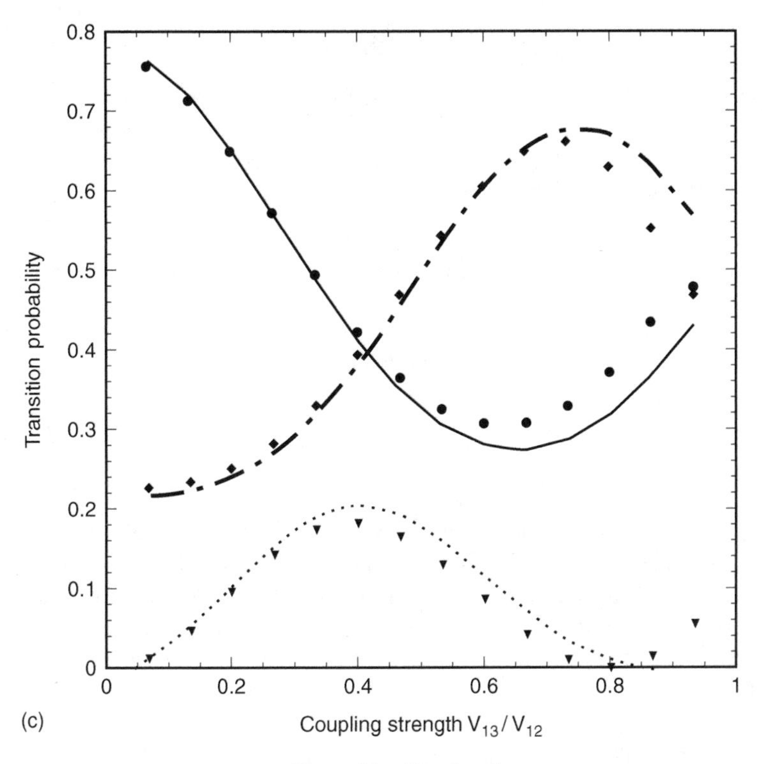


**Figure 16.** *(Continued)*

by Carroll and Hioe [43], Ostrovsky and Nakamura [38] obtained the general solution for the $N$-level case where the Hamiltonian is given by

$$H_{0,0} = 0 \tag{6.95}$$

$$H_{i,i} = \beta_j t \tag{6.96}$$

$$H_{j,0} = H_{0j} = V_j (j \neq 0) \tag{6.97}$$

and

$$H_{j,k} = 0 (k \neq 0, j \neq 0) \tag{6.98}$$

where

$$\ldots \beta_{-3} < \beta_{-2} < \beta_{-1} < \beta_0 < \beta_1 < \beta_2 < \ldots \tag{6.99}$$

The probability $P_{ij}$ of the transition from the diabatic state $i$ to the diabatic state $j$ is exactly obtained. For example, $P_{00}$ is given by

$$P_{00} = \left[1 - \prod_{n>0} p_n - \prod_{n<0} p_n\right]^2 \tag{6.100}$$

where $p_j$ is given by the LZ formula, namely,

$$p_j = \exp\left[-\frac{\pi V_j^2}{|\beta_j|}\right] \tag{6.101}$$

The general transition probability $P_{ij}$ for arbitrary $i$ and $j$ can also be written in terms of $p_k$ and $1 - p_k$, and the physical interpretation can be provided [42]. It is interesting to note that the total transition probability for both the bowtie and Demkov–Osherov model can be simply given in terms of a product of the two-state LZ probability. This, however, does not hold for general multilevel problems because of the interference effect and deviation from the linear model. As described in the Section VI.C, the present theory works well for the general case, since it not only contains the interference effect but also can deal with curved potentials.

## VII. NEW WAY OF CONTROLLING MOLECULAR PROCESSES BY TIME-DEPENDENT EXTERNAL FIELDS

### A. Basic Theory

In this section, being based on the developments in the theory of nonadiabatic transition discussed above, we propose a new idea [46, 47] to control nonadiabatic processes so that an overall transition probability to any specified state in a multichannel curve crossing system becomes unity. This can be realized by periodically sweeping the external field in time at the crossing point. By periodically changing the field, either field strength or the frequency, we can use not only the nonadiabatic transition probability for one passage of crossing point but also the various phases and the number of sweeping periods as control parameters. By taking a simple two-state curve crossing as a function of time (see Fig. 17), we explain and formulate our basic idea. It should be emphasized that the theory proposed here can be applied to general multichannel problems as far as the crossings are relatively well separated.

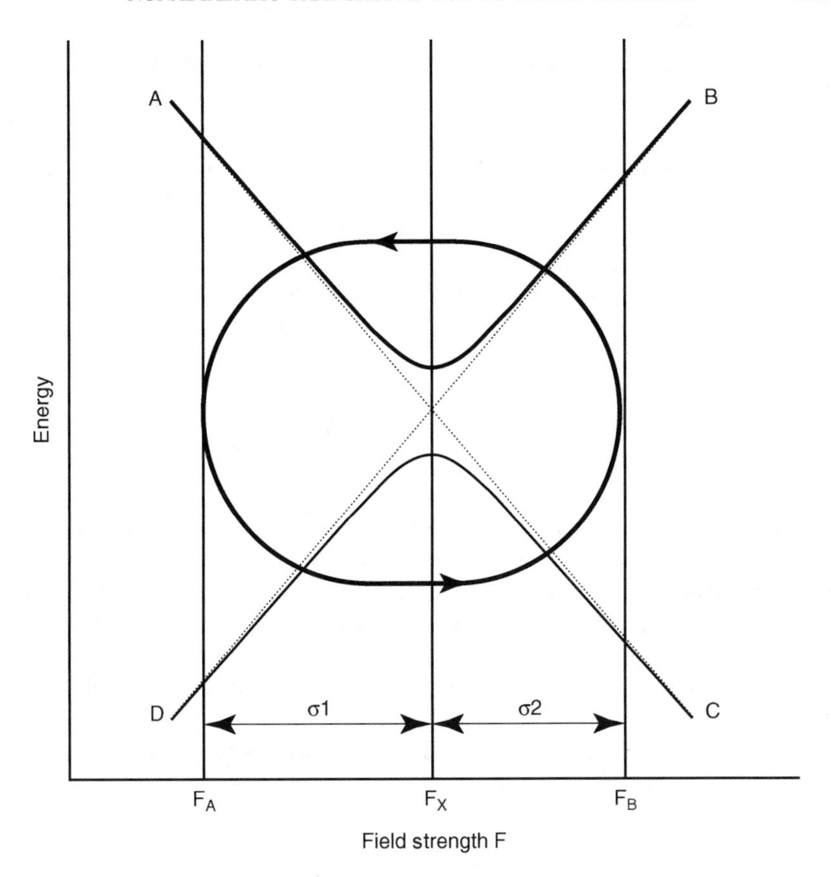


**Figure 17.** Schematic two diabatic (dotted lines) and two adiabatic (solid lines) potentials. External field oscillates between $F_a$ and $F_b$, striding the avoided crossing point $F_x$. The phase $\sigma_1$ ($\sigma_2$) can be controlled by changing $F_a(F_b)$.

The transition matrix $I$, which describes the transition from $F_a$ to $F_b$ (see Fig. 17), is given by

$$I = \begin{bmatrix} \sqrt{1-p}e^{i(\phi_s+\sigma_1/2+\sigma_2/2)} & \sqrt{p}e^{i(\sigma_0-\sigma_1/2+\sigma_2/2)} \\ -\sqrt{p}e^{-i(\sigma_0-\sigma_1/2+\sigma_2/2)} & \sqrt{1-p}e^{-i(\phi_s+\sigma_1/2+\sigma_2/2)} \end{bmatrix} \tag{7.1}$$

while the transpose of this matrix, $I^t$, describes the backward transition from $F_b$ to $F_a$. Here, $p$ represents the nonadiabatic transition probability by one passage of the crossing point $F_x$, $\phi_s$ is the Stokes phase, and $\sigma_0$, $\sigma_1$, and $\sigma_2$ are the phase factors which are defined, respectively, by

$$\sigma_0 = \mathrm{Re}\left(\int_{F_x}^{F_*} \Delta E(F)\frac{dt}{dF}dF\right) = \mathrm{Re}\left[\int_{t_x}^{t_*} \Delta E(t)dt\right] \tag{7.2}$$

$$\sigma_1 = \int_{F_a}^{F_x} \Delta E(F)\frac{dt}{dF}dF = \int_{t_a}^{t_x} \Delta E(t)dt \tag{7.3}$$

and

$$\sigma_2 = \int_{F_x}^{F_b} \Delta E(F)\frac{dt}{dF}dF = \int_{t_x}^{t_b} \Delta E(t)dt \tag{7.4}$$

where $\Delta E(F)$ is the adiabatic energy difference at the field parameter $F$, $F_*$ is the solution of $\Delta E(F_*) = 0$, and $F_x$ is the field parameter corresponding to the diabatic crossing point. The time $t_\alpha$ for $\alpha = a$, $b$, $x$, and $*$ is the time at which $F(t_\alpha) = F_\alpha$ is satisfied. Since $\Delta E \neq 0$ on the real axis, $F_*$ and $t_*$ are complex numbers. Note that the compact analytical expressions of these quantities are available even for the cases where the two diabatic potentials tangentially touch each other ($F_b = F_x$) or avoid crossing ($F_b < F_x$).

The final overall transition matrix $T_n$ after $n$ periods of oscillation between $F_a$ and $F_b$ is expressed as

$$T_n = T^n \tag{7.5}$$

where $T$ is the transition matrix for one period, which is given by

$$T \equiv I^t I = \begin{bmatrix} \{p + (1-p)e^{2i\psi}\}e^{-i\sigma} & -2i\sqrt{p(1-p)}\sin\psi \\ -2i\sqrt{p(1-p)}\sin\psi & \{p + (1-p)e^{-2i\psi}\}e^{i\sigma} \end{bmatrix} \tag{7.6}$$

with $\psi \equiv \phi_s + \sigma_0 + \sigma_2$ and $\sigma \equiv 2\sigma_0 + \sigma_2 - \sigma_1$.

In the case of $n$ and half periods of traversing the crossing point, the overall transition matrix is given by

$$T_{n+1/2} = I(I^t I)^n = IT^n \tag{7.7}$$

Note that the adiabatic potentials, and thus the parameters $p$, $\psi$, and $\sigma_i (i = 0 \sim 2)$, are dependent on the external field. Roughly speaking, the nonadiabatic transition probability $p$, the Stokes phase $\phi_s$, and the phase $\sigma_0$ are dependent on the local functionality of the adiabatic potentials around the crossing point, namely, the sweep velocity ($dF/dt$) of the external field at the crossing point; while, the phase factors $\sigma_1$ and $\sigma_2$ are dependent on the global functionality of the adiabatic potentials in the range ($F_a$, $F_b$) of the field. We try to find conditions for the parameters ($n, p, \psi, \sigma_i$ ($i = 1 \sim 2$)) to satisfy

$$P_{12}^{(n)} \equiv |(T_n)_{12}|^2 = 0 \text{ or } 1 \tag{7.8}$$

or

$$P_{12}^{(n+1/2)} \equiv |(T_{n+1/2})_{12}|^2 = 0 \text{ or } 1 \tag{7.9}$$

By using the Lagrange–Sylvester formula, we obtain

$$T_n = T^n = \frac{\lambda_+\lambda_-(\lambda_-^{n-1} - \lambda_+^{n-1})}{\lambda_+ - \lambda_-}E + \frac{\lambda_+^n - \lambda_-^n}{\lambda_+ - \lambda_-}T \tag{7.10}$$

where $E$ is the unit matrix and $\lambda_\pm$ are the eigenvalues of $T$, which are given by

$$\lambda_\pm = e^{\pm i\xi} \tag{7.11}$$

where

$$\cos\xi = (1-p)\cos(2\psi - \sigma) + p\cos(\sigma) \tag{7.12}$$

The unitarity of the matrix $T$ requires $\xi$ to be real. Equation (7.12) implies that the nonadiabatic transition probability $p$ should satisfy

$$\frac{1 - |\cos\xi|}{2} \leq p \leq \frac{1 + |\cos\xi|}{2} \tag{7.13}$$

Then, the requirements of Eq. (7.8) lead, respectively, to

$$\begin{aligned} P_{12}^{(n)} &= \left|\frac{\lambda_+^n - \lambda_-^n}{\lambda_+ - \lambda_-}T_{12}\right|^2 \\ &= 4\frac{\sin^2(n\xi)}{\sin^2\xi}p(1-p)\sin\psi = 0 \end{aligned} \tag{7.14}$$

or

$$P_{12}^{(n)} = 4\frac{\sin^2(n\xi)}{\sin^2\xi}p(1-p)\sin^2\psi = 1 \tag{7.15}$$

In the case of Eq. (7.14), we simply have the condition $\sin(n\xi) = 0$ or $\sin\psi = 0$. It is more interesting and worthwhile to consider Eq. (7.15). If $P_{12}^{(n)} = 1$ holds for $n$ periods, then $P_{12}^{(2n)} = 0$ must be satisfied for $2n$ periods; thus Eq. (7.15) for $n$ may be divided into the following two conditions:

$$\sin(2n\xi) = 0 \tag{7.16}$$

and

$$4p(1-p)\sin^2\psi = \sin^2\xi \tag{7.17}$$

Since $\sin(n\xi) \neq 0$ in general because of Eq. (7.15), Eq. (7.16) leads to

$$\sin^2(n\xi) = 1 \tag{7.18}$$

This equation determines $\xi$ for a given $n$, and Eq. (7.17) gives a condition for $p$ and $\psi$ for a given $\xi$. The phase $\sigma$ may be determined from Eq. (7.12). Equation (7.17) implies that $p$ must satisfy the following condition:

$$p(1-p) \geq \frac{1}{4}\sin^2\xi \tag{7.19}$$

which is the same as Eq. (7.13).

The above analysis can be summarized as follows: (1) For a given system, the nonadiabatic transition probability $p$ is estimated as a function of the external field. (2) From Eq. (7.18), $\xi$ is determined for an appropriately specified $n$. If $p$ is not in the range of Eq. (7.13), the external field (mainly the sweep velocity) and/or $n$ should be modified so that this condition is satisfied. (3) The phase $\psi$ is controlled by changing $\sigma_1$ to satisfy Eq. (7.17), while $\sigma$ is controlled by changing $\sigma_2$ to satisfy Eq. (7.12). These can be realized by adjusting the oscillation period $n$ and the range $(F_a, F_b)$ of the field. When the above procedure is completed, then we can achieve $P_{12}^{(n)} = 1$. The required range of $p$ as a function of $\xi$ given by Eq. (7.13) can be easily known and be used to search for appropriate conditions for the parameters.

Let us next consider $n$ and half-periods of oscillation of the external field [see Eq. (7.9)]. This case together with the $n$ period case discussed above is useful to treat general multilevel problems, since this enables us to follow any specified path from any initial state to any desirable final state. If either one of the conditions of Eq. (7.9) is satisfied for $n$ and half-periods, then $P_{12}^{(2n+1)} = 0$ must be satisfied for $2n + 1$ periods. Then, from Eq. (7.14), we have

$$\sin^2[(2n+1)\xi] = 0 \tag{7.20}$$

Now, the condition $P_{12}^{(n+1/2)} = 0$ can be explicitly expressed as

$$4(1-p)\sin^2(\psi - \sigma) = \frac{\sin^2\xi}{\sin^2(n\xi)} \tag{7.21}$$

which tells

$$(1-p) \geq \frac{\sin^2\xi}{4\sin^2(n\xi)} \tag{7.22}$$

On the other hand, the condition $P_{12}^{(n+1/2)} = 1$ can be reduced to the following equation:

$$4p\sin^2(\psi - \sigma) = \frac{\sin^2\xi}{\sin^2(n\xi)} \tag{7.23}$$

which implies

$$p \geq \frac{\sin^2\xi}{4\sin^2(n\xi)} \tag{7.24}$$

From Eqs. (7.20), (7.22), and (7.24), we end up with the same condition for the range of $p$ as Eq. (7.13).

Search for the best condition of the parameters can be done in the same way as in the $n$-period case. In the case of the requirement $P_{12}^{(n+1/2)} = 0$, Eq. (7.18) in the above step (2) should be replaced by Eqs. (7.20), and (7.17) has to be replaced by Eq. (7.21). In the case of $P_{12}^{(n+1/2)} = 1$, on the other hand, Eqs. (7.20) and (7.23) take the place of Eqs. (7.18) and (7.17) in the procedure.

An example of multilevel crossing is shown in Figure 18, which is taken from the quantum tunneling of the magnetization of $Mn_{12}Ac$ in a time-dependent magnetic field [67, 97, 98]. Here, we consider the lowest three adiabatic states 1–3, and demonstrate control of the nonadiabatic processes by our idea presented above. The Hamiltonian is taken from Eq. (1) of [97]. In Figure 18, the energy is scaled by the anisotropy energy $D$. Figure 19 shows the time evolution of the state probability from the point "a" on state 1 to "b" on state 3 via two avoided crossings A and B. At the avoided crossing A(B) four-(five-) period oscillation of the field is applied. This is shown in Figure 19($a$). The probability of the state 1($P_1$) becomes zero after the four periods, as is seen in Figure 19($b$). The probability $P_2$ reaches unity when $P_1$ becomes zero, and after five periods at B it becomes zero [Fig. 19($c$)] at which time $P_3$ reaches unity [Fig. 19($d$)]. Figure 20 demonstrates another path from "a" on state 1 to "c" on state 3

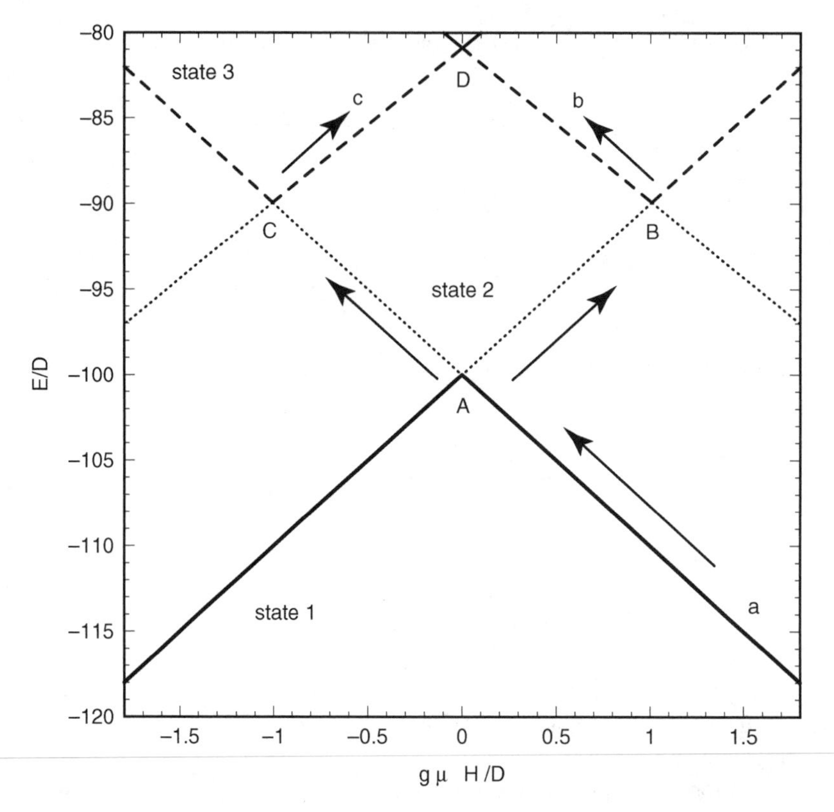


**Figure 18.** Adiabatic spin states as a function of an external magnetic field, $g\mu_\beta H$: three lowest levels (scaled by the anisotropy energy D) are shown, where $g$ is the Landé $g$ factor, $\mu_B$ is the Bohr magneton, and $H$ is the magnetic field. The nonadiabatic probabilities $p$ at avoided crossings A–C are 0.039, 0.977, and 0.977, respectively.

via two avoided crossings A and C. In this case, we have applied four and half (five)-periods of oscillation of the field at the avoided crossing A(C). This example clearly demonstrates that we can choose any path to reach any specified final state with unity probability.

Figures 19 and 20 are the results of a numerical solution of the coupled Schrödinger equations; but we have confirmed that the semiclassical theory developed by the present authors [44] based on the present new theory for the time-dependent nonadiabatic transition gives the results almost indistinguishable from Figure 19(*b*)–(*d*) and Figure 20(*b*)–(*d*) except for the humps and dips that appear when the probability jumps abruptly. This guarantees that

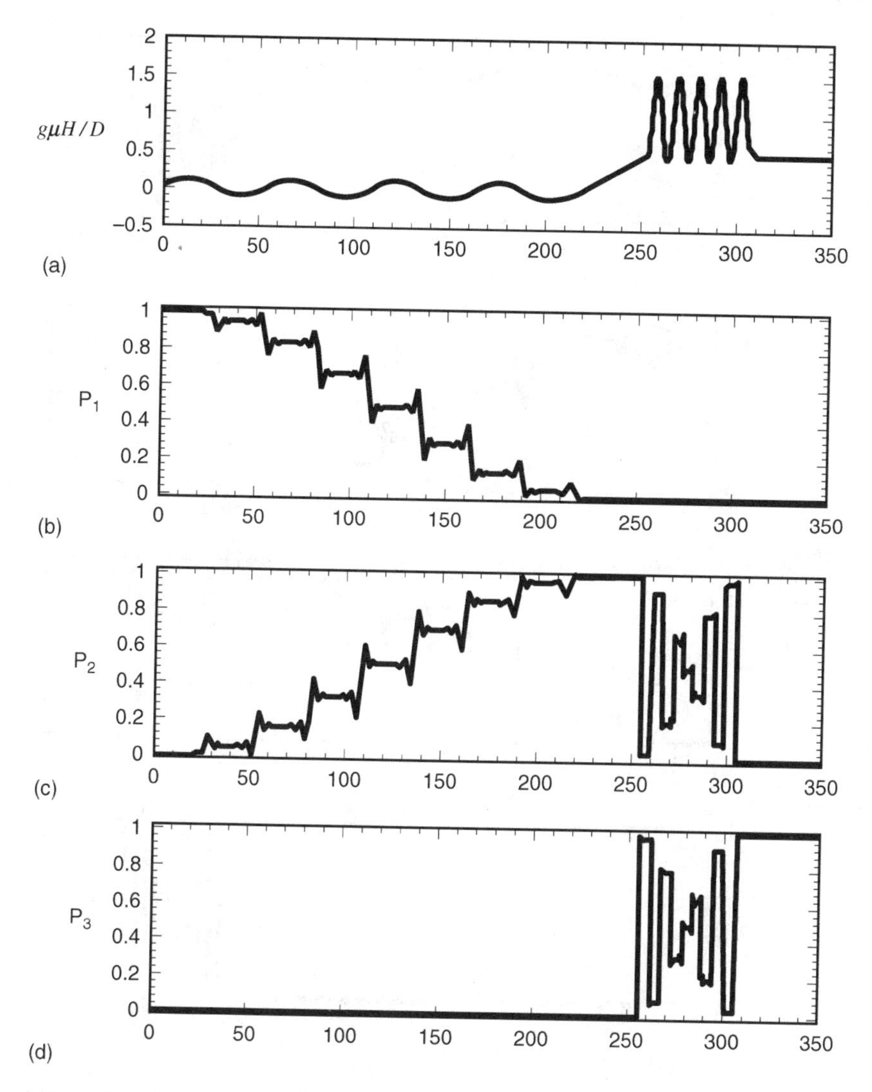


**Figure 19.** Controlled nonadiabatic processes, starting from "$a$" on state 1 and ending at "$b$" on state 3 via avoided crossings A and B (see Fig. 18). ($a$) Variation of the external magnetic field as a function of time, $tD/\hbar$. The first four-period oscillation around $g\mu_\beta H/D = 0$ corresponds to the control at the avoided crossing A. The second five-period oscillation around $g\mu_\beta H/D = 1.0$ corresponds to the control at B. ($b$) Time evolution of the probability $P_1$ for the system to be staying on the state 1. ($c$) Time evolution of $P_2$. ($d$) Time evolution of $P_3$.

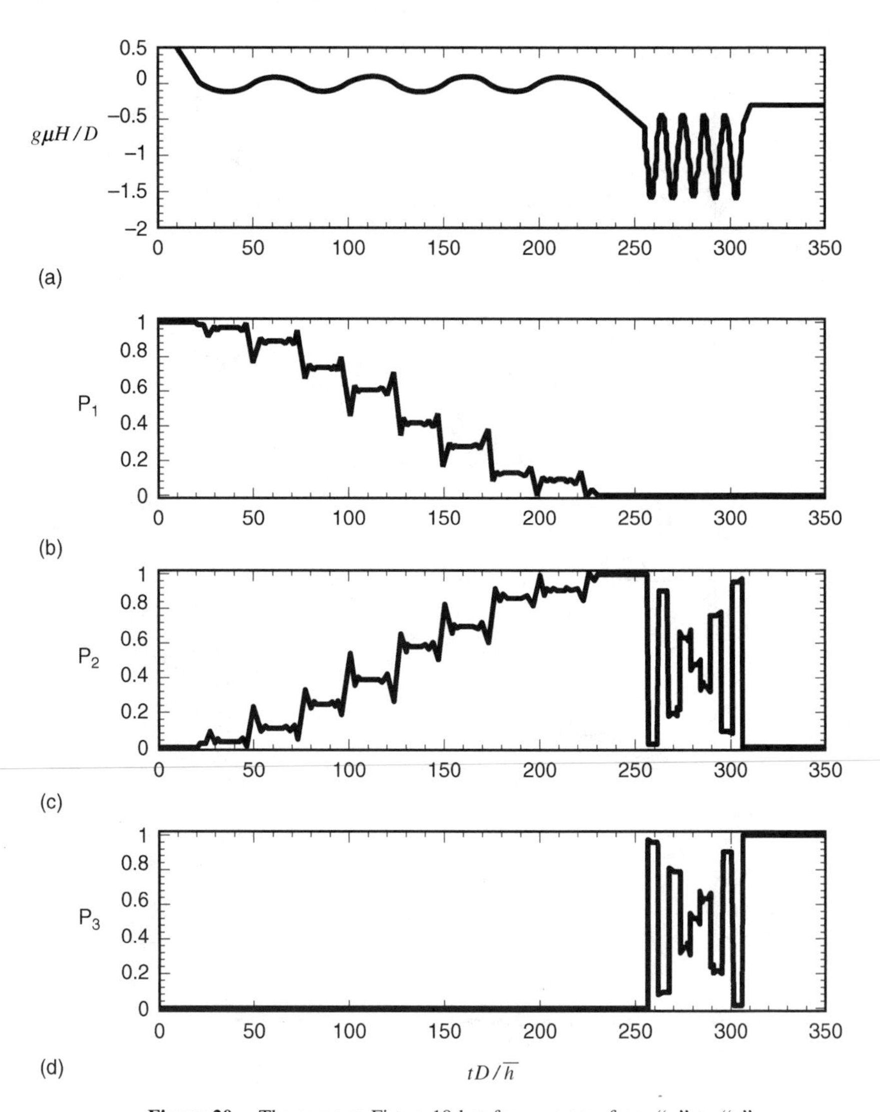


**Figure 20.** The same as Figure 19 but for a process from "$a$" to "$c$".

we do not have to solve multichannel coupled equations numerically, and that we can formulate all necessary conditions of control analytically.

## B. Control by Laser Field

With the help of nonadiabatic Floquet theory [45], our control theory is applicable to controlling by a laser field. In this case, the frequency and/or the intensity can be the adiabatic parameter(s) to sweep. Sweeping the frequency induces the LZ type of nonadiabatic transition, while sweeping the intensity induces the RZ type. Utilizing such nonadiabatic transition several times, unit transition probability can be realized with a small laser intensity.

Let us take, as an example, a 1D model of a laser-induced ring-puckering isomerization of trimethylenimine, which was discussed by Sugawara and Fujimura [99]. This problem may be reduced to a double well problem in which the left (right) well corresponds to the isomer A (B), and the isomerization from A to B occurs through tunneling from the left well to the right well [see Fig. 21(*a*)]. We try to control this isomerization with use of the various types of laser pulses. All the parameters to determine the potential system are taken from [99].

The Floquet-state diagram as a function of laser frequency $\omega$ ($cm^{-1}$) with constant intensity [$I = 0.1(TW/cm^2)$] is shown in Fig. 21(*b*). A lot of avoided crossings appear, where the energy gap is proportional to the laser intensity $I$ and the square of the transition dipole moment between the corresponding two states. We can treat each avoided crossing separately unless the laser intensity is extremely strong and avoided crossings overlap with each other.

Nonadiabatic transitions among the Floquet states induced by the variation of intensity and/or frequency can be described by the nonadiabatic Floquet theory [45, 100], and we can employ the various analytical theories of nonadiabatic transition to analyze them. Next, demonstrate the control of vibrational transitions and isomerization numerically with use of the various theoretical schemes.

### 1. *Landau–Zener Type of Nonadiabatic Transition*

If the system has a clear avoided crossing as a function of the adiabatic parameter, the LZS type transition can be utilized. The Hamiltonian of the simplest linear potential model of LZ, as is well known, is given by

$$H_{LZ} = \begin{bmatrix} \alpha_1 t & V \\ V & \alpha_2 t \end{bmatrix} \tag{7.25}$$

where $V$ is the constant diabatic coupling and $t$ is the time. In the case of a laser, this model corresponds to the constant intensity and linear sweeping of the frequency. This model is good enough to explain qualitative features of our control scheme. The nonadiabatic transition occurs at $t = 0$ with the transition probability $p$ given by

$$p_{LZ} = \exp\left(-2\pi\frac{V^2}{\hbar\alpha}\right) = \exp\left(-2\pi\frac{I\varepsilon^2}{\hbar^2|l-m|\dot{\omega}}\right) \tag{7.26}$$

where $\alpha \equiv |\alpha_1 - \alpha_2|$; $I$ and $\dot{\omega}$ are the laser intensity and the sweep velocity of the frequency at the avoided crossing, respectively; $l$ and $m$ (can be negative) are the

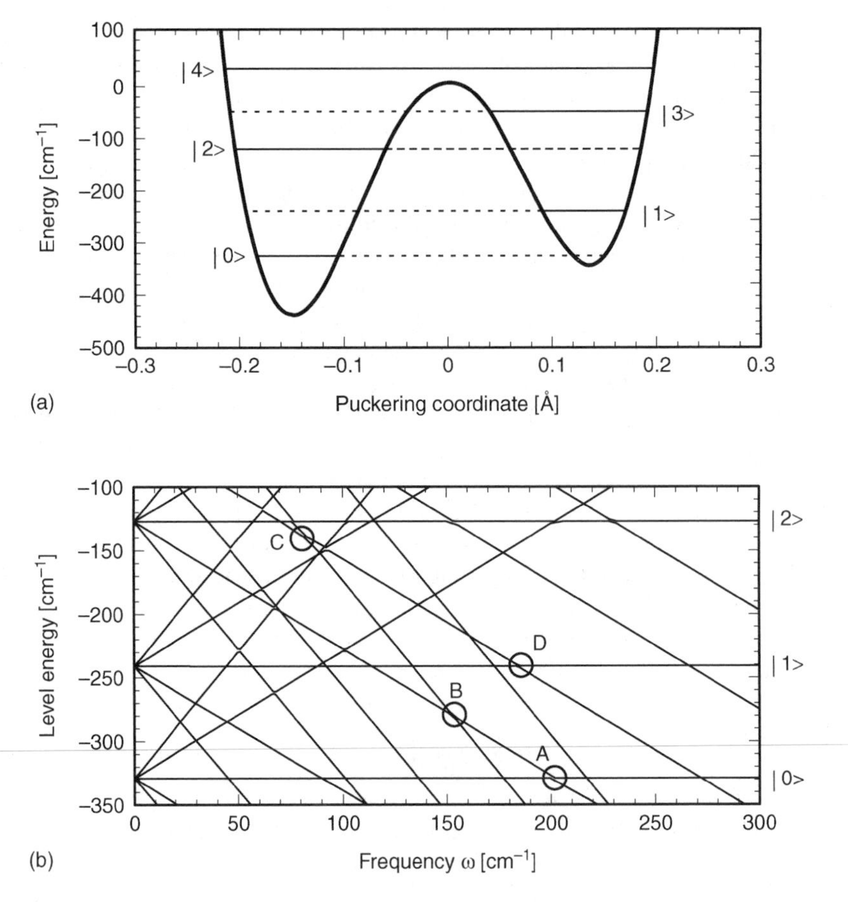


**Figure 21.** (*a*) A double well potential model of ring-puckering isomerization of trimethylenimine (see [99]). The origin of potential energy is taken at the barrier top. The horizontal solid line represents the main portion of the eigenfunction. (*b*) Floquet-state diagram, that is, vibrational levels in (*a*) as a function of laser frequency $\omega(\mathrm{cm}^{-1})$ at fixed intensity $[I = 0.1(\mathrm{TW/cm^2})]$. The gap at each avoided crossing is proportional to the transition dipole moment between the corresponding two states.

photon numbers; and $\varepsilon$ is the dipole matrix element. The conventional adiabatic passage requires large laser intensity and small sweep velocity to make $p_{\mathrm{LZ}}$ very small. For instance, the adiabatic passage with $p_{\mathrm{LZ}} \leq 0.001$ requires

$$1.0994\, \hbar\dot{\omega}/\varepsilon \leq I \tag{7.27}$$

For the values $\varepsilon = 1.0$ (Åe) and $\dot{\omega} = 0.5$ ($\mathrm{cm^{-1}/ps}$), $I$ must be $> 1.0$ ($\mathrm{TW/cm^2}$).

In order to accomplish the passage in a reasonably short timescale with relatively large $\dot{\omega}$, a large intensity $I$ or a large number of oscillation is required. If $p_{\mathrm{LZ}} = 0.5$ can be attained without difficulty, then one period of oscillation enables us to achieve exactly zero or unit final transition probability. The required intensity in this case is given by

$$I = 0.1103\hbar\dot{\omega}/\varepsilon \tag{7.28}$$

Namely, one period of oscillation requires the intensity by one order smaller compared to the case of one passage for the same $\varepsilon$ and $\dot{\omega}$. Furthermore, 10 periods of oscillation require the following condition [see Eq. (7.15)]:

$$0.982 \times 10^{-3}\hbar\dot{\omega}/\varepsilon \leq I \leq 0.810\hbar\dot{\omega}/\varepsilon \tag{7.29}$$

This requires only 1000 of the intensity required in the case of one passage. Many periods of oscillation, however, requires high accuracy of $p$ and phases. When $p$ is large, it is sensitive to the error in the exponent that is proportional to the intensity. In the case of adiabatic passage, on the other hand, $p$ is relatively stable against the error in the exponent, since the exponent is large. When $p \simeq 0.5$, 15% of error in the exponent yields the fluctuation in the range $0.45 < p_{\mathrm{LZ}} < 0.65$.

So far, our discussion is based on the simple model Eq. (7.25). For finding the actual parameters, however, it is much better to use the new semiclassical theory based on the quadratic model because this theory is applicable to the general functionality of $\omega$, even if the two diabatic potentials touch each other or avoid crossing.

Let us first consider the vibrational transition $|0> \rightarrow |2>$ via the avoided crossing A in Figure 21($b$) with use of the LZS type curve crossing model. In this case, the complete control can be achieved by one period of oscillation with reasonable values of the laser intensity and the sweep velocity. Figure 22($b$) and ($c$) show the frequency and intensity as a function of time. The frequency is taken to be a quadratic function of time, that is, $\omega(t) = at^2 + b$, and the analytical theory for quadratic model has been used. The resonance frequency $\omega_X$ corresponding to the avoided crossing is $\omega_X = 202.6\ \mathrm{cm}^{-1}$. Two functional forms, constant(solid line) and $4A^2\mathrm{sech}^2(\beta t)/\varepsilon^2$ (dash line), are assumed for the laser intensity [Fig. 22($c$)]. The frequencies shown in Figure 22($b$) are the solutions of our control theory corresponding to the intensities given in Figure 22($c$). Figure 22($a$) shows the time variation of the transition probability for the process $|0> \rightarrow |2>$ with use of the avoided crossing A in Figure 21($b$). Nonadiabatic transitions occur twice, and each time $p = 0.5$ is achieved. Unit transition probability is finally realized in the two cases. In the case of the intensity of pulse shape [dash line in Fig. 22($c$)], not only the frequency but also the area of

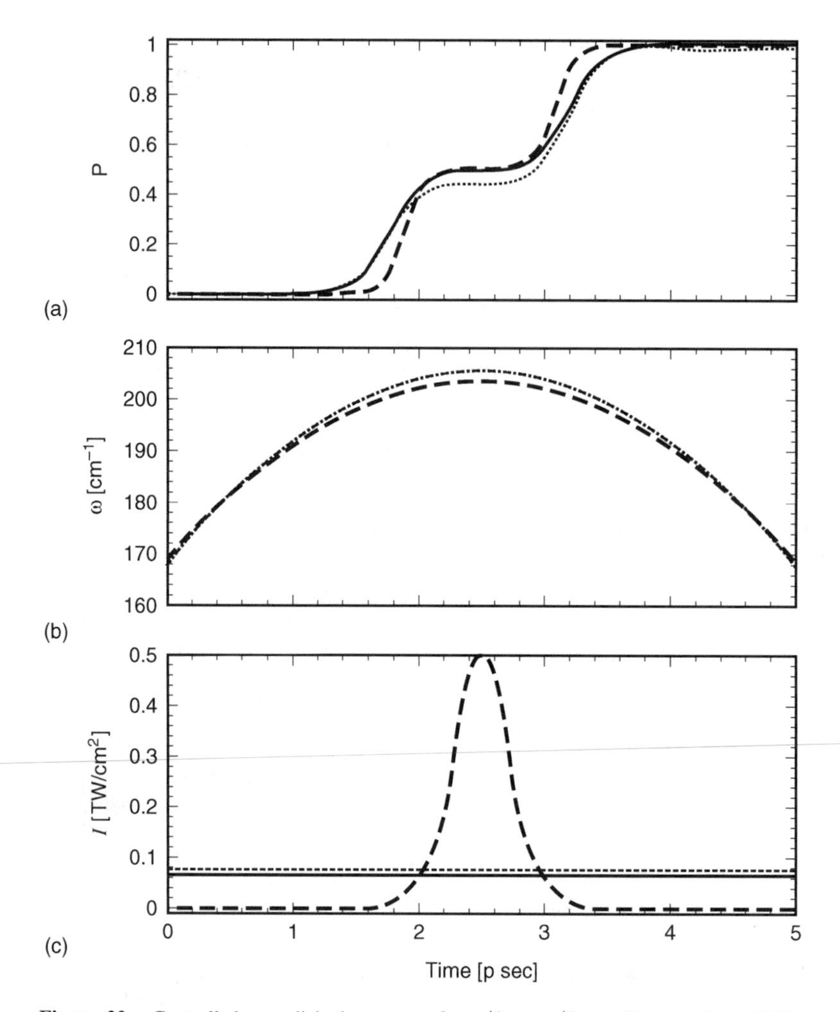


**Figure 22.** Controlled nonadiabatic process from $|0>$ to $|2>$ with use of the LZS type nonadiabatic transition at the avoided crossing "A" at 202.6 $cm^{-1}$ in Figure 21(*b*). (*a*) Time evolution of the transition probability. (*b*) Variation of laser frequency as a function of time. (*c*) Variation of laser intensity as a function of time. Solid line is the case of constant intensity and quadratic variation of frequency, and dash line corresponds to pulse shape intensity and quadratic variation of frequency. In both cases, complete control is attained. The dotted line shows the sensitivity to the constant shift of frequency. The final probability is $\sim 0.975$ in this case.

the intensity pulse contribute to the phase. Thus the corresponding frequency in Figure 22(*b*) is slightly smaller than the solid line. In this LZS type of nonadiabatic transition, functionality of the intensity is not important, but the intensity at the avoided crossing is critical. This is the reason why the difference in frequency is so small in the two cases. The dotted line is to show the effects of intensity variation on the final result. Although the nonadiabatic transition probability $p$ is reduced to 0.45 by a small shift in intensity, the final transition probability is not as bad, since the phase is accurate.

If we sweep the frequency more than once at the avoided crossing, we can naturally achieve the final unit transition probability with smaller intensity $I$, but it takes a longer time. It should be noted that the required peak intensity for the case of constant intensity is smaller than that required by a $\pi$ pulse (discussion will be made later in relation to Fig. 24). Note that the constant intensity can of course be cut off outside the transition region.

For a transition between two states with a small transition moment $\varepsilon$, a larger intensity or a smaller sweep velocity is required to satisfy $p = 0.5$ [see Eq. (7.26)]. This means that the direct isomerization from $|0>$ to $|1>$ requires a very large intensity or a very long transition time (very slow sweeping). For the isomerization, it is thus better to use an indirect process that is composed of the transitions of relatively large transition moments [99]. Note that the square of the transition moment for the direct process $|0> \rightarrow |1>$ is about four orders of magnitude smaller than that for $|0> \rightarrow |2>$. Figure 23 shows an example of such indirect isomerization: $|0> \rightarrow |2> \rightarrow |4> \rightarrow |3> \rightarrow |1>$, where four pulses are applied corresponding to these four transitions. That is to say, the first pulse achieves the complete transition $|0> \rightarrow |2>$, and the second one does $|2> \rightarrow |4>$, and so on. The corresponding avoided crossings are designated as A–D in Figure 21($b$). To obtain Figure 23($a$), 12 Floquet states are taken into account. Here, we have used exactly the same shape of $\omega(t)$ at four avoided crossings. This can be done by simply multiplying a certain constant to the intensity $I$.

Since the essential qualitative features such as the characteristics of various types of nonadiabatic transitions that we want to address here do not depend on the transitions, we consider the transition $|0> \rightarrow |2>$ for a while.

### 2. *Rosen–Zener Type of Nonadiabatic Transition*

The Hamiltonian of the Rosen–Zener–Demkov model in the diabatic representation is given by

$$H_{\mathrm{RZ}} = \begin{bmatrix} \Delta/2 & A\exp(\beta t) \\ A\exp(\beta t) & -\Delta/2 \end{bmatrix} \tag{7.30}$$

This model describes the process with constant frequency and exponentially rising intensity in the case of a laser. The nonadiabatic transition occurs at $t = \log(\Delta/A)/\beta$, and the transition probability $p$ is given by

$$p_{\mathrm{RZ}} = \frac{1}{1 + \exp(2\delta)} \tag{7.31}$$

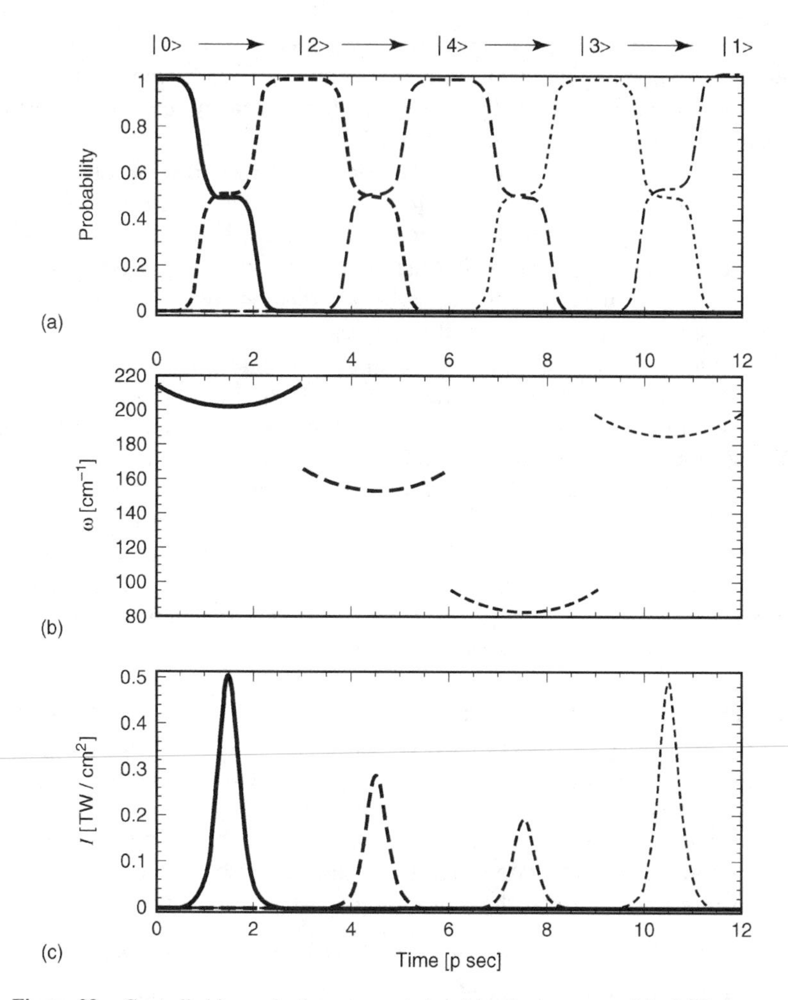


**Figure 23.** Controlled isomerization process induced by the sequence of the LZS type transitions, $|0> \to |2> \to |4> \to |3> \to |1>$. The corresponding avoided crossings are designated as A–D in Figure 21(*b*). The corresponding resonance frequencies are 202.6, 153.6, 82.6, and 185.2 $cm^{-1}$, respectively. (*a*) Time evolution of the probability. At each stage complete transition is attained. Twelve Floquet states are taken into account for the calculation. The time variations of frequency and intensity are shown in (*b*) and (*c*), respectively.

where

$$\delta \equiv \frac{\pi\Delta}{2\hbar\beta} \tag{7.32}$$

As is clearly seen from Eqs. (7.31) and (7.32), the range of $p$ is $0 \leq p \leq 0.5$, and $p \to 0$ when $\Delta/\beta \to \infty$; while $p \to 0.5$ when $\Delta/\beta \to 0$. Note that $p_{RZ}$ does depend only on $\Delta/\beta$ and not on the laser intensity $A$, and that $p \simeq 0.5$ can be achieved even with very small intensity (small $A$) in short time with large $\beta$. Thus, by adjusting the phases, we can achieve the unit final transition probability with use of one laser pulse of the shape $E(t) = 2A\text{sech}(\beta t)/\varepsilon$, namely, by one period of oscillation. This process of one pulse with $\Delta = 0$ corresponds to the $\pi$ pulse. The condition $\Delta = 0$ leads to $p_{RZ} = 0.5$ and thus $P_{12}^{(n)}$ given by Eq. (7.15) with $n = 1$ reduces to $\sin^2\psi = 1$, which coincides with the condition of the area of $\pi$ pulse. Note that our theory is quite general, and that the condition for unit final transition probability can be attained for any frequency, if we use more than one pulse.

The sensitivity of $p_{RZ}$ against an error in $\delta$ is largest when $p_{RZ} = 0.5$, and decreases as $p_{RZ}$ decreases. In the case of $\pi$ pulse ($\Delta = 0$), a constant shift of the frequency $\Delta$ from zero yields a relatively large effect on $p_{RZ}$ when $\beta$ or the pulse width is small. If the pulse width is, say, several picoseconds, about several reciprocal centimeters of shift in $\Delta$ yields $p_{RZ} \simeq 0.45$. It should be noted, however, that a small fluctuation of $\Delta$ in time (not a constant shift) causes a large error in the nonadiabatic transition probability $p$.

Figure 24 is an example of the so-called $\pi$ pulse (solid lines) with parameters chosen so that the overall transition time between the corresponding Floquet states becomes the same order as that in Figure 22. Again, nonadiabatic transitions occur twice with the transition probability $p = 0.5$, and the final overall transition probability is controlled to be unity with the use of the phase accumulated between the two transitions. In the case of the $\pi$ pulse, the condition of $p = 0.5$ is attained by fixing the frequency at the resonance frequency $\omega_X$, or the frequency at the avoided crossing. The phase condition is satisfied by adjusting the area of the intensity pulse. These conditions are relatively simple and seem to be easily realized compared to the LZS case.

It should be noted that a small variation of the frequency could cause unexpected transitions and errors if the frequency is close to the resonance. Dashed and dotted lines in Figure 24 demonstrate the sensitivity of the $\pi$-pulse method to frequency variations. The dashed line shows the effects of a constant shift of the frequency from the resonance. As discussed above, frequency can contain a constant shift error up to several reciprocal centimeters, if the phase due to the intensity variation is accurate, and the intensity can have a 10% error, and if the frequency is exactly at the resonance. More shocking is the large effect of a very small time-dependent fluctuation in frequency, as is demonstrated by the dotted line. This small fluctuation induces unexpected curve crossing type nonadiabatic transitions between the closely lying states effectively, and causes a big effect in the final result, as seen in Figure 24($a$). In order to avoid this instability, it might be worthwhile to use the off-resonant case explicitly. In the case of off-resonance, the nonadiabatic transition probability $p$ is $< 0.5$, and more than one pulse, that is, more than one period of oscillation, is required. Figure 25 shows an example of two pulses. The frequency shift is kept constant at 5 $\text{cm}^{-1}$ (solid line) and the various parameters are chosen so that the transition time is

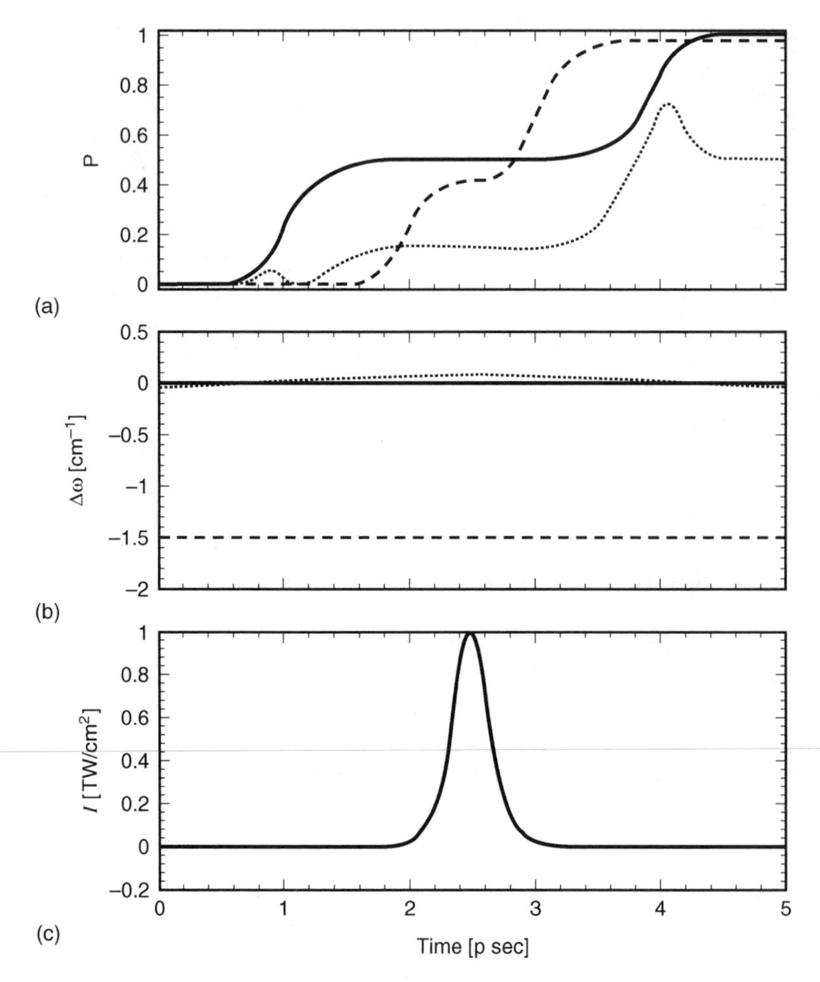


**Figure 24.** The same as Figure 22 for the case of $\pi$ pulse. (*a*) Time evolution of the transition probability. Time variations of frequency and intensity are shown in (*b*) and (*c*), respectively. The parameter $\Delta\omega$ represents the shift from the resonance frequency $\omega_X = 202.6\ \mathrm{cm}^{-1}$. The dashed line shows the case of constant shift in frequency. The solid line shows the case of exact one. The dotted line shows the large effect of small time variation of the frequency around resonance. Note that the small fluctuation of frequency in (*b*) gives a large effect on the transition probability [dotted line in (*a*)].

the same order as that in Figures 22 and 24. As seen from this figure in comparison with Figure 24, this scheme is very effective. The nonadiabatic transition probability for one passage is $\sim 0.2$, and after four nonadiabatic transitions the final transition probability reaches unity. This can be achieved

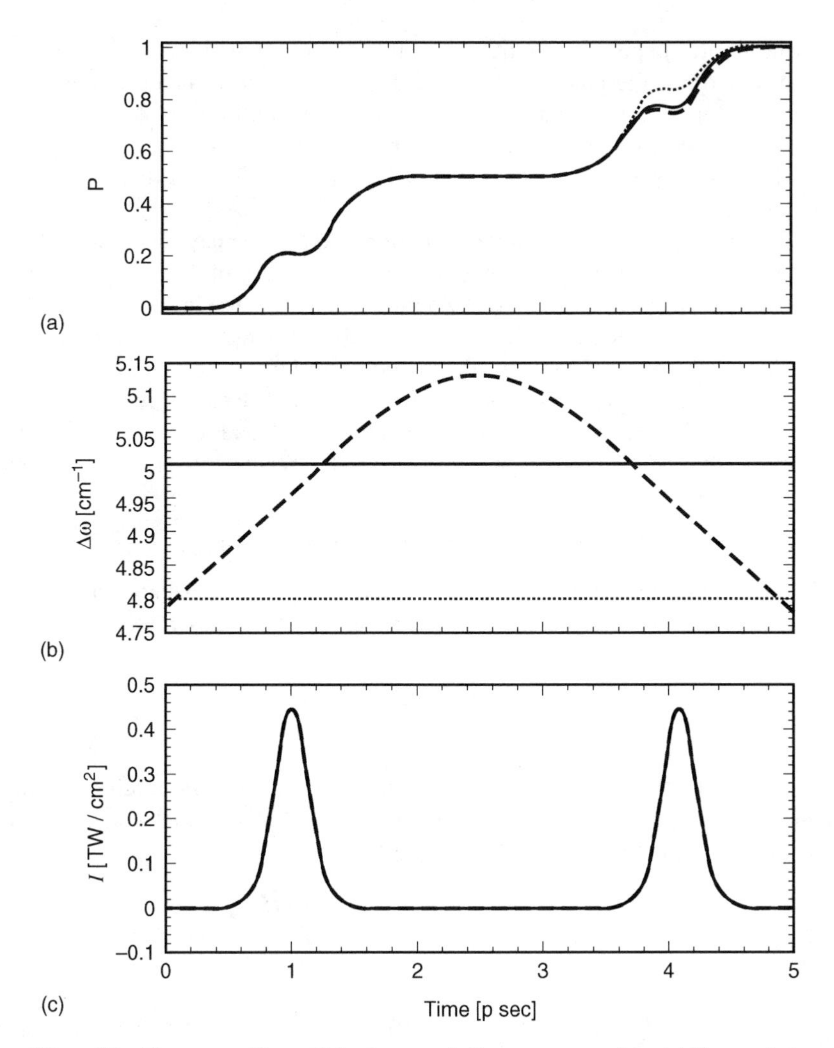


**Figure 25.** The same as Figure 22 for the case of off-resonant two pulses. (*a*) Time evolution of the probability. Time variations of frequency and intensity are shown in (*b*) and (*c*), respectively. The parameter $\Delta\omega$ represents the shift from the resonance frequency $\omega_X = 202.6\ \text{cm}^{-1}$. The solid line is the case of complete control. Dashed and dotted lines demonstrate the stability of the method against time variation of frequency(dashed line) and the constant shift in frequency(dotted line).

with a smaller peak intensity compared to the $\pi$ pulse. This is because the necessary phase can be accumulated not only by the intensity but also by the frequency because of the off-resonance. In the case of off-resonance, however, we should adjust not only the height and shape of the pulse $I(t)$ but also the interval of two pulses, because the nonadiabatic transition probability $p_{\text{RZ}}$

depends on the exponent of $I(t)$, that is, $\beta$ in Eq. (7.30) and the interval of two pulses determine the phase $\sigma_2$ (see Fig. 17). It might be difficult to adjust the shape of $I(t)$ accurately, but the control by off-resonant pulses is very attractive. It requires a small peak intensity. The stability against frequency fluctuation is also satisfactory, as demonstrated by the dashed and dotted lines in Figure 25.

We can choose either the LZS (oscillation of the frequency) or the RZD type (oscillation of the intensity) depending on the availability of the laser. One thing we should keep in mind is that the LZS type requires a large intensity to achieve $p = 0.5$, while the RZD (including the $\pi$ pulse) requires a large intensity to satisfy the phase condition. In other words, the LZS does not require large intensity to satisfy the phase condition and the RZD does not require that to satisfy $p = 0.5$. Thus we may think of a hybrid of LZS and RZD that enables us to achieve $p = 0.5$ by changing the intensity and to accumulate enough phase by changing the frequency.

### 3. *Exponential Type of Nonadiabatic Transition*

The well-known model that contains time variations in both diabatic energy and diabatic coupling is the exponential model [6, 31, 32]. The Hamiltonian of this model is given by

$$H_{\text{exp}} = \begin{bmatrix} U_1 + V_1 e^{-\beta t} & V e^{-\beta t} \\ V e^{-\beta t} & U_2 + V_2 e^{-\beta t} \end{bmatrix} \tag{7.33}$$

In the case of a laser, this model describes the process of exponentially changing intensity and frequency with the same exponent. The nonadiabatic transition probability in this model is given by [6, 32]

$$p_{\text{exp}} = \exp(-\pi\delta_2)\sinh(\pi\delta_1)/\sinh[\pi(\delta_1 + \delta_2)] \tag{7.34}$$

where

$$\delta_1 = \frac{U_1 - U_2}{2\hbar\beta}\left[1 + \frac{(V_1 - V_2)/2V}{\sqrt{1 + (V_1 - V_2/2V)^2}}\right] \tag{7.35}$$

and

$$\delta_2 = \frac{U_1 - U_2}{2\hbar\beta}\left[1 - \frac{(V_1 - V_2)/2V}{\sqrt{1 + (V_1 - V_2/2V)^2}}\right] \tag{7.36}$$

In the same way as in the RZD case, we can achieve $p = 0.5$ within a relatively short time with a small intensity. The necessary phase, on the other hand, may be accumulated by changing the intensity as well as the frequency. This means that a large intensity is not necessary to accumulate the necessary phase, and we can achieve unit transition probability with a small intensity by one period of oscillation (one pulse). The control scheme with use of this exponential model may provide one of the most effective ones (rapid transition with small intensity). A model with different exponents for the diabatic energy and the diabatic coupling would be more versatile and useful, although there is no analytical theory available yet. It should, however, be noted that the transition probability $p_{\mathrm{exp}}$ is rather sensitive to the functionalities of both intensity and frequency as a function of time, and a small experimental error might affect the final overall transition probability. Comparative studies on the effectiveness and the stability of the various control schemes described here are made hereafter together with the presentation of numerical examples.

Figure 26 (solid line) shows the control of the transition $|0> \to |2>$ by one pulse (period) with two nonadiabatic transitions of the exponential type ($p = 0.5$). It should be noted that the required intensity for the same order of transition time as before is quite small and the frequency is not necessary to be kept at resonance; while, as demonstrated before, the LZS type requires the larger intensity and RZD requires the frequency to be close to resonance to achieve $p = 0.5$. Due to the off-resonance, small fluctuation of $\omega$ does not cause any appreciable errors as in the case of the $\pi$ pulse. The necessary phase can be accumulated by means of both frequency and intensity, which is the reason why the required intensity can be so small. Dashed and dotted lines in Figure 26 show the stability of the method against the variations of intensity and frequency. Thus, the exponential model may provide quite an efficient control method compared to LZS and RZD.

By taking the vibrational transition $|0> \to |2>$, we have explained the characteristics of various types of nonadiabatic transitions and proved that the exponential model presents the most effective way of control. As mentioned before, this does not depend on the transitions and thus is also true for the tunneling transition $|0> \to |1>$. Figure 27 demonstrates this. As is well known, it is far better to use the detour $|0> \to |2> \to |4> \to |3> \to |1>$ in the same way as in Figure 27 than to take the direct path $|0> \to |1>$, since the dipole moment for the latter is about two orders of magnitude smaller than that of $|0> \to |2>$. As is seen in Figure 27, the required intensity is much smaller than that in Figure 23. If the laser intensity could be kept constant, however, we could further reduce the necessary maximum intensity, as was demonstrated in Figure 22. Figure 28 shows this type of control of the isomerization.

A more general model, like an exponential model with different exponents for frequency and intensity, for which no analytical theory is unfortunately available yet, is expected to provide a more efficient scheme. Note, however, that the exponential model in general requires accurate shaping of both intensity and

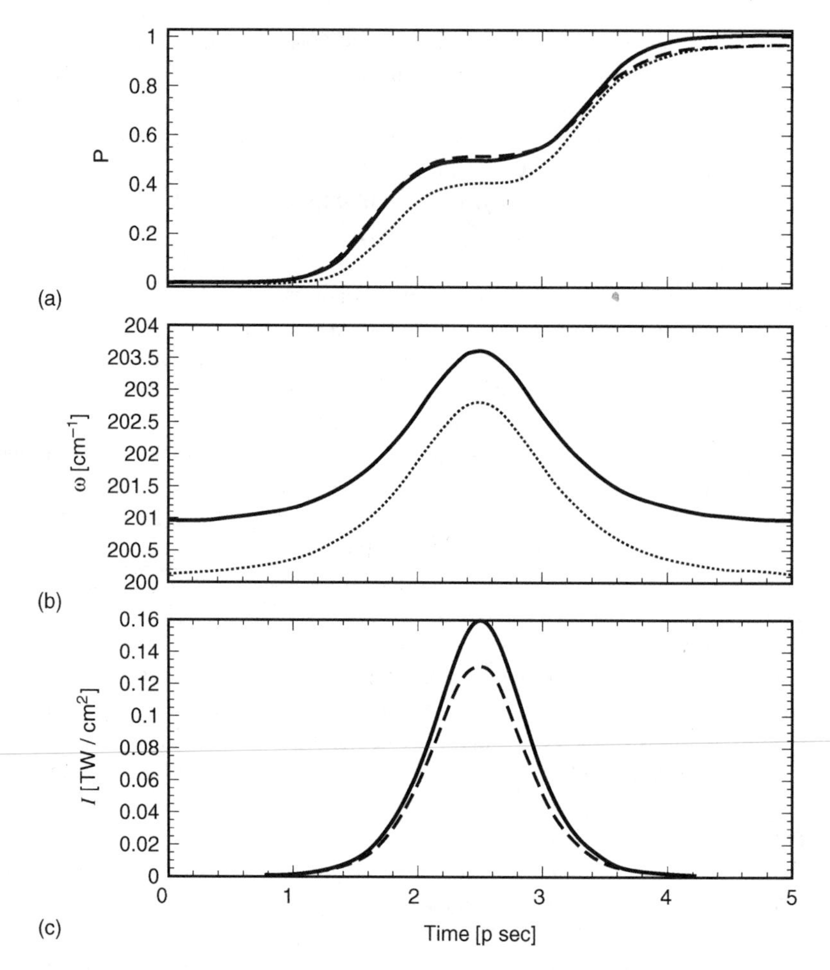


**Figure 26.** The same as Figure 25 for the case of exponential model. The solid line is the case of complete control. Frequency is swept around the avoided crossing "A" at $\omega_X = 202.6$ cm$^{-1}$. Dashed and dotted lines demonstrate the stability of the method against the error in the exponent of intensity(dashed line) and the constant shift in frequency from the solid line(dotted line).

frequency pulses, since the nonadiabatic transition probability $p_{\text{exp}}$ depends explicitly on the exponents of both intensity and frequency.

So far, we have discussed several control schemes with the help of analytical theories. We can choose one of them depending on the molecular process and availability of lasers. If accurate pulse shaping of both intensity and frequency is possible, the scheme with the exponential type of nonadiabatic transition is the best. If pulse shaping is available for intensity, but not for frequency, the RZD type works relatively well. In this case, the off-resonant RZD type is

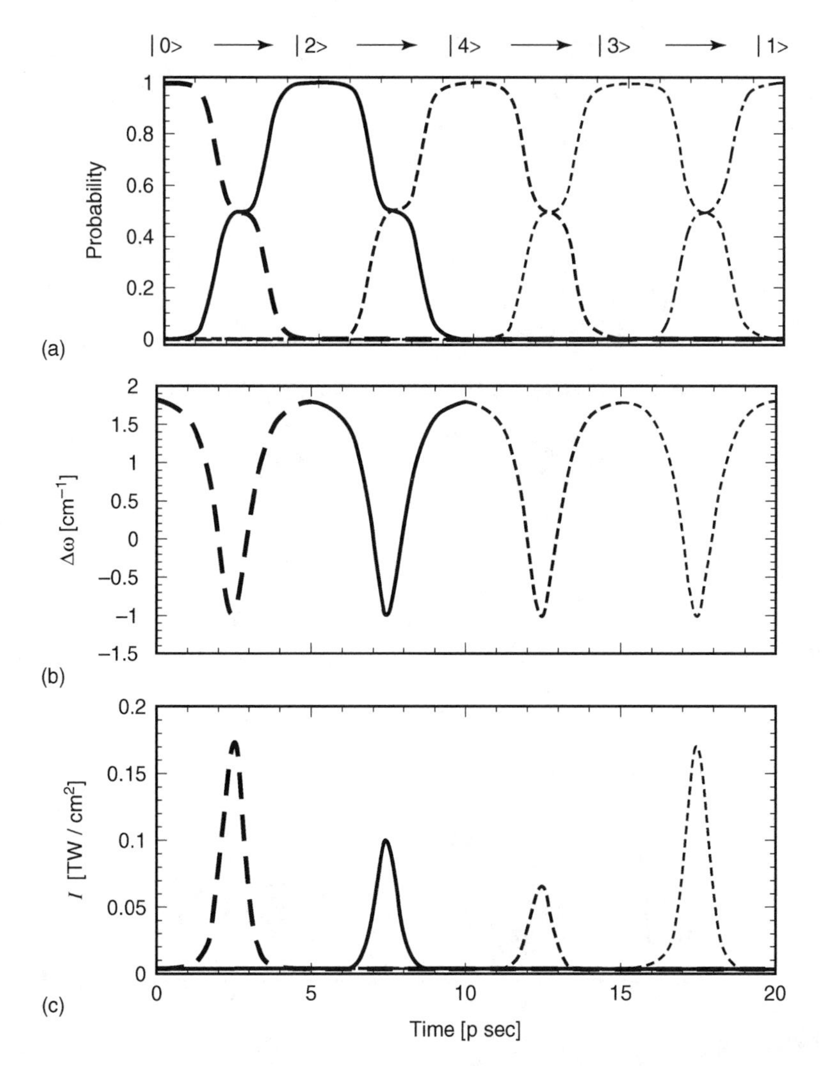


**Figure 27.** The same as Figure 23, that is, control of the isomerization process $|0> \rightarrow |2> \rightarrow |4> \rightarrow |3> \rightarrow |1>$. The exponential model is employed. Twelve Floquet states are taken into account to obtain (*a*). It should be noted that the intensity is reduced by $\sim$ a factor of 3 compared to Figure 23. In (*b*), the frequency shift $\Delta\omega$ from the resonance is shown instead of $\omega$ itself in order to clearly show its exponential form. The shape is taken to be the same for the four transitions.

recommended. If accurate shaping of intensity is not attainable, the resonant RZD ($\pi$ pulse) or the LZS methods are recommended.

Generally speaking, molecular processes in a laser field can be considered to be a sequence of nonadiabatic transitions and adiabatic propagation as before,

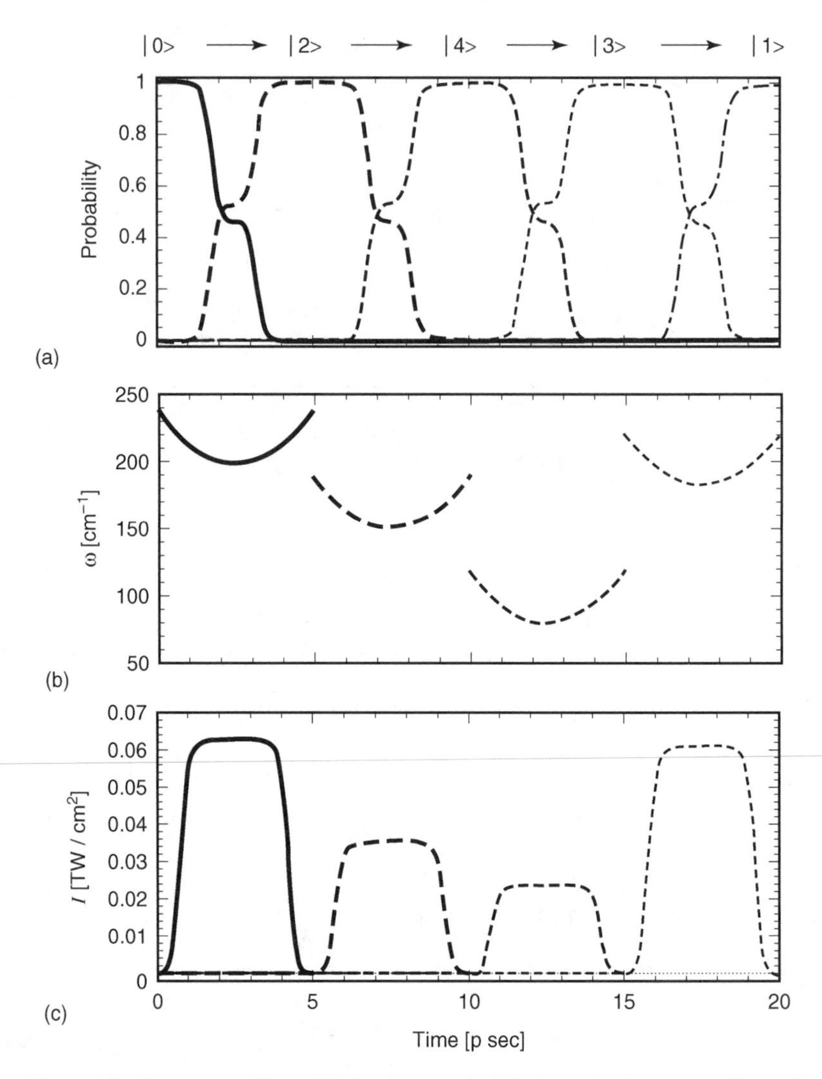


**Figure 28.** The same as Figure 23, that is, control of the isomerization process $|0>\rightarrow|2>\rightarrow|4>\rightarrow|3>\rightarrow|1>$. The constant intensity and the quadratic variation of frequency are utilized.

and we can treat the whole control problem analytically, if the nonadiabatic transitions are separated from each other in time. This indicates that we may construct a control scheme for a general shape of pulse, even if no analytical theory for each nonadiabatic transition is available. Once we know the phase change and the probability change by one pulse (one period of oscillation), we can design an efficient way of control based on the analytical scheme. The optimum conditions for one pulse, namely, the conditions to satisfy $\sin^2\psi = 1$ and $p = 0.5$, may be found numerically or even experimentally.

In the present work, we have focused our attention mainly on the laser case, in which the diabatic coupling (energy) depends on the intensity (frequency) under the nonadiabatic Floquet formalism. By sweeping the adiabatic parameter(s) periodically, the unit transition probability can be achieved due to the interference effect among different passages of the nonadiabatic transition point. One pulse (or one period of oscillation) requires the nonadiabatic transition probability $p$ for one passage to be 0.5. The required range of $p$ becomes wider as the number of oscillation increases. The nonadiabatic transition can be any type such as LZS, RZD, and the type of exponential model. Even a transition for which no analytical theory is available might be used.

It is known that $I$ (laser intensity) $\simeq 10$ TW/cm$^{-2}$ is the upper limit for us to safely focus on the vibrational transitions [10]. The maximum intensity required in some of the processes discussed in the present work is 1.0 TW/cm$^{-2}$. The best way of control based on our new idea requires only 60 $\simeq$ 170 GW/cm$^{-2}$ maximum intensity, two orders of magnitude smaller than the upper limit.

Since nonadiabatic transitions play important roles in various molecular processes in external fields, our idea of controlling nonadiabatic transitions would give a versatile control scheme of molecular processes. Magnetic resonance, ESR, or NMR can be another interesting example. With the help of analytical theories of time-dependent nonadiabatic transitions, various possibilities can be discussed. One can choose an appropriate one depending on the availability and quality of the lasers. The physical conditions of our controlling scheme are very clear, and the analysis can be done analytically without any heavy computations.

## VIII. FUTURE PERSPECTIVES

In this chapter, we have presented an overview of the newly established theory for time-independent two-state curve crossing problems. The basic two-state theory is simple enough for the general users to utilize, and can still work well over the entire range of energy and coupling strength. The theory is further demonstrated to be applicable to multichannel and multidimensional problems. The time-dependent version of the theory has also been presented, and a new idea of controlling molecular processes by time-dependent external fields was discussed. These semiclassical theories are completely established in adiabatic representation, and there is no need for any diabatization and for any information about nonadiabatic coupling. In contrast, the semiclassical theory for the noncurve crossing case has not been complete yet; especially no theory is available for energies lower than the transition region. This is a big challenge for the future. Formulation of a unified theory that can cover both crossing and noncrossing cases in a unified way would be a very challenging subject.

If we consider the fact that the LZS type potential surface crossings, including conical intersections, most commonly appear and play crucial roles in various real physical and chemical dynamic processes, it is very important and worthwhile to develop practically useful accurate methodologies to deal with multi-dimensional nonadiabatic transition problems. As was discussed in Section IV, the new semiclassical theory of nonadiabatic transition can be easily incorporated into the various frameworks such as the TSH method [27, 30], the semiclassical propagation method based on the IVR [28], and the CFGWP method [29]. Accurate formulas not only for nonadiabatic transition probability but also for all necessary phases are available in present semiclassical theory; and even when the surface crossing is located in the classically forbidden region, the accurate treatment is possible with the present semiclassical theory. In the case of conical intersection, the geometrical phase [102] should just be added to the phase along each trajectory.

Controlling molecular processes by time-dependent external fields is one of the important modern subjects in chemical dynamics. Again, nonadiabatic transitions play a crucial role there. We have proposed a new idea of controlling nonadiabatic transitions that is summarized in Section VII. Various generalizations and applications of this idea can be thought of and should be carried out.

## Acknowledgments

The present work is partly supported by a Grant-in-Aid for Scientific Research on Priority Area "Molecular Physical Chemistry" and also by a Research grant 10440179 from the Ministry of Education, Science, Culture and Sports of Japan.

## References

1. L. D. Landau, *Phys. Z. Sowjetunion* **2**, 46 (1932).
2. C. Zener, *Proc. R. Soc. Ser. A* **137**, 696 (1932).
3. E. C. G. Stueckelberg, *Helv. Phys. Acta* **5**, 369 (1932).
4. N. Rosen and C. Zener, *Phys. Rev.* **40**, 502 (1932).
5. M. S. Child, *Molecular Collision Theory*, Academic, London, 1974; *Semiclassical Mechanics with Molecular Applications*, Oxford University Press, 1991.
6. E. E. Nikitin and S. Ya. Umanskii, *Theory of Slow Atomic Collisions*, Springer, Berlin, 1984.
7. D. S. F. Crothers, *Adv. Phys.* **20**, 405 (1971).
8. H. Nakamura, *Int. Rev. Phys. Chem.* **10**, 123 (1991).
9. J. C. Tully, *Nonadiabatic Processes in Molecular Collision*, Part B, W. H. Miller, Ed., Plenum, New York, 1976, p. 217.
10. M. Baer, in *Theory of Chemical Reaction Dynamics*, Vol II, M. Baer, Ed., CRC Press, Boca Roton, FL, 1985, p. 219.

11. E. E. Nikitin, *Ann. Rev. Phys. Chem.* **50**, 1 (1999).
12. C. Zhu and H. Nakamura, *J. Math. Phys.* **33**, 2697 (1992).
13. C. Zhu, H. Nakamura, N. Re, and V. Aquilanti, *J. Chem. Phys.* **97**, 1892 (1992).
14. C. Zhu and H. Nakamura, *J. Chem. Phys.* **97**, 8497 (1992).
15. C. Zhu and H. Nakamura, *J. Chem. Phys.* **98**, 6208 (1993).
16. C. Zhu and H. Nakamura, *J. Chem. Phys.* **101**, 4855 (1994).
17. C. Zhu and H. Nakamura, *J. Chem. Phys.* **102**, 7448 (1995).
18. C. Zhu and H. Nakamura, *J. Chem. Phys.* **109**, 4689 (1998).
19. C. Zhu and H. Nakamura, *J. Chem. Phys.* **101**, 10630 (1994).
20. C. Zhu and H. Nakamura, *J. Chem. Phys.* **108**, 7501 (1998).
21. H. Nakamura and C. Zhu, *Comments Atomic Mol. Phys.* **32**, 249 (1996).
22. H. Nakamura, in *Dynamics of Molecules and Chemical Reactions*, R. E. Wyatt and J. Z. H. Zhang, Eds., Marcel Dekker, 1996, Chapter 12.
23. H. Nakamura, "Complete solutions of the Landau–Zener–Stueckelberg curve crossing problems and their generalizations and applications." XXI–ICPEAC (Sendai, 1999) invited talk.
24. C. Zhu and H. Nakamura, *Chem. Phys. Lett.* **258**, 342 (1996); *J. Chem. Phys.* **106**, 2599 (1997).
25. C. Zhu and H. Nakamura, *Chem. Phys. Lett.* **274**, 205 (1997); *J. Chem. Phys.* **107**, 7839 (1997).
26. C. Zhu, H. Nakamura, and K. Nobusada, *Phys. Chem. Chem. Phys.* **2**, 557 (2000).
27. J. C. Tully and R. K. Preston, *J. Chem. Phys.* **55**, 562 (1971).
28. W. H. Miller, *J. Chem. Phys.* **53**, 3578 (1970); X. Sun and W. H. Miller, *J. Chem. Phys.* **106**, 3578 (1997).
29. E. J. Heller, *J. Chem. Phys.* **94**, 2723 (1991); Wolton and Manolopoulos, *Mol. Phys.* **84**, 961 (1996).
30. J. C. Tully, *J. Chem. Phys.* **93**, 1061 (1990).
31. V. I. Osherov and H. Nakamura, *J. Chem. Phys.* **105**, 2770 (1996).
32. V. I. Osherov, V. G. Ushakov, and H. Nakamura, *Phys. Rev. A*, **57**, 2672 (1998).
33. V. I. Osherov and H. Nakamura, *Phys. Rev. A*, **59**, 2486 (1999).
34. L. Pichl, V. I. Osherov, and H. Nakamura, *J. Phys. A: Math. Gen.* **33**, 3361 (2000).
35. Yu. N. Demkov and V. I. Osherov, *Sov. Phys. JETP* **26**, 916 (1968).
36. Yu. N. Demkov and M. Kunike, *Vestn. Leningr. Univ. Fi. Ckemia.* **16**, 39 (1969).
37. Yu. N. Demkov and V. N. Ostrovsky, *J. Phys. B* **28**, 403 (1995).
38. V. N. Ostrovskii and H. Nakamura, *J. Phys. A* **30**, 6939 (1997).
39. S. Toshev, *Phys. Lett.* **196**, 170 (1987).
40. Yu. N. Demkov and V. N. Ostrovsky, *Zero-Range Potentials and Their Applications in Atomic Physics*, Plenum, New York, 1988.
41. S. Brundobler and V. Elser, *J. Phys. A* **26**, 1211 (1993).
42. Yu. N. Demkov and V. N. Ostrovsky, *Phys. Rev. A* (2000).
43. C. E. Carroll and F. T. Hioe, *J. Phys. A* **19**, 1151 (1986).
44. Y. Teranishi and H. Nakamura, *J. Chem. Phys.* **107**, 1904 (1997).
45. S. I. Chu, *Adv. Chem. Phys.* **73**, 739 (1989).
46. Y. Teranishi and H. Nakamura, *Phys. Rev. Lett.* **81**, 2032 (1998).
47. Y. Teranishi and H. Nakamura, *J. Chem. Phys.* **111**, 1415 (1999).

48. S. S. Shaik and P. C. Hiberty, in *Theoretical Models of Chemical Bonding*, Part 4, Z. B. Maksic, Ed., Springer-Verlag, Berlin, 1991, p. 269.
49. D. DeVault, *Quantum-Mechanical Tunneling in Biological Systems*, Cambridge University Press, Cambridge, 1984.
50. J. R. Bolton, N. Mataga, and G. Mclendon, Eds., Electron Transfer in Inorganic, Organic and Biological Systems, Advances in Chemistry Series 228, American Chemical Society, Washington, DC, 1991.
51. A. Yoshimori and M. Tsukada, Eds., *Dynamical Processes and Ordering on Solid Surfaces*, Springer-Verlag, Berlin, 1985.
52. F. O. Goodman and H. Nakamura, *Prog. Surface Sci.* **50**, 389 (1995).
53. R. H. Bartram and A. M. Stoneham, *Solid State Comm.* **21**, 1325 (1978).
54. M. N. Kabler and R. T. Williams, *Phys. Rev.* **18**, 1948 (1978).
55. Y. Kayanuma and K. Nasu, *Solid State Comm.* **27**, 1371 (1978).
56. R. Engleman, *Non-Radiative Decay of Ions And Molecules in Soilds*, North-Holland, Amsterdam, The Netherlands, 1979.
57. A. Thiel, *J. Phys. G.* **16**, 867 (1990).
58. J. Zakrzewski and M. Kus, *Phys. Rev. Lett.* **67**, 2749 (1991).
59. R. J. Gordon and S. A. Rice, *Adv. Chem. Phys.* **48**, 601 (1997).
60. E. Ben-Jacob and Y. Gefen, *Phys. Lett.* **108A**, 289 (1985).
61. M. Buttiker, Y. Imry and R. Landauer, *Phys. Lett.* **96A**, 523 (1983).
62. R. Landauer and M. Buttiker, *Phys. Rev. Lett.* **54**, 2049 (1986).
63. Y. Gefen, E. Ben-Jacob, and A. O. Caldeira, *Phys. Rev. B* **36**, 2770 (1987).
64. Y. Gefen and D. J. Thouless, *Phys. Rev. Lett.* **59**, 1572 (1987).
65. G. Blatter and D. A. Browne, *Phys. Rev. B* **37**, 3856 (1988).
66. K. Mullen, E. Ben-Jacob, Y. Gefen and Z. Schuss, *Phys. Rev. Lett.* **62**, 2543 (1989).
67. L. Thomas et al., Nature (London) **383**, 145 (1996).
68. B. Schwarzchild, *Phys. Today*, June 17, 1986; J. N. Bahcall and M. H. Pinsonneault, *Rev. Mod. Phys.* **64**, 885 (1992).
69. L. Pichl, H. Nakamura and H. Deguchi, Proceedings of the V-th Joint Conference on Information Sciences (Atlantic city, N.J., March, 2000).
70. Y. Kami and E. E. Nikitin, *J. Chem. Phys.* **100**, 2027 (1994).
71. A. Bárány, *J. Phys.* **13**, 147 (1980).
72. A. Bárány, Institute of Theoretical Physics, Uppsala University, Sweden, Report No. 25 (1979).
73. J. B. Delos and W. R. Thorson, *Phys. Rev. A* **6**, 728 (1972).
74. J. Heading, *An Introduction to Phase Integral Methods*, Methuer, London, 1962.
75. P. V. Coveney, M. S. Child, and A. Bárány, *J. Phys. B* **18**, 4557 (1985).
76. H. Nakamura, *J. Chem. Phys.* **97**, 256 (1992).
77. S. Nanbu, H. Nakamura and F. O. Goodman, *J. Chem. Phys.* **107**, 5445 (1997).
78. H. Nakamura, *J. Chem. Phys.* **110**, 10253 (1999).
79. H. Nakamura, *J. Chem. Phys.* **87**, 4031 (1987).
80. H. Nakamura, *Adv. Chem. Phys. LXXXII, State-Selected and State-to-State Ion Molecule Reaction Dynamics*, Part 2: Theory, M. Baer an C-Y. Ng, Eds., Wiley, New York, 1992, p. 243.

81. H. J. Seaton, *Rep. Prog. Phys.* **46**, 167 (1983).
82. B. K. Homer and M. S. Child, *Faraday Discuss. Chem. Soc.* **75**, 831 (1983).
83. P. S. Julienne and M. J. Krauss, *J. Mol. Spectrosc.* **56**, 270 (1975).
84. J. C. Tully, in *Modern Methods for Multidimensional Dynamics Computations in Chemistry*, D. L. Thompson, Ed., World Scientific Publishing, New York, 1998, p. 34.
85. W. H. Miller, *Adv. Chem. Phys.* **25**, 69 (1974); **30**, 77 (1975); M. V. Berry and K. E. Mount, *Rep. Prog. Phys.* **35**, 315 (1972); V. P. Maslov and M. V. Fedoriuk, *Semi-Classical Approximation in Quantum Mechanics*, Reidel, Boston, 1981.
86. A. Kupperman and P. G. Hipes, *J. Chem. Phys.* **84**, 5962 (1986); J. M. Launay and B. Lepetit, *Chem. Phys. Lett.* **144**, 346 (1998); G. C. Schatz, D. Sokolovski and J. N. L. Connor, in Advanced Molecular Vibrations and Collision Dynamics, J. M. Bowman, Ed., JAI Press, London, 1994. Vol. 2B, p. 1; D. E. Manolopoulos and D. C. Clary, in *Annual Reports on the Progress of Chemistry*, Section C, Vol. 86, 1989, p. 95.
87. R. T. Pack and G. A. Parker, *J. Chem. Phys.* **87**, 3888 (1987); **90**, 3511 (1989); G. A. Parker and R. T. Pack, **98**, 6883 (1992).
88. V. Aquilanti, S. Cavalli and D. D. Fazio, *J. Chem. Phys.* **109**, 3792 (1998).
89. O. I. Tolstikhin and H. Nakamura, *J. Chem. Phys.* **108**, 8899 (1998).
90. K. Nobusada, O. I. Tolstikhin and H. Nakamura, *J. Chem. Phys.* **108**, 8992 (1998).
91. K. Nobusada, O. I. Tolstikhin and H. Nakamura, *J. Phys. Chem.* **102**, 9445 (1998).
92. C. Zhu, *J. Phys. A* **29**, 4159 (1996).
93. C. Zhu, *J. Chem. Phys.* **105**, 4159 (1996).
94. V. I. Osherov and A. I. Voronin, *Phys. Rev. A* **49**, 2672 (1994).
95. Y. Luke, *Mathematical Functions and Their Approximations*, Academic, New York, 1975.
96. D. Kleppner, M. G. Littman, and M. L. Zimmerman, *Rydberg States of Atoms and Molecules*, R. F. Stebbings and F. B. Dunning, Eds., Cambridge University Press, 1983.
97. I. Ya. Korenblit and E. F. Shender, *JETP* **48**, 937 (1987).
98. J. R. Friedman, M. P. Sarachik, J. Tejada and R. Ziolo, *Phys. Rev. Lett.* **76**, 3830 (1996).
99. M. Sugawara and Y. Fujimura, *J. Chem. Phys.* **100**, 5646 (1994).
100. T. S. Ho and S. I. Chu, *Chem. Phys. Lett.* **141**, 315 (1987).
101. A. D. Bandrauk, Ed., *Molecules in Laser Fields*, Marcel Dekker, New York, 1994.
102. A. Kuppermann, in *Dynamics of Molecules and Chemical Reactions*, R. E. Wyatt and J. Z. H. Zhang, Eds., Marcel Dekker, 1996, p. 411.

# MULTIDIMENSIONAL RAMAN SPECTROSCOPY

JOHN T. FOURKAS*

*Eugene F. Merkert Chemistry Center, Boston College, Chestnut Hill, MA*

## CONTENTS



## I. INTRODUCTION

Raman spectroscopy is a powerful technique for investigating the dynamics, structure and interactions of molecules. Because Raman spectroscopy is sensitive to vibrations that alter the polarizability of a molecule, this technique is complementary to infrared absorption, which is sensitive to vibrations that alter the dipole moment of a molecule. In addition, since the polarizability of a

*Camille and Henry Dreyfus New Faculty Fellow, Sloan Research Fellow, Research Corporation Cottrell Scholar, and Camille Dreyfus Teacher-Scholar.

*Advances in Chemical Physics, Volume 117*, Edited by I. Prigogine and Stuart A. Rice.
ISBN 0-471-40542-6 

molecule is a tensor quantity, it is possible to perform polarization-selective Raman experiments that eliminate particular contributions to the spectrum or that yield additional information about intermolecular or intramolecular structure.

In its most common implementations, Raman spectroscopy is a one-dimensional technique. In other words, a Raman spectrum is obtained by scanning one frequency (or, in time-domain Raman spectroscopy, a Raman response function is obtained by scanning a single delay time). As a result, despite the selectivity available from controlling the experimental polarizations, it is often not possible to make an unambiguous separation of the effects that contribute to the shape of a given Raman line.

The problem in identifying the individual contributions to the Raman line shape is entirely analogous to the difficulty of distinguishing the different couplings or broadening mechanisms in nuclear magnetic resonance (NMR) spectra. NMR spectroscopists solve this problem by employing complex pulse sequences that can suppress specific contributions to a spectrum selectively [1]. Sequences that involve multiple pulses also involve multiple delay times between pulses, which allows for the construction of multidimensional spectra. An appropriate multidimensional representation of such an experiment can provide a readily-interpretable depiction of molecular aspects such as the spins that are coupled to one another and the strengths of these couplings.

Multidimensional techniques have the potential to be every bit as revolutionary for optical spectroscopies as they have been for NMR. Multidimensional optical techniques have to date found their greatest use in electronic spectroscopies [1–5]. The implementation of multidimensional vibrational techniques is more technically challenging, and so fewer examples of multidimensional Raman and IR spectroscopies currently exist. However, a large amount of effort is being expended on the development of these techniques, as well as on the theory that will allow for the interpretation of multidimensional Raman spectra. In this chapter, we will review the experimental and theoretical advances that have been made in recent years in multidimensional Raman spectroscopy.

## II. ONE-DIMENSIONAL RAMAN SPECTROSCOPY

In discussing multidimensional Raman spectroscopy, it is helpful first to review the salient features of conventional, 1D Raman spectroscopy [6]. Spontaneous Raman spectroscopy is the most common example of a 1D Raman technique. However, since all multidimensional Raman techniques are necessarily coherent spectroscopies, we will focus on coherent techniques in discussing 1D Raman spectroscopy.

Before considering the theoretical details of coherent 1D Raman spectroscopy, we will give a brief description of its experimental implementation.

Coherent anti-Stokes Raman scattering (CARS) spectroscopy is perhaps the most common example of a 1D coherent Raman technique. In this spectroscopy, two ultrashort pulses of laser light (femtoseconds to picoseconds in duration) arrive simultaneously at the sample of interest. The center frequencies of the two pulses ($\omega_1$ and $\omega_2$), which are generally in the visible range, are adjusted such that the difference between them matches the fundamental frequency of a Raman-active vibrational mode in the sample. These "pump" pulses thereby initiate a coherent oscillation of the chosen vibrational mode. This coherence is monitored some time later by a "probe" pulse of center frequency $\omega_3$, which causes the sample to emit a signal pulse of center frequency $\omega_1 - \omega_2 + \omega_3$ in a unique spatial direction. By scanning the delay time between the pump and probe pulses, the decay of the vibrational coherence can be measured.

Although both CARS and spontaneous Raman measurements are formally equivalent in information content [7], each spectroscopy has distinct advantages and disadvantages. Coherent anti-Stokes Raman spectroscopy is a background-free technique, since the signal propagates in a unique direction. The coherent nature of CARS also allows this spectroscopy to be performed readily in small sample volumes. The polarizations of each of the four beams in a CARS experiment can be controlled separately, making polarization-selective experiments straightforward [8]. On the other hand, CARS spectroscopy requires at least two ultrashort lasers, at least one of which must be tunable. In addition, because CARS experiments involve laser beams of different colors, care must be taken to ensure that the experimental geometry meets the appropriate phase-matching criteria. This problem becomes all the more difficult if one wishes to be able to tune the frequencies of the laser pulses.

In Sections II A and B, we will cover the fundamental physics of the coherent 1D Raman spectroscopy of intramolecular and intermolecular vibrational modes, and we will discuss how polarization control can be used to enhance or suppress different spectral contributions in both cases.

### A. Intramolecular Vibrations

In discussing CARS, we will confine ourselves to the case in which the bandwidth of the laser pulses is considerably greater than the width of the vibrational line of interest, but is narrow enough compared to the anharmonicity of the vibrational mode that we need only consider transitions between the ground and first excited vibrational states. We consider a phase-matching geometry for which the signal wavevector is given by $\mathbf{k}_s = \mathbf{k}_1 - \mathbf{k}_2 + \mathbf{k}_3$. We denote by $E_n$ and $\boldsymbol{\eta}_n$ the envelope and polarization vector of pulse $n$, respectively. Note that the pulses need not interact in numerical order, so we will denote the time of the $n$th interaction by $t_n$.

We begin by considering how the first two pulses create a Raman coherence. The first interaction creates an electronic coherence between the ground electronic state and an excited electronic state that is far off resonance. The second interaction turns this electronic coherence into a vibrational coherence between the ground and first excited vibrational states of the electronic ground state. Since these pulses are time coincident, there are two possible ways to excite the same vibrational coherence with them (see Fig. 1). Integrating over the possible interaction times for each pulse for the first pathway (a) we find that the time dependence of the Raman coherence created at a single vibrational frequency $\omega_{g'g}$ is given by

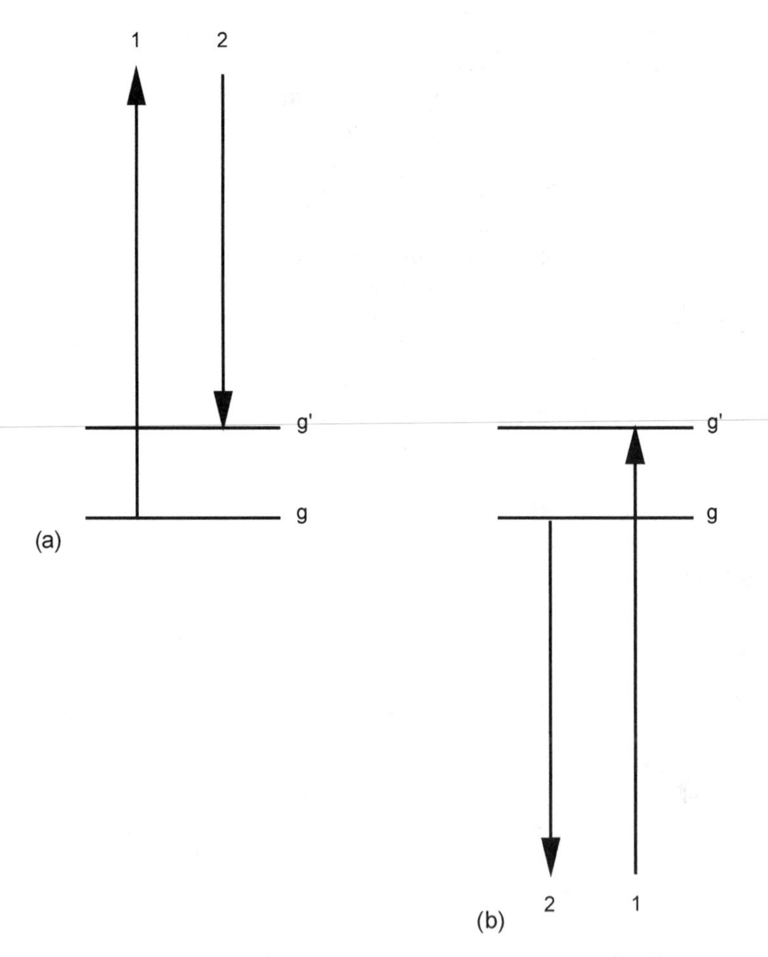


**Figure 1.** The two vibrationally resonant pathways through which two laser beams (1 and 2) can create a Raman coherence at frequency $\omega_1 - \omega_2$. The vibrational ground state is $g$ and the first vibrational excited state is $g'$.

$$R_a(\omega_{g'g};t_3) \propto e^{-i\mathbf{k}_1\cdot\mathbf{r}}e^{i\mathbf{k}_2\cdot\mathbf{r}}\sum_{\{e\}}(\boldsymbol{\mu}_{eg}\cdot\boldsymbol{\eta}_1)(\boldsymbol{\mu}_{g'e}\cdot\boldsymbol{\eta}_2)$$
$$\times\int_{-\infty}^{t_3}dt_2\int_{-\infty}^{t_2}dt_1E_1(t_1)E_2(t_2)e^{-i\omega_1t_1}e^{i\omega_2t_2}e^{-i\Omega_{eg}(t_2-t_1)}e^{-i\Omega_{g'g}(t_3-t_2)}, \quad (2.1)$$

where the sum is over all possible electronic excited states $e, g$ and $g'$ are the ground and an excited vibrational state in the ground electronic state, $\boldsymbol{\mu}_{ab}$ is the transition dipole moment between states $a$ and $b$, and the material Bohr frequencies are given by $\Omega_{ab} = \omega_{ab} - i\Gamma_{ab}$, where $\Gamma_{ab}$ is a phenomenological exponential damping constant that includes the effects of population relaxation and pure dephasing. Since the electronic states are far off-resonance, we can assume that the time delay between the first and second interactions is fleetingly small, such that the arguments of the electric-field envelopes are identical at the time of the interaction, so we can write

$$R_a(\omega_{g'g};t_3) \propto e^{-i\mathbf{k}_1\cdot\mathbf{r}}e^{i\mathbf{k}_2\cdot\mathbf{r}}\sum_{\{e\}}(\boldsymbol{\mu}_{eg}\cdot\boldsymbol{\eta}_1)(\boldsymbol{\mu}_{g'e}\cdot\boldsymbol{\eta}_2)$$
$$\times\int_{-\infty}^{t_3}dt_2\int_{-\infty}^{t_2}dt_1E_1(t_2)E_2(t_2)e^{-i\omega_1t_1}e^{i\omega_2t_2}e^{-i\Omega_{eg}(t_2-t_1)}e^{-i\Omega_{g'g}(t_3-t_2)} \quad (2.2)$$

Performing the integration over $t_1$ then yields

$$R_a(\omega_{g'g};t_3) \propto -ie^{-i\mathbf{k}_1\cdot\mathbf{r}}e^{i\mathbf{k}_2\cdot\mathbf{r}}\sum_{\{e\}}\frac{(\boldsymbol{\mu}_{eg}\cdot\boldsymbol{\eta}_1)(\boldsymbol{\mu}_{g'e}\cdot\boldsymbol{\eta}_2)}{\Omega_{eg}-\omega_1}$$
$$\times\int_{-\infty}^{t_3}dt_2E_1(t_2)E_2(t_2)e^{-i(\omega_1-\omega_2)t_2}e^{-i\Omega_{g'g}(t_3-t_2)} \quad (2.3)$$

In pathway (b), pulse 2 interacts before pulse 1. Going through the same steps leads to

$$R_b(\omega_{g'g};t_3) \propto ie^{-i\mathbf{k}_1\cdot\mathbf{r}}e^{i\mathbf{k}_2\cdot\mathbf{r}}\sum_{\{e\}}\frac{(\boldsymbol{\mu}_{eg}\cdot\boldsymbol{\eta}_2)(\boldsymbol{\mu}_{g'e}\cdot\boldsymbol{\eta}_1)}{\Omega_{eg}+\omega_2}$$
$$\times\int_{-\infty}^{t_3}dt_2E_1(t_2)E_2(t_2)e^{-i(\omega_1-\omega_2)t_2}e^{-i\Omega_{g'g}(t_3-t_2)} \quad (2.4)$$

As these pathways are indistinguishable, we must add their amplitudes together, which gives

$$R(\omega_{g'g};t_3) \propto ie^{-i\mathbf{k}_1\cdot\mathbf{r}}e^{i\mathbf{k}_2\cdot\mathbf{r}}\sum_{\{e\}}\left[\frac{(\boldsymbol{\mu}_{eg}\cdot\boldsymbol{\eta}_2)(\boldsymbol{\mu}_{g'e}\cdot\boldsymbol{\eta}_1)}{\omega_{eg}+\omega_2+i\Gamma_{eg}} - \frac{(\boldsymbol{\mu}_{eg}\cdot\boldsymbol{\eta}_1)(\boldsymbol{\mu}_{g'e}\cdot\boldsymbol{\eta}_2)}{\omega_{eg}-\omega_1+i\Gamma_{eg}}\right] \times \int_{-\infty}^{t_3} dt_2 E_1(t_2)E_2(t_2)e^{-i(\omega_1-\omega_2)t_2}e^{-i\Omega_{g'g}(t_3-t_2)} \tag{2.5}$$

Note that the integral is now independent of the electronic state. The quantity in the summation is generally written as

$$\sum_{\{e\}}\left[\frac{(\boldsymbol{\mu}_{eg}\cdot\boldsymbol{\eta}_2)(\boldsymbol{\mu}_{g'e}\cdot\boldsymbol{\eta}_1)}{\omega_{eg}+\omega_2+i\Gamma_{eg}} - \frac{(\boldsymbol{\mu}_{eg}\cdot\boldsymbol{\eta}_1)(\boldsymbol{\mu}_{g'e}\cdot\boldsymbol{\eta}_2)}{\omega_{eg}-\omega_1+i\Gamma_{eg}}\right] \equiv \boldsymbol{\eta}_1\cdot\tilde{\boldsymbol{\pi}}\cdot\boldsymbol{\eta}_2 \tag{2.6}$$

where $\tilde{\boldsymbol{\pi}}$ is the molecular polarizability tensor, which is also often denoted by $\tilde{\boldsymbol{\alpha}}$. Note that the polarizability tensor depends explicitly on $\omega_1$ and $\omega_2$, although this fact is generally safely ignored.

Pulse 3 arrives at some delay time $\tau$ after pulses 1 and 2 (Fig. 2). If we restrict our consideration to delay times that are significantly longer than the pulse

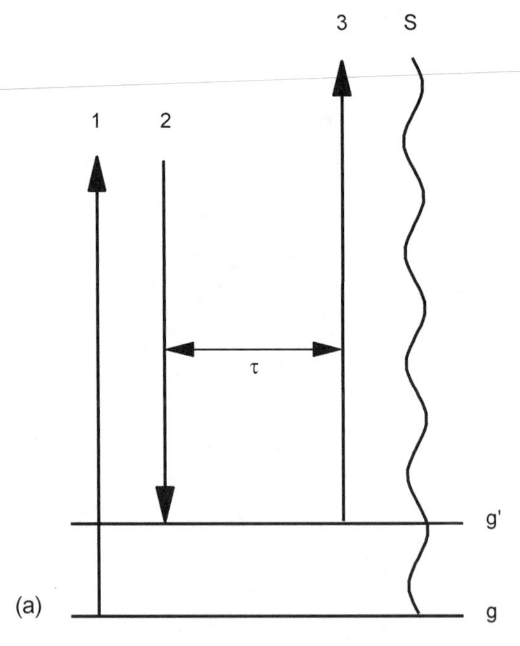


**Figure 2.** A representative wave-mixing pathway for CARS spectroscopy. Beams 1 and 2 create a Raman coherence that is allowed to propagate for time $\tau$ before being probed by pulse 3.

lengths, then the response at a single vibrational frequency can be approximated as

$$R(\omega_{g'g};t_s) \propto e^{-i(\mathbf{k}_1-\mathbf{k}_2+\mathbf{k}_3)\cdot\mathbf{r}}e^{-i(\omega_3+\omega_{g'g})t_s}e^{-i(\omega_1-\omega_2+\omega_3-\omega_{g'g})\tau} \times (\boldsymbol{\eta}_1\cdot\tilde{\boldsymbol{\pi}}\cdot\boldsymbol{\eta}_2)(\boldsymbol{\eta}_3\cdot\tilde{\boldsymbol{\pi}}\cdot\boldsymbol{\eta}_s)e^{-\Gamma_{g'g}\tau} \tag{2.7}$$

where the subscript $s$ denotes the signal. Within this set of approximations, we expect a CARS signal that decays as a single exponential with a time constant that depends on the dephasing rate. However, three modifications should be made to this simple picture. First, we must integrate over the inhomogeneous distribution of oscillator frequencies. This integration is straightforward, and is formally equivalent to including the Fourier transform (FT) of the inhomogeneity in the Raman line shape in the decay (indeed, the CARS decay can be thought of as a Raman free-induction decay, and is thus directly proportional to the FT of the full Raman line shape). Second, we should acknowledge that $\omega_{g'g}$ can depend on time, since the environment around a molecule in a liquid is ever-changing. This effect is taken into account by integrating the instantaneous frequency $\omega_{g'g}$ from time 0 to time $\tau$ inside the argument of the second exponent in Eq. (2.7). Kubo line shape theory [9] is commonly used to calculate how the time variation of this frequency affects the observed Raman coherence decay. Third, we must also recognize that $\tilde{\boldsymbol{\pi}}$ is also time dependent; at any given instant, its value depends on the value of the vibrational coordinate, the local environment, and the molecular orientation. We will consider the importance of each of these factors in turn.

The basis of Raman spectroscopy is that the polarizability tensor of a molecule depends on the values of the molecule's vibrational coordinates $q_i$. The dependence of $\tilde{\boldsymbol{\pi}}$ on the coordinates is in general complicated and difficult to calculate. As a result, the coordinate dependence is generally expressed in terms of a Taylor-series expansion:

$$\begin{aligned}\tilde{\boldsymbol{\pi}}(q_1,\ldots,q_N) &= \tilde{\boldsymbol{\pi}}^{(0)} + \sum_i \left(\frac{\partial\tilde{\boldsymbol{\pi}}}{\partial q_i}\right)_0 q_i + \frac{1}{2}\sum_{i,j}\left(\frac{\partial^2\tilde{\boldsymbol{\pi}}}{\partial q_i\partial q_j}\right)_0 q_iq_j + \cdots \\ &= \tilde{\boldsymbol{\pi}}^{(0)} + \sum_i \tilde{\boldsymbol{\pi}}_i^{(1)} q_i + \frac{1}{2}\sum_{i,j}\tilde{\boldsymbol{\pi}}_{ij}^{(2)} q_iq_j + \cdots\end{aligned} \tag{2.8}$$

where $\tilde{\boldsymbol{\pi}}^{(0)}$ is the polarizability in the equilibrium geometry and the derivatives are evaluated at this same point. We denote the polarizations of beams 1–3 and the signal beam by $\alpha$, $\beta$, $\gamma$, and $\delta$, respectively. If $\omega_1 - \omega_2$ is set to be resonant with the vibrational frequency of mode $i$, then to first approximation CARS is sensitive to the correlation function

$$\langle \tilde{\boldsymbol{\pi}}^{(1)}_{i,\gamma\delta}(t)\tilde{\boldsymbol{\pi}}^{(1)}_{i,\alpha\beta}(0)q_i(\tau)q_i(0)\rangle \tag{2.9}$$

Within this so-called Placzek approximation [10], a mode is Raman active if the first derivative of the polarizability with respect to the mode evaluated at the equilibrium geometry is not zero. Note that although the vibrational modes themselves are time dependent, we have also specifically made the $\tilde{\boldsymbol{\pi}}^{(1)}$ tensor elements in Eq. (2.9) time dependent. The time dependence of these elements accounts for the fact that time-dependent dipole/induced-dipole interactions with neighboring molecules that do not depend on the modes $q$ can influence the effective polarizability tensor of a molecule and the fact that molecular reorientation affects the elements of the polarizability tensor. In intramolecular Raman spectroscopy, the former effect is usually a minor one, whereas the latter effect can play an important role in determining the line shape. Furthermore, so long as the vibrational and reorientational dynamics are reasonably decoupled, Eq. (2.9) can be factorized into the product of the polarizability correlation function and the vibrational correlation function.

It is convenient to separate the molecular polarizability tensor into its isotropic and anisotropic components, $\alpha$ and $\beta$. These components are given by

$$\alpha = \tfrac{1}{3}Tr(\tilde{\boldsymbol{\pi}}) \tag{2.10}$$

and

$$\beta = \sqrt{\frac{1}{2}\left[(\pi_{xx} - \pi_{yy})^2 + (\pi_{xx} - \pi_{zz})^2 + (\pi_{yy} - \pi_{zz})^2\right]} \tag{2.11}$$

Since the trace of a tensor is rotationally invariant and since the trace of the derivative of a tensor with respect to a coordinate is equal to the derivative of the trace, the isotropic portion of the polarizability is not affected by molecular reorientation. On the other hand, reorientation does change the anisotropic portion of the derivative of the polarizability tensor.

Eperiments performed with different polarization conditions are sensitive to the isotropic and anisotropic correlation functions to different extents [8, 11]. The three most important cases are the depolarized $(xyxy)$ correlation function

$$C_{\text{dep}}(\tau) = \tfrac{1}{15}\langle\beta(\tau)\beta(0)\rangle \tag{2.12}$$

the isotropic ($xxmm$, where $m$ denotes the magic angle, 54.7°) correlation function

$$C_{\text{iso}}(\tau) = 3\langle\alpha(\tau)\alpha(0)\rangle \tag{2.13}$$

and the polarized $(xxxx)$ correlation function

$$C_{\text{pol}}(\tau) = \langle \alpha(\tau)\alpha(0) \rangle + \tfrac{4}{45} \langle \beta(\tau)\beta(0) \rangle \tag{2.14}$$

Note that the depolarized correlation function depends only on $\beta$ and the isotropic correlation function depends only on $\alpha$.

In spontaneous Raman spectroscopy, it is often much easier to measure the polarized and depolarized spectra than it is to measure the isotropic spectrum. However, the polarized correlation function depends on both $\alpha$ and $\beta$, and so the isotropic correlation function must be derived from the polarized and depolarized correlation functions via the relation

$$C_{\text{iso}}(\tau) = 3C_{\text{pol}}(\tau) - 4C_{\text{dep}}(\tau) \tag{2.15}$$

In CARS and other coherent Raman spectroscopies, the direct measurement of the isotropic response or spectrum is much more straightforward.

A common means of measuring the relative strengths of the isotropic and anisotropic responses in spontaneous Raman spectroscopy is the depolarization ratio $\rho$, which is the ratio of the depolarized and polarized spectra [6]. As can be seen from Eqs. (2.12) and (2.14), the depolarization ratio for a completely depolarized mode is 0.75, while that for a completely isotropic mode (which is often called a polarized mode) is 0. The magnitude of the depolarization ratio of a mode depends on its symmetry. In particular, a completely depolarized mode must change the trace of the polarizability symmetrically about its equilibrium point; an example would be the in-plane ring distortion mode of benzene. On the other hand, a completely polarized mode cannot change the anisotropy of the polarizability tensor; an example would be the breathing mode of carbon tetrachloride.

Since the isotropic correlation function is not affected by molecular reorientation, for modes that are not completely depolarized the isotropic response can be measured to isolate the vibrational correlation function. Assuming that the vibrational and reorientational dynamics are separable, for a mode that is partly polarized dividing the depolarized response by the isotropic response yields the second-rank, single-molecule orientational correlation function.

### B. Intermolecular Vibrations

Liquids are dense enough that free rotation and translation of molecules is not possible. Instead, frustration of these degree of freedom leads to pseudooscillatory rotational and translational motions of a molecule within the potential energy cage formed by its nearest neighbors. In most liquids, these intermolecular modes are confined to frequencies $< 200\,\text{cm}^{-1}$. A significant subset of the intermolecular modes of a liquid affect its polarizability, and can therefore be probed via Raman transitions.

The spontaneous technique that is used to probe Raman-active intermolecular modes is called Rayleigh-wing or low-frequency Raman scattering [12]. A number of coherent time-domain techniques have also been developed to study these modes. Such time-domain techniques are often classified as optical Kerr effect (OKE) spectroscopies [13, 14].

While OKE spectroscopy has a considerable amount in common with the CARS spectroscopy of high-frequency intramolecular modes, there are also some significant differences. First, the bandwidth of the laser pulses that are commonly used to perform OKE spectroscopy is broad enough to cover the entire intermolecular spectrum; thus, it is not necessary to employ two or more lasers that are tuned to different wavelengths. In fact, a single beam can be used to create the initial coherence, which can be probed by a second beam. Second, the intermolecular modes are of low enough frequency to have substantial excited-state populations, such that downward vibrational transitions contribute to the signal in addition to the usual upward transitions. Furthermore, the existence of significant excited-state populations implies that anharmonicity can play an important role in intermolecular spectra.

The intermolecular spectrum is related to the time correlation function of the many-body polarizability of a liquid, $\tilde{\mathbf{\Pi}}$. The many-body polarizability is affected by both single-molecule and collective effects in the liquid. Single-molecule scattering arises from the oscillation of individual molecules due to frustrated motions. Collective (or interaction-induced) scattering, on the other hand, is the result of dipole/induced-dipole interactions between molecules. It is instructive at this point to consider the properties of both types of scattering in some detail.

The contribution of single-molecule effects to the many-body polarizability is found by summing over the individual molecular polarizability tensors:

$$\tilde{\mathbf{\Pi}}_{\mathrm{sm}}(t) = \sum_{i=1}^{N} \tilde{\boldsymbol{\pi}}_i(t) \tag{2.16}$$

As in the case of intramolecular Raman spectroscopy, the leading contribution to the intermolecular Raman spectrum arises from the first derivative of the many-body polarizability with respect to each of the modes, which we denote $\tilde{\mathbf{\Pi}}_i^{(1)}$. Translational motion of a molecule does not affect its polarizability tensor, so single-molecule scattering must arise purely from frustrated rotations (librations) of molecules. Since the trace of the polarizability tensor of a molecule is rotationally invariant, only the anisotropic portion of the molecular polarizability contributes to the single-molecule portion of the intermolecular spectrum. As a result, single-molecule scattering is always completely depolarized. Of course, the intensity of the single-molecule scattering per unit volume increases with increasing density.

Interaction-induced (II) scattering occurs when the laser-induced dipole moment on one molecule induces a dipole moment on another molecule. Interaction-induced scattering was first discovered in noble gas fluids, and as such is often attributed to density fluctuations even in molecular liquids. However, dipole/induced-dipole interactions are also modulated by the same frustrated rotational translational motions that lead to single-molecule scattering. Consequently, there can be a strong (and generally negative) cross-term between single-molecule and orientational II scattering [15]. Neither orientational nor translational fluctuations are guaranteed to preserve the trace of $\tilde{\mathbf{\Pi}}$, so II scattering need not be completely depolarized (and, indeed, the isotropic OKE spectrum arises entirely from II scattering). However, isotropic II scattering does require that the environment around each molecule be significantly anisotropic. That this criterion is rarely met is attested to by the fact that the depolarization ratio for Rayleigh-wing scattering rarely deviates significantly from 0.75 in liquids. Finally, we should note that three-body effects tend to cancel two-body effects in II scattering, which means that the II scattering intensity generally decreases with increasing liquid density [16].

There has been a considerable amount of debate as to whether depolarized Rayleigh-wing scattering in liquids composed of molecules with anisotropic polarizabilities is dominated by single-molecule or II effects, or, if II effects do dominate, whether orientational or translational motions play a larger role [16–19]. A recent simulation of $CS_2$ suggested that the depolarized Rayleigh-wing spectrum of this liquid is strongly dominated by single-molecule scattering, and that orientational and translational motions contribute nearly equally to the isotropic spectrum [20]. Since II effects should be the strongest in highly polarizable liquids, it seems likely that these results hold true for most molecular liquids.

The Raman excitation step in OKE spectroscopy also acts to create a partial alignment of the liquid molecules. Consequently, the depolarized OKE decay also contains a contribution from orientational diffusion. Unlike the depolarized intramolecular Raman spectrum, which is sensitive to the second-rank single-molecule orientational correlation function, the OKE decay is sensitive to the second-rank collective orientational correlation function [21]. For a symmetric-top molecule, the single-molecule and collective orientational correlation times are related by

$$\tau_{\mathrm{coll}} = \frac{g_2}{j_2}\tau_{\mathrm{sm}} \tag{2.17}$$

where $g_2$ and $j_2$ are the static and dynamic orientational pair correlation parameters, respectively. The latter parameter generally takes on a value near unity for liquids [21], whereas $g_2$, which measures the degree of orientational ordering, is $\geq 1$. Thus, the collective orientational correlation time is always greater than or equal to the single-molecule orientational correlation time.

Figure 3(*a*) illustrates typical OKE decays for simple liquids, which were obtained in benzene and $CS_2$ at room temperature [22]. The long-time exponential tails in these decays arise from orientational diffusion. If the reorientational tails are removed from the decays, a second exponential decay is revealed [Fig. 3(*b*)]. Although this so-called intermediate exponential [23] decay is a common feature in the OKE response of simple liquids, its source is not yet understood. At even earlier delay times, the OKE response is notably nonexponential. It has recently been demonstrated that timescale of the intermediate response is approximately four times faster than that for orientational diffusion for a large number of liquids, which has led to the suggestion that the intermediate response arises from structural fluctuations [22]. A FT deconvolution procedure [24] can also be used to convert the OKE decays into Bose–Einstein-corrected Rayleigh-wing spectra. The spectra corresponding to the decays in Figure 3(*a*) and (*b*) are shown in Figure 4(*a*) and (*b*), respectively.

## III. RAMAN-ECHO SPECTROSCOPY

As discussed above, the CARS decay is the FT of the Raman line shape, and therefore contains information identical to that in the corresponding spontaneous Raman spectrum. In particular, the CARS decay is affected by both homogeneous and inhomogeneous broadening. Although the distinction between these types of broadening is perhaps not so clear in liquids as it is in other media, it is still desirable to be able to effect a separation of the contributions of relatively slow and relatively fast fluctuations to the Raman line shape.

The problem of separating the homogeneous and inhomogeneous contributions to the line shape was solved long ago in NMR spectroscopy with the advent of the spin–echo pulse sequence [1, 25]. This sequence starts by using a $\pi/2$ pulse to create a spin coherence that is allowed to evolve for delay time $\tau$, at which point a spin of frequency $\omega$ in the rotating frame has picked up a phase of $\omega\tau$ radians. A $\pi$ pulse is then used to reverse the sign of the coherence. After another time period of duration $\tau$, the net phase of the spin has returned to zero. Thus, all of the spins in the system rephase at a total delay time after the first pulse of $2\tau$, removing the effects of inhomogeneous broadening. The large, coherent magnetization in the sample at time $2\tau$ is the spin echo. Thus, by monitoring the strength of the echo as a function of $\tau$, it is possible to isolate the portion of the coherence decay that arises from homogeneous processes.

It was realized subsequently that a similar idea could be used to rephase inhomogeneity in electronic coherences [26]. This technique, which is called the photon echo, has found considerable use in the spectroscopy of materials such as low-temperature glasses [27]. More recently, photon echoes have become a powerful tool for the study of electronic dephasing of dye molecules in

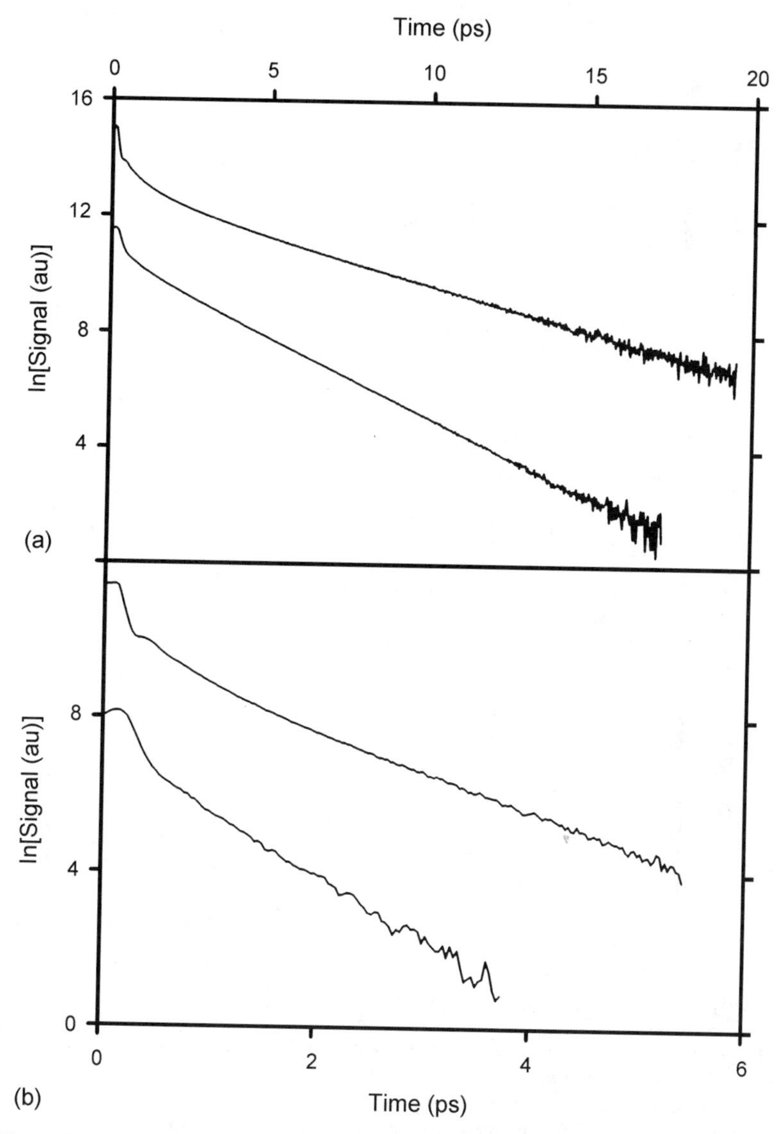


**Figure 3.** Room temperature OKE data for bulk benzene (upper trace in each panel) and $CS_2$ (lower trace in each panel). Panel (*a*) shows the complete OKE decays whereas panel (*b*) shows decays with the orientational diffusion component removed. The decays in each panel have been offset for clarity. (Adapted from [22].)

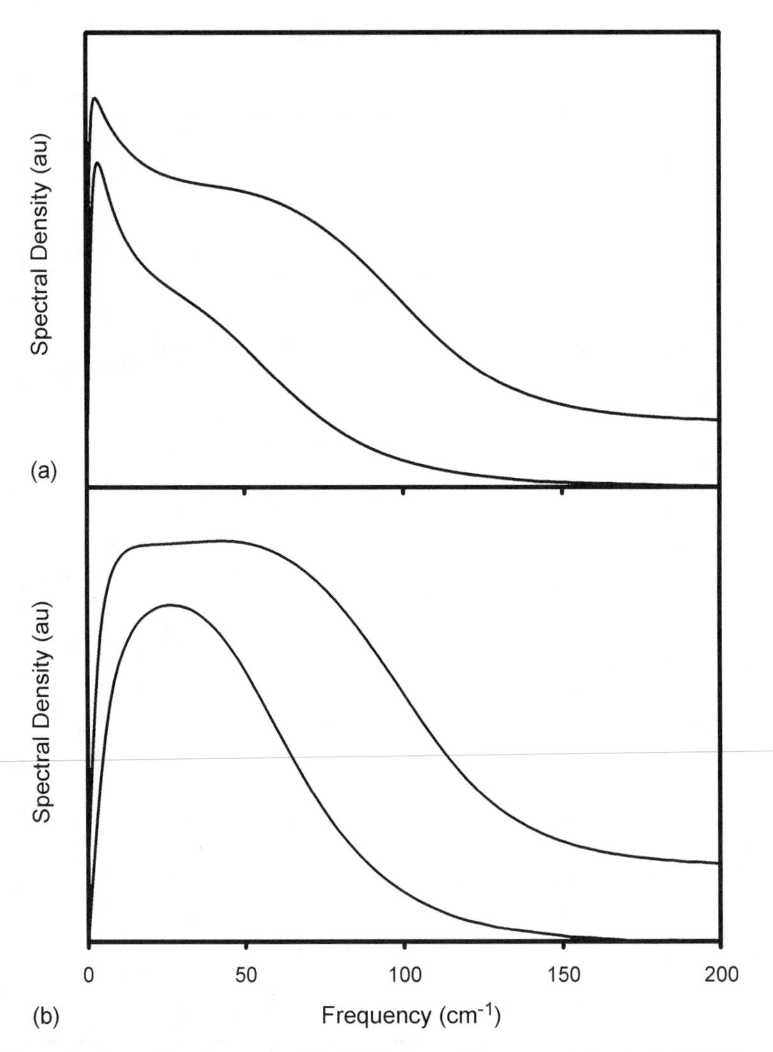


**Figure 4.** Bose–Einstein corrected Rayleigh-wing spectra corresponding to the OKE decays in Figure 3. The upper trace in each panel is for benzene and the lower trace is for $CS_2$. Panel (*a*) is the spectra calculated from the complete OKE decays and panel (*b*) is the spectra from the decays from which the orientational diffusion component has been subtracted. The spectra in each panel have been offset for clarity.

solution [28]. Several years after the advent of the photon echo, it was suggested that optical echo experiments could also be performed via Raman transitions rather than electronic transitions [29]. In this section, we will discuss this so-called Raman-echo spectroscopy.

### A. Background

The basic scheme for the Raman echo is shown in Figure 5. The technique generally involves five laser pulses of at least two different colors. Two time-coincident pulses of light (1 and 2) are used to create a Raman coherence at frequency $\omega_{g'g}$ that is allowed to evolve for time $\tau_1$, after which the response for a single vibrational frequency is given by

$$R(\omega_{g'g}; \tau_1) \propto ie^{-i(\mathbf{k}_1-\mathbf{k}_2)\cdot\mathbf{r}} e^{-i\omega_{g'g}\tau_1} (\boldsymbol{\eta}_1 \cdot \tilde{\boldsymbol{\pi}}(0) \cdot \boldsymbol{\eta}_2) e^{-\Gamma_{g'g}\tau_1} \tag{3.1}$$

where we have assumed that $\tau_1$ is appreciably longer than the pulse length. At this point, another pulse pair (3 and 4) is incident upon the sample. Each pulse in the second pair interacts with the system twice, reversing the coherence to be at frequency $\omega_{gg'}$. This coherence is allowed to evolve for a second delay time $\tau_2$, after which the response for a single vibrational frequency is given by

$$\begin{aligned} R(\omega_{g'g}; \tau_1 + \tau_2) \propto\ & e^{-i(\mathbf{k}_1-\mathbf{k}_2+2\mathbf{k}_3-2\mathbf{k}_4)\cdot\mathbf{r}} e^{-i\omega_{g'g}(\tau_1-\tau_2)} \\ & \times (\boldsymbol{\eta}_1 \cdot \tilde{\boldsymbol{\pi}}(0) \cdot \boldsymbol{\eta}_2)(\boldsymbol{\eta}_3 \cdot \tilde{\boldsymbol{\pi}}(\tau_1) \cdot \boldsymbol{\eta}_4)^2 e^{-\Gamma_{g'g}(\tau_1+\tau_2)} \end{aligned} \tag{3.2}$$

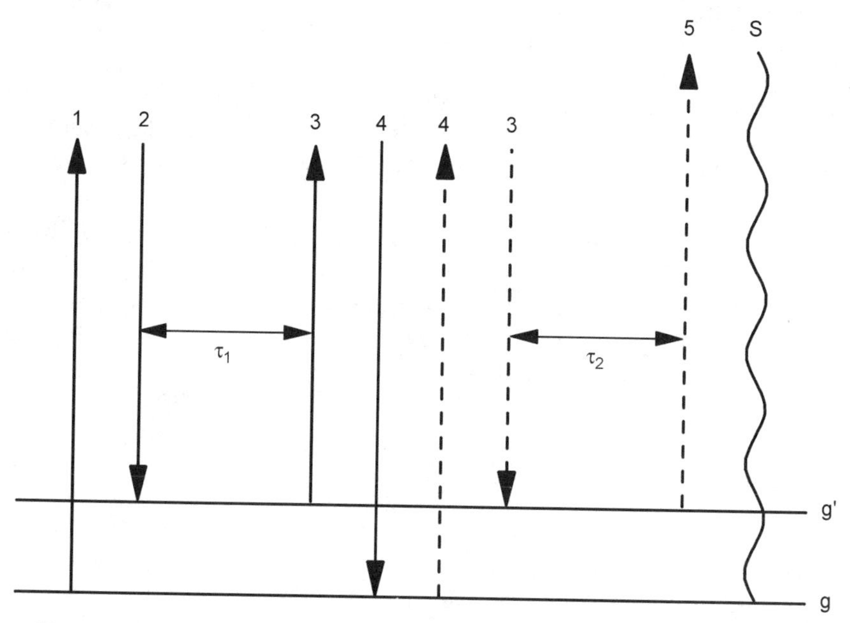


**Figure 5.** A representative wave-mixing pathway for Raman-echo spectroscopy. Beams 1 and 2 create a Raman coherence that is allowed to propagate for time $\tau_1$. Beams 3 and 4 then reverse the frequency of the coherence. After a second propagation time $\tau_2$ the coherence is probed by pulse 5.

Note that the coherence has a net phase of $\omega_{g'g}(\tau_1 - \tau_2)$, whereas the degree of damping is proportional to $\tau_1 + \tau_2$. In addition, the factor $[\boldsymbol{\eta}_3 \cdot \tilde{\boldsymbol{\pi}}(\tau_1) \cdot \boldsymbol{\eta}_4]$ is squared, since there are two Raman interactions with pulses 3 and 4 after the first delay.

A final pulse (5) interacts with the sample at a delay time $\tau_2$ after the second pulse pair, and a signal beam is generated by the scattering of this probe pulse off of any remaining Raman coherence, yielding

$$R(\omega_{g'g}; t_s) \propto e^{-i(\mathbf{k}_1 - \mathbf{k}_2 + 2\mathbf{k}_3 - 2\mathbf{k}_4 + \mathbf{k}_5)\cdot\mathbf{r}} e^{-i\lfloor\omega_5(\tau_1+\tau_2)+\omega_{g'g}(\tau_1-\tau_2)\rfloor} \times [(\boldsymbol{\eta}_1 \cdot \tilde{\boldsymbol{\pi}}(0) \cdot \boldsymbol{\eta}_2)(\boldsymbol{\eta}_3 \cdot \tilde{\boldsymbol{\eta}}(\tau_1) \cdot \boldsymbol{\eta}_4)]^2 [\boldsymbol{\eta}_5 \cdot \tilde{\boldsymbol{\pi}}(\tau_1 + \tau_2) \cdot \boldsymbol{\eta}_s] e^{-\Gamma_{g'g}(\tau_1+\tau_2)} \tag{3.3}$$

As was the case for CARS, the response must now be integrated over all oscillator frequencies, and time-dependent oscillator frequencies must be introduced if needed. Notice that while the damping of the signal depends on $\tau_1 + \tau_2$, the coherence frequency at $t_s$ depends on $\tau_1 - \tau_2$. Thus, in a sample with a large degree of inhomogeneous broadening an echo will occur, and the maximum signal will occur when $\tau_1 = \tau_2$. On the other hand, if the line broadening is predominantly homogeneous, echo-like behavior will not be observed. In this case, the maximum signal will instead be observed at $\tau_2 = 0$.

It is evident from Eq. (3.3) that the Raman-echo signal will depend on the polarizability of the system at three different times. In fact, since the second pulse pair is responsible for two Raman transitions, we need consider a four-time correlation function [30] for which two of the times are identical:

$$C_{\alpha\beta\gamma\delta\varepsilon\zeta}(\tau_1, \tau_2) = \langle \tilde{\boldsymbol{\pi}}_{\varepsilon\zeta}(\tau_1 + \tau_2)\tilde{\boldsymbol{\pi}}_{\gamma\delta}(\tau_1)\tilde{\boldsymbol{\pi}}_{\gamma\delta}(\tau_1)\tilde{\boldsymbol{\pi}}_{\alpha\beta}(0)\rangle \tag{3.4}$$

where $\alpha, \beta, \gamma, \delta, \varepsilon$, and $\zeta$ are the polarizations of pulses $1, 2, 3, 4, 5$, and the signal, respectively. We can again make the Placzek approximation [10] and factorize the correlation function, leading to

$$\begin{aligned} C_{\alpha\beta\gamma\delta\varepsilon\zeta}(\tau_1, \tau_2) =& \langle \tilde{\boldsymbol{\pi}}^{(1)}_{\varepsilon\zeta}(\tau_1 + \tau_2)\tilde{\boldsymbol{\pi}}^{(1)}_{\gamma\delta}(\tau_1)\tilde{\boldsymbol{\pi}}^{(1)}_{\gamma\delta}(\tau_1)\tilde{\boldsymbol{\pi}}^{(1)}_{\alpha\beta}(0)\rangle \\ & \times \langle q(\tau_1 + \tau_2)q(\tau_1)q(\tau_1)q(0)\rangle \end{aligned} \tag{3.5}$$

Note also that as opposed to CARS spectroscopy, which relies on a third-order nonlinearity, the Raman echo depends on a seventh-order nonlinearity. Thus, the Raman echo is a much more difficult technique to use than is CARS. However, the Raman echo has been used successfully to study dynamics in solid- and gas-phase systems and, more recently, in liquids. In Sections III B and C, we will review the results of these experiments.

### B. Experiments in Solids and Gases

Although the Raman echo was first proposed in 1968 [29], performing an experiment relying on seventh-order nonlinearity presented enough difficulties that the first observation of this phenomenon was not made until 1976. Hu et al. [31] used a two-color nanosecond dye laser to pump Raman spin-flip transitions of bound donors in CdS in a magnetic field. A cavity-dumped argon ion laser was used to probe the spin coherences. In samples with dilute impurities, their data show a clear echo in the appropriate unique spatial direction, demonstrating that there is significant inhomogeneity in the system. Upon increasing the donor concentration the echo disappears, which is indicative of rapid loss of phase memory due to interactions between donors.

The first successful gas-phase Raman-echo experiments were reported in 1982 by Leung et al. [32, 33]. They used nanosecond dye-laser pulses to drive 7793-$cm^{-1}$ Raman transitions between the $6P_{1/2}$ ground state and the $6P_{3/2}$ excited state of atomic thallium vapor. By employing laser frequencies that were only slightly detuned from the $6P_{1/2}$-$7S_{1/2}$ and $6P_{3/2}$-$7S_{1/2}$ transitions, they were able to produce sizable echo signals [32]. The observed echo strength was found to scale as expected with the square of the number density of thallium atoms, although saturation effects were observed at high densities and were attributed to loss of phase matching and nonnegligible population transfer in the gas [32]. In addition, Raman-echo studies in the presence of significant buffer gas pressures revealed that phase-interrupting collisions are the primary mechanism of pure dephasing in thallium gas, whereas velocity-changing collisions play a negligible role [33]. This latter result can be explained by the fact that the considerable mass of a thallium atom prevents it from experiencing large velocity changes in single collisions, even when the collision partner is as large as xenon. In further support of this idea, the phase-changing collision cross-sections for various noble buffer gases as calculated from the thallium Raman-echo data were found to be in good agreement with theoretical values [33].

In the same year, Langelaar et al. [34] reported the first Raman-echo data obtained in a molecular gas. These initial experiments, which were performed using nanosecond laser pulses, involved the ground and first excited vibrational states of $N_2$ [34]. As in the case of thallium vapor [32], the echo intensity was found to increase quadratically with density before saturating [34]. These experiments were followed up by an additional study of the same system employing picosecond laser pulses [35]. The maximum signal intensity in these experiments was weak (100 aJ per pulse), but it was strong enough to monitor the echo decay out to $>100$ ps [35].

### C. Experiments in Liquids

It is considerably more challenging to perform Raman-echo experiments on liquids than on gases or on dilute impurities in solids. Beyond the problem of the Raman echo depending on a seventh-order nonlinearity, a number of other difficulties can arise in liquids. First, the frequency dependence of the dispersion in liquids is large enough that meeting the appropriate phase-matching criterion for a Raman echo can become a serious issue. Second, the interactions in liquids are strong enough and the dynamic timescale fast enough that the degree of inhomogeneity is necessarily far less than that in gases. Consequently, Raman-echo experiments in liquids must be performed with shorter laser pulses than can be used in gases, but the pulses must still be long enough that they do not excite transitions from the first to the second vibrational excited state of the mode of interest. For these and other reasons, early attempts to use the Raman echo to study vibrational dynamics in liquids met with failure [36]. Indeed, it was argued that so many other undesired processes would occur at the intensities of light needed to observe Raman-echo signals in liquids that the experiment was impossible [36].

These practical difficulties were overcome by Vanden Bout et al. [37], who reported the first Raman-echo signals from a liquid in 1991. The subject of this groundbreaking experiment was the symmetric methyl stretch of acetonitrile, which was chosen in part because it has a strongly polarized Raman transition, such that reorientational effects will not influence the echo signal significantly. Representative Raman-echo data obtained by scanning $\tau_2$ for various fixed values of $\tau_1$ are shown in Figure 6. For comparison, the solid lines show the shape of the Raman free-induction decay (FID) under the same conditions. Note that there is no sign of any peak at $\tau_1 = \tau_2$ in the Raman-echo data; indeed, for every value of $\tau_1$ the echo data are indistinguishable from the FID data. On this basis, it was concluded that the Raman spectrum of the symmetric methyl stretch of acetonitrile is homogeneously broadened [37].

Following this initial application of Raman-echo spectroscopy to the study of intramolecular vibrational dynamics in liquids, Yoshihara and co-workers [38, 39] also successfully used this technique to study the C≡N stretching mode of benzonitrile and the S—H stretch of ethyl mercaptan [40]. In both cases, the Raman line widths were found to be dominated entirely by homogeneous dephasing [38–40]. In fact, to date there are no Raman-echo data that show evidence of any significant inhomogeneous broadening in a pure liquid at ambient temperature.

Berg and co-workers [30, 41] used Schweizer and Chandler's [42] theory of vibrational dephasing in liquids to suggest an explanation for the absence of any appreciable inhomogeneity in intramolecular Raman lines in pure liquids. Based on the idea that relatively slow fluctuations in attractive forces lead to vibrational dephasing, Schweitzer and Chandler [42] predicted that the inhomogeneous line width $\Delta$ can be approximated by the relation

$$\Delta^2 = \frac{\chi(\delta\omega)^2}{N} \tag{3.6}$$

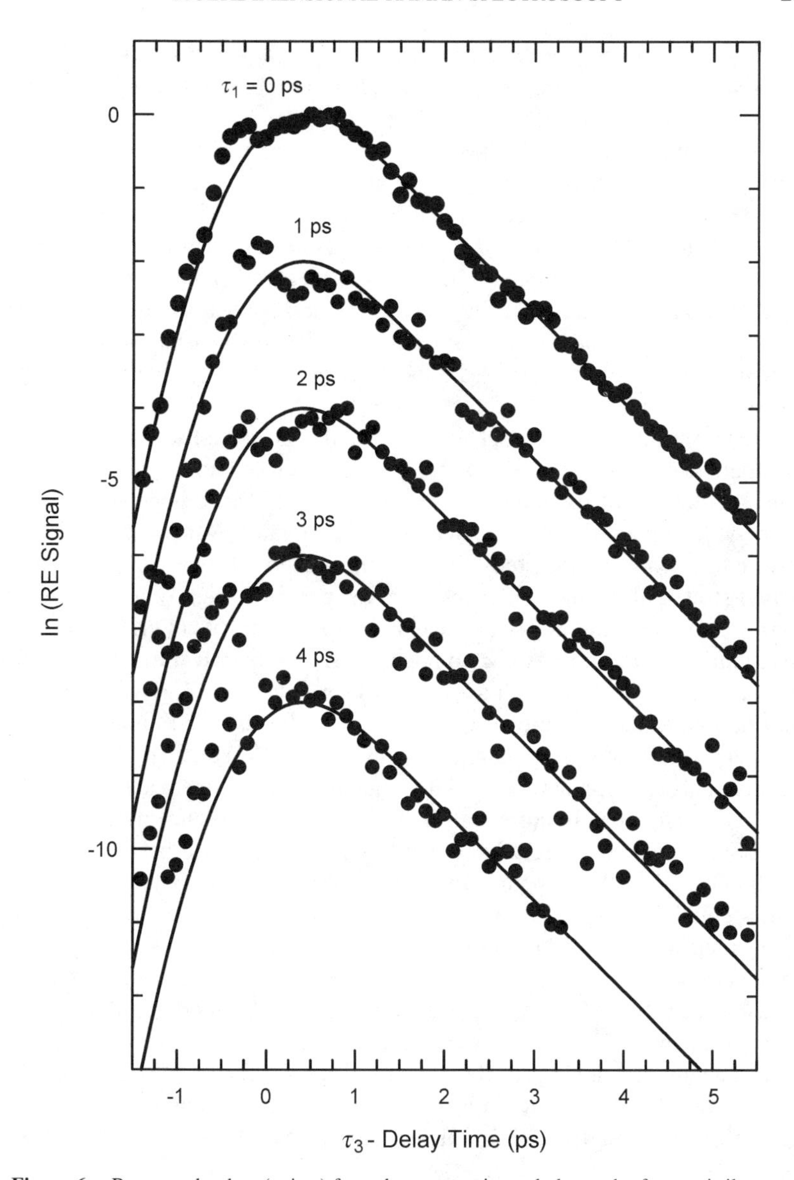


**Figure 6.** Raman-echo data (points) from the symmetric methyl stretch of acetonitrile at room temperature. The lines are the corresponding Raman free-induction decays. Note that the echo decay along $\tau_2$ matches the free-induction decay for all values of $\tau_1$, indicating that there is no appreciable inhomogeneous broadening in the Raman spectrum of this mode. (Adapted from [37] copyright (1991) by the American Physical Society.)

where $\chi$ is the isothermal compressibility of the liquid, $\delta\omega$ is the vibrational frequency shift upon going from the gas phase to the liquid phase, and $N$ is the number of molecules interacting directly with any given vibrating molecule. Initial attempts to apply this theory to Raman spectra of liquids such as acetonitrile suggested that significant inhomogeneous broadening should exist [43, 44]. Muller et al. pointed out that because it is molecules in the first solvation shell that lead to vibrational dephasing, one should use the considerably smaller value of the compressibility of this shell rather than that of the bulk liquid in evaluating Eq. (3.6). In essence, the first solvent shell is such a stable structure that the number of molecules in it is relatively constant, which does not allow for enough fluctuation of the local density to lead to significant inhomogeneous broadening.

Even if the inhomogeneity of a vibrational line is insignificant in a pure liquid, one might expect to find considerable inhomogeneous broadening in the same line upon vitrification of the liquid, since static structural inhomogeneity is a characteristic feature of glasses [45]. With this idea in mind, Vanden Bout et al. [46], performed a comparative study of the Raman-echo behavior of the symmetric methyl stretch of ethanol-1,1-$d_2$ at room temperature, slightly below the glass-transition temperature (80 K), and deeply in the glassy state (12 K). As expected, this Raman line was found to be homogeneously broadened at room temperature. However, the same was found to be the case even at 12 K. These results suggest that intramolecular vibrational redistribution (IVR) is rapid enough to play a dominant role in the relaxation of the symmetric methyl stretch of ethanol, allowing this mode to exchange energy with other modes on a time scale of 300 fs [46]. If this interpretation is correct, IVR is likely to contribute more strongly to vibrational relaxation in relatively small molecules than had been believed.

Muller et al. [30], recognized that even if slow density fluctuations in pure liquids could not engender enough inhomogeneity to be observed in a Raman-echo experiment, concentration fluctuations in binary mixtures had the possibility of providing a considerably greater degree of inhomogeneous broadening. Indeed, broadening of Raman lines upon mixing had been observed in numerous binary systems [47–49]. To search for concentration-fluctuation-induced inhomogeneous broadening, a 50% binary mixture of methyl iodide in $CDCl_3$ was chosen [30], based on strong evidence for inhomogeneous broadening of the symmetric methyl stretch in this system [47, 48]. As shown in Figure 7, the

**Figure 7.** Raman-echo data (points) from the symmetric methyl stretch of methyl iodide in a 50% mixture in $CDCl_3$ at room temperature. The lines are fit to various models for the decays, and the model that incorporates fast and intermediate modulation of the vibrational frequency (dotted lines) clearly works best. The echo decay along $\tau_2$ changes significantly as a function of $\tau_1$, demonstrating that inhomogeneous broadening does make a significant contribution to the width of this Raman mode. (Adapted from [30].)

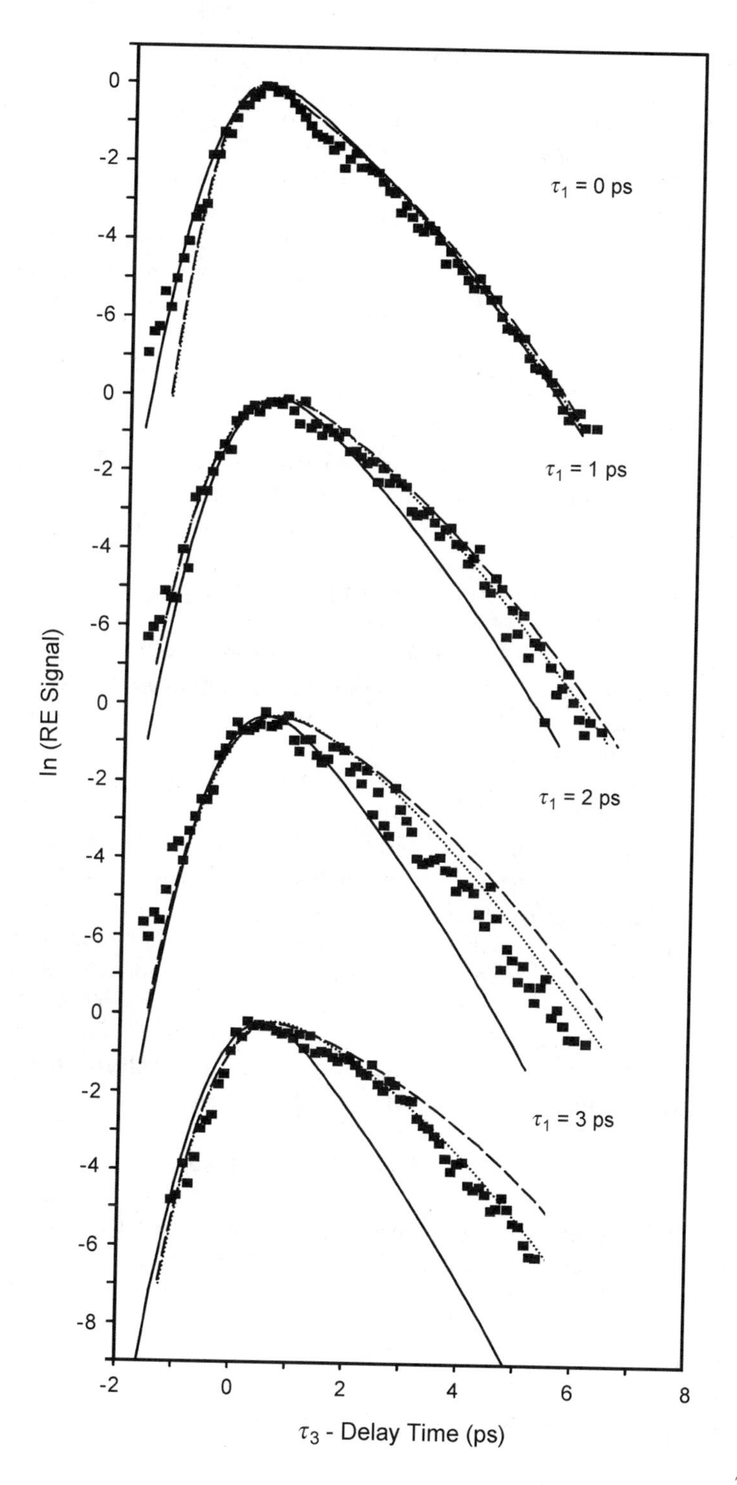

0
-2
-4
-6
0
-2
-4
-6
0
-2
-4
-6
0
-2
-4
-6
-8
-2
0
2
4
6
8
ln (RE Signal)
$\tau_1$ = 0 ps
$\tau_1$ = 1 ps
$\tau_1$ = 2 ps
$\tau_1$ = 3 ps
$\tau_3$ - Delay Time (ps)

shape of the Raman-echo decay along $\tau_2$ does change significantly as $\tau_1$ is increased, and for large values of $\tau_1$ the decay differs considerably from the FID. These data represented the first unambiguous Raman-echo evidence for inhomogeneous vibrational broadening in a liquid. To interpret their results, Muller et al. [42] extended Schweitzer–Chandler theory to binary mixtures, finding in analogy to Eq. (3.6) that the inhomogeneous broadening in such systems is given by

$$\Delta^2 = \frac{\chi(x_A \delta\omega_A + x_B \delta\omega_B)^2}{N} + \frac{x_A x_B (\delta\omega_A - \delta\omega_B)^2}{N} \tag{3.7}$$

where $x_A$ and $x_B$ are the mole fractions of the two components and $\delta\omega_A$ and $\delta\omega_B$ are the gas-to-liquid frequency shifts of the vibration of interest in the pure liquid (A) and in the partner liquid (B). The first term in Eq. (3.7) again describes the role of density fluctuations, whereas the second term arises from concentration fluctuations. So long as the gas-to-liquid frequency shifts of the two components differ considerably, the second term in Eq. (3.7) has the potential to be considerably larger than the first term. In essence, concentration fluctuations can play a large role in inhomogeneous broadening because while the number of molecules in the first solvent shell cannot vary considerably, the composition of this shell can. If the vibrational frequency depends strongly on this composition (i.e., if $|\delta\omega_A - \delta\omega_B|$ is large), then inhomogeneous broadening can be significant.

Based on these Raman-echo experiments, the inhomogeneous broadening of the symmetric methyl stretch of methyl iodide in $CDCl_3$ was determined to be $5.15\ \mathrm{cm}^{-1}$. Assuming that the 10.1-$\mathrm{cm}^{-1}$ frequency shift of the methyl stretch in going from pure methyl iodide to a dilute solution in chloroform is a good approximation of $|\delta\omega_A - \delta\omega_B|$, application of Eq. (3.7) implies that around any given methyl iodide molecule there are, on average, 5.4 other molecules that perturb its symmetric methyl stretch [30]. This number is in good agreement with the number of molecules thought to be in the first solvent shell around methyl iodide in the pure liquid [30].

## IV. FIFTH-ORDER RAMAN SPECTROSCOPY

Rayleigh-wing spectra tend to be broad and rather featureless, making it difficult to assess the degree of inhomogeneity in Raman-active intermolecular vibrational modes. Although numerous curve-fitting procedures have been employed with Rayleigh-wing spectra in an effort to address these issues, it remains difficult to make an unambiguous determination of the relative importance of different broadening mechanisms (and, indeed, different scattering mechanisms) in such spectra. Obtaining spectra of pure liquids at different state points has been of some help in this regard [22, 50–55], as has studying mixtures of liquids [56–58]. Nonetheless, it remains impossible to make an unequivocal identification of the phenomena that determine the shape of the Rayleigh-wing spectrum of any liquid composed of anisotropic molecules.

In 1993, Tanimura and Mukamel [59] proposed a new, higher order spectroscopic technique with the ability to rephase inhomogeneity in Raman-active intermolecular vibrations. As with the Raman echo, this spectroscopy is two dimensional. Two time-coincident laser pulses create a coherence in Raman-active intermolecular modes. After the coherence has evolved for a time $\tau_1$, another two time-coincident pulses alter the coherence. Time $\tau_2$ later a final pulse probes the coherence and generates a signal in a unique spatial direction. Fifth-order spectroscopy has attracted a considerable amount of experimental [60–70] and theoretical [71–92] attention since that time. Performing the experiments such that the desired data are not contaminated by unwanted artifacts has proven more difficult than had been thought initially, but these problems now appear to have been resolved. In this section, we will review the theoretical and experimental progress that has been made in understanding fifth-order spectroscopy.

### A. Background

All of the techniques discussed above need involve only a pair of vibrational energy levels, and each Raman transition can arise through the first-derivative terms in the polarizability. However, fifth-order spectroscopy relies on the existence of some sort of nonlinearity, either in the coordinate dependence of the polarizability [59] or in the vibrational potential, and explicitly involves at least three vibrational energy levels [75]. In the former case, the lowest order contribution to the fifth-order signal arises from two interactions via $\tilde{\mathbf{\Pi}}_i^{(1)}$ and one via $\tilde{\mathbf{\Pi}}_{ii}^{(2)}$, while in the latter case there are three interactions via $\tilde{\mathbf{\Pi}}_i^{(1)}$ but the cubic anharmonicity $g_{iii}^{(3)}$ plays a role in one of these interactions. These nonlinearities can also couple two different modes together rather than being within a single mode [78]. The lowest order contribution when there is polarizability coupling between the modes comes from one interaction via $\tilde{\mathbf{\Pi}}_i^{(1)}$, one via $\tilde{\mathbf{\Pi}}_j^{(1)}$ and one via $\tilde{\mathbf{\Pi}}_{ij}^{(2)}$. In the case of anharmonic coupling, the lowest order contribution arises from two interactions via $\tilde{\mathbf{\Pi}}_i^{(1)}$ and one via $\tilde{\mathbf{\Pi}}_j^{(1)}$, with the assistance of $g_{iij}^{(3)}$.

Representative pathways that can generate fifth-order signal involving a single mode are illustrated in Figure 8 [72, 82, 85]. Note that each pathway involves either a zero-quantum or a two-quantum transition, which is why either $\tilde{\mathbf{\Pi}}_{ii}^{(2)}$ or $g_{iii}^{(3)}$ must be involved in the lowest order contribution to the signal. The first type of pathway, which begins with a zero-quantum transition, does not contribute to the signal [59]. The second type of pathway, which begins with a two-quantum transition, can contribute to this signal, but only when the damping depends on quantum number [82]. In the absence of time-dependent oscillator frequencies,

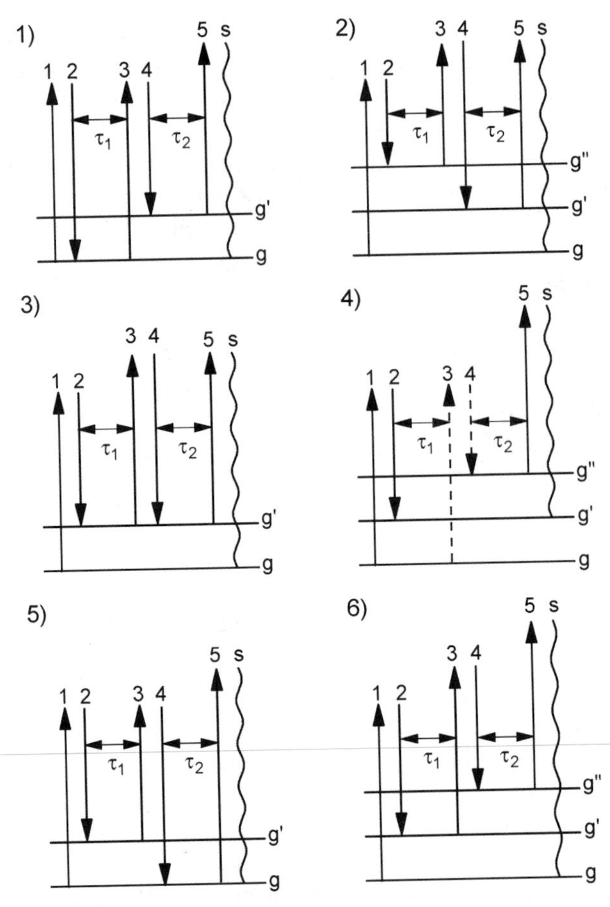


**Figure 8.** Representative examples of the six types of wave-mixing pathways that contribute to the signal in fifth-order spectroscopy. Beams 1 and 2 create a Raman coherence that is allowed to propagate for time $\tau_1$. Beams 3 and 4 then modify the coherence. After a second propagation time $\tau_2$ the coherence is probed by pulse 5.

its contribution to the fifth-order response for a single mode is given by [82,85].

$$
\begin{aligned}
R^{(5)}_{2,\alpha\beta\gamma\delta\varepsilon\zeta}(\omega_{g'g};\tau_1,\tau) \propto \frac{i\hbar^2}{8\omega^2} D(2;\tau_1,\tau_2)\coth\left(\frac{\hbar\omega}{2k_BT}\right)\cos\left[\omega_{g'g}(2\tau_1+\tau_2)\right] \\
\times \left[\left\langle \tilde{\mathbf{\Pi}}^{(1)}_{\varepsilon\zeta}(\tau_1+\tau_2)\tilde{\mathbf{\Pi}}^{(1)}_{\gamma\delta}(\tau_1)\tilde{\mathbf{\Pi}}^{(2)}_{\alpha\beta}(0)\right\rangle \right. \\
\left. - \frac{1}{3\omega^2}\left\langle \tilde{\mathbf{\Pi}}^{(1)}_{\varepsilon\zeta}(\tau_1+\tau_2)\tilde{\mathbf{\Pi}}^{(1)}_{\gamma\delta}(\tau_1)\tilde{\mathbf{\Pi}}^{(1)}_{\alpha\beta}(0)g^{(3)}_{iii}(0)\right\rangle\right]
\end{aligned}
\tag{4.1}
$$

where $\alpha, \beta, \gamma, \delta, \varepsilon$, and $\zeta$ are the polarizations of the five input beams and the signal beam, respectively, and $D(2;\tau_1,\tau_2)$ is a model-dependent damping function. The next two types of pathway involve zero- and two-quantum transitions at the second interaction, respectively, and have contributions to the response of a single mode given by

$$R^{(5)}_{3,\alpha\beta\gamma\delta\varepsilon\zeta}(\omega_{g'g};\tau_1,\tau) \propto -\frac{i\hbar^2}{8\omega^2}D(3;\tau_1,\tau_2)\cos\left[\omega_{g'g}(\tau_1+\tau_2)\right] \times \left[\langle\tilde{\mathbf{\Pi}}^{(1)}_{\varepsilon\zeta}(\tau_1+\tau_2)\tilde{\mathbf{\Pi}}^{(2)}_{\gamma\delta}(\tau_1)\tilde{\mathbf{\Pi}}^{(1)}_{\alpha\beta}(0)\rangle - \frac{1}{3\omega^2}\langle\tilde{\mathbf{\Pi}}^{(1)}_{\varepsilon\zeta}(\tau_1+\tau_2)\tilde{\mathbf{\Pi}}^{(1)}_{\gamma\delta}(\tau_1)\tilde{\mathbf{\Pi}}^{(1)}_{\alpha\beta}(0)g^{(3)}_{iii}(\tau_1)\rangle\right] \tag{4.2}$$

and

$$R^{(5)}_{4,\alpha\beta\gamma\delta\varepsilon\zeta}(\omega_{g'g};\tau_1,\tau) \propto \frac{i\hbar^2}{8\omega^2}D(4;\tau_1,\tau_2)\cos\left[\omega_{g'g}(\tau_1-\tau_2)\right] \times \left[\langle\tilde{\mathbf{\Pi}}^{(1)}_{\varepsilon\zeta}(\tau_1+\tau_2)\tilde{\mathbf{\Pi}}^{(2)}_{\gamma\delta}(\tau_1)\tilde{\mathbf{\Pi}}^{(1)}_{\alpha\beta}(0)\rangle - \frac{1}{3\omega^2}\langle\tilde{\mathbf{\Pi}}^{(1)}_{\varepsilon\zeta}(\tau_1+\tau_2)\tilde{\mathbf{\Pi}}^{(1)}_{\gamma\delta}(\tau_1)\tilde{\mathbf{\Pi}}^{(1)}_{\alpha\beta}(0)g^{(3)}_{iii}(\tau_1)\rangle\right] \tag{4.3}$$

The final two types of pathway involve zero- and two-quantum transitions at the final step, and have contributions to the response of a single mode given by

$$R^{(5)}_{5,\alpha\beta\gamma\delta\varepsilon\zeta}(\omega_{g'g};\tau_1,\tau) \propto \frac{i\hbar^2}{8\omega^2}D(5;\tau_1,\tau_2)\cos\left[\omega_{g'g}\tau_1\right] \times \left[\langle\tilde{\mathbf{\Pi}}^{(2)}_{\varepsilon\zeta}(\tau_1+\tau_2)\tilde{\mathbf{\Pi}}^{(1)}_{\gamma\delta}(\tau_1)\tilde{\mathbf{\Pi}}^{(1)}_{\alpha\beta}(0)\rangle - \frac{1}{3\omega^2}\langle\tilde{\mathbf{\Pi}}^{(1)}_{\varepsilon\zeta}(\tau_1+\tau_2)\tilde{\mathbf{\Pi}}^{(1)}_{\gamma\delta}(\tau_1)\tilde{\mathbf{\Pi}}^{(1)}_{\alpha\beta}(0)g^{(3)}_{iii}(\tau_1+\tau_2)\rangle\right] \tag{4.4}$$

and

$$R^{(5)}_{6,\alpha\beta\gamma\delta\varepsilon\zeta}(\omega_{g'g};\tau_1,\tau) \propto -\frac{i\hbar^2}{8\omega^2}D(6;\tau_1,\tau_2)\cos\left[\omega_{g'g}(\tau_1+2\tau_2)\right] \times \left[\langle\tilde{\mathbf{\Pi}}^{(2)}_{\varepsilon\zeta}(\tau_1+\tau_2)\tilde{\mathbf{\Pi}}^{(1)}_{\gamma\delta}(\tau_1)\tilde{\mathbf{\Pi}}^{(1)}_{\alpha\beta}(0)\rangle - \frac{1}{3\omega^2}\langle\tilde{\mathbf{\Pi}}^{(1)}_{\varepsilon\zeta}(\tau_1+\tau_2)\tilde{\mathbf{\Pi}}^{(1)}_{\gamma\delta}(\tau_1)\tilde{\mathbf{\Pi}}^{(1)}_{\alpha\beta}(0)g^{(3)}_{iii}(\tau_1+\tau_2)\rangle\right] \tag{4.5}$$

Note that the argument of the cosine in pathway 4 is proportional to $\tau_1 - \tau_2$. This type of term therefore leads to rephasing when the two delay times are equal. This rephasing is made possible because the initial coherence at $\omega_{g'g}$ that evolves for time $\tau_1$ is then converted into a coherence of frequency $\omega_{g'g''}$ (which is equal to $\omega_{gg'}$) for the second time period. Thus, fifth-order spectroscopy is similar to the tri-level echo in electronic spectroscopy [93]. However, unlike in the case of the Raman echo and other nonlinear techniques, the pathways that do not have echo-like behavior cannot be removed using phase-matching tricks, and so the echo necessarily competes with these other processes in the signal.

The fact that the fifth-order signal arises from an odd number of Raman transitions plays a major role in determining the scattering mechanisms that can contribute to this spectroscopy [85]. In essence, single-molecule scattering can only contribute strongly to even invariants of the many-body polarizability tensor (such as the anisotropy). In OKE spectroscopy, single-molecule scattering makes no contribution whatsoever to the isotropic portion of the polarizability tensor (i.e., the trace), and so its dominant contribution arises from the anisotropy [20, 85, 94]. The fifth-order signal arises from a cubic invariant (which is weak for single-molecule scattering) and from cubic products involving $\alpha$ and $\beta^2$ (the first of which is zero for single-molecule scattering). Furthermore, for microscopic symmetry reasons the isotropic portion of $\tilde{\mathbf{\Pi}}^{(1)}$ for II scattering is generally quite small, so the strongest product of invariants that can contribute to the fifth-order signal is $\alpha(\tilde{\mathbf{\Pi}}^{(2)})\beta^2(\tilde{\mathbf{\Pi}}^{(1)}, \tilde{\mathbf{\Pi}}^{(1)})$. While the first term in this invariant must arise from II scattering, the second can come from single-molecule scattering. Thus, even though the OKE signal is probably generally dominated by single-molecule effects, the fifth-order signal must be dominated by II scattering or its cross term with single-molecule scattering [85, 95]. The strength of the $\alpha(\tilde{\mathbf{\Pi}}^{(2)})\beta^2(\tilde{\mathbf{\Pi}}^{(1)}, \tilde{\mathbf{\Pi}}^{(1)})$ can be used to advantage in developing polarization conditions that selectively enhance the echo-like pathways and their partners in which the zero-quantum transition comes second [85,95]. However, the flip side of this coin is that because the third- and fifth-order signals arise from different scattering mechanisms, there is no good way to use third-order data as a constraint for fitting fifth-order data [85].

It has also been shown that if the dominant nonlinearity that produces fifth-order signal arises from coupling between modes, the signatures of coupling through the polarizability and coupling through anharmonicity are significantly different [66,78]. In particular, polarizability coupling tends to wash out any sort of echo-like behavior, as the coupled modes can have considerably different frequencies [87]. On the other hand, anharmonic effects can most efficiently couple modes with frequencies that are related harmonically, which can potentially lead to echo-like features for $\tau_1 = 2\tau_2$ and $2\tau_1 = \tau_2$ rather than $\tau_1 = \tau_2$.

### B. Intermolecular Vibrations

The application of fifth-order spectroscopy to the study of liquids has been fraught with experimental difficulties that may only now have been resolved. One of the major problems is that the fifth-order signal relies on either the cubic anharmonicity or the second derivative of the polarizability of a system, and therefore the signal is so weak that it is easily swamped by signals arising from undesired processes. We will briefly discuss the experiments that have been reported to date in this light.

The first reported observation of fifth-order signal in a liquid was by Tominaga and Yoshihara [60], who concluded that at room temperature the intermolecular spectrum of $CS_2$ is characterized by at most a small amount of inhomogeneity. Further studies by these authors supported their initial conclusions [62], and an additional study of $CS_2$ dissolved in alkanes suggested that the degree of inhomogeneity in the intermolecular spectrum is, if anything, reduced upon dilution [63].

After these initial studies, Steffen and Duppen [65] used a new experimental geometry to reinvestigate the fifth-order signal of $CS_2$. Their data differ significantly from those reported previously, and they concluded that there is no inhomogeneous broadening in the intermolecular spectrum of this liquid. However, they recognized subsequently that in the geometry they employed, the fifth-order signal could be contaminated by other third- and fifth-order processes [64, 83]. When they repeated their experiments in a new geometry their results were similar to those of Tominaga and Yoshihara [64,83].

Tokmakoff and Fleming [66] developed another experimental geometry that was designed to minimize any contamination of the fifth-order signal by slightly phase-mismatched third-order signals. By using this geometry, they obtained fifth-order data on $CS_2$ using a variety of different polarization combinations with an exceptional signal/noise ratio. Surprisingly, neither these data nor others obtained at different temperatures [69] or in different liquids [68] could be explained by existing models of fifth-order spectroscopy.

Ulness et al. [96] subsequently pointed out that the fifth-order signal could be contaminated by cascaded third-order signals. In cascading events, the electric field generated by one third-order event acts as a source field for a second third-order event (Fig. 9). The possibility of serial cascading events, in which the field from one third-order event acts as a pump field for a second event (upper diagram in Fig. 9) was considered by Tominaga and Yoshihara [62], who were able to use phase-matching conditions to rule out such contributions to the fifth-order signal. However, parallel cascading events (lower diagrams in Fig. 9), in which the field from one third-order event acts as the probe field for a second event, are more difficult to avoid. Indeed, Blank et al. [66, 68, 69] showed that

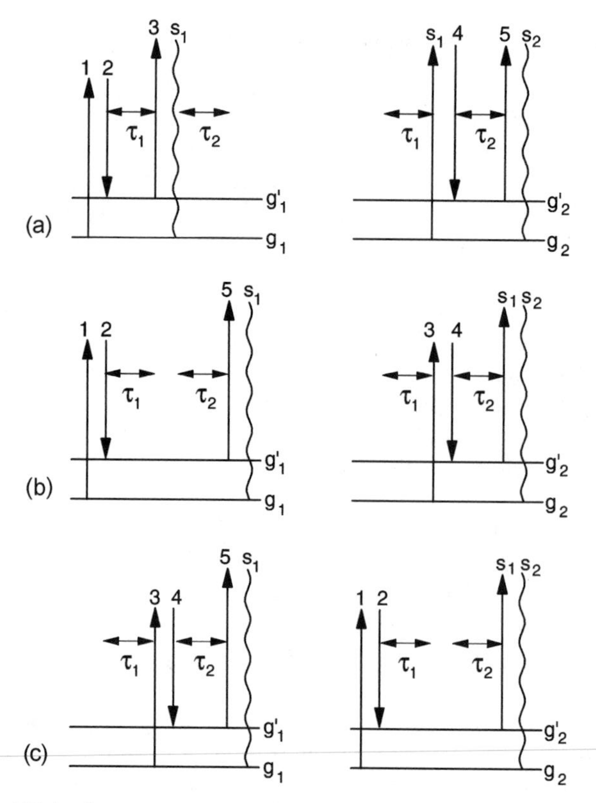


**Figure 9.** Third-order cascading events that can contaminate the fifth-order intermolecular signal. Pathway (*a*) is a serial cascade, in which the signal from one third-order process acts as a pump field for a second third-order process. Pathways (*b*) and (*c*) are parallel cascading events, in which the signal field from one third-order process acts as a probe field for a second third-order process.

the intermolecular fifth-order data from the Fleming group could be explained entirely in terms of parallel third-order cascades [70].

Since the realization that parallel cascades can dominate the third-order signal, this problem has received both theoretical [97] and experimental [98] attention. Blank et al. [98] designed a phase-matching geometry that minimizes the effects of cascaded third-order signals. By combining this geometry with a polarization scheme that also minimizes cascading, they have obtained what appear to be uncontaminated fifth-order $CS_2$ data [see Fig. 10(*a*)] [98]. These data do seem to be consistent with theoretical models for fifth-order spectroscopy. Any inhomogeneity in the intermolecular modes in these data should show up as a ridge along $\tau_1 = \tau_2$. The cut of the data taken along this diagonal [Fig. 10(*b*)] decays to zero within 400 fs, which suggests that the intermolecular modes of this liquid are essentially completely homogeneously broadened.

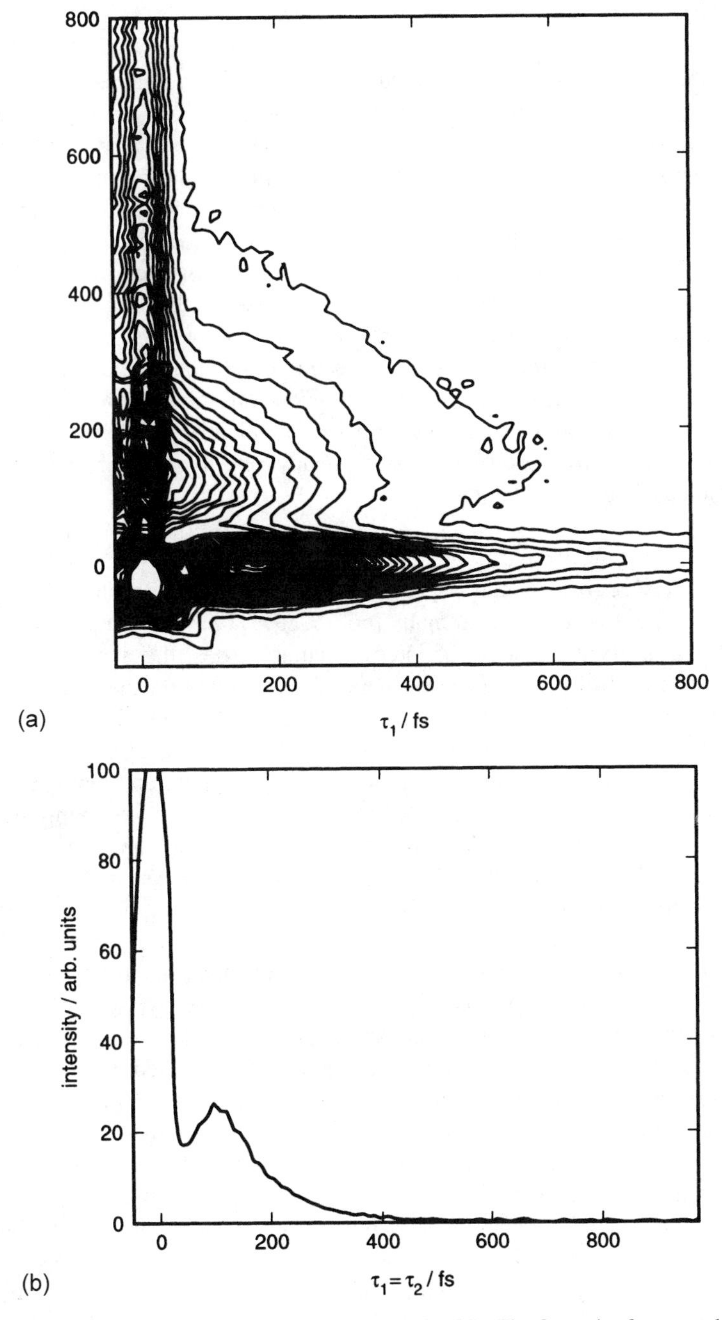


**Figure 10.** (*a*) Room temperature fifth-order data for $CS_2$. The first pair of pump pulses was polarized at the magic angle, and the remaining pulses and the signal were $z$-polarized. (*b*) A cut through the data along $\tau_1 = \tau_2$ decays rapidly, indicating that there is no significant inhomogeneity present in the Raman-active intermolecular modes of this liquid. (Adapted from [98].)

Further analysis of these fifth-order data and others obtained on the same liquid at other polarization conditions and temperatures should demonstrate relatively conclusively whether or not the Raman-active intermolecular modes of simple liquids are characterized by any significant inhomogeneity. The mass and the moment of inertia of $CS_2$ are quite large for the size of this molecule. Both of these features of $CS_2$ tend to slow down the timescale of pseudooscillatory intermolecular dynamics, and thus favor the existence of long-lived inhomogeneity. If such inhomogeneity cannot be seen in this liquid, it is unlikely to be found in any liquid within its normal temperature range.

Ma and Stratt [99] recently used a molecular dynamics simulation to compute the fifth-order Raman spectrum of liquid xenon. The *zzzzzz* fifth-order response derived from this simulation shows no sign of a ridge along $\tau_1 = \tau_2$, further suggesting that it will be difficult to observe any long-lived inhomogeneity in the intermolecular modes of liquids.

### C. Intramolecular Vibrations

Fifth-order spectroscopy can also be performed on intramolecular vibrations in liquids, either in the time domain [66, 75, 78, 91] or the frequency domain [90, 92, 100]. As in the case of intermolecular modes, the appearance of the intramolecular fifth-order spectrum depends sensitively on the source of nonlinearity (in this case, in the coupling among the modes) [67, 75, 78]. In the absence of any coupling between modes, a two-dimensional vibrational spectrum will consist of peaks along the diagonals ($\omega_1 = \omega_2$) and along a second-harmonic diagonal ($2\omega_1 = \omega_2$) [66]. On the other hand, coupling between modes, either through polarizability or anharmonicity, can produce additional cross-peaks. Furthermore, the relative strengths of cross-peaks in different quadrants of the spectrum can be used to deduce the relative importance of polarizability and anharmonicity in coupling two modes [66]. Thus, in principle this technique allows the coupling strengths and mechanisms between different Raman-active modes of a molecule to be measured directly.

Tokmakoff et al. [66] reported the first application of fifth-order spectroscopy to the study of intramolecular modes in liquids. They studied the Raman-active modes of chloroform, carbon tetrachloride, and a mixture of these two liquids out to a frequency of approximately 600 $\text{cm}^{-1}$ (Fig. 11). Cross-peaks are evident in the spectra of both liquids, but show up most strongly for carbon tetrachloride. For this liquid, particularly strong cross-peaks are seen between the $\nu_1$ and $\nu_2$ modes at 460 and 219 $\text{cm}^{-1}$, presumably due to Fermi-resonance coupling between these vibrations. In chloroform, the only strong coupling appear to be between the $\nu_6$ asymmetric bending mode at 262 $\text{cm}^{-1}$ and the $\nu_3$ symmetric bending mode at 368 $\text{cm}^{-1}$. Theoretical calculations suggest that these two modes have a relatively large cubic anharmonic coupling coefficient but relatively weak polarizability coupling coefficients [91].

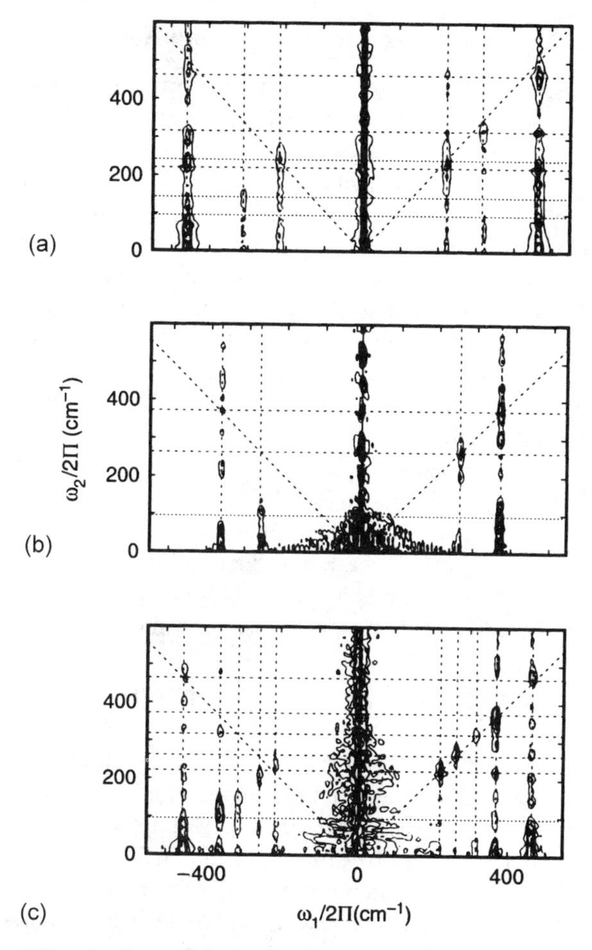


**Figure 11.** Fifth-order Raman spectra of carbon tetrachloride (*a*), chloroform (*b*), and an equimolar mixture of the two liquids (*c*). Note that in each case significant cross-peaks are evident. (Adapted from [67] copyright (1997) by the American Physical Society.)

Interestingly, the fifth-order data for the liquid mixture show cross-peaks between chloroform and carbon tetrachloride modes. These cross-peaks might plausibly arise through intermolecular polarizability coupling or even intermolecular Fermi resonances. However, it is also possible that these cross-peaks (and some, but not all, of those in the pure liquids) arise from cascaded scattering. Further experiments in the newly designed experimental geometry that minimizes cascades [98] will be needed to ascertain which of these peaks arise from true fifth-order scattering.

### D. Fifth-Order Spectroscopy with Noisy Light

Rather than using ultrafast laser pulses to perform time-resolved nonlinear experiments, it is also possible to employ nanosecond broadband laser pulses that have a short coherence time. Such noisy light has been exploited to make numerous time-resolved measurements such as in CARS spectroscopy [101–110]. Noisy-light CARS has proven to be a useful technique that is in many senses complementary to more conventional CARS techniques.

By the same token, it has also proven possible to use incoherent light to perform a version of fifth-order spectroscopy. Such techniques have been investigated in some detail both theoretically [96, 111, 112] and experimentally [113–117]. As might be expected, the same time of cascading pathways that have plagued coherent-light fifth-order spectroscopy also can contaminate spectra obtained using noisy light [96, 111, 114]. Kirkwood et al. [114] have devised a clever scheme to test whether a given spectrum arises from a true fifth-order process or from third-order cascades. Unfortunately, it appears that cascades dominate all of the data that have been obtained to date [113, 114]. It remains possible that the same sort of phase matching and polarization that have been used to remove cascades from the coherent fifth-order signal [98] will be similarly useful in the noisy-light version of this experiment.

## V. OVERTONE DEPHASING SPECTROSCOPY

Taking their inspiration from fifth-order intermolecular spectroscopy, Tominaga and Yoshihara [118–120] recognized that in addition to probing the dephasing of intramolecular fundamentals, two-color CARS experiments also could be used to probe the dephasing of overtones of intramolecular modes. In its simplest form [Fig. 12(*a*)], two pump pulses (1 and 2) are allowed to interact with the vibrational mode of interest twice. The frequencies of these pulses are adjusted such that $2\omega_1 - 2\omega_2 = \omega_{g''g}$. So long as the vibrational mode of interest is not extremely anharmonic, these pump interactions can create a sizeable coherence between the vibrational ground state and the second vibrational excited state. This coherence can be probed by a third, pump pulse some delay time $\tau$ later. The scattered signal with wave vector $\mathbf{k}_s = 2\mathbf{k}_1 - 2\mathbf{k}_2 + \mathbf{k}_3$ and frequency $\omega_s = 2\omega_1 - 2\omega_2 + \omega_3$ is then detected. In analogy to Eq. (2.7),

$$R(\omega_{g''g}; t_s) \propto e^{-i(2\mathbf{k}_1 - 2\mathbf{k}_2 + \mathbf{k}_3)\cdot\mathbf{r}} e^{-i(\omega_3 + \omega_{g''g})t_s} e^{-i(2\omega_1 - 2\omega_2 + \omega_3 - \omega_{g''g})\tau} \times (\boldsymbol{\eta}_1 \cdot \tilde{\boldsymbol{\pi}} \cdot \boldsymbol{\eta}_2)(\boldsymbol{\eta}_1 \cdot \tilde{\boldsymbol{\pi}} \cdot \boldsymbol{\eta}_2)(\boldsymbol{\eta}_3 \cdot \tilde{\boldsymbol{\pi}} \cdot \boldsymbol{\eta}_s) e^{-\Gamma_{g''g}\tau} \quad (5.1)$$

is the response at a single vibrational frequency. Thus, this technique measures the free–induction decay of the overtone coherence. Similarly, by tuning the pump lasers such that $3\omega_1 - 3\omega_2 = \omega_{g'''g}$ and detecting signal with wavevector

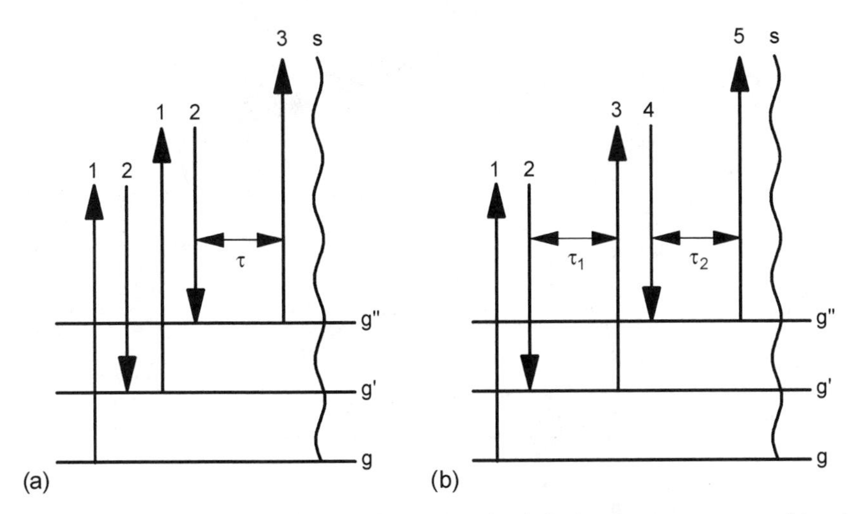


**Figure 12.** Representative pathways for overtone free-induction decay spectroscopy (*a*) and two-dimensional overtone dephasing spectroscopy (*b*).

$\mathbf{k}_s = 3\mathbf{k}_1 - 3\mathbf{k}_2 + \mathbf{k}_3$ and frequency $\omega_s = 3\omega_1 - 3\omega_2 + \omega_3$, the dephasing of a coherence between the ground vibrational state and the third excited vibrational state can be probed, and so on [120].

Tominaga and Yoshihara [120] used such a technique to probe the C—H stretching dynamics of chloroform and the C—D stretching dynamics of its deuterated analog [118,120]. In the rapid-modulation limit, one expects an exponential coherence decay that is proportional to the square of the difference in quantum number in the two states in the coherence. In agreement with this prediction, exponential coherence decays were observed, although the quantum-number dependence was found to be sub-quadratic [118, 120]. Thus, it is possible that population processes dominate the dephasing in this system. Gayathri et al. [121] also suggested, based on an extension of Kubo theory [9], that biphasic microscopic friction could lead to the observed subquadratic quantum-number dependence of the dephasing time.

By adding a second dimension to the experiment [Fig. 12(*b*)], it is possible to probe the correlations in energy fluctuations among the different quantum levels of the mode of interest [119]. In this spectroscopy, a coherence of between the ground and first excited vibrational states is allowed to propagate for time $\tau_1$, after which it is turned into a coherence between the ground and second excited vibrational states that is then probed some time $\tau_2$ later. The response is then proportional to the correlation function [119]

$$R^{(5)}(\tau_1, \tau_2) \propto \left\langle \exp\left( -i \int_0^{\tau_1} \Delta\omega_{g'g}(t_1)dt_1 - i \int_{\tau_1}^{\tau_2} \Delta\omega_{g''g}(t_2)dt_2 \right) \right\rangle \tag{5.2}$$

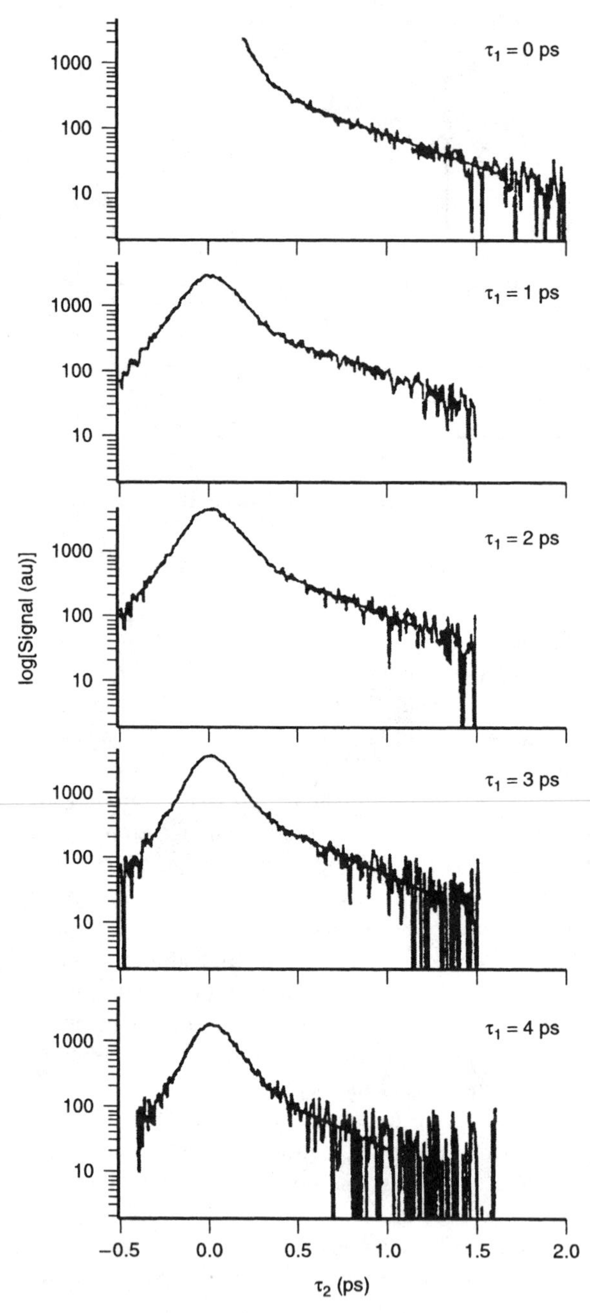


**Figure 13.** Two-dimensional overtone dephasing data for the C—D stretch of $CDCl_3$ at room temperature. Note that the decay time along $\tau_2$ decreases with increasing $\tau_1$. (Adapted from [119] copyright (1997) by the American Physical Society.)

In the simplest picture, the fluctuations during the two time periods would be completely correlated, and the decay along $\tau_2$ would not depend on the value of $\tau_1$. However, it is also possible that the coupling of the different oscillator states to the environment is different, due to anharmonicity or differences in oscillator displacement, for example.

Two-dimensional overtone dephasing data for the C—D stretch of chloroform at room temperature are shown in Fig. 13. As can be seen clearly in this figure, the decay time along $\tau_2$ becomes shorter as $\tau_1$ is increased. This result suggests that there is at least some correlated, slow modulation that contributes to line broadening, except that in this case one would expect to observe Gaussian rather than exponential decays [119]. Although reorientation can contribute to the signal with the polarizations that were used in these experiments, all of the effects of reorientation should appear along the first delay time rather than the second [85, 112]. More experiments are needed to clarify the origin of this unexpected behavior.

## VI. FREQUENCY-RESOLVED OPTICAL KERR EFFECT SPECTROSCOPY

The final type of multidimensional Raman spectroscopy that we will discuss is significantly different from the others that we have considered, in that it involves a time and a frequency dimension rather than two time or two frequency dimensions. Ziegler and co-workers [123, 124] reported experiments in which they have detected OKE signal as a function of the detuning from the frequency of the carrier of the probe pulse. By employing a probe pulse that is longer than the pump pulse (and therefore has a correspondingly narrower bandwidth) and then detecting the time-dependent birefringence of the sample within a narrow frequency range about a desired detuning, it is possible to obtain information about the time evolution of individual frequency components within the Rayleigh-wing spectrum. If the broadening of the intermolecular spectrum is entirely homogeneous, then the appearance of the dispersed birefringence signal should be independent of the detuning from the carrier frequency. On the other hand, if inhomogeneous broadening is present, the temporal behavior of the dispersed birefringence will depend on the detuning. The form of the signal as a function of detuning can thus be used as a means of assessing the degree of inhomogeneity in the intermolecular spectrum of a liquid [124].

Although no results on inhomogeneity in the Rayleigh-wing spectrum of liquids have been reported to date, this technique appears to be a promising one. Unlike data from fifth-order spectroscopy, data from this technique have the advantage of being able to be compared directly to conventional OKE data. In addition, this technique does not rely on higher order nonlinearities, which means that signals are much larger than for fifth-order spectroscopy (even though spectrally resolving the signal reduces its intensity to some extent). Finally, this technique is not susceptible to the effects of cascaded nonlinearities.

## VII. FUTURE DIRECTIONS

Multidimensional Raman spectroscopy is still in its infancy. The experiments and techniques described here have only just begun to scratch the surface of the information about vibrational dynamics that multidimensional Raman spectroscopy will eventually be able to provide. Because the techniques described here depend on multiple nonresonant interactions, the signals in many cases are necessarily weak, and the techniques are potentially fraught with significant technical difficulties. It appears that most if not all of these problems are well on their way to being overcome, however. As these techniques become easier to use, they will be applied to far greater range of systems. In addition, new techniques such as seventh-order intermolecular spectroscopy may actually prove easier to implement than techniques such as fifth-order spectroscopy [97]. With good fortune and hard work, the days in which multidimensional Raman spectroscopy routinely provides the kind of detailed information available from multidimensional NMR spectroscopies may be in the not too distant future.

## Acknowledgments

I would like to thank Andreas Albrecht, Mark Berg, David Blank, Minheang Cho, Brian Loughnane, Bob Murry, Alessandra Scodinu, Andrea Tokmakoff, and Keisuke Tominaga for their assistance in the preparation of this article. Part of the work described here was supported by the National Science Foundation, grant CHE-9501598.

## References

1. R. Benn and H. Gunther, *Agnew. Chem. Int. Ed. Eng.* **22**, 350 (1983).
2. W. P. de Boeij, M. S. Pshenichnikov, and D. A. Wiersma, *Chem. Phys. Lett.* **238**, 1 (1995).
3. W. P. de Boeij, M. S. Pshenichnikov, and D. A. Wiersma, *Chem. Phys. Lett.* **253**, 53 (1996).
4. G. R. Fleming, T. Joo, and M. Cho, *Adv. Chem. Phys.* **101**, 141 (1997).
5. S. A. Passino, Y. Nagasawa, T. Joo, and G. R. Fleming, *J. Phys. Chem. A* **101**, 725 (1997).
6. J. A. Koningstein, *Introduction to the Theory of the Raman Effect*, Reidel, Dordrecht, The Netherlands, 1972.
7. S. Mukamel, *Principles of Nonlinear Optical Spectroscopy*, Oxford University Press, New York, 1995, p. 543.
8. B. Dick, *Chem. Phys.* **113**, 131 (1987).
9. R. Kubo, in *Fluctutation, Relaxation and Resonance in Magnetic Systems*, D. T. Haar, Ed., Oliver and Boyd, Edinburgh, 1961.
10. G. Placzek, in *Handbuch der Radiologie*, E. Marx, Ed., Vol. VI, Pt. 2. Academische Verlagsgesellschaft, Leipzig, 1934.

11. A. Tokmakoff, *J. Chem. Phys.* **105**, 1 (1996).
12. B. J. Berne and R. Pecora, *Dynamic Light Scattering*, Wiley, New York, 1976.
13. R. Righini, *Science* **262**, 1386 (1993).
14. S. Kinoshita, Y. Kai, T. Ariyoshi, and Y. Shimada, *Int. J. Mod. Phys. B* **10**, 1229 (1996).
15. T. Keyes, D. Kivelson, and J. P. McTague, *J. Chem. Phys.* **55**, 4096 (1971).
16. A. Patkowski, W. Steffen, H. Nilgens, E. W. Fischer, and R. Pecora, *J. Chem. Phys.* **106**, 8401 (1997).
17. H. Z. Cummins, G. Li, W. Du, R. M. Pick, and C. Dreyfus, *Phys. Rev. E* **53**, 896 (1996).
18. K. L. Ngai, G. Floudas, and A. K. Rizos, *J. Chem. Phys.* **106**, 6957 (1997).
19. R. Torre, P. Bartolini, and R. M. Pick, *Phys. Rev. E* **57**, 1912 (1998).
20. R. L. Murry, J. T. Fourkas, W.-X. Li, and T. Keyes, *Phys. Rev. Lett.* **83**, 3550 (1999).
21. D. Kivelson and P. A. Madden, *Annu. Rev. Phys. Chem.* **31**, 523 (1980).
22. B. J. Loughnane, A. Scodinu, R. A. Farrer, and J. T. Fourkas, *J. Chem. Phys.* **111**, 2686 (1999).
23. D. McMorrow and W. T. Lotshaw, *Chem. Phys. Lett.* **178**, 69 (1991).
24. D. McMorrow and W. T. Lotshaw, *J. Phys. Chem.* **95**, 10395 (1991).
25. E. L. Hahn, *Phys. Rev.* **80**, 580 (1950).
26. N. A. Kurnit, I. D. Abella, and S. R. Hartmann, *Phys. Rev. Lett.* **13**, 567 (1964).
27. C. A. Walsh, M. Berg, L. R. Narasimhan, and M. D. Fayer, *Acc. Chem. Res.* **20**, 120 (1987).
28. G. R. Fleming and M. Cho, *Annu. Rev. Phys. Chem.* **47**, 109 (1996).
29. S. R. Hartmann, *IEEE J. Quantum Electron.* **QE-4**, 802 (1968).
30. L. J. Muller, D. Vanden Bout, and M. Berg, *J. Chem. Phys.* **99**, 810 (1993).
31. P. Hu, S. Geschwind, and T. M. Jedju, *Phys. Rev. Lett.* **37**, 1357 (1976).
32. K. P. Leung, T. W. Mossberg, and S. R. Hartmann, *Opt. Commun.* **43**, 145 (1982).
33. K. P. Leung, T. W. Mossberg, and S. R. Hartmann, *Phys. Rev. A* **25**, 3097 (1982).
34. J. Langelaar, D. Bebelaar, and J. D. W. van Voorst, *Appl. Phys. B* **28**, 274 (1982).
35. V. Bruckner, E. A. J. M. Bente, J. Langelaar, D. Bebelaar, and J. D. W. Van Voorst, *Opt. Commun.* **51**, 49 (1984).
36. M. Muller, K. Wynne, and J. D. W. Van Voorst, *Chem. Phys.* **128**, 549 (1988).
37. D. Vanden Bout, L. J. Muller, and M. Berg, *Phys. Rev. Lett.* **67**, 3700 (1991).
38. R. Inaba, K. Tominaga, M. Tasumi, K. A. Nelson, and K. Yoshihara, *Chem. Phys. Lett.* **211**, 183 (1993).
39. K. Yoshihara, R. Inaba, H. Okamoto, M. Tasumi, K. Tominaga, and K. A. Nelson, in *Femtosecond Reaction Dynamics*, D. Wiersma, Ed., North-Holland, Amsterdam, The Netherlands, 1994.
40. K. Tominaga, R. Inaba, T. J. Kang, Y. Naitoh, K. A. Nelson, M. Tasumi, and K. Yoshihara, in *Proceedings of the XIV International Conference on Raman Spectroscopy* N.-T. Yu and X.-Y. Li, Eds., Wiley, New York, 1994.
41. M. Berg and D. Vanden Bout, *Acc. Chem. Res.* **30**, 65 (1997).
42. K. S. Schweizer and D. Chandler, *J. Chem. Phys.* **76**, 2296 (1982).
43. S. M. George and C. B. Harris, *J. Chem. Phys.* **77**, 4781 (1982).
44. S. M. George, A. L. Harris, M. Berg, and C. B. Harris, *J. Chem. Phys.* **80**, 83 (1984).
45. J. T. Fourkas, D. Kivelson, U. Mohanty, and K. A. Nelson, in *Supercooled Liquids: Advances and Novel Applications*, J. T. Fourkas, D. Kivelson, U. Mohanty, and K. A. Nelson, Eds., ACS Books, Washington, DC, 1997, p. 2.

46. D. Vanden Bout, J. E. Freitas, and M. Berg, *Chem. Phys. Lett.* **229**, 97 (1994).
47. G. Doege, R. Arndt, H. Buhl, and G. Bettermann, *Z. Naturforsch. A* **35A**, 468 (1980).
48. E. W. Knapp and S. F. Fischer, *J. Chem. Phys.* **76**, 4730 (1982).
49. S. Bratos and G. Tarjus, *Phys. Rev. A* **32**, 2431 (1985).
50. S. Ruhman, B. Kohler, A. G. Joly, and K. A. Nelson, *Chem. Phys. Lett.* **141**, 16 (1987).
51. S. Ruhman and K. A. Nelson, *J. Chem. Phys.* **94**, 859 (1991).
52. B. Kohler and K. A. Nelson, *J. Phys. Chem.* **96**, 6532 (1992).
53. B. J. Loughnane, R. A. Farrer, and J. T. Fourkas, in *Ultrafast Phenomena X*, P. F. Barbara, J. G. Fujimoto, W. H. Knox, and W. Zinth, Eds., Springer-Verlag, Berlin, 1996, p. 304.
54. R. A. Farrer, B. J. Loughnane, L. A. Deschenes, and J. T. Fourkas, *J. Chem. Phys.* **106**, 6901 (1997).
55. W. T. Lotshaw, D. McMorrow, N. Thantu, J. S. Melinger, and R. Kitchenham, *J. Raman Spectrosc.* **26**, 571 (1995).
56. C. Kalpouzos, D. McMorrow, W. T. Lotshaw, and G. A. Kenney-Wallace, *Chem. Phys. Lett.* **155**, 240 (1989).
57. D. McMorrow, N. Thantu, J. S. Melinger, S. K. Kim, and W. T. Lotshaw, *J. Phys. Chem.* **100**, 10389 (1996).
58. T. Steffen, N. A. C. M. Meinders, and K. Duppen, *J. Phys. Chem. A* **102**, 4213 (1998).
59. Y. Tanimura and S. Mukamel, *J. Chem. Phys.* **99**, 9496 (1993).
60. K. Tominaga and K. Yoshihara, *Phys. Rev. Lett.* **74**, 3061 (1995).
61. K. Tominaga, G. P. Keogh, Y. Naitoh, and K. Yoshihara, *J. Raman Spectrosc.* **26**, 495 (1995).
62. K. Tominaga and K. Yoshihara, *J. Chem. Phys.* **104**, 4419 (1996).
63. K. Tominaga and K. Yoshihara, *J. Chem. Phys.* **104**, 1159 (1996).
64. T. Steffen and K. Duppen, *J. Chem. Phys.* **106**, 3854 (1997).
65. T. Steffen and K. Duppen, *Phys. Rev. Lett.* **76**, 1224 (1996).
66. A. Tokmakoff and G. R. Fleming, *J. Chem. Phys.* **106**, 2569 (1997).
67. A. Tokmakoff, M. J. Lang, D. S. Larsen, G. R. Fleming, V. Chernyak, and S. Mukamel, *Phys. Rev. Lett.* **79**, 2702 (1997).
68. A. Tokmakoff, M. J. Lang, D. S. Larsen, and G. R. Fleming, *Chem. Phys. Lett.* **272**, 48 (1997).
69. A. Tokmakoff, M. J. Lang, X. J. Jordanides, and G. R. Fleming, *Chem. Phys.* **233**, 231 (1998).
70. D. A. Blank, L. J. Kaufman, and G. R. Fleming, *J. Chem. Phys.* **111**, 3105 (1999).
71. S. Palese, J. T. Buontempo, L. Schilling, W. T. Lotshaw, Y. Tanimura, S. Mukamel, and R. J. D. Miller, *J. Phys. Chem.* **98**, 12466 (1994).
72. V. Khidekel and S. Mukamel, *Chem. Phys. Lett.* **240**, 304 (1995).
73. V. Chernyak and M. S., *J. Chem. Phys.* **108**, 5812 (1998).
74. W. M. Zhang, V. Chernyak, and S. Mukamel, *J. Chem. Phys.* **110**, 5011 (1999).
75. K. Okumura and Y. Tanimura, *J. Chem. Phys.* **107**, 2267 (1997).
76. K. Okumura and Y. Tanimura, *J. Chem. Phys.* **106**, 1687 (1997).
77. K. Okumura and Y. Tanimura, *Chem. Phys. Lett.* **277**, 159 (1997).
78. K. Okumura and Y. Tanimura, *Chem. Phys. Lett.* **278**, 175 (1997).
79. Y. Tanimura and K. Okumura, *J. Chem. Phys.* **106**, 2078 (1997).
80. M. Cho, K. Okumura, and Y. Tanimura, *J. Chem. Phys.* **108**, 1326 (1998).
81. Y. Tanimura, *Chem. Phys.* **233**, 217 (1998).

82. T. Steffen, J. T. Fourkas, and K. Duppen, *J. Chem. Phys.* **105**, 7364 (1996).
83. T. Steffen and K. Duppen, *Chem. Phys. Lett.* **273**, 47 (1997).
84. T. Steffen and K. Duppen, *Chem. Phys. Lett.* **290**, 229 (1998).
85. R. L. Murry and J. T. Fourkas, *J. Chem. Phys.* **107**, 9726 (1997).
86. R. L. Murry and F. J. T., *J. Chem. Phys.* **109**, 7913 (1998).
87. S. Saito and I. Ohmine, *J. Chem. Phys.* **108**, 240 (1998).
88. T. Keyes and J. T. Fourkas, *J. Chem. Phys.* **112**, 287 (2000).
89. M. Cho, *J. Chem. Phys.* **109**, 6227 (1998).
90. M. Cho, *J. Chem. Phys.* **109**, 5327 (1998).
91. S. Hahn, K. Park, and M. Cho, *J. Chem. Phys.* **111**, 4121 (1999).
92. J. H. Ivanecky III and J. C. Wright, *J. Chem. Phys.* **206**, 437 (1993).
93. T. Mossberg, A. Flusberg, R. Kachru, and S. R. Hartmann, *Phys. Rev. Lett.* **39**, 1523 (1977).
94. R. L. Murry, J. T. Fourkas, W.-X. Li, and T. Keyes, *J. Chem. Phys.* **110**, 10410 (1999).
95. R. L. Murry, J. T. Fourkas, W.-X. Li, and T. Keyes, *J. Chem. Phys.* **110**, 10423 (1999).
96. D. J. Ulness, J. C. Kirkwood, and A. C. Albrecht, *J. Chem. Phys.* **108**, 3897 (1998).
97. M. Cho, D. A. Blank, J. Sung, K. Park, S. Hahn, and G. R. Fleming, *J. Chem. Phys.* **112**, 2082 (2000).
98. D. A. Blank, L. J. Kaufman, and G. R. Fleming, *J. Chem. Phys.* **113**, 771 (2000).
99. A. Ma and R. M. Stratt, *Phys. Rev. Lett.* **85**, 1004 (2000).
100. M. Cho, *J. Chem. Phys.* **106**, 7550 (1997).
101. M. A. Dugan and A. C. Albrecht, *Phys. Rev. A* **43**, 3922 (1991).
102. M. A. Dugan and A. C. Albrecht, *Phys. Rev. A* **43**, 3877 (1991).
103. M. J. Stimson, D. J. Ulness, and A. C. Albrecht, *Chem. Phys. Lett.* **263**, 185 (1996).
104. D. J. Ulness and A. C. Albrecht, *Phys. Rev. A* **53**, 1081 (1996).
105. D. C. DeMott, D. J. Ulness, and A. C. Albrecht, *Phys. Rev. A* **55**, 761 (1997).
106. D. J. Ulness, M. J. Stimson, J. C. Kirkwood, and A. C. Albrecht, *J. Phys. Chem. A* **101**, 4587 (1997).
107. D. J. Ulness, J. C. Kirkwood, M. J. Stimson, and A. C. Albrecht, *J. Chem. Phys.* **107**, 7127 (1997).
108. M. J. Stimson, D. J. Ulness, and A. C. Albrecht, *Chem. Phys.* **222**, 17 (1997).
109. A. Kummrow, A. Lau, and H. G. Ludewig, *Opt. Commun.* **107**, 137 (1994).
110. A. Lau, A. Kummrow, M. Pfeiffer, and S. Woggon, *J. Raman Spectrosc.* **25**, 607 (1994).
111. J. C. Kirkwood, A. C. Albrecht, and D. J. Ulness, *J. Chem. Phys.* **111**, 253 (1999).
112. M. Pfeiffer and A. Lau, *J. Chem. Phys.* **108**, 4159 (1998).
113. J. C. Kirkwood, D. J. Ulness, A. C. Albrecht, and M. J. Stimson, *Chem. Phys. Lett.* **293**, 417 (1998).
114. J. C. Kirkwood, A. C. Albrecht, D. J. Ulness, and M. J. Stimson, *J. Chem. Phys.* **111**, 272 (1999).
115. A. Lau, M. Pfeiffer, and A. Kummrow, *Chem. Phys. Lett.* **263**, 435 (1996).
116. A. Lau, M. Pfeiffer, V. Kozich, and F. Tschirschwitz, *J. Chem. Phys.* **108**, 4173 (1998).
117. A. Lau, M. Pfeiffer, V. Kozich, and A. Kummrow, *Laser Chem.* **19**, 35 (1999).
118. K. Tominaga and K. Yoshihara, *Phys. Rev. Lett.* **76**, 987 (1996).
119. K. Tominaga and K. Yoshihara, *Phys. Rev. A* **55**, 831 (1997).

120. K. Tominaga and K. Yoshihara, *J. Phys. Chem. A* **102**, 4222 (1998).
121. N. Gayathri, S. Bhattacharyya, and B. Bagchi, *J. Chem. Phys.* **107**, 10381 (1997).
122. A. Tokmakoff, *J. Chem. Phys.* **105**, 13 (1996).
123. S. Constantine, Y. Zhou, J. Morais, and L. D. Ziegler, *J. Phys. Chem. A* **101**, 5456 (1997).
124. Y. Zhou, S. Constantine, S. Harrel, and L. D. Ziegler, *J. Chem. Phys.* **110**, 5893 (1999).

# BIREFRINGENCE AND DIELECTRIC RELAXATION IN STRONG ELECTRIC FIELDS

J. L. DÉJARDIN and YU. P. KALMYKOV

*Centre d'Etudes Fondamentales, Université de Perpignan, 52 Avenue de Villeneuve, 66860 Perpignan Cedex, France*

P. M. DÉJARDIN

*Department of Applied Mathematics and Theoretical Physics, The Queen's University of Belfast, Belfast BT7 1NN, Northern Ireland*

## CONTENTS



*Advances in Chemical Physics, Volume 117*, Edited by I. Prigogine and Stuart A. Rice.
ISBN 0-471-40542-6 

# I. INTRODUCTION

During the last decades, considerable progress has been achieved in the theoretical study of the rotational Brownian motion of particles (or molecules) and relaxation processes in condensed matter, which arises from the application of external stimuli such as electric or magnetic fields [1–7]. Among many physical phenomena entering this area of research, one can mention dielectric and Kerr effect relaxation of liquids and liquid crystals as well as magnetic relaxation of single domain (superparamagnetic) particles. All these phenomena can be described in the context of the model of the rotational Brownian motion of a particle both interacting with the thermal environment and subjected to a field of force. The theoretical treatment of these problems is based upon the Langevin equation [8] and/or the Fokker–Planck equation [9] approaches. The Fokker–Planck equation (which is a partial differential equation) describes the time evolution of the orientational distribution function of a particle in configuration (or phase) space. The Langevin equation is a stochastic vector differential equation for angular variables. The Fokker–Planck equation can be directly derived from the Langevin equation by calculating the drift and the diffusion coefficients. However, the statistical averages can be calculated both from the Fokker–Planck and the Langevin equations so that both approaches are completely equivalent.

In this chapter, our analysis will be essentially focused on dielectric and Kerr effect relaxation in strong external electric fields. At the end, we shall present two examples illustrating how to apply the same method to other physical phenomena, such as rotational diffusion of molecules in a mean-field potential and nonlinear magnetic relaxation of superparamagnets. In any case and regardless of the physical system under consideration, the mathematical approaches are very similar insofar as they all involve the rotational Brownian motion of a particle in an external potential. The solution of either the Fokker–Planck equation or the Langevin equation is always reduced to that of solving an infinite hierarchy of differential-recurrence equations for statistical averages describing the dynamic behavior of appropriate physical quantities [8, 9]. In order to solve this hierarchy, we used the matrix continued fraction expansion for the Green's function, which allows us to obtain solutions in closed form. When the recurrence relation is a three-term one, it is possible to obtain exact analytical expressions for the relaxation functions in terms of ordinary continued fractions and for the relaxation times in terms of known mathematical functions. However, in the majority of cases such a situation does not occur because the number of terms in the underlying recurrence relations is generally $> 3$ as, for example, in the theoretical treatment of dielectric and Kerr effect relaxation of an assembly of both polar and anisotropically polarizable molecules. Nevertheless, as first shown by Risken [9], and then extensively elaborated by Coffey et al. [8], it is always possible to reduce a multiterm scalar recurrence relation to a three-term matrix one (tridiagonal form) so that the solution can be given in terms of matrix continued fractions. Here, we shall use this powerful method for the calculation of nonlinear responses of systems of particles acted on by external stimuli, and compare the results whenever possible to those obtained by experimental data. Moreover, it is worth mentioning that the algorithms employed in such an approach are very effective, hence allowing us to evaluate the physical quantities of interest and thus going well beyond the previous approaches available. Finally, we shall restrict ourselves mainly to the low-frequency region where inertial effects may be completely ignored and the Fokker–Planck equation in phase space may be reduced to the Smoluchowski equation in configuration space only (the components of the angular momentum may be considered here as fast variables) [2, 4, 5, 9]. It could be possible naturally to include inertia but the mathematical task becomes more and more intricate and this aspect at the present time is in progress. In this way, we will show how to take account of inertial effects in the context of the extended rotational diffusion model ($J$ diffusion) [10, 11] whose dynamics is described by the generalized Langevin equation containing a memory kernel. In this introduction, we have underlined the essential points of our approach for evaluating the nonlinear responses inherent in relaxation processes in the framework of the rotational Brownian motion.

The contents of the remaining parts of this chapter can be detailed as follows. In Section II, we recall the main results of the previous treatments of nonlinear dielectric and Kerr effect relaxation in order to measure the real progress made since then by using matrix continued fraction techniques. In Section III, we present the theoretical basis for the treatment of nonlinear relaxation processes in the context of the rotational diffusion model (noninertial limit) in the presence of strong electric fields. Two approaches are described in detail, one is based on the Langevin equation and the second on the Smoluchowski equation. We show how to solve infinite hierarchies of three-term or multiterm differential equations for the moments (ensemble averages of the spherical harmonics) using the scalar or matrix continued fraction approach. In Section IV, we apply the above procedure to a truly nonlinear problem, namely, the transient dielectric relaxation and dynamic Kerr effect of polar and polarizable molecules subjected to a strong direct current (dc) electric field whose magnitude and direction may be suddenly changed. In Section V, we give analytical solutions for the dielectric and Kerr effect relaxation of polar and/or polarizable molecules in a strong dc bias field superimposed on a weak alternating current (ac) electric field. In particular, we evaluate exact expressions for the correlation functions in the frequency domain and for the relaxation times. Section VI is devoted to the calculation of the nonlinear ac stationary response of polar and polarizable molecules to an ac field of arbitrary strength. This is done both for dielectric relaxation and electric birefringence. Pronounced nonlinear effects due to the high ac field and the coupling of the dc and ac fields are shown in diagrams of the frequency behavior of the in-phase and out-of-phase components of the electric polarization and birefringence. In Section VII, we consider the nonlinear effects arising from the first and second hyperpolarizabilities on the Kerr effect relaxation response of an assembly of polar and anisotropically polarizable molecules in a strong dc electric field. The problem of linear response in dielectric and Kerr effect relaxation of a classical ensemble of linear polar molecules, acted on by a strong constant electric field with account of inertial effects, is considered in Section VIII. The application of the matrix continued fraction method to other physical phenomena is demonstrated in Section IX for two examples. The first concerns the rotational diffusion of polar molecules in a mean-field potential. The second refers to the study of the nonlinear magnetic relaxation of superparamagnetic particles. The appendixes contain a detailed account of some equations presenting some difficulty in their derivation in order to allow the reader an easier understanding of the mathematical developments.

## II. PREVIOUS TREATMENTS OF THE NONLINEAR DIELECTRIC AND BIREFRINGENCE RELAXATION

We shall consider a dielectric liquid consisting of an assembly of electrically noninteracting molecules (very dilute system) acted on by an electric field $\mathbf{E}$. Hence, collective effects may be ignored, and the particles pertaining to this assembly can be considered as effecting independent rotational motions. This allows us to reduce the problem to the orientational motion of an individual molecule. The starting equation commonly used for a theoretical analysis of these problems was the Fokker–Planck equation for the probability distribution function $W$ of orientations of Brownian particles in configuration space when inertial effects are neglected [1]. On assuming that all the particles are identical and axially symmetric, and denoting by $V$ the potential energy of the particle in an external field $\mathbf{E}(t)$, this equation (also called the Smoluchowski equation) can be written down in a general form as [12]

$$\frac{1}{D}\frac{\partial}{\partial t}W(\vartheta,\varphi,\psi,t) = \frac{1}{kT}\mathrm{div}[W(\vartheta,\varphi,\psi,t)\mathrm{grad}V(\vartheta,\varphi,\psi,t)] + \Delta W(\vartheta,\varphi,\psi,t) \tag{2.1}$$

where $D$ is the rotational diffusion coefficient with respect to the axis that is perpendicular to the axis of symmetry; $\Delta$ is the Laplace operator in terms of the Euler angles $\vartheta(t), \varphi(t), \psi(t)$ [13]; $k$ is the Boltzmann constant; and $T$ is the absolute temperature. The Smoluchowski equation (2.1) is a projection of the complete Fokker–Planck equation in phase space onto the coordinate subspace [9]. This restriction amounts to implicitly assuming that the angular momentum relaxes rapidly compared with the angular orientational variables. Equations similar to Eq. (2.1) may also be used for the evaluation of the nonlinear response of other physical systems. In particular, an analogous Smoluchowski equation can be applied to the calculation of the dielectric response of nematic liquid crystals [14], where $V = V_0 + V_1$ ($V_0$, and $V_1$ have the meaning as a mean-field potential and as the field dependent part, respectively). A very similar equation may be applied to the nonlinear magnetic response of an assembly of single domain ferromagnetic particles (superparamagnetism), where $V$ now has the meaning as the free energy per unit volume (see Section IX.B).

When a dielectric liquid comprised of polar and anisotropically polarizable molecules is acted on by an external stimulus such as an electric field $\mathbf{E}$, for example, it becomes birefringent, acquiring the same properties as a uniaxial crystal. If one assumes that the molecule under study carries a permanent dipole moment $\boldsymbol{\mu}$ directed along its symmetry axis and is anisotropically polarizable (i.e., has an induced dipole moment) and if the magnitude of the electric field is not too high, then, the total dipole moment is given by

$$\mathbf{m} = \boldsymbol{\mu} + \hat{\alpha}\mathbf{E} \tag{2.2}$$

where $\hat{\alpha}$ is the molecular polarizability tensor of second rank. In Eq. (2.2), we have assumed that the induced dipole is linearly dependent on the electric field. When this condition is not fulfilled, it is necessary to include nonlinear terms that can be characterized by the hyperpolarizability tensors $\hat{\beta}$ (third order) and $\hat{\gamma}$ (fourth order): This question will be the subject matter of Section VII. So, if Eq. (2.2) is valid and if the geometric axes of the molecule are chosen in such a manner that they coincide with those of the polarizability tensor, the orientational potential energy $V$ will be $\vartheta$ dependent only and given by [2,4]:

$$V(\vartheta, t) = -\mu E(t)\cos\vartheta - \frac{1}{2}\Delta\alpha E^2(t)\cos^2\vartheta - \frac{1}{2}\alpha_\perp E^2(t) \tag{2.3}$$

where $\Delta\alpha = \alpha_\parallel - \alpha_\perp$ is the difference between the principal electric polarizabilities parallel and perpendicular to the symmetry axis, the direction of which makes the angle $\vartheta$ with the electric field $\mathbf{E}$ at time $t$. After taking account of all these considerations, the noninertial Fokker–Planck equation (2.1) reduces to the well-known one-dimensional (1D) rotational diffusion equation, namely [12],

$$2\tau_D \frac{\partial}{\partial t} W(\vartheta, t) = \frac{1}{\sin\vartheta} \frac{\partial}{\partial\vartheta} \left[ \sin\vartheta \left( \frac{\partial}{\partial\vartheta} W(\vartheta, t) + \frac{W(\vartheta, t)}{kT} \frac{\partial}{\partial\vartheta} V(\vartheta, t) \right) \right] \tag{2.4}$$

where $\tau_D = (2D)^{-1}$ is the Debye relaxation time for isotropic diffusion. The distribution function $W(\vartheta, t)$ may be expanded as a series of Legendre polynomials $P_n(\cos\vartheta)$ of order $n$ as follows:

$$W(\vartheta, t) = \sum_{n=0}^{\infty} a_n(t) P_n(\cos\vartheta) \tag{2.5}$$

to give the following infinite hierarchy of differential-recurrence equations [2, 4, 12]

$$\begin{aligned} \frac{2\tau_D}{n(n+1)} \frac{d}{dt} f_n(t) &+ \left[ 1 - \frac{2\sigma(t)}{(2n-1)(2n+3)} \right] f_n(t) \\ &= \frac{\zeta(t)}{(2n+1)} [f_{n-1}(t) - f_{n+1}(t)] \\ &\quad + \frac{2\sigma(t)}{(2n+1)} \left[ \frac{(n-1)}{(2n-1)} f_{n-2}(t) - \frac{(n+2)}{(2n+3)} f_{n+2}(t) \right] \end{aligned} \tag{2.6}$$

where

$$\zeta(t) = \frac{\mu E(t)}{kT} \qquad \sigma(t) = \frac{\Delta\alpha}{2kT} E^2(t) \tag{2.7}$$

and $f_n(t)$ are the expectation values of the $P_n(\cos\vartheta)$ such that

$$\begin{aligned} f_n(t) = \langle P_n(\cos\vartheta) \rangle(t) &= \int_0^{\pi} P_n(\cos\vartheta) W(\vartheta, t) \sin\vartheta d\vartheta \\ &= \frac{1}{2n+1} \frac{a_n(t)}{a_0} \end{aligned} \tag{2.8}$$

The physical quantities, which are interesting from an experimental point of view and appropriate to dielectric and Kerr effect relaxation, are the electric polarization $P(t)$ and the electric birefringence function $K(t)$ defined, respectively, by

$$P(t) = b_1 + N_0\mu \langle \cos\vartheta \rangle(t) = b_1 + N_0\mu \langle P_1(\cos\vartheta) \rangle(t) \tag{2.9}$$

$$K(t) = b_2 \frac{2\pi N_0}{\bar{n}} (\alpha_{\parallel}^0 - \alpha_{\perp}^0) \langle P_2(\cos\vartheta)\rangle(t) \tag{2.10}$$

where $N_0$ denotes the number of molecules per unit volume, $\alpha_{\parallel}^0$ and $\alpha_{\perp}^0$ are the components of the optical polarizability due to the electric field (optical frequency) of the light beam passing through the liquid medium, and $\bar{n}$ is the mean refractive index. The coefficients $b_1$ and $b_2$ depend on the particle depolarization factors and the dielectric permeability of the solution. In our consideration, we will assume that these coefficients do not depend on the frequency of the electric field (in dielectric relaxation) and on the light frequency (in Kerr effect relaxation). For simplicity, below we will assume that $b_1 = 0$ and $b_2 = 1$. Here, the internal field effects are not taken into account. This means that the effects of long-range torques due to the connection between the average moments and the Maxwell fields are not taken into account. A treatment of these effects for the static nonlinear dielectric increment has been given, for example, in [15, 16]. However, for the dynamic nonlinear response, the account of the internal field effects is a very difficult problem. Nevertheless, these effects may be ignored for diluted systems in first approximation. Thus, the theory developed here is relevant to situations where dipole–dipole interactions have been eliminated by means of suitable extrapolation of data to infinite dilution [15].

Theoretical approaches to the analysis of the dielectric relaxation and dynamic Kerr effect usually started (see, e.g., [2, 12]) with the solution of the infinite hierarchy of recurrence relations for the moments of Eq. (2.6) for various problems such as the rise and decay transients, the response in ac and pulse electric fields, and so on (see, e.g., [17–24] and references cited therein). Attempts to calculate the *nonlinear response* by solving this hierarchy have been made by many authors but, were mainly calculated by applying the perturbation theory (e.g., [12, 17, 21–23]) where the potential energy of a Brownian particle in electric fields is less than the thermal energy. Morita [25] and Morita and Watanabe [26] generalized the perturbation approach and proposed a formal theory of nonlinear response arising from the transient and stationary processes for systems whose dynamics is governed by the Smoluchowski equation. They expanded the Green's function for the unperturbed state in terms of appropriate orthogonal functions. They showed that this Green's function is sufficient to calculate the nonlinear behavior of the distribution function perturbed by a strong external field. They found also that in the stationary state the nonlinear response function can be expressed in terms of integrals of products of infinite matrices whose elements are composed of correlation functions in the absence of perturbations. However, this formalism is very difficult to apply to the calculation of transient responses in nonlinear dielectric and Kerr effect relaxation in the presence of high dc and ac fields. Indeed, the only analytical expression for the birefringence they obtained was derived for a weak ac superimposed on a weak dc bias field. Moreover, the theory [25,26] may *only* be used for the calculation of the nonlinear response of a system where there is a small parameter. A further development of this approach was given by Déjardin [4]. The nonlinear response to an electric field of *arbitrary strength* is an intrinsically nonlinear problem as small parameters are absent. In this case, the application of the perturbation theory is no longer possible. Nevertheless, some progress has been achieved on applying analytical [27–33] and numerical methods (e.g., [18,34–36]). Below, we present in two parts the main results obtained in previous treatments of transient and ac responses in dielectric and Kerr effect relaxation.

### A. Transient Responses in High Fields

Transient relaxation arises when the magnitude and/or the direction of a dc electric field suddenly changes at the moment $t = 0$ from $\mathbf{E}_{\mathrm{I}}$ to $\mathbf{E}_{\mathrm{II}}$. We are interested in the relaxation of the system starting from an equilibrium state I with the distribution function $W_{\mathrm{I}}\,(t \leq 0)$ to another equilibrium state II with the distribution function $W_{\mathrm{II}}\,(t \to \infty)$. The following problems have been under investigation: the rise (step-on) and decay (step-off) transients in linear and nonlinear regimes [2,18,20,27,28], the response in pulse [2,4,17,22], rapidly reversing [2,17,27] and rapidly rotating [2,37] dc electric fields. The majority of the results were obtained for the step-on response by taking $\mathbf{E}_{\mathrm{I}} = 0$, $\mathbf{E}_{\mathrm{II}} = \mathbf{E}_0$ and the dimensionless parameters $\sigma(t)$ and $\zeta(t)$ in Eq. (2.6) constant. For the step-on response, one can evaluate the relaxation functions $f_n(t)$ by

$$f_n(t) = \langle P_n(\cos\vartheta)\rangle(\infty) - \langle P_n(\cos\vartheta)\rangle(t) \tag{2.11}$$

obeying the same Eq. (2.6). First, we shall consider two particular cases, where either $\sigma(t) = 0$ and $\zeta(t) \neq 0$ or $\zeta(t) = 0$ and $\sigma(t) \neq 0$, before tackling the general case $\sigma(t) \neq 0$ and $\zeta(t) \neq 0$. These cases are the simplest from a theoretical point of view. Other quantities of interest are the one-sided Fourier transforms (FT)

$$\tilde{f}_n(i\omega) = \int_0^\infty f_n(t)e^{-i\omega t}dt$$

of the relaxation functions $f_n(t)$ and the relaxation times $\tau_n$ defined as the area under the *normalized* relaxation function $f_n(t)$, namely

$$\tau_n = \int_0^\infty \frac{f_n(t)}{f_n(0)}dt = \frac{\tilde{f}_n(0)}{f_n(0)} \tag{2.12}$$

It should be noted that in some cases $f_n(0)$ may be equal to zero so that Eq. (2.12) can no longer be used. In such cases, it is sufficient to evaluate the area $\Pi_n$ under the relaxation functions $f_n(t)$:

$$\Pi_n = \int_0^\infty f_n(t)dt = \tilde{f}_n(0) \tag{2.13}$$

From an experimental point of view, the determination of the relaxation times gives interesting information about molecular parameters like the rotational diffusion and friction coefficients themselves linked to the dimensions of the molecules. However, it is obvious that more insight into the molecular dynamics can be gained from the relaxation spectra defined as the one-sided FT of the normalized relaxation functions, namely,

$$\chi_n(\omega) = \frac{\tilde{f}_n(i\omega)}{f_n(0)} \tag{2.14}$$

In particular, the analysis of $\chi_n(\omega)$ allows one to evaluate all the characteristic frequencies of the relaxation process.

### 1. *Step-On Response: Polar Molecules*

For purely polar molecules, one can put $\sigma(t) = 0$ in Eq. (2.6). Since the electric field applied to the molecules is constant, we can take the Laplace transform of this equation, which yields

$$[2s\tau_D + n(n+1)]\tilde{f}_n(s) = \frac{\xi n(n+1)}{2n+1}[\tilde{f}_{n-1}(s) - \tilde{f}_{n+1}(s)] + 2\tau_D f_n(0) \tag{2.15}$$

where

$$\tilde{f}_n(s) = \int_0^\infty f_n(t)e^{-st}dt$$
$$\xi = \mu E_0/kT \tag{2.16}$$

and the $f_n(0)$ values are the initial conditions calculated from the Maxwell–Boltzmann distribution function. Watanabe and Morita [2] examined this problem starting from recurrence equations for the Fourier–Laplace transforms of the coefficients $a_n(t)$ [see Eq. (2.8)]. They did not solve the corresponding three-term recurrence relation (2.15) in terms of a continued fraction, just calculating the Laplace transforms of the first two Legendre polynomials. Their results are summarized in Eqs. (2.17) and (2.18)

$$\int_0^\infty \langle P_1(\cos\vartheta)\rangle(t)e^{-st}dt = \frac{S_1(s)}{s} \tag{2.17}$$

$$\int_0^\infty \langle P_2(\cos\vartheta)\rangle(t)e^{-st}dt = \frac{1}{s}\left[1 - \frac{3}{\xi}(1 + s\tau_D)S_1(s)\right] \tag{2.18}$$

where the continued fraction $S_n(s)$ is solution of the homogeneous Eq. (2.15) $[f_n(0) = 0]$ and given by

$$S_n(s) = \frac{\xi}{(2(2n+1)/n(n+1))s\tau_D + 2n + 1 + \xi S_{n+1}(s)} \tag{2.19}$$

Equations (2.17)–(2.19) yield an exact solution of the problem under consideration for an arbitrary strength of the dc electric field $\mathbf{E}_0$.

### 2. *Step-On Response: Nonpolar Polarizable Molecules*

After putting $\zeta(t) = 0$ in Eq. (2.6) and applying the Laplace transform to this equation, one has a three-term recurrence relation:

$$\begin{aligned}\left[\frac{2\tau_D s}{n(n+1)} + 1 - \frac{2\sigma}{(2n-1)(2n+3)}\right]\tilde{f}_n(s) + \frac{2\sigma(n+2)\tilde{f}_{n+2}(s)}{(2n+1)(2n+3)} \\ = \frac{2\tau_D f_n(0)}{n(n+1)} + \frac{2\sigma(n-1)}{4n^2-1}\tilde{f}_{n-2}(s)\end{aligned} \tag{2.20}$$

where

$$\sigma = \frac{\Delta\alpha}{2kT}E_0^2 \tag{2.21}$$

Watanabe and Morita [2] obtained an exact solution of Eq. (2.20) for the Laplace transform of the second Legendre polynomial in terms of a continued fraction, namely,

$$\int_0^\infty \langle P_2(\cos\vartheta)\rangle(t)e^{-st}dt = \frac{S_2(s)}{s} \tag{2.22}$$

where the continued fraction $S_n(s)$, which is the solution of the homogeneous Eq. (2.20), is defined as

$$S_n(s) = \frac{2\sigma(n-1)/(4n^2-1)}{(2s\tau_D/n(n+1)) + 1 - (2\sigma/(2n-1)(2n+3)) + (2\sigma(n+2)/(2n+1)(2n+3))S_{n+2}(s)} \tag{2.23}$$

When both the permanent and the induced dipole moments are taken into account (general case of polar and polarizable molecules), the differential recurrence Eq. (2.6) contains five terms. Perturbation solutions of this problem for small fields have been obtained on many occasions (see, e.g., [4, 12, 23, 38]). For treating this problem in moderate fields, Watanabe and Morita [2] used a matrix approach involving tedious mathematical manipulations, and the results they obtained can be compared to the perturbation solution. In particular, they gave the Laplace transform of $\langle P_2(\cos\vartheta)\rangle$ suitable to electric birefringence up to $E^4$. However, this problem can be solved exactly in terms of matrix continued fractions as will be shown in Section IV.

### 3. *Relaxation Times for Build-up Processes*

Watanabe and Morita [2] tried to obtain analytical expressions for the relaxation times in the two limiting cases $\sigma = 0$ and $\xi = 0$. They only derived approximate expressions valid up to $\xi = 3$ for pure permanent dipole moments and up to $\sigma = 3.25$ for pure induced dipole orientation. As is obvious from their definition [Eq. (2.12)], in order to evaluate the relaxation times $\tau_n$, it is necessary to have the exact expressions for the relaxation functions $f_n(t)$ solutions of the inhomogeneous recurrence relations. Eskin [27] and Aizenberg and Eskin [28] were able to obtain exact expressions for the relaxation time $\tau_2$ of the electric birefringence of a solution of symmetric-top macromolecules diluted in a nonpolar solvent and acted on by an electric field emitted in the form of either a rectangular pulse of constant amplitude $\mathbf{E}_0$ (step-on response) or a reverse pulse (rapidly reversing dc field when at $t < 0$ the solution is at equilibrium in the field $\mathbf{E}_0$ whose direction is suddenly changed at $t = 0$ in the opposite direction). They restricted themselves to two particular cases $\sigma = 0$ (polar nonpolarizable molecules) and $\xi = 0$ (anisotropically nonpolar molecules) and evaluated in closed form the area under the normalized relaxation function $f_2(t)$ defined by Eq. (2.13). In our notation, their results may be formulated as follows. For polarizable molecules ($\xi = 0$), the birefringence relaxation time for the step-on response is [31]

$$\begin{aligned}\tau_2 &= \frac{1}{f_2(0)}\int_0^\infty f_2(t)dt\\ &= \frac{45\pi\tau_D}{64}\sum_{n=1}^{\infty}\left(-\frac{\sigma^2}{4}\right)^{n-1}\frac{(4n+1)\Gamma(2n)M^2(n+1/2,2n+3/2,\sigma)}{(2n+1)\Gamma^2(2n+3/2)M(1/2,3/2,\sigma)M(3/2,7/2,\sigma)}\end{aligned} \tag{2.24}$$

where $\Gamma(z)$ is the gamma function [39] and $M(a, b, z)$ is the confluent hypergeometric (Kummer) function defined as [39]

$$M(a,b,z) = 1 + \frac{a}{b}\frac{z}{1!} + \frac{a(a+1)}{b(b+1)}\frac{z^2}{2!} + \frac{a(a+1)(a+2)}{b(b+1)(b+2)}\frac{z^3}{3!} + \cdots \tag{2.25}$$

Here, we have taken into account that

$$\begin{aligned} f_2(0) &= \langle P_2(\cos\vartheta)\rangle(\infty) = \frac{1}{\int_0^\pi e^{\sigma\cos^2\vartheta}\sin\vartheta d\vartheta}\int_0^\pi P_2(\cos\vartheta)e^{\sigma\cos^2\vartheta}\sin\vartheta d\vartheta \\ &= \frac{2\sigma M(3/2,7/2,\sigma)}{15M(1/2,3/2,\sigma)} \end{aligned}$$

For small values of $\sigma$, Aizenberg and Eskin [28] obtained

$$\frac{\tau_2}{\tau_D} = \frac{1}{3}\left[1 + \frac{2}{21}\sigma - \frac{212}{11{,}025}\sigma^2 + O(\sigma^3)\right] \tag{2.26}$$

In the case where $\sigma \to \infty$ ($\sigma > 0$), they showed that

$$\frac{\tau_2}{\tau_D} \sim \frac{3}{2\sigma} \tag{2.27}$$

It can also be shown that for $\sigma \to -\infty$ ($\sigma < 0$) [31]

$$\frac{\tau_2}{\tau_D} \sim \frac{3(1-\ln 2)}{|\sigma|} \tag{2.28}$$

For polar molecules ($\sigma = 0$), the results obtained by Eskin [27] for the build-up process can be presented as follows:

$$\begin{aligned} \tau_2 &= \frac{\tilde{f}_2(0)}{f_2(0)} = \frac{3\tau_D}{\xi I_{1/2}(\xi)}\left[I_{3/2}(\xi) + \frac{2}{\xi I_{5/2}(\xi)}\sum_{n=2}^{\infty}(-1)^n\frac{(2n+1)}{n(n+1)}I_{n+1/2}^2(\xi)\right] \\ &= \frac{3\tau_D I_{1/2}(\xi)}{\xi I_{5/2}(\xi)}\left[\coth(\xi) - \frac{3}{\xi} - \frac{4}{\sinh^2(\xi)}\int_0^1 \ln(x)\sinh(2\xi x)dx\right] \end{aligned} \tag{2.29}$$

Here, we have used

$$f_2(0) = \left(\int_0^\pi P_2(\cos\vartheta)e^{\xi\cos\vartheta}\sin\vartheta d\vartheta\right)\left(\int_0^\pi e^{\xi\cos\vartheta}\sin\vartheta d\vartheta\right)^{-1} = \frac{I_{5/2}(\xi)}{I_{1/2}(\xi)} \tag{2.30}$$

where $I_\nu(z)$ is the modified Bessel function of the first kind [39]. The asymptotic limit of $\tau_2$ as $\xi \to \infty$ is given by [27]

$$\frac{\tau_2}{\tau_D} \sim \frac{3}{\xi} \tag{2.31}$$

Eskin [27] also calculated the area under the relaxation function $f_2(t)$ [which is $\tilde{f}_2(0)$ in our notations] for a rapidly reversing dc field, where $\mathbf{E}_\mathrm{I} = \mathbf{E}_0$, $\mathbf{E}_\mathrm{II} = -\mathbf{E}_0$ [in this case $f_2(0) = 0$ and the definition of the relaxation time (2.12) loses its physical sense]. He obtained [27]

$$\begin{aligned}\tilde{f}_2(0) &= \frac{6\tau_D}{\xi I_{1/2}^2(\xi)}\left[I_{3/2}(\xi)I_{5/2}(\xi) - \frac{1}{\xi}\sum_{n=1}^{\infty}\frac{(4n+3)}{(n+1)(2n+1)}I_{2n+3/2}^2(\xi)\right] \\ &= \frac{6\tau_D}{\xi}\left\{\coth(\xi) - \frac{1}{\xi} - \frac{2}{\sinh^2(\xi)}\int_0^1 \ln(x)[\sinh(2\xi x)\right. \\ &\qquad \left. - \sinh(2\xi - 2\xi x) + \sinh(2\xi - 4\xi x)]dx\right\}\end{aligned} \tag{2.32}$$

The asymptotic behavior of $\tau_2$ [Eq. (2.29)] as $\xi \to \infty$ is given by [27]

$$\tilde{f}_2(0) \sim \frac{6\tau_D}{\xi}\left(1 - \frac{\ln\xi}{\xi}\right)$$

Several approximate equations for the relaxation times $\tau_1$ and $\tau_2$ characterizing the step-on response were also obtained by Morita [2, 29] on using the effective eigenvalue approach [40]. In particular, Morita [29] showed that if the condition $3t/\tau_D \gg 1$ is fulfilled, the behavior of the averaged first Legendre polynomial in the step-on response of rigid dipolar molecules is described by a simple analytical equation

$$\langle P_1(\cos\vartheta)\rangle(t) = L(\xi)\left[1 - e^{-t/\tau_1^{ef}}\right] \tag{2.33}$$

where

$$\tau_1^{ef}/\tau_D = 3L(\xi)/\xi \tag{2.34}$$

and

$$L(z) = \coth(z) - 1/z \tag{2.35}$$

is the Langevin function.

## B. The ac Field Responses

### 1. *Dielectric Response*

As long as the orientational potential energy resulting from the application of an ac field to an ensemble of molecules is far less than the thermal energy, linear response theory works well. Debye [41] in 1913 was the first to apply this approach to the calculation of the after-effect response of the electric polarization of polar molecules following the sudden removal of a constant electric field and showed that this response allowed him to obtain the dynamic response of the molecules to an ac field. Both responses are described by the same relaxation function. In the presence of strong ac fields, linear response theory is no longer applicable, and the theoretical treatment of the problem becomes more complicated. Nowadays, experimental devices using electric fields of relatively high amplitudes are available (see, e.g., [18, 20, 21, 42–44]) so that it is important to tackle these nonlinear aspects theoretically in view of a correct interpretation of data, and hence of a better understanding of molecular relaxation mechanisms in dielectric fluids. Coffey and Paranjape [45] proposed an extension of the Debye theory to obtain the field and the frequency dependence of the relative permittivity of polar liquids in the case of either a strong ac field $E(t) = E_1 \cos \omega t$ or a strong dc bias field superimposed on a weak ac field in the same direction $E(t) = E_0 + E_1 \cos \omega t$. Their calculation provides analytic expressions for the dielectric relaxation as far as terms in $E_1^3$. Further progress in the calculation of the ac response was achieved by Morita [25] and Morita and Watanabe [26]. As already mentioned, they proposed a general formal perturbation theory of nonlinear response arising from the transient and stationary processes. Using a perturbation theory based on Morita's approach, Déjardin et al. [4, 5, 46–50] found analytic formulas for the nonlinear response in Kerr effect relaxation that are restricted to second order in the field strength. With a similar approach [46], Déjardin also obtained the harmonic components of the electric polarization valid up to the third order of the electric field by highlighting the nonlinear coupling effect due to the dual action of the dc $\mathbf{E}_0$ and ac $\mathbf{E}_1(t)$ fields on molecules both polar and anisotropically polarizable. By comparing these results with the exact solutions that we shall develop in Sections V and VI, we summarize the results of [4, 5, 45–50]. These results were obtained for the stationary regime (the system has removed all the transient effects so that we consider its behavior a long time after the electric field has been switched on). The nonlinear response of the electric polarization is therefore given by

$$P(t) = P_{st}(\omega) + \sum_{j=1}^{3} P_j'(\omega) \cos(j\omega t) + P_j''(\omega) \sin(j\omega t) \tag{2.36}$$

where $P_{st}(\omega)$ is a time independent but frequency-dependent component only due to the presence of the dc bias field $E_0$, and $P_j'(\omega)$ and $P_j''(\omega)$ are the real and imaginary parts of the complex electric polarization $P_j(\omega)$ of rank $j$ defined by $P_j(\omega) = P'_j(\omega) + iP''_j(\omega)$. On setting

$$\xi_1 = \frac{\mu E_1}{kT} \qquad \xi = \frac{\mu E_0}{kT} \qquad \kappa = \frac{\Delta\alpha}{\mu^2}kT \qquad \chi_0^1(\omega) = \frac{3P_{st}(\omega)}{\xi}$$
$$\chi_1^1(\omega) = \frac{3P_1(\omega)}{\xi_1} \qquad \chi_2^1(\omega) = \frac{90P_2(\omega)}{\xi_1^2\xi} \qquad \chi_3^1(\omega) = \frac{180P_3(\omega)}{\xi_1^3} \tag{2.37}$$

we have [46–48]:

$$\chi_0^1(\omega) = 1 + \frac{\xi^2}{15}\left[\kappa(2 + \xi_1^2/\xi^2) - 1\right] - \xi_1^2\frac{3 + (7\omega^2\tau_D^2/9) + \kappa(-1 + 5\omega^2\tau_D^2/3)}{30(1 + \omega^2\tau_D^2)(1 + \omega^2\tau_D^2/9)} \tag{2.38}$$

for the steady-state component,

$$\begin{aligned}\mathrm{Re}\{\chi_1^1(\omega)\} &= \frac{1}{1 + \omega^2\tau_D^2} - \xi_1^2\frac{(27 - 13\omega^2\tau_D^2) + \kappa(-54 + 20\omega^2\tau_D^2 + 14\omega^4\tau_D^4)}{60(1 + \omega^2\tau_D^2)^2(9 + 4\omega^2\tau_D^2)} \\ &\quad - \xi^2\frac{(27 + \omega^2\tau_D^2 - 2\omega^4\tau_D^4) - \kappa(54 + 14\omega^2\tau_D^2 + 8\omega^4\tau_D^4)}{15(1 + \omega^2\tau_D^2)^2(9 + \omega^2\tau_D^2)}\end{aligned} \tag{2.39}$$

$$\begin{aligned}&\mathrm{Im}\{\chi_1^1(\omega)\} \\ &= \omega\tau_D\left[\frac{1}{1 + \omega^2\tau_D^2} - \frac{\xi_1^2}{30}\frac{(21 + \omega^2\tau_D^2) + \kappa\left[(-75/2) - (7\omega^2\tau_D^2/2) + 4\omega^4\tau_D^4\right]}{(1 + \omega^2\tau_D^2)^2(9 + 4\omega^2\tau_D^2)}\right. \\ &\quad \left. - \frac{\xi^2}{15}\frac{(42 + 25\omega^2\tau_D^2 + \omega^4\tau_D^4) - \kappa(75 + 32\omega^2\tau_D^2 + 5\omega^4\tau_D^4)}{(1 + \omega^2\tau_D^2)^2(9 + \omega^2\tau_D^2)}\right]\end{aligned} \tag{2.40}$$

for the first harmonic components (varying at the fundamental frequency $\omega$ of the alternating field),

$$\begin{aligned}&\mathrm{Re}\{\chi_2^1(\omega)\} \\ &= -\frac{81 - 153\omega^2\tau_D^2 - 62\omega^4\tau_D^4 - 8\omega^6\tau_D^6 + \kappa(-162 + 144\omega^2\tau_D^2 + 58\omega^4\tau_D^4 - 8\omega^6\tau_D^6)}{3(1 + 4\omega^2\tau_D^2)(1 + \omega^2\tau_D^2)(4 + 4\omega^2\tau_D^2)(1 + \omega^2\tau_D^2/9)}\end{aligned} \tag{2.41}$$

$$\mathrm{Im}\{\chi_2^1(\omega)\} = -2\omega\tau_\mathrm{D} \frac{126 + 44\omega^2\tau_\mathrm{D}^2 + 8\omega^4\tau_\mathrm{D}^4 - \kappa(225 + 130\omega^2\tau_\mathrm{D}^2 + 29\omega^4\tau_\mathrm{D}^4 + 4\omega^6\tau_\mathrm{D}^6)}{3(1 + 4\omega^2\tau_\mathrm{D}^2)(1 + \omega^2\tau_\mathrm{D}^2)(9 + 4\omega^2\tau_\mathrm{D}^2)(1 + \omega^2\tau_\mathrm{D}^2/9)} \tag{2.42}$$

for the second harmonic components in $2\omega$, and

$$\mathrm{Re}\{\chi_3^1(\omega)\} = -\frac{3 - 17\omega^2\tau_\mathrm{D}^2 + \kappa[-6 + 20\omega^2\tau_\mathrm{D}^2 + (3\omega^4\tau_\mathrm{D}^4/2)]}{3(1 + \omega^2\tau_\mathrm{D}^2)[1 + (4\omega^2\tau_\mathrm{D}^2/9)](1 + 9\omega^2\tau_\mathrm{D}^2)} \tag{2.43}$$

$$\mathrm{Im}\{\chi_3^1(\omega)\} = -\omega\tau_\mathrm{D} \frac{14 - 6\omega^2\tau_\mathrm{D}^2 - \kappa(25 + 5\omega^2\tau_\mathrm{D}^2)}{3(1 + \omega^2\tau_\mathrm{D}^2)[1 + (4\omega^2\tau_\mathrm{D}^2/9)](1 + 9\omega^2\tau_\mathrm{D}^2)} \tag{2.44}$$

for the third harmonic components in $3\omega$. The appearance of the steady-state and $2\omega$-components in dielectric relaxation is a direct consequence of the nonlinear nature of the response when a constant field is simultaneously impressed on an alternating field. This is characterized by cross-terms between permanent and induced dipole moments. On setting $\kappa = 0$ and $\xi = 0$ in Eqs. (2.38)–(2.44), the results so obtained coincide exactly with those of Coffey and Paranjape [45]. The dimensionless parameter $\kappa$ in Eq. (2.37) measures the contribution of the induced dipole moment with respect to that of the permanent one: For *polar nonpolarizable* molecules $\kappa = 0$, while for *nonpolar polarizable* molecules $\kappa = \pm\infty$.

### 2. *Kerr Effect Response*

We now recall the analytical expressions suitable to Kerr effect relaxation and obtained up to second order in the electric field strength [49, 50]. As the time evolution of the electric birefringence is proportional to the expectation value of the second Legendre polynomial, the stationary solution may be represented as follows:

$$\begin{aligned}\langle P_2(\cos\vartheta)\rangle(t) = \langle P_2(\cos\vartheta)\rangle_\mathrm{st} + \sum_{j=1}^{2} \langle P_2(\cos\vartheta)\rangle_j'(\omega)\cos(j\omega t) \\ + \langle P_2(\cos\vartheta)\rangle_j''(\omega)\sin(j\omega t)\end{aligned} \tag{2.45}$$

where $\langle P_2(\cos\vartheta)\rangle_\mathrm{st}$ is a constant term containing two completely uncoupled parts. One is frequency independent while the other is frequency dependent due to the presence of the permanent moment. The terms $\langle P_2(\cos\vartheta)\rangle_j'(\omega)$ and $\langle P_2(\cos\vartheta)\rangle_j''(\omega)$ are the real and imaginary parts of the complex orientational factor $\langle P_2(\cos\vartheta)\rangle_j(\omega)$ of the electric birefringence. Equations for the reduced quantities,

$$\chi_0^2(\omega) = 15\frac{\langle P_2(\cos\vartheta)\rangle_{\mathrm{st}}}{\xi^2 + \xi_1^2/2} \qquad \chi_1^2(\omega) = \frac{15}{2}\frac{\langle P_2(\cos\vartheta)\rangle_1(\omega)}{\xi_1\xi}$$
$$\chi_2^2(\omega) = 30\frac{\langle P_2(\cos\vartheta)\rangle_2(\omega)}{\xi_1^2} \tag{2.46}$$

are given by

$$\chi_0^2(\omega) = \kappa + \frac{1}{1+2(\xi/\xi_1)^2}\left[\frac{1}{1+\omega^2\tau_{\mathrm{D}}^2} + 2(\xi/\xi_1)^2\right] \tag{2.47}$$

for the steady component,

$$\chi_1^2(\omega) = \frac{(1-\omega^2\tau_{\mathrm{D}}^2/3)/2 + (\kappa+1/2)(1+\omega^2\tau_{\mathrm{D}}^2) + i\omega\tau_{\mathrm{D}}\left[1+(\kappa+1/2)(1+\omega^2\tau_{\mathrm{D}}^2)\right]/3}{(1+\omega^2\tau_{\mathrm{D}}^2/9)(1+\omega^2\tau_{\mathrm{D}}^2)} \tag{2.48}$$

for the first complex harmonic component in $\omega$,

$$\chi_2^2(\omega) = \frac{(1-2\omega^2\tau_{\mathrm{D}}^2/3) + \kappa(1+\omega^2\tau_{\mathrm{D}}^2) + i\omega\tau_{\mathrm{D}}\left[5+2\kappa(1+\omega^2\tau_{\mathrm{D}}^2)\right]/3}{(1+\omega^2\tau_{\mathrm{D}}^2/9)(1+\omega^2\tau_{\mathrm{D}}^2)} \tag{2.49}$$

for the second complex harmonic component in $2\omega$. Equations (2.48) and (2.49) are in full agreement with those obtained by Thurston and Bowling [23] who considered the ac response only and by Morita and Watanabe [26] for an ac field combined with a dc bias field.

On the other hand, Raikher et al. [35, 36] considered the problem of the dynamic Kerr effect of rigid noninteracting dipolar molecules in a strong ac field by solving numerically Eq. (2.6) for $\sigma = 0$ in terms of Fourier coefficients depending on the angular frequency $\omega$ as well as on the ac field amplitude $\xi_1$. They were also able to give analytic expressions for $\langle P_2(\cos\vartheta)\rangle(t)$ in two limiting cases: (1) for the quasistatic regime corresponding to very low frequencies ($\omega\tau_{\mathrm{D}} \ll 1$), it is assumed that the distribution function is close to equilibrium, so that the response may be expressed in terms of the Langevin function [35, 36]

$$\langle P_2(\cos\vartheta)\rangle(t) = 1 - 3\frac{L(\xi_1\cos\omega t)}{\xi_1\cos\omega t} \tag{2.50}$$

(2) in the high-frequency limit ($\omega\tau_D \gg 1$) and for strong fields ($\xi_1 \gg 1$), the noninertial rotational motion of the molecule can be described by the dynamic equation

$$2\tau_D\dot{\vartheta} + (\xi_1 \cos \omega t) \sin \vartheta = 0 \tag{2.51}$$

representing the balance between viscous and field-induced torques. Equation (2.51) has an analytical solution

$$\cos \vartheta(t) = 2\left(1 + \exp[-2s(t)] \tan^2\left[\frac{\vartheta(0)}{2}\right]\right)^{-1} - 1 \tag{2.52}$$

where $s(t) = (\xi_1/2\omega\tau_D) \sin \omega t$ and $\vartheta(0)$ represents the initial orientation of the molecule. The solution (2.52) allows one to write down $P_2(\cos \vartheta)$, namely,

$$P_2(\cos \vartheta)(t) = \frac{3}{2}\left[\frac{1 - \exp[-2s(t)] \tan^2(\vartheta(0)/2)}{1 + \exp[-2s(t)] \tan^2(\vartheta(0)/2)}\right]^2 - \frac{1}{2} \tag{2.53}$$

Assuming that at time $t = 0$, all the molecules are completely oriented at random, one obtains [35, 36]

$$\langle P_2(\cos \vartheta)\rangle(t) = 1 - 3s(t)L[s(t)]/ \sinh^2 s(t) \tag{2.54}$$

Equation (2.54) is in full agreement with Eq. (4.73) obtained by Watanabe and Morita [2]. Similarly, one can obtain from Eq. (2.51) the expectation value of the first Legendre polynomial [51]

$$\langle P_1(\cos \vartheta)\rangle(t) = \coth s(t) - s(t)/ \sinh^2 s(t) \tag{2.55}$$

Till then, we had briefly discussed the main results drawn from previous theoretical studies in dielectric relaxation and dynamic Kerr effect, by restricting ourselves to those suitable for further consideration in Sections IV–VI. The interested reader can find more examples, discussions, and complete bibliography in reviews [2, 5] and in the books [1, 4].

## III. ORIENTATIONAL RELAXATION PROCESSES IN THE PRESENCE OF STRONG ELECTRIC FIELDS: THE ROTATIONAL DIFFUSION MODEL

### A. The Langevin Equation Approach

Before proceeding, we first consider the general aspects of the problem of the three-dimensional (3D) rotational Brownian motion of a particle (macromolecule) acted on by an external electric field $\mathbf{E}(t)$. Denoting by $\mathbf{u}(t)$ a unit vector through the center of mass of the particle, we can write the equation of motion for the rate of change of $\mathbf{u}(t)$ and the angular velocity $\boldsymbol{\omega}(t)$ of the particle [1, 52]

$$\frac{d\mathbf{u}(t)}{dt} = \boldsymbol{\omega}(t) \times \mathbf{u}(t) \tag{3.1}$$

$$\hat{\mathbf{I}}\frac{d\boldsymbol{\omega}(t)}{dt} + \varsigma\boldsymbol{\omega}(t) = \mathbf{m}(t) \times \mathbf{E}(t) + \boldsymbol{\lambda}(t) \tag{3.2}$$

where $\hat{\mathbf{I}}$ is the inertia tensor of the molecule, $\mathbf{m}$ is the total dipole moment that takes into account the permanent dipole moment of the molecule as well as effects due to the polarizability and hyperpolarizability, $\varsigma\boldsymbol{\omega}(t)$ is the damping torque due to Brownian movement ($\varsigma = kT/D$ is the rotational friction coefficient, assumed to be a scalar for simplicity) and $\boldsymbol{\lambda}(t)$ is the white noise driving torque, again due to Brownian movement, so that $\boldsymbol{\lambda}(t)$ has the following properties:

$$\overline{\lambda_i(t)} = 0$$

$$\overline{\lambda_i(t_1)\lambda_j(t_2)} = 2kT\varsigma\delta_{ij}\delta(t_1 - t_2)$$

Here the overbar means a statistical average over an ensemble of Brownian particles that *all* start at time $t$ with the *same* angular velocity $\boldsymbol{\omega}$ and orientation $\mathbf{u}$ [8]; $\delta_{ij}$ is Kronecker's delta; indexes $i, j = 1, 2, 3$ correspond to the Cartesian axes $X$, $Y$, $Z$ of the laboratory coordinate system $OXYZ$; and $\delta(t)$ is the Dirac delta function. Equation (3.1) is a purely kinematic relation with no particular reference either to the Brownian movement or to the shape of the particle, while Eq. (3.2) is the Euler–Langevin equation [52]. The term $\mathbf{m}(t) \times \mathbf{E}$ in Eq. (3.2) is the torque acting on the molecule. This torque can be expressed in terms of the potential energy $V(\mathbf{u}, t)$ of the molecule in the electric field $\mathbf{E}$ as a function of the components of the vector $\mathbf{u}$, namely,

$$\mathbf{m} \times \mathbf{E} = -\mathbf{u} \times \frac{\partial}{\partial \mathbf{u}} V \tag{3.3}$$

where

$$\frac{\partial}{\partial \mathbf{u}} = \mathbf{i}\frac{\partial}{\partial u_X} + \mathbf{j}\frac{\partial}{\partial u_Y} + \mathbf{k}\frac{\partial}{\partial u_Z}$$

$\mathbf{i}$, $\mathbf{j}$, and $\mathbf{k}$ are the unit vectors along the axes $X$, $Y$, and $Z$ respectively, $u_X$, $u_Y$, and $u_Z$ are the Cartesian components of the unit vector $\mathbf{u}(t)$. The field $\mathbf{E}$ in Eq. (3.3) may include externally applied ac and dc fields.

The noninertial limit (or the Debye approximation) occurs when the inertia term in Eq. (3.2) is neglected. In this limit, the angular velocity vector may be immediately obtained from Eqs. (3.2) and (3.3) as [31]

$$\boldsymbol{\omega}(t) = \varsigma^{-1}\left[\boldsymbol{\lambda}(t) - \mathbf{u} \times \frac{\partial}{\partial \mathbf{u}} V\right] \tag{3.4}$$

On combining Eqs. (3.1) and (3.4) one obtains [8, 31]

$$\varsigma \frac{d\mathbf{u}(t)}{dt} = -\frac{\partial}{\partial \mathbf{u}} V + \mathbf{u}(t)\left[\mathbf{u}(t) \cdot \frac{\partial}{\partial \mathbf{u}} V\right] + \boldsymbol{\lambda}(t) \times \mathbf{u}(t) \tag{3.5}$$

Equation (3.5) is a vector stochastic differential equation. It contains multiplicative noise terms given by the components of the vector product $\boldsymbol{\lambda}(t) \times \mathbf{u}(t)$. This poses an interpretation problem for this equation as discussed in [8] and [9]. As far as we are concerned, here we shall use the Stratonovich definition [53] of the average of the multiplicative noise term, as that definition always constitutes the mathematical idealization of the physical stochastic process of orientational relaxation in the noninertial limit [8]. Thus, it is not necessary to transform Eq. (3.5) into Itô equations (e.g., [54]), so that we avoid complicated mathematical manipulations.

We recall [9] that on taking the set of the Langevin equations for $N$ stochastic variables $\{\boldsymbol{\xi}(t)\} = \{\xi_1(t), \xi_2(t), \ldots, \xi_N(t)\}$:

$$\frac{d}{dt}\xi_i(t) = h_i(\{\boldsymbol{\xi}(t)\}, t) + g_{ij}(\{\boldsymbol{\xi}(t)\}, t)\Gamma_j(t) \qquad (i, j = 1, \ldots, N) \tag{3.6}$$

with

$$\begin{aligned} \overline{\Gamma_i(t)} &= 0 \\ \overline{\Gamma_i(t)\Gamma_j(t')} &= 2D\delta_{ij}\delta(t - t') \end{aligned} \tag{3.7}$$

and interpreting them as Stratonovich equations, that the averaged equations for the sharp values $\xi_i(t) = x_i$ at time $t$ are [8, 9]

$$\frac{dx_i}{dt} = \lim_{\tau \to 0} \frac{\overline{\xi_i(t+\tau) - x_i}}{\tau} = h_i(\{\mathbf{x}\}, t) + Dg_{kj}(\{\mathbf{x}\}, t)\frac{\partial}{\partial x_k} g_{ij}(\{\mathbf{x}\}, t) \qquad (i, j = 1, \ldots, N) \tag{3.8}$$

where $\xi_i(t+\tau)(\tau > 0)$ is the solution of Eq. (3.6) with the initial conditions $\xi_i(t) = x_i$. In Eqs. (3.6) and (3.8), the summation over $j$ and $k$ is understood

(Einstein's notation). The proof of Eq. (3.8) can be found elsewhere (see [9], pp. 54, 55). In like manner, we can prove that the averaged equation for an arbitrary differentiable function $f(\{\xi\})$ has the following form (see Appendix A):

$$\begin{aligned}\frac{df(\{\mathbf{x}\})}{dt} &= \lim_{\tau\to 0}\frac{\overline{f(\{\boldsymbol{\xi}(t+\tau)\}) - f(\{\mathbf{x}\})}}{\tau} \\ &= h_i(\{\mathbf{x}\},t)\frac{\partial}{\partial x_i}f(\{\mathbf{x}\}) + Dg_{kj}(\{\mathbf{x}\},t)\frac{\partial}{\partial x_k}\left[g_{ij}(\{\mathbf{x}\},t)\frac{\partial}{\partial x_i}f(\{\mathbf{x}\})\right]\end{aligned} \tag{3.9}$$

where summation over $i$, $j$, and $k$ is also understood. For the problem under consideration, according to Eq. (3.5), $x_k = u_k$ $(k = X, Y, Z)$ and the tensorial elements of $g_{ik}$ are given by

$$\begin{array}{lll} g_{XX} = 0 & g_{XY} = u_Z/\varsigma & g_{XZ} = -u_Y/\varsigma \\ g_{YX} = -u_Z/\varsigma & g_{YY} = 0 & g_{YZ} = u_X/\varsigma \\ g_{ZX} = u_Y/\varsigma & g_{ZY} = -u_X/\varsigma & g_{ZZ} = 0 \end{array} \tag{3.10}$$

As the methods of ordinary analysis are applicable to mathematical transformations of Stratonovich stochastic differential equations [53, 54], one can readily obtain from Eq. (3.5) the stochastic equation for any function $f(\mathbf{u})$, namely,

$$\begin{aligned}\frac{d}{dt}f[\mathbf{u}(t)] &= \dot{\mathbf{u}}(t)\cdot\frac{\partial}{\partial\mathbf{u}}f[\mathbf{u}(t)] \\ &= \varsigma^{-1}[\boldsymbol{\lambda}(t)\times\mathbf{u}(t)]\cdot\frac{\partial}{\partial\mathbf{u}}f[\mathbf{u}(t)] \\ &\quad - \varsigma^{-1}\left[\frac{\partial}{\partial\mathbf{u}}V - \mathbf{u}(t)\left(\mathbf{u}(t)\cdot\frac{\partial}{\partial\mathbf{u}}V\right)\right]\cdot\frac{\partial}{\partial\mathbf{u}}f[\mathbf{u}(t)]\end{aligned} \tag{3.11}$$

The right-hand side of Eq. (3.11) consists of two terms, namely, the deterministic drift and the noise-induced (or spurious) drift. Let us first evaluate the noise-induced drift [the first term in the right-hand side of Eq. (3.11)] on averaging it over an ensemble of Brownian particles, that all start at time $t$ with the same orientation $\mathbf{u}$. According to Eqs. (3.9) and (3.10), it can be written as follows [55, 56]:

$$
\begin{aligned}
&\overline{\varsigma^{-1}[\boldsymbol{\lambda}(t)\times\mathbf{u}(t)]\cdot\frac{\partial}{\partial\mathbf{u}}f[\mathbf{u}(t)]} \\
&\quad = kT\varsigma\left\{g_{kj}\frac{\partial}{\partial u_k}\left[g_{ij}\frac{\partial}{\partial u_Z}f(\mathbf{u})\right]+g_{kj}\frac{\partial}{\partial u_k}\left[g_{ij}\frac{\partial}{\partial u_X}f(\mathbf{u})\right]\right. \\
&\qquad\left.+g_{kj}\frac{\partial}{\partial u_k}\left[g_{ij}\frac{\partial}{\partial u_Y}f(\mathbf{u})\right]\right\}
\end{aligned}
\tag{3.12}
$$

On considering the first term in the right-hand side of Eq. (3.12) and taking account of Eq. (3.10), we have

$$
\begin{aligned}
Dg_{kj}\frac{\partial}{\partial u_k}\left[g_{ij}\frac{\partial}{\partial u_Z}f(\mathbf{u})\right] &= \frac{kT}{\varsigma}\left[(1-u_Z^2)\frac{\partial^2}{\partial u_Z^2}-u_Xu_Z\frac{\partial^2}{\partial u_X\partial u_Z}\right. \\
&\qquad\left.-u_Yu_Z\frac{\partial^2}{\partial u_Y\partial u_Z}-2u_Z\frac{\partial}{\partial u_Z}\right]f(\mathbf{u})
\end{aligned}
\tag{3.13}
$$

The other two terms can be readily obtained by means of a cyclic permutation of indexes $X\to Y$, $Y\to Z$, $Z\to X$, which leads to

$$
\begin{aligned}
&\overline{\varsigma^{-1}[\boldsymbol{\lambda}(t)\times\mathbf{u}(t)]\cdot\frac{\partial}{\partial\mathbf{u}}f[\mathbf{u}(t)]} \\
&\quad = \frac{kT}{\varsigma}\left[(1-u_X^2)\frac{\partial^2}{\partial u_X^2}+(1-u_Y^2)\frac{\partial^2}{\partial u_Y^2}+(1-u_Z^2)\frac{\partial^2}{\partial u_Z^2}\right. \\
&\qquad -2\left(u_Yu_X\frac{\partial^2}{\partial u_Y\partial u_X}+u_Xu_Z\frac{\partial^2}{\partial u_X\partial u_Z}+u_Yu_Z\frac{\partial^2}{\partial u_Y\partial u_Z}\right. \\
&\qquad\left.\left.+u_X\frac{\partial}{\partial u_X}+u_Y\frac{\partial}{\partial u_Y}+u_Z\frac{\partial}{\partial u_Z}\right)\right]f(\mathbf{u}) \\
&\quad = \frac{kT}{\varsigma}\Delta f(\mathbf{u})
\end{aligned}
\tag{3.14}
$$

where $\Delta$ is the Laplacian. We remark also that $u_X$, $u_Y$, $u_Z$ in Eqs. (3.12)–(3.14) and $u_X(t), u_Y(t), u_Z(t)$ in Eqs. (3.5) and (3.11) have different meanings, namely, $u_X(t)$, $u_Y(t)$, $u_Z(t)$ in Eqs. (3.5) and (3.11) are stochastic variables while $u_X$, $u_Y$, $u_Z$ in Eq. (3.12)–(3.14) are the sharp (definite) values $u_k(t)=u_k$ at time $t$ [8,9]. Thus, $\mathbf{u}$ is the sharp value of the stochastic variable $\mathbf{u}(t)$ at the moment of averaging $t$. Instead of using different symbols for the two quantities, we have distinguished the sharp values at time $t$ from the stochastic variables by deleting the time argument as in [8] and [9].

Let us now consider the deterministic drift term. We have [55]

$$
\begin{aligned}
&\overline{-\varsigma^{-1}\left\{\frac{\partial}{\partial \mathbf{u}}V[\mathbf{u}(t),t]-\mathbf{u}(t)\left[\mathbf{u}(t)\cdot\frac{\partial}{\partial \mathbf{u}}V[\mathbf{u}(t),t]\right]\right\}\cdot\frac{\partial}{\partial \mathbf{u}}f[\mathbf{u}(t)]} \\
&\quad = (2\varsigma)^{-1}\{V(\mathbf{u},t)\Delta f(\mathbf{u})+f(\mathbf{u})\Delta V(\mathbf{u},t)-\Delta[V(\mathbf{u},t)f(\mathbf{u})]\}
\end{aligned}
\tag{3.15}
$$

Thus on combining Eqs. (3.14) and (3.15) we obtain [55]

$$
2\tau_{\mathrm{D}}\dot{f}(\mathbf{u}) = \frac{1}{2kT}[V(\mathbf{u},t)\Delta f(\mathbf{u})+f(\mathbf{u})\Delta V(\mathbf{u},t)-\Delta(V(\mathbf{u},t)f(\mathbf{u}))]+\Delta f(\mathbf{u})
\tag{3.16}
$$

where $\tau_{\mathrm{D}} = \varsigma/(2kT) = 1/(2D)$ is the Debye relaxation time. As we have already mentioned, all the quantities in Eq. (3.16) are in general functions of $u_k$, which are themselves random variables with the probability density function $W$ such that $W du_k$ is the probability of finding $u_k$ in the interval $(u_k,\ u_k + du_k)$.

Below, we will confine ourselves to the consideration of symmetric top molecules, where the vector $\mathbf{u}$ is directed along the axis of symmetry of the molecule. For this case, the quantities suitable for dielectric and Kerr effect relaxation are averages involving the spherical harmonics $Y_{l,m}$ defined as [57]

$$
\begin{aligned}
Y_{l,m}(\vartheta,\varphi) &= (-1)^m\sqrt{\frac{(2l+1)(l-m)!}{4\pi(l+m)!}}e^{im\varphi}P_l^m(\cos\vartheta) \\
Y_{l,-m} &= (-1)^m Y_{l,m}^* \; (m>0)
\end{aligned}
\tag{3.17}
$$

where $P_l^m(x)$ is the associated Legendre function [57], the asterisk denotes the complex conjugate, and $\vartheta$ and $\varphi$ are the polar and azimuthal angles of spherical polar coordinates, respectively.

Thus, Eq. (3.16) yields

$$
2\tau_{\mathrm{D}}\dot{Y}_{l,m} = \frac{1}{2kT}[V\Delta Y_{l,m}+Y_{l,m}\Delta V-\Delta(VY_{l,m})]+\Delta Y_{l,m}
\tag{3.18}
$$

where the Laplace operator $\Delta$ is expressed in terms of $\vartheta$ and $\varphi$ as follows [57]

$$
\Delta = \frac{1}{\sin\vartheta}\frac{\partial}{\partial\vartheta}\left(\sin\vartheta\frac{\partial}{\partial\vartheta}\right)+\frac{1}{\sin^2\vartheta}\frac{\partial^2}{\partial\varphi^2}
$$

On using the known relationships [57, 58]

$$\Delta Y_{l,m} = -l(l+1)Y_{l,m} \tag{3.19}$$

$$Y_{l,m}Y_{l_1,m_1} = \sqrt{\frac{(2l+1)(2l_1+1)}{4\pi}} \sum_{\substack{l_2=|l-l_1| \\ \Delta l_2=2}}^{l+l_1} \frac{\langle l,0,l_1,0|l_2,0\rangle\langle l,m,l_1,m_1|l_2,m+m_1\rangle}{\sqrt{2l_2+1}} Y_{l_2,m+m_1} \tag{3.20}$$

($\langle l_1, m_1, l_2, m_2|l, m\rangle$ are the Clebsch–Gordan coefficients [57, 58]) one can show [59] that for any potential $V$, which can be expanded in terms of spherical harmonics as

$$V = \sum_{R,S} v_{R,S} Y_{R,S}$$

Eq. (3.18) can be presented as

$$\tau_D \dot{Y}_{l,m} = \sum_{r,s} e_{l,m,l+r,m+s} Y_{l+r,m+s} \tag{3.21}$$

where [59]

$$\begin{aligned} e_{l,m,l+r,m+s} = & -\frac{l(l+1)}{2}\delta_{r,0}\delta_{s,0} + (-1)^m \frac{\sqrt{(2l+1)(2l+2r+1)}}{8kT} \\ & \times \sum_{R=1}^{\infty} v_{R,s} \frac{r(2l+r+1) - R(R+1)}{\sqrt{\pi(2R+1)}} \langle l,0,l+r,0|R,0\rangle \\ & \times \langle l,m,l+r,-m-s|R,-s\rangle \end{aligned} \tag{3.22}$$

In order to obtain equations for the moments $\langle Y_{l,m}\rangle$, which govern the relaxation dynamics of the system, we must also average Eq. (3.21) over $W$ [8]. Thus, we obtain

$$\tau_D \frac{d}{dt}\langle Y_{l,m}\rangle(t) = \sum_{r,s} e_{l,m,l+r,m+s}\langle Y_{l+r,m+s}\rangle(t) \tag{3.23}$$

where

$$\langle Y_{l,m}\rangle(t) = \int_0^{2\pi}\int_0^{\pi} Y_{l,m}(\vartheta,\varphi)W(\vartheta,\varphi,t)\sin\vartheta\, d\vartheta\, d\varphi$$

However, if the system under consideration is in equilibrium, all averages are either constant or zero. Thus, first we need to construct from Eq. (3.21) a set of differential-recurrence equations for equilibrium correlation functions (for details, see [8, 56]). In the absence of ac external fields, the system is in equilibrium with the Boltzmann distribution function $W_0$ given by

$$W_0(\vartheta, \varphi) = \frac{1}{Z} \exp\left[ -\frac{V(\vartheta, \varphi)}{kT} \right] \tag{3.24}$$

where $Z$ is the partition function. Thus, on multiplying Eq. (3.21) by the function $Y^*_{L,m}(0)$, and averaging the resulting equation over the equilibrium distribution function $W_0$ at the instant $t = 0$, we obtain

$$\tau_D \frac{d}{dt} c_{l,m}(t) = \sum_{s,r} e_{l,m,l+r,m+s} c_{l+r,m+s}(t) \tag{3.25}$$

where the equilibrium correlation functions $c_{l,m}(t)$ are defined by

$$c_{l,m}(t) = \langle Y^*_{L,m}[\vartheta(0), \varphi(0)] Y_{l,m}[\vartheta(t), \varphi(t)] \rangle_0$$

and $\langle\ \rangle_0$ designates the equilibrium average over the distribution $W_0$.

### B. The Noninertial Fokker–Planck (Smoluchowski) Equation Approach

In Section III.A, we derived an equation for the matrix elements $e_{l,m,l',m'}$ of the moment system in the context of the Langevin equation approach. Here, we shall show how the same results can be obtained from the Fokker–Planck equation. In the context of the noninertial rotational diffusion model, the dynamics of a Brownian symmetric-top particle in a potential obeys the Smoluchowski equation (2.1) for the probability density distribution $W$ of orientations of the particle in configuration space, which can be written in the form [14]

$$\begin{aligned} \tau_D \frac{\partial}{\partial t} W &= L_{FP} W \\ &= \frac{1}{2} \Delta W + \frac{1}{4kT} [\Delta(WV) + W\Delta V - V\Delta W] \end{aligned} \tag{3.26}$$

where $L_{FP}$ is the Fokker–Planck operator. In general, the Smoluchowski equation (3.26) can be formally solved by means of an expansion of the distribution function $W$ in spherical harmonics. In such an approach, the problem can also be reduced to the solution of an infinite system of multiterm differential-recurrence

equations for the moments [the expectation values of the spherical harmonics $\langle Y_{l,m}\rangle(t)$].
We look for a solution of the Smoluchowski equation (3.26) as [59]:

$$W(\theta, \varphi, t) = \Psi(\theta, \varphi, t)\Psi^*(\theta, \varphi, t) \tag{3.27}$$

where $\Psi(\theta, \varphi, t)$ is given by

$$\Psi(\theta, \varphi, t) = \sum_{l,m} f_{l,m}(t) Y_{l,m}(\theta, \varphi) \tag{3.28}$$

The normalization condition for $W(\theta, \varphi, t)$ is

$$\int_0^{2\pi} d\varphi \int_0^{\pi} \sin\theta d\theta W(\theta, \varphi, t) = \sum_{l,m} |f_{l,m}(t)|^2 = 1 \tag{3.29}$$

The representation (3.27) has an advantage that it is unnecessary to apply additional conditions to the distribution function in order for it to be physically meaningful (e.g., $W$ should be positive and real). Moreover, the direct quantum mechanical analogy is obvious because the function $W$ is now similar to the probability density $|\Psi|^2$ ($\Psi$ is the wave function), which obeys the continuity equation ([60], p. 75):

$$\frac{\partial|\Psi|^2}{\partial t} + \operatorname{div}\mathbf{j} = 0 \tag{3.30}$$

where $\mathbf{j}$ is the probability density current.

On substituting Eq. (3.27) into Eq. (3.26), we can now obtain the moment system by means of the transformation [59]

$$\begin{aligned}
\tau_D \frac{d}{dt}\langle Y_{l,m}\rangle(t) &= \tau_D \int_0^{2\pi} d\varphi \int_0^{\pi} \sin\theta d\theta Y_{l,m} \dot{W} \\
&= \sum_{\substack{l',l'' \\ m',m''}} f_{l',m'}(t) f^*_{l'',m''}(t) \int_0^{2\pi} d\varphi \int_0^{\pi} \sin\theta d\theta Y_{l,m} L_{\mathrm{FP}}(Y_{l',m'} Y^*_{l'',m''}) \\
&= \sum_{\substack{l',l'',l''' \\ m',m'',m'''}} \sqrt{\frac{(2l'+1)(2l''+1)}{4\pi(2l'''+1)}} \langle l', 0, l'', 0 | l''', 0\rangle \langle l', m', l'', m'' | l''', m'''\rangle \\
&\qquad \times f_{l',m'}(t) f^*_{l'',m''}(t) \int_0^{2\pi} d\varphi \int_0^{\pi} \sin\theta d\theta Y_{l,m} L_{\mathrm{FP}} Y^*_{l''',m'''}
\end{aligned}$$

or

$$\tau_D \frac{d}{dt}\langle Y_{l,m}\rangle(t) = \sum_{l',m'} d_{l',m',l,m}\langle Y_{l',m'}\rangle(t) \tag{3.31}$$

where

$$d_{l',m',l,m} = \int_0^{2\pi} d\varphi \int_0^{\pi} \sin\theta d\theta Y_{l,m} L_{\mathrm{FP}} Y^*_{l',m'} \tag{3.32}$$

are the matrix elements of the Fokker–Planck operator and

$$\begin{aligned}\langle Y_{l,m}\rangle(t) &= \int_0^{2\pi}\int_0^{\pi} Y_{l,m}(\theta,\varphi)W(\theta,\varphi,t)\sin\theta d\theta d\varphi \\ &= \sum_{\substack{l',l''\\ m',m''}} \sqrt{\frac{(2l+1)(2l'+1)}{4\pi(2l''+1)}}\langle l,0,l',0|l'',0\rangle\langle l,m,l',m'|l'',m''\rangle f_{l',m'}(t) f^*_{l'',m''}(t)\end{aligned}$$

Here, we have used Eq. (3.20) and that [57]

$$\begin{aligned}&\int_0^{2\pi} d\varphi \int_0^{\pi} \sin\theta d\theta Y_{l,m} Y_{l',m'} Y^*_{l'',m''} \\ &\qquad = \sqrt{\frac{(2l+1)(2l'+1)}{4\pi(2l''+1)}}\langle l,0,l',0|l'',0\rangle\langle l,m,l',m'|l'',m''\rangle\end{aligned}$$

On using Eqs. (3.19), (3.20), and (3.26), we obtain [59]

$$\begin{aligned}d_{l',m\pm s,l,m} &= \int_0^{2\pi} d\varphi \int_0^{\pi} \sin\theta d\theta Y_{l,m} L_{\mathrm{FP}} Y^*_{l',m\pm s} \\ &= -\frac{l(l+1)\delta_{l,l'}\delta_{s,0}}{2} + (-1)^m \frac{1}{8kT}\sqrt{\frac{(2l+1)(2l'+1)}{\pi}} \\ &\quad \times \sum_{r=s}^{\infty} v_{r,\pm s}\frac{[l'(l'+1) - r(r+1) - l(l+1)]}{\sqrt{2r+1}} \\ &\quad \times \langle l,0,l',0|r,0\rangle\langle l,m,l',-m\mp s|r,\mp s\rangle\end{aligned} \tag{3.33}$$

On comparing Eqs. (3.22) and (3.33), one can readily prove that [59]

$$d_{l',m',l,m} = e_{l,m,l',m'} \tag{3.34}$$

Equation (3.34) demonstrates clearly that both moment systems given by Eqs. (3.23) and (3.31) are in complete agreement. This also proves the equivalence between the Langevin and the Fokker–Plank equation approaches.

In the cases when the potential $V$ does not depend on the azimuthal angle $\varphi$, so that it can be expanded in terms of spherical harmonics $Y_{R,S}$ with $S = 0$ only, namely,

$$V = \sum_R v_R Y_{R,0}$$

it is more convenient to use the recurrence equation for the expectation values of the Legendre polynomials $\langle P_l(\cos\vartheta)\rangle(t)$ instead of that for $\langle Y_{l,m}\rangle(t)$. On using the known relation [57],

$$Y_{l,0}(\vartheta,\varphi) = \sqrt{\frac{2l+1}{4\pi}} P_l(\cos\vartheta)$$

one can obtain from Eqs. (3.31) and (3.33)

$$\tau_D \frac{d}{dt}\langle P_l\rangle(t) = \sum_{l'} a_{l',l}\langle P_{l'}\rangle(t) \tag{3.35}$$

where

$$\begin{aligned} a_{l',l} &= \sqrt{\frac{2l'+1}{2l+1}} d_{l',0,l,0} \\ &= -\frac{l(l+1)\delta_{l,l'}}{2} + \frac{(2l'+1)}{8kT}\sum_{r=1}^{\infty} v_r \frac{[l'(l'+1) - r(r+1) - l(l+1)]}{\sqrt{\pi(2r+1)}}\langle l,0,l',0|r,0\rangle^2 \end{aligned} \tag{3.36}$$

For example, one can readily derive Eq. (2.6) for the potential given by Eq. (2.3) from Eqs. (3.33) and (3.36).

## C. ORDINARY AND MATRIX CONTINUED FRACTION APPROACH TO THE SOLUTION OF INFINITE HIERARCHIES OF DIFFERENTIAL-RECURRENCE EQUATIONS OF MOMENT SYSTEMS

As was shown above, the moment equations that are generated by averaging the nonlinear Langevin equation or the accompanying Fokker–Planck equation take the form of multiterm differential-recurrence relations between the set of statistical averages describing the dynamical behavior of appropriate physical quantities. If the recurrence relation between the averages is a three-term one, its solution may be expressed as an infinite continued fraction [8, 9]. In the majority of problems, the Langevin or Fokker–Planck equations do not lead to scalar three-term recurrence relations. Hence, the method based on the conversion of the recurrence relation to an ordinary continued fraction no longer applies. Examples of this are those problems that involve diffusion in phase space and diffusion in configuration space where the form of the potential is such as to give rise to a five or higher order term recurrence relation. These difficulties may, however, be circumvented since there is a method of converting a multiterm scalar recurrence relation to one of a three-term matrix [9]. Thus, multiterm recurrence relations may be solved in terms of matrix continued fractions [8, 9]. We shall confine our discussion to matrix continued fractions, as the scalar continued fraction constitutes simply a special case of the matrix one.

A vector tridiagonal recurrence relation, in the notation of Risken [9], may be written down in general form as

$$\tau_D \frac{d}{dt}\mathbf{C}_p(t) = \mathbf{Q}_p^- \mathbf{C}_{p-1}(t) + \mathbf{Q}_p \mathbf{C}_p(t) + \mathbf{Q}_p^+ \mathbf{C}_{p+1}(t) \tag{3.37}$$

where the $\mathbf{C}_p(t)$ are column vectors, $\mathbf{Q}_p^-$, $\mathbf{Q}_p^+$, and $\mathbf{Q}_p$ are, in a general case, time-independent noncommutative matrices, and $\tau_D$ is the characteristic relaxation time (for the problems under consideration $\tau_D$ is the Debye relaxation time for the isotropic diffusion). A general method of solution of Eq. (3.37) in terms of continued fractions was described by Risken [9] and later extended by Coffey et al. [8]. The last approach has the merit of being simpler than the previously available solution. The solution of such a recurrence relation may be accomplished as follows [8]. On taking the Laplace transform of Eq. (3.37), we have

$$(s\tau_D \mathbf{I} - \mathbf{Q}_p)\tilde{\mathbf{C}}_p(s) = \tau_D \mathbf{C}_p(0) + \mathbf{Q}_p^- \tilde{\mathbf{C}}_{p-1}(s) + \mathbf{Q}_p^+ \tilde{\mathbf{C}}_{p+1}(s) \tag{3.38}$$

($\mathbf{I}$ is the unit matrix). Let us seek the solution of Eq. (3.38) in the form

$$\tilde{\mathbf{C}}_p(s) = \mathbf{S}_p(s)\tilde{\mathbf{C}}_{p-1}(s) + \mathbf{q}_p(s) \tag{3.39}$$

where the matrix continued fraction $\mathbf{S}_p(s)$ satisfies the homogeneous Eq. (3.38) with $\mathbf{C}_p(0) = 0$, namely,

$$\mathbf{S}_p(s) = [s\tau_D \mathbf{I} - \mathbf{Q}_p - \mathbf{Q}_p^+ \mathbf{S}_{p+1}(s)]^{-1} \mathbf{Q}_p^- \tag{3.40}$$

or

$$\mathbf{S}_p(s) = \cfrac{\mathbf{I}}{\tau_D s\mathbf{I} - \mathbf{Q}_p - \mathbf{Q}_p^+ \cfrac{\mathbf{I}}{\tau_D s\mathbf{I} - \mathbf{Q}_{p+1} - \mathbf{Q}_{p+1}^+ \cfrac{\mathbf{I}}{\tau_D s\mathbf{I} - \mathbf{Q}_{p+2} - \cdots} \mathbf{Q}_{p+2}^-} \mathbf{Q}_{p+1}^-} \mathbf{Q}_p^- \tag{3.41}$$

(the fraction lines designate the matrix inversions). The infinite matrix continued fraction $\mathbf{S}_p(s)$ represents the *complementary solution* of the recurrence relation. The *particular solution* may be found as follows. We have from Eqs. (3.38)–(3.40)

$$(s\tau_D\mathbf{I} - \mathbf{Q}_p)\mathbf{q}_p(s) - \mathbf{Q}_p^+[\mathbf{S}_{p+1}(s)\mathbf{q}_p(s) + \mathbf{q}_{p+1}(s)] = \tau_D\mathbf{C}_p(0) \tag{3.42}$$

hence we have an expression for $\mathbf{q}_p(s)$, namely,

$$\begin{aligned} \mathbf{q}_p(s) &= \tau_D(s\tau_D\mathbf{I} - \mathbf{Q}_p - \mathbf{Q}_p^+\mathbf{S}_{p+1}(s))^{-1}[\mathbf{C}_p(0) + \mathbf{Q}_p^+\mathbf{q}_{p+1}(s)] \\ &= \tau_D\mathbf{S}_p(s)(\mathbf{Q}_p^-)^{-1}[\mathbf{C}_p(0) + \mathbf{Q}_p^+\mathbf{q}_{p+1}(s)] \end{aligned} \tag{3.43}$$

The exact solution of the recurrence relation (3.43) is given by [8]

$$\mathbf{q}_p(s) = \tau_D\mathbf{S}_p(s)(\mathbf{Q}_p^-)^{-1}\left[\mathbf{C}_p(0) + \sum_{k=1}^{\infty}\left(\prod_{n=1}^{k}\mathbf{Q}_{p+n-1}^+\mathbf{S}_{p+n}(s)(\mathbf{Q}_{p+n}^-)^{-1}\right)\mathbf{C}_{p+k}(0)\right] \tag{3.44}$$

Thus, on using Eqs. (3.39) and (3.44), one has [8]

$$\begin{aligned} \tilde{\mathbf{C}}_p(s) &= \mathbf{S}_p(s)\tilde{\mathbf{C}}_{p-1}(s) \\ &\quad + \tau_D\mathbf{S}_p(s)(\mathbf{Q}_p^-)^{-1}\left[\mathbf{C}_p(0) + \sum_{k=1}^{\infty}\left(\prod_{n=1}^{k}\mathbf{Q}_{p+n-1}^+\mathbf{S}_{p+n}(s)(\mathbf{Q}_{p+n}^-)^{-1}\right)\mathbf{C}_{p+k}(0)\right] \end{aligned} \tag{3.45}$$

Equation (3.45) constitutes the solution of Eq. (3.38) rendered as a sum of products of matrix continued fractions in the $s$ domain.

Thus, the *exact solution* for the Laplace transform of $\mathbf{C}_1(t)$ of Eq. (3.37) with $\mathbf{C}_0(t) = \mathbf{0}$ is given by [8]

$$\tilde{\mathbf{C}}_1(s) = \tau_D[s\tau_D\mathbf{I} - \mathbf{Q}_1 - \mathbf{Q}_1^+\mathbf{S}_2(s)]^{-1} \times \left[\mathbf{C}_1(0) + \sum_{k=2}^{\infty}\left(\prod_{n=2}^{k}\mathbf{Q}_{n-1}^+[s\tau_D\mathbf{I} - \mathbf{Q}_n - \mathbf{Q}_n^+\mathbf{S}_{n+1}(s)]^{-1}\right)\mathbf{C}_k(0)\right] \tag{3.46}$$

In many physical applications, the initial conditions $\mathbf{C}_p(0)$ can be expressed in terms of equilibrium (stationary in the general case) averages, which are in fact the equilibrium (stationary) solutions of the corresponding Fokker–Planck equation, or deduced equivalently from the vector recurrence relation (3.37) [9]:

$$\mathbf{Q}_p^-\mathbf{C}_{p-1}^0 + \mathbf{Q}_p\mathbf{C}_p^0 + \mathbf{Q}_p^+\mathbf{C}_{p+1}^0 = \mathbf{0} \tag{3.47}$$

This has a tridiagonal form, so that it is possible to express the equilibrium averages $\mathbf{C}_p^0$ in terms of the matrix continued fraction $\mathbf{S}_p(0)$ from Eq. (3.41). In order to accomplish this, let us consider Eq. (3.47) for $p = 1$. We have

$$\mathbf{Q}_1^-\mathbf{C}_0^0 + \mathbf{Q}_1\mathbf{C}_1^0 + \mathbf{Q}_1^+\mathbf{C}_2^0 = \mathbf{0} \tag{3.48}$$

or equivalently

$$[\mathbf{Q}_1^- + \mathbf{Q}_1\mathbf{S}_1(0) + \mathbf{Q}_1^+\mathbf{S}_2(0)\mathbf{S}_1(0)]\mathbf{C}_0^0 = \mathbf{0} \tag{3.49}$$

where the column vector $\mathbf{C}_0^0$ is given by

$$\mathbf{C}_0^0 = \begin{pmatrix} C_0^1 \\ C_0^2 \\ \vdots \\ C_0^p \end{pmatrix} \tag{3.50}$$

We may always choose an element of $\mathbf{C}_0^0$, for example, $C_0^p = 1$, because of the normalization condition [9]. Equation (3.47) is then a set of inhomogeneous linear equations hence the other components $C_0^1, C_0^2, \ldots, C_0^{p-1}$ and thus the vector $\mathbf{C}_0^0$ can be determined. We can then find the other vectors $\mathbf{C}_p^0$ by successive iteration

$$\mathbf{C}_p^0 = \mathbf{S}_p(0)\mathbf{S}_{p-1}(0)\ldots\mathbf{S}_1(0)\mathbf{C}_0^0 \tag{3.51}$$

and hence calculate $\mathbf{C}_p(0)$.

Equations (3.46) and (3.51) are important in view of their applications both to linear and nonlinear responses. For the particular case of a scalar three-term recurrence relation, Eq. (3.37) may be written down as [8, 9]

$$\tau_D \frac{d}{dt} C_n(t) = q_n^- C_{n-1}(t) + q_n C_n(t) + q_n^+ C_{n+1}(t) \qquad n = 1, 2, 3, \ldots \tag{3.52}$$

where the $C_n(t)$ are the appropriate relaxation functions, and the $q_n^-$, $q_n^+$, and $q_n$ are time-independent coefficients. Thus, according to Eq. (3.46), the *exact solution* of Eq. (3.52) with $C_0(t) = 0$ for the Laplace transform of $C_1(t)$ is given by

$$\tilde{C}_1(s) = \frac{\tau_D}{\tau_D s - q_1 - q_1^+ S_2(s)} \left\{ C_1(0) + \sum_{n=2}^{\infty} \left[ \prod_{k=2}^{n} \frac{q_{k-1}^+ S_k(s)}{q_k^-} \right] C_n(0) \right\} \tag{3.53}$$

where the infinite continued fraction $S_n(s)$ is defined as

$$S_n(s) = \cfrac{q_n^-}{\tau_D s - q_n - \cfrac{q_n^+ q_{n+1}^-}{\tau_D s - q_{n+1} - \cfrac{q_{n+1}^+ q_{n+2}^-}{\tau_D s - q_{n+2} - \cdots}}} \tag{3.54}$$

## D. RELAXATION TIME FOR NONLINEAR RESPONSE OF A BROWNIAN PARTICLE SUBJECTED TO A STRONG STEP EXTERNAL FIELD: ANALYTICAL SOLUTIONS FOR ONE-DIMENSIONAL MODELS

A system initially in an equilibrium (stationary) state and suddenly disturbed by an external stimulus (e.g., by applying a step external field) will evolve to a new equilibrium (stationary) state. Presently, a satisfactory theory is available for linear response only where the energy of the system arising from the external stimulus is much lower than the thermal energy [9, 61]. Here we need only *linear* (in the external stimulus) deviations of the expectation value of the dynamical variable of interest in the stationary state in order to evaluate the generalized susceptibility and/or response functions in terms of the appropriate equilibrium (stationary) correlation functions. Linear response theory is widely used for the interpretation of nonequilibrium phenomena such as dielectric and magnetic relaxation, conductivity problems, and so on.

First, we shall summarize the principal results of *linear response theory* ([9], Chapter 7) for systems where the dynamics obey 1D Fokker–Planck equations for the distribution function $W(x,t)$ of a variable $x$, such that

$$\frac{\partial}{\partial t}W = L_{\mathrm{FP}}W \tag{3.55}$$

Thus let us consider the Fokker–Planck operator $L_{\mathrm{FP}}$ of a system subject to a *small* perturbing force $F(t)$. On account of this $L_{\mathrm{FP}}$ may be represented as [9]

$$\begin{aligned} L_{\mathrm{FP}} &= \frac{\partial}{\partial x}\left[D^{(2)}(x)e^{-V(x)+B(x)F(t)}\frac{\partial}{\partial x}e^{V(x)-B(x)F(t)}\right] \\ &= L^{0}_{\mathrm{FP}}(x) + L_{\mathrm{ext}}(x)F(t) \end{aligned} \tag{3.56}$$

with

$$L^{0}_{\mathrm{FP}}(x) = \frac{\partial}{\partial x}\left[D^{(2)}(x)e^{-V(x)}\frac{\partial}{\partial x}e^{V(x)}\right] \qquad L^{0}_{\mathrm{FP}}(x)W_0(x) = 0 \tag{3.57}$$

$$L_{\mathrm{ext}}(x) = \frac{\partial}{\partial x}(D^{(2)}(x)B'(x)) \qquad B'(x) = \frac{d}{dx}B \tag{3.58}$$

where $L^{0}_{\mathrm{FP}}(x)$ is the Fokker–Planck operator in the *absence* of the perturbation, $W_0$ is the equilibrium (stationary) distribution function, $V$ is called a generalized (effective) potential [9], $D^{(2)}(x)$ is the diffusion coefficient, and $B(x)$ denotes a dynamical quantity. The step-off and step-on relaxation functions (when, on the one hand, a small constant force $F_1$ is suddenly switched off and on the other hand switched on at time $t = 0$, respectively, a statistical equilibrium having been achieved prior to the imposition of the stimulus in both instances) for a dynamic variable $A(x)$ are then

$$\begin{aligned} \langle A\rangle^{\mathrm{off}}(t) - \langle A\rangle_0 &= F_1 C_{AB}(t) \\ \langle A\rangle^{\mathrm{on}}(t) - \langle A\rangle_0 &= F_1[C_{AB}(0) - C_{AB}(t)] \qquad (t > 0) \end{aligned} \tag{3.59}$$

where the quantity

$$\begin{aligned} C_{AB}(t) &= \langle A(x(0))B(x(t))\rangle_0 - \langle A\rangle_0\langle B\rangle_0 \\ &= \int_{x_1}^{x_2}[A(x) - \langle A\rangle_0]\,e^{L^{0}_{\mathrm{FP}}t}[B(x) - \langle B\rangle_0]W_0(x)dx \end{aligned} \tag{3.60}$$

is the *equilibrium* (stationary) correlation function, and the symbols $\langle\,\rangle$ and $\langle\,\rangle_0$ designate the statistical averages over $W$ and $W_0$, respectively, with $x$ defined in the range $x_1 < x < x_2$. Furthermore, the spectrum of $\langle A \rangle(t)$ (ac response) is

$$\langle A \rangle_\omega = F_\omega \left[ C_{AB}(0) - i\omega \int_0^\infty C_{AB}(t) e^{-i\omega t} dt \right] \tag{3.61}$$

where $\langle A \rangle_\omega$ and $F_\omega$ are the Fourier components of $\langle A \rangle(t)$ and $F(t)$, respectively. We remark that Eqs. (3.59) and (3.61) are particular examples of Kubo's linear response theory [61]. Moreover, an exact integral formula [62, 63] exists for the correlation time $\tau_A$ [defined as the area under the curve of the normalized autocorrelation function $C_{AA}(t)$ as is apparent from Eq. (3.60) for $A = B$], namely [9],

$$\begin{aligned} \tau_A &= \frac{1}{C_{AA}(0)} \int_0^\infty C_{AA}(t) dt \\ &= \frac{1}{C_{AA}(0)} \int_{x_1}^{x_2} \frac{1}{D^{(2)}(x) W_0(x)} \left[ \int_{x_1}^{x} [A(x') - \langle A \rangle_0] W_0(x') dx' \right]^2 dx \end{aligned} \tag{3.62}$$

The derivation of an expression for the correlation time when $A \neq B$ is given in Appendix B.

In contrast, nonlinear response theory has been much less well developed because of its inherent mathematical–physical complexity. The calculation of the nonlinear response even for systems described by a single coordinate is a difficult task as there is no longer *any* connection between the step-on and the step-off responses and the ac response because the response now depends on the precise nature of the stimulus—as no *unique* response function valid for all stimuli unlike linear response exists. Nonlinear dielectric relaxation and dynamic Kerr effect are natural examples of the application of nonlinear response theory. As we have already mentioned in Section II, the results that have been obtained have mainly emerged either by perturbation theory or by numerical simulations. However, a few exact analytical solutions of particular problems dealing with nonlinear step responses exist (e.g., [27–32]). Following [32], we will now show that it is possible to derive an exact general equation in terms of an integral [similar to Eq. (3.62)] for the *nonlinear step response* relaxation time of a system governed by a 1D Fokker–Planck equation (3.55) just as in linear response.

We consider the 1D Brownian movement of a particle subject to a potential $V(x)$ and we assume that the relaxational dynamics of the particle obeys the Fokker–Planck equation (3.55). Let us suppose that at time $t = 0$ the value of the generalized potential $V$ is suddenly changed from $V_\mathrm{I}$ to $V_\mathrm{II}$ (e.g., by applying a strong external field or by a change in some parameter characterizing the system). We are interested in the relaxation of the system starting from an equilibrium (stationary) state I with the distribution function $W_{\mathrm{I}}(x)$, which evolves under the action of the stimulus to another equilibrium (stationary) state II with the distribution function $W_{\mathrm{II}}(x)$. Our goal is to evaluate the relaxation time $\tau_A$ of a dynamical variable $A$. This problem is intrinsically nonlinear because we assume that changes in the magnitude of the potential are now significant. Thus the concept of relaxation functions and relaxation times must be used rather than correlation functions and correlation times.

We define the normalized relaxation function $f_A(t)$ of a dynamical variable $A$ by

$$f_A(t) = \begin{cases} \dfrac{\langle A\rangle(t) - \langle A\rangle_{\mathrm{II}}}{\langle A\rangle_{\mathrm{I}} - \langle A\rangle_{\mathrm{II}}} & t > 0 \\ 1, & t \le 0 \end{cases} \tag{3.63}$$

where $\langle A\rangle_{\mathrm{I}}$ and $\langle A\rangle_{\mathrm{II}}$ are equilibrium (stationary) averages defined as

$$\langle A\rangle_{\mathrm{I}} = \int_{x_1}^{x_2} A(x)W_{\mathrm{I}}(x)dx \qquad \langle A\rangle_{\mathrm{II}} = \int_{x_1}^{x_2} A(x)W_{\mathrm{II}}(x)dx \tag{3.64}$$

and $\langle A\rangle(t)$ is the time-dependent average

$$\langle A\rangle(t) = \int_{x_1}^{x_2} A(x)W(x,t)dx \tag{3.65}$$

The relaxation time $\tau_A$ defined as the area under the curve of $f_A(t)$ at $t > 0$ is then given by

$$\tau_A = \int_0^\infty f_A(t)dt = \lim_{s\to 0}\int_0^\infty e^{-st}f_A(t)dt = \tilde{f}_A(0) \tag{3.66}$$

where $\tilde{f}_A(s)$ is the Laplace transform of $f_A(t)$. On interchanging the orders of integration over $x$ and $t$ in Eq. (3.66), we have

$$\tau_A = \frac{1}{\langle A\rangle_{\mathrm{I}} - \langle A\rangle_{\mathrm{II}}}\int_{x_1}^{x_2} [A(x) - \langle A\rangle_{\mathrm{II}}]\widetilde{W}(x,0)dx \tag{3.67}$$

where

$$\widetilde{W}(x,0) = \lim_{s\to 0}\widetilde{W}(x,s) \tag{3.68}$$

and

$$\widetilde{W}(x,s) = \int_0^\infty W(x,t)e^{-st}dt \tag{3.69}$$

The quantity $\tilde{W}(x,0)$ in Eq. (3.67) can be calculated analytically as follows. On using the final value theorem of Laplace transformation [39], namely,

$$\lim_{s\to 0} s\widetilde{W}(x,s) = \lim_{t\to\infty} W(x,t) = W_{\mathrm{II}}(x) \tag{3.70}$$

and on taking into account Eqs. (3.57) and (3.58), we obtain from Eq. (3.55) at $t > 0$

$$W_{\mathrm{II}}(x) - W_{\mathrm{I}}(x) = \frac{d}{dx}\left[D^{(2)}(x)\left(\frac{d}{dx}\widetilde{W}(x,0) + \widetilde{W}(x,0)\frac{d}{dx}V_{\mathrm{II}}(x)\right)\right] \tag{3.71}$$

The solution of Eq. (3.71) is

$$\widetilde{W}(x,0) = W_{\mathrm{II}}(x)\int_{x_1}^{x} \frac{\Phi(y)dy}{D^{(2)}(y)W_{\mathrm{II}}(y)} \tag{3.72}$$

where

$$\Phi(y) = \int_{x_1}^{y} [W_{\mathrm{II}}(z) - W_{\mathrm{I}}(z)]dz \tag{3.73}$$

Thus, we obtain from Eqs. (3.67) and (3.72)

$$\tau_A = \frac{1}{\langle A\rangle_{\mathrm{I}} - \langle A\rangle_{\mathrm{II}}}\int_{x_1}^{x_2} [A(x) - \langle A\rangle_{\mathrm{II}}]W_{\mathrm{II}}(x)\int_{x_1}^{x} \frac{\Phi(y)}{D^{(2)}(y)W_{\mathrm{II}}(y)}dydx$$

so that, on integration by parts,

$$\tau_A = \frac{1}{\langle A\rangle_{\mathrm{II}} - \langle A\rangle_{\mathrm{I}}}\int_{x_1}^{x_2} \frac{\Phi(x)\Psi(x)}{D^{(2)}(x)W_{\mathrm{II}}(x)}dx \tag{3.74}$$

where

$$\Psi(x) = \int_{x_1}^{x} [A(y) - \langle A\rangle_{\mathrm{II}}]W_{\mathrm{II}}(y)dy \tag{3.75}$$

Equation (3.74) is an exact equation for the nonlinear step response relaxation time, which is analogous to Eq. (3.62) for the linear response.

The above results are directly applicable to the noninertial rotational Brownian motion of a particle in an external *uniaxial* potential $V$. Here, the relevant Fokker–Planck equation for the distribution function $W$ of the orientations of the particle is given by Eq. (2.4), which using as new variable $x = \cos\vartheta$ reads

$$2\tau_D \frac{\partial}{\partial t} W(x,t) = \frac{\partial}{\partial x}\left[(1-x^2)\left(\frac{\partial}{\partial x} W(x,t) + \frac{W(x,t)}{kT}\frac{\partial}{\partial x} V_{II}(x)\right)\right] \qquad (t > 0) \tag{3.76}$$

In this case, $W_I(x)$ and $W_{II}(x)$ are the Maxwell–Boltzmann distribution functions

$$W_I(x) = e^{-V_I(x)/kT}/Z_I,\ W_{II}(x) = e^{-V_{II}(x)/kT}/Z_{II} \tag{3.77}$$

$$x_1 = -1 \qquad x_2 = 1 \tag{3.78}$$

$$D^{(2)}(x) = \frac{1-x^2}{2\tau_D} \tag{3.79}$$

$Z_I$ and $Z_{II}$ are the partition functions in the states I and II. Thus, Eq. (3.74) becomes

$$\tau_A = \frac{2\tau_D}{\langle A\rangle_{II} - \langle A\rangle_I}\int_{-1}^{1}\frac{e^{V_{II}(z)/kT}\Phi(z)\Psi(z)dz}{1-z^2} \tag{3.80}$$

where

$$\Phi(z) = \int_{-1}^{z}\left[Z_{II}^{-1}e^{-V_{II}(y)/kT} - Z_I^{-1}e^{-V_I(y)/kT}\right]dy \tag{3.81}$$

$$\Psi(z) = \int_{-1}^{z}[A(x) - \langle A\rangle_{II}]e^{-V_{II}(x)/kT}dx \tag{3.82}$$

Equation (3.80) can be used to calculate the relaxation time of the nonlinear dielectric and dynamic Kerr effect transient responses of systems consisting of permanently polar and polarizable molecules. The orientational relaxation of molecules here is governed by the Fokker–Planck equation (3.76) with the uniaxial potential $V$ given by Eq. (2.3), which can be presented as

$$\frac{V_N(x)}{kT} = -\sigma_N(x^2 + 2h_N x) \tag{3.83}$$

where $N = \mathrm{I}$ for the initial state and $N = \mathrm{II}$ for the final state, and

$$h_N = \xi_N/2\sigma_N \qquad \xi_N = \frac{\mu E_N}{kT} \qquad \sigma_N = \frac{\Delta\alpha}{2kT}E_N^2 \tag{3.84}$$

When $\xi_N = 0$, the potential given by Eq. (3.83) is a double-well symmetrical one and has a barrier at $\vartheta = (\pi/2)$ corresponding to a maximum; the height relative to the minima at $\vartheta = 0$ and $\vartheta = \pi$ is thus equal to $\sigma_N$. The potential becomes asymmetrical for $\xi_N \neq 0$, and the double well structure disappears at $h_N = h_s = 1$.

The quantities of greatest interest in the nonlinear response of these systems are the relaxation times $\tau_n (n = 1, 2)$ of the relaxation functions $f_1(t)$ and $f_2(t)$ of the first and second Legendre polynomials, namely,

$$f_1(t) = \langle P_1(\cos\vartheta)\rangle(t) - \langle P_1(\cos\vartheta)\rangle_{\mathrm{II}} \tag{3.85}$$

and

$$f_2(t) = \langle P_2(\cos\vartheta)\rangle(t) - \langle P_2(\cos\vartheta)\rangle_{\mathrm{II}} \tag{3.86}$$

The relaxation functions $f_1(t)$ and $f_2(t)$ govern the dielectric relaxation and the dynamic Kerr effect, respectively. The distribution functions in the equilibrium states I and II are given by

$$W_N(z) = e^{\sigma_N(2h_N z + z^2)}/Z_N \qquad (N = \mathrm{I, II}) \tag{3.87}$$

where

$$Z_N = \frac{1}{2}\sqrt{\frac{\pi}{\sigma_N}}e^{-\sigma_N h_N^2}\{\mathrm{erf}\,i[(1+h_N)\sqrt{\sigma_N}] + \mathrm{erf}\,i[(1-h_N)\sqrt{\sigma_N}]\} \tag{3.88}$$

and

$$\mathrm{erf}\,i(x) = \frac{2}{\sqrt{\pi}}\int_0^x e^{t^2}\,dt \tag{3.89}$$

is the error function of imaginary argument [39]. We have from Eqs. (3.74),

(3.87), and (3.88)

$$\tau_n = \frac{2\tau_D}{\langle P_n \rangle_{II} - \langle P_n \rangle_I} \int_{-1}^{1} \frac{\Phi(z)\Psi_n(z)e^{-\sigma_{II}z^2 - \xi_{II}z}dz}{1 - z^2} \tag{3.90}$$

where

$$\begin{aligned}
\Phi(z) &= \int_{-1}^{z} [W_{II}(z') - W_I(z')]dz' \\
&= \frac{\pi^{1/2}e^{-\sigma_{II}h_{II}^2}}{2\sigma_{II}^{1/2}Z_{II}}\{\text{erf}\, i[(z + h_{II})\sqrt{\sigma_{II}}\,] + \text{erf}\, i[(1 - h_{II})\sqrt{\sigma_{II}}\,]\} \\
&\quad - \frac{\pi^{1/2}e^{-\sigma_I h_I^2}}{2\sigma_I^{1/2}Z_I}\{\text{erf}\, i[(z + h_I)\sqrt{\sigma_I}\,] + \text{erf}\, i[(1 - h_I)\sqrt{\sigma_I}\,]\}
\end{aligned} \tag{3.91}$$

$$\begin{aligned}
\Psi_1(z) &= \int_{-1}^{z} [P_1(z') - \langle P_1 \rangle_{II}]e^{\sigma_{II}(z'^2 + 2h_{II}z')}dz' \\
&= \frac{1}{2\sigma_{II}}[e^{\sigma_{II}(z^2 + 2h_{II}z)} - e^{\sigma_{II}(1 - 2h_{II})}] - e^{\sigma_{II}(1 - h_{II}^2)}\frac{\pi^{1/2}\sinh(2\sigma_{II}h_{II})}{2\sigma_{II}^{3/2}Z_{II}} \\
&\quad \times \{\text{erf}\, i[(z + h_{II})\sqrt{\sigma_{II}}\,] + \text{erf}\, i[(1 - h_{II})\sqrt{\sigma_{II}}\,]\}
\end{aligned} \tag{3.92}$$

$$\begin{aligned}
\Psi_2(z) &= \int_{-1}^{z} [P_2(z') - \langle P_2 \rangle_{II}]e^{\sigma_{II}(z'^2 + 2h_{II}z')}dz' \\
&= \frac{3}{4\sigma_{II}}\Bigg\{e^{\sigma_{II}(z^2 + 2h_{II}z)}(z - h_{II}) + e^{\sigma_{II}(1 - 2h_{II})}(1 + h_{II}) \\
&\qquad - e^{\sigma_{II}(1 - h_{II}^2)}\frac{\pi^{1/2}[\cosh(2\sigma_{II}h_{II}) - h_{II}\sinh(2\sigma_{II}h_{II})]}{\sigma_{II}^{1/2}Z_{II}} \\
&\qquad \times [\text{erf}i[(z + h_{II})\sqrt{\sigma_{II}}\,] + \text{erf}i[(1 - h_{II})\sqrt{\sigma_{II}}\,]]\Bigg\}
\end{aligned} \tag{3.93}$$

$$\langle P_1 \rangle_N = \frac{e^{\sigma_N}\sinh(2\sigma_N h_N)}{\sigma_N Z_N} - h_N \tag{3.94}$$

$$\langle P_2 \rangle_N = \frac{3e^{\sigma_N}[\cosh(2\sigma_N h_N) - h_N\sinh(2\sigma_N h_N)]}{2\sigma_N Z_N} + \frac{3h_N^2}{2} - \frac{3}{4\sigma_N} - \frac{1}{2} \tag{3.95}$$

First, we shall compare the foregoing results with those previously quoted in Section II.A.3. Aizenberg and Eskin [28] and Déjardin et al. [30] by solving the infinite hierarchy of differential-recurrence relations for the averaged spherical harmonics, obtained the exact analytical solution for the relaxation time $\tau_2$ for the nonlinear rise transient of the dynamic Kerr effect for *nonpolar polarizable* molecules, which is given by Eq. (2.24). In our notation, the relaxation time $\tau_2$ for the model considered in [27, 30], and [31] is given by Eq. (3.90) for $n = 2$ with

$$h_{\mathrm{I}} = 0 \qquad \sigma_{\mathrm{I}} = 0 \qquad h_{\mathrm{II}} = 0 \qquad \sigma_{\mathrm{II}} = \sigma \tag{3.96}$$

where according to Eqs. (3.91)–(3.95)

$$\Phi(z) = \frac{\operatorname{erf} i(\sqrt{\sigma}z)}{2\operatorname{erf} i(\sqrt{\sigma})} - \frac{z}{2} \tag{3.97}$$

$$\Psi_2(z) = \frac{3}{4\sigma}\left[ze^{\sigma z^2} - e^{\sigma}\frac{\operatorname{erf} i(\sqrt{\sigma}z)}{\operatorname{erf} i(\sqrt{\sigma})}\right] \tag{3.98}$$

$$\langle P_2\rangle_{\mathrm{II}} = \frac{3}{4\sigma}\left[\frac{2e^{\sigma}\sqrt{\sigma}}{\sqrt{\pi}\operatorname{erf} i(\sqrt{\sigma})} - 1\right] - \frac{1}{2} \tag{3.99}$$

so that

$$\tau_2 = \frac{3\tau_{\mathrm{D}}}{4\sigma\langle P_2\rangle_{\mathrm{II}}}\int_{-1}^{1}\left\{z\frac{\operatorname{erf} i(\sqrt{\sigma}z)}{\operatorname{erf} i(\sqrt{\sigma})}[1 + e^{\sigma(1-z^2)}] - e^{\sigma(1-z^2)}\frac{\operatorname{erf} i^2(\sqrt{\sigma}z)}{\operatorname{erf} i^2(\sqrt{\sigma})} - z^2\right\}\frac{dz}{1-z^2} \tag{3.100}$$

The result of the calculation of the correlation time $\tau_2$ from Eq. (3.100) is *in complete agreement* with that predicted by Eq. (2.24) (e.g., for $\sigma = 5$ we obtained $\tau_2/\tau_{\mathrm{D}} = 0.281269\ldots$ in both representations).

We have also borne our attention to another particular case, when a strong constant field is suddenly switched-on at $t = 0$ on a system of *polar and nonpolarizable* molecules. This corresponds to the following values of the parameters:

$$\xi_{\mathrm{I}} = 0 \qquad \xi_{\mathrm{II}} = \xi \qquad \sigma_{\mathrm{I}} = \sigma_{\mathrm{II}} = 0 \tag{3.101}$$

All the quantities in Eq. (3.90) can be expressed in terms of elementary functions, namely,

$$\langle P_1 \rangle_{\mathrm{I}} = \langle P_2 \rangle_{\mathrm{I}} = 0 \qquad \langle P_1 \rangle_{\mathrm{II}} = \coth \xi - \frac{1}{\xi} \qquad \langle P_2 \rangle_{\mathrm{II}} = 1 - \frac{3}{\xi}\left[\coth \xi - \frac{1}{\xi}\right] \tag{3.102}$$

$$\Phi(z) = \frac{1}{2}\left[\frac{e^{\xi z} - e^{-\xi}}{\sinh \xi} - z - 1\right] \tag{3.103}$$

$$\Psi_1(z) = e^{\xi z}(z - \coth \xi) + e^{-\xi}(1 + \coth \xi) \tag{3.104}$$

$$\Psi_2(z) = \frac{3}{2\xi}\left[e^{\xi z}\left(z^2 - \frac{2z}{\xi} - 1 + \frac{2\coth \xi}{\xi}\right) - \frac{2e^{-\xi}}{\xi}(1 + \coth \xi)\right] \tag{3.105}$$

Thus, the exact equations for the relaxation times $\tau_1$, $\tau_2$ are [33]:

$$\tau_1 = \frac{2\tau_{\mathrm{D}}}{\xi \coth \xi - 1}\left[1 - \frac{\mathrm{Cinh}(2\xi)}{\sinh^2 \xi}\right] \tag{3.106}$$

for dielectric relaxation, and

$$\tau_2 = \frac{\tau_{\mathrm{D}}}{1 - \xi \coth \xi + \xi^2/3}\left[\xi \coth \xi - 3 + 2\frac{\mathrm{Cinh}(2\xi)}{\sinh^2 \xi}\right] \tag{3.107}$$

for Kerr effect relaxation, where

$$\mathrm{Cinh}(z) = \int_0^z \frac{\cosh t - 1}{t}\, dt$$

is the integral hyperbolic cosine [39]. A similar analytical solution for the relaxation time $\tau_2$ for the nonlinear rise transient of the dynamic Kerr effect for *polar nonpolarizable* molecules was also obtained by Eskin [27] [cf. Eq. (2.29)]. As well as in the previous case the results yielded by Eqs. (3.107) and (2.29) coincide.

In [31], we also evaluated the nonlinear dielectric and birefringence rise transients for a more general model of both *polar and polarizable* molecules (the solution was obtained in terms of matrix continued fractions). The model corresponds to

$$h_{\mathrm{I}} = 0 \qquad \sigma_{\mathrm{I}} = 0 \qquad h_{\mathrm{II}} = h \qquad \sigma_{\mathrm{II}} = \sigma \tag{3.108}$$

Here we have also found a complete agreement between the relaxation times yielded by both solutions. Equation (3.90) can also be applied to other nonlinear problems of the dynamic birefringence and dielectric relaxation considered by Morita and Watanabe [2]. In particular, Eq. (3.90) for

$$h_{\mathrm{I}} = -h_{\mathrm{II}} \qquad \sigma_{\mathrm{II}} = \sigma_{\mathrm{I}} = \sigma$$

yields the nonlinear dielectric relaxation time for a transient process where a homogeneous electric field $\mathbf{E}_0$ applied to a system of polar and polarizable particles for a time of sufficient duration to allow the system to reach the equilibrium state for $t < 0$, is suddenly reversed at $t = 0$. Furthermore, Eq. (3.90) (for $h_{\mathrm{I}} \neq h_{\mathrm{II}}$, $\sigma_{\mathrm{II}} \neq \sigma_{\mathrm{I}}$) gives the exact solution for the nonlinear birefringence and dielectric relaxation times when a strong homogeneous electric field $\mathbf{E}_0$ is suddenly applied to the system where a Maxwell–Boltzmann distribution of the orientations of particles has been established by another homogeneous electric field $\mathbf{E}'_0$.

## IV. NONLINEAR TRANSIENTS IN DIELECTRIC RELAXATION AND DYNAMIC KERR EFFECT ARISING FROM SUDDEN CHANGES OF A STRONG DC ELECTRIC FIELD

The goal of this section is to present a general theory for the transient dynamic birefringence and nonlinear dielectric relaxation response of polar and anisotropically polarizable particles (macromolecules) dissolved in nonpolar solvents when both magnitude and direction of the dc field may suddenly be changed. We shall demonstrate that the theory developed contains as particular cases all the results previously obtained for various particular transient relaxation problems caused by sudden changes of external fields and is in agreement with available experimental data.

### A. Formulation and Solution of the Problem

Let us suppose that *both magnitude and direction* of the dc field are suddenly changed at time $t = 0$ from $\mathbf{E}_{\mathrm{I}}$ to $\mathbf{E}_{\mathrm{II}}$. As already described in Section II.A, in order to evaluate this transient response, one needs to consider the relaxation of a system of particles (macromolecules) diluted in a nonpolar solvent starting from an equilibrium state I with the distribution function

$$W_{\mathrm{I}} = e^{-V_{\mathrm{I}}/kT}/Z_{\mathrm{I}} \ (t \leq 0) \tag{4.1}$$

and evolving to another equilibrium state II with the distribution function

$$W_{\mathrm{II}} = e^{-V_{\mathrm{II}}/kT}/Z_{\mathrm{II}} \ (t \to \infty) \tag{4.2}$$

On neglecting effects due to the hyperpolarizability of the molecule, the reduced potential energy $V_N/kT$ is given by [55]

$$\frac{V_N}{kT} = -\xi_N \cos \Xi_N - \sigma_N \cos^2 \Xi_N \qquad (N = \mathrm{I},\ \mathrm{II}) \tag{4.3}$$

where $\xi_N$, $\sigma_N$ are determined by Eq. (3.84) and $\Xi_N$ is the angle between the vectors $\mathbf{u}$ and $\mathbf{E}_N$ as shown in Figure 1. The problem is intrinsically nonlinear

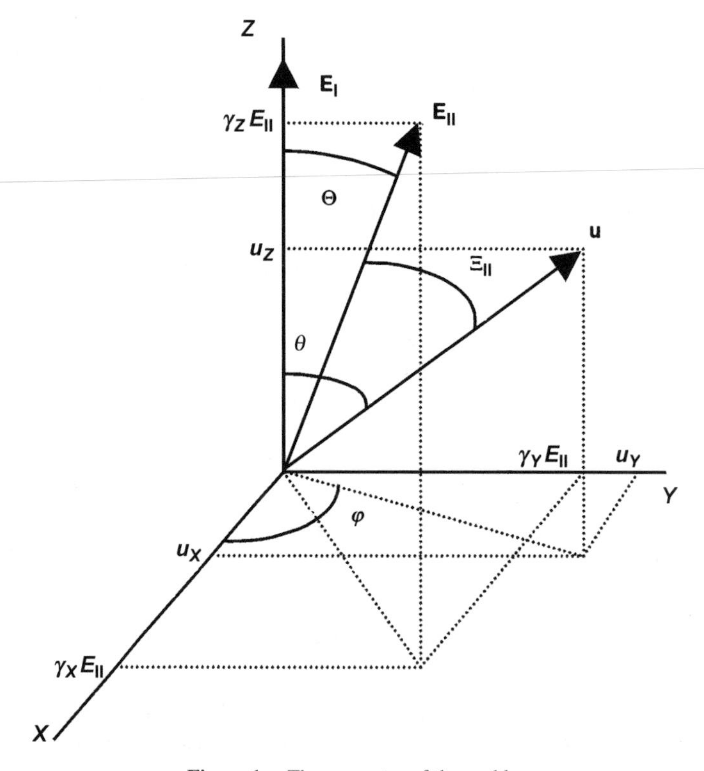


**Figure 1.** The geometry of the problem.

because it is assumed that changes both in the magnitude and in the direction of the dc field are significant (the particular case of only rigid polar molecules was considered in [64]).

The approach of the evaluation of the expectation values of the spherical harmonics developed in Section III can be directly applied to the present problem. Equation (3.23) yields the infinite hierarchy of differential-recurrence relations for the moments $\langle Y_{l,m}\rangle(t)$, namely,

$$\tau_D \frac{d}{dt}\langle Y_{l,m}\rangle(t) = \sum_{r,s} e_{l,m,l+r,m+s}\langle Y_{l+r,m+s}\rangle(t) \tag{4.4}$$

where the coefficients $e_{l,m,l+r,m+s}$ are defined by Eq. (3.22). For the potential given by Eq. (4.3), we have

$$\frac{V_N}{kT} = \sum_{R=1}^{2}\sum_{S=-R}^{R} v_{R,S}^N Y_{R,S} - \frac{\sigma_N}{3} \tag{4.5}$$

where

$$v_{1,0}^N = -\sqrt{\frac{4\pi}{3}}\xi_N \gamma_Z^N \tag{4.6}$$

$$v_{1,1}^N = -(v_{1,-1}^N)^* = \sqrt{\frac{2\pi}{3}}\xi_N(\gamma_X^N - i\gamma_Y^N) \tag{4.7}$$

$$v_{2,0}^N = -\sqrt{\frac{4\pi}{45}}\sigma_N\left[3(\gamma_Z^N)^2 - 1\right] \tag{4.8}$$

$$v_{2,1}^N = -(v_{2,-1}^N)^* = \sqrt{\frac{8\pi}{15}}\sigma_N\gamma_Z^N(\gamma_X^N - i\gamma_Y^N) \tag{4.9}$$

$$v_{2,2}^N = (v_{2,-2}^N)^* = -\sqrt{\frac{2\pi}{15}}\sigma_N(\gamma_X^N - i\gamma_Y^N)^2 \tag{4.10}$$

Here,

$$\gamma_X^N = \sin\Theta_N\cos\Phi_N \qquad \gamma_Y^N = \sin\Theta_N\sin\Phi_N \qquad \gamma_Z^N = \cos\Theta_N \tag{4.11}$$

are the direction cosines of $\mathbf{E}_N$ in the coordinate system $OXYZ$ (see Fig. 1). Without loss of generality, we will suppose that the field $\mathbf{E}_{\mathrm{I}}$ is directed along the $Z$ axis. Thus, below we will set $\Theta_{\mathrm{I}} = 0$ and $\Theta_{\mathrm{II}} = \Theta$. For the problem in question, it is convenient to introduce the relaxation functions $c_{n,m}(t)$ defined as

$$c_{n,m}(t) = \langle Y_{n,m}\rangle(t) - \langle Y_{n,m}\rangle_{\text{II}} \tag{4.12}$$

Thus, we have the following 21 term differential-recurrence relations:

$$\tau_{\text{D}}\frac{d}{dt}c_{n,m}(t) = \sum_{r=-2}^{2}\sum_{s=-2}^{2} e_{n,m,n+r,m+s}c_{n+r,m+s}(t) \tag{4.13}$$

where the coefficients $e_{n,m,n+r,m+s}$ are given by

$$\begin{aligned} e_{n,m,n+r,m+s} = {} & -\frac{n(n+1)}{2}\delta_{s,0}\delta_{r,0} + \frac{(-1)^m\sqrt{(2n+1)(2n+2r+1)}}{8} \\ & \times \Bigg\{ v_{1,s}\left[\frac{r(2n+r+1)-2}{\sqrt{3\pi}}\right] \\ & \quad \times \langle n,0,n+r,0|1,0\rangle\langle n,m,n+r,-m-s|1,-s\rangle \\ & \quad + v_{2,s}\left[\frac{r(2n+r+1)-6}{\sqrt{5\pi}}\right] \\ & \quad \times \langle n,0,n+r,0|2,0\rangle\langle n,m,n+r,-m-s|2,-s\rangle \Bigg\} \end{aligned} \tag{4.14}$$

with $e_{n,m,n\,\pm\,1,m\,\pm\,2} \equiv 0$. In order to derive Eq. (4.13), we have used the fact that the equilibrium averages $\langle Y_{n,m}\rangle_N (N = \text{I, II})$ satisfy the recurrence relation:

$$\sum_{r=-2}^{2}\sum_{s=-2}^{2} e_{n,m,n+r,m+s}\langle Y_{n+r,m+s}\rangle_N = 0 \tag{4.15}$$

which follows from Eq. (4.4) in the stationary regime.

Thus, on solving Eq. (4.13) one is able to evaluate the transient responses of the electric polarization $P(t)$ and the birefringence function $K(t)$, which are conveniently described by the normalized relaxation functions,

$$\begin{aligned} f_1(t) &= \frac{P(t) - P_{\text{II}}}{P_{\text{I}} - P_{\text{II}}} \\ &= \frac{\gamma_Z^{\text{II}}c_{1,0}(t) - \sqrt{2}\text{Re}\{(\gamma_X^{\text{II}} - i\gamma_Y^{\text{II}})c_{1,1}(t)\}}{\gamma_Z^{\text{II}}c_{1,0}(0) - \sqrt{2}\text{Re}\{(\gamma_X^{\text{II}} - i\gamma_Y^{\text{II}})c_{1,1}(0)\}} \end{aligned} \tag{4.16}$$

and

$$
\begin{aligned}
f_2(t) &= \frac{K(t) - K_{\mathrm{II}}}{K_{\mathrm{I}} - K_{\mathrm{II}}} \\
&= \frac{[3(\gamma_Z^{\mathrm{II}})^2 - 1]c_{2,0}(t) + \sqrt{6}\mathrm{Re}\{(\gamma_X^{\mathrm{II}} - i\gamma_Y^{\mathrm{II}})^2 c_{2,2}(t)\} - 2\sqrt{6}\gamma_Z^{\mathrm{II}}\mathrm{Re}\{(\gamma_X^{\mathrm{II}} - i\gamma_Y^{\mathrm{II}})c_{2,1}(t)\}}{[3(\gamma_Z^{\mathrm{II}})^2 - 1]c_{2,0}(0) + \sqrt{6}\mathrm{Re}\{(\gamma_X^{\mathrm{II}} - i\gamma_Y^{\mathrm{II}})^2 c_{2,2}(0)\} - 2\sqrt{6}\gamma_Z^{\mathrm{II}}\mathrm{Re}\{(\gamma_X^{II} - i\gamma_Y^{\mathrm{II}})c_{2,1}(0)\}}
\end{aligned}
\tag{4.17}
$$

respectively. Other quantities of interest are the integral relaxation times $\tau_n$ defined either by Eq. (2.12) or by (2.13).

The formal matrix continued fraction method to the solution of the recurrence equations, such as Eq. (4.13), where two indices vary, was suggested in [9]. However, in practice it is rather inconvenient, as one must use matrices of infinite dimension. Below, we shall use a more refined approach to the solution of Eq. (4.13) recently suggested in [65] and [66] for the solution of similar recurrence equations, so that it is possible to reduce the computational task to operations involving matrices of finite dimensions.

Following [55] let us introduce a vector $\mathbf{C}_n(t)$, consisting of $8n$ elements:

$$
\mathbf{C}_n(t) = \begin{pmatrix} c_{2n,-2n}(t) \\ c_{2n,-2n+1}(t) \\ \vdots \\ c_{2n,2n}(t) \\ c_{2n-1,-2n+1}(t) \\ c_{2n-1,-2n+2}(t) \\ \vdots \\ c_{2n-1,2n-1}(t) \end{pmatrix}
\tag{4.18}
$$

Then, Eq. (4.13) can be transformed in a matrix three-term differential-recurrence Eq. (3.37), namely,

$$
\tau_{\mathrm{D}} \frac{d}{dt}\mathbf{C}_n(t) = \mathbf{Q}_n^{-}\mathbf{C}_{n-1}(t) + \mathbf{Q}_n\mathbf{C}_n(t) + \mathbf{Q}_n^{+}\mathbf{C}_{n+1}(t) \qquad n = 1, 2, 3, \ldots \tag{4.19}
$$

where

$$
\mathbf{C}_0(t) = \mathbf{0} \tag{4.20}
$$

and

$$\mathbf{C}_1(t) = \begin{pmatrix} c_{2,-2}(t) \\ c_{2,-1}(t) \\ c_{2,0}(t) \\ c_{2,1}(t) \\ c_{2,2}(t) \\ c_{1,-1}(t) \\ c_{1,0}(t) \\ c_{1,1}(t) \end{pmatrix} \tag{4.21}$$

The matrices $\mathbf{Q}_n$, $\mathbf{Q}_n^+$, $\mathbf{Q}_n^-$ in Eq. (4.19) are given by

$$\mathbf{Q}_n = \begin{pmatrix} \mathbf{X}_{2n} & \mathbf{W}_{2n} \\ \mathbf{Y}_{2n-1} & \mathbf{X}_{2n-1} \end{pmatrix}$$

$$\mathbf{Q}_n^+ = \begin{pmatrix} \mathbf{Z}_{2n} & \mathbf{Y}_{2n} \\ \mathbf{O} & \mathbf{Z}_{2n-1} \end{pmatrix} \tag{4.22}$$

$$\mathbf{Q}_n^- = \begin{pmatrix} \mathbf{V}_{2n} & \mathbf{O} \\ \mathbf{W}_{2n-1} & \mathbf{V}_{2n-1} \end{pmatrix}$$

where

$$\mathbf{Y}_n = -\frac{n}{n+2}\mathbf{W}_{n+1}^{\dagger} \qquad \mathbf{Z}_n = -\frac{n}{n+3}\mathbf{V}_{n+2}^{\dagger} \tag{4.23}$$

and $\mathbf{O}$ is the zero matrix of appropriate dimension. Here the symbol † denotes the Hermitian conjugation (transposition and complex conjugation). Thus, the matrices $\mathbf{Q}_n$, $\mathbf{Q}_n^+$, and $\mathbf{Q}_n^-$ can be expressed in terms of the submatrices $\mathbf{W}_l$, $\mathbf{X}_l$ and $\mathbf{V}_l$, which are defined as

$$\mathbf{W}_l = \begin{pmatrix} w_{l,-l}^+ & 0 & 0 & \dots & 0 & 0 \\ w_{l,-l+1} & w_{l,-l+1}^+ & 0 & \dots & 0 & 0 \\ w_{l,-l+2}^- & w_{l,-l+2} & w_{l,-l+2}^+ & \dots & 0 & 0 \\ \vdots & \vdots & \vdots & \ddots & \vdots & \vdots \\ 0 & 0 & 0 & \dots & w_{l,l-2} & w_{l,l-2}^+ \\ 0 & 0 & 0 & \dots & w_{l,l-1}^- & w_{l,l-1} \\ 0 & 0 & 0 & \dots & 0 & w_{l,l}^- \end{pmatrix} \tag{4.24}$$

$$\mathbf{X}_l = \begin{pmatrix} x_{l,-l} & x^{+}_{l,-l} & x^{++}_{l,-l} & 0 & \ddots & 0 & 0 \\ x^{-}_{l,-l+1} & x_{l,-l+1} & x^{+}_{l,-l+1} & x^{++}_{l,-l+1} & \ddots & 0 & 0 \\ x^{--}_{l,-l+2} & x^{-}_{l,-l+2} & x_{l,-l+2} & x^{+}_{l,-l+2} & \ddots & \vdots & \vdots \\ 0 & \ddots & \ddots & \ddots & \ddots & 0 & 0 \\ \vdots & \ddots & \ddots & \ddots & \ddots & x^{+}_{l,l-2} & x^{++}_{l,l-2} \\ 0 & \ddots & 0 & x^{--}_{l,l-1} & x^{-}_{l,l-1} & x_{l,l-1} & x^{+}_{l,l-1} \\ 0 & \cdots & 0 & 0 & x^{--}_{l,l} & x^{-}_{l,l} & x_{l,l} \end{pmatrix} \tag{4.25}$$

$$\mathbf{V}_l = \begin{pmatrix} v^{++}_{l,-l} & 0 & \cdots & 0 & 0 \\ v^{+}_{l,-l+1} & v^{++}_{l,-l+1} & \ddots & 0 & 0 \\ v_{l,-l+2} & v^{+}_{l,-l+2} & \ddots & \ddots & \vdots \\ v^{-}_{l,-l+3} & v_{l,-l+1} & \ddots & v^{++}_{l,l-5} & 0 \\ v^{--}_{l,-l+4} & v^{-}_{l,-l+4} & \ddots & v^{+}_{l,l-4} & v^{++}_{l,l-4} \\ 0 & v^{--}_{l,-l+5} & \ddots & v_{l,l-3} & v^{+}_{l,l-3} \\ \vdots & \ddots & \ddots & v^{-}_{l,l-2} & v_{l,l-2} \\ 0 & 0 & \ddots & v^{--}_{l,l-1} & v^{-}_{l,l-1} \\ 0 & 0 & \cdots & 0 & v^{--}_{l,l} \end{pmatrix} \tag{4.26}$$

and have dimensions $(2l+1) \times (2l-1)$, $(2l+1) \times (2l+1)$, and $(2l+1) \times (2l-3)$, respectively. The elements of the submatrices $\mathbf{W}_l$, $\mathbf{X}_l$, and $\mathbf{V}_l$ are given by

$$w_{n,m} = \frac{\gamma_Z^N \xi_N (n+1)}{2} \sqrt{\frac{n^2 - m^2}{(2n-1)(2n+1)}}$$

$$w^{+}_{n,m} = -(w^{-}_{n,-m})^* = \frac{(\gamma_X^N - i\gamma_Y^N)\xi_N (n+1)}{4} \sqrt{\frac{(n-m-1)(n-m)}{(2n-1)(2n+1)}}$$

$$x_{n,m} = \frac{\sigma_N[3(\gamma_Z^N)^2 - 1][n(n+1) - 3m^2]}{2(2n-1)(2n+3)} - \frac{n(n+1)}{2}$$

$$x_{n,m}^- = -(x_{n,-m}^+)^* = -\frac{3\sigma_N(2m-1)\gamma_Z^N(\gamma_X^N + i\gamma_Y^N)}{2(2n-1)(2n+3)}\sqrt{(n+1-m)(n+m)}$$

$$x_{n,m}^{--} = (x_{n,-m}^{++})^*$$

$$= -\frac{3\sigma_N(\gamma_X^N + i\gamma_Y^N)^2}{4(2n-1)(2n+3)}\sqrt{(n-m+1)(n-m+2)(n+m-1)(n+m)}$$

$$v_{n,m} = \frac{\sigma_N(n+1)}{2(2n-1)}[3(\gamma_Z^N)^2 - 1]\sqrt{\frac{[n^2 - m^2][(n-1)^2 - m^2]}{(2n+1)(2n-3)}}$$

$$v_{n,m}^- = -(v_{n,-m}^+)^*$$

$$= -\frac{(n+1)\sigma_N\gamma_Z^N(\gamma_X^N + i\gamma_Y^N)}{(2n-1)}\sqrt{\frac{(n^2 - m^2)(n+m-2)(n+m-1)}{(2n-3)(2n+1)}}$$

$$v_{n,m}^{--} = (v_{n,-m}^{++})^*$$

$$= \frac{(n+1)\sigma_N(\gamma_X^N + i\gamma_Y^N)^2}{4(2n-1)}\sqrt{\frac{(n+m-3)(n+m-2)(n+m-1)(n+m)}{(2n-3)(2n+1)}}$$

The column vector $\mathbf{C}_1(t)$ [Eq. (4.21)] contains all the $c_{l,m}(t)$ that are necessary for the calculation of the relaxation functions $f_1(t)$ and $f_2(t)$ from Eqs. (4.16) and (4.17). On taking the Laplace transform of Eq. (4.19), we have already seen that the *exact* solution for the Laplace transform $\tilde{\mathbf{C}}_1(s)$ in terms of *matrix continued fractions* is given by Eq. (3.46), namely,

$$\tilde{\mathbf{C}}_1(s) = \tau_D[\tau_D s\mathbf{I} - \mathbf{Q}_1 - \mathbf{Q}_1^+ \mathbf{S}_2^{\mathrm{II}}(s)]^{-1} \times \left\{\mathbf{C}_1(0) + \sum_{n=2}^{\infty}\left(\prod_{k=2}^{n}\mathbf{Q}_{k-1}^+[\tau_D s\mathbf{I} - \mathbf{Q}_k - \mathbf{Q}_k^+\mathbf{S}_{k+1}^{\mathrm{II}}(s)]^{-1}\right)\mathbf{C}_n(0)\right\} \tag{4.27}$$

where $\mathbf{S}_n^{\mathrm{II}}(s)$ is the infinite matrix continued fraction defined by Eq. (3.14), namely,

$$\mathbf{S}_n^{\mathrm{II}}(s) = \cfrac{\mathbf{I}}{\tau_D s\mathbf{I} - \mathbf{Q}_n - \mathbf{Q}_n^+ \cfrac{\mathbf{I}}{\tau_D s\mathbf{I} - \mathbf{Q}_{n+1} - \mathbf{Q}_{n+1}^+ \cfrac{\mathbf{I}}{\tau_D s\mathbf{I} - \mathbf{Q}_{n+2} \ddots}\mathbf{Q}_{n+2}^-}\mathbf{Q}_{n+1}^-}\mathbf{Q}_n^- \tag{4.28}$$

and the matrices $\mathbf{Q}_n$, $\mathbf{Q}_n^+$, and $\mathbf{Q}_n^-$ are evaluated for the parameters $\xi_N$, $\sigma_N$, $\gamma_i^N$ in the final state (the state II here). The dimensions of the matrices $\mathbf{Q}_n$, $\mathbf{Q}_n^+$, and

$\mathbf{Q}_n^-$ are accordingly equal to $8n \times 8n$, $8n \times 8(n+1)$, and $8n \times 8(n-1)$. The exception is

$$\mathbf{Q}_1^- = \begin{pmatrix} \mathbf{V}_2 \\ \mathbf{W}_1 \end{pmatrix}$$

which degenerates to a column vector of dimension 8.

The vectors of the initial conditions $\mathbf{C}_n(0)$ in Eq. (4.27) can be also calculated with the help of the matrix-continued fractions on noting that the components $c_{n,m}(0)$ of $\mathbf{C}_n(0)$ are given by

$$c_{n,m}(0) = \langle Y_{n,m} \rangle_{\mathrm{I}} - \langle Y_{n,m} \rangle_{\mathrm{II}} \tag{4.29}$$

We transform Eq. (4.15) to matrix recurrence relations to obtain

$$\mathbf{Q}_n^- \mathbf{R}_{n-1}^N + \mathbf{Q}_n \mathbf{R}_n^N + \mathbf{Q}_n^+ \mathbf{R}_{n+1}^N = 0 \qquad n = 1, 2, 3, \ldots \tag{4.30}$$

where

$$\mathbf{R}_n^N = \begin{pmatrix} \langle Y_{2n,-2n} \rangle_N \\ \langle Y_{2n,-2n+1} \rangle_N \\ \vdots \\ \langle Y_{2n,2n} \rangle_N \\ \langle Y_{2n-1,-2n+1} \rangle_N \\ \langle Y_{2n-1,-2n+2} \rangle_N \\ \vdots \\ \langle Y_{2n-1,2n-1} \rangle_N \end{pmatrix} \tag{4.31}$$

and the matrices $\mathbf{Q}_n$, $\mathbf{Q}_n^+$, and $\mathbf{Q}_n^-$ are evaluated for the parameters $\xi_N$, $\sigma_N$, $\gamma_i^N$ in the corresponding state. The solution of Eq. (4.30) is given by

$$\mathbf{R}_n^N = \mathbf{S}_n^N(0)\mathbf{R}_{n-1}^N = \frac{1}{\sqrt{4\pi}} \prod_{k=1}^{n} \mathbf{S}_k^N(0) \tag{4.32}$$

where

$$\mathbf{S}_n^N(0) = [-\mathbf{Q}_n - \mathbf{Q}_n^+ \mathbf{S}_{n+1}^N(0)]^{-1} \mathbf{Q}_n^- \tag{4.33}$$

Thus, the initial conditions $\mathbf{C}_n(0)$ are given by

$$\mathbf{C}_n(0) = \frac{1}{\sqrt{4\pi}} \left[ \prod_{k=1}^{n} \mathbf{S}_k^{\mathrm{I}}(0) - \prod_{k=1}^{n} \mathbf{S}_k^{\mathrm{II}}(0) \right] \qquad n = 1, 2, 3, \ldots \tag{4.34}$$

In particular, for $n = 1$ we have

$$\mathbf{C}_1(0) = \frac{1}{\sqrt{4\pi}}[\mathbf{S}_1^{\mathrm{I}}(0) - \mathbf{S}_1^{\mathrm{II}}(0)] \tag{4.35}$$

By putting $s = i\omega$ in Eq. (4.27), we are now able to calculate the one-sided Fourier transforms of the relaxation functions $f_1(t)$ and $f_2(t)$. Moreover, by using Eq. (2.13), we can also calculate the relaxation times $\tau_1$ and $\tau_2$. The matrix continued fraction solution [Eq. (4.27)] that we have obtained is very convenient for the purpose of computation. All the matrix continued fractions and series involved converge very rapidly, thus 8–10 downward iterations in calculating these continued fractions and 8–10 terms in the series are enough to estimate the spectrum $\tilde{\mathbf{C}}_1(i\omega)$ at an accuracy not $< 6$ significant digits in the majority of cases. Having determined $\tilde{\mathbf{C}}_1(i\omega)$, we are now able to calculate from Eqs. (4.16)–(4.18) all the quantities of interest. It should be noted that for a given value of $\xi$, the two angles $\Theta$ (polar) and $\Phi$ (azimuthal) define the direction cosines of the applied field $\mathbf{E}_{\mathrm{II}}$ in general. However, due to the symmetry properties the solution is independent on the angle $\Phi$, so that we may set $\Phi = 0$ in the calculations.

Let us first calculate the transient responses when a strong dc field $\mathbf{E}$ is suddenly rotated at an angle $\Theta$, which leads one to consider that only the direction of the field is changed, that is, $\xi_{\mathrm{I}} = \xi_{\mathrm{II}}$. The rapidly rotating field method was introduced by Watanabe and Morita [2]. In comparison with other methods (such as rise transient or rapidly reversing field methods) this method has the advantage of obtaining a larger value for the birefringence. Previously, the theory of this method was developed only for *nonpolarizable* molecules. The approach developed here allows us to calculate the relaxation time and the spectra of the relaxation functions for both *polar and polarizable molecules.* Some results of these calculations are shown in Figures 2–7. The calculations were carried out for

$$\sigma_N = \frac{\xi_N^2}{2}\kappa \qquad \text{with} \qquad \kappa = 1$$

(we recall that the ratio $\kappa$ [Eq. (2.37)] characterizes the relative effect of the induced dipole moment with respect to the permanent one and can vary from $-\infty$ to $+\infty$ [2, 4, 23]). The evolution of the relaxation time $\tau_1 = \tilde{f}_1(0)$ of the electric polarization as a function of the angle $\Theta$ and the dimensionless parameter $\xi = \xi_{\mathrm{I}} = \xi_{\mathrm{II}}$ (which characterizes the strength of the dc field) is illustrated by a surface plot in Figure 2. For $\Theta \approx 0$, the relaxation time $\tau_1 \approx \tau_{\mathrm{D}}$ for $\xi \ll 1$ (small fields), while $\tau_1 \sim \tau_{\mathrm{D}}/\xi$ for high values of $\xi$. For small values of $\Theta$, $\tau_1$ decreases monotonically to zero with increasing $\xi$. However, as $\Theta \to \pi$ (rever-

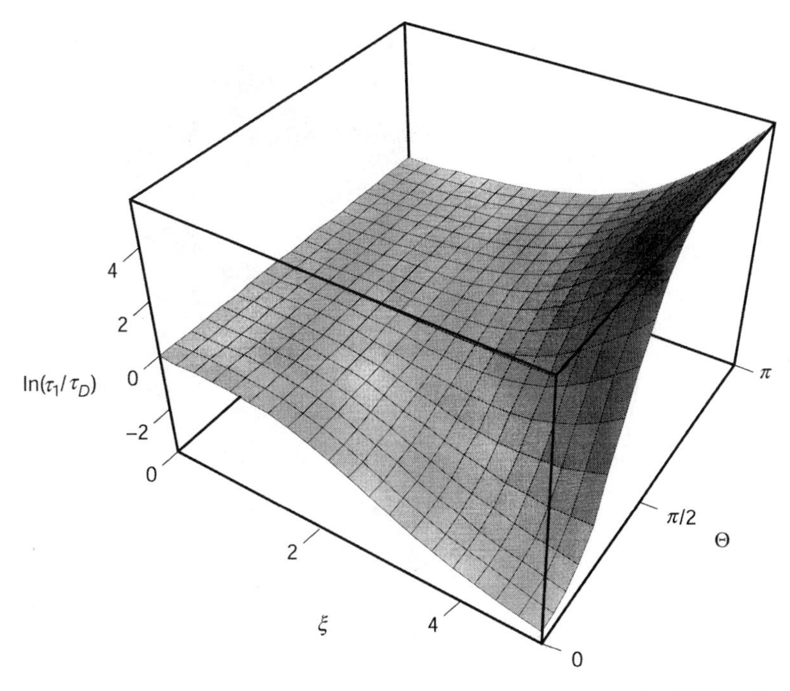


**Figure 2.** The $\ln(\tau_1/\tau_D)$ as a function of $\xi$ and $\Theta$ for $\kappa = 1$ (a rapidly rotating field $\xi_I = \xi_{II} = \xi$, $0.01 \leq \xi \leq 5$).

sing field) one may notice that $\tau_1$ increases with increasing $\xi$. A similar plot of the Kerr effect relaxation time $\tau_2 = \tilde{f}_2(0)$ is shown in Figure 3. For $\Theta \approx 0$, $\tau_2 \approx \tau_D/3$ for $\xi \ll 1$ and $\tau_2 \sim \tau_D/\xi$ for $\xi \gg 1$. Just as for $\tau_1$, for a fixed value of $\Theta \approx 0$ the relaxation time $\tau_2$ decreases monotonically to zero with increasing $\xi$. For a reversing field ($\Theta \to \pi$), the relaxation time $\tau_2$ tends to infinity. However, such behavior is only due to the definition of the relaxation time $\tau_2$ given by Eq. (2.12), as in this case ($\xi_I = \xi_{II}$) $\langle P_2 \rangle_I = \langle P_2 \rangle_{II}$ for $\Theta = \pi$. Thus, a more suitable quantity characterizing this particular case would be the area under the unnormalized birefringence relaxation function, which is not equal to zero for $\Theta = \pi$.

The real and imaginary parts of the spectra of the one-sided FT of the relaxation functions $f_1(t)$ and $f_2(t)$ are illustrated in Figures 4–7. As one can see in these figures, the dispersion curves have a very complicated behavior. It is clearly seen (Figs. 5 and 7) that in the vicinity of $\Theta = 0$ corresponding to the step-on (or step-off) field response, two relaxation processes appear in these spectra. One (slow) Arrhenius-like process describes the overbarrier reversal of the molecule in the potential (4.3), while the second one describes the fast

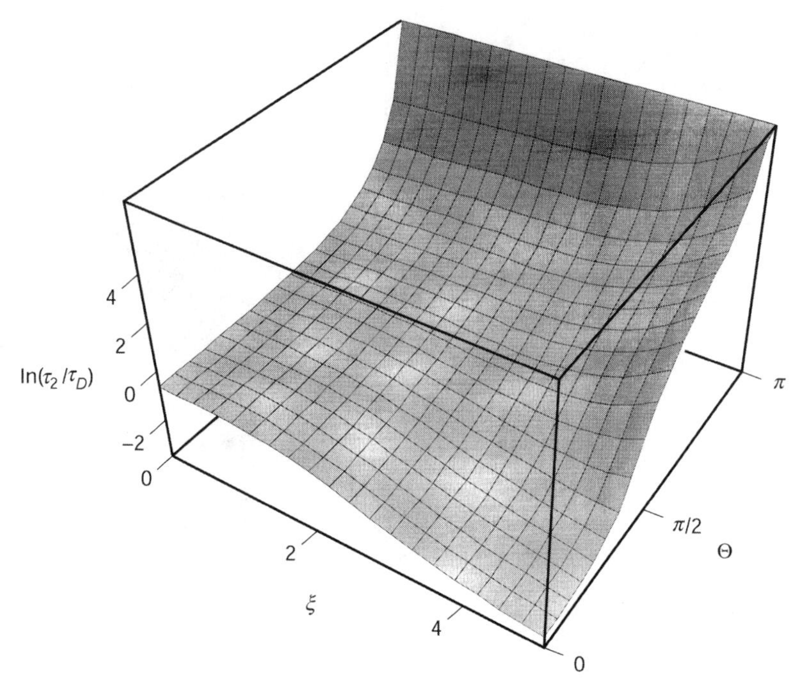


**Figure 3.** The $\ln(\tau_2/\tau_D)$ as a function of $\xi$ and $\Theta$ for $\kappa = 1$ (a rapidly rotating field ($\xi_I = \xi_{II} = \xi$, $0.01 \leq \xi \leq 5$).

relaxation inside the wells located at the minima of the potential energy, namely, at $\vartheta = 0$ and $\vartheta = \pi$. On increasing $\Theta$, although both processes continue to exist, the amplitude of the slow process enhances and progressively masks the high-frequency relaxation. Such a behavior implies that for the rapidly rotating field the relaxation functions $f_1(t)$ and $f_2(t)$ appropriate to polar and anisotropically polarizable molecules may not be approximated by a single exponential in contrast to the Debye-like behavior encountered when only permanent moments are taken into account [64]. This difference arises mainly from the double-well structure of the potential energy, Eq. (4.3), considered in this instance.

We considered above the transient behavior for a rapidly rotating field assuming that there is no change in the strength of the field. Similar calculations can also be carried out when the strength of the field may be different in the states I and II. The last case covers all possible situations for transient relaxation at sudden changes of the external field. If we suppose that only the strength but not the direction of the external field does change, then the calculations can be considerably simplified as in this case, the dynamics of the system is governed by a five term differential-recurrence Eq. (2.6). This corresponds to $\Theta = 0$

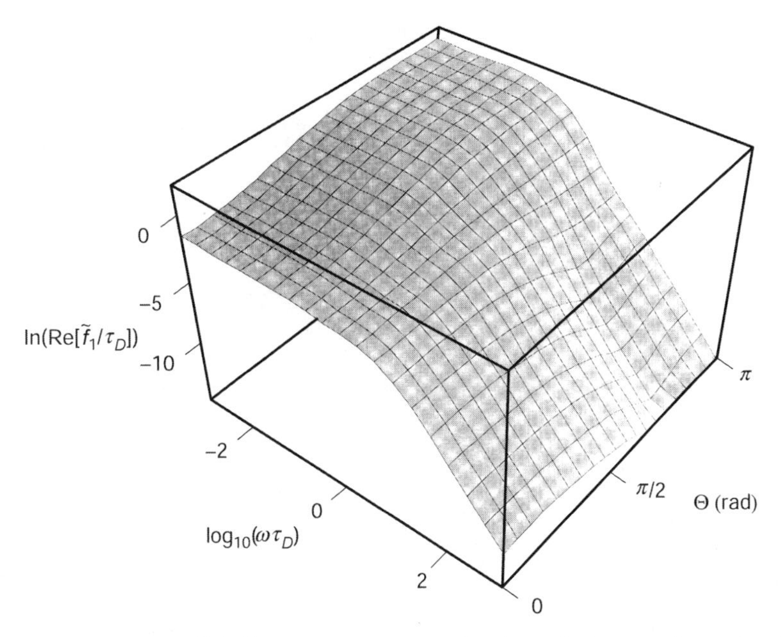


**Figure 4.** The $\ln[\mathrm{Re}(\tilde{f}_1/\tau_\mathrm{D})]$ as a function of $\log_{10}(\omega\tau_\mathrm{D})$ and $\Theta$ for a rapidly rotating field ($\xi_\mathrm{I} = \xi_\mathrm{II} = 3$ and $\kappa = 1$).

(step-on and/or step-off response) and $\Theta = \pi$ (suddenly reversing field response) for arbitrary values of $E_\mathrm{I}$ and $E_\mathrm{II}$. The details of the matrix continued fraction approach to the solution of all these problems will be given below. Moreover, for step-on and/or step-off ($\Theta = 0$) and suddenly reversing fields ($\Theta = \pi$) the relaxation times $\tau_1$ and $\tau_2$ can be calculated in integral form (3.90), which is a direct consequence of the nonlinear transient response theory developed in [32] for systems whose dynamics are governed by a 1D Fokker–Planck equation. Equation (3.90) provides us with an independent check of the matrix continued fraction solution as well as allowing us to evaluate the relaxation times $\tau_1$ and $\tau_2$ readily for various particular cases.

Thus, in the context of the noninertial rotational diffusion model, the transient nonlinear dielectric relaxation and dynamic Kerr effect responses of an ensemble of noninteracting polar and polarizable molecules in a strong dc field, when both the magnitude and the direction of the dc field may suddenly be changed, can be evaluated from Eq. (4.27) in terms of matrix-continued fractions. The theory contains as particular cases all the results previously obtained for various particular transient relaxation problems such as transient responses on step-on, step-off, suddenly reversing or suddenly rotating fields. However, in these cases the theory can be considerably simplified.

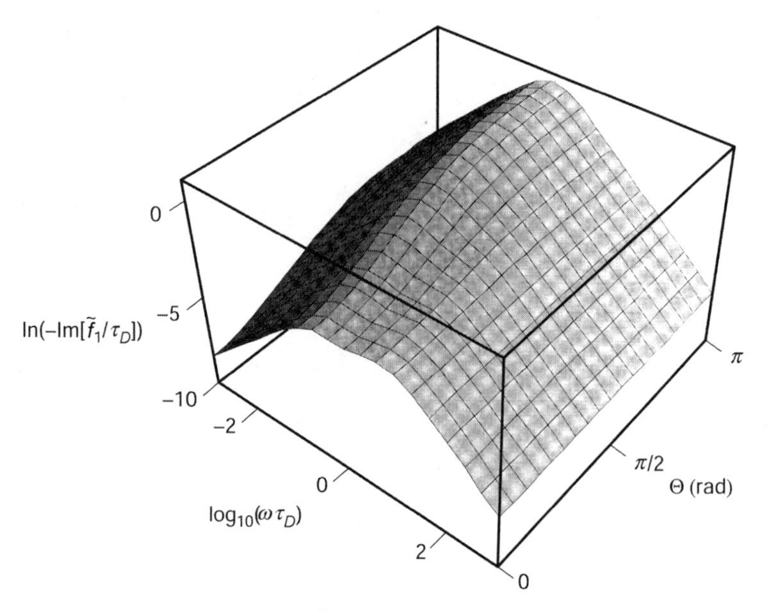


**Figure 5.** The $\ln[-\mathrm{Im}(\tilde{f}_1/\tau_\mathrm{D})]$ as a function of $\log_{10}(\omega\tau_\mathrm{D})$ and $\Theta$.

### B. Exact Solutions for Transient Kerr Effect and Nonlinear Dielectric Relaxation in a Strong dc Field Suddenly Switched on

In Section IV.A, we have presented a general treatment of transient relaxation when both magnitude and direction of the dc field are suddenly changed at time $t = 0$ from $\mathbf{E}_\mathrm{I}$ to $\mathbf{E}_\mathrm{II}$. Now, let us consider a particular case of this problem, namely; suppose that a strong constant electric field $\mathbf{E}_\mathrm{II} = \mathbf{E}_0$ (applied along the $Z$ axis) is suddenly switched on at time $t = 0$. In this case, the theory can be considerably simplified on noting that the vector $(\partial/\mu\partial\mathbf{u})V$ has only a $Z$ component, if the geometric axes of the molecule are chosen to be coincident with those of the molecular polarizability tensor, namely,

$$\frac{\partial}{\mu\partial\mathbf{u}}V = -\left[E_0 + \frac{\Delta\alpha}{\mu}E_0^2 u_z\right]\mathbf{k} \tag{4.36}$$

Having switched on the electric field $\mathbf{E}_0$, the system will tend as $t \to \infty$ to a new equilibrium state with the Boltzmann distribution function

$$W_0(\vartheta) = Z^{-1}\exp\left(-\frac{V}{kT}\right) = Z^{-1}\exp(\xi\cos\vartheta + \sigma\cos^2\vartheta)$$

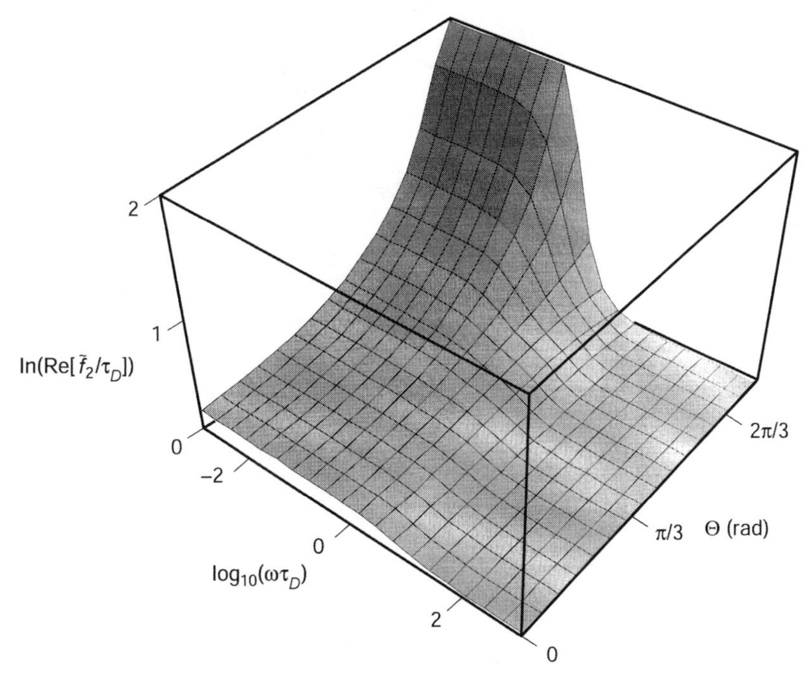


**Figure 6.** The $\mathrm{Re}(\tilde{f}_2/\tau_\mathrm{D})$ as a function of $\log_{10}(\omega\tau_\mathrm{D})$ and $\Theta$.

where

$$\xi = \frac{\mu E_0}{kT} \qquad \sigma = \frac{\Delta\alpha}{2kT}E_0^2 \tag{4.37}$$

The relaxation process under consideration will be described by the hierarchy of five-term differential-recurrence relations [Eq. (2.6)] for the relaxation functions $f_n(t)$ given by Eq. (2.11), where one must put $\zeta(t) = \xi$ and $\sigma(t) = \sigma$. Then, Eq. (2.6) can be transformed into the matrix three-term differential-recurrence equation (3.37) if we arrange it as follows [31]

$$\tau_\mathrm{D}\frac{d}{dt}\begin{pmatrix} f_{2n-1}(t) \\ f_{2n}(t) \end{pmatrix} = \mathbf{Q}_n^-\begin{pmatrix} f_{2n-3}(t) \\ f_{2n-2}(t) \end{pmatrix} + \mathbf{Q}_n\begin{pmatrix} f_{2n-1}(t) \\ f_{2n}(t) \end{pmatrix} + \mathbf{Q}_n^+\begin{pmatrix} f_{2n+1}(t) \\ f_{2n+2}(t) \end{pmatrix} \tag{4.38}$$

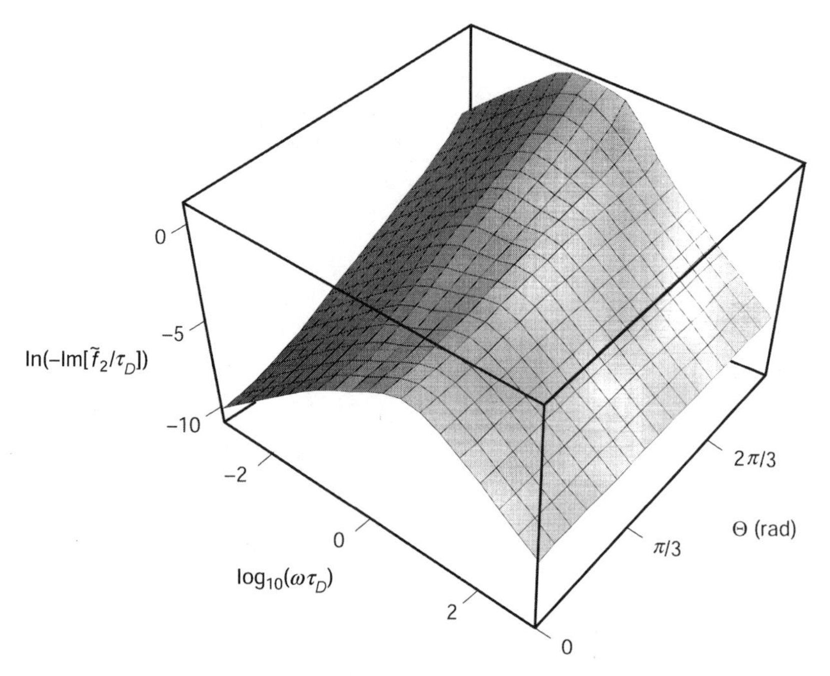


**Figure 7.** The $\ln[-\mathrm{Im}(\tilde{f}_2/\tau_\mathrm{D})]$ as a function of $\log_{10}(\omega\tau_\mathrm{D})$ and $\Theta$.

where

$$\mathbf{Q}_n^- = \begin{pmatrix} \dfrac{4\sigma n(n-1)(2n-1)}{(4n-1)(4n-3)} & \dfrac{\xi n(2n-1)}{4n-1} \\ 0 & \dfrac{2\sigma n(4n^2-1)}{16n^2-1} \end{pmatrix} \tag{4.39}$$

$$\mathbf{Q}_n = \begin{pmatrix} n(2n-1)\left[\dfrac{2\sigma}{(4n-3)(4n+1)} - 1\right] & -\dfrac{\xi n(2n-1)}{4n-1} \\ \dfrac{\xi n(2n+1)}{4n+1} & n(2n+1)\left[\dfrac{2\sigma}{(4n-1)(4n+3)} - 1\right] \end{pmatrix} \tag{4.40}$$

$$\mathbf{Q}_n^+ = \begin{pmatrix} -\dfrac{2\sigma n(4n^2-1)}{16n^2-1} & 0 \\ -\dfrac{\xi n(2n+1)}{4n+1} & -\dfrac{4\sigma n(n+1)(2n+1)}{(4n+1)(4n+3)} \end{pmatrix} \tag{4.41}$$

On applying the general method of solution of the matrix three-term recurrence equation (Section III.C), we have a solution for the Laplace transform $\tilde{\mathbf{C}}_1(s)$ in terms of matrix continued fractions

$$\begin{pmatrix} \tilde{f}_1(s) \\ \tilde{f}_2(s) \end{pmatrix} = \tau_D[\tau_D s\mathbf{I} - \mathbf{Q}_1 - \mathbf{Q}_1^+\mathbf{S}_2(s)]^{-1} \times \left\{ \mathbf{C}_1(0) + \sum_{n=2}^{\infty}\prod_{k=2}^{n} \mathbf{Q}_{k-1}^+ \mathbf{S}_k(s)(\mathbf{Q}_k^-)^{-1}\mathbf{C}_n(0) \right\} \quad (4.42)$$

where $\mathbf{I}$ is the $2 \times 2$ identity matrix, $\mathbf{Q}_n$, $\mathbf{Q}_n^{\pm}$ are the $2 \times 2$ matrices given in Eqs. (4.39)–(4.41), and $\mathbf{S}_n(s)$ is the matrix continued fraction defined by Eq. (3.41). The initial value vectors

$$\mathbf{C}_n(0) = \begin{pmatrix} f_{2n-1}(0) \\ f_{2n}(0) \end{pmatrix} = \begin{pmatrix} \langle P_{2n-1}(\cos\vartheta)\rangle_0 \\ \langle P_{2n}(\cos\vartheta)\rangle_0 \end{pmatrix} \quad (4.43)$$

may be evaluated from the recurrence relation

$$\left[1 - \frac{2\sigma}{(2n-1)(2n+3)}\right]\langle P_n\rangle_0 = \frac{\xi}{2n+1}[\langle P_{n-1}\rangle_0 - \langle P_{n+1}\rangle_0] + \left[\frac{2\sigma(n-1)}{(2n-1)(2n+1)}\langle P_{n-2}\rangle_0 - \frac{2\sigma(n+2)}{(2n+1)(2n+3)}\langle P_{n+2}\rangle_0\right] \quad (4.44)$$

where the first three members of the hierarchy are [31]

$$\langle P_0\rangle_0 = 1 \quad (4.45)$$

$$\langle P_1\rangle_0 = \frac{1}{\sqrt{\sigma}[(\coth\xi + 1)D(\sqrt{\sigma} + \xi/2\sqrt{\sigma}) + (\coth\xi - 1)D(\sqrt{\sigma} - \xi/2\sqrt{\sigma})]} - \frac{\xi}{2\sigma} \quad (4.46)$$

$$\langle P_2 \rangle_0 = \frac{3(\coth\xi - \xi/2\sigma)}{2\sqrt{\sigma}[(\coth\xi + 1)D(\sqrt{\sigma} + \xi/2\sqrt{\sigma}) + (\coth\xi - 1)D(\sqrt{\sigma} - \xi/2\sqrt{\sigma})]} + \frac{3\xi^2}{8\sigma^2} - \frac{3}{4\sigma} - \frac{1}{2} \tag{4.47}$$

Here, $D(x)$ is the Dawson's integral defined by [39]

$$D(x) = \frac{\sqrt{\pi}}{2} e^{-x^2} \operatorname{erf} i(x) = e^{-x^2} \int_0^x e^{t^2} dt$$

Equation (4.44) follows from Eq. (2.6). It should be noted, however, that this upward iteration is unstable and one should use it with caution. A much more efficient method consists in evaluating the initial value vectors $\mathbf{C}_n(0)$ with the aid of matrix continued fractions. Let us transform Eq. (4.44) to the matrix form

$$\mathbf{Q}_n^- \begin{pmatrix} \langle P_{2n-3}(\cos\vartheta)\rangle_0 \\ \langle P_{2n-2}(\cos\vartheta)\rangle_0 \end{pmatrix} + \mathbf{Q}_n \begin{pmatrix} \langle P_{2n-1}(\cos\vartheta)\rangle_0 \\ \langle P_{2n}(\cos\vartheta)\rangle_0 \end{pmatrix} + \mathbf{Q}_n^+ \begin{pmatrix} \langle P_{2n+1}(\cos\vartheta)\rangle_0 \\ \langle P_{2n+2}(\cos\vartheta)\rangle_0 \end{pmatrix} = \mathbf{0}$$

or

$$\mathbf{Q}_n^- \mathbf{C}_{n-1}(0) + \mathbf{Q}_n \mathbf{C}_n(0) + \mathbf{Q}_n^+ \mathbf{C}_{n+1}(0) = \mathbf{0} \tag{4.48}$$

The solution of Eq. (4.48) is then given by

$$\mathbf{C}_n(0) = \mathbf{S}_n(0)\mathbf{C}_{n-1}(0) = \mathbf{S}_n(0)\mathbf{S}_{n-1}(0)\ldots\mathbf{S}_1(0)\begin{pmatrix} 0 \\ 1 \end{pmatrix} \quad (n = 1, 2, \ldots) \tag{4.49}$$

In particular, for $n = 1$ we have

$$\mathbf{C}_1(0) = \begin{pmatrix} f_1(0) \\ f_2(0) \end{pmatrix} = \begin{pmatrix} \langle P_1(\cos\vartheta)\rangle_0 \\ \langle P_2(\cos\vartheta)\rangle_0 \end{pmatrix} = \mathbf{S}_1(0)\begin{pmatrix} 0 \\ 1 \end{pmatrix} \tag{4.50}$$

We can now evaluate $\tilde{f}_n(0) = \tau_n \langle P_n \rangle_0 / \tau_D$ from Eqs. (2.13), (4.42), and (4.49) (see Figs. 8 and 9). In these Figures, we have plotted $\tau_1 \langle P_1 \rangle_0 / \tau_D$ and $\tau_2 \langle P_2 \rangle_0 / \tau_D$ rather than the relaxation times $\tau_1$ and $\tau_2$ as $f_n(0) = \langle P_n \rangle_0$ may become zero at some values of the model parameters.

We remark that the solution in the form of Eq. (4.42) is mainly needed for the calculation of the relaxation times. The Laplace transforms of $\langle P_1(\cos\vartheta)\rangle(t)$ and $\langle P_2(\cos\vartheta)\rangle(t)$ have a simpler representation as

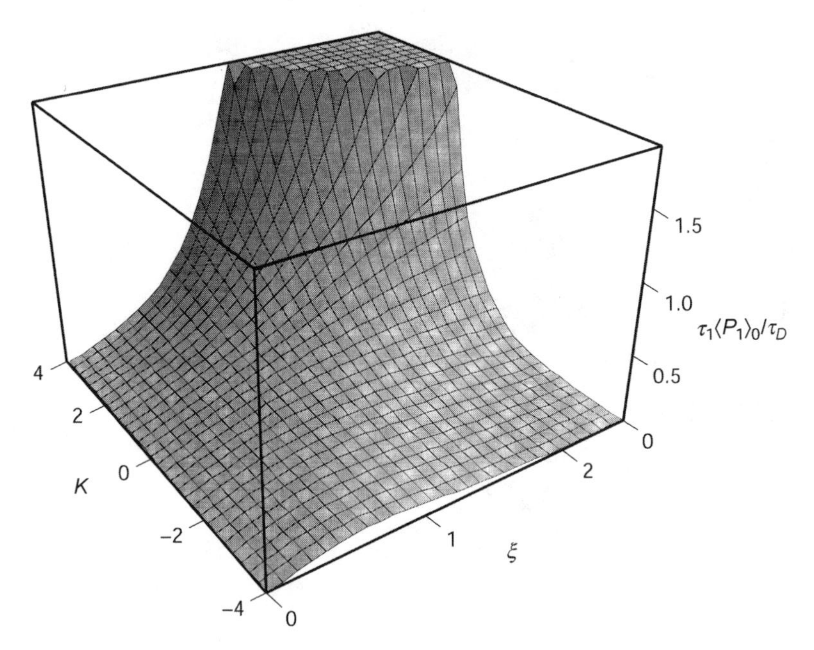


**Figure 8.** The $\tau_1\langle P_1\rangle_0/\tau_D$ as a function of $\xi$ and $\kappa$ at $0.01 < \xi < 5$ and $-4 < \kappa < 4$.

$$\begin{pmatrix} \int_0^\infty \langle P_1(\cos\vartheta)\rangle(t)e^{-st}dt \\ \int_0^\infty \langle P_2(\cos\vartheta)\rangle(t)e^{-st}dt \end{pmatrix} = s^{-1}\mathbf{S}_1(s)\begin{pmatrix} 0 \\ 1 \end{pmatrix} \tag{4.51}$$

On taking into account Eqs. (2.11), (4.49), and (4.50), we can simplify Eq. (4.42) as follows

$$\begin{pmatrix} \tilde{f}_1(s) \\ \tilde{f}_2(s) \end{pmatrix} = s^{-1}[\mathbf{S}_1(0) - \mathbf{S}_1(s)]\begin{pmatrix} 0 \\ 1 \end{pmatrix} \tag{4.52}$$

Thus, in order to calculate the nonlinear dielectric and dynamic Kerr effect step-on responses we simply need to evaluate the matrix continued fraction $\mathbf{S}_1(s)$. Note also that the matrix continued fraction $\mathbf{S}_1(0)$ can be represented in closed form as

$$\mathbf{S}_1(0) = \begin{pmatrix} 0 & \langle P_1(\cos\vartheta)\rangle_0 \\ 0 & \langle P_2(\cos\vartheta)\rangle_0 \end{pmatrix} \tag{4.53}$$

where $\langle P_1\rangle_0$ and $\langle P_2\rangle_0$ are given by Eqs. (4.46) and (4.47).

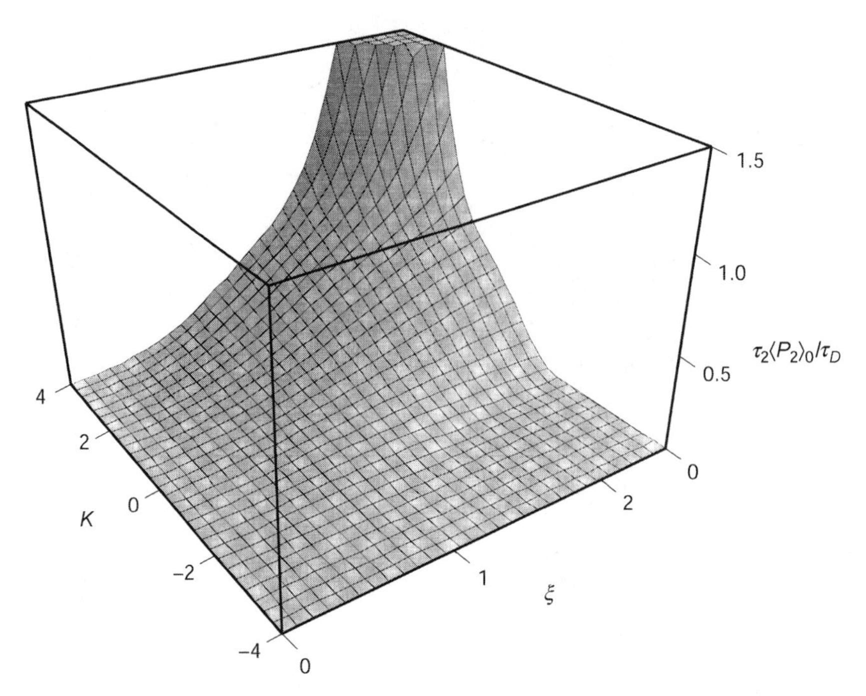


**Figure 9.** The $\tau_2\langle P_2\rangle_0/\tau_D$ as a function of $\xi$ and $\kappa$ at $0.01 < \xi < 5$ and $-4 < \kappa < 4$.

### *1. Evaluation of the Step-On Response for $\xi = 0$*

This is a nonlinear problem concerning the rise transient and relaxation time of the induced dipole Kerr effect, which was investigated by Eskin [27] and more recently by Déjardin et al. [30]. However, below we reexamine this problem and propose another presentation of the solution. For $\xi = 0$, Eq. (4.38) can be considerably simplified and reduced to a three-term recurrence relation (2.20). Following the method described in Section III.C, we seek the solution of Eq. (2.20) in the form

$$\tilde{f}_n(s) = \tilde{f}_{n-2}S_n(s) + q_n(s) \tag{4.54}$$

where the continued fraction $S_n(s)$ defined by Eq. (2.23) is the solution of the homogeneous Eq. (2.20) [with $f_n(0) = 0$]. As we mentioned in Section II, Watanabe and Morita [2] obtained the solution of the homogeneous equation (2.20) for $\langle P_2(\cos\vartheta)\rangle(t)$, which is Eq. (2.22). However, in order to evaluate the Kerr effect relaxation time one must obtain a solution of the inhomogeneous

Eq. (2.20). On substituting Eq. (4.54) into Eq. (2.20) and using Eq. (2.22), we have

$$q_n = \left[\frac{\tau_D}{\sigma} a_n f_n(0) - b_n q_{n+2}\right] S_n(s)$$

where

$$a_n = \frac{4n^2 - 1}{n(n^2 - 1)} \qquad b_n = \frac{(n+2)(2n-1)}{(n-1)(2n+3)}$$

Thus

$$\tilde{f}_n(s) = \left\{\tilde{f}_{n-2}(s) + \frac{\tau_D}{\sigma} a_n f_n(0) - b_n q_{n+2}\right\} S_n(s) \tag{4.55}$$

Since all the relaxation functions with odd indexes are equal to zero, we can solve Eq. (4.55) for even $n$ only. In particular, for $n = 2$, we have

$$\tilde{f}_2(s) = \left\{\frac{5\tau_D}{2\sigma} f_2(0) - \frac{12}{7} q_4\right\} S_2(s) \tag{4.56}$$

We obtain by induction [31]

$$\begin{aligned}\tilde{f}_2(s) &= \frac{3\tau_D}{2\sigma} \sum_{n=1}^{\infty} (-1)^{n+1} f_{2n}(0) a_{2n} \prod_{k=1}^{n} b_{2k-2} S_{2k}(s) \\ &= \frac{3\sqrt{\pi}\tau_D}{8\sigma} \sum_{n=1}^{\infty} (-1)^{n+1} \frac{(4n+1)\Gamma(n)}{\Gamma(n+3/2)} f_{2n}(0) \prod_{k=1}^{n} S_{2k}(s)\end{aligned} \tag{4.57}$$

The Kerr effect relaxation time $\tau_2$ is given by Eq. (2.12), so that from Eq. (4.56) we have

$$\tau_2 = \frac{3\sqrt{\pi}\tau_D}{8\sigma f_2(0)} \sum_{n=1}^{\infty} (-1)^{n+1} \frac{(4n+1)\Gamma(n)}{\Gamma(n+3/2)} f_{2n}(0) \prod_{k=1}^{n} S_{2k}(0) \tag{4.58}$$

The continued fraction $S_n(0)$ can be expressed [8] as a ratio of confluent hypergeometric functions $M(a, b, z)$ [Eq. (2.25)]. This is accomplished by noting that $S_n(0)$ from Eq. (2.23) can be rearranged to yield after some algebra [8]

$$1 - S_n(0) = \cfrac{1}{1 + \cfrac{\cfrac{2\sigma(n-1)}{(2n+1)(2n-1)}}{1 - \cfrac{2\sigma(n+2)}{(2n+1)(2n+3)}[1 - S_{n+2}(0)]}} \tag{4.59}$$

On comparing Eq. (4.59) with the continued fraction ([67], p. 347)

$$\frac{M(a+1,b+1,z)}{M(a,b,z)} = \cfrac{1}{1 - \cfrac{\cfrac{z(b-a)}{b(b+1)}}{1 + \cfrac{z(a+1)}{(b+1)(b+2)}\cfrac{M(a+2,b+3,z)}{M(a+1,b+2,z)}}} \tag{4.60}$$

we can see that the fraction (4.59) is identical to (4.60) if

$$z = -\sigma, \quad a = n/2, \qquad \text{and} \qquad b = n - \tfrac{1}{2}$$

Thus we can write

$$S_n(0) = 1 - \frac{M(1+(n/2), n+(1/2), -\sigma)}{M((n/2), n-(1/2), -\sigma)} \tag{4.61}$$

On using the known relations [39]

$$M(a,b,z) = e^z M(b-a, b, -z) \tag{4.62}$$

and

$$M(a, b-1, z) - M(a,b,z) = \frac{az}{b(b-1)} M(a+1, b+1, z) \tag{4.63}$$

we can rearrange Eq. (4.61) as follows [8]:

$$S_n(0) = \frac{2(n-1)\sigma}{(4n^2-1)} \frac{M(((n+1)/2), n+(3/2), \sigma)}{M(((n-1)/2), n-(1/2), \sigma)} \tag{4.64}$$

On the other hand, one can note that the equilibrium averages $\langle P_n(\cos\vartheta)\rangle(\infty)$ satisfy Eq. (4.44) for $\xi = 0$, which has the solution

$$S_n(0) = \frac{\langle P_n(\cos\vartheta)\rangle(\infty)}{\langle P_{n-2}(\cos\vartheta)\rangle(\infty)} \tag{4.65}$$

Thus, we have

$$f_{2n}(0) = \langle P_{2n}(\cos\vartheta)\rangle(\infty) = S_{2n}(0)S_{2n-2}(0)\ldots S_2(0) = \prod_{k=1}^{n} S_{2k}(0) \tag{4.66}$$

which using Eq. (4.64) yields

$$f_{2n}(0) = \frac{\sigma^n\Gamma(n+(1/2))M(n+(1/2),2n+(3/2),\sigma)}{2\Gamma(2n+(3/2))M(1/2,3/2,\sigma)} \tag{4.67}$$

Thus on taking Eqs. (4.66) and (4.67) into account, we obtain from Eq. (4.58) the *exact analytical* solution Eq. (2.24) for the relaxation time $\tau_2$. The behavior of $\tau_2$ as a function of $\sigma$ is shown in Figure 10 for $\sigma$ values in the range $-20 < \sigma < 20$. For positive $\sigma$ values, the Kerr effect relaxation time passes through a maximum at a certain value of $\sigma$ before decreasing monotonically to zero with increasing $\sigma$ as a result already obtained by Watanabe and Morita [2] and Déjardin et al. [30]. For negative $\sigma$ values, $\tau_2$ decays monotonically to zero with increasing $|\sigma|$. Moreover, one can see that the asymptotic Eqs. (2.27)

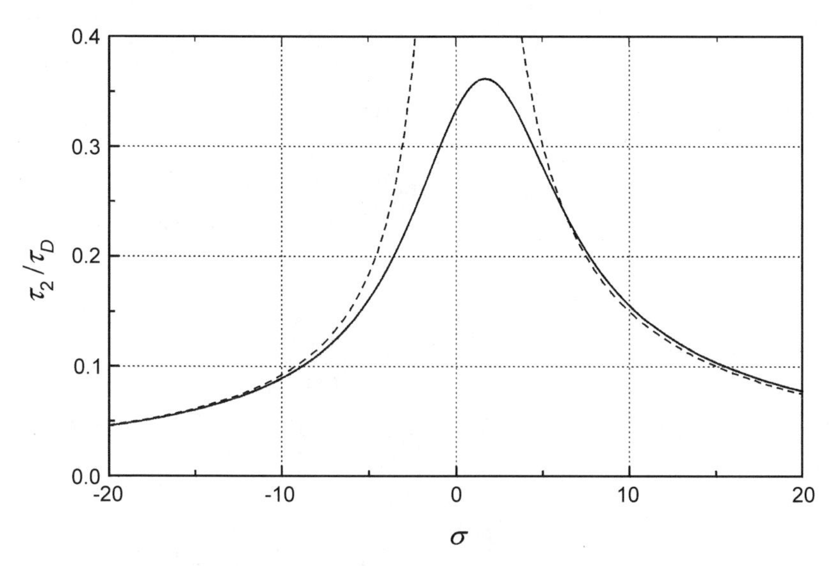


**Figure 10.** The relaxation time $\tau_2$ as a function of $\sigma$ (solid line). The dashed lines are the asymptotes given by Eqs. (2.27) and (2.28).

and (2.28) closely fit the exact solution for $\sigma > 6$ and $\sigma < -10$. The surprising increase in $\tau_2$ at intermediate electric fields was also predicted from a numerical solution of Eq. (2.20) and observed experimentally by Tolles [18].

### 2. *Evaluation of the Relaxation Times and Spectra for* $\sigma = 0$

If the contribution of the induced dipole moment is negligible in comparison with that of the permanent dipole moment, the dynamics of the system is described by Eq. (2.15), where the initial conditions are given by [31]

$$f_n(0) = \int_0^{\pi} P_n(\cos\vartheta) W_0(\vartheta) \sin\vartheta d\vartheta = \frac{I_{n+1/2}(\xi)}{I_{1/2}(\xi)} \tag{4.68}$$

Here, $W_0(\vartheta)$ is the equilibrium Boltzmann distribution function

$$W_0(\vartheta) = \frac{\xi}{4\pi \sinh(\xi)} \exp(\xi \cos\vartheta) \tag{4.69}$$

The exact analytic solution of a scalar three-term recurrence Eq. (2.15) is in like manner [31] (see Section III.C):

$$\tilde{f}_n(s) = \tilde{f}_{n-1}(s) S_n(s) + \frac{2\tau_D}{\xi} \sum_{k=1}^{\infty} (-1)^{k+1} \frac{(2n+2k-1)}{(n+k-1)(n+k)} f_{n+k-1}(0) \prod_{m=n}^{n+k-1} S_m(s) \tag{4.70}$$

In particular, for $n = 1$ and $n = 2$, we have

$$\tilde{f}_1(s) = \frac{2\tau_D}{\xi} \sum_{n=1}^{\infty} (-1)^{n+1} \frac{(2n+1)}{n(n+1)} f_n(0) \prod_{k=1}^{n} S_k(s) \tag{4.71}$$

$$\begin{aligned} \tilde{f}_2(s) &= \tilde{f}_1(s) S_2(s) + \frac{2\tau_D}{\xi} \sum_{n=2}^{\infty} (-1)^n \frac{(2n+1)}{n(n+1)} f_n(0) \prod_{k=2}^{n} S_k(s) \\ &= \frac{3\tau_D}{\xi} f_1(0) S_1(s) S_2(s) + \frac{2\tau_D}{\xi} (1 - S_1(s) S_2(s)) \sum_{n=2}^{\infty} (-1)^n \frac{(2n+1)}{n(n+1)} f_n(0) \prod_{k=2}^{n} S_k(s) \end{aligned} \tag{4.72}$$

where the continued fraction $S_n(s)$ is given by Eq. (2.19). If we use the definition of the relaxation times Eq. (2.12), we have from Eqs. (4.71) and (4.72)

$$\tau_1 = \frac{2\tau_D}{\xi} \sum_{n=1}^{\infty} (-1)^{n+1} \frac{(2n+1) f_n(0)}{n(n+1) f_1(0)} \prod_{k=1}^{n} S_n(0) \tag{4.73}$$

$$\tau_2 = \tau_1 + \frac{2\tau_D}{\xi} \sum_{n=2}^{\infty} (-1)^{n} \frac{(2n+1) f_n(0)}{n(n+1) f_2(0)} \prod_{k=2}^{n} S_k(0) \tag{4.74}$$

where

$$S_n(0) = \frac{\xi}{2n+1+\xi S_{n+1}(0)} = \frac{I_{n+1/2}(\xi)}{I_{n-1/2}(\xi)} \tag{4.75}$$

since the modified Bessel functions $I_\nu(z)$ satisfy the recurrence relation [39]

$$I_{\nu-1}(z) - I_{\nu+1}(z) = \frac{2\nu}{z} I_\nu(z) \tag{4.76}$$

which can be represented as the continued fraction

$$\frac{I_\nu(z)}{I_{\nu-1}(z)} = \frac{z}{2\nu + z(I_{\nu+1}(z)/I_\nu(z))} \tag{4.77}$$

Note that the equilibrium averages $\langle P_n(\cos\vartheta)\rangle(\infty)$ satisfy Eq. (4.44) for $\sigma = 0$ whose solution is

$$S_n(0) = \frac{\langle P_n(\cos\vartheta)\rangle(\infty)}{\langle P_{n-1}(\cos\vartheta)\rangle(\infty)} \tag{4.78}$$

leading to

$$f_n(0) = \langle P_n(\cos\vartheta)\rangle(\infty) = S_n(0) S_{n-1}(0) \ldots S_1(0) = \prod_{k=1}^{n} S_k(0) \tag{4.79}$$

which, on taking into account Eq. (4.75), reduces to Eq. (4.68). On substituting Eqs. (4.75) and (4.79) into Eqs. (4.71) and (4.72), we obtain

$$\tau_1 = \frac{2\tau_D}{\xi I_{1/2}(\xi) I_{3/2}(\xi)} \sum_{n=1}^{\infty} (-1)^{n+1} \frac{(2n+1)}{n(n+1)} I^2_{n+(1/2)}(\xi) \tag{4.80}$$

$$\tau_2 = \tau_1 + \frac{2\tau_D}{\xi I_{3/2}(\xi) I_{5/2}(\xi)} \sum_{n=2}^{\infty} (-1)^n \frac{(2n+1)}{n(n+1)} I^2_{n+(1/2)}(\xi) \tag{4.81}$$

On using the known equality [68]

$$\sum_{n=1}^{\infty} (-1)^{n+1} \frac{(2n+1)}{n(n+1)} I^2_{n+(1/2)}(\xi) = 1 + \frac{2\xi}{\sinh^2\xi} \int_0^1 \ln(x) \sinh(2\xi x) dx \tag{4.82}$$

one can show that Eqs. (4.80) and (4.81) coincide with Eqs. (3.106) and (3.107).

In the limit $\xi \to 0$ on using the Taylor expansion [39]

$$I_\nu(z) = \left(\frac{z}{2}\right)^{\nu} \sum_{k=0}^{\infty} \frac{(z/2)^{2k}}{k!\Gamma(k+\nu+1)} \tag{4.83}$$

we have

$$\frac{\tau_1}{\tau_D} \approx 1 - \frac{4}{45}\xi^2 + \frac{89}{9450}\xi^4 + O(\xi^6) \tag{4.84}$$

$$\frac{\tau_2}{\tau_D} \approx \frac{4}{3} - \frac{13}{126}\xi^2 + \frac{673}{66150}\xi^4 + O(\xi^6) \tag{4.85}$$

In the opposite limit $\xi \to \infty$, on using the asymptotic expansion [39]

$$I_\nu(z) \sim \frac{e^z}{\sqrt{2\pi z}} \left[1 + O\left(\frac{1}{z}\right)\right] \tag{4.86}$$

we find

$$\frac{\tau_1}{\tau_D} \sim \frac{2}{\xi} \tag{4.87}$$

The behavior of the relaxation time $\tau_2$ at large $\xi$ is given by Eq. (2.31). The relaxation times $\tau_1$ and $\tau_2$ as functions of $\xi$ are shown in Figure 11. Both $\tau_1$ and $\tau_2$ decrease monotonically to zero with increasing $\xi$. One can see in this figure that the asymptotic Eqs. (4.87) and (2.31) closely fit the exact solution for $\xi > 5$. The present problem has also been considered by Watanabe and Morita [2]. However, they did not give exact expressions for the relaxation times, neither did they obtained the solution of Eq. (2.15) in terms of continued fractions [Eqs. (2.17) and (2.18)].

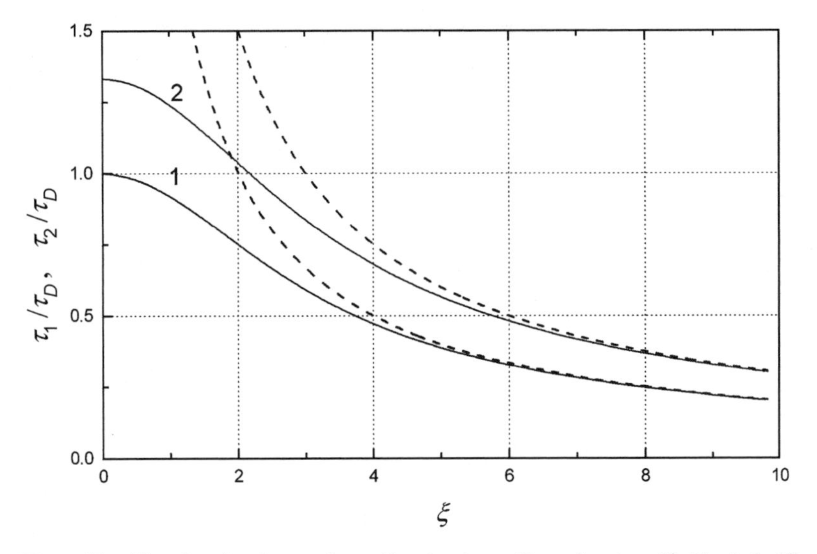


**Figure 11.** The relaxation times $\tau_1$ (curve 1) and $\tau_2$ (curve 2) as a function of $\xi$. The dashed lines are the asymptotes (dashed lines) given by Eqs. (4.88) and (2.31).

### 3. *Evaluation of the Relaxation Spectra and Discussion*

The behavior of the real and imaginary parts of the one-sided Fourier transforms of the normalized relaxation functions defined as

$$\chi_n(\omega) = \frac{\tilde{f}_n(i\omega)}{\tau_D f_n(0)} \tag{4.88}$$

is shown in Figs. 12–15. Here, the spectra evaluated from the exact solutions given by Eqs. (4.42), (4.57), (4.71), and (4.72), are compared with the Debye spectrum

$$\chi_{Dn}(\omega) = \frac{\tau_n/\tau_D}{1 + i\omega\tau_n} \tag{4.89}$$

where the $\tau_n$ are the relaxation times calculated from Eqs. (2.12), (2.29), (4.42), (2.24), (4.80), and (4.81). Equation (4.89) corresponds to the representation of the relaxation functions $f_n(t)$ by a purely exponential

$$f_n(t) = f_n(0)e^{-t/\tau_n} \tag{4.90}$$

It is apparent from Figs. 12–15 that Lorentzian behavior is obtained for the spectra $\chi_1(\omega)$ for arbitrary $\xi$ and at $\sigma \approx 0$ and $\chi_2(\omega)$ for arbitrary $\sigma$ and $\xi = 0$.

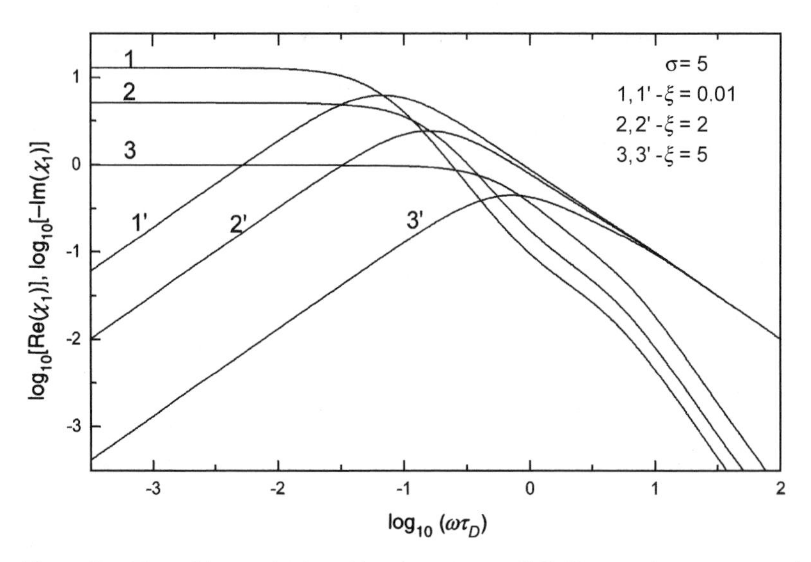


**Figure 12.** The real (curves 1,2,3) and imaginary (curves $1', 2', 3'$) parts of the spectrum $\chi_1(\omega)$ of the normalized relaxation function at $\sigma = 5$. Curves $1, 1'$; $2, 2'$; and $3, 3'$ correspond to $\xi = 0.01$, 1, and 5, respectively.

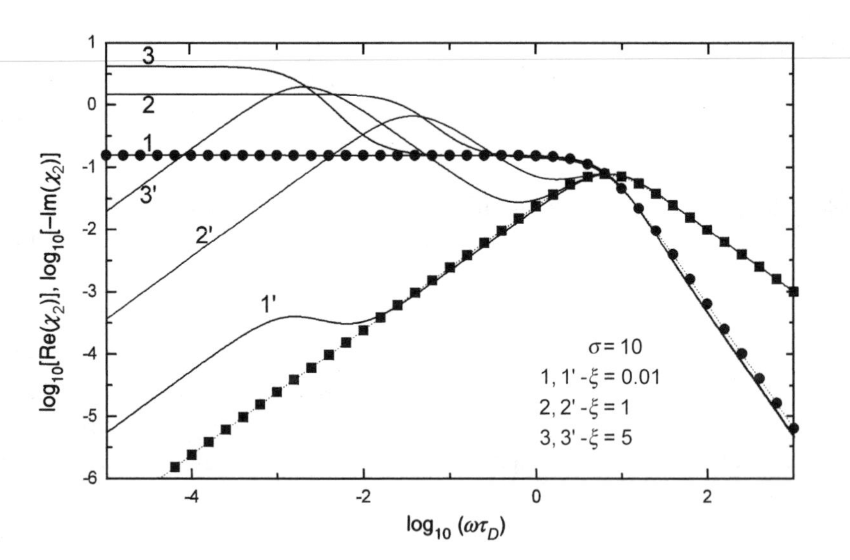


**Figure 13.** The real (curves 1,2,3) and imaginary (curves $1', 2', 3'$) parts of the spectrum $\chi_2(\omega)$ of the normalized relaxation function at $\sigma = 10$. Curves $1, 1'$; $2, 2'$; and $3, 3'$ correspond to $\xi = 0.01$, 1, and 5, respectively. Filled circles and squares are the real and imaginary parts of the spectrum $\chi_{D2}(\omega)$ of the normalized exponential relaxation function with the relaxation time $\tau_2$ from Eq. (2.24).

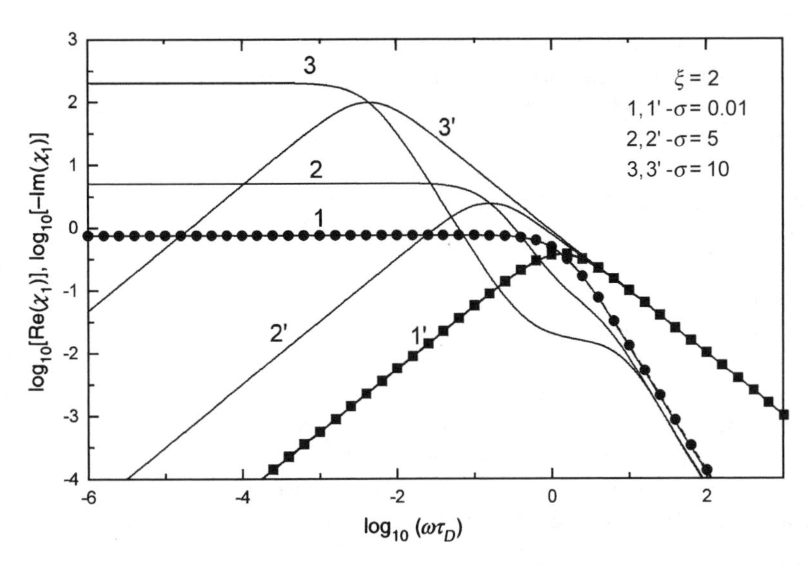


**Figure 14.** The real (curves 1,2,3) and imaginary (curves $1', 2', 3'$) parts of the spectrum $\chi_1(\omega)$ of the normalized relaxation function at $\xi = 2$. Curves $1, 1', 2, 2'$ and $3, 3'$ correspond to $\sigma = 0.01$, 5, and 10, respectively. Filled circles and squares are the real and imaginary parts of the spectrum $\chi_{D1}(\omega)$ of the normalized exponential relaxation function with the relaxation time $\tau_1$ from Eq. (4.80).

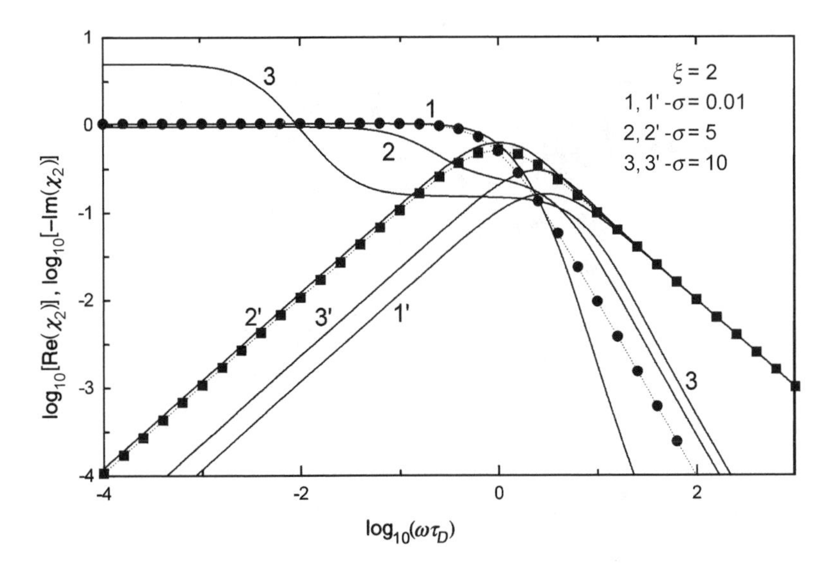


**Figure 15.** The real (curves 1,2,3) and imaginary (curves $1', 2', 3'$) parts of the spectrum $\chi_2(\omega)$ of the normalized relaxation function at $\xi = 2$. Curves $1, 1', 2, 2'$ and $3, 3'$ correspond to $\sigma = 0.01$, 5, and 10, respectively. Filled circles and squares are the real and imaginary parts of the spectrum $\chi_{D2}(\omega)$ of the normalized exponential relaxation function with the relaxation time $\tau_2$ from Eq. (4.81).

Thus, in these cases alone the relaxation functions $f_1(t)$ and $f_2(t)$ can be approximated by a single exponential. In all other cases the decay of the relaxation functions $f_n(t)$ has a more complicated behavior. This may be explained as follows. The relaxation dynamics in the potential given by Eq. (2.3) is determined by two relaxation processes. One relaxation (activation) process governs the crossing of the potential barrier between two positions of equilibrium by a current of molecules. Another process describes orientational relaxation inside the wells. In the case of nonpolarizable molecules ($\sigma = 0$), when the potential (2.3) transforms to a single well cosine potential, we observe one relaxation process only, namely, the reorientation of the molecule inside the well.

In the literature, we have found only a few experimental data [18, 20, 69], which can be used for checking the nonlinear theory for transient responses. Tolles and co-workers [18, 69] presented experimental results for the step-on nonlinear relaxation time $\tau_2$ of *nonpolar polarizable* zinc oxide particles as a function of an applied electric field. The comparison of the theory with these data is given in Figure 16. Here $\tau_2$ [Eq. (3.100)] is plotted as a function of

$$A = 2.4E_{\mathrm{II}}\sqrt{\frac{\varepsilon_0\tau_D}{\eta}\left(\frac{\ln(2l/d) - 1.57 - 7\{[\ln(2l/d)]^{-1} - 0.28\}^2}{3.49\ln(2l/d) - 1.84}\right)} \tag{4.91}$$

where $\varepsilon_0$ is the permittivity of free space, $\eta$ is the viscosity of the fluid, $l/d$

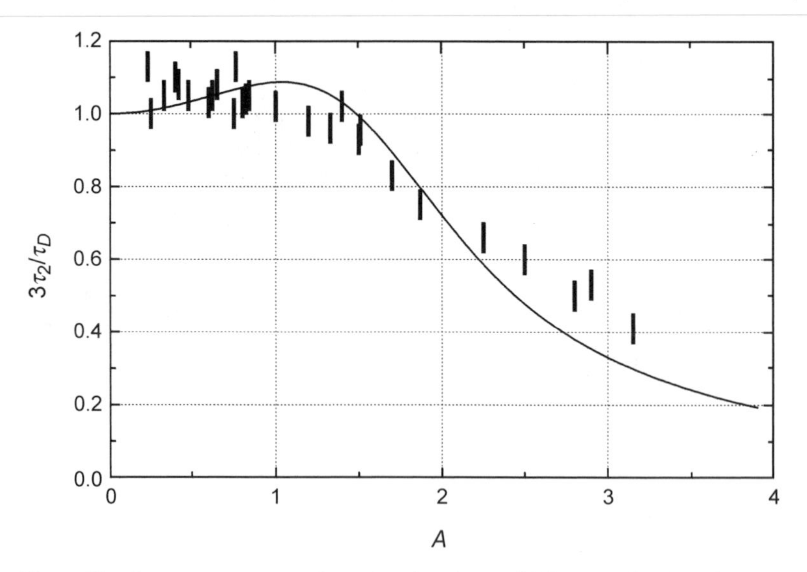


**Figure 16.** Step-on response transient relaxation time (solid line) as a function of $A$ given by Eq. (3.100). Bars are the experimental data from [18].

is the length-to-diameter ratio, taken to be 15 for these experiments. The agreement of the theory with the experimental data [18, 69] is good. The theory also agrees in all respects with the numerical solutions of the Smoluchowski equation obtained in [18].

Some measurements of transient and steady-state electric birefringence of rodlike macromolecules $[\text{helical}\,(\text{Lys}\cdot\text{HBr})_n]$ in a methanol–water mixture were made by Kikuchi [20]. For this system the electrooptical response exhibits the characteristics of a pure-induced dipole orientation mechanism (the contribution from the permanent dipole moment is $< 5\%$). The theoretical prediction for the steady state birefringence

$$K_{\text{II}} = K_s \langle P_2 \rangle_{\text{II}}$$

where $\langle P_2 \rangle_{\text{II}}$ is given by Eq. (3.99), is in accordance with clearly noticed experimental observations of upward deviations from Kerr's law. Unfortunately, the author did not present experimental results for the field dependence of the transient relaxation time. Therefore, the detailed comparison with this experiment cannot be carried out.

# V. STEADY-STATE RESPONSE ARISING FROM A WEAK AC ELECTRIC FIELD SUPERIMPOSED ON A STRONG DC BIAS FIELD

## A. Analytical Solutions for the Dynamic Kerr Effect: Linear Response of Polar and Polarizable Molecules to a Weak ac Electric Field Superimposed on a Strong dc Bias Field

### 1. *Linear Response Theory for the Dynamic Kerr Effect*

We shall illustrate the possibilities of linear response theory by evaluating the *linear* dynamic Kerr effect for *polar and polarizable* molecules in a small ac electric field superimposed on a strong dc bias field. First, we shall calculate the after-effect solution following the sudden removal of a small electric field $\mathbf{E}_1(t)$ applied in a direction parallel to a strong dc bias field $\mathbf{E}_0$. Then, we shall give the expressions for the spectra of the birefringence function and the corresponding relaxation time in terms of matrix continued fractions [70]. Also, exact analytical expressions for the Kerr effect relaxation time are derived in terms of integrals and compared with the matrix continued fraction result. The last part is devoted to the representation of the spectra of the real and imaginary parts of the complex birefringence function.

In the presence of an ac field $\mathbf{E}_1(t)$ superimposed on a dc bias field $\mathbf{E}_0$, and in the simplest case of a symmetric top molecule acted on by the total field

$\mathbf{E}_0 + \mathbf{E}_1(t)$ applied along the $Z$ axis of the laboratory frame, the orientational potential energy of the molecule is given by [cf. Eq. (2.3)]

$$V = -\mu[E_0 + E_1(t)]\cos\vartheta - \frac{1}{2}\Delta\alpha[E_0 + E_1(t)]^2\cos^2\vartheta \tag{5.1}$$

In the linear approximation in $E_1(t)$, which amounts to neglecting the quadratic term in Eq. (5.1), we have

$$V = -\mu E_0\cos\vartheta - \frac{1}{2}\Delta\alpha E_0^2\cos^2\vartheta - [\mu\cos\vartheta + \Delta\alpha E_0\cos^2\vartheta]E_1(t) \tag{5.2}$$

The principal results of linear response theory for systems whose dynamics obey the 1D Fokker–Planck equation for the distribution function $W$ of a variable $x$ have been discussed in Section III.D [Eqs. (3.59)–(3.61)]. Their application to the linear dynamic Kerr response is straightforward. Insofar as we are concerned here with the problem of the noninertial rotational Brownian motion of a particle in an external potential $V$, the relevant Fokker–Planck equation for the distribution function $W$ of $x = \cos\vartheta$ is given by Eq. (3.76), where $V$ is given by Eq. (5.2). Thus, we can simply use Eqs. (3.59)–(3.61) by putting

$$A(x) = P_2(x) \tag{5.3}$$

$$\langle A\rangle(t) = \langle P_2\rangle(t) \tag{5.4}$$

and

$$B(x) = [\mu x + \Delta\alpha E_0 x^2]/kT \tag{5.5}$$

so that the step-off, step-on and ac response solutions are, respectively, given by

$$\langle P_2(\cos\vartheta)\rangle^{\text{off}}(t) - \langle P_2(\cos\vartheta)\rangle_0 = \frac{\mu E_1}{kT} f_K(t) \tag{5.6}$$

$$\langle P_2(\cos\vartheta)\rangle^{\text{on}}(t) - \langle P_2(\cos\vartheta)\rangle_0 = \frac{\mu E_1}{kT}[f_K(0) - f_K(t)] \tag{5.7}$$

and

$$\langle P_2(\cos\vartheta)\rangle_\omega = \frac{\mu E_\omega}{kT}[f_K(0) - i\omega\tilde{f}_K(i\omega)] \tag{5.8}$$

where

$$f_K(t) = f_1(t) + \rho f_2(t) \tag{5.9}$$

$$\rho = \frac{2\Delta\alpha E_0}{3\mu} \tag{5.10}$$

and the *equilibrium* correlation functions $f_n(t)$ are defined as

$$f_n(t) = \langle P_2(\cos\vartheta(0))P_n(\cos\vartheta(t))\rangle_0 - \langle P_2(\cos\vartheta)\rangle_0\langle P_n(\cos\vartheta)\rangle_0 \qquad (n = 1, 2) \tag{5.11}$$

Here, $\tilde{f}_K(i\omega)$ is the one-sided FT of $f_K(t)$, namely,

$$\tilde{f}_K(i\omega) = \int_0^\infty f_K(t)e^{-i\omega t}dt$$

Moreover, another quantity that can be measured experimentally, is the Kerr effect relaxation time $\tau_K$ defined as the area under the normalized relaxation function, namely,

$$\tau_K = \int_0^\infty \frac{f_K(t)}{f_K(0)}dt = \frac{\tilde{f}_1(0) + \rho\tilde{f}_2(0)}{f_1(0) + \rho f_2(0)} \tag{5.12}$$

From Eqs. (5.6)–(5.10), it follows that for pure polar molecules the dynamic Kerr response functions depend on the cross-correlation function $f_1(t)$ of the first two Legendre polynomials. For pure induced dipole moments, one simply needs to calculate the autocorrelation function $f_2(t)$ of the second Legendre polynomial.

The recurrence equation for the equilibrium correlation functions $f_n(t)$ calculated in the absence of the ac field $\mathbf{E}_1(t)$ are given by an equation similar to Eq. (2.6), which now becomes (see for details [70])

$$\begin{aligned}
&\frac{2\tau_D}{n(n+1)}\frac{d}{dt}f_n(t) + \left[1 - \frac{2\sigma}{(2n-1)(2n+3)}\right]f_n(t) \\
&\quad = \frac{\xi}{2n+1}[f_{n-1}(t) - f_{n+1}(t)] \\
&\quad + 2\sigma\left[\frac{(n-1)}{(2n-1)(2n+1)}f_{n-2}(t) - \frac{(n+2)}{(2n+1)(2n+3)}f_{n+2}(t)\right] \qquad n = 1, 2, \ldots
\end{aligned} \tag{5.13}$$

where $\xi$ and $\sigma$ are given by Eq. (4.37).

### 2. *Evaluation of the Dynamic Kerr Effect Response for* $\xi = 0$

This is a linear problem concerning the rise and decay transient and relaxation time of the induced dipole Kerr effect. For $\xi = 0$, Eq. (5.13) is considerably simplified and reduced to a three-term differential-recurrence relation. We have

$$\begin{aligned} &\frac{2\tau_D}{n(n+1)}\frac{d}{dt}f_n(t) + \left[1 - 2\sigma\frac{1}{(2n-1)(2n+3)}\right]f_n(t) \\ &\quad = 2\sigma\left[\frac{(n-1)}{(2n-1)(2n+1)}f_{n-2}(t) - \frac{(n+2)}{(2n+1)(2n+3)}f_{n+2}(t)\right] \end{aligned} \tag{5.14}$$

with $f_0(t) = 0$. Since the reduced potential $V$ at equilibrium is of the form $\sigma\cos^2\vartheta$, only the even Legendre polynomials will contribute to the initial conditions. These are [70]

$$\begin{aligned} f_{2n}(0) &= \langle P_2(\cos\vartheta)P_{2n}(\cos\vartheta)\rangle_0 - \langle P_2(\cos\vartheta)\rangle_0\langle P_{2n}(\cos\vartheta)\rangle_0 \\ &= \frac{3(n+1)(2n+1)}{(4n+3)(4n+1)}\langle P_{2n+2}(\cos\vartheta)\rangle_0 + \frac{3n(2n-1)}{16n^2-1}\langle P_{2n-2}(\cos\vartheta)\rangle_0 \\ &\quad + \frac{2n(2n+1)}{(4n+3)(4n-1)}\langle P_{2n}(\cos\vartheta)\rangle_0 - \langle P_2(\cos\vartheta)\rangle_0\langle P_{2n}(\cos\vartheta)\rangle_0 \end{aligned} \tag{5.15}$$

where the equilibrium averages $\langle P_{2n}(\cos\vartheta)\rangle_0$ are given by Eq. (4.67), namely,

$$\langle P_{2n}(\cos\vartheta)\rangle_0 = \frac{\sigma^n\Gamma(n+1/2)M(n+1/2, 2n+3/2, \sigma)}{2\Gamma(2n+3/2)M(1/2, 3/2, \sigma)}$$

The method of the solution of Eq. (5.14) in terms of continued fractions has been described in detail in Section IV.B.1. For the present problem, we have found that [70]

$$\begin{aligned} \tilde{f}_{2n}(s) &= \tilde{f}_{2n-2}(s)S_{2n}(s) + \frac{\tau_D(4n-1)\Gamma(n+1/2)}{4\sigma(2n-1)\Gamma(n+1)}\sum_{k=0}^{\infty}(-1)^k \\ &\quad \times \frac{(4n+1+4k)\Gamma(n+k)}{\Gamma(k+n+3/2)}f_{2n+2k}(0)\prod_{m=0}^{k}S_{2n+2m}(s) \end{aligned} \tag{5.16}$$

where $S_n(s)$ is given by Eq. (2.23). In particular, for $n = 1$ we have the desired quantity

$$\tilde{f}_2(s) = \frac{3\sqrt{\pi}\tau_D}{8\sigma}\sum_{k=0}^{\infty}(-1)^k\frac{(4k+5)\Gamma(k+1)}{\Gamma(k+5/2)}f_{2k+2}(0)\prod_{m=0}^{k}S_{2m+2}(s) \tag{5.17}$$

The Kerr effect relaxation time $\tau_K$, defined from Eq. (5.12) is

$$\tau_K = \frac{3\sqrt{\pi}\tau_D}{8\sigma f_2(0)}\sum_{n=1}^{\infty}(-1)^{n+1}\frac{(4n+1)\Gamma(n)}{\Gamma(n+3/2)}f_{2n}(0)\prod_{k=1}^{n}S_{2k}(0) \tag{5.18}$$

Noting that

$$S_n(0) = \frac{\langle P_n(\cos\vartheta)\rangle_0}{\langle P_{n-2}(\cos\vartheta)\rangle_0} \tag{5.19}$$

is given by Eq. (4.64) and on using Eq. (4.66), we can find the exact analytical solution for the relaxation time [70]

$$\tau_K = \frac{3\sqrt{\pi}\tau_D}{8}\sum_{n=0}^{\infty}(-\sigma)^n\frac{f_{2n+2}(0)\Gamma(n+1)M(n+3/2,2n+7/2,\sigma)}{f_2(0)(n+3/2)\Gamma(2n+5/2)M(1/2,3/2,\sigma)} \tag{5.20}$$

In the limit $\sigma \to 0$, on using the Taylor expansion (2.25) of Kummer's function $M(a,b,z)$, we obtain from Eq. (5.20)

$$\frac{\tau_K}{\tau_D} \approx \frac{1}{3} + \frac{2}{63}\sigma - \frac{284}{33075}\sigma^2 + O(\sigma^3) \tag{5.21}$$

For $\sigma = 0$, it is apparent that $\tau_K = \tau_D/3$, which is the usual birefringence relaxation time obtained in the case of very weak fields. Unfortunately, in the opposite limit $\sigma \to \infty$, the asymptotic expansion is not available presently.

### 3. *Evaluation of the Dynamic Kerr Effect Response for* $\sigma = 0$

If the contribution of the induced dipole moment is negligible in comparison with that of the permanent dipole moment, we have, on putting $\sigma = 0$ in Eq. (5.13),

$$2\tau_D\frac{d}{dt}f_n(t) + n(n+1)f_n(t) = \frac{\xi n(n+1)}{(2n+1)}[f_{n-1}(t) - f_{n+1}(t)] \tag{5.22}$$

The initial conditions for $f_n(t)$ in Eq. (5.22) are given by [70]

$$\begin{aligned} f_n(0) &= \langle P_2(\cos\vartheta)P_n(\cos\vartheta)\rangle_0 - \langle P_2(\cos\vartheta)\rangle_0\langle P_n(\cos\vartheta)\rangle_0 \\ &= \frac{3(n+1)(n+2)}{2(2n+1)(2n+3)}\frac{I_{n+5/2}(\xi)}{I_{1/2}(\xi)} + \frac{3n(n-1)}{2(2n+1)(2n-1)}\frac{I_{n-3/2}(\xi)}{I_{1/2}(\xi)} \\ &\quad + \frac{n(n+1)}{(2n+3)(2n-1)}\frac{I_{n+1/2}(\xi)}{I_{1/2}(\xi)} - \frac{I_{5/2}(\xi)}{I_{1/2}(\xi)}\frac{I_{n+1/2}(\xi)}{I_{1/2}(\xi)} \end{aligned} \tag{5.23}$$

where we have used that [cf. Eq. (4.68)]

$$\langle P_n(\cos\vartheta)\rangle_0 = \frac{I_{n+1/2}(\xi)}{I_{1/2}(\xi)} \tag{5.24}$$

The solution of Eq. (5.22) has been given in Section IV.B.1. Thus, on using the results of that section, we recall that the Laplace transform of the correlation function $f_1(t)$ is [70]

$$\tilde{f}_1(s) = \frac{2\tau_D}{\xi}\sum_{n=1}^{\infty}(-1)^{n+1}\frac{(2n+1)}{n(n+1)}f_n(0)\prod_{k=1}^{n}S_k(s) \tag{5.25}$$

where the continued fraction $S_n(s)$ is given by Eq. (2.19). Using the definition of the relaxation time [Eq. (5.12)] we have from Eq. (5.25)

$$\tau_K = 2\frac{\tau_D}{\xi}\sum_{n=1}^{\infty}(-1)^{n+1}\frac{(2n+1)f_n(0)}{n(n+1)f_1(0)}\prod_{k=1}^{\infty}S_k(0) \tag{5.26}$$

On using Eq. (4.75), Eq. (5.26) can be presented as

$$\tau_K = \frac{2\tau_D}{\xi I_{1/2}(\xi)f_1(0)}\sum_{n=1}^{\infty}(-1)^{n+1}\frac{(2n+1)}{n(n+1)}f_n(0)I_{n+1/2}(\xi) \tag{5.27}$$

In the limit $\xi \to 0$ on using the Taylor expansion of $I_\nu(z)$ [Eq. (4.83)], we have

$$\frac{\tau_K}{\tau_D} \approx \frac{5}{6} - \frac{23}{252}\xi^2 + \frac{328}{33075}\xi^4 + O(\xi^6) \tag{5.28}$$

The leading term of Eq. (5.28) is in agreement with the results previously derived by Ullman [71] who considered the problem of the longitudinal dielectric relaxation is a strong dc bias field. In the opposite limit $\xi \to \infty$, on using the asymptotic expansion of $I_\nu(z)$, we find that

$$\tau_K \sim \tau_D/\xi \tag{5.29}$$

which means that for high $\xi$ values, $\tau_K$ follows an hyperbolic law.

### 4. *Evaluation of the Relaxation Functions and Relaxation Times in the General Case*

As we have already shown in Section IV.B, Eq. (5.13) can be transformed [8] into the matrix three-term differential-recurrence equation (3.37). We recall that the exact solution for the Laplace transform $\widetilde{\mathbf{C}}_1(s)$ in terms of matrix continued fractions is therefore

$$\begin{pmatrix} \tilde{f}_1(s) \\ \tilde{f}_2(s) \end{pmatrix} = \tau_D[\tau_D s\mathbf{I} - \mathbf{Q}_1 - \mathbf{Q}_1^+\mathbf{S}_2(s)]^{-1}\left\{\mathbf{C}_1(0) + \sum_{n=2}^{\infty}\prod_{k=2}^{n}\mathbf{Q}_{k-1}^+\mathbf{S}_k(s)(\mathbf{Q}_k^-)^{-1}\mathbf{C}_n(0)\right\} \tag{5.30}$$

where the $2 \times 2$ matrices $\mathbf{Q}_n, \mathbf{Q}_n^{\pm}$ are given by Eqs. (4.39)–(4.41), the matrix continued fraction $\mathbf{S}_n(s)$ is defined by Eq. (3.41), and the initial value vectors are

$$\begin{aligned} \mathbf{C}_n(0) &= \begin{pmatrix} f_{2n-1}(0) \\ f_{2n}(0) \end{pmatrix} \\ &= \begin{pmatrix} \langle P_2(\cos\vartheta)P_{2n-1}(\cos\vartheta)\rangle_0 - \langle P_2(\cos\vartheta)\rangle_0\langle P_{2n-1}(\cos\vartheta)\rangle_0 \\ \langle P_2(\cos\vartheta)P_{2n}(\cos\vartheta)\rangle_0 - \langle P_2(\cos\vartheta)\rangle_0\langle P_{2n}(\cos\vartheta)\rangle_0 \end{pmatrix} \end{aligned} \tag{5.31}$$

which may be evaluated in terms of matrix continued fractions as [70]

$$\mathbf{C}_n(0) = [\mathbf{X}_n + \mathbf{Y}_n\mathbf{S}_n(0) + \mathbf{Z}_n\mathbf{S}_{n+1}(0)\mathbf{S}_n(0)]\mathbf{S}_{n-1}(0)\mathbf{S}_{n-2}(0)\ldots\mathbf{S}_1(0)\begin{pmatrix} 0 \\ 1 \end{pmatrix} \tag{5.32}$$

where

$$\mathbf{X}_n = \begin{pmatrix} \dfrac{3(n-1)(2n-1)}{(4n-1)(4n-3)} & 0 \\ 0 & \dfrac{3n(2n-1)}{16n^2-1} \end{pmatrix} \tag{5.33}$$

$$\mathbf{Y}_n = \begin{pmatrix} \frac{2n(2n-1)}{(4n+1)(4n-3)} - \langle P_2(\cos\vartheta)\rangle_0 & 0 \\ 0 & \frac{2n(2n+1)}{(4n-1)(4n+3)} - \langle P_2(\cos\vartheta)\rangle_0 \end{pmatrix} \tag{5.34}$$

$$\mathbf{Z}_n = \begin{pmatrix} \frac{3n(2n+1)}{16n^2-1} & 0 \\ 0 & \frac{3(n+1)(2n+1)}{(4n+3)(4n+1)} \end{pmatrix} \tag{5.35}$$

Here, we have used Eqs. (4.49), (4.53), and

$$\begin{aligned} f_n(0) = {} & \frac{3(n+2)(n+1)}{2(2n+3)(2n+1)}\langle P_{n+2}(\cos\vartheta)\rangle_0 + \frac{3n(n-1)}{2(4n^2-1)}\langle P_{n-2}(\cos\vartheta)\rangle_0 \\ & + \frac{n(n+1)}{(2n+3)(2n-1)}\langle P_n(\cos\vartheta)\rangle_0 - \langle P_2(\cos\vartheta)\rangle_0\langle P_n(\cos\vartheta)\rangle_0 \end{aligned} \tag{5.36}$$

From Eqs. (5.30)–(5.36), we can now evaluate the relaxation time $\tau_K$ from Eq. (5.12) and the one-sided FT $\chi_K(\omega)$ of the normalized birefringence function such as

$$\chi_K(\omega) = -\int_0^\infty \dot{f}_K(t)e^{-i\omega t}dt = f_K(0) - i\omega\tilde{f}_K(i\omega) \tag{5.37}$$

representing the linear ac response of the electric birefringence.

### 5. *Integral Representation for the Correlation Time*

We recall (see Section III.D and Appendix B) that for a stochastic system the dynamics of which obeys the one-variable Fokker–Planck equation, the correlation time $\tau_{AB}$ of the equilibrium *correlation* function $C_{AB}(t) = \langle A[x(0)]B[x(0)]\rangle_0 - \langle A[x(0)]\rangle_0\langle B[x(0)]\rangle_0$ of the dynamic variables $A(x)$ and $B(x)$ is given by Eq. (A28) in Appendix B. For the problem of noninertial rotational Brownian motion of a dipolar particle in an external potential $V$, the relevant Fokker–Planck equation for the distribution function $W$ of the orientations of the particles is given by Eq. (3.76) and the variables $A(x)$ and $B(x)$ are defined by Eqs. (5.4) and (5.5). On using Eqs. (A28)–(A30) in Appendix B, we have an analytical expression (in terms of an integral) for the linear dynamic Kerr effect relaxation time, namely,

$$\tau_K = \frac{2\tau_D}{Z[f_1(0)+\rho f_2(0)]}\int_{-1}^{1}\frac{e^{-\xi z-\sigma z^2}\Phi(z)[\Psi(z)+\rho\Phi(z)]dz}{1-z^2} \tag{5.38}$$

where the functions $Z, f_1(0), f_2(0)$,

$$\Psi(z) = \int_{-1}^{z}[P_1(z') - \langle P_1\rangle_0]e^{\xi z'+\sigma z'^2}dz'$$

and

$$\Phi(z) = \int_{-1}^{z}[P_2(z') - \langle P_2\rangle_0]e^{\xi z'+\sigma z'^2}dz'$$

can be evaluated with the help of Eqs. (3.88), (3.91)–(3.95), and (5.11). For particular cases $\xi = 0$ and $\sigma = 0$, Eq. (5.38) can be considerably simplified. For the pure induced dipole moment Kerr effect ($\xi = 0$), we obtain [70]

$$\tau_K = \frac{2\tau_D\sqrt{\sigma}}{\sqrt{\pi}\mathrm{erf}i(\sqrt{\sigma})f_2(0)}\int_{-1}^{1}\left[ze^{\sigma z^2} - \frac{e^{\sigma}\mathrm{erf}i(z\sqrt{\sigma})}{\mathrm{erf}i(\sqrt{\sigma})}\right]^2\frac{e^{-\sigma z^2}dz}{1-z^2} \tag{5.39}$$

where

$$f_2(0) = \frac{9e^{\sigma}}{4\sqrt{\pi\sigma}\mathrm{erf}i(\sqrt{\sigma})}\left[1 - \frac{e^{\sigma}}{\sqrt{\pi\sigma}\mathrm{erf}i(\sqrt{\sigma})}\right] + \frac{9}{8\sigma^2}\left[1 - \frac{\sqrt{\sigma}e^{\sigma}}{\sqrt{\pi}\mathrm{erf}i(\sqrt{\sigma})}\right] \tag{5.40}$$

While, for the pure permanent dipole moment Kerr effect ($\sigma = 0$), we have [70]

$$\tau_K = \frac{\tau_D\xi}{f_1(0)\sinh\xi}\int_{-1}^{1}\frac{e^{-\xi z}\Phi(z)\Psi(z)dz}{1-z^2} \tag{5.41}$$

where

$$f_1(0) = \frac{3}{\xi}\left[\coth^2\xi + \frac{\coth\xi}{\xi} - 1 - \frac{2}{\xi^2}\right] \tag{5.42}$$

$$\Psi(z) = \xi^{-1}[ze^{\xi z} + e^{-\xi} - \coth\xi(e^{\xi z} - e^{-\xi})] \tag{5.43}$$

$$\Phi(z) = \frac{3}{2\xi}\left[e^{\xi z}\left(z^2 - \frac{2z}{\xi} - 1 + \frac{2\coth\xi}{\xi}\right) - \frac{2e^{-\xi}}{\xi}(1+\coth\xi)\right] \tag{5.44}$$

The results of the calculations for the correlation time $\tau_K$ that we have obtained in Sections V.A.2–4 in terms of continued fractions [Eqs. (5.20), (5.26) and (5.12), (5.30)] are in full agreement with those predicted by Eqs. (5.38)–(5.44).

### 6. *Discussion of the Results*

The behavior of the relaxation time $\tau_K$ from Eq. (5.39) as a function of $\sigma$ for the pure induced dipole moment Kerr effect is shown in Figure 17 for $\sigma$ values in the range $-20 < \sigma < 20$. For positive $\sigma$ values, this Kerr-effect relaxation time passes through a maximum at a certain value of $\sigma$ before decreasing monotonically to zero with increasing $\sigma$ [72]. For negative $\sigma$ values, $\tau_K$ decays monotonically to zero with increasing $|\sigma|$. The behavior of the relaxation time $\tau_K$ from Eq. (5.41) as a function of $\xi$ for the pure permanent dipole moment Kerr effect is shown in Figure 18. We remark that $\tau_K$ decreases monotonically to zero with increasing $\xi$ values. The asymptotic expansion for $\tau_K$ given by Eq. (5.29) also closely fits the exact solution for $\xi > 10$. In Figure 19, the logarithm of the relaxation time $\tau_K$ is plotted in three dimensions as a function of the parameters $\xi$ and $\sigma$ when both types of moments (permanent and induced) are taken into account. As is apparent from this figure, the relaxation process follows activation law behavior in some definite range of the parameters

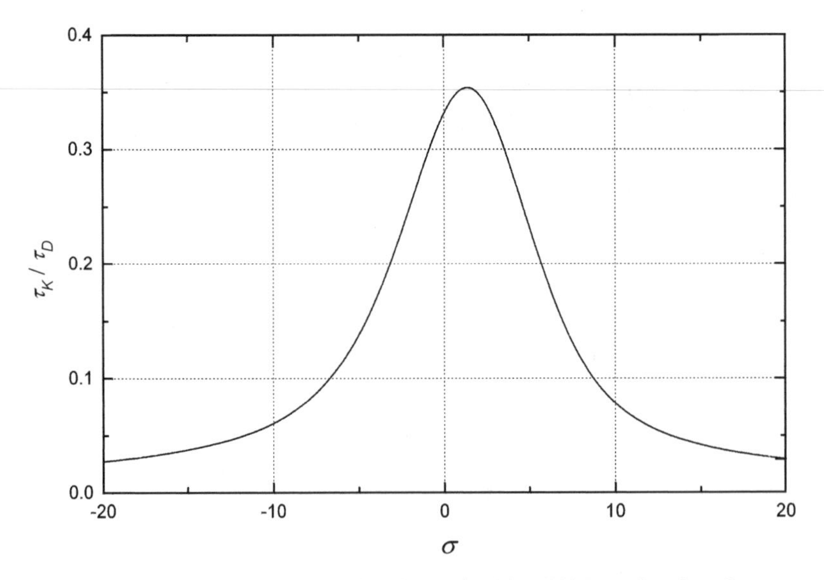


**Figure 17.** Plot of the relaxation time $\tau_K/\tau_D$ [Eq. (5.39)] as a function of $\sigma$.

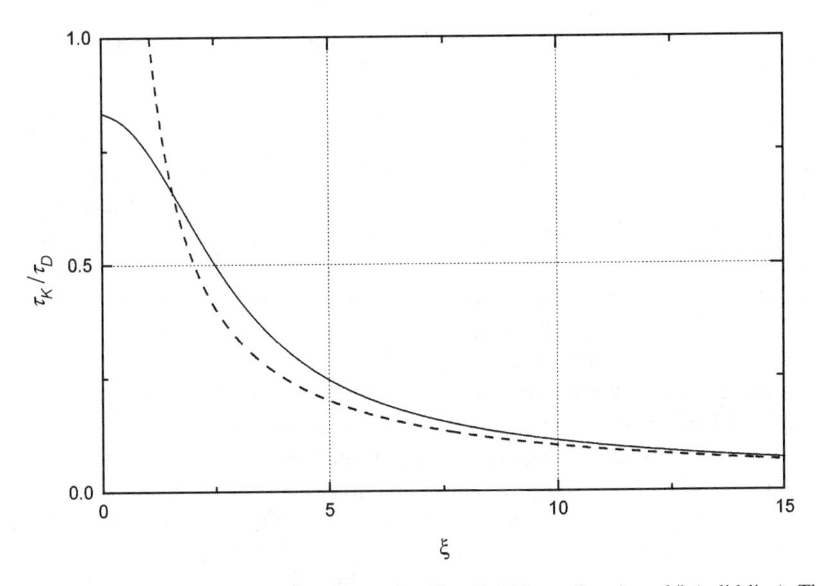


**Figure 18.** Plot of the relaxation time $\tau_K/\tau_D$ [Eq. (5.41)] as a function of $\xi$ (solid line). The dashed line is the asymptotic dependence of $\tau_K$ given by Eq. (5.29).

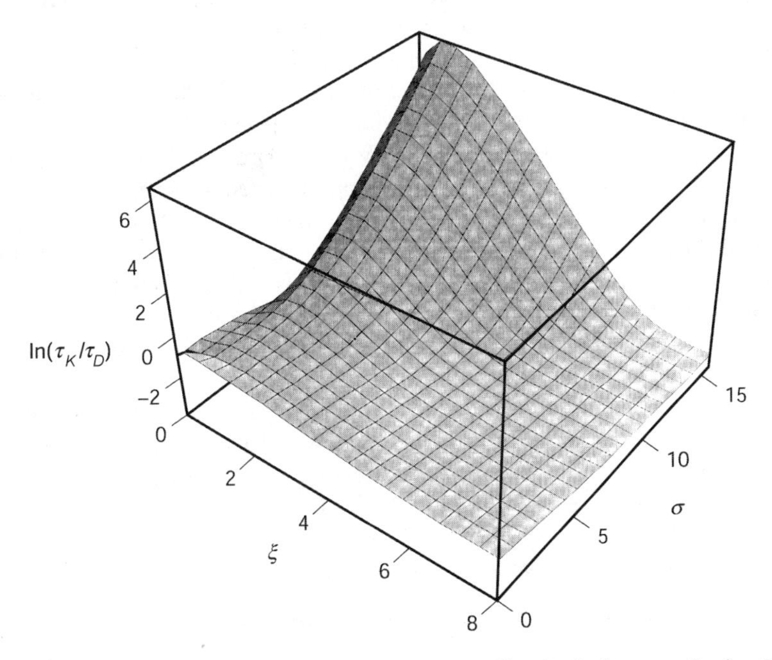


**Figure 19.** Plot of the relaxation time $\tau_K$ as a function of $\xi$ and $\sigma$ in the ranges $0 < \xi < 8$ and $0 < \sigma < 16$.

$\xi$ and $\sigma$ (i.e., the exponential increase of $\tau_K$ with increasing $\sigma$). This may be explained as follows. In the absence of the ac field the relaxation dynamics in the potential given by Eq. (5.2) with $E_1(t) = 0$ (which has in general two potential wells) is determined by two relaxation processes. One relaxation (activation) process governs the crossing of the potential barrier between two positions of equilibrium. Another (fast) process describes orientational relaxation inside the wells. In the case of nonpolarizable molecules ($\sigma = 0$), when the potential transforms to the single well potential $\xi \cos \vartheta$, we observe intrawell relaxation modes only. The activation process is responsible for relaxation, which is determined by the expectation value of the first Legendre polynomial (e.g., for dielectric relaxation). The appearance of this process in the Kerr effect response is due to the coupling between the first and second Legendre polynomials. However, with increasing $\xi$ the potential (5.2) with $E_1(t) = 0$ becomes more and more asymmetrical and the activation process is suppressed due to the depletion of the upper well [73]. This depletion occurs at values of the constant electric field $\mathbf{E}_0$, which are considerably smaller than the value of a critical field at which the double well structure of the potential disappears (this critical field is given by $\xi/(2\sigma) = 1$). A similar result has been observed for the linear response in magnetic relaxation of single domain ferromagnetic particles with high anisotropy barriers in the presence of a strong constant magnetic field following an infinitesimal change in that field [74].

The behavior of the real and imaginary parts of the spectra $\chi_K(\omega)$ of the birefringence function [Eq. (5.37)] is shown in Figures 20–23. It is apparent from these figures that Debye-like behavior is obtained for the spectra $\chi_K(\omega)$ for arbitrary $\xi$ and at $\sigma \approx 0$ and for arbitrary $\sigma$ and large $\xi$. Thus, in these cases alone the relaxation function $f_K(t)$ can be approximated by a single exponential. In all other cases, the decay of the relaxation functions $f_K(t)$ has a more complicated behavior and is determined by the cross-correlation function $f_1(t)$ of the first two Legendre polynomials and the autocorrelation function $f_2(t)$ of the second Legendre polynomial. The deviation from the Debye spectrum is still more pronounced in the Cole–Cole diagrams representing the variation of $-\mathrm{Im}\{\chi_K(\omega)/\chi_K(0)\}$ vs. $\mathrm{Re}\{\chi_K(\omega)/\chi_K(0)\}$, where one can see considerable deviations from the usual Debye semicircle (see Figs. 22 and 23). In particular, in Figure 22, for $\xi = 2$ (fixed value), it is shown that two small arcs manifest themselves in the very low frequency region with increasing $\sigma$. A similar behavior may be observed in Figure 23 for $\sigma = 5$ (fixed value) when $\xi = 1$. Moreover, all these Cole–Cole plots have their maxima below 0.5, the value obtained with a pure Debye semicircle.

### B. Steady-State Response: Perturbation Solution

By using the perturbation theory, we shall extend here the approach of Section V.A to evaluating the ac stationary solution of the dynamic birefringence and

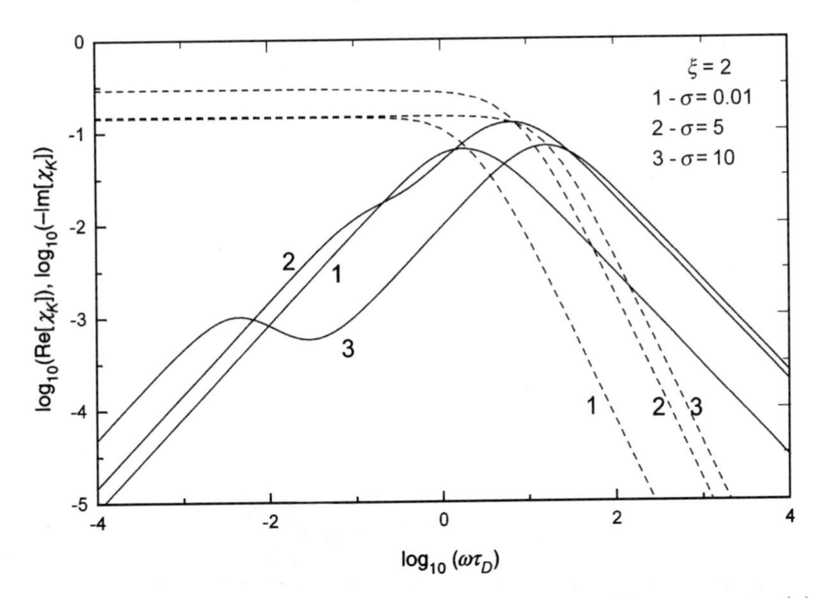


**Figure 20.** The real (dashed lines) and imaginary (solid lines) parts of the spectrum of $\chi_K(\omega)$ at $\xi = 2$. Curves 1, 2, and 3 correspond to $\sigma = 0.01$, 5, and 10, respectively.

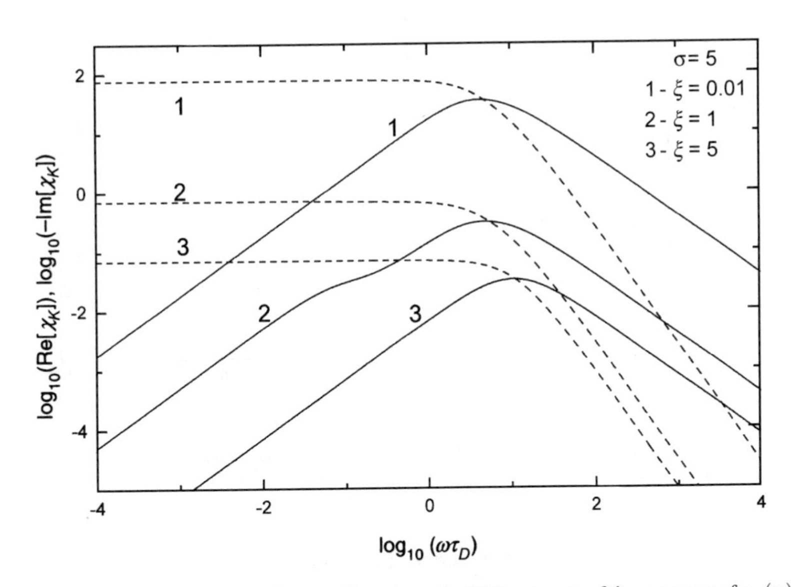


**Figure 21.** The real (dashed lines) and imaginary (solid lines) parts of the spectrum of $\chi_K(\omega)$ for $\sigma = 5$. Curves 1, 2, and 3 correspond to $\xi = 0.01$, 1, and 5, respectively.

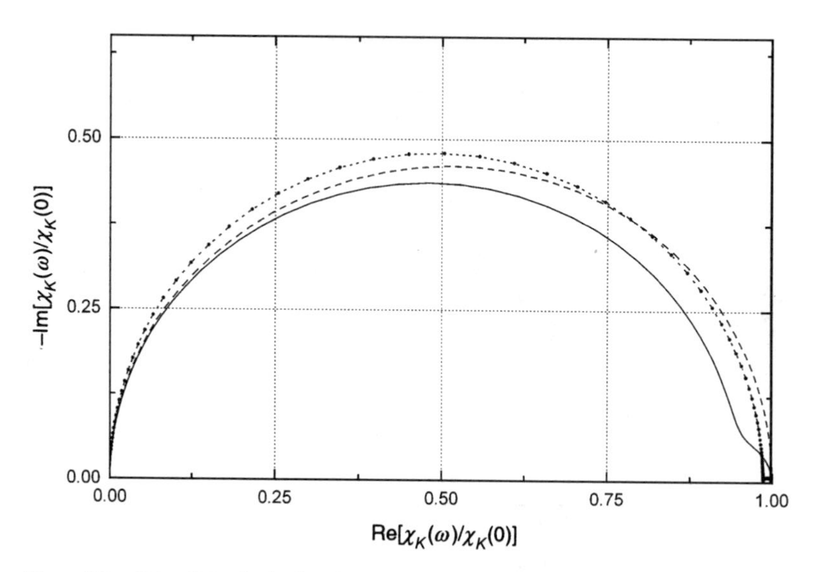


**Figure 22.** Cole–Cole plot for $\xi = 2$. The parameter $\sigma = 0.01$ (dashed line), $\sigma = 5$ (solid line), and $\sigma = 10$ (dotted line).

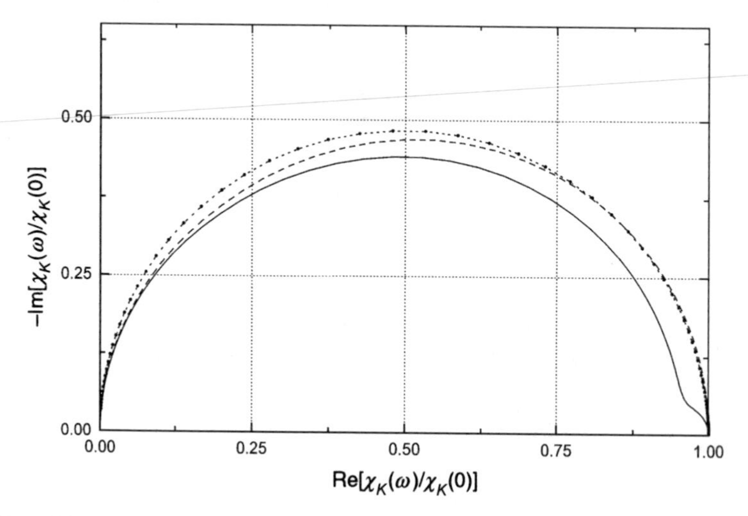


**Figure 23.** Cole–Cole plot for $\sigma = 5$. The parameter $\xi = 0.01$ (dashed line), $\xi = 1$ (solid line), and $\xi = 5$ (dotted line).

dielectric relaxation up to second order in the ac electric field strength [75, 76]. Here, as above, we assume that the rotational Brownian motion in superimposed dc $\mathbf{E}_0$ and ac $\mathbf{E}_1(t)$ fields of polar and anisotropically polarizable symmetric-top molecules dissolved in a nonpolar solvent is described by the noninertial

Fokker–Planck equation (3.76) with $V(\vartheta, t)$ given by Eq. (5.2) [both $\mathbf{E}_0$ and $\mathbf{E}_1(t)$ are assumed to be oriented along the $Z$ axis]. Moreover, we will suppose that the amplitude of the ac field $\mathbf{E}_1(t)$ is small so that

$$\xi_1 = \frac{\mu E_1}{kT} \ll 1 \tag{5.45}$$

is a small parameter, and the effects due to the hyperpolarizability are neglected. On expanding $W(\vartheta, t)$ as a series of Legendre polynomials $P_n(\cos\vartheta)$, one has the set of differential-recurrence relations (2.6) (e.g., [2, 4]), where now

$$\zeta(t) = \frac{\mu[E_0 + E_1(t)]}{kT} \tag{5.46}$$

$$\sigma(t) = \frac{\Delta\alpha}{2kT}[E_0 + E_1(t)]^2 \tag{5.47}$$

and

$$f_n(t) = \langle P_n(\cos\vartheta)\rangle(t) \tag{5.48}$$

As the time-dependent quantities appropriate to the dielectric and Kerr effect relaxation [the electric polarization and birefringence functions $P(t)$ and $K(t)$] are proportional to $f_1(t)$ and $f_2(t)$ and as we are solely concerned with the stationary ac responses, we must calculate the stationary state solution of Eq. (2.6) in the presence of the small ac electric field $E_1(t)$ [in order to deduce the corresponding relaxation spectra $\tilde{f}_n(\omega)$ for $n = 1, 2$]. This ac field $E_1(t)$ may be regarded as a small perturbation so that we shall expand $f_n(t)$ as follows:

$$f_n(t) = f_n^{(0)} + \xi_1 f_n^{(1)}(t) + \xi_1^2 f_n^{(2)}(t) = f_n^{(0)} + \xi_1 e^{i\omega t}\tilde{f}_n^{(1)}(\omega) + \xi_1^2 e^{i2\omega t}\tilde{f}_n^{(2)}(2\omega) \tag{5.49}$$

where the superscripts represent the order of the perturbation, and terms higher than $\xi_1^2$ are discarded. Hence, we shall calculate exactly $\tilde{f}_n^{(1)}(\omega)$ and $\tilde{f}_n^{(2)}(2\omega)$ corresponding to first- and second-order perturbations, respectively. Before proceeding, we introduce the following reduced parameters

$$\zeta(t) = \xi + \xi_1 e(t) \tag{5.50}$$

$$\sigma(t) = \sigma + \tfrac{3}{2}\rho\xi_1 e(t) + \tfrac{1}{2}\kappa\xi_1^2 e^2(t) \tag{5.51}$$

where

$$e(t) = e^{i\omega t} \tag{5.52}$$

and the dimensionless parameters $\xi$, $\sigma$, $\kappa$, and $\rho$ have already been defined in Eqs. (2.21), (2.37) and (5.10). Next, we first evaluate the equilibrium $f_n^{(0)}$ [for $E_1(t) = 0$] and first-order $\tilde{f}_n^{(1)}(\omega)$ solutions of Eq. (2.6) for any $n$, which are *all necessary* for calculating the second-order solution $\tilde{f}_n^{(2)}(\omega)$. In particular, we shall show that the first-order solution $\tilde{f}_n^{(1)}(\omega)$ agrees fully with the results of Section V.A based on linear response theory.

### 1. Equilibrium and First-Order Solutions in Terms of Matrix Continued Fractions

In the absence of any time-dependent perturbing field, the relaxation functions are time-independent equilibrium averages so that

$$f_n(t) = f_n^{(0)} = \langle P_n(\cos\vartheta)\rangle_0 \tag{5.53}$$

and Eq. (2.6) leads to the five-term recurrence relation Eq. (4.44), which can be transformed into a homogeneous matrix three-term recurrence relation as follows:

$$\mathbf{Q}_n^- \mathbf{C}_{n-1}^{(0)} + \mathbf{Q}_n \mathbf{C}_n^{(0)} + \mathbf{Q}_n^+ \mathbf{C}_{n+1}^{(0)} = \mathbf{0} \tag{5.54}$$

where

$$\mathbf{C}_n^{(0)} = \begin{pmatrix} \langle P_{2n-1}\rangle_0 \\ \langle P_{2n}\rangle_0 \end{pmatrix} \tag{5.55}$$

is a column vector containing the equilibrium averages of the Legendre polynomials of the odd and even order and $\mathbf{Q}_n^-$, $\mathbf{Q}_n$, and $\mathbf{Q}_n^+$ are $2 \times 2$ matrices defined by Eqs. (4.39)–(4.41). As we have already shown in Section IV.B, Eq. (5.54) can be solved in terms of matrix continued fractions [cf. Eq. (4.49)]

$$\mathbf{C}_n^{(0)} = \left[\prod_{k=1}^{n} \mathbf{S}_k(0)\right] \begin{pmatrix} 0 \\ 1 \end{pmatrix} \tag{5.56}$$

where the matrix continued fraction $\mathbf{S}_n(0)$ is defined by Eq. (3.41), where $s = 0$.

Now, if we restrict ourselves to the first order of perturbation, we can seek all the $f_n(t)$ in the form

$$f_n(t) = f_n^{(0)} + \xi_1 f_n^{(1)}(t) \tag{5.57}$$

Moreover, Eq. (5.51) simplifies to yield

$$\sigma(t) = \sigma + \frac{3\rho\xi_1 e(t)}{2} \tag{5.58}$$

On substituting Eqs. (5.50), (5.57), and (5.58) into Eq. (2.6) and equating terms with the first power of $\xi_1$ we obtain the set of the differential-recurrence relations for $f_n^{(1)}(t)$, namely,

$$\begin{aligned}
&\frac{2\tau_D}{n(n+1)}\frac{d}{dt}f_n^{(1)}(t)+\left[1-\frac{2\sigma}{(2n-1)(2n+3)}\right]f_n^{(1)}(t)\\
&=\frac{\xi}{2n+1}\left[f_{n-1}^{(1)}(t)-f_{n+1}^{(1)}(t)\right]+2\sigma\left[\frac{(n-1)}{(2n-1)(2n+1)}f_{n-2}^{(1)}(t)-\frac{(n+2)}{(2n+1)(2n+3)}f_{n+2}^{(1)}(t)\right]\\
&\quad+3\rho e(t)\left[\frac{f_n^{(0)}}{(2n-1)(2n+3)}+\frac{(n-1)}{(2n-1)(2n+1)}f_{n-2}^{(0)}-\frac{(n+2)}{(2n+1)(2n+3)}f_{n+2}^{(0)}\right]
\end{aligned} \tag{5.59}$$

Denoting by $\tilde{f}_n^{(1)}(\omega)$ the relaxation functions in the frequency domain, we set $f_n^{(1)}(t) = \tilde{f}_n^{(1)}(\omega)e^{i\omega t}$ and $e(t) = e^{i\omega t}$. Then, Eq. (5.59) can be arranged as an inhomogeneous matrix three-term recurrence relation to yield

$$i\omega\tau_D \tilde{\mathbf{C}}_n^{(1)}(\omega) = \mathbf{Q}_n^- \tilde{\mathbf{C}}_{n-1}^{(1)}(\omega) + \mathbf{Q}_n \tilde{\mathbf{C}}_n^{(1)}(\omega) + \mathbf{Q}_n^+ \tilde{\mathbf{C}}_{n+1}^{(1)}(\omega) + \mathbf{C}_n^{(1)} \tag{5.60}$$

where

$$\tilde{\mathbf{C}}_n^{(1)}(\omega) = \begin{pmatrix} \tilde{f}_{2n-1}^{(1)}(\omega) \\ \tilde{f}_{2n}^{(1)}(\omega) \end{pmatrix} \tag{5.61}$$

$$\mathbf{C}_n^{(1)} = \mathbf{q}_n^- \mathbf{C}_{n-1}^{(0)} + \mathbf{q}_n \mathbf{C}_n^{(0)} + \mathbf{q}_n^+ \mathbf{C}_{n+1}^{(0)} \tag{5.62}$$

and

$$\mathbf{q}_n^- = \begin{pmatrix} \dfrac{6\rho n(2n-1)(n-1)}{(4n-3)(4n-1)} & \dfrac{n(2n-1)}{4n-1} \\ 0 & \dfrac{3\rho n(4n^2-1)}{16n^2-1} \end{pmatrix} \tag{5.63}$$

$$\mathbf{q}_n = \begin{pmatrix} \dfrac{3\rho n(2n-1)}{(4n-3)(4n+1)} & -\dfrac{n(2n-1)}{4n-1} \\ \dfrac{n(2n+1)}{4n+1} & \dfrac{3\rho n(2n+1)(2n-1)}{(4n-1)(4n+3)} \end{pmatrix} \tag{5.64}$$

$$\mathbf{q}_n^+ = \begin{pmatrix} -\dfrac{3\rho n(4n^2-1)}{16n^2-1} & 0 \\ -\dfrac{n(2n+1)}{4n+1} & -\dfrac{6\rho n(2n+1)(n+1)}{(4n+1)(4n+3)} \end{pmatrix} \tag{5.65}$$

The last term in the right-hand side of Eq. (5.60) is frequency-independent and can be calculated from Eq. (5.56).

On noting that $\tilde{\mathbf{C}}_0^{(1)}(\omega) = \mathbf{0}$, the first harmonic components $\tilde{f}_1^{(1)}(\omega)$ and $\tilde{f}_2^{(1)}(\omega)$ varying at the fundamental frequency now follow directly from Eq. (3.46), namely,

$$\tilde{\mathbf{C}}_1^{(1)}(\omega) = \begin{pmatrix} \tilde{f}_1^{(1)}(\omega) \\ \tilde{f}_2^{(1)}(\omega) \end{pmatrix} = [i\omega\tau_D\mathbf{I} - \mathbf{Q}_1 - \mathbf{Q}_1^+\tilde{\mathbf{S}}_2(\omega)]^{-1} \times \left\{ \mathbf{C}_1^{(1)} + \sum_{k=2}^{\infty} \left[ \prod_{m=2}^{k} \mathbf{Q}_{m-1}^+ \tilde{\mathbf{S}}_m(\omega)(\mathbf{Q}_m^-)^{-1} \right] \mathbf{C}_k^{(1)} \right\} \tag{5.66}$$

where $\tilde{\mathbf{S}}_n(\omega) = \mathbf{S}_n(i\omega)$ and the continued fraction $\mathbf{S}_n(s)$ is defined by Eq. (3.41). Having determined $\tilde{\mathbf{C}}_1^{(1)}(\omega)$, we can evaluate from the recurrence Eq. (5.60) $\tilde{\mathbf{C}}_n^{(1)}(\omega)$ for any $n$. The exact solution for $\tilde{f}_1^{(1)}(\omega)$ and $\tilde{f}_2^{(1)}(\omega)$ [Eq. (5.66)] differs in mathematical form from that obtained in Section V.A by using linear response theory [Eq. (5.30)]. However, the numerical calculations demonstrate that both solutions are *in full agreement*.

### 2. *Exact Second-Order Perturbation Solutions*

Having determined $f_n^{(0)}$ and $\tilde{f}_n^{(1)}(\omega)$ for all $n$, we may seek a second-order perturbation solution of Eq. (2.6) in the form of Eq. (5.49). On substituting

Eqs. (5.49)–(5.51) into Eq. (2.6), and equating the terms in $\xi_1^2$, we have

$$\begin{aligned}
\frac{2\tau_D}{n(n+1)}\frac{d}{dt}f_n^{(2)}(t) + \left[1 - \frac{2\sigma}{(2n-1)(2n+3)}\right]f_n^{(2)}(t) &= \frac{\xi}{2n+1}\left[f_{n-1}^{(2)}(t) - f_{n+1}^{(2)}(t)\right] \\
&+ 2\sigma\left[\frac{(n-1)}{(2n-1)(2n+1)}f_{n-2}^{(2)}(t) - \frac{(n+2)}{(2n+1)(2n+3)}f_{n+2}^{(2)}(t)\right] \\
&+ \frac{e(t)}{2n+1}\left[f_{n-1}^{(1)}(t) - f_{n+1}^{(1)}(t)\right] \\
&+ 3\rho e(t)\left[\frac{f_n^{(1)}(t)}{(2n-1)(2n+3)} + \frac{(n-1)}{(2n-1)(2n+1)}f_{n-2}^{(1)}(t)\right. \\
&\qquad \left. - \frac{(n+2)}{(2n+1)(2n+3)}f_{n+2}^{(1)}(t)\right] \\
&+ \kappa e^2(t)\left[\frac{f_n^{(0)}}{(2n-1)(2n+3)} + \frac{(n-1)}{(2n-1)(2n+1)}f_{n-2}^{(0)}\right. \\
&\qquad \left. - \frac{(n+2)}{(2n+1)(2n+3)}f_{n+2}^{(0)}\right]
\end{aligned} \tag{5.67}$$

Thus, the time evolution of the second-order relaxation functions $f_n^{(2)}(t)$ depends on all lower order functions, namely, the zero-order $f_n^{(0)}$ and the first-order $f_n^{(1)}(t)$ functions. Setting $f_n^{(2)}(t) = \tilde{f}_n^{(2)}(2\omega)e^{2i\omega t}$, $f_n^{(1)}(t) = \tilde{f}_n^{(1)}(\omega)e^{i\omega t}$, and $e(t) = e^{i\omega t}$ in Eq. (5.67), we may derive the inhomogeneous matrix three-term recurrence relation for the second harmonic components in $2\omega$, which can be written in a manner similar to the first-order one as follows:

$$2i\omega\tau_D\tilde{\mathbf{C}}_n^{(2)}(2\omega) = \mathbf{Q}_n^-\tilde{\mathbf{C}}_{n-1}^{(2)}(2\omega) + \mathbf{Q}_n\tilde{\mathbf{C}}_n^{(2)}(2\omega) + \mathbf{Q}_n^+\tilde{\mathbf{C}}_{n+1}^{(2)}(2\omega) + \mathbf{C}_n^{(2)} \tag{5.68}$$

where

$$\tilde{\mathbf{C}}_n^{(2)}(2\omega) = \begin{pmatrix} \tilde{f}_{2n-1}^{(2)}(2\omega) \\ \tilde{f}_{2n}^{(2)}(2\omega) \end{pmatrix} \tag{5.69}$$

$$\mathbf{C}_n^{(2)} = \mathbf{q}_n^-\tilde{\mathbf{C}}_{n-1}^{(1)}(\omega) + \mathbf{q}_n\tilde{\mathbf{C}}_n^{(1)}(\omega) + \mathbf{q}_n^+\tilde{\mathbf{C}}_{n+1}^{(1)}(\omega) + \mathbf{K}_n^-\mathbf{C}_{n-1}^{(0)} + \mathbf{K}_n\mathbf{C}_n^{(0)} + \mathbf{K}_n^+\mathbf{C}_{n+1}^{(0)} \tag{5.70}$$

and $\mathbf{K}_n^-$, $\mathbf{K}_n$, and $\mathbf{K}_n^+$ are $2 \times 2$ diagonal matrices given by

$$\mathbf{K}_n^- = \begin{pmatrix} \dfrac{2\kappa n(2n-1)(n-1)}{(4n-3)(4n-1)} & 0 \\ 0 & \dfrac{\kappa n(4n^2-1)}{16n^2-1} \end{pmatrix} \tag{5.71}$$

$$\mathbf{K}_n = \begin{pmatrix} \dfrac{\kappa n(2n-1)}{(4n-3)(4n+1)} & 0 \\ 0 & \dfrac{\kappa n(2n+1)}{(4n-1)(4n+3)} \end{pmatrix} \tag{5.72}$$

$$\mathbf{K}_n^+ = \begin{pmatrix} -\dfrac{\kappa n(4n^2-1)}{(16n^2-1)} & 0 \\ 0 & -\dfrac{2\kappa n(2n+1)(n+1)}{(4n+1)(4n+3)} \end{pmatrix} \tag{5.73}$$

We remark that the $\mathbf{C}_n^{(2)}$ in Eq. (5.70) may be calculated with the help of Eqs. (5.56) and (5.66) and so represent known quantities. However, $\mathbf{C}_n^{(2)}$ is now frequency-dependent since it contains the $\tilde{f}_n^{(1)}(\omega)$. Equation (5.68) can be solved in terms of matrix continued fractions just as (5.66) above to yield

$$\begin{pmatrix} \tilde{f}_1^{(2)}(2\omega) \\ \tilde{f}_2^{(2)}(2\omega) \end{pmatrix} = [2i\omega\tau_D \mathbf{I} - \mathbf{Q}_1 - \mathbf{Q}_1^+ \tilde{\mathbf{S}}_2(2\omega)]^{-1} \times \left\{ \mathbf{C}_1^{(2)} + \sum_{k=2}^{\infty} \left[ \prod_{m=2}^{k} \mathbf{Q}_{m-1}^+ \tilde{\mathbf{S}}_m(2\omega)(\mathbf{Q}_m^-)^{-1} \right] \mathbf{C}_k^{(2)} \right\} \tag{5.74}$$

Equation (5.74) is the *exact* solution in terms of matrix continued fractions for the second-order perturbation spectra $\tilde{f}_1^{(2)}(2\omega)$ of the dielectric relaxation and $\tilde{f}_2^{(2)}(2\omega)$ of the birefringence function.

It is worth noting that, for $\xi = 0$ and $\sigma = 0$ the results may be radically simplified since Eqs. (5.59) and (5.67) reduce to *scalar three-term* differential-recurrence relations that may be solved in terms of *ordinary* continued fractions. For $\xi = 0$ (purely induced moments), the first- and the second-order solutions of

Eqs. (5.59) and (5.67) are given by [75,76]:

$$\tilde{f}_2^{(\alpha)}(\omega) = \frac{3\sqrt{\pi}}{8\sigma}\sum_{k=0}^{\infty}(-1)^k \frac{(4k+5)\Gamma(k+1)}{\Gamma(k+5/2)} f_{2k+2}^{(\alpha)} \prod_{m=0}^{k} S_{2m+2}(i\alpha\omega) \qquad (\alpha = 1,2) \tag{5.75}$$

where the continued fraction $S_n(s)$ is defined by Eq. (2.23),

$$\begin{aligned} f_{2n}^{(1)} = 4\sqrt{\sigma}\bigg[ & \frac{n(2n+1)}{(4n+3)(4n-1)}\langle P_{2n}(\cos\vartheta)\rangle_0 - \frac{2n(n+1)(2n+1)}{(4n+3)(4n+1)} \\ & \times \langle P_{2n+2}(\cos\vartheta)\rangle_0 + \frac{n(4n^2-1)}{16n^2-1}\langle P_{2n-2}(\cos\vartheta)\rangle_0 \bigg] \end{aligned} \tag{5.76}$$

$$\begin{aligned} f_{2n}^{(2)} = 4\sqrt{\sigma}\bigg[ & \frac{n(2n+1)}{(4n+3)(4n-1)}\tilde{f}_{2n}^{(1)}(\omega) - \frac{2n(n+1)(2n+1)}{(4n+3)(4n+1)}\tilde{f}_{2n+2}^{(1)}(\omega) \\ & + \frac{n(4n^2-1)}{16n^2-1}\tilde{f}_{2n-2}^{(1)}(\omega)\bigg] \\ + 2\bigg[ & \frac{n(2n+1)}{(4n+3)(4n-1)}\langle P_{2n}(\cos\vartheta)\rangle_0 - \frac{2n(n+1)(2n+1)}{(4n+3)(4n+1)} \\ & \times \langle P_{2n+2}(\cos\vartheta)\rangle_0 + \frac{n(4n^2-1)}{16n^2-1}\langle P_{2n-2}(\cos\vartheta)\rangle \bigg] \end{aligned} \tag{5.77}$$

and $\langle P_{2n}\rangle_0$ is given by Eq. (4.67). In this case, the perturbation parameter $\xi_1 = E_1\sqrt{\Delta\alpha/2kT}$. If the contribution of the induced dipole moment is negligible in comparison with that of the permanent dipole moment, that is, $\sigma = 0$, we have [75, 76]:

$$\tilde{f}_1^{(\alpha)}(\alpha\omega) = \frac{2}{\xi}\sum_{k=1}^{\infty}(-1)^{k+1}\frac{(2k+1)}{k(k+1)} f_k^{(\alpha)} \prod_{m=1}^{k} S_m(i\alpha\omega) \tag{5.78}$$

$$\begin{aligned} \tilde{f}_2^{(\alpha)}(\alpha\omega) = & \frac{3\tilde{f}_1^{(\alpha)} S_1(i\alpha\omega)S_2(i\alpha\omega)}{\xi} + \frac{2[1 - S_1(i\alpha\omega)S_2(i\alpha\omega)]}{\xi} \\ & \times \sum_{k=2}^{\infty}(-1)^{k+1}\frac{(2k+1)}{k(k+1)} f_k^{(\alpha)} \prod_{m=2}^{k} S_m(i\alpha\omega) \end{aligned} \tag{5.79}$$

where $\alpha = 1$ and 2, for the first- and the second-order response, respectively, and the continued fraction $S_n(s)$ is given by Eq. (2.19),

$$f_n^{(1)} = \frac{n(n+1)}{(2n+1)} [\langle P_{n-1} \rangle_0 - \langle P_{n+1} \rangle_0] \tag{5.80}$$

$$f_n^{(2)} = \frac{n(n+1)}{(2n+1)} \left[ \tilde{f}_{n-1}^{(1)}(\omega) - \tilde{f}_{n+1}^{(1)}(\omega) \right] \tag{5.81}$$

and $\langle P_n(\cos \vartheta) \rangle_0$ is defined by Eq. (4.68).

### 3. Discussion of the Results

In order to illustrate all the results obtained for the nonlinear electrooptical responses, we have plotted diagrams showing the typical features of the real and imaginary parts of the complex birefringence and dielectric response functions. In Figures 24–27, we present dispersion and absorption plots of the birefringence relaxation spectra for first and second orders in the ac electric field strength. Various values of the electrical parameters $\xi_0$ and $\sigma_0$ are proposed in order to highlight the essential features of these spectra. The evolution of $\tilde{f}_2^{(1)}(\omega)$ and $\tilde{f}_2^{(2)}(2\omega)$ has the usual relaxational (Debye-like) behavior with char-

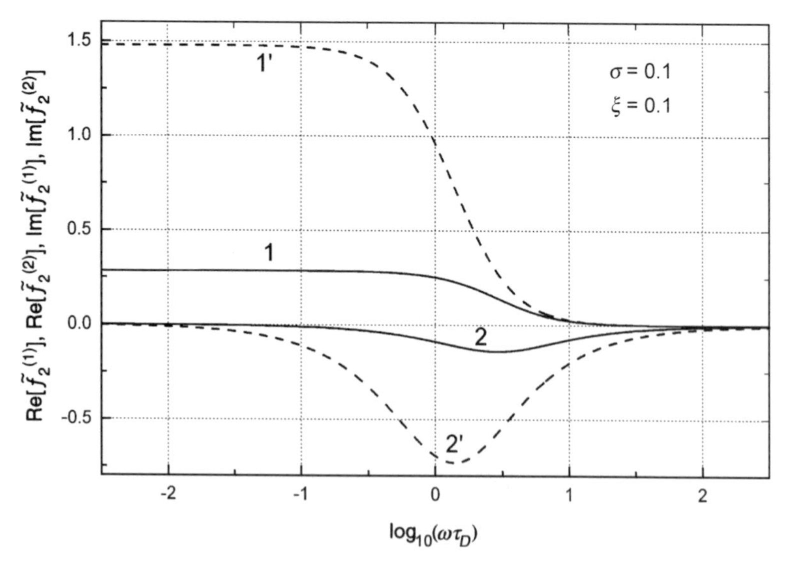


**Figure 24.** The real $(1, 1')$ and imaginary $(2, 2')$ parts of $\tilde{f}_2^{(1)}$ (solid lines) and $\tilde{f}_2^{(2)}(2\omega)$ (dashed lines) for $\sigma = 0.1$ and $\xi = 0.1$.

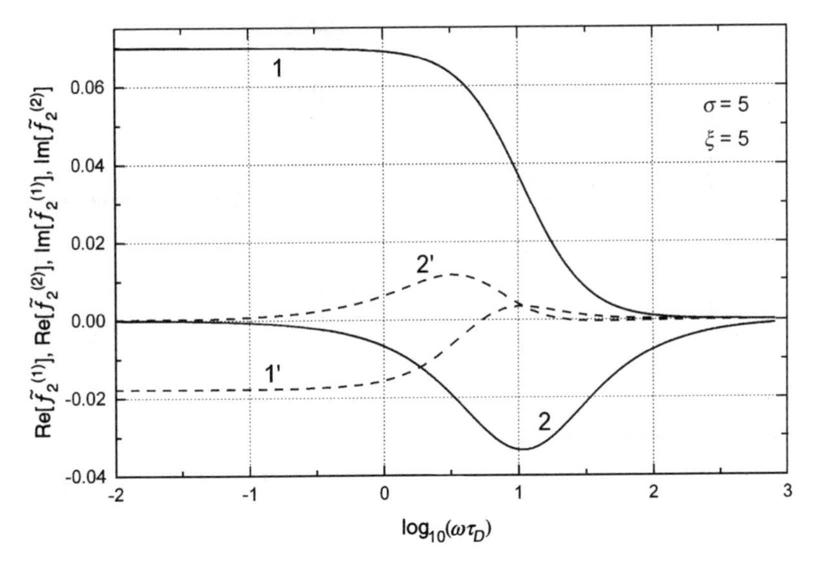


**Figure 25.** The same as in Figure 24 for $\sigma = 5$ and $\xi = 5$.

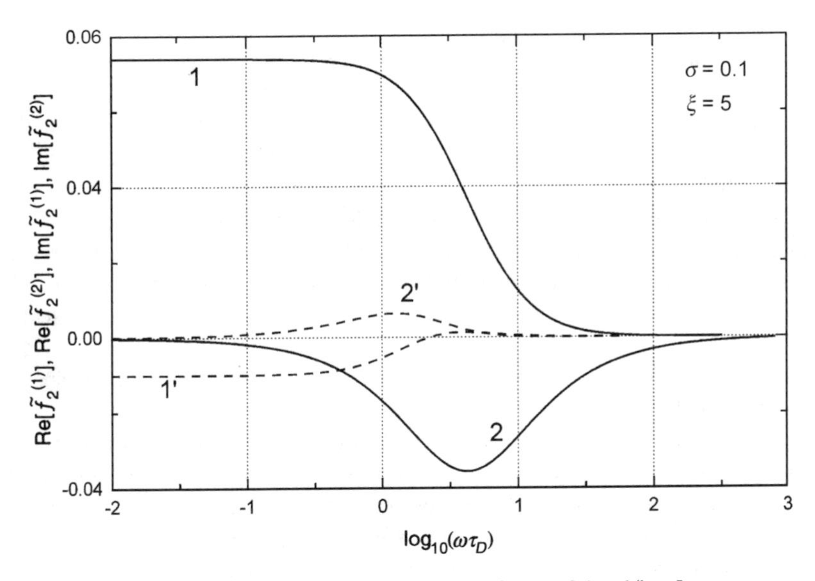


**Figure 26.** The same as in Figure 24 for $\sigma = 0.1$ and $\xi = 5$.

acteristic times $\sim \tau_D/3$ and $\sim 2\tau_D/3$, respectively when $\xi$ and $\sigma$ are small (Fig. 24). As $\xi$ and $\sigma$ simultaneously increase, the amplitude of the second-order response decreases (in absolute value) and, as shown in Figure 25, at relatively large values of $\xi$ and $\sigma$ (equal to 5, say), there is an inversion of the sign between

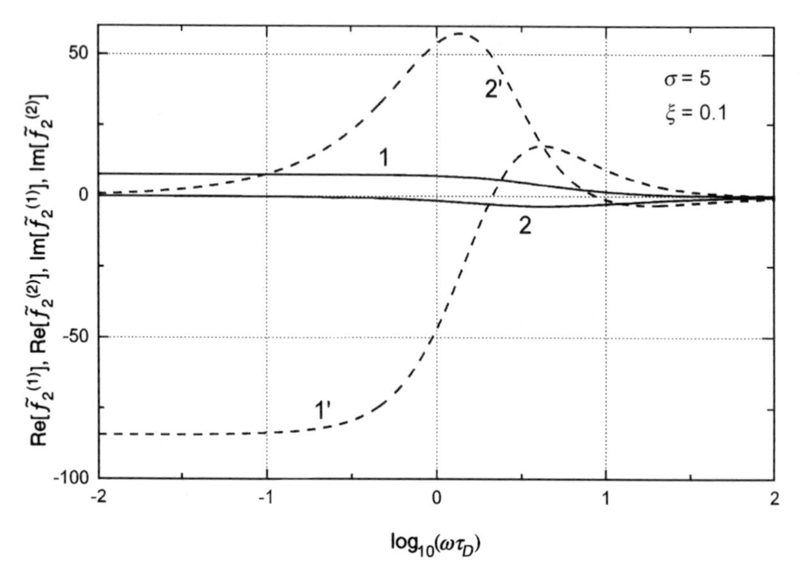


**Figure 27.** The same as in Figure 24 for $\sigma = 5$ and $\xi = 0.1$.

the first- and second-order spectra and different qualitative behavior of the second-order spectrum. In particular, the real part of $\tilde{f}_2^{(2)}(2\omega)$ changes sign at a frequency $\sim 2/\tau_D$. This tendency is maintained for large $\xi$ values and small $\sigma$ values (Fig. 26). In the reverse situation, where $\sigma$ is $\gg \xi$ (Fig. 27), the amplitude for the second-order prevails in absolute value over that of the first-order. As we have already mentioned, all the first-order plots coincide exactly with those that we have independently calculated by using linear response theory developed in Section V.A. Regarding the second-order relaxation spectra, the nonlinear effects arise essentially due to the coupling of the ac and dc electric fields.

In Figures 28–31, the dispersion plots of the dielectric relaxation spectra obtained for the first and second order in the ac field strength are presented for the same values of parameters $\xi$ and $\sigma$. Various values of the electrical parameters $\xi$ and $\sigma$ are proposed in order to emphasize the behavior of the spectra. In all cases, one can see the effects of the nonlinearity. If $\xi$ and $\sigma$ are small (0.1, say) the first- and second-order spectra have the usual Debye-like behavior with characteristic times $\sim \tau_D$ and $\sim 2\tau_D$, respectively (Fig. 28). For large values of $\xi$ and $\sigma$ (equal to 5, say, see Fig. 29), there is an inversion of the sign and a different qualitative behavior of the second-order components, namely: the real and imaginary parts of $\tilde{f}_1^{(2)}(2\omega)$ change sign at some particular frequencies. This is also the case for large values of $\xi$ and small ones of $\sigma$ (Fig. 30). In the opposite limit, when $\sigma$ is $\gg \xi$ (Fig. 31), the second-order response is

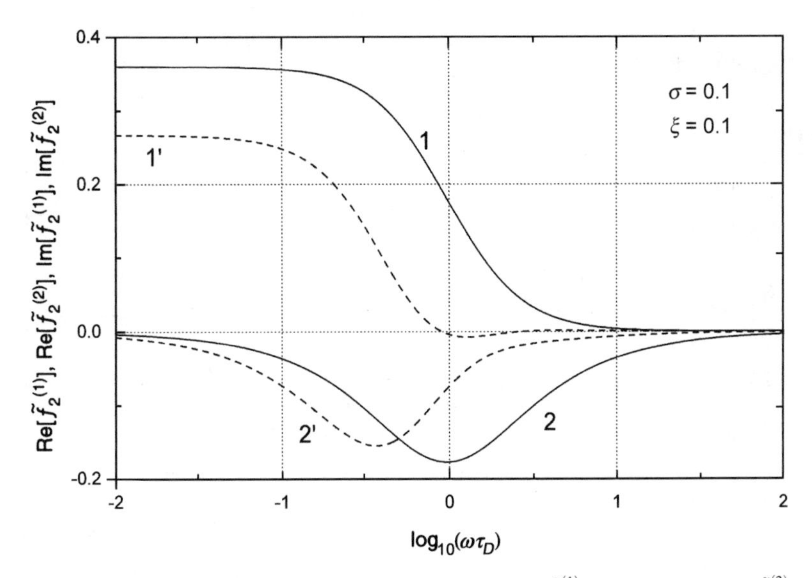


**Figure 28.** The real $(1, 1')$ and the imaginary $(2, 2')$ parts of $\tilde{f}_1^{(1)}(\omega)$ (solid lines) and $\tilde{f}_1^{(2)}(2\omega)$ (dashed lines) for $\sigma = 0.1$ and $\xi = 0.1$.

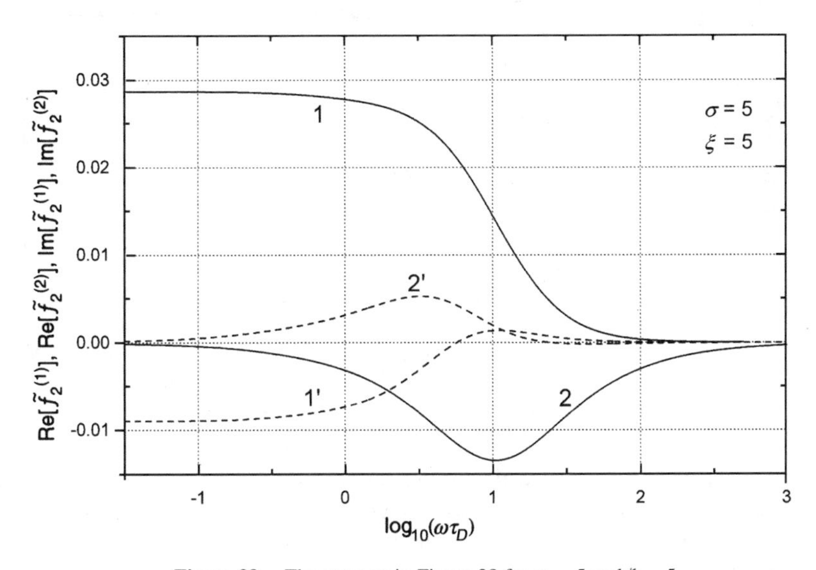


**Figure 29.** The same as in Figure 28 for $\sigma = 5$ and $\xi = 5$.

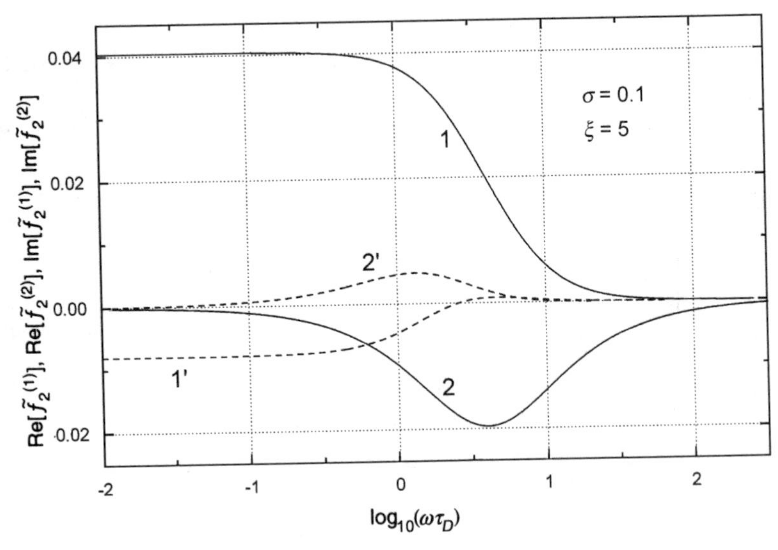


**Figure 30.** The same as in Figure 28 for $\sigma = 0.1$ and $\xi = 5$.

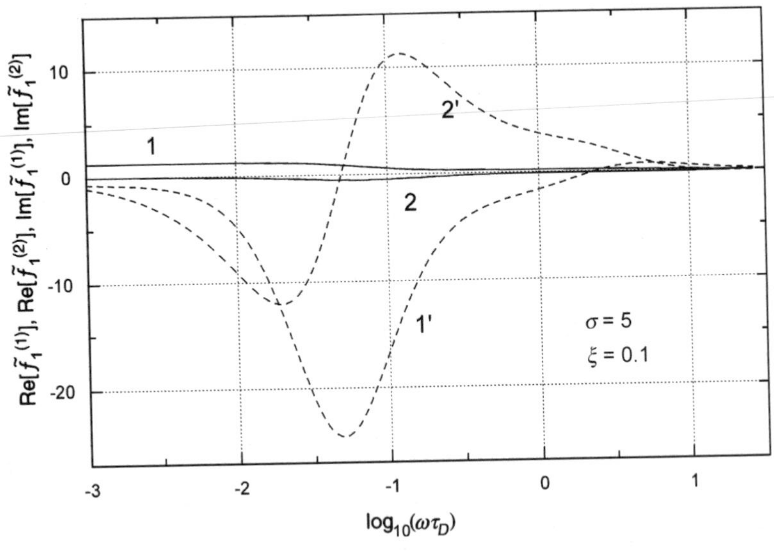


**Figure 31.** The same as in Figure 28 for $\sigma = 5$ and $\xi = 0.1$.

completely prevailed, due to the strong induced dipole moment. Both the real and the imaginary parts have two extrema (which can be evaluated). We remark in this case that both parts of the second-order spectrum are, in absolute value, much more significant than the first-order ones.

In the low dc bias field limit (i.e., for $\xi \ll 1$), the results obtained are in agreement with the perturbation solutions, which have been discussed in Section II. For dielectric response, the results predicted by the perturbation theory have been recently checked experimentally [42–44], as the first- and second-order responses can always be separated, and so measured. We remark that here we have confined ourselves to the study of the nonlinear ac response up to second-order in the ac electric field strength only. However, the approach we have developed may also be applied in like manner to the evaluation of higher order perturbation expansion terms of the nonlinear response. It is also possible to use such an approach to calculating the dielectric response with account of hyper-polarizability effects in order to see how the spectra are modified in the frequency domain.

## VI. NONLINEAR STATIONARY RESPONSES IN STRONG AC AND DC ELECTRIC FIELDS

The perturbation procedure developed in Section V is inapplicable to the treatment of nonlinear response in high ac fields. It is the purpose of this section to show how our approach can also be applied to the calculation of the nonlinear ac stationary response of systems of rigid polar molecules to an ac field of arbitrary strength. This approach [77, 78] is, in some respects, analogous to those used in [79] for the calculation of the harmonic mixing in a cosine potential and in [80] for the evaluation of the mean beat frequency of the dithered-ring-laser gyroscope. However, the model used here and the solution so obtained differ from those of [79] and [80]. Moreover, our solution has the merit of being considerably simpler than those previously available.

### A. Nonlinear Dielectric Relaxation of Rigid Polar Molecules in Superimposed ac and dc Electric Fields

Here, we shall consider the nonlinear dielectric relaxation of an assembly of *rigid polar* symmetric top particles (macromolecules) dissolved in a nonpolar solvent and acted on by strong external superimposed dc $\mathbf{E}_0$ and ac $\mathbf{E}_1(t) = \mathbf{E}_1 \cos \omega t$ electric fields. Each particle contains a rigid electric dipole $\boldsymbol{\mu}$ directed along the axis of symmetry. Let us also suppose, for simplicity, that both $\mathbf{E}_0$ and $\mathbf{E}_1$ are directed along the $Z$ axis of the laboratory coordinate system and that effects due to the anisotropy of the polarizability of the particles can be neglected. Then, the noninertial rotational Brownian motion of the particles may be described by the Smoluchowski equation (2.4), where the orientational potential energy of the molecule is given by

$$V(\vartheta, t) = -\mu[E_0 + E_1(t)] \cos \vartheta \tag{6.1}$$

The problem we want to solve is intrinsically nonlinear because we assume that the magnitudes of both ac and dc fields are large enough so that the energy of the molecule in these fields may be comparable or higher than $kT$. The dynamics of the system is governed by the following set of differential-recurrence equations [see Eq. (2.6) in which $\sigma = 0$]

$$\tau_D \frac{d}{dt} f_n(t) + \frac{n(n+1)}{2} f_n(t) = \zeta(t) \frac{n(n+1)}{2(2n+1)} [f_{n-1}(t) - f_{n+1}(t)] \tag{6.2}$$

where $f_n(t)$ denotes the expectation value of the Legendre polynomial of order $n$ [Eq. (2.8)] and $\zeta(t)$ is a dimensionless field parameter that may be separated into two parts as follows:

$$\zeta(t) = \xi + \xi_1 \cos \omega t \tag{6.3}$$

Here, the dimensionless parameters $\xi$ and $\xi_1$ are defined by Eq. (2.37). Our goal is to evaluate the ac stationary response of the electric polarization $P(t)$ proportional to $f_1(t)$.

The noninertial rotational Brownian motion in the presence of the ac electric field $\mathbf{E}_1(t) = \mathbf{E}_1 \cos \omega t$ is a nonstationary Markovian process where the symmetry under the time translation is retained in a discrete time transformation $t \to t + 2\pi/\omega$ only [81]. Since we are solely concerned with the stationary ac response, which is independent of the initial conditions, we need only to calculate the solution of Eq. (6.2) corresponding to the stationary state. To accomplish this, one may seek all the $f_n(t)$ in the form

$$f_n(t) = \sum_{k=-\infty}^{\infty} F_k^n(\omega) e^{ik\omega t} \tag{6.4}$$

As all the $f_n(t)$ are real, the Fourier amplitudes $F_k^n$ satisfy the following condition:

$$F_{-k}^n = (F_k^n)^* \tag{6.5}$$

where the asterisk denotes the complex conjugate. On substituting Eq. (6.4) into Eq. (6.2), we have the following recurrence relations for the Fourier amplitudes $F_k^n(\omega)$, namely;

$$\begin{aligned} z_{n,k}(\omega) F_k^n(\omega) &- 2\xi[F_k^{n-1}(\omega) - F_k^{n+1}(\omega)] \\ &- \xi_1[F_{k-1}^{n-1}(\omega) + F_{k+1}^{n-1}(\omega) - F_{k-1}^{n+1}(\omega) - F_{k+1}^{n+1}(\omega)] = 0 \end{aligned} \tag{6.6}$$

where

$$z_{n,k}(\omega) = 2(2n+1)\left[1 + \frac{2i\omega\tau_D k}{n(n+1)}\right] \tag{6.7}$$

The solution of Eq. (6.6) can be obtained in terms of matrix continued fractions as follows. Let us introduce the column vectors $\mathbf{C}_n(\omega)$ and $\mathbf{C}_0$:

$$\mathbf{C}_n(\omega) = \begin{pmatrix} \vdots \\ F^n_{-2}(\omega) \\ F^n_{-1}(\omega) \\ F^n_0(\omega) \\ F^n_1(\omega) \\ F^n_2(\omega) \\ \vdots \end{pmatrix} \quad \text{and} \quad \mathbf{C}_0(\omega) = \mathbf{C}_0 = \begin{pmatrix} \vdots \\ 0 \\ 0 \\ 1 \\ 0 \\ 0 \\ \vdots \end{pmatrix} \tag{6.8}$$

[As is obvious from its definition, the vector $\mathbf{C}_1$ contains all the Fourier amplitudes of $f_1(t)$, which are necessary for obtaining the ac nonlinear dielectric response.] Then, the *seven-term* recurrence Eq. (6.6) can be transformed into the *matrix three-term* recurrence equations

$$\mathbf{Q}_n(\omega)\mathbf{C}_n(\omega) + \mathbf{Y}\mathbf{C}_{n+1}(\omega) = \mathbf{Y}\mathbf{C}_{n-1}(\omega) \qquad n = 1, 2, 3, \ldots \tag{6.9}$$

where $\mathbf{Y}$ and $\mathbf{Q}_n(\omega)$ are tridiagonal and diagonal infinite matrices, respectively, defined as

$$\mathbf{Y} = \begin{pmatrix} \ddots & \vdots & \vdots & \vdots & \vdots & \vdots & \iddots \\ \cdots & 2\xi & \xi_1 & 0 & 0 & 0 & \cdots \\ \cdots & \xi_1 & 2\xi & \xi_1 & 0 & 0 & \cdots \\ \cdots & 0 & \xi_1 & 2\xi & \xi_1 & 0 & \cdots \\ \cdots & 0 & 0 & \xi_1 & 2\xi & \xi_1 & \cdots \\ \cdots & 0 & 0 & 0 & \xi_1 & 2\xi & \cdots \\ \iddots & \vdots & \vdots & \vdots & \vdots & \vdots & \ddots \end{pmatrix} \tag{6.10}$$

and

$$\mathbf{Q}_n(\omega) = \begin{pmatrix} \ddots & \vdots & \vdots & \vdots & \vdots & \vdots & ⋰ \\ \cdots & z_{n,-2}(\omega) & 0 & 0 & 0 & 0 & \cdots \\ \cdots & 0 & z_{n,-1}(\omega) & 0 & 0 & 0 & \cdots \\ \cdots & 0 & 0 & z_{n,0}(\omega) & 0 & 0 & \cdots \\ \cdots & 0 & 0 & 0 & z_{n,1}(\omega) & 0 & \cdots \\ \cdots & 0 & 0 & 0 & 0 & z_{n,2}(\omega) & \cdots \\ ⋰ & \vdots & \vdots & \vdots & \vdots & \vdots & \ddots \end{pmatrix} \tag{6.11}$$

where $z_{n,k}(\omega)$ is given by Eq. (6.7). Insofar as we are interested in the determination of $\mathbf{C}_1(\omega)$ only, the infinite system of Eq. (6.9) can readily be solved in terms of matrix continued fractions. Thus, we obtain

$$\mathbf{C}_1(\omega) = \mathbf{S}(\omega)\mathbf{C}_0 \tag{6.12}$$

where the infinite matrix continued fraction $\mathbf{S}(\omega)$ is given by

$$\mathbf{S}(\omega) = \frac{\mathbf{I}}{\mathbf{Q}_1(\omega) + \mathbf{Y}\dfrac{\mathbf{I}}{\mathbf{Q}_2(\omega) + \mathbf{Y}\dfrac{\mathbf{I}}{\mathbf{Q}_3(\omega) + \cdots}\mathbf{Y}}\mathbf{Y}}\mathbf{Y} \tag{6.13}$$

and $\mathbf{I}$ is the identity matrix of infinite dimension. Having determined the column vector $\mathbf{C}_1(\omega)$ from Eq. (6.12), one can calculate the stationary ac response function $f_1(t)$ suitable for dielectric relaxation, which may be presented as follows:

$$f_1(t) = F_0^1(\omega) + 2\sum_{k=1}^{\infty} \mathrm{Re}\{F_k^1(\omega)e^{ik\omega t}\} \tag{6.14}$$

The $F_0^1(\omega)$ in the right-hand side of Eq. (6.14) is a time-independent, but frequency-dependent term. This frequency dependence is due to the coupling effect of the dc bias $\mathbf{E}_0$ and ac $\mathbf{E}_1(t)$ fields. In the absence of the dc bias field, that is, for $\xi = 0$, the series (6.14) contains only the odd components of $F_k^n$ (all the even components are equal to zero) and reduces to

$$f_1(t) = 2\sum_{k=1}^{\infty} \mathrm{Re}\{F_{2k-1}^1(\omega)e^{i(2k-1)\omega t}\} \tag{6.15}$$

TABLE I
The Low and High-Frequency Behavior of the Fourier Components $F_n^k(\omega)$ for $\xi = 0$

| | $\omega\tau_D \ll 1, \xi_1 \ll 1$ | $\omega\tau_D \to \infty, \xi_1$ any |
|---|---|---|
| $F_{-1}^1 = (F_1^1)^*$ | $\approx \frac{\xi_1}{6}[1 + i\omega\tau_D + O(\omega^2\tau_D^2)] + O(\xi_1^3)$ | $\sim \frac{\xi_1}{6\omega^2\tau_D^2}(1 + i\omega\tau_D)$ |
| $F_{-3}^1 = (F_3^1)^*$ | $\approx -\frac{\xi_1^3}{360}\left[1 + \frac{14}{3}i\omega\tau_D + O(\omega^2\tau_D^2)\right] + O(\xi_1^5)$ | $\sim \frac{\xi_1^3}{720\omega^4\tau_D^4}\left(\frac{17}{6} + i\omega\tau_D\right)$ |
| $F_{-5}^1 = (F_5^1)^*$ | $\approx \frac{\xi_1^5}{15{,}120}\left[1 + \frac{41}{4}i\omega\tau_D + O(\omega^2\tau_D^2)\right] + O(\xi_1^7)$ | $\sim \frac{\xi_1^5}{80{,}640\omega^6\tau_D^6}\left(\frac{299}{60} + i\omega\tau_D\right)$ |
| $F_{-7}^1 = (F_7^1)^*$ | $\approx -\frac{\xi_1^7}{604{,}800}\left[1 + \frac{266}{15}i\omega\tau_D + O(\omega^2\tau_D^2)\right] + O(\xi_1^9)$ | $\sim \frac{\xi_1^7}{9{,}676{,}800\omega^8\tau_D^8}\left(\frac{3179}{420} + i\omega\tau_D\right)$ |

As far as the calculation of the infinite matrix continued fraction $\mathbf{S}$ in Eq. (6.13) is concerned, we approximated it by some matrix continued fraction of finite order (by putting $\mathbf{Q}_n$ at some $n = N$). At the same time we confined the dimensions of the matrices $\mathbf{Q}_n$ and $\mathbf{Y}$ to some finite number $M$. Both $N$ and $M$ depend on the field parameters $\xi$, $\xi_1$ and on the number of harmonics to be determined. They must be chosen taking account of the desired degree of accuracy of the calculation. For example, for the calculation of $F_k^1(\omega)$ up to $k = 7$ and for $\xi$ and $\xi_1$ up to 20, the dimensions of $\mathbf{Q}_n$ and $\mathbf{Y}$ need not exceed 50, and 15–20 iterations in calculating $\mathbf{S}$ are enough to arrive at an accuracy of not $< 6$ significant digits in the majority of cases.

Let us first of all consider the main features of the ac nonlinear response in the absence of the dc bias field, that is, for $\xi = 0$. The low $(\omega \to 0)$ and high $(\omega \to \infty)$ frequency asymptotic behavior of the Fourier components $F_k^1(\omega)$ may be evaluated from the recurrence relation (6.6). These asymptotic estimates for $F_k^1(\omega)$ at $k = 1, 3, 5$, and 7 are summarized in Table I. Equations for the low-frequency behavior of $F_k^1(\omega)$ presented in Table I were obtained by using the perturbation expansion of $F_k^1(\omega)$ in powers of $\xi_1$ so that they are valid for $\xi_1 \ll 1$ only. The high-frequency asymptotic expansions of $F_k^1(\omega)$ were derived by assuming arbitrary $\xi_1$, thus they are applicable to any $\xi_1$. This may be explained by the fact that in the limit $\omega \to \infty$ the dipole polarization plays no role as the dipoles are "frozen" due to viscosity of the solution and they have no time to follow the changes of the ac field independent of the strength of the ac field.

The results of the calculation of the real and imaginary parts of the normalized nonlinear harmonic components of the electric polarization varying in $\omega$, $3\omega$, and $5\omega$, namely,

$$\chi_1^1(\omega) = 6F_1^{1*}(\omega)/\xi_1, \qquad \chi_3^1(\omega) = 360F_3^{1*}(\omega)/\xi_1^3 \qquad \text{and}$$

$$\chi_5^1(\omega) = 15{,}120F_5^{1*}(\omega)/\xi_1^5$$

are presented in Figures 32–37. [The normalization was chosen in order to satisfy the condition $|\chi_n^1(0)| = 1$ at $\xi_1 \to 0$.] The spectra of $\mathrm{Re}\{\chi_1^1(\omega)\}$ (dispersion) and $\mathrm{Im}\{\chi_1^1(\omega)\}$ (absorption) and the corresponding Cole–Cole diagram of the first harmonic component are shown in Figures 32 and 33. Here, it is clearly seen how the relaxation spectrum of $\chi_1^1(\omega)$ at $\xi_1 \ll 1$ (linear response) is transformed to the nonlinear response spectrum in high fields: with increasing of $\xi_1$ the absorption and dispersion curves are shifted to higher frequencies with decreasing of the amplitude due to the saturation. A saturation level seems to be reached at $\xi_1 \sim 5$, where all the Fourier components $F_k^1(\omega)$ become comparable in the order of magnitude (see Table I). Moreover, one can note that the half-width of the spectra $\mathrm{Im}\{\chi_1^1(\omega)\}$ enlarges (Fig. 32) as $\xi_1$ increases. Defining the phase angle $\theta_1$ between in-phase and out-of-phase components of $\chi_1^1(\omega)$ as

$$\theta_1 = \arctan\left(\mathrm{Im}\{\chi_1^1(\omega)\}/\mathrm{Re}\{\chi_1^1(\omega)\}\right)$$

one can remark (Fig. 33) that the asymptotic limit of $\theta_1(\omega \to \infty)$ is equal to $\pi/2$ regardless of the value of $\xi_1$. This is so because neither the polarizability nor the inertia of the molecules was taken into account here. The account of the induced moment and the inertia effect contributions to the response may lead to different behavior of $\theta_1$.

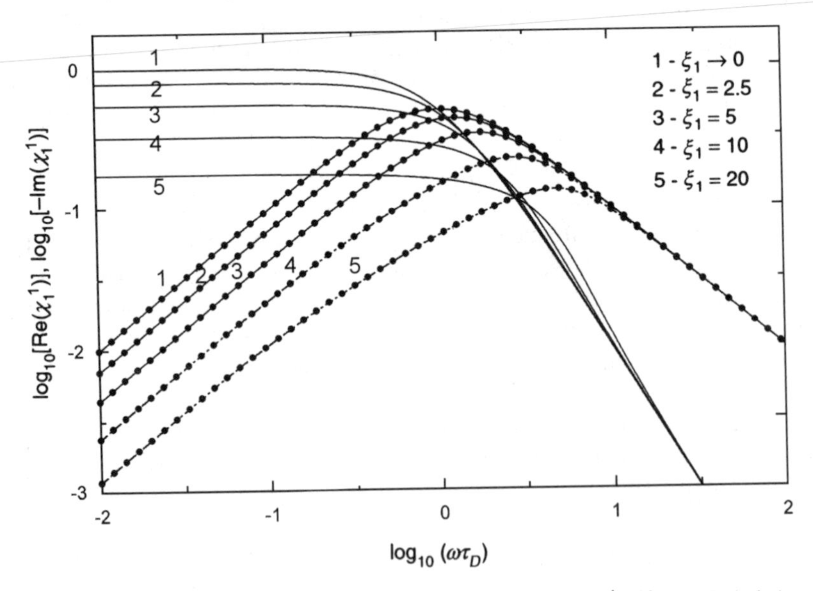


**Figure 32.** The $\log_{10}\mathrm{Re}\{\chi_1^1(\omega)\}$ (solid lines) and the $\log_{10}\mathrm{Im}\{\chi_1^1(\omega)\}$ (filled circles) as a function of $\log_{10}(\omega\tau_D)$ for various values of $\xi_1$. Curves 1 correspond to the linear response. Note that all the curves merge in a single asymptote in the high frequency region ($\omega \to \infty$).

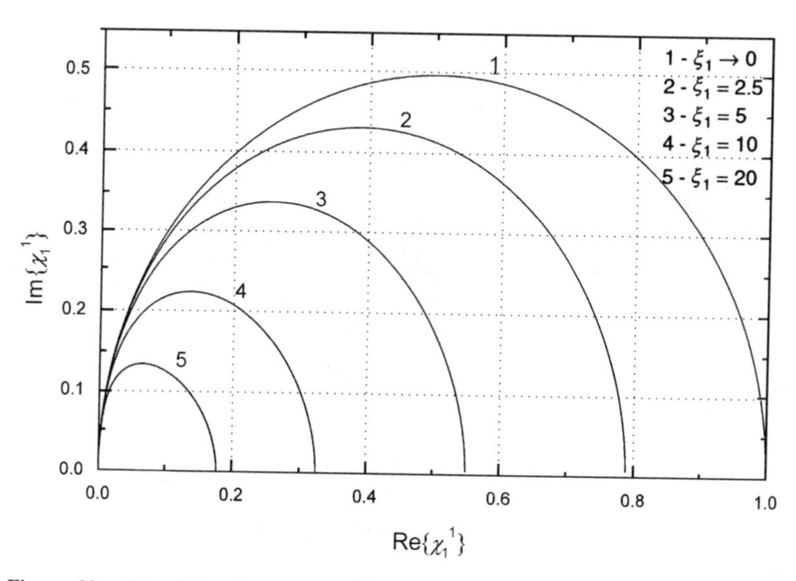


**Figure 33.** Cole–Cole diagram for $\chi_1^1(\omega)$ at various values of $\xi_1$. Curve 1 (semicircle) corresponds to the linear response.

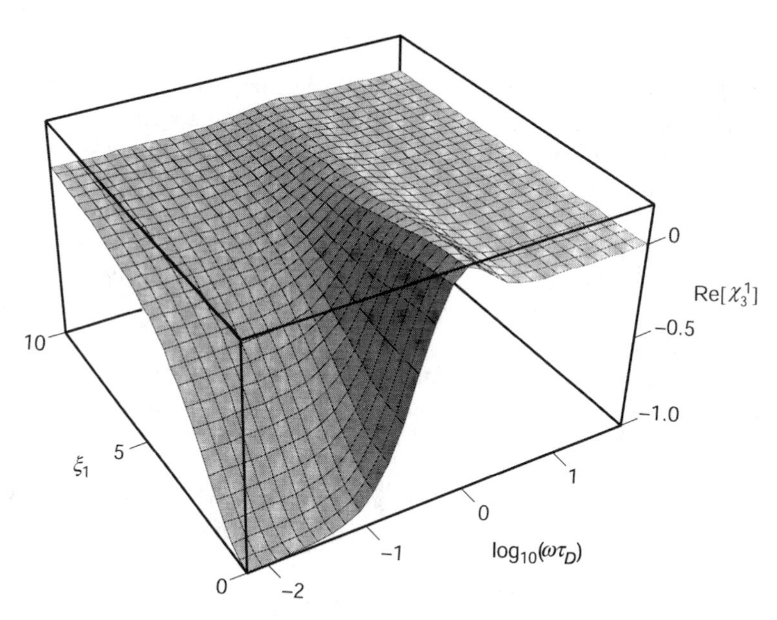


**Figure 34.** The $\mathrm{Re}\{\chi_3^1\}$ (third harmonic component) as a function of $\log_{10}(\omega\tau_D)$ and $\xi_1$.

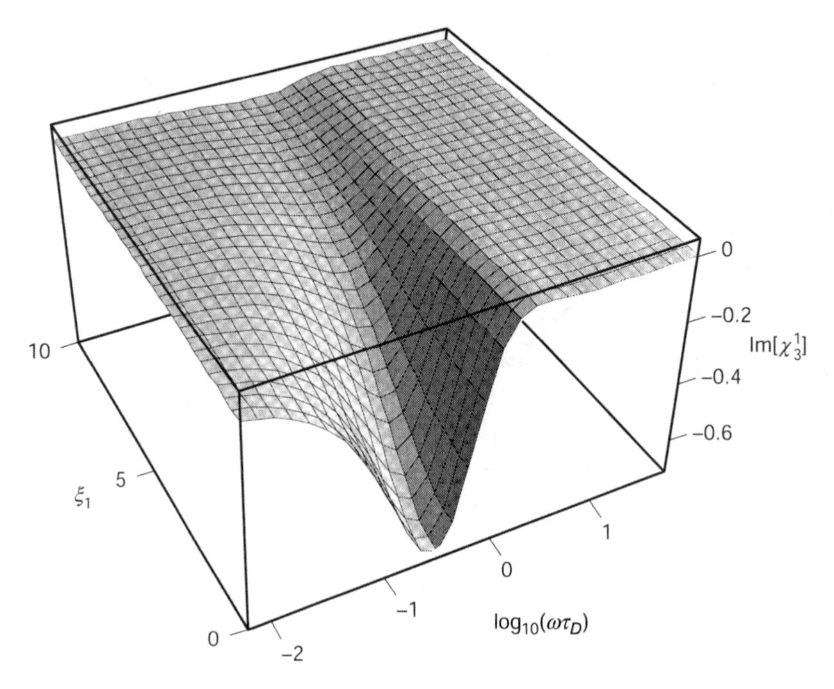


**Figure 35.** The same as in Figure 34 for $\mathrm{Im}\{\chi_3^1\}$.

The frequency behavior of the third harmonic component $\chi_3^1(\omega)$ is shown in Figures 34 and 35. For $\xi_1 \ll 1$, the real part of the $3\omega$-component (Fig. 34) starts from $-1$ at low frequencies, then reaches a positive maximum at $\omega\tau_D \approx 0.77$ before decreasing monotonically to 0 when $\omega$ tends to $\infty$. The spectrum becomes more and more flattened as $\xi_1$ increases. The imaginary part of the $3\omega$ component (Fig. 35) passes through a negative minimum at $\omega\tau_D \approx 0.26$ for $\xi_1 \ll 1$. This minimum is shifted to higher frequencies and its absolute value decreases with increasing $\xi_1$. An analogous behavior in the frequency domain (as that presented for the harmonic of rank 3) can be observed for $\chi_5^1(\omega)$ (fifth harmonic) with the exception that now both the real and imaginary parts are positive. As expected, the increasing of the ac field strength results in the saturation of all the Fourier components considered. All higher harmonics may be investigated in a similar way.

At small ac fields ($\xi_1 \leq 0.5$) and at $\xi = 0$, the results of our calculations are in full agreement with the perturbation solution for the first and third harmonics previously obtained by Coffey and Paranjape [45], namely,

$$\chi_1^1(\omega) = \frac{1 + i\omega\tau_D}{1 + \omega^2\tau_D^2} - \xi_1^2 \frac{27 - 13\omega^2\tau_D^2 + i\omega\tau_D(42 + 2\omega^2\tau_D^2)}{60(1 + \omega^2\tau_D^2)(9 + 4\omega^2\tau_D^2)} + O(\xi_1^4) \quad (6.16)$$

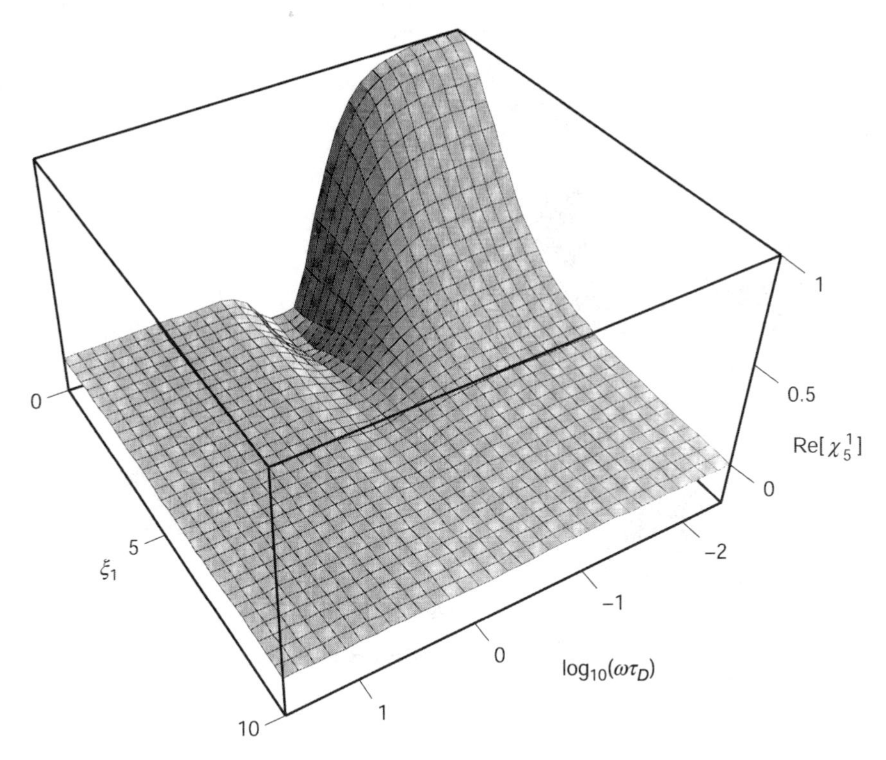


**Figure 36.** The Re$\{\chi_5^1\}$ (fifth harmonic component) as a function of $\log_{10}(\omega\tau_D)$ and $\xi_1$.

$$\chi_3^1(\omega) = -3\frac{3 - 17\omega^2\tau_D^2 + i\omega\tau_D(14 - 6\omega^2\tau_D^2)}{(1+\omega^2\tau_D^2)(9+4\omega^2\tau_D^2)(1+9\omega^2\tau_D^2)} + O(\xi_1^2) \tag{6.17}$$

One can readily see that the asymptotic estimates for $F_1^1(\omega)$ and $F_3^1(\omega)$ presented in Table I agree in all respects with Eqs. (6.16) and (6.17). For $\xi \neq 0$ similar, but more complicated equations [Eqs. (2.39), (2.40) and (2.43), (2.44)] have already been discussed in Section II.

In order to demonstrate how the dc bias field $\mathbf{E}_0$ affects the ac nonlinear response, we present here, as an example, the first harmonic component $\chi_1^1(\omega)$ as a function of the bias field parameter $\xi$. The main features of this dc bias field effect are shown in Figures 38 and 39, where the real and imaginary parts of $\chi_1^1(\omega)$ are plotted as functions of $\xi$ and $\xi_1$ for $\omega\tau_D = 1$, and in Figures 40 and 41, where the spectra of Re$\{\chi_1^1\}$ and Im$\{\chi_1^1\}$ are presented for $0 \leq \xi \leq 10$ and $\xi_1 = 5$. As one can see in these figures, the nonlinear effects arising from the increasing of the amplitude $\xi_1$ of the ac field coupled with the dc bias field $\mathbf{E}_0$

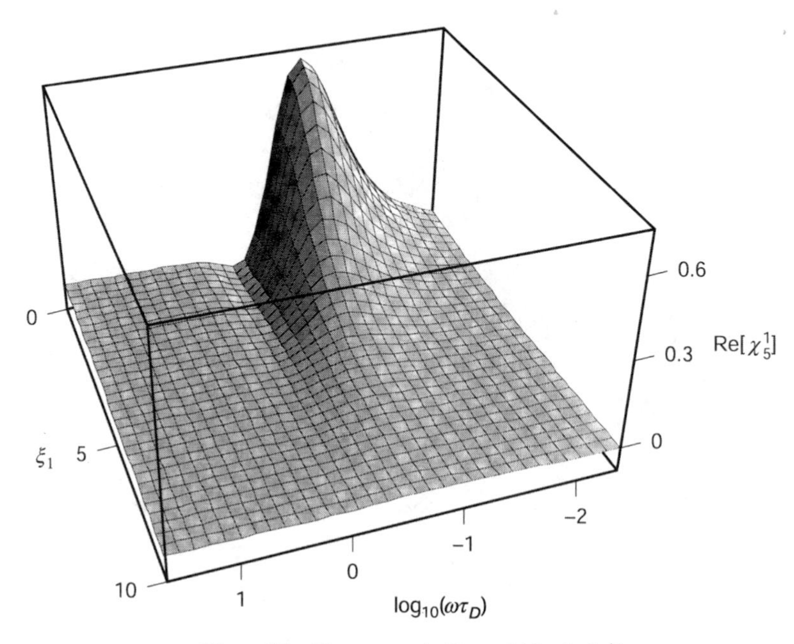


**Figure 37.** The same as in Figure 34 for $\mathrm{Im}\{\chi_5^1\}$.

are very similar to those when $\mathbf{E}_0$ is set equal to zero (cf. Figs. 32 and 33). However, the increase of the bias field parameter $\xi$ results in a further decrease of the response and in its shift to higher frequencies. The half-width of $\mathrm{Im}\{\chi_1^1\}$ enlarges with increasing $\xi$ as well. One can also see in Figures 38–41 that the decrease of $\mathrm{Re}\{\chi_1^1\}$ and $\mathrm{Im}\{\chi_1^1\}$ with increasing the amplitude of the dc bias field is several times more than that due to the ac field. For a small ac field ($\xi_1 \ll 1$) superimposed to a strong dc bias field ($\xi \gg 1$), our results are in complete agreement with those obtained in Section V, where the ac nonlinear response has been investigated by using the perturbation approach. The above conclusions in relation with these dc field effects can also be applied to higher harmonics.

### B. Nonlinear Dielectric Relaxation and Birefringence in Strong Superimposed ac and dc Bias Electric Fields: Polar and Polarizable Molecules

In Section VI.A, we have been able to calculate the steady-state ac nonlinear response of an ensemble of rigid polar molecules in strong superimposed ac and dc bias fields in terms of a matrix continued fraction [Eqs. (6.12) and (6.13)]. The calculation proceeded by expanding all the moments (relaxation functions)

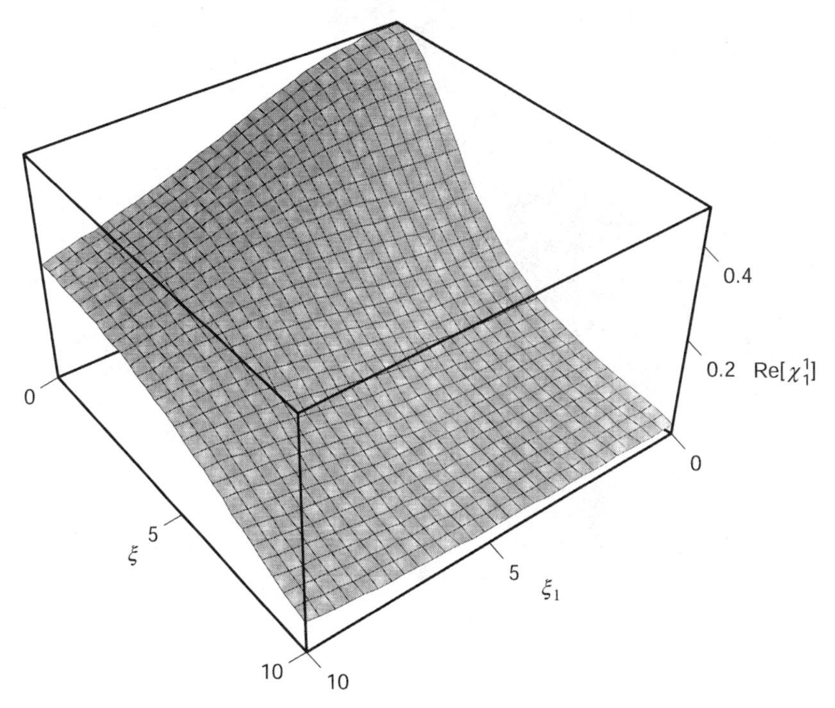


**Figure 38.** The $\mathrm{Re}\{\chi_1^1\}$ as a function of $\xi$ and $\xi_1$ for $\omega\tau_D = 1$.

of the infinite hierarchy of the coupled equations, which describe the ac nonlinear response of the system, as a Fourier series in the time domain. Hence, another infinite hierarchy of recurrence equations for the Fourier components of the relaxation functions was derived. The exact solution of this hierarchy in terms of a matrix continued fraction was obtained that allowed us to evaluate the stationary ac nonlinear dielectric response. The method of the solution that we have presented is quite general and can also be used for calculating stationary solutions of analogous nonlinear response problems, where time-dependent stimuli in high ac external fields are considered in the context of the Brownian motion of a particle in an external potential. For example, the approach presented can also be applied to the calculation of the nonlinear impedance of a microwave-driven Josephson junction [82, 83] (as known, the dynamics of these systems is governed by very similar recurrence equations). Here, we shall elaborate the approach to the calculation of the dynamic Kerr effect responses of *polar* and *anisotropically polarizable* molecules in a high ac electric field superimposed to a strong dc bias field. Although the system under consideration and

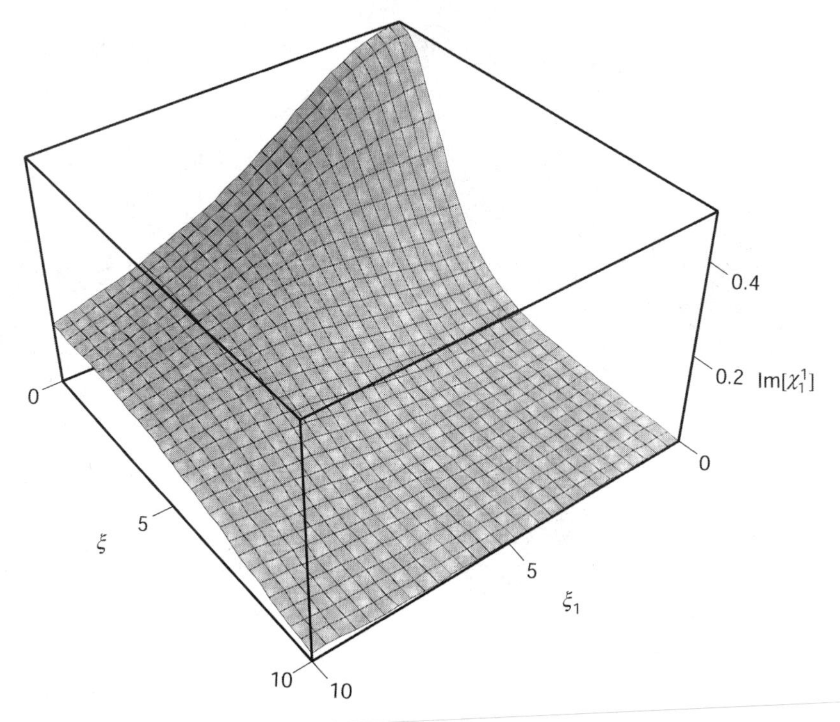


**Figure 39.** The same as in Figure 38 for $\text{Im}\{\chi_1^1\}$.

the quantities of interest differ from those of Section VI.A, the calculation can be accomplished in a similar manner so that the ac stationary nonlinear Kerr effect and ac dielectric responses can be evaluated for arbitrary ac and dc bias field amplitudes.

Let us consider an assembly of *polar and anisotropically polarizable* symmetric top molecules acted on by strong superimposed external dc $\mathbf{E}_0$ and ac $\mathbf{E}_1(t) = \mathbf{E}_1 \cos \omega t$ electric fields. We shall suppose for the sake of simplicity that both $\mathbf{E}_0$ and $\mathbf{E}_1$ are in the same direction along the $Z$ axis of the laboratory coordinate system. Then, the noninertial rotational Brownian motion of the molecules dissolved in a nonpolar solvent (noninteracting molecules) may be described by the Smoluchowski equation (2.4) for the probability distribution function $W(\vartheta, t)$ of the molecular orientations where the orientational potential energy is given by Eq. (5.1). We shall also assume that the nonlinear effects due to the hyperpolarizability may be neglected. By assuming that the magnitudes of both ac and dc fields are large enough so that the energy of the molecule in these fields can be comparable or higher than the thermal energy $kT$, one is faced with an intrinsically nonlinear problem we shall solve as follows. On expanding

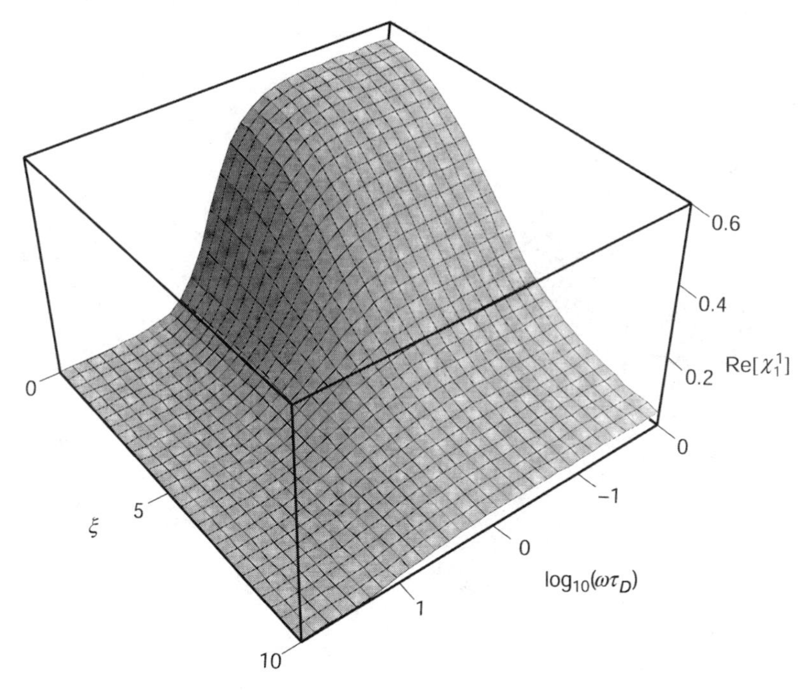


**Figure 40.** The $\mathrm{Re}\{\chi_1^1\}$ as a function of $\log_{10}(\omega\tau_D)$ and $\xi$ for $\xi_1 = 5$ (strong nonlinear regime).

$W(\vartheta, t)$ as a series of Legendre polynomials $P_n(\cos\vartheta)$, we have from Eq. (2.4) the following set of differential-recurrence equations for the relaxation functions $f_n(t)$ [cf. Eq. (2.6)]:

$$\tau_D \frac{d}{dt} f_n(t) + \frac{n(n+1)}{2} f_n(t) = \zeta(t)[a_n f_{n-1}(t) + b_n f_{n+1}(t)] + \sigma(t)[c_n f_{n-2}(t) + d_n f_n(t) + g_n f_{n+2}(t)] \tag{6.18}$$

where

$$a_n = -b_n = \frac{n(n+1)}{2(2n+1)} \qquad d_n = \frac{n(n+1)}{(2n-1)(2n+3)} \tag{6.19}$$

$$c_n = \frac{n(n+1)(n-1)}{(2n-1)(2n+1)} \qquad g_n = -\frac{n(n+1)(n+2)}{(2n+1)(2n+3)} \tag{6.20}$$

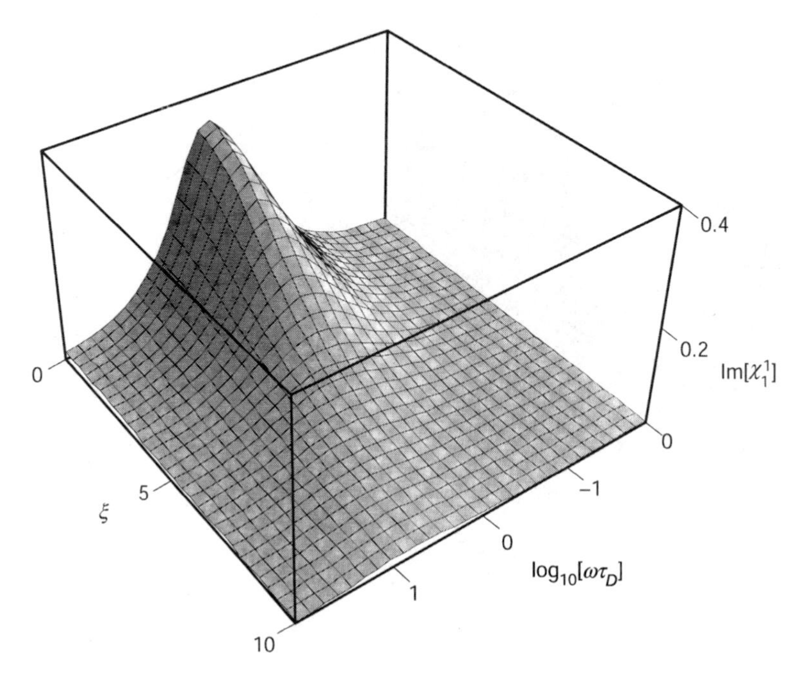


**Figure 41.** The same as in Figure 40 for $\mathrm{Im}\{\chi_1^1\}$.

$$\zeta(t) = \xi + \xi_1 \cos \omega t \tag{6.21}$$

$$\sigma(t) = \sigma_0 + \kappa\xi\xi_1 \cos \omega t + \tfrac{1}{4}\kappa\xi_1^2 \cos 2\omega t \tag{6.22}$$

$$\sigma_0 = \kappa(\xi_1^2 + 2\xi^2)/4 \tag{6.23}$$

and the parameters $\xi_1$, $\xi$, and $\kappa$ are defined by Eqs. (2.37). Our goal is to evaluate the electric polarization $P(t)$ and the birefringence function $K(t)$. As in Section VI.A, in order to evaluate the ac response corresponding to the stationary state, which is independent of the initial conditions, we may seek all the $f_n(t)$ in the form of a Fourier series like Eq. (6.4), namely,

$$f_n(t) = \sum_{k=-\infty}^{\infty} F_k^n(\omega)e^{ik\omega t} = F_0^n(\omega) + 2\sum_{k=1}^{\infty} \mathrm{Re}\{F_k^n(\omega)e^{ik\omega t}\} \tag{6.24}$$

On substituting Eq. (6.24) into Eq. (6.18), we can obtain the following multiterm recurrence equations for the Fourier amplitudes $F_k^n(\omega)$

$$
\begin{aligned}
&[2\sigma_0 d_n - 2i\omega\tau_D k - n(n+1)]F_k^n(\omega) + 2\xi a_n[F_k^{n-1}(\omega) - F_k^{n+1}(\omega)] \\
&\quad + \xi_1 a_n[F_{k-1}^{n-1}(\omega) + F_{k+1}^{n-1}(\omega) - F_{k-1}^{n+1}(\omega) - F_{k+1}^{n+1}(\omega)] \\
&\quad + 2\sigma_0[c_n F_k^{n-2}(\omega) + g_n F_k^{n+2}(\omega)] + \kappa\xi_1\xi\{c_n[F_{k-1}^{n-2}(\omega) + F_{k+1}^{n-2}(\omega)] \\
&\quad + d_n[F_{k-1}^{n}(\omega) + F_{k+1}^{n}(\omega)] + g_n[F_{k-1}^{n+2}(\omega) + F_{k+1}^{n+2}(\omega)]\} \\
&\quad + \frac{\kappa\xi_1^2}{4}\{c_n[F_{k-2}^{n-2}(\omega) + F_{k+2}^{n-2}(\omega)] + d_n[F_{k-2}^{n}(\omega) + F_{k+2}^{n}(\omega)] \\
&\quad + g_n[F_{k-2}^{n+2}(\omega) + F_{k+2}^{n+2}(\omega)]\} = 0
\end{aligned}
\tag{6.25}
$$

The *exact* solution of Eq. (6.25) can be expressed in terms of a matrix continued fraction as follows. Let us introduce the column vectors $\mathbf{C}_n$ and $\mathbf{R}$ given by

$$
\mathbf{C}_n(\omega) = \begin{pmatrix} \mathbf{c}_{2n}(\omega) \\ \mathbf{c}_{2n-1}(\omega) \end{pmatrix}
\tag{6.26}
$$

and

$$
\mathbf{R} = \begin{pmatrix} \mathbf{r}_2 \\ \mathbf{r}_1 \end{pmatrix}
\tag{6.27}
$$

where the subvectors $\mathbf{c}_k$ and $\mathbf{r}_k$ of infinite dimension are given by

$$
\mathbf{c}_n(\omega) = \begin{pmatrix} \vdots \\ F_{-2}^n(\omega) \\ F_{-1}^n(\omega) \\ F_0^n(\omega) \\ F_1^n(\omega) \\ F_2^n(\omega) \\ \vdots \end{pmatrix} \qquad \mathbf{r}_1 = \begin{pmatrix} \mathbf{0} \\ 0 \\ \xi_1/3 \\ 2\xi/3 \\ \xi_1/3 \\ 0 \\ \mathbf{0} \end{pmatrix} \qquad \text{and} \qquad \mathbf{r}_2 = \begin{pmatrix} \mathbf{0} \\ \kappa\xi_1^2/10 \\ 2\kappa\xi\xi_1/5 \\ 4\sigma_0/5 \\ 2\kappa\xi\xi_1/5 \\ \kappa\xi_1^2/10 \\ \mathbf{0} \end{pmatrix}
\tag{6.28}
$$

Here the symbol $\mathbf{0}$ designates a zero column vector of infinite dimension. Thus, the *21-term scalar* recurrence Eq. (6.25) can be transformed into the *3-term matrix* recurrence equation as follows:

$$
\mathbf{Q}_1\mathbf{C}_1(\omega) + \mathbf{Q}_1^+\mathbf{C}_2(\omega) = -\mathbf{R}
\tag{6.29}
$$

$$\mathbf{Q}_n^- \mathbf{C}_{n-1}(\omega) + \mathbf{Q}_n \mathbf{C}_n(\omega) + \mathbf{Q}_n^+ \mathbf{C}_{n+1}(\omega) = \mathbf{0} \qquad n = 2, 3, \ldots \tag{6.30}$$

where the matrices $\mathbf{Q}_n^-$, $\mathbf{Q}_n$, and $\mathbf{Q}_n^+$ are defined as

$$\mathbf{Q}_n^- = \begin{pmatrix} c_{2n}\mathbf{X} & \mathbf{O} \\ a_{2n-1}\mathbf{Y} & c_{2n-1}\mathbf{X} \end{pmatrix}$$

$$\mathbf{Q}_n = \begin{pmatrix} \mathbf{Z}_{2n} & a_{2n}\mathbf{Y} \\ -a_{2n-1}\mathbf{Y} & \mathbf{Z}_{2n-1} \end{pmatrix}$$

$$\mathbf{Q}_n^+ = \begin{pmatrix} g_{2n}\mathbf{X} & -a_{2n}\mathbf{Y} \\ \mathbf{O} & g_{2n-1}\mathbf{X} \end{pmatrix}$$

Here $a_n$, $c_n$, and $g_n$ are given by Eqs. (6.19) and (6.20), $\mathbf{O}$ is the zero matrix of infinite dimension, and the five-diagonal submatrices $\mathbf{X}$ and $\mathbf{Z}_n$ are defined as

$$\mathbf{X} = \kappa \begin{pmatrix} \ddots & \vdots & \vdots & \vdots & \vdots & \vdots & \iddots \\ \cdots & \xi^2 + \xi_1^2/2 & \xi_1\xi & \xi_1^2/4 & 0 & 0 & \cdots \\ \cdots & \xi_1\xi & \xi^2 + \xi_1^2/2 & \xi_1\xi & \xi_1^2/4 & 0 & \cdots \\ \cdots & \xi_1^2/4 & \xi_1\xi & \xi^2 + \xi_1^2/2 & \xi_1\xi & \xi_1^2/4 & \cdots \\ \cdots & 0 & \xi_1^2/4 & \xi_1\xi & \xi^2 + \xi_1^2/2 & \xi_1\xi & \cdots \\ \cdots & 0 & 0 & \xi_1^2/4 & \xi_1\xi & \xi^2 + \xi_1^2/2 & \cdots \\ \iddots & \vdots & \vdots & \vdots & \vdots & \vdots & \ddots \end{pmatrix}$$

$$\mathbf{Z}_n = \kappa d_n \begin{pmatrix} \ddots & \vdots & \vdots & \vdots & \vdots & \vdots & \iddots \\ \cdots & z_{n,-2} & \xi_1\xi & \xi_1^2/4 & 0 & 0 & \cdots \\ \cdots & \xi_1\xi & z_{n,-1} & \xi_1\xi & \xi_1^2/4 & 0 & \cdots \\ \cdots & \xi_1^2/4 & \xi_1\xi & z_{n,0} & \xi_1\xi & \xi_1^2/4 & \cdots \\ \cdots & 0 & \xi_1^2/4 & \xi_1\xi & z_{n,1} & \xi_1\xi & \cdots \\ \cdots & 0 & 0 & \xi_1^2/4 & \xi_1\xi & z_{n,2} & \cdots \\ \iddots & \vdots & \vdots & \vdots & \vdots & \vdots & \ddots \end{pmatrix}$$

and the tridiagonal submatrix $\mathbf{Y}$ is given by Eq. (6.10). Here,

$$z_{n,m} = \xi^2 + \frac{\xi_1^2}{2} - \frac{2i\omega\tau_D m + n(n+1)}{\kappa d_n} \qquad (m = -\infty, \ldots, -1, 0, 1, \ldots, \infty).$$

The system of Eqs. (6.29) and (6.30) can readily be solved for $\mathbf{C}_1(\omega)$ in terms of matrix continued fractions. One has

$$\mathbf{C}_1(\omega) = \cfrac{\mathbf{I}}{-\mathbf{Q}_1 - \mathbf{Q}_1^+ \cfrac{\mathbf{I}}{-\mathbf{Q}_2 - \mathbf{Q}_2^+ \cfrac{\mathbf{I}}{-\mathbf{Q}_3 - \cdots}\mathbf{Q}_3^-}\mathbf{Q}_2^-}\mathbf{R} \tag{6.31}$$

As is obvious from its definition [Eqs. (6.26) and (6.28)], the vector $\mathbf{C}_1(\omega)$ contains all the Fourier amplitudes of $f_1(t)$ and $f_2(t)$, which are necessary to evaluate the nonlinear dielectric and Kerr effect relaxation responses. The matrix continued fraction solution [Eq. (6.31)] we have obtained is very convenient for the purpose of computation. The calculations have shown that the results are stable by choosing the dimension of the subvectors and submatrices involved equal to 20–30. Moreover, the matrix continued fraction involved converges very rapidly, thus 15–20 downward iterations in evaluating the continued fraction in Eq. (6.31) are enough to arrive at not less than six significant digits in the majority of cases.

Having determined $F_k^n(\omega)$, one can calculate the stationary ac response functions $f_1(t)$ and $f_2(t)$. Below, we shall confine ourselves to the evaluation of the ac response for $\xi = 0$ only, because the response to a weak ac field in the presence of a strong dc bias field has already been investigated in detail in Section V and the effect of increasing the strength of the ac field coupled with the dc bias field $\mathbf{E}_0$ is similar to that when $\mathbf{E}_0$ is set equal to zero. In the absence of the dc bias field, that is, for $\xi = 0$, the series (6.24) contains only the odd components for $f_1(t)$ and the even components for $f_2(t)$, namely,

$$f_1(t) = 2\sum_{k=1}^{\infty} \mathrm{Re}\{F_{2k-1}^1(\omega)e^{i(2k-1)\omega t}\} \tag{6.32}$$

$$f_2(t) = F_0^2(\omega) + 2\sum_{k=1}^{\infty} \mathrm{Re}\{F_{2k}^2(\omega)e^{i2k\omega t}\} \tag{6.33}$$

The steady-state component $F_0^2(\omega)$ in the right-hand side of Eq. (6.33) is a time-independent, but frequency-dependent term mainly due to the permanent dipole moment contribution to the Kerr effect relaxation process.

In order to illustrate the nonlinear effects induced by the ac field, we have plotted the spectra of the real and imaginary parts of the normalized Fourier components of the polarization and the birefringence functions. We have chosen to consider, as examples, the normalized first and third harmonics of the electric polarization defined by

$$\chi_1^1(\omega) = \frac{6F_1^{1*}(\omega)}{\xi_1} \qquad \text{and} \qquad \chi_3^1(\omega) = \frac{360F_3^{1*}(\omega)}{\xi_1^3} \tag{6.34}$$

and the steady-state component and the second harmonic of the Kerr effect given by

$$\chi_0^2(\omega) = \frac{30F_0^2(\omega)}{\xi_1^2} \qquad \text{and} \qquad \chi_2^2(\omega) = \frac{60F_2^{2*}(\omega)}{\xi_1^2} \tag{6.35}$$

The normalization was chosen in order to satisfy the condition $|\chi_n^k(0)| = 1$ at $\xi_1 \to 0$ for $\kappa = 0$ and $\xi = 0$. Three-dimensional plots of the spectra $\chi_k^n(\omega)$ vs. $\log_{10}(\omega\tau_D)$ and $\kappa$ are shown in Figures 42–48 for the ac field parameter $\xi_1 = 2.5$ corresponding to a strong nonlinear regime. As one can see in these figures,

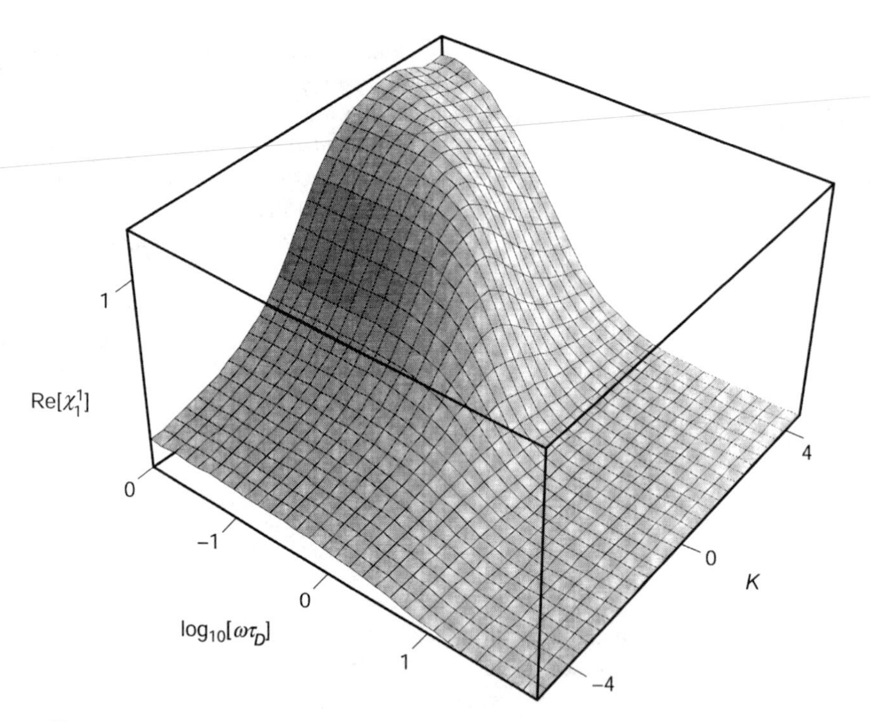


**Figure 42.** The real part of $\chi_1^1(\omega)$ (first harmonic component of the electric polarization) as a function of $\log_{10}(\omega\tau_D)$ and $\kappa$ for $\xi_1 = 2.5$ and $\xi = 0$.

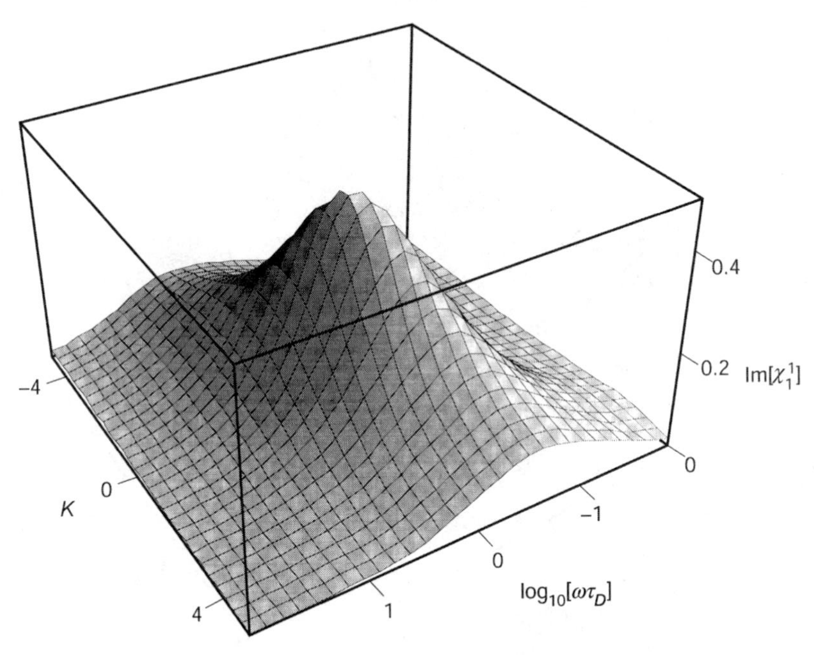


**Figure 43.** The imaginary part of $\chi_1^1(\omega)$ as a function of $\log_{10}(\omega\tau_D)$ and $\kappa$ for $\xi_1 = 2.5$ and $\xi = 0$.

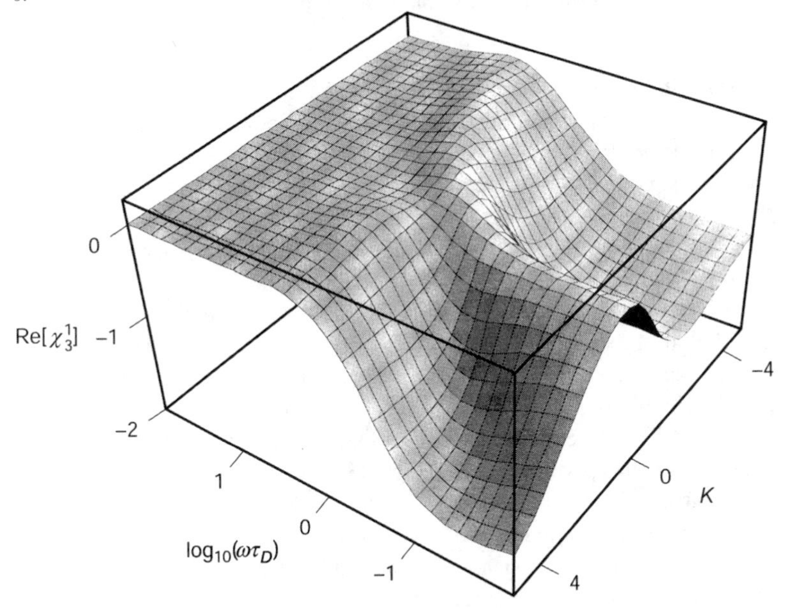


**Figure 44.** The same as in Figure 42 for $\mathrm{Re}\{\chi_3^1(\omega)\}$.

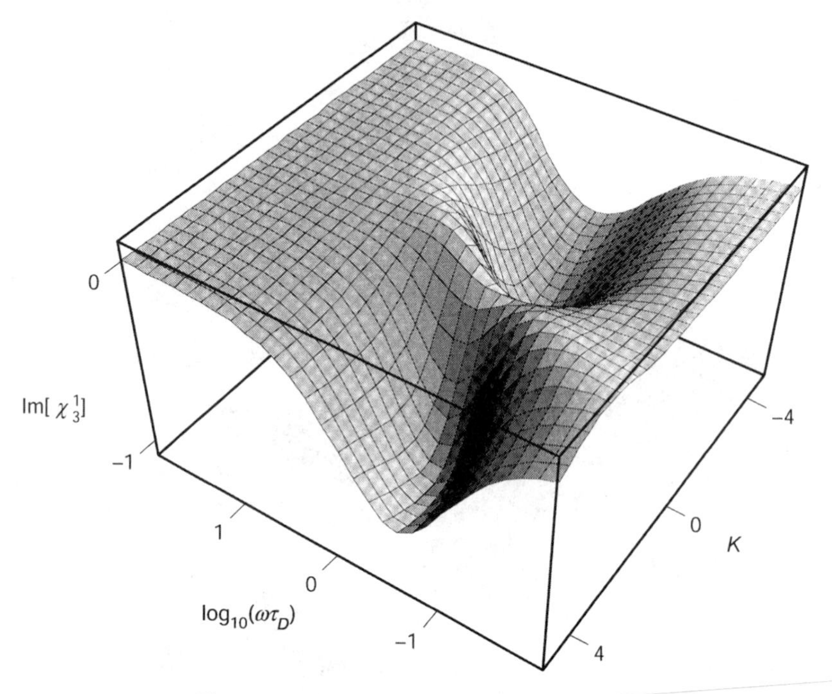


**Figure 45.** The same as in Figure 43 for $\mathrm{Im}\{\chi_3^1(\omega)\}$.

the ratio of the permanent and induced dipole components given by the parameter κ crucially affects the ac response both for nonlinear dielectric and Kerr effect relaxation. Both the shape of the spectra and the sign of the components $\chi_k^n(\omega)$ may vary with κ. Figures 42 and 43 show the first harmonic component of the electric polarization. As long as the parameter $|\kappa| < 0.5$, the spectra have a typical relaxation (Debye-type) behavior, namely, the shape of the curves $\mathrm{Re}\{\chi_1^1(\omega)\}$ and $\mathrm{Im}\{\chi_1^1(\omega)\}$ being that of Lorentzian profiles. Here, the contribution of the permanent dipole moment to the response is dominant or comparable with that of the induced moment. For $|\kappa| > 0.5$, the plots are more and more distorted indicating strong nonlinear effects due to the prominence of the induced dipole moments on the field-off (permanent) ones. The evolution of $\mathrm{Re}\{\chi_1^1(\omega)\}$ (Fig. 45) consists in a continuous decrease to zero with increasing ω. The imaginary part of $\chi_1^1(\omega)$ (Fig. 43) has a maximum whose position depends on κ. With increasing $|\kappa|$, the half-width of $\mathrm{Im}\{\chi_1^1(\omega)\}$ enlarges and the spectrum shifts to higher or to lower frequencies depending on the sign of κ. In Figures 44 and 45, the spectra of the real and imaginary parts of the third harmonic component of the electric polarization are presented. As one can see in these figures, the behavior of $\mathrm{Re}\{\chi_3^1(\omega)\}$ and $\mathrm{Im}\{\chi_3^1(\omega)\}$ is very complicated

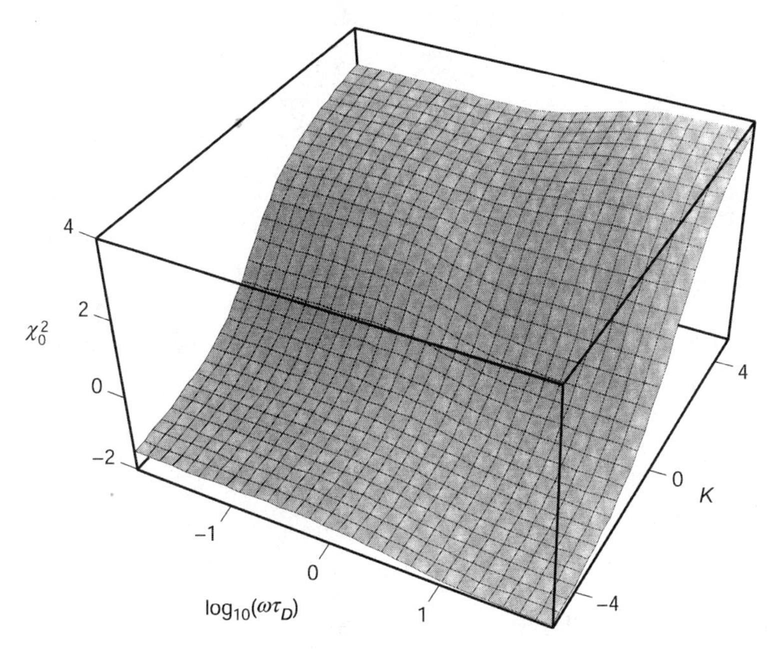


**Figure 46.** Steady-state component $\chi_0^2(\omega)$ (steady-state component of the Kerr effect) as a function of $\log_{10}(\omega\tau_D)$ and $\kappa$ for $\xi_1 = 2.5$ and $\xi = 0$.

and denotes strong interaction between the induced and permanent dipole moments; the values taken by $\mathrm{Re}\{\chi_3^1(\omega)\}$ and $\mathrm{Im}\{\chi_3^1(\omega)\}$ are for the most part negative. The steady-state component $\chi_0^2(\omega)$ of the electric birefringence is shown in Figure 46. As already mentioned, this component is not time dependent but is only frequency dependent. For $-\infty < \kappa < 2$, the curves are decreasing with increasing $\omega$, while for $\kappa > 2$, $\chi_0^2(\omega)$ is increasing when $\omega$ increases. Figures 47 and 48 present the spectra of the real and imaginary parts of the second harmonic component of the birefringence function. For $-1 < \kappa < \infty$, the amplitude of $\mathrm{Re}\{\chi_2^2(\omega)\}$ increases as $\kappa$ increases and $\mathrm{Re}\{\chi_2^2(\omega)\}$ is monotonically decaying to zero at $\omega \to \infty$, $\mathrm{Im}\{\chi_2^2(\omega)\}$ being positive and having a maximum at $\omega\tau_D \sim 1$. For $\kappa < -1$, $\mathrm{Re}\{\chi_2^2(\omega)\}$ starts from negative values before reaching zero when $\omega \to \infty$; $\mathrm{Im}\{\chi_2^2(\omega)\}$ has now a negative minimum at $\omega\tau_D \sim 3$. All higher harmonics may be investigated in a similar way.

The effect of increasing the ac field amplitude $\xi_1$ is demonstrated in Figures 49–55, where $\mathrm{Re}\{\chi_k^n(\omega)\}$ and $\mathrm{Im}\{\chi_k^n(\omega)\}$ [Eqs. (6.34) and (6.35)] vs. $\xi_1$ are presented. Here, the results of the calculation are also compared with analytical solutions for $\chi_k^n(\omega)$ derived previously in the limit of a weak ac field, $\xi_1 \to 0$, with the help of the perturbation theory [these analytical solutions are given by

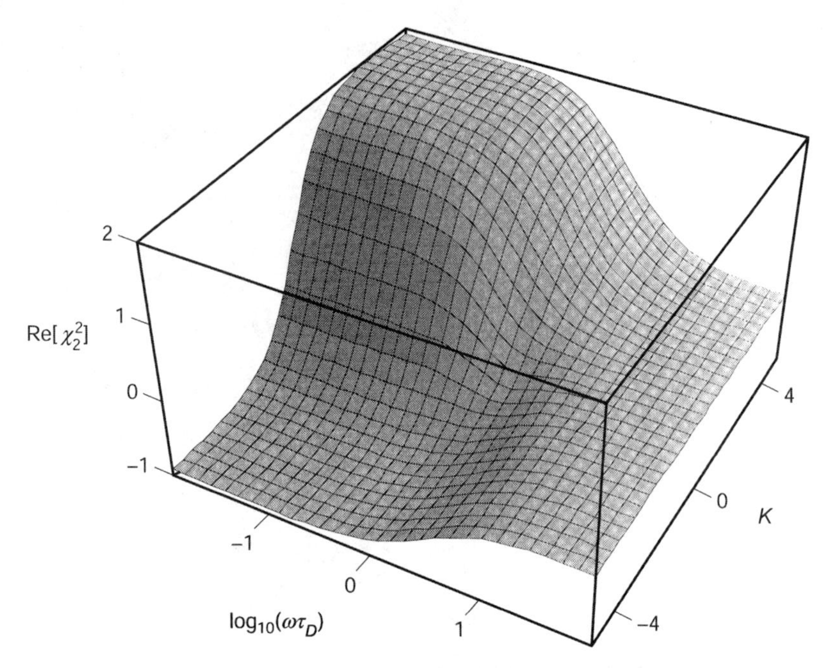


**Figure 47.** The same as in Figure 42 for $\text{Re}\{\chi_2^2(\omega)\}$ (second harmonic component of the Kerr effect).

Eqs. (2.39), (2.40), (2.43), (2.44), (2.47), and (2.49)]. They are shown in Figures 49–55 by stars. As one can see in these figures, the results of the exact calculations of $\chi_1^1(\omega)$, $\chi_0^2(\omega)$, and $\chi_2^2(\omega)$ are in agreement with the perturbation solution even for $\xi_1 \sim 1$. However, for higher harmonics, for example, for $\chi_3^1(\omega)$ shown in Figures 51 and 52, the difference between the perturbation and exact solutions becomes more pronounced and both solutions agree only for $\xi_1 \ll 1$. Concerning the nonlinear dielectric responses $\chi_3^1(\omega)$, one observes an inversion of sign at high values of $\xi_1$ before reaching the saturation. This inversion of sign is a nonlinear effect due to the influence of the induced dipole moments (the effect is absent for $\kappa = 0$). Given a value of $\kappa$, the maximum of $\text{Im}\{\chi_1^1(\omega)\}$ shifts to lower frequencies as $\xi_1$ increases. As for $\chi_0^2(\omega)$ and $\chi_2^2(\omega)$, one can also note a saturation effect and a shift of these spectra to higher frequencies with increasing $\xi_1$. The increase of the ac field strength results in the saturation of other Fourier components as well.

### C. Nonstationary ac Response

In Sections VI.A and B, we have applied the matrix continued fraction method for the evaluation of the ac *stationary* response in high ac electric fields.

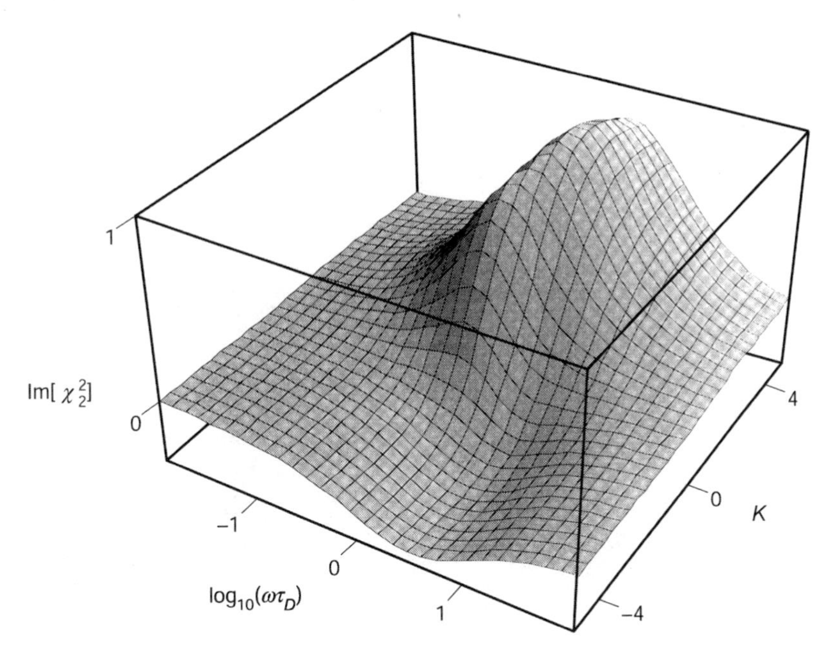


**Figure 48.** The same as in Figure 43 for $\mathrm{Im}\{\chi_2^2(\omega)\}$.

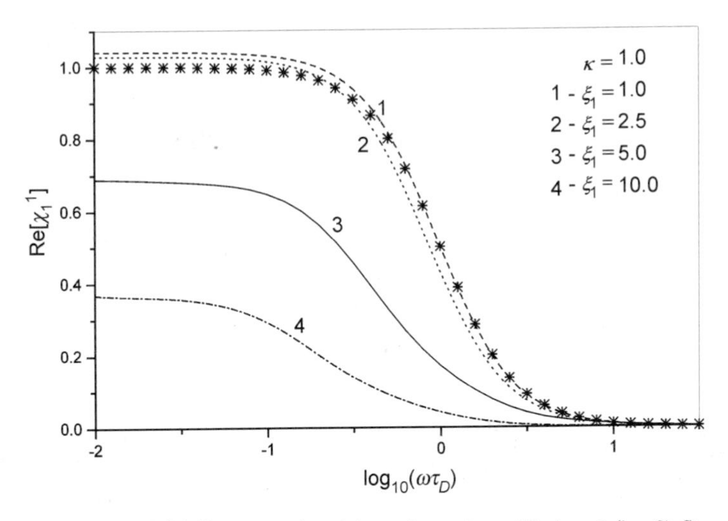


**Figure 49.** The $\mathrm{Re}\{\chi_1^1(\omega)\}$ vs. $\log_{10}(\omega\tau_D)$ for various values of $\xi_1$ ($\kappa = 1, \xi = 0$). Stars are the calculation in the limit $\xi \to 0$ (linear response).

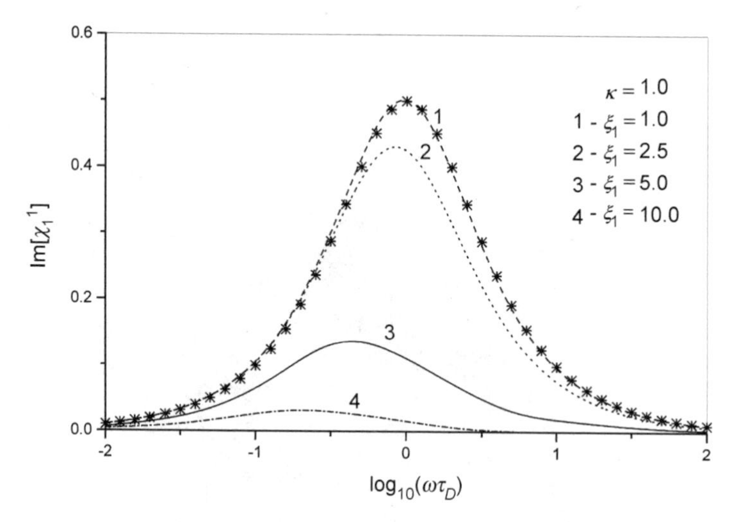


**Figure 50.** The same as in Figure 49 for $\mathrm{Im}\{\chi_1^1(\omega)\}$.

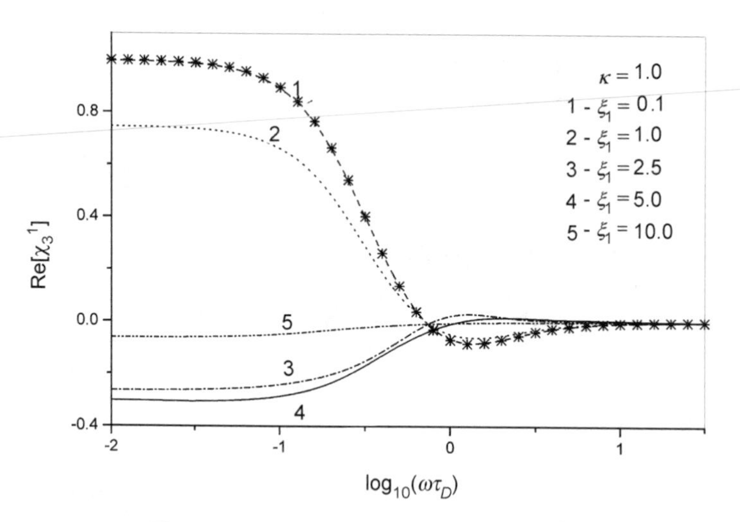


**Figure 51.** The same as in Figure 49 for $\mathrm{Re}\{\chi_3^1(\omega)\}$.

However, the method can also be generalized for the treatment of nonstationary response in ac fields [9, 81]. Here, we briefly discuss this problem on the example of the nonstationary response of a system of rigid polar molecules in an ac electric field $E(t) = E_1 \sin(\omega t)$. In this case, the Smoluchowski equation (2.4) for the distribution function $W$ of the orientations of the molecules may

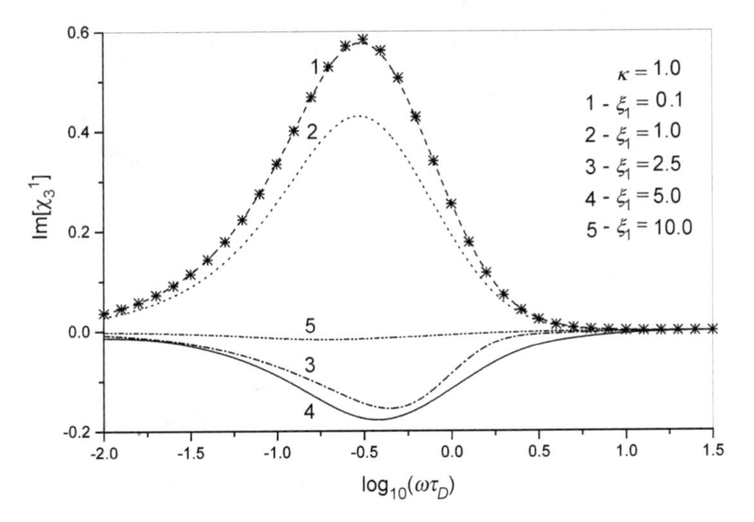


**Figure 52.** The same as in Figure 49 for $\mathrm{Im}\{\chi_3^1(\omega)\}$.

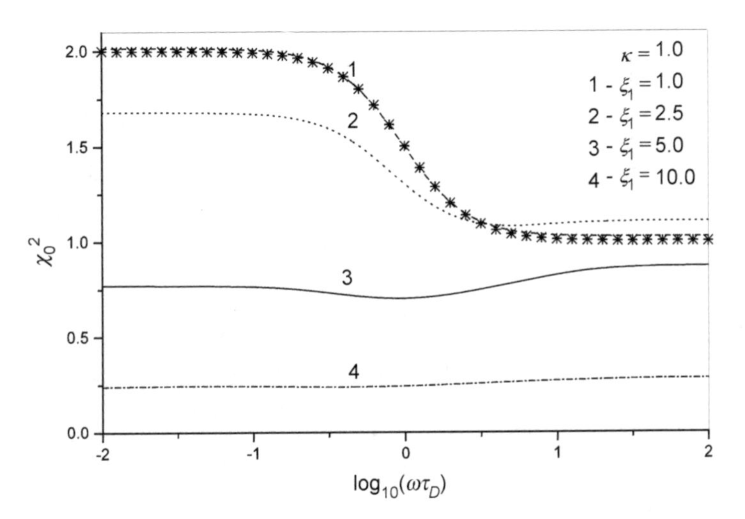


**Figure 53.** The same as in Figure 49 for $\chi_0^2(\omega)$.

be written as

$$
\begin{aligned}
\frac{\partial}{\partial t} W(x,t) &= L_{\mathrm{FP}}(x,t) W(x,t) \\
&= \left[L_{\mathrm{FP}}^0(x) + \sin(\omega t) L_\xi(x)\right] W(x,t)
\end{aligned}
\tag{6.36}
$$

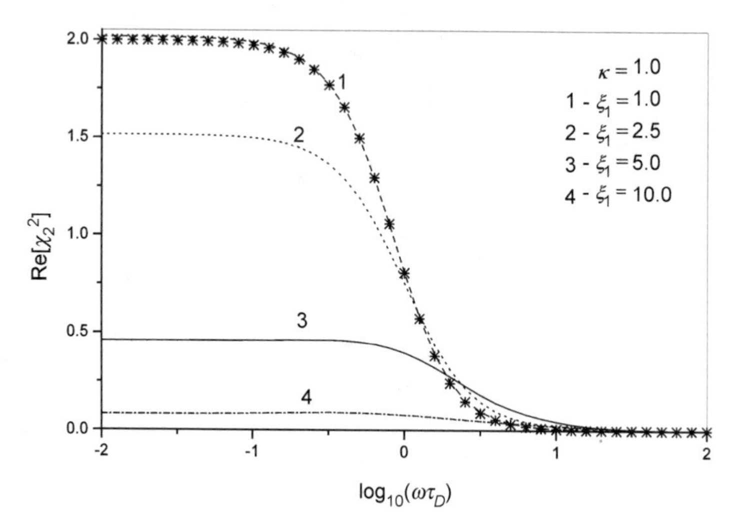


**Figure 54.** The same as in Figure 49 for $\mathrm{Re}\{\chi_2^2(\omega)\}$.

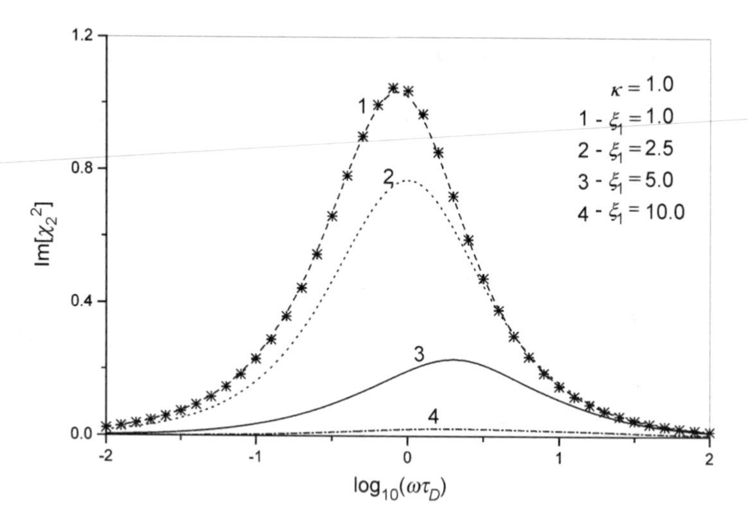


**Figure 55.** The same as in Figure 49 for $\mathrm{Im}\{\chi_2^2(\omega)\}$.

where $x = \cos\vartheta$,

$$L_{\mathrm{FP}}^{0}(x) = D\frac{\partial}{\partial x}\left[(1-x^2)\frac{\partial}{\partial x}\right] \tag{6.37}$$

is the unperturbed Fokker–Planck operator and

$$L_{\xi}(x) = -D\xi_1 \left[(1 - x^2)\frac{\partial}{\partial x} - 2x\right] \tag{6.38}$$

is the operator due to the interaction of the ac electric field with the rigid dipolar molecules. As the Fokker–Planck operator $L_{FP}(x, t)$ is periodic in time with the period $T = 2\pi/\omega$, Eq. (6.36) may be solved by applying Floquet's theorem ([39], p. 727), that is, by seeking the solution in the form of a superposition of Floquet's modes $w_{\lambda_M}(x, t)$ [9, 81], namely,

$$W(x, t) = \sum_{M=1}^{\infty} e^{-\lambda_M t} w_{\lambda_M}(x, t) \tag{6.39}$$

where $w_{\lambda_M}(x, t)$ and $\lambda_M$ are the eigenfunctions and eigenvalues determined from the following equation

$$\left[\frac{\partial}{\partial t} - L_{FP}(x, t)\right] w_{\lambda_M}(x, t) = \lambda_M w_{\lambda_M}(x, t) \tag{6.40}$$

On taking into account that all the $w_{\lambda_M}(x, t)$ are periodic functions in time with the period $T = 2\pi/\omega$, that is,

$$w_{\lambda_M}(x, t) = w_{\lambda_M}(x, t + T) \tag{6.41}$$

one has [81]

$$\begin{aligned} e^{-\lambda_M t} w_{\lambda_M}(x, t) &= e^{-(\lambda_M + im\omega)t} w_{\lambda_M}(x, t) e^{im\omega t} \\ &= e^{-\Lambda_M t} W_{\lambda_M}(x, t) \end{aligned} \tag{6.42}$$

where

$$\Lambda_M = \lambda_M + im\omega \qquad m = 0, \pm 1, \pm 2, \ldots \tag{6.43}$$

and

$$W_{\lambda_M}(x, t) = w_{\lambda_M}(x, t) e^{im\omega t} = W_{\lambda_M}(x, t + T)$$

Thus, the Floquet eigenvalues can be defined only mod $(i\omega)$ [81].

On expanding $w_{\lambda_M}(x,t)$ in a Fourier series as

$$w_{\lambda_M}(x,t) = \sum_{p=-\infty}^{+\infty} w_{\lambda_M}^{(p)}(x) e^{ip\omega t} \tag{6.44}$$

and on substituting Eq. (6.44) into Eq. (6.36), we obtain

$$2[ip\omega - \lambda_M - L_{\mathrm{FP}}^0(x)] w_{\lambda_M}^{(p)}(x) = -iL_\xi(x)\left[w_{\lambda_M}^{(p-1)}(x) - w_{\lambda_M}^{(p+1)}(x)\right] \tag{6.45}$$

On expanding further the $w_{\lambda_M}^{(p)}(x)$ as a series of Legendre polynomials, namely,

$$w_{\lambda_M}^{(p)}(x) = \sum_{p=0}^{\infty} {}^M a_p^m P_m(x) \tag{6.46}$$

and on noting that [39]

$$\frac{d}{dx}\left[(1-x^2)\frac{d}{dx}P_n(x)\right] = -n(n+1)P_n(x) \tag{6.47}$$

we have the following recurrence relation for the coefficients ${}^M a_p^m$ $(m = 0, 1, 2, \ldots, \infty)$

$$\sum_{m=0}^{\infty}\left\{2(2n+1)\left[\frac{2ip\omega\tau_{\mathrm{D}} - \lambda_M}{n(n+1)} + 1\right]\delta_{n,m}\, {}^M a_p^m \right.$$
$$\left. - \xi_1(\delta_{n,m+1} - \delta_{n,m-1})\left({}^M a_{p-1}^m - {}^M a_{p+1}^m\right)\right\} = 0 \tag{6.48}$$

or

$$\sum_{m=0}^{\infty} \mathbf{Q}_p^{nm}\, {}^M a_p^m + (\mathbf{Q}_p^+)^{nm}\, {}^M a_{p+1}^m + (\mathbf{Q}_p^-)^{nm}\, {}^M a_{p-1}^m = 0 \tag{6.49}$$

where the matrix elements of the matrices $\mathbf{Q}_p$, $\mathbf{Q}_p^+$, and $\mathbf{Q}_p^-$ of infinite dimension are given by

$$\begin{aligned}
(\mathbf{Q}_p)^{nm} &= 2(2n+1)\left[\frac{2ip\omega\tau_{\mathrm{D}} - \lambda_M}{n(n+1)} + 1\right]\delta_{m,n} \\
(\mathbf{Q}_p^+)^{nm} &= -(\mathbf{Q}_p^-)^{nm} = \xi_1\left(\delta_{n,m+1} - \delta_{n,m-1}\right)
\end{aligned} \tag{6.50}$$

All the eigenvalues $\lambda_M$ and all the coefficients ${}^M a_p^m$ can be evaluated [9] from Eq. (6.49) by using the matrix continued fraction approach. By putting $\lambda_M = 0$ corresponding to the stationary state response in the ac regime (asymptotic value of the distribution function), Eq. (6.49) leads to a tridiagonal vector recurrence relation of exactly the same form as Eq. (6.6). A similar consideration can be used for the nonstationary response of a system of polar and polarizable molecules to an ac electric field. In order to accomplish this, one should add to the right-hand side of Eq. (6.36) the perturbation operator $L_\sigma(x,t)$ due to the interaction of the ac electric field $E(t) = E_1 \sin(\omega t)$ with induced dipoles, namely,

$$L_\sigma(x,t) = \frac{1}{2}\sin^2(\omega t) D\kappa\xi_1^2 \left[(3x^2 - 1) - x(1 - x^2)\frac{\partial}{\partial x}\right] \tag{6.51}$$

Furthermore, in the context of Floquet's approach one can readily take into account the effect of the dc bias field. Thus, one can apply the matrix continued fraction method for the evaluation of the *nonstationary* response in superimposed high ac and dc electric fields, as well. The interested reader can find more details on the application of Floquet's approach to the solution of the Fokker–Planck equation in review [81].

## VII. EFFECTS DUE TO HYPERPOLARIZABILITIES OF MOLECULES ON KERR EFFECT RELAXATION

### A. Nonlinear Step-On Response

Until now, all the effects due to hyperpolarizability of the molecules have been ignored. However, the theory developed in previous sections can be generalized to take into account these effects. Equation (2.2), which we have used explicitly many times, may not be applicable to high field strengths, because the induced dipole moment can no longer be considered as linear with the field. This aspect has been discussed and treated theoretically by many authors (see, e.g. [84–89]). If $i, j, k, l$ are four indexes used for representing Cartesian components along the three directions of space ($i, j, k, l = 1, 2, 3$), we have instead

$$m_i = \mu_i + \alpha_{ij}E_j + \frac{1}{2}\beta_{ijk}E_jE_k + \frac{1}{3!}\gamma_{ijkl}E_jE_kE_l + \cdots$$

where summation over repeated indexes is understood, $\alpha_{ij}$ is the linear polarizability and $\beta_{ijk}$ and $\gamma_{ijkl}$ are the components of the first and second hyperpolarizabilities (components of third and fourth rank tensors, respectively). For molecules with axial symmetry, the only useful tensorial components of

$\beta_{ijk}$ and $\gamma_{ijkl}$ are [85]

$$\beta_{113} = \beta_{223}, \beta_{333} \qquad \gamma_{1111} = \gamma_{2222} = 3\gamma_{1122} \qquad \gamma_{1133} = \gamma_{2233}, \gamma_{3333}$$

These tensorial components may also be regarded as the successive partial derivatives of the $m_i$ component of the total dipole moment with respect to those of the electric field components, namely,

$$\alpha_{ij} = \frac{\partial m_i}{\partial E_j}$$

$$\beta_{ijk} = \frac{\partial^2 m_i}{\partial E_j \partial E_k} = \frac{\partial}{\partial E_k}\alpha_{ij}$$

$$\gamma_{ijkl} = \frac{\partial^3 m_i}{\partial E_j \partial E_k \partial E_l} = \frac{\partial}{\partial E_l}\beta_{ijk}$$

Next, we shall assume that the polarizability tensors are symmetric in all pairs of indexes, which is true for constant fields and remains valid for time-dependent fields whose frequencies lie far from absorption bands [90]. We also assume that the liquid consists of an assembly of axially symmetric and noninteracting molecules acted on by an electric field. This reduces considerably the number of tensorial components and allows us to take into account the rotational Brownian motion of a single molecule. The orientational potential energy of such a molecule compelled to rotate in 3D space is thus given by [91, 92]

$$\begin{aligned} V = &-\mu E(t)\cos\vartheta - \frac{\Delta\alpha E^2(t)}{2}\cos^2\vartheta - \frac{\alpha_\perp E^2(t)}{2} \\ &- \frac{\beta_{113}E^3(t)}{2}\cos\vartheta - \frac{\Delta\beta}{6}E^3(t)\cos^3\vartheta - \frac{1}{12}(3\gamma_{1133} - \gamma_{1111})E^4(t)\cos^2\vartheta \\ &- \frac{\Delta\gamma}{24}E^4(t)\cos^4\vartheta - \frac{\gamma_{1111}}{24}E^4(t) - \cdots \end{aligned} \tag{7.1}$$

where $\vartheta$, $\mu$, $\alpha_\perp$, and $\Delta\alpha$ have the same meaning as before,

$$\Delta\beta = \beta_{333} - 3\beta_{113}$$

is the first hyperpolarizability anisotropy, and

$$\Delta\gamma = \gamma_{3333} - 6\gamma_{1133} + \gamma_{1111}$$

may be regarded as the second hyperpolarizability anisotropy. The study of these hyperpolarizabilities has received a continuing interest in development of nonlinear optics for useful practical applications. This is the case, for example, of the second harmonic generation of laser beams obtained by using new organic and polymeric materials having high nonlinear optical coefficient rather than standard inorganic crystals [93]. This effect is quadratic and based on first-order hyperpolarizability. An other important application may be found in telecommunications by optical fibers where amplifiers [94] are conceived as all-optical systems (in this case both the first and second hyperpolarizability play an important role). In Section IV, we have derived exact analytical equations for the nonlinear response and related integral relaxation time arising from the sudden application of an electric field for an assembly of polar and polarizable molecules in terms of matrix continued fractions. It is the purpose of this section to extend this matrix continued fraction approach to the calculation of nonlinear dielectric and Kerr responses and relaxation times to see how the effects due to the hyperpolarizabilities modify the results obtained above. A previous study of this problem has been accomplished by Lee et al. [34] using the Runge–Kutta method to solve the infinite hierarchy for the averaged Legendre polynomials for this particular problem numerically. However, they restricted themselves to pure permanent moments or pure polarizable molecules, and analyzed the Kerr response in the time domain, although not emphasizing the nonlinear relaxation behavior and not having calculated the relaxation time. Nevertheless, as the nonlinear relaxation behavior is better shown from relaxation spectra, we shall proceed below to the evaluation of the dielectric and Kerr nonlinear responses (as with our method both are obtained through one calculation only) when hyperpolarizabilities are included and focus our attention more especially on the Kerr response (the dielectric response could be investigated in a similar way).

So, we concentrate on the transient step-on response by supposing that a strong constant electric field $\mathbf{E}_0$ (applied along the $Z$ axis) is suddenly switched on at time $t = 0$ (the formulation of this problem is given in Section IV). The potential given by Eq. (7.1) can be equivalently expressed in terms of spherical $Y_{n,m}(\vartheta, \varphi)$ as follows:

$$V(\vartheta, t) = \sum_{R=1}^{4} v_R(t) Y_{R,0}(\vartheta, \varphi) - \frac{a_0}{2} E^2(t) - \frac{c_0}{24} E^4(t) \tag{7.2}$$

where

$$v_1(t) = -2\sqrt{\frac{\pi}{3}}\left(\mu E(t) + \frac{3}{10}\beta_{113} E^3(t)\right) \tag{7.3}$$

$$v_2(t) = -\frac{2}{3}\sqrt{\frac{\pi}{5}}\left(\Delta\alpha E^2(t) + \frac{3\gamma_{1133} - \gamma_{1111}}{21}E^4(t)\right) \tag{7.4}$$

$$v_3(t) = -\frac{2}{15}\sqrt{\frac{\pi}{7}}\Delta\beta E^3(t) \tag{7.5}$$

$$v_4(t) = -\frac{16\sqrt{\pi}}{105}\Delta\gamma E^4(t) \tag{7.6}$$

Here,

$$a_0 = \tfrac{1}{3}(\alpha_{33} + 2\alpha_{11})$$

is the mean polarizability and

$$c_0 = \frac{1}{15}(8\gamma_{1111} + 12\gamma_{1133} + 3\gamma_{3333})$$

is the mean second hyperpolarizability [89].

Thus, the infinite hierarchy of equations for the moments $\langle P_n \rangle(t)$, which is the basis of the solution of the problem, is given by Eqs. (3.35) and (3.36) that now become

$$\tau_D \frac{d}{dt}\langle P_n(\cos\vartheta)\rangle(t) = \sum_{m=-4}^{4} a_{n+m,n}\langle P_{n+m}(\cos\vartheta)\rangle(t) \tag{7.7}$$

where

$$\begin{aligned} a_{m,n} = & -\frac{n(n+1)\delta_{n,m}}{2} + \frac{2(m+n)+1}{8kT} \\ & \times \left\{ \sum_{r=1}^{4} v_r(t) \frac{[m(2n+m+1) - r(r+1)]}{\sqrt{\pi(2r+1)}} \langle n, 0, n+m, 0 | r, 0\rangle^2 \right\} \end{aligned}$$

and the $v_r(t)$ are defined by Eqs. (7.3)–(7.6). On using an explicit representation for the Clebsch–Gordan coefficients $\langle n, 0, m, 0 | r, 0\rangle$ and on taking into account that the equilibrium ensemble averages $\langle P_n(\cos\vartheta)\rangle_0 = \langle P_n(\cos\vartheta)\rangle(\infty)$ obey a similar recurrence equation, namely,

$$\sum_{m=-4}^{4} a_{m,n}\langle P_m(\cos\vartheta)\rangle_0 = 0$$

one can derive from Eq. (7.7) the following differential recurrence relations for the relaxation function $f_n(t) = \langle P_n(\cos\vartheta)\rangle(\infty) - \langle P_n(\cos\vartheta)\rangle(t)$ [95]

$$\begin{aligned}
&\frac{2\tau_D}{n(n+1)}\frac{d}{dt}f_n(t) + \left[1 - \frac{2(\sigma+H_I^A)}{(2n-1)(2n+3)} - \frac{12H_I(n^2+n-3)}{(2n-3)(2n-1)(2n+3)(2n+5)}\right]f_n(t) \\
&\quad = \frac{\xi+H_P^A}{2n+1}[f_{n-1}(t) - f_{n+1}(t)] + \frac{3H_P(n^2+n-3)}{2n+1} \\
&\quad \times \left[\frac{f_{n-1}(t)}{(2n-3)(2n+3)} - \frac{f_{n+1}(t)}{(2n-1)(2n+5)}\right] \\
&\quad + 2(\sigma+H_I^A)\left[\frac{n-1}{(2n-1)(2n+1)}f_{n-2}(t) - \frac{n+2}{(2n+1)(2n+3)}f_{n+2}(t)\right] \\
&\quad + 4H_I\left[\frac{(n-1)(2n^2-n-9)}{(2n-5)(2n-1)(2n+1)(2n+3)}f_{n-2}(t)\right. \\
&\qquad \left. - \frac{(n+2)(2n^2+5n-6)}{(2n-1)(2n+1)(2n+3)(2n+7)}f_{n+2}(t)\right] \\
&\quad + 3H_P\left[\frac{(n-1)(n-2)}{(2n-3)(2n-1)(2n+1)}f_{n-3}(t)\right. \\
&\qquad \left. - \frac{(n+2)(n+3)}{(2n+1)(2n+3)(2n+5)}f_{n+3}(t)\right] \\
&\quad + 4H_I\left[\frac{(n-1)(n-2)(n-3)}{(2n-5)(2n-3)(2n-1)(2n+1)}f_{n-4}(t)\right. \\
&\qquad \left. - \frac{(n+2)(n+3)(n+4)}{(2n+1)(2n+3)(2n+5)(2n+7)}f_{n+4}(t)\right]
\end{aligned} \tag{7.8}$$

where the parameters $\xi$ and $\sigma$ are defined by Eq. (4.37), and the parameters $H_P^A$, $H_P$, $H_I^A$, and $H_I$ are given by

$$H_P^A = \frac{\beta_{113}E_0^3}{2kT} \tag{7.9}$$

$$H_P = \frac{\Delta\beta E_0^3}{6kT} \tag{7.10}$$

$$H_I^A = \frac{(3\gamma_{1133} - \gamma_{1111})E_0^4}{12kT} \tag{7.11}$$

$$H_I = \frac{\Delta\gamma E_0^4}{24kT} \tag{7.12}$$

These are dimensionless parameters reflecting the contribution of each term of the potential energy with respect to thermal energy. As can be seen from Eq. (7.8), the nonlinear effect due to the electric field strength can be characterized by two effective dipole moments, which in terms of reduced parameters write down:

$$\xi^* = \xi + H_P^A \tag{7.13}$$

for the permanent moment and

$$\sigma^* = \sigma + H_I^A \tag{7.14}$$

for the induced moment.

Equation (7.8) forms an infinite hierarchy of nine-term recurrence relations, which can be transformed into a matrix three-term recurrence Eq. (3.37), where $\mathbf{C}_n(t)$ is a column vector defined by

$$\mathbf{C}_n(t) = \begin{pmatrix} f_{4n-3}(t) \\ f_{4n-2(t)} \\ f_{4n-1}(t) \\ f_{4n}(t) \end{pmatrix} \qquad n = 1, 2, 3, \ldots \tag{7.15}$$

and $\mathbf{Q}_n^-$, $\mathbf{Q}_n$, and $\mathbf{Q}_n^+$ are three $4 \times 4$ matrices such that

$$\mathbf{Q}_n^- = \begin{pmatrix} 4\alpha_{4n-3} & 3\beta_{4n-3} & 2\gamma_{4n-3} & \delta_{4n-3} \\ 0 & 4\alpha_{4n-2} & 3\beta_{4n-2} & 2\gamma_{4n-2} \\ 0 & 0 & 4\alpha_{4n-1} & 3\beta_{4n-1} \\ 0 & 0 & 0 & 4\alpha_{4n} \end{pmatrix} \tag{7.16}$$

$$\mathbf{Q}_n = \begin{pmatrix} \varepsilon_{4n-3} & -\varphi_{4n-3} & -2\psi_{4n-3} & -3\varpi_{4n-3} \\ \delta_{4n-2} & \varepsilon_{4n-2} & -\varphi_{4n-2} & -2\psi_{4n-2} \\ 2\gamma_{4n-1} & \delta_{4n-1} & \varepsilon_{4n-1} & -\varphi_{4n-1} \\ 3\beta_{4n} & 2\gamma_{4n} & \delta_{4n} & \varepsilon_{4n} \end{pmatrix} \tag{7.17}$$

$$\mathbf{Q}_n^+ = \begin{pmatrix} -4\Theta_{4n-3} & 0 & 0 & 0 \\ -3\varpi_{4n-2} & -4\Theta_{4n-2} & 0 & 0 \\ -2\psi_{4n-1} & -3\varpi_{4n-1} & -4\Theta_{4n-1} & 0 \\ -\varphi_{4n} & -2\psi_{4n} & -3\varpi_{4n} & -4\Theta_{4n} \end{pmatrix} \tag{7.18}$$

and

$$\alpha_n = \frac{n(n+1)(n-1)(n-2)(n-3)H_I}{2(2n-5)(2n-3)(2n-1)(2n+1)} \tag{7.19}$$

$$\beta_n = \frac{n(n+1)(n-1)(n-2)H_P}{2(2n-3)(2n-1)(2n+1)} \tag{7.20}$$

$$\gamma_n = \frac{n(n+1)(n-1)\sigma^*}{2(2n-1)(2n+1)} + \frac{n(n+1)(n-1)(2n^2-n-9)H_I}{(2n-5)(2n-1)(2n+1)(2n+3)} \tag{7.21}$$

$$\delta_n = \frac{n(n+1)\xi^*}{2(2n+1)} + \frac{n(n+1)(n^2+n-3)H_P}{2(2n-3)(2n+1)(2n+3)} \tag{7.22}$$

$$\varepsilon_n = \frac{n(n+1)}{2}\left[\frac{12(n^2+n-3)H_I}{(2n-3)(2n-1)(2n+1)(2n+3)} + \frac{2\sigma^*}{(2n-1)(2n+3)} - 1\right] \tag{7.23}$$

$$\varphi_n = \frac{n(n+1)\xi^*}{2(2n+1)} + \frac{n(n+1)(n^2+n-3)H_P}{2(2n-1)(2n+1)(2n+5)} \tag{7.24}$$

$$\psi_n = \frac{n(n+1)(n+2)\sigma^*}{2(2n+1)(2n+3)} + \frac{n(n+1)(n+2)(2n^2+5n-6)H_I}{(2n-1)(2n+1)(2n+3)(2n+7)} \tag{7.25}$$

$$\varpi_n = \frac{n(n+1)(n+2)(n+3)H_P}{2(2n+1)(2n+3)(2n+5)} \tag{7.26}$$

$$\Theta_n = \frac{n(n+1)(n+2)(n+3)(n+4)H_I}{2(2n+1)(2n+3)(2n+5)(2n+7)} \tag{7.27}$$

As has been described in Section III.C, we can readily obtain the exact solution for the Laplace transform $\tilde{\mathbf{C}}_1(s)$ [95], namely,

$$\begin{pmatrix} \tilde{f}_1(s) \\ \tilde{f}_2(s) \\ \tilde{f}_3(s) \\ \tilde{f}_4(s) \end{pmatrix} = \tau_D[\tau_D s\mathbf{I} - \mathbf{Q}_1 - \mathbf{Q}_1^+ \mathbf{S}_2(s)]^{-1}$$

$$\times \left\{ \mathbf{C}_1(0) + \sum_{n=2}^{\infty} \prod_{k=2}^{n} \mathbf{Q}_{k-1}^+ \mathbf{S}_k(s)(\mathbf{Q}_k^-)^{-1}\mathbf{C}_n(0) \right\} \quad (7.28)$$

where $\mathbf{S}_n(s)$ is the matrix continued fraction given by Eq. (3.41) with the matrices $\mathbf{Q}_n^-$, $\mathbf{Q}_n$, and $\mathbf{Q}_n^+$ defined by Eqs. (7.16)–(7.27), and $\mathbf{I}$ is now the $4 \times 4$ identity matrix. The $\mathbf{C}_n(0)$ is a column vector containing the initial conditions, which can be evaluated at equilibrium (see Section IV for details). Thus, we have

$$\mathbf{C}_n(0) = \begin{pmatrix} \langle P_{4n-3} \rangle_0 \\ \langle P_{4n-2} \rangle_0 \\ \langle P_{4n-1} \rangle_0 \\ \langle P_{4n} \rangle_0 \end{pmatrix} = \prod_{k=1}^{n} \mathbf{S}_k(0) \begin{pmatrix} 0 \\ 0 \\ 0 \\ 1 \end{pmatrix} \quad (7.29)$$

We recall that $\langle P_n \rangle_0 = \langle P_n(\cos\vartheta) \rangle(\infty)$ denotes the equilibrium average value of the $n$th Legendre polynomial in the presence of the dc electric field. For $n = 1$, Eq. (7.29) reduces to

$$\mathbf{C}_1(0) = \mathbf{S}_1(0) \begin{pmatrix} 0 \\ 0 \\ 0 \\ 1 \end{pmatrix} \quad (7.30)$$

Just as in Section IV.B, the matrix continued fraction $\mathbf{S}_1(0)$ can be represented in closed form as

$$\mathbf{S}_1(0) = \begin{pmatrix} 0 & 0 & 0 & \langle P_1(\cos\vartheta) \rangle_0 \\ 0 & 0 & 0 & \langle P_2(\cos\vartheta) \rangle_0 \\ 0 & 0 & 0 & \langle P_3(\cos\vartheta) \rangle_0 \\ 0 & 0 & 0 & \langle P_4(\cos\vartheta) \rangle_0 \end{pmatrix}$$

On using Eq. (7.9) and that

$$\begin{pmatrix} \int_0^\infty \langle P_1(\cos\vartheta)\rangle(t)e^{-st}dt \\ \int_0^\infty \langle P_2(\cos\vartheta)\rangle(t)e^{-st}dt \\ \int_0^\infty \langle P_3(\cos\vartheta)\rangle(t)e^{-st}dt \\ \int_0^\infty \langle P_4(\cos\vartheta)\rangle(t)e^{-st}dt \end{pmatrix} = s^{-1}\mathbf{S}_1(s)\begin{pmatrix} 0 \\ 0 \\ 0 \\ 1 \end{pmatrix} \tag{7.31}$$

Eq. (7.28) can be simplified to yield

$$\tilde{\mathbf{C}}_1(s) = s^{-1}[\mathbf{S}_1(0) - \mathbf{S}_1(s)]\begin{pmatrix} 0 \\ 0 \\ 0 \\ 1 \end{pmatrix} \tag{7.32}$$

Equation (7.32) clearly shows that the step-on response may be calculated from the sole matrix continued fraction $\mathbf{S}_1(s)$. In what follows, we focus on the Kerr effect step-on response only. The Kerr effect relaxation time which is an interesting physical quantity from an experimental point of view is

$$\tau_K = \tilde{f}_2(0)/f_2(0) \tag{7.33}$$

Another way of obtaining $\tau_K$ consists of using an integral representation of this relaxation time as described in detail in Section III.D. On applying Eq. (3.80) to the problem under consideration, we have

$$\tau_K = \frac{2\tau_D}{\langle P_2\rangle_0}\int_{-1}^{1}\frac{\exp[V(z)/kT]\phi(z)\psi_2(z)}{1-z^2}dz \tag{7.34}$$

where

$$\psi_2(z) = \int_{-1}^{z}[P_2(x) - \langle P_2(x)\rangle_0]\exp[-V(x)/kT]dx \tag{7.35}$$

$$\phi(z) = \frac{\int_{-1}^{z}\exp[-V(y)/kT]dy}{\int_{-1}^{1}\exp[-V(y)/kT]dy} - \frac{(z+1)}{2} \tag{7.36}$$

Equations (7.33) and (7.34), although differing in form, yield the same numerical results. Equation (7.34) is a general formula used to evaluate the nonlinear integral relaxation time.

In Figure 56, we present the plots of $\tilde{f}_2(0) = \tau_K \langle P_2 \rangle_0 / \tau_D$ (the area under the relaxation function) as a function of $\sigma^*$ and $H_I$, which corresponds to the case of purely induced dipole moments (nonpolar molecules, $\xi^* = 0$ and $H_P = 0$). The same result would be obtained for the optical birefringence since the atomic polarization no longer plays any role in this case. For purely permanent moments, the behavior of $\tilde{f}_2(0) = \tau_K \langle P_2 \rangle_0 / \tau_D$ as a function of $\xi^*$ and $H_P$ is illustrated in Figure 57. Figure 58 represents the general case showing the hyperpolarizability effects when permanent and induced dipole moments are both taken into account. The spectra of the real and imaginary parts of the relaxation function $\tilde{f}_2(\omega)$ are shown in Figures 59 and 60. One may remark that the activation process becomes more important as $H_I$ increases (the effective polarizability anisotropy increases). On the other hand, the increase of $H_P$ first compensates and then destroys the activation process because of the disappearance of the bistable structure of the potential $V(\vartheta, t)$ when the odd power terms in $\cos\vartheta$ become dominant. Thus, the $[\gamma]$ effect makes the activation process more important while the $[\beta]$ effect tends to reinforce the fast relaxation processes in such a manner that the low-frequency shoulder disappears in the spectra.

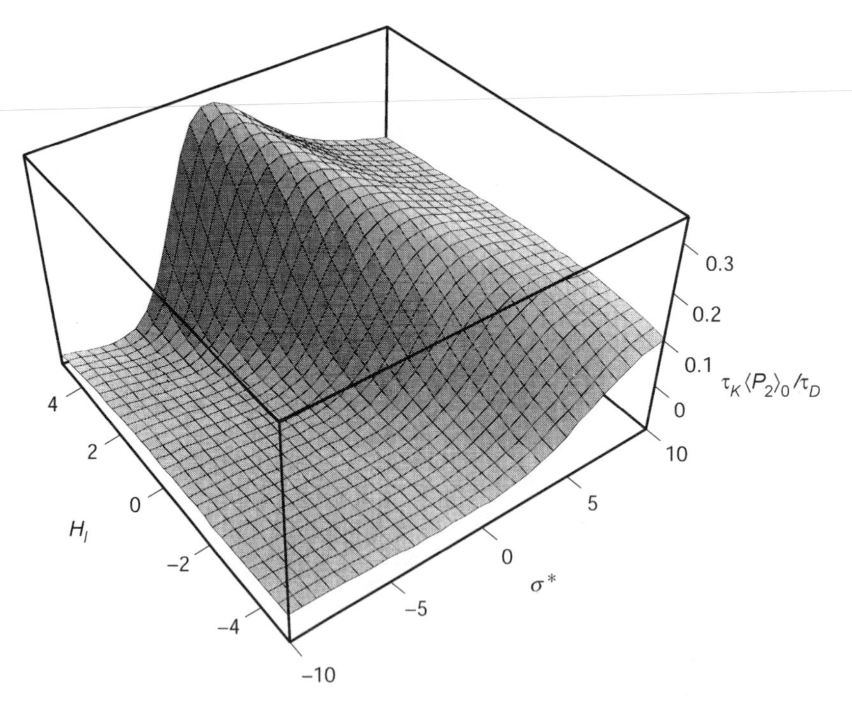


**Figure 56.** The $\tau_K \langle P_2 \rangle_0 / \tau_D$ as a function of $\sigma^*$ and $H_I$ for $\xi^* = 0$ and $H_P = 0$.

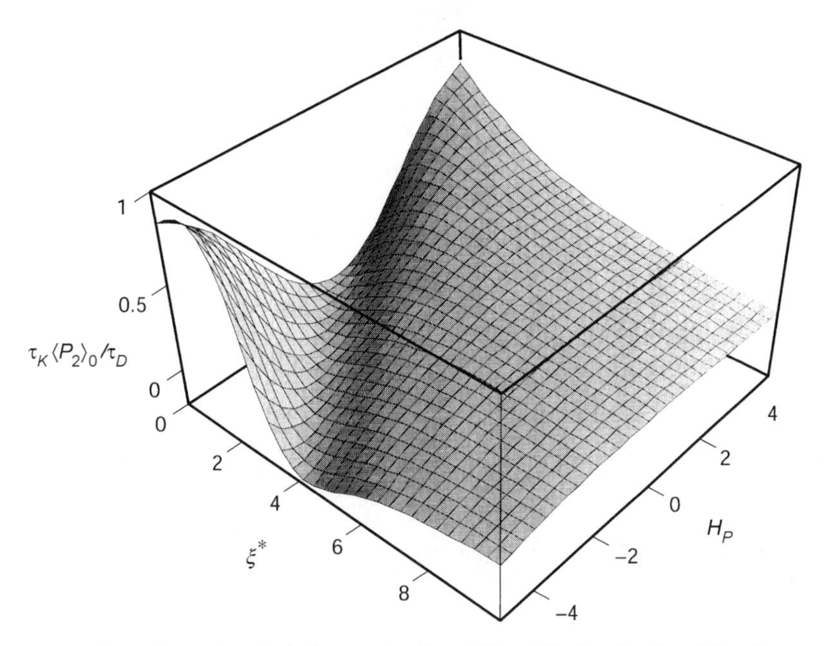


**Figure 57.** The $\tau_K \langle P_2 \rangle_0 / \tau_D$ as a function of $\xi^*$ and $H_P$ for $\sigma^* = 0$ and $H_I = 0$.

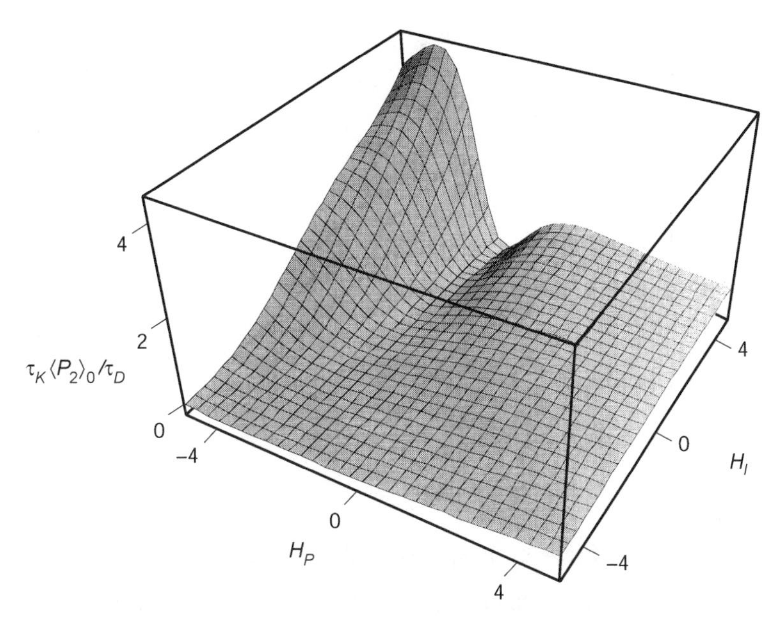


**Figure 58.** The $\tau_K \langle P_2 \rangle_0 / \tau_D$ as a function of $H_I$ and $H_P$ for $\sigma^* = 2$ and $\xi^* = 2$.

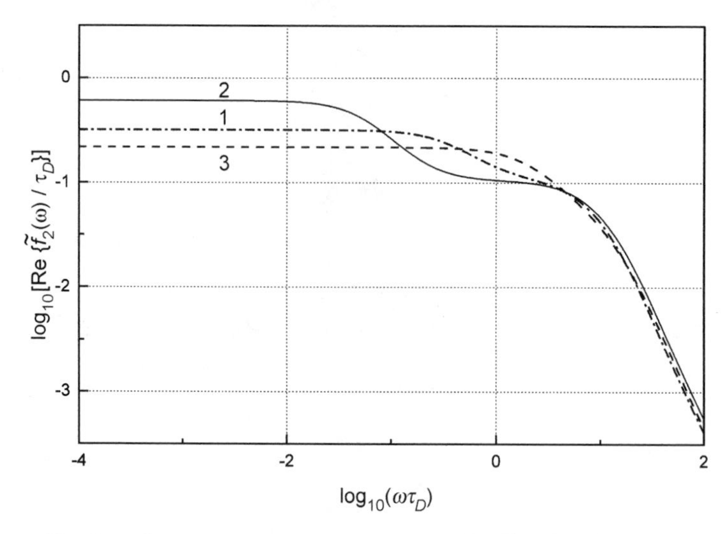


**Figure 59.** Kerr effect nonlinear step-on response: $\log_{10}[\mathrm{Re}\{\tilde{f}_2(\omega)\}/\tau_\mathrm{D}]$ for different values of $H_P$ and $H_I$. Curve 1: $H_P = H_I = 0$; curve 2: $H_P = 1, H_I = 4$; curve 3: $H_P = 5, H_I = 1$. $\xi^*$ and $\sigma^*$ are chosen arbitrarily equal to 10.

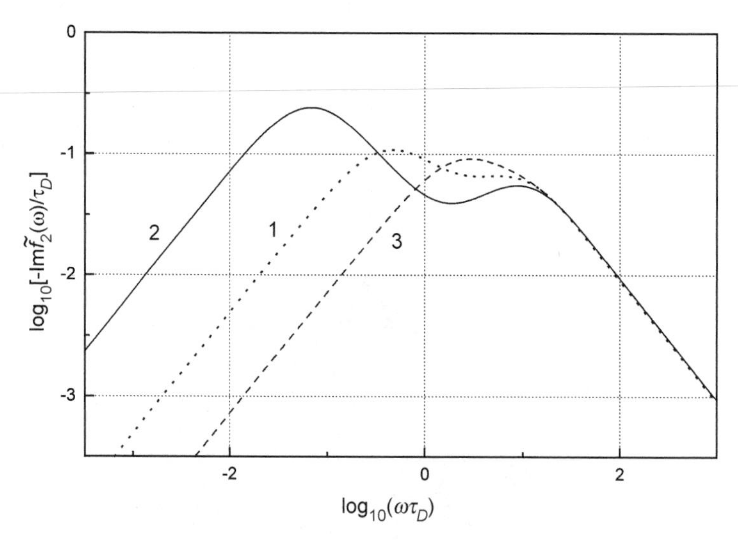


**Figure 60.** The same as in Figure 59 for $\log_{10}[-\mathrm{Im}\{\tilde{f}_2(\omega)\}/\tau_\mathrm{D}]$.

## B. Linear ac Response and After Effect Solution

As another application of the model, one can consider the linear response to an ac field of small amplitude. The case of ignoring hyperpolarizability effects has been treated in Section V.A. As in that section, we will suppose that the molecules are subjected to the influence of two electric fields, namely, a strong dc electric field $E_0$ superimposed on a weak field $E_1(t)$. In order to study the linear dynamic Kerr effect, we shall determine the aftereffect solution of the rotational diffusion equation following a small but abrupt change in the small field $E_1$ at $t = 0$. Since we are interested in the Kerr effect decay process, the aftereffect birefringence relaxation function is

$$f_K(t) = \langle P_2(\cos\vartheta)\rangle(t) - \langle P_2(\cos\vartheta)\rangle_0 \tag{7.37}$$

so that we are required to evaluate Eq. (7.37) in the linear approximation in the field $E_1$. On noting that in this approximation

$$E = E_0 + E_1$$

$$E^2 = E_0^2 + 2E_0E_1 + O(E_1^2)$$

$$E^3 = E_0^3 + 3E_0^2E_1 + O(E_1^2)$$

$$E^4 = E_0^4 + 4E_0^3E_1 + O(E_1^2)$$

we have [95]

$$f_K(t) = (\Lambda + 3\Phi)f_1(t) + (\rho^* + \Psi)f_2(t) + 2\Phi f_3(t) + \frac{2\Psi f_4(t)}{5} \tag{7.38}$$

where

$$\Lambda = 1 + \frac{3H_P^A}{\xi} \qquad \rho^* = \frac{4\sigma}{3\xi} + \frac{8H_I^A}{3\xi} \qquad \Phi = \frac{3H_P}{5\xi} \qquad \Psi = \frac{16H_1}{7\xi}$$

and $f_n(t)$ are the equilibrium correlation functions:

$$f_n(t) = \langle P_2[\cos\vartheta(0)]P_n[\cos\vartheta(t)]\rangle_0 - \langle P_2[\cos\vartheta(0)]\rangle_0\langle P_n[\cos\vartheta(0)]\rangle_0 \tag{7.39}$$

where the subscript "0" indicates an equilibrium ensemble average (at $E_1 = 0$). It is easy to verify that the correlation functions (7.39) are solutions of Eq. (7.8) so that the same as (in Section VII.A) matrix continued fraction method can be again applied. The only change concerns the initial conditions that can be expressed as follows:

$$\mathbf{C}_n(0) = \begin{pmatrix} f_{4n-3}(0) \\ f_{4n-2}(0) \\ f_{4n-1}(0) \\ f_{4n}(0) \end{pmatrix} = [\mathbf{X}_n + \mathbf{Y}_n\mathbf{S}_n(0) + \mathbf{Z}_n\mathbf{S}_{n+1}(0)\mathbf{S}_n(0)]\mathbf{S}_{n-1}(0)\cdots\mathbf{S}_1(0)\begin{pmatrix} 0 \\ 0 \\ 0 \\ 1 \end{pmatrix} \tag{7.40}$$

where $\mathbf{X}_n$, $\mathbf{Y}_n$, and $\mathbf{Z}_n$ are three square $(4 \times 4)$ matrix defined by

$$\mathbf{X}_n = \begin{pmatrix} 0 & 0 & c_{4n-3} & 0 \\ 0 & 0 & 0 & c_{4n-2} \\ 0 & 0 & 0 & 0 \\ 0 & 0 & 0 & 0 \end{pmatrix} \tag{7.41}$$

$$\mathbf{Y}_n = \begin{pmatrix} b_{4n-3} & 0 & a_{4n-3} & 0 \\ 0 & b_{4n-2} & 0 & a_{4n-2} \\ c_{4n-1} & 0 & b_{4n-1} & 0 \\ 0 & c_{4n} & 0 & b_{4n} \end{pmatrix} \tag{7.42}$$

$$\mathbf{Z}_n = \begin{pmatrix} 0 & 0 & 0 & 0 \\ 0 & 0 & 0 & 0 \\ a_{4n-1} & 0 & 0 & 0 \\ 0 & a_{4n} & 0 & 0 \end{pmatrix} \tag{7.43}$$

and

$$a_n = \frac{3(n+1)(n+2)}{2(2n+1)(2n+3)} \tag{7.44}$$

$$b_n = \frac{n(n+1)}{(2n-1)(2n+3)} - \langle P_2 \rangle_0 \tag{7.45}$$

$$c_n = \frac{3n(n-1)}{2(2n-1)(2n+1)} \tag{7.46}$$

Thus, we can determine from Eq. (5.37) the relaxation spectrum of $\chi_K(\omega)$ corresponding to the ac response. Then, we can also obtain the relaxation time, which is

$$\tau_K = \frac{\tilde{f}_K(0)}{f_K(0)} \tag{7.47}$$

On the other hand,

$$\begin{aligned}\tilde{f}_K(0) &= (\Lambda + 3\Phi)\tau_1 f_1(0) + (\rho^* + \Psi)\tau_2 f_2(0) \\ &\quad + 2\Phi\tau_3 f_3(0) + \frac{2\Psi\tau_4 f_4(0)}{5}\end{aligned} \tag{7.48}$$

where according to linear response theory [61] (see Appendix B), the correlation times $\tau_n (n = 1, 2, 3, 4)$ for every correlation function $f_n(t)$ are given by

$$\tau_n = \frac{2\tau_D}{Zf_n(0)} \int_{-1}^{1} \frac{e^{V(z)/kT}\psi_2(z)\phi_n(z)dz}{1-z^2} \tag{7.49}$$

Here, $\psi_2(z)$ is given by Eq. (7.35) and

$$Z = \int_{-1}^{1} e^{-V(z')/kT} dz'$$

$$\phi_n(z) = \int_{-1}^{z} [P_n(z') - \langle P_n \rangle_0] e^{-V(z')/kT} dz'$$

Thus

$$\tau_K = \frac{(\Lambda + 3\Phi)\tau_1 f_1(0) + (\rho^* + \Psi)\tau_2 f_2(0) + 2\Phi\tau_3 f_3(0) + 2\Psi\tau_4 f_4(0)/5}{(\Lambda + 3\Phi) f_1(0) + (\rho^* + \Psi) f_2(0) + 2\Phi f_3(0) + 2\Psi f_4(0)/5} \tag{7.50}$$

Here, we have given two examples of the application of our approach to the problem under consideration when the effect due to hyperpolarizability of the molecules are taken into account. However, many other nonlinear responses treated in this chapter, as the second-order perturbation solution, the response in high ac field, and so on, can be evaluated in a similar manner.

## VIII. INERTIAL EFFECTS IN DIELECTRIC AND BIREFRINGENCE RELAXATION IN A STRONG DC ELECTRIC FIELD: LINEAR RESPONSE

So far, we have dealt with dielectric and Kerr effect relaxation, we always treated these phenomena by ignoring the inertia of the molecules. In doing so, the solutions obtained for the dielectric and Kerr effect relaxation spectra are not valid in the high-frequency range. In order to account for inertial effects, one should solve the inertial Langevin equation or the corresponding Fokker–Planck equation [8, 9, 52]. Many authors have done this for rotation in two dimensions, either exactly or by perturbation expansions (for the treatment of the dynamic Kerr effect and dielectric relaxation, see, e.g., [8, 96]). In general and for nonlinear dynamics (anharmonic potentials, e.g.), this leads to solve the so-called Brinkman's hierarchy [97], which governs the dynamics of the expansion coefficients of the conditional probability density (Green's function of the Fokker–Planck equation) on a proper orthogonal set of functions. More precisely, when the expression for the potential energy is periodic (cosine potential), Brinkman's hierarchy [for a two-dimensional (2D) problem] becomes a doubly infinite set of differential recurrence relations whose solution can only be carried out by numerical methods. As an example, one can quote the matrix continued fraction procedure [9], whose efficiency is proved when the barrier height parameter (potential energy) is not too high. Unfortunately, Risken reported in his book [9] that when large barrier height parameter values are chosen, the size of the matrices involved becomes so large that the numerical problem becomes intractable, especially when the low-damping limit is considered. We add the argument that in view of considering 3D rotational motion, solving the Fokker–Planck equation becomes a formidably difficult task, even when considering the needle model [1, 8] because the mathematics involved in order to solve this equation are much more complicated.

However, it must be noted that the Fokker–Planck–Langevin model is one of the various equivalent existing kinetic models (the word "equivalent" just means that all these models yield the same qualitative results). These models are often termed as extended rotational diffusional models [10, 11] because they imply a generalized concept of diffusion, when the free rotational motion of the molecule between collisions is possible and the collisions may be considered as "weak" as well as "strong". In addition, the main feature of these models is that the collisions for these models are assumed instantaneous, namely, the average time between collisions is much larger than the duration of molecular collisions. Among these models, we mention the Kilson–Storer model [11], the

Bhatnagar–Gross–Krook (BGK) model [98, 99], and the $J$ and $M$ diffusion models [10, 11]. Although for the first time introduced by Gross [100] and Sack [101, 102] as a particular case of the BGK model [98], the $J$-diffusion model has been recovered independently by Gordon [103]. Its applications to the study of rotational diffusion in fluids have been extensively reviewed by McClung [10]. In particular, McClung discussed this model in terms of the Mori–Zwanzig memory function approach [104], which is equivalent to that of the kinetic equation, and having the advantage of being much easier to handle. Later on, the $J$-diffusion model was further elaborated on by Leicknam et al. [105, 106] in order to apply it to the calculation of orientational correlation functions of asymmetric top molecules. More recently, Kalmykov [107, 108] applied the $J$-diffusion model in the context of the memory function formalism to the calculation of the dielectric susceptibility of linear molecules in a uniform external dc field. This latter approach has been very recently extended by Déjardin [109] to the linear dynamic Kerr effect response of polar molecules to a small ac field in a strong external field. The main purpose of this section is to apply the $J$-diffusion model in order to calculate the linear dielectric susceptibilities and the Kerr effect response of polar linear molecules in superimposed weak ac and strong dc bias fields. The method presented here has the advantage that the mathematical treatment is relatively simpler than solving the inertial Fokker–Planck equation for rotation in three dimensions, and allows us to obtain analytical expressions rather easily. For the sake of simplicity, below we restrict ourselves to the case where the molecules have a permanent dipole moment only, but before proceeding, it is necessary to recall some generalities about extended rotational diffusion models as well as the main equations of the formalism.

## A. Extended Rotational Diffusion Model in a Strong dc Electric Field

### 1. General Equations

The extended rotational diffusion for free motion and in a field of force was discussed elsewhere [10, 11, 108], and thus we briefly describe the model here. We assume, as before, that a rigid linear molecule may be assimilated to a classical rotator, whose permanent moment is $\mu$ and moment of inertia is $I$. The molecule rotates in a constant external field $\mathbf{E}_0$. The rotation of the dipole is interrupted by instantaneous collisions, namely, the duration of a collision is much shorter than the average time between collisions $\tau$. The collisions occur at random times governed by a Poisson law and randomize both the direction and magnitude of the angular momentum $\mathbf{J}$, where the random values of $\mathbf{J}$ are governed by a Boltzmann law. It should be emphasized that the term "collision", in the context of the $J$-diffusion model, does not mean a bimolecular collision as is presumed to occur in dilute gases. A "collision" is an idealized event wherein the molecule experiences an impulsive torque [10]. The memory function formalism [104] is very easily adapted to the treatment of such a problem. We recall that the starting point of this formalism is the generalized Langevin equation for a dynamic variable $A(t)$ [104], namely,

$$\frac{d}{dt}A(t) = -i\Omega A(t) - \int_0^t K_0(t-t')A(t')dt' + \Gamma(t) \tag{8.1}$$

where $K_0(t)$ is the memory function and $\Omega$ is a time-independent coefficient, which is equal to zero for the problems in question. The memory function $K_0(t)$ is defined as the correlation function of the random "force" $\Gamma(t)$. Thus, the equilibrium correlation function $C(t) = \langle A(0)A(t)\rangle_0$ obeys the following Volterra integrodifferential equation [104]:

$$\frac{d}{dt}C(t) = -\int_0^t K_0(t-t')C(t')dt' \tag{8.2}$$

Here, we have assumed that $\Omega = 0$. One can readily solve Eq. (8.2) with the help of a one-sided FT:

$$\tilde{C}(\omega) = \frac{C(0)}{-i\omega + \tilde{K}_0(\omega)} \tag{8.3}$$

where the tilde denotes the one-sided FT, namely,

$$\tilde{f}(\omega) = \int_0^\infty f(t)e^{i\omega t}dt$$

By inspection of Eq. (8.3), it is clear that if one knows the memory function or its FT, then one is able to evaluate the frequency spectrum of the correlation function $C(t)$.

We apply this formalism to the extended rotational diffusion model cited above. We formulate the model just as the $J$-diffusion model of Gordon [103]. As in the context of the $J$ diffusion, it is assumed that a molecule rotates freely between successive collisions and that the collisions are governed by a Poisson law. This simply means that between two collisions, the memory kernel of Eq. (8.2) is the *free rotational* memory function, and the correlation function is the free rotational correlation function. By "free" it is meant that the rotation of the molecule between collisions is determined by deterministic equations of motion. Keeping in mind that the free rotational memory function is the memory function for the molecule that has experienced no collision, and that the memory function for the molecule that has experienced a collision is zero, the memory friction for the ensemble is

$$K_0(t) = K_0^c(t)e^{-t/\tau} \tag{8.4}$$

where $K_0^c(t)$ is the memory function for free rotors (where the superscript "c" means "collisionless"). The free rotation correlation function $C_c(t)$ obeys the equation:

$$\frac{d}{dt}C_c(t) = -\int_0^t K_0^c(t-t')C_c(t')dt' \tag{8.5}$$

so that by using Eqs. (8.2)–(8.5) we are able to express the correlation function of interest in terms of the free rotational correlation function $C_c(t)$. Thus, one has

$$\tilde{C}(\omega) = \frac{\tilde{C}_c(\omega + i/\tau)}{1 - \tilde{C}_c(\omega + i/\tau)/[\tau C_c(0)]} \tag{8.6}$$

According to linear response theory the generalized susceptibility is then given by [104]:

$$\chi(\omega) \propto C(0) + i\omega\tilde{C}(\omega) \tag{8.7}$$

In the following, we will use Eqs. (8.6) and (8.7) in order to work out the linear longitudinal and transverse dynamic susceptibilities, the Kerr effect linear response, and the relaxation times of an ensemble of rigid linear molecules under the influence of a strong dc external field $\mathbf{E}_0$.

### 2. *Solution of the Equation of Motion for the Free Rotation*

Before proceeding, we give some generalities concerning the *free rotational* motion of a linear molecule in space in the presence of a dc external electric field $\mathbf{E}_0$ (which is directed along the $Z$ axis of the laboratory coordinate system $OXYZ$). In the laboratory frame, the equations of motion of the molecule are those of a rigid rotator, namely,

$$\frac{d}{dt}\boldsymbol{\mu}(t) = \boldsymbol{\Omega}(t) \times \boldsymbol{\mu}(t) \tag{8.8}$$

$$I\frac{d}{dt}\boldsymbol{\Omega}(t) = \boldsymbol{\mu}(t) \times \mathbf{E}_0 \tag{8.9}$$

where $I$ is the moment of inertia of the molecule, $\boldsymbol{\mu}(t)$ is the dipole moment vector, and $\boldsymbol{\Omega}(t)$is the angular velocity vector. On differentiating Eq. (8.8) with respect to time, we have

$$\frac{d^2}{dt^2}\boldsymbol{\mu}(t) = \left[\frac{d}{dt}\boldsymbol{\Omega}(t)\right] \times \boldsymbol{\mu}(t) + \boldsymbol{\Omega}(t) \times \left[\frac{d}{dt}\boldsymbol{\mu}(t)\right] \tag{8.10}$$

On using Eqs. (8.8) and (8.9) and the triple vector cross-product formula, Eq. (8.10) reduces to

$$\frac{d^2}{dt^2}\boldsymbol{\mu}(t) = -\boldsymbol{\Omega}^2(t)\boldsymbol{\mu}(t) + I^{-1}[\mu^2\mathbf{E}_0 - \boldsymbol{\mu}(t)(\boldsymbol{\mu}(t)\cdot\mathbf{E}_0)] \tag{8.11}$$

The vector Eq. (8.11) can be equivalently presented as [108]

$$\frac{d^2}{dt^2}u_{\parallel}(t) + \Omega^2(t)u_{\parallel}(t) = \frac{\mu E_0}{I}[1 - u_{\parallel}^2(t)] \tag{8.12}$$

$$\frac{d^2}{dt^2}u_{\perp}(t) + \Omega^2(t)u_{\perp}(t) = -\frac{\mu E_0}{I}u_{\perp}(t)u_{\parallel}(t) \tag{8.13}$$

where

$$u_{\parallel}(t) = \cos\vartheta(t) \qquad \text{and} \qquad u_{\perp}(t) = \sin\vartheta(t)\cos\varphi(t)$$

are the parallel and perpendicular components of the unit vector $\mathbf{u}(t) = \boldsymbol{\mu}(t)/\mu$ on $\mathbf{E}_0$. Here, $\vartheta$ and $\varphi$ are the polar and azimuthal angles, respectively. On noting that the Hamiltonian $H$ of the molecule

$$H = \frac{p_\vartheta^2}{2I} + \frac{p_\varphi^2}{2I\sin^2\vartheta} - \mu E_0\cos\vartheta = \frac{I\Omega^2}{2} - \mu E_0\cos\vartheta \tag{8.14}$$

is a constant of motion, we can obtain from Eqs. (8.12)–(8.14):

$$\frac{d^2}{dt'^2}u_{\parallel}(t') + hu_{\parallel}(t') = \xi[1 - 3u_{\parallel}^2(t')]/2 \tag{8.15}$$

$$\frac{d^2}{dt'^2}u_{\perp}(t') + hu_{\perp}(t') = -3\xi u_{\perp}(t')u_{\parallel}(t')/2 \tag{8.16}$$

Here, $p_\vartheta = I\dot{\vartheta}$, $p_\varphi = I\dot{\varphi}\sin^2\vartheta$, $\Omega^2 = \dot{\vartheta}^2 + \dot{\varphi}^2\sin^2\vartheta$, and we have introduced the normalized quantities:

$$t' = \frac{t}{\eta} \qquad h = \frac{H}{kT} \qquad r^2 = \frac{p_\varphi^2}{2IkT} \qquad \eta^2 = \frac{I}{2kT} \qquad \xi = \frac{\mu E_0}{kT} \tag{8.17}$$

Equation (8.15) can be readily integrated to yield [108]:

$$\left[\frac{d}{dt'}u_\parallel(t')\right]^2 = [h + \xi u_\parallel(t')][1 - u_\parallel^2(t')] - r^2 = \Phi[u_\parallel(t')] \tag{8.18}$$

In the right-hand side of Eq. (8.18), $\Phi[u_\parallel(t')]$ is a third-order polynomial in $u_\parallel(t)$. Thus, $\Phi[u_\parallel(t')]$ has three roots, two of which are in the interval $[-1, 1]$ (we denote them $e_1$ and $e_2, e_1 > e_2$) and the third one, $e_3$, is $\leq -1$ [108]. For numerical purposes, these roots may be expressed in trigonometric form. Thus we have [108]

$$e_1 = 2p\cos(\psi/3) - h/3\xi \tag{8.19}$$

$$e_2 = -2p\cos[(\psi + \pi)/3] - h/3\xi \tag{8.20}$$

$$e_3 = -2p\cos[(\psi - \pi)/3] - h/3\xi \tag{8.21}$$

where

$$p = \frac{1}{3}[3 + h^2/\xi^2]^{1/2} \tag{8.22}$$

$$\psi = \arccos\left\{p^{-3}\left[\frac{h}{3\xi} - \left(\frac{h}{3\xi}\right)^3 - \frac{r^2}{2\xi}\right]\right\} \tag{8.23}$$

Equation (8.18) with the initial condition $u_\parallel(0) = \cos\vartheta(0)$ is then solved [108] in terms of the Jacobian doubly periodic function $\mathrm{sn}(u|m)$ [39]:

$$u_\parallel(t') = e_1 - (e_1 - e_2)\mathrm{sn}^2(t'\sqrt{\xi(e_1 - e_3)/4} + \delta|m) \tag{8.24}$$

where

$$m = \frac{e_1 - e_2}{e_1 - e_3} \tag{8.25}$$

$$\delta = \int_0^{u'} [(1 - mx^2)(1 - x^2)]^{-1/2} dx \tag{8.26}$$

$$u' = \sqrt{[e_1 - u_{\|}(0)]/(e_1 - e_2)} \tag{8.27}$$

One may now solve Eq. (8.16), which, after insertion of Eq. (8.24), has the form of Lamé's equation [110], namely,

$$\frac{d^2}{dx^2} u_{\perp}(x) = [A + 6m\mathrm{sn}^2(x|m)]u_{\perp}(x) \tag{8.28}$$

where

$$A = -\frac{4h + 6\xi e_1}{\xi(e_1 - e_3)} \tag{8.29}$$

$$x = t'\sqrt{\xi(e_1 - e_3)/4} + \delta \tag{8.30}$$

The solution of Lamé's equation (8.28) in terms of Jacobian $\Theta$ functions [39] is given in [110]. However, an alternative method can be used here [106]. This method consists of expressing $u_{\perp}(t')$ in terms of $u_{\|}(t')$ given by Eq. (8.24) and $\varphi(t')$. The latter can be easily evaluated on noting that

$$p_{\varphi} = I \sin^2\vartheta \frac{d\varphi}{dt} \tag{8.31}$$

is a constant of motion (since $\varphi$ is a cyclic coordinate). Thus, we have

$$\begin{aligned}\frac{d}{dt'}\varphi(t') &= \frac{r}{\sin^2\vartheta(t')} \\ &= \frac{r}{2}\left[\frac{1}{1 - \cos\vartheta(t')} + \frac{1}{1 + \cos\vartheta(t')}\right] \\ &= \frac{r}{2}\left[\frac{1}{1 - e_1 + (e_1 - e_2)\mathrm{sn}^2(x|m)} + \frac{1}{1 + e_1 - (e_1 - e_2)\mathrm{sn}^2(x|m)}\right]\end{aligned} \tag{8.32}$$

where $x$ is defined by Eq. (8.30). On integrating this differential equation, we have

$$\varphi(t') = \varphi(0) + \frac{r}{\sqrt{\xi(e_1 - e_3)}} \int_\delta^x \frac{dz}{1 - e_1 + (e_1 - e_2)\mathrm{sn}^2(z|m)} + \frac{r}{\sqrt{\xi(e_1 - e_3)}} \int_\delta^x \frac{dz}{1 + e_1 - (e_1 - e_2)\mathrm{sn}^2(z|m)} \tag{8.33}$$

On noting that $u_\perp(t')$ can be written as

$$u_\perp(t') = \sin\vartheta(t')\cos\varphi(t') = \sqrt{[1 - \cos\vartheta(t')][1 + \cos\vartheta(t')]}\cos\varphi(t') \tag{8.34}$$

we have

$$\begin{aligned} u_\perp(t') = {} & \{(1 - e_1^2)[1 - m\mathrm{sn}^2(ib|m)\mathrm{sn}^2(x|m)] \\ & \times [1 - m\mathrm{sn}^2(ia|m)\mathrm{sn}^2(x|m)]\}^{1/2} \\ & \times \cos\left[\varphi(0) + \frac{r}{(1 - e_1)\sqrt{\xi(e_1 - e_3)}} \int_\delta^x \frac{dz}{1 - m\mathrm{sn}^2(ia|m)\mathrm{sn}^2(z|m)} \right. \\ & \left. + \frac{r}{(1 + e_1)\sqrt{\xi(e_1 - e_3)}} \int_\delta^x \frac{dz}{1 - m\mathrm{sn}^2(ib|m)\mathrm{sn}^2(z|m)}\right] \end{aligned} \tag{8.35}$$

where

$$\mathrm{sn}^2(ia|m) = \frac{e_1 - e_3}{e_1 - 1} \qquad \mathrm{sn}^2(ib|m) = \frac{e_1 - e_3}{e_1 + 1} \tag{8.36}$$

The parameters $a$ and $b$ can be calculated from equations:

$$a = F\left[\arctan\left(\sqrt{\frac{e_1 - e_3}{1 - e_1}}\right) \middle| 1 - m\right] \tag{8.37}$$

$$b = iK(m) + F\left[\arctan\left(\sqrt{\frac{-1 - e_3}{1 + e_2}}\right) \middle| 1 - m\right] \tag{8.38}$$

where $K(m)$ is the complete elliptic integral of the first kind [39]. Next, on using the formulas [39, 111]

$$1 - m\mathrm{sn}^2(\alpha|m)\mathrm{sn}^2(u|m) = \Theta^2(0)\frac{\Theta(u+\alpha)\Theta(u-\alpha)}{\Theta^2(u)\Theta^2(\alpha)} \tag{8.39}$$

$$\int_0^u \frac{dz}{1 - m\mathrm{sn}^2(\alpha|m)\mathrm{sn}^2(z|m)} = u + \frac{\mathrm{sn}(\alpha|m)}{\mathrm{cn}(\alpha|m)\mathrm{dn}(\alpha|m)}\left[u\frac{\Theta'(\alpha)}{\Theta(\alpha)} + \frac{1}{2}\ln\frac{\Theta(u-\alpha)}{\Theta(u+\alpha)}\right] \tag{8.40}$$

$$\frac{\Theta'(\alpha)}{\Theta(\alpha)} = \frac{H'(\alpha)}{H(\alpha)} - \frac{\mathrm{cn}(\alpha|m)\mathrm{dn}(\alpha|m)}{\mathrm{sn}(\alpha|m)} \tag{8.41}$$

Eq. (8.35) can be transformed as follows:

$$\begin{aligned} u_\perp(t') = {} & \frac{(1-e_1^2)^{1/2}\Theta^2(0)}{2\Theta(ia)\Theta(ib)} \\ & \times \left\{ e^{i[\varphi(0)+\lambda t']}\frac{\Theta(x+ia)\Theta(x+ib)}{\Theta^2(x)}\left[\frac{\Theta(\delta-ia)\Theta(\delta-ib)}{\Theta(\delta+ia)\Theta(\delta+ib)}\right]^{1/2} \right. \\ & \left. + e^{-i[\varphi(0)+\lambda t']}\frac{\Theta(x-ia)\Theta(x-ib)}{\Theta^2(x)}\left[\frac{\Theta(\delta+ia)\Theta(\delta+ib)}{\Theta(\delta-ia)\Theta(\delta-ib)}\right]^{1/2} \right\} \end{aligned} \tag{8.42}$$

where $\Theta(u)$ and $H(u)$ are the Jacobian theta-functions [39, 111] and

$$\lambda = i\frac{\sqrt{\xi(e_1 - e_3)}}{2}\left[\frac{H'(ia)}{H(ia)} + \frac{H'(ib)}{H(ib)}\right] \tag{8.43}$$

Here, we have used the following equalities [39, 108]:

$$\frac{(1-e_1)^{-1}r\mathrm{sn}(ia|m)}{\sqrt{\xi(e_1-e_3)}\mathrm{cn}(ia|m)\mathrm{dn}(ia|m)} = \sqrt{\frac{r^2}{\xi(e_1-1)(e_2-1)(e_3-1)}} = i \tag{8.44}$$

$$\frac{(1+e_1)^{-1}r\mathrm{sn}(ib|m)}{\sqrt{\xi(e_1-e_3)}\mathrm{cn}(ib|m)\mathrm{dn}(ib|m)} = \sqrt{\frac{r^2}{\xi(e_1+1)(e_2+1)(e_3+1)}} = i \tag{8.45}$$

$$\mathrm{sn}^2(u|m) + \mathrm{cn}^2(u|m) = 1$$

$$m\mathrm{sn}^2(u|m) + \mathrm{dn}^2(u|m) = 1$$

Having determined the solution of the deterministic Eqs. (8.15) and (8.16), we can now apply the memory function formalism to the calculation of orientational correlation functions in the context of the $J$-diffusion model.

## B. Application to Dielectric Relaxation and Linear Kerr Effect Responses

### 1. Dielectric Response

In the zero wavevector limit, the tensor of the complex dielectric susceptibility $\chi_{ij}(\omega) = \chi'_{ij}(\omega) + i\chi''_{ij}(\omega)$ is diagonal and has only two independent components, namely, perpendicular $\chi_\perp(\omega) = \chi_{XX}(\omega) = \chi_{YY}(\omega)$, and parallel, $\chi_\|(\omega) = \chi_{ZZ}(\omega)$, to the $Z$ axis. For a system of noninteracting molecules these components are given by

$$\chi_\gamma(\omega) = \frac{\mu^2 N_0}{kT}\left[C_{1,\gamma}(0) + i\omega\int_0^\infty C_{1,\gamma}(t)e^{i\omega t}dt\right] \qquad (\gamma = \|, \perp) \tag{8.46}$$

where $N_0$ is the concentration of molecules and

$$C_{1,\|}(t) = \langle u_\|(0)u_\|(t)\rangle_0 - \langle u_\|(0)\rangle_0^2 \tag{8.47}$$

and

$$C_{1,\perp}(t) = \langle u_\perp(0)u_\perp(t)\rangle_0 \tag{8.48}$$

The evaluation of the longitudinal $[\chi_\|(\omega)]$ and transverse $[\chi_\perp(\omega)$ components of the complex susceptibility in the context of the $J$-diffusion model requires, according to Eqs. (8.6) and (8.46), the knowledge of the Fourier–Laplace transforms

$$\tilde{C}^c_{1,\gamma}(\omega + i/\tau) = \int_0^\infty C^c_{1,\gamma}(t)e^{i(\omega+i/\tau)t}dt$$

of the free rotation correlation functions $C^c_{1,\|}(t)$ and $C^c_{1,\perp}(t)$. The correlation functions $C^c_{1,\|}(t)$ and $C^c_{1,\perp}(t)$ are also defined by Eqs. (8.47) and (8.48), where $u_\|(t)$ and $u_\perp(t)$ are given by Eqs. (8.24) and (8.42), respectively, and the statistical averages are defined as follows:

$$\langle(\cdots)\rangle_0 = Z^{-1}\int_0^{2\pi}\int_0^{\pi}\int_{-\infty}^{+\infty}\int_{-\infty}^{+\infty}(\cdots)\exp(-h)d\vartheta(0)d\varphi(0)dp_\vartheta(0)dp_\varphi(0) \quad (8.49)$$

where the partition function $Z$ is given by

$$Z = 8\pi^2 IkT \sinh(\xi)/\xi$$

However, in order to evaluate the statistical averages in Eq. (8.49), it is more convenient to change the variables $\{\vartheta, \varphi, p_\vartheta, p_\varphi\}$ into the variables $\{\delta, \varphi, h, r\}$. The Jacobian of this transformation is given by [108]

$$\frac{\partial(\vartheta, \varphi, p_\vartheta, p_\varphi)}{\partial(\delta, \varphi, h, r)} = \frac{2IkT}{\sqrt{\xi(e_1 - e_3)}} \quad (8.50)$$

Let us first calculate the Fourier–Laplace transform of $C^c_{1,\|}(t)$. On using Eq. (8.24) and the Fourier series [110]

$$\mathrm{sn}^2(u|m) = \frac{1}{m}\left[1 - \frac{E(m)}{K(m)}\right] - \frac{2\pi^2}{mK^2(m)}\sum_{n=1}^{\infty}\frac{nq^n}{1-q^{2n}}\cos\left[\frac{n\pi u}{K(m)}\right] \quad (8.51)$$

where $q = \exp[-\pi K(1-m)/K(m)]$ and $E(m)$ is the complete elliptic integral of the second kind [39], we obtain, after an analytic integration over $\delta$, $\varphi$, and $t'$ and some algebra [108]:

$$\tilde{C}^c_{1,\|}(\omega + i/\tau) = \frac{iC^c_{1,\|}(0)}{\omega + i/\tau} - \int_{-\xi}^{\infty} e^{-h}\int_0^{l_m}\sum_{n=1}^{\infty} S_n^{1,\|}\left[\frac{\eta(\omega + i/\tau)}{\sqrt{\xi(e_1 - e_3)}}\middle| m\right] dr dh \quad (8.52)$$

Here,

$$S_n^{1,\|}(u|m) = \frac{i\eta\pi^5(e_1 - e_3)n^4}{u\sinh(\xi)K^3(m)\sinh^2[\pi nK(1-m)/K(m)][\pi^2n^2 - 4u^2K^2(m)]} \quad (8.53)$$

$$l_m = \frac{1}{3}\left[8h + \frac{2(h^2 + 3\xi^2)}{h + (h^2 + 3\xi^2)^{1/2}}\right]^{1/2} \quad (8.54)$$

and

$$C^c_{1,\parallel}(0) = \langle \cos^2\vartheta\rangle_0 - \langle \cos\vartheta\rangle_0^2 = 1 - \coth^2\xi + \frac{1}{\xi^2} \tag{8.55}$$

On using Eqs. (8.6) and (8.7), one can calculate the normalized longitudinal susceptibility given by

$$\chi_1^{\parallel}(\omega) = C^c_{1,\parallel}(0) + \frac{i\omega\tilde{C}^c_{1,\parallel}(\omega + i/\tau)}{1 - \tilde{C}^c_{1,\parallel}(\omega + i/\tau)/\tau C^c_{1,\parallel}(0)} \tag{8.56}$$

The calculation of the transverse susceptibility can be accomplished in a similar manner. On using the Fourier series (see Appendix C)

$$\begin{aligned}\frac{\Theta(u+\alpha)\Theta(u+\beta)}{\Theta^2(u)} = & \frac{\pi}{2K(m)H'^2(0)} \\ & \times \sum_{n=-\infty}^{\infty} \frac{H(\beta)H'(\alpha) + H(\alpha)H'(\beta) - i\pi n H(\alpha)H(\beta)/K(m)}{\sin\left(\pi[\alpha+\beta+2inK(1-m)]/2K(m)\right)} \\ & \times \exp\left[\frac{in\pi u}{K(m)}\right]\end{aligned} \tag{8.57}$$

one obtains (by integrating analytically over $\delta$, $\varphi$, and $t'$) the spectrum of the transverse correlation function $\tilde{C}^c_{1,\perp}(\omega + i/\tau)$ [108]:

$$\tilde{C}^c_{1,\perp}(\omega + i/\tau) = \frac{iC^c_{1,\perp}(0)}{(\omega + i/\tau)} - \int_{-\xi}^{\infty} e^{-h} \int_0^{l_m} \sum_{n=-\infty}^{\infty} S_n^{1,\perp}\left[\frac{\eta(\omega + i/\tau)}{\sqrt{\xi(e_1 - e_3)}}\middle| m\right] dr dh \tag{8.58}$$

where

$$S_n^{1,\perp}(u|m) = \frac{i\eta\pi(e_1 - e_3)B_n^4}{u\sinh(\xi)K(m)\sinh^2[a + b + \pi nK(1-m)/K(m)](B_n^2 - u^2)} \tag{8.59}$$

$$B_n = \frac{n\pi}{2K(m)} + \frac{i}{2}\left[\frac{H'(ia)}{H(ia)} + \frac{H'(ib)}{H(ib)}\right]$$

$$C^c_{1,\perp}(0) = L(\xi)/\xi \tag{8.60}$$

and $L(\xi)$ is the Langevin function Eq. (2.35). Thus the normalized transverse susceptibility $\chi_{1,\perp}(\omega)$ is given by

$$\chi_{1,\perp}(\omega) = C^c_{1,\perp}(0) + \frac{i\omega\tilde{C}^c_{1,\perp}(\omega + i/\tau)}{1 - \tilde{C}^c_{1,\perp}(\omega + i/\tau)/\tau C^c_{1,\perp}(0)} \tag{8.61}$$

Equations (8.56) and (8.61) are the general solution for the dielectric response and they are valid for any values of $\xi$ and $\tau$. In the absence of the constant field (i.e., in the limit $\mathbf{E}_0 \to 0$), the results obtained for $\chi_{1,\|}(\omega)$ and $\chi_{1,\perp}(\omega)$ coincide precisely with those of Gordon's $J$-diffusion model for linear molecules [11]

$$\chi_{1,\|}(\omega) = \chi_{1,\perp}(\omega) = \frac{1 + i(\omega + i/\tau)\tilde{C}_{FR}(\omega + i/\tau)}{3[1 - \tilde{C}_{FR}(\omega + i/\tau)/\tau]} \tag{8.62}$$

where

$$\tilde{C}_{FR}(\omega + i/\tau) = -i\eta^2(\omega + i/\tau)\exp[-\eta^2(\omega + i/\tau)^2]E_1[-\eta^2(\omega + i/\tau)^2]$$

and

$$E_1(z) = \int_z^\infty \frac{\exp(-t)}{t}\,dt$$

is the first-order exponential integral function [39].

### 2. *Linear Kerr Effect Response*

Now, let us consider the *linear* Kerr effect response of rigid polar molecules to a small ac field on which is superimposed a strong dc field. As Section V.A showed, the relevant correlation function is in this case:

$$C_2(t) = \langle \cos\vartheta(0)P_2(\cos\vartheta(t))\rangle_0 - \langle \cos\vartheta(0)\rangle_0\langle P_2(\cos\vartheta(0))\rangle_0 \tag{8.63}$$

The Kerr effect response can be worked out just as in the dielectric case. We obtain, on using Eq. (8.51) and the Fourier series:

$$\operatorname{sn}^4(u|m) = \frac{1}{3m^2}\left[2 + m - 2(1+m)\frac{E(m)}{K(m)}\right] + \frac{\pi^2}{3m^2K^2(m)}\sum_{n=1}^{\infty}\left[\frac{n^2\pi^2}{K^2(m)} - 4(m+1)\right] \times \frac{nq^n}{1-q^{2n}}\cos\left[\frac{n\pi u}{K(m)}\right] \tag{8.64}$$

[the derivation of Eq. (8.64) is given in Appendix C] the one-sided FT of the free rotation correlation function $C_2^c(t)$. Thus, we have [109]

$$\tilde{C}_2^c(\omega + i/\tau) = \frac{iC_2^c(0)}{\omega + i/\tau} - \int_{-\xi}^{\infty} e^{-h} \int_0^{l_m} \sum_{n=1}^{\infty} S_n^2 \left[ \frac{\eta(\omega + i/\tau)}{\sqrt{\xi(e_1 - e_3)}} \middle| m \right] drdh \tag{8.65}$$

where

$$S_n^2(u|m) = \frac{i\eta\pi^5 n^4 (e_1 - e_3)\{3e_1 - (e_1 - e_3)[m + 1 - n^2\pi^2/4K^2(m)]\}}{uK^3(m) \sinh^2[\pi n K(1-m)/K(m)][\pi^2 n^2 - 4K^2(m)u^2]} \tag{8.66}$$

$$C_2^c(0) = \frac{3\coth\xi}{\xi} + \frac{3\coth\xi}{\xi^2} - \frac{6}{\xi^3} - \frac{3}{\xi} \tag{8.67}$$

The normalized *linear* Kerr effect response is given by

$$\chi_K(\omega) = C_2^c(0) + \frac{i\omega\tilde{C}_2^c(\omega + i/\tau)}{1 - \tilde{C}_2^c(\omega + i/\tau)/\tau C_2^c(0)} \tag{8.68}$$

### 3. *Spectra and Relaxation Times*

Equations (8.52), (8.56), (8.59), (8.61), (8.65), and (8.68) allow us to calculate the dielectric susceptibilities as well as the linear Kerr effect response to an ac field when the molecule is subjected to a strong external field. The advantage of using these expressions is that the series involved are very rapidly convergent, so that only three to seven terms in the series in each expression (8.52), (8.59), and (8.65) are necessary to ensure six digits of accuracy in the majority of cases.

We turn to the discussion of dielectric and birefringence relaxation spectra. In Figures 61–66, the imaginary parts of the longitudinal susceptibility, transverse susceptibility and linear Kerr effect response, for various values of the parameters $\eta/\tau$ and $\xi$ are represented. The longitudinal dielectric and Kerr response exhibit inertia-corrected Debye behavior for large values of the parameter $\eta/\tau$. Conversely, when this parameter is small, these spectra exhibit two absorption bands. One crosses a maximum at $\omega \sim \tau^{-1}$ and represents the Debye absorption. A second absorption band appears at high frequencies, located approximately at $\omega_{\|} = 2\omega_E$, where

$$\omega_E = \sqrt{\frac{\mu E_0}{I}} \tag{8.69}$$

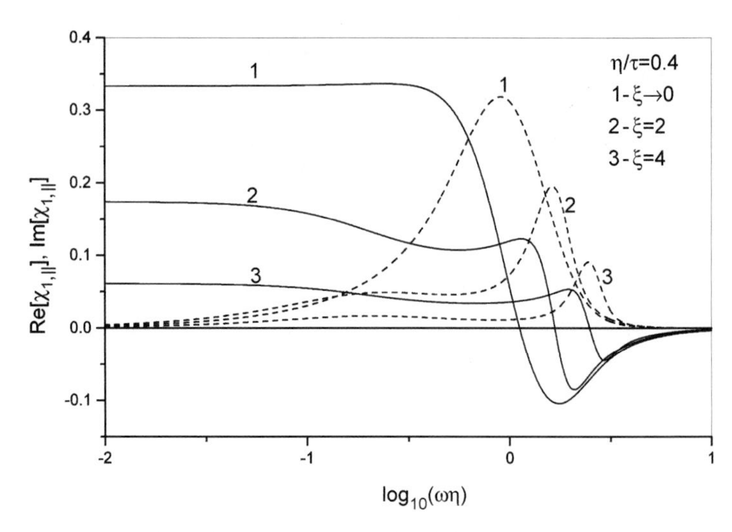


**Figure 61.** The $\mathrm{Re}[\chi_{1,\|}(\omega)]$(solid lines) and $\mathrm{Im}[\chi_{1,\|}(\omega)]$ as functions of $\log_{10}(\omega\eta)$. The curves are plotted for $\eta/\tau = 0.4$ and $\xi = 0$ (curves 1), $\xi = 2$ (curves 2) and $\xi = 4$ (curves 3).

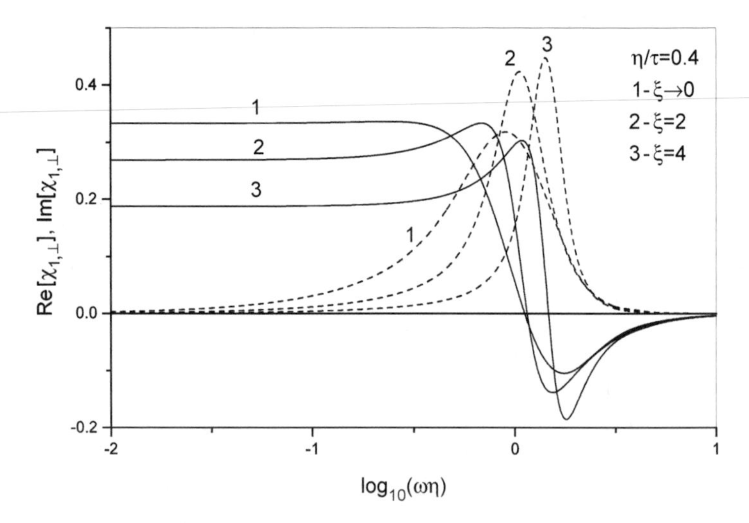


**Figure 62.** The same as Figure 61 for the transverse component $\chi_{1,\perp}(\omega)$.

as expected from the parametric oscillation of the dipole [9]. In the case of transverse dielectric relaxation, there is no relaxation band since the average projection of the dipole moment vector onto any axis perpendicular to $\mathbf{E}_0$ is equal to zero. There is a high-frequency band only located at $\omega_\perp = \omega_E$. The real

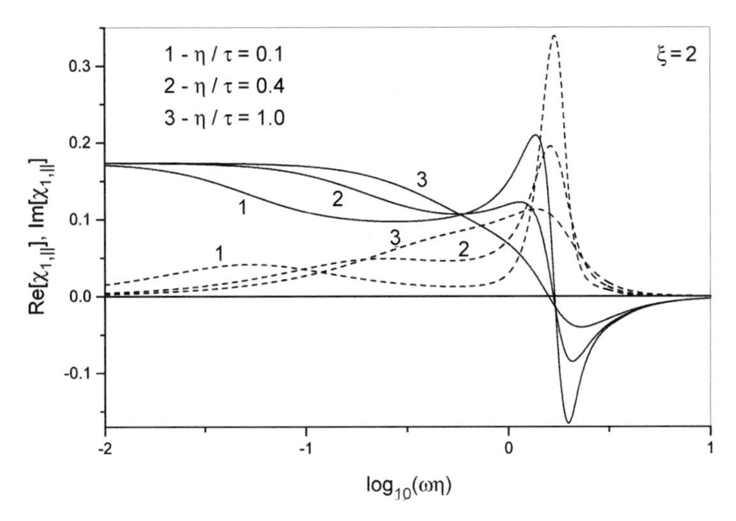


**Figure 63.** The $\mathrm{Re}[\chi_{1,\|}(\omega)]$ (solid lines) and $\mathrm{Im}[\chi_{1,\|}(\omega)]$ as functions of $\log_{10}(\omega\eta)$. The curves are plotted for $\xi = 2$ and $\eta/\tau = 0.1$ (curves 1), $\eta/\tau = 0.4$ (curves 2) and $\eta/\tau = 1.0$ (curves 3).

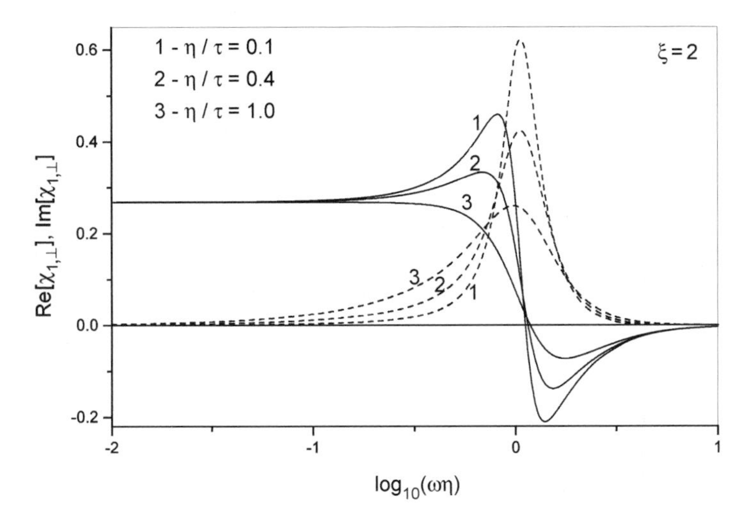


**Figure 64.** The same as Figure 63 for the transverse component $\chi_{1,\perp}(\omega)$.

parts of the spectra have inertia-corrected Debye behavior for frequent collisions and exhibit a negative part, typical of inertial effects.

We can now focus our attention on some properties of the model. It is certainly of interest to have asymptotic expansions for the spectra and relaxation times. The relaxation time is defined as before, as the area under the curve of the normalized relaxation function (here, this is the correlation time since the

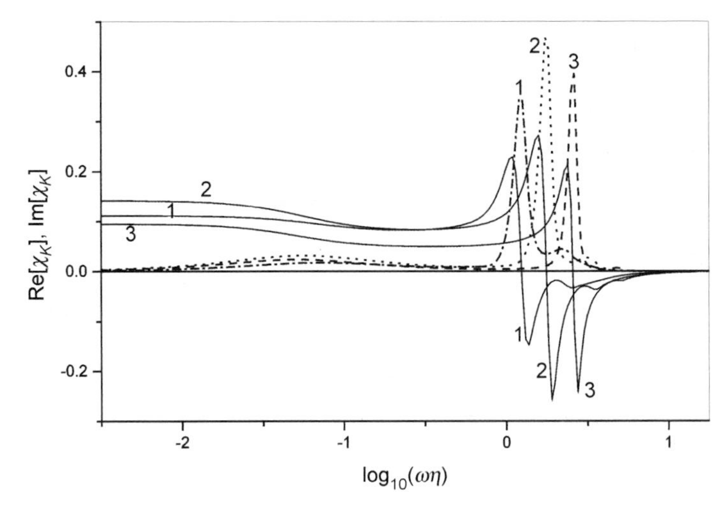


**Figure 65.** The Re[$\chi_K(\omega)$] (solid lines) and Im[$\chi_K(\omega)$] as functions of $\log_{10}(\omega\eta)$. The curves are plotted for $\xi = 1$ and $\eta/\tau = 0.1$ (curves 1), $\eta/\tau = 0.2$ (curves 2) and $\eta/\tau = 0.8$ (curves 3).

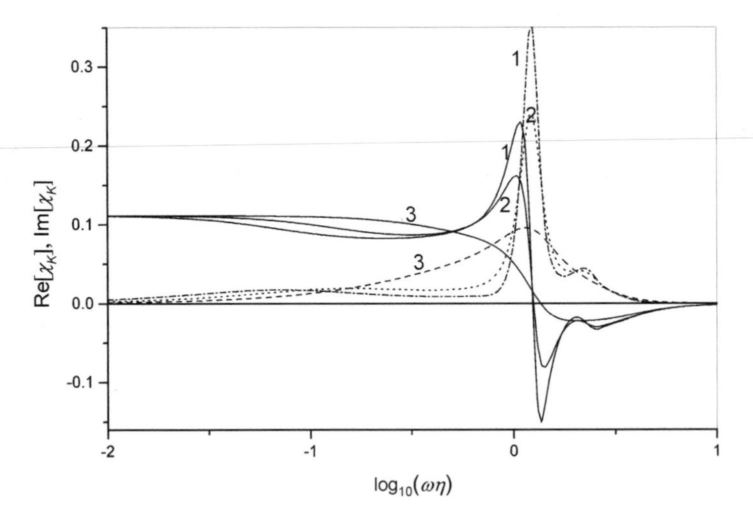


**Figure 66.** The same as Figure 5 for $\eta/\tau = 0.1$ (rare collision limit) and $\xi = 1$ (curves 1), $\xi = 2$ (curves 2), and $\xi = 3$ (curves 1).

relaxation function is a correlation function). Thus we have

$$\tau_\gamma = \frac{\tilde{C}_\gamma(0)}{C_\gamma(0)} = \frac{\tilde{C}^c_\gamma(i/\tau)/C^c_\gamma(0)}{1 - \tilde{C}^c_\gamma(i/\tau)/\tau C^c_\gamma(0)} \qquad \gamma = 1, \|; 1, \perp, \text{ or } 2 \tag{8.70}$$

On the other hand, in the low-frequency limit, all the responses may be reduced to a Debye one. We have

$$\chi_\gamma(\omega) \propto \frac{1}{1 - i\omega\tau_\gamma} \tag{8.71}$$

Moreover, in the frequent collision regime ($\tau \to 0$), the free correlation function spectrum for any case considered here may be expanded as

$$\tilde{C}^c_\gamma(z) = i\frac{C^c_\gamma(0)}{z} - i\frac{\ddot{C}^c_\gamma(0)}{z^3} + O(z^{-5}) \qquad z \to \infty \tag{8.72}$$

since $C^c_\gamma(t)$ is a real and even function of time. On using the equations of motion (8.12) and (8.13) and the equipartition theorem, one finds easily that [108, 109]

$$\ddot{C}^c_{1,\parallel}(0) = -L(\xi)/(\xi\eta^2) \tag{8.73}$$

$$\ddot{C}^c_{1,\perp}(0) = -\frac{1}{2\eta^2}[1 - L(\xi)/\xi] \tag{8.74}$$

$$\ddot{C}^c_2(0) = \frac{3}{\eta^2\xi^2}[3L(\xi) - \xi] \tag{8.75}$$

Thus, according to Eqs. (8.70) and (8.72), the relaxation times are given by

$$\tau_\gamma = -\frac{C^c_\gamma(0)}{\tau\ddot{C}^c_\gamma(0)} \tag{8.76}$$

Equation (8.76) determines the relaxation times for any field strength in the frequent collision regime. In that regime, on defining the Debye relaxation time as $\tau_D = I/(2\tau kT)$ [11] and on using explicit expressions for $C^c_\gamma(0)$ and $\ddot{C}^c_\gamma(0)$, one can obtain [108]

$$\tau_{1,\parallel} = \tau_D \frac{1 + \xi^2 - \xi^2 \coth^2 \xi}{\xi \coth \xi - 1} \tag{8.77}$$

$$\tau_{1,\perp} = 2\tau_D \frac{\xi \coth \xi - 1}{1 + \xi^2 - \xi \coth \xi} \tag{8.78}$$

for dielectric relaxation, and [109]

$$\tau_2 = \tau_D \frac{\xi \coth \xi + \xi^2 \coth^2 \xi - 2 - \xi^2}{3 + \xi^2 - 3\xi \coth \xi} \tag{8.79}$$

for Kerr effect relaxation. In the high field limit ($\xi \to \infty$), all these relaxation times follow a hyperbolic law

$$\tau_\gamma \sim \tau_D/\xi \tag{8.80}$$

In the low-field limit ($\xi \to 0$), we have [108]

$$\tau_{1,\parallel} = \tau_D \left[1 - \frac{2}{15}\xi^2 + O(\xi^4)\right] \tag{8.81}$$

$$\tau_{1,\perp} = \tau_D \left[1 - \frac{1}{10}\xi^2 + O(\xi^4)\right] \tag{8.82}$$

(for dielectric relaxation) and [109]

$$\tau_2 = \tau_D \left[\frac{2}{3} - \frac{4}{63}\xi^2 + \frac{44}{6615}\xi^4 + O(\xi^6)\right] \tag{8.83}$$

(for the birefringence relaxation time $\tau_2 = \tau_K$).

In the rare collision limit ($\tau \to \infty$), the correlation functions $C^c_{1,\parallel}(t)$ and $C^c_2(t)$ do not vanish after long periods of time as a consequence of the conservation of the projection of $u_\parallel(t)$ on the direction of the dc electric field. After long periods of time, we thus have

$$\tilde{C}^c_\gamma(\omega + i/\tau) = \frac{C^c_\gamma(\infty)}{-i\omega + 1/\tau} \tag{8.84}$$

and the relaxation times become

$$\tau_\gamma = \tau \frac{\Delta_\gamma}{1 - \Delta_\gamma} \tag{8.85}$$

where

$$\Delta_\gamma = \frac{C^c_\gamma(\infty)}{C^c_\gamma(0)} \tag{8.86}$$

In order to evaluate $C_{\gamma}^{c}(\infty)$ in Eq. (8.86), we recall that the correlation functions $C_{1,\parallel}^{c}(t)$ and $C_{2}^{c}(t)$ may be split in two parts: a time-dependent and a constant one. After having noticed that $P_1(u_\parallel) = u_\parallel$, $P_2(u_\parallel) = (3u_\parallel^2 - 1)/2$, where $u_\parallel(t) = \cos\vartheta(t)$ is given by Eq. (8.24), we have

$$\langle \cos\vartheta(0)\cos\vartheta(\infty)\rangle_0 = \langle (e_1 - A)^2\rangle_0 \tag{8.87}$$

$$\langle \cos\vartheta(\infty) P_2(\cos\vartheta(0))\rangle_0 = \frac{3}{2}\langle (e_1^2 - 2e_1 A + B - 1)(e_1 - A)\rangle_0 \tag{8.88}$$

where

$$A = (e_1 - e_3)\left[1 - \frac{E(m)}{K(m)}\right] \tag{8.89}$$

$$B = \frac{(e_1 - e_3)^2}{3}\left[2 + m - 2(1+m)\frac{E(m)}{K(m)}\right] \tag{8.90}$$

and $m$ is given by Eq. (8.25). We have Eqs. (8.87) and (8.88) because the time-dependent parts of the correlation functions $C_{1,\parallel}^{c}(t)$ and $C_{2}^{c}(t)$ vanish after averaging over $\delta$ at $t \to \infty$. Thus, one has

$$C_{1,\parallel}^{c}(\infty) = \langle\cos\vartheta(0)\cos\vartheta(\infty)\rangle_0 - \langle\cos\vartheta(0)\rangle_0^2 = \langle (e_1 - A)^2\rangle_0 - (\coth\xi - 1/\xi)^2 \tag{8.91}$$

$$\begin{aligned} C_{2}^{c}(\infty) &= \langle\cos\vartheta(\infty) P_2(\cos\vartheta(0))\rangle_0 - \langle\cos\vartheta(0)\rangle_0 \langle P_2(\cos\vartheta(0))\rangle_0 \\ &= \frac{3}{2}\langle (e_1^2 - 2e_1 A + B - 1)(e_1 - A)\rangle_0 \\ &\quad + \frac{3}{\xi^3} + \frac{1}{\xi} - \coth\xi - \frac{6\coth\xi}{\xi^2} + \frac{3\coth^2\xi}{\xi} \end{aligned} \tag{8.92}$$

Equation (8.85) stipulates that the relaxation rates are inversely proportional to the time between collisions. This behavior is analogous to that of the Kramers formula [112] for chemical reaction rates in the low-damping regime, as in this case the rate is proportional to the collision frequency. We note also a similarity between the Kramers rate at moderate-to-high damping and the results obtained for the relaxation times in the frequent collision limit, as those relaxation times are proportional to the time between collisions, that is, inversely proportional to the friction parameter. The effect of the dc external field on the relaxation times

$\tau_{1,\|}$ and $\tau_K = \tau_2$ is shown in Figures 67 and 68. Here, it is clearly seen that these relaxation times exhibit a frictional crossover just as in the Kramers theory of reaction rates [112]. When the external field is increased, then this crossover region is shifted to lower $\tau/\eta$.

It is of interest to make a comparison of results given by the *J*-diffusion and Fokker–Planck model. This may be accomplished in the high damping regime only, because exact solutions for rotation in three dimensions including inertia are not presently available for the Fokker–Planck model. We recall (see Section V) that the asymptotic expansion of the birefringence correlation time obtained in the low field limit is given by Eq. (5.28). This expression obviously differs from Eq. (8.83) although giving the same qualitative behavior. This may be explained as follows. The solution of the Fokker–Planck equation for the *linear response* of a system of polar molecules in a constant dc field is always obtained by solving the hierarchy of differential recurrence relations (5.22) for the equilibrium correlation functions $f_n(t)$, namely,

$$2\tau_D \frac{d}{dt} f_n(t) + n(n+1) f_n(t) = \frac{\xi n(n+1)}{(2n+1)} [f_{n-1}(t) - f_{n+1}(t)] \tag{8.93}$$

where

$$f_n(t) = \langle \cos \vartheta(0) P_n[\cos \vartheta(t)] \rangle_0 - \langle \cos \vartheta \rangle_0 \langle P_n(\cos \vartheta) \rangle_0 \tag{8.94}$$

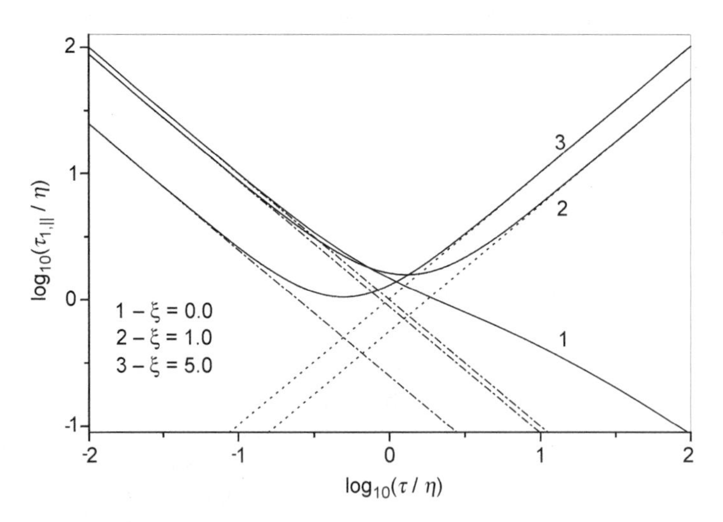


**Figure 67.** The $\log_{10}(\tau_{1,\|}/\eta)$ vs. $\log_{10}(\tau/\eta)$. Comparison of the exact Eq. (8.70) (solid lines) for $\xi = 0$ (curves 1) $\xi = 1$ (curves 2) and $\xi = 5$ (curves 3) with Eq. (8.85) (dotted lines) for rare collisions ($\tau \to \infty$) and with Eq. (8.76) (dot dashed lines) for frequent collisions ($\tau \to 0$).

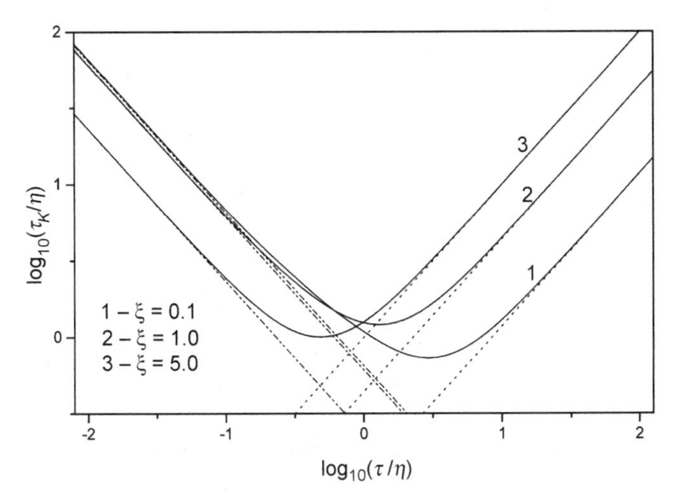


**Figure 68.** The $\log_{10}(\tau_K/\eta)$ vs. $\log_{10}(\tau/\eta)$. Comparison of the exact Eq. (8.70) (solid lines) for $\xi = 0.1$ (curves 1) $\xi = 1$ (curves 2) and $\xi = 5$ (curves 3) with Eq. (8.85) (dotted lines) for rare collisions ($\tau \to \infty$) and with Eq. (8.66) (dot dashed lines) for frequent collisions ($\tau \to 0$).

is for the longitudinal dielectric relaxation and

$$f_n(t) = \langle P_2[\cos\vartheta(0)]P_n[\cos\vartheta(t)]\rangle_0 - \langle P_2(\cos\vartheta)\rangle_0\langle P_n(\cos\vartheta)\rangle_0 \tag{8.95}$$

is for the dynamic Kerr effect [cf. Eq. (2.11)]. Equation (8.93) allows one to evaluate the effective relaxation time, which is the reciprocal of the initial slope of the normalized correlation function ([40] and references quoted therein). The effective relaxation time $\tau_{\text{ef}}$ of the correlation function $C(t)$, which is defined as [40]

$$\tau_{\text{ef}} = -\frac{C_1(0)}{(d/dt)C_1(t)|_{t=0}} \tag{8.96}$$

is the time constant associated with the initial slope of the correlation function. It also contains the contributions from all the eigenvalues just as the relaxation time. The behavior of the correlation time is sometimes similar to that of the effective relaxation time [40]. However, for some models these quantities may differ considerably in certain ranges of the model parameters, particularly if an activation process is involved and in this case $\tau_{\text{ef}}$ gives information on the initial decay of the correlation function only. Thus, we have from Eqs. (8.93) and (8.96):

$$\frac{\tau_{\text{ef}}}{\tau_{\text{D}}} = -\frac{f_1(0)}{\dot{f}_1(0)} = \left[1 + \frac{\xi f_2(0)}{3f_1(0)}\right]^{-1} \tag{8.97}$$

where

$$f_n(0) = \frac{n+1}{2n+1}\frac{I_{n+3/2}(\xi)}{I_{1/2}(\xi)} + \frac{n}{2n+1}\frac{I_{n-1/2}(\xi)}{I_{1/2}(\xi)} - \frac{I_{3/2}(\xi)}{I_{1/2}(\xi)}\frac{I_{n+1/2}(\xi)}{I_{1/2}(\xi)} \tag{8.98}$$

for the longitudinal dielectric relaxation and

$$\begin{aligned} f_n(0) = {} & \frac{3(n+1)(n+2)}{2(2n+1)(2n+3)}\frac{I_{n+5/2}(\xi)}{I_{1/2}(\xi)} + \frac{3n(n-1)}{2(4n^2-1)}\frac{I_{n-3/2}(\xi)}{I_{1/2}(\xi)} \\ & + \left[\frac{n(n+1)}{(2n+3)(2n-1)} - \frac{I_{5/2}(\xi)}{I_{1/2}(\xi)}\right]\frac{I_{n+1/2}(\xi)}{I_{1/2}(\xi)} \end{aligned} \tag{8.99}$$

for the dynamic Kerr effect [cf. Eq. (5.23)]. On using Eqs. (8.97)–(8.99) and the representation of the modified Bessel functions $I_{\nu+1/2}(\xi)$ in terms of the hyperbolic functions [39], one can readily obtain that Eq. (8.97) predicts the expressions (8.77) and (8.79) for the relaxation times $\tau_{1,\parallel}$ and $\tau_2$. This means that in the limit of frequent collisions, the $J$-diffusion model yields a result that coincides exactly with the effective relaxation time obtained in the context of the Fokker–Planck–Langevin model. This is due to the two different collision mechanisms underlying these models. Indeed, in both models, the collisions are assumed instantaneous, but the main difference between them is that the molecule undergoes both a damping torque and a white noise torque permanently in the Fokker–Planck–Langevin model while the rotation is free between collisions in the $J$-diffusion model. This difference between $\tau_K^{\mathrm{FP}}$ and $\tau_{\mathrm{ef}}$ ($\approx 15\%$) exists only for $\xi < 1$ and tends to disappear by increasing the strength of the constant field. The small difference (several %) between the two models exists for the dielectric relaxation time as well [8, 40]. One may expect that the behavior of the relaxation functions for the $J$-diffusion model is very similar to that for other extended rotational diffusion models [11], so that it does not really matter which model should be used for qualitative estimations of the linear response.

It should be mentioned that the imaginary parts of the linear response spectra $\mathrm{Im}[\chi_{1,\parallel}(\omega)]$, $\mathrm{Im}[\chi_{1,\perp}(\omega)]$, and $\mathrm{Im}[\chi_K(\omega)]$ satisfy the following sum rules [108, 109]:

$$\int_0^\infty \omega\,\mathrm{Im}[\chi_{1,\parallel}(\omega)]d\omega = \frac{\pi kT}{I\xi}L(\xi) \tag{8.100}$$

$$\int_0^\infty \omega\,\mathrm{Im}[\chi_{1,\perp}(\omega)]d\omega = \frac{\pi kT}{2I}\left(1 - \frac{L(\xi)}{\xi}\right) \tag{8.101}$$

$$\int_0^\infty \omega \, \mathrm{Im}[\chi_K(\omega)]d\omega = \frac{3\pi kT}{I\xi^2}[3L(\xi) - \xi] \tag{8.102}$$

The solutions we have presented concern two particular problems for linear dielectric and Kerr effect responses. The next step to accomplish would be to include the effect of the induced dipole moment in order to make the model more realistic from the point of view of the field dependence of the spectrum. This case will include a barrier-crossing phenomenon, and one may expect that an additional relaxation process will appear at low frequencies. The relaxation time may then have an exponential behavior, like in Kramers's theory of chemical reaction rates [112]. On the other hand, for the evaluation of the Kerr effect response, one could also take into account the finite duration of collisions, which means that a second-order memory function would be needed, as well as another timescale for characterizing the duration of collisions [113].

# IX. APPLICATIONS TO RELATED RELAXATION PHENOMENA

## A. Rotational Diffusion in a Mean-Field Potential

### 1. *Application to Dielectric Relaxation of Linear Molecules in a Cubic Potential*

The noninertial rotational Brownian motion of a particle in a mean-field potential arises in a variety of problems. Examples can be found in rotational dynamics of impurities in molecular crystals and orientational relaxation of molecules in nematic liquid crystals (e.g. [7, 14, 114–118]). This problem bears a close resemblance to that of magnetic relaxation of single domain ferromagnetic particles as we shall see in Section IX.B. Although the mean-field approximation has a restricted area of applicability, as it ignores local order effects, the model nevertheless is easily visualized and moreover allows us to carry out quantitative evaluations of the physical parameters. The rotational dynamics of the particle in the mean potential is described by the noninertial Fokker–Planck (Smoluchowski) equation (2.1) for the distribution function $W$ of the orientations of a unit vector $\mathbf{u}$, fixed in the molecule [1, 3], where now $V = V_0 + V_1$ and $V_0$ and $V_1$ having the meaning of a mean-field potential and of the field-dependent part, respectively (in the absence of external electric fields $V = V_0$). As already seen in Section III, the Smoluchowski equation (2.1) can be formally solved by means of an expansion of the distribution function $W$ in spherical harmonics. In such an approach, the problem is reduced to the solution of an infinite system of multiterm differential-recurrence equations for the moments [the expectation values of the spherical harmonics $\langle Y_{l,m}\rangle(t)$]. This system may, however, be solved if one uses the matrix continued fraction method. Here, this method is used to calculate the low-frequency spectrum of the susceptibility $\chi(\omega)$ and the relaxation time $\tau$ of a system of noninteracting dipole particles in a cubic field. The problem of the orientational relaxation in a cubic field has been considered in a few papers. Brot and Darmon [119] numerically treated a $J$-diffusion model of a linear rotator in a cubic potential. Dynamics of the linear molecule in a potential of cubic symmetry has also been considered by De Raedt et al. [117, 120] in the context of the Langevin model. However, the above authors confined themselves to the study of high-frequency dynamics and did not investigate in detail the long-time behavior associated with the jumps between potential wells. This long-time behavior in the noninertial limit of the rotational diffusion in a cubic potential was treated either in the discrete orientation approximation or only in the high barrier limit (see, e.g. [112] and references cited therein). Although the use of these approximations considerably simplifies the analysis, the results obtained have, however, a restricted area of applicability. Both approaches are not applied when the potential energy $V$ is of the order of magnitude of the thermal energy $kT$ and do not allow us to study high-frequency (intrawell) relaxation processes. Recently, this problem was reexamined by Déjardin and Kalmykov [121]. Here, we follow their solution.

Our calculations are based on linear (in the amplitude of an external electric field) response theory. In order to evaluate $\chi(\omega)$ and $\tau$, one must first calculate the equilibrium dipole correlation function $C(t)$, which is more conveniently accomplished directly from the underlying Langevin equation rather than from the Smoluchowski equation. Hence, we shall use the results already presented in Section III.A, particularly those given by Eqs. (3.5), (3.16), (3.18), (3.21), and (3.22). Let us now specialize the problem to the calculation of the dielectric relaxation in a cubic potential. We shall use the following representation for the potential energy, namely,

$$\begin{aligned} V &= \frac{K}{4}(\sin^4\vartheta \sin^2 2\varphi + \sin^2 2\vartheta) \\ &= v_{4,0}Y_{4,0} + v_{4,4}Y_{4,4} + v_{4,-4}Y_{4,-4} + \frac{K}{5} \end{aligned} \tag{9.1}$$

where

$$v_{4,0} = -\frac{2K}{15}\sqrt{\pi} \qquad v_{4,4} = -\frac{K}{15}\sqrt{\frac{10\pi}{7}} \qquad v_{4,-4} = -\frac{K}{15}\sqrt{\frac{10\pi}{7}} \tag{9.2}$$

$K$ is the anisotropy constant, which may have both positive and negative values. For $K > 0$, the potential (9.1) has 6 minima, 8 maxima, and 12 saddle points, for example, in the directions [100] (minimum), [111] (maximum), and [110] (saddle point), accordingly. Both the saddle energy and the potential barrier height are equal to $\sigma$, where

$$\sigma = \frac{K}{4kT} \tag{9.3}$$

is the dimensionless anisotropy parameter. If $K < 0$, the maxima and minima are interchanged so that the barrier height and the saddle energy are equal to $|\sigma|/3$ and $-|\sigma|$, respectively [121].

By using Eqs. (9.1) and (9.2) in Eq. (3.21), one can obtain the 15-term differential-recurrence equation [121]

$$\tau_{\mathrm{D}} \dot{Y}_{l,m} = \sum_{s=-1}^{1} \sum_{r=-2}^{2} d_{l,m,l+2r,m+4s} Y_{l+2r,m+4s} \tag{9.4}$$

where

$$\begin{aligned} d_{l,m,l+2r,m+4s} = & -\frac{l(l+1)}{2} \delta_{r,0}\delta_{s,0} + (-1)^m v_{4,4s} \frac{\sqrt{(2l+1)(2l+4r+1)}}{12\sqrt{\pi}} \\ & \times [r(2l+2r+1) - 10] \\ & \times \langle l, 0, l+2r, 0|4, 0\rangle \langle l, m, l+2r, -m-4s|4, -4s\rangle \end{aligned} \tag{9.5}$$

(the $d_{n,m,r,s}$, which are necessary in the calculations, are listed explicitly in Appendix D). In the absence of external fields, the system is in equilibrium with the Boltzmann distribution function $W_0$ given by

$$W_0(\vartheta, \varphi) = \frac{1}{Z} \exp\left[-\frac{V(\vartheta, \varphi)}{kT}\right]$$

Thus, on multiplying Eq. (9.4) by $\cos \vartheta(0)$, and averaging the resulting equation over the equilibrium distribution function $W_0$ at the instant $t = 0$, we obtain

$$\tau_{\mathrm{D}} \frac{d}{dt} c_{l,m}(t) = \sum_{s=-1}^{1} \sum_{r=-2}^{2} d_{l,m,l+2r,m+4s} c_{l+2r,m+4s}(t) \tag{9.6}$$

where the equilibrium correlation functions $c_{l,m}(t)$ are defined by

$$c_{l,m}(t) = \left\langle \cos \vartheta(0) Y_{l,m}(t) \right\rangle_0 \tag{9.7}$$

and $\langle\ \rangle_0$ designates the equilibrium average at $t = 0$.

### 2. *Evaluation of the Complex Susceptibility in Terms of Matrix Continued Fractions*

Before proceeding, we first recall the principal results of *linear response theory*. The application of this theory to the present problem states that the decay of the dielectric polarization $P(t)$ of a system of noninteracting particles, when a small constant external field $\mathbf{E}_1[(\boldsymbol{\mu}\cdot\mathbf{E}_1)/kT \ll 1]$ applied along the $Z$ axis has been switched off at time $t = 0$ is

$$P(t) = \chi_S E_1 C(t) \tag{9.8}$$

where

$$\chi_S = \frac{N_0\mu^2}{kT}\sqrt{\frac{4\pi}{3}}c_{1,0}(0) = \frac{N_0\mu^2}{3kT}$$

is the static susceptibility, $N_0$ is the number of dipolar particles per unit volume, and

$$C(t) = \frac{\langle \cos\vartheta(0)\cos\vartheta(t)\rangle_0}{\langle \cos^2\vartheta(0)\rangle_0} = \frac{c_{1,0}(t)}{c_{1,0}(0)} \tag{9.9}$$

is the normalized dipole autocorrelation function. Here, we have taken into account that for the cubic anisotropy $\langle \cos^2\vartheta\rangle_0 = \frac{1}{3}$ due to the symmetry properties. Thus, the complex susceptibility $\chi(\omega)$ is given by

$$\chi(\omega) = \chi'(\omega) - i\chi''(\omega) = \chi_S[1 - i\omega\tilde{C}(i\omega)] \tag{9.10}$$

where the tilde denotes the one-sided FT, namely,

$$\tilde{C}(i\omega) = \int_0^\infty C(t)e^{-i\omega t}dt$$

Another quantity of interest is the correlation time $\tau$ which is given by

$$\tau = \int_0^\infty C(t)dt = \tilde{C}(0) = \frac{\tilde{c}_{1,0}(0)}{c_{1,0}(0)} \tag{9.11}$$

By following [121], Eq. (9.6) can be transformed into the matrix three-term differential-recurrence equation

$$\tau_D\frac{d}{dt}\mathbf{C}_n(t) = \mathbf{Q}_n^-\mathbf{C}_{n-1}(t) + \mathbf{Q}_n\mathbf{C}_n(t) + \mathbf{Q}_n^+\mathbf{C}_{n+1}(t) \qquad n = 1, 2, 3, \ldots \tag{9.12}$$

where the matrices $\mathbf{Q}_n^-$, $\mathbf{Q}_n$, $\mathbf{Q}_n^+$ are explicitly defined in Appendix C and the column vector $\mathbf{C}_n(t)$ is given by

$$\mathbf{C}_n(t) = \begin{pmatrix} \mathbf{c}_{4n}(t) \\ \mathbf{c}_{4n-1}(t) \\ \mathbf{c}_{4n-2}(t) \\ \mathbf{c}_{4n-3}(t) \end{pmatrix} \tag{9.13}$$

This vector consists of four column subvectors $\mathbf{c}_{4n-i}(t)$ such that

$$\mathbf{c}_{4n-i}(t) = \begin{pmatrix} c_{4n-i,-4(n-1+\delta_{i0})}(t) \\ c_{4n-i,-4(n-2+\delta_{i0})}(t) \\ \vdots \\ c_{4n-i,4(n-1+\delta_{i0})}(t) \end{pmatrix} \qquad i = 0, 1, 2, 3 \tag{9.14}$$

In particular

$$\mathbf{C}_0(t) = \mathbf{0} \quad \text{and} \quad \mathbf{C}_1(t) = \begin{pmatrix} C_{4,-4}(t) \\ C_{4,0}(t) \\ C_{4,4}(t) \\ C_{3,0}(t) \\ C_{2,0}(t) \\ C_{1,0}(t) \end{pmatrix} \tag{9.15}$$

On applying the general method of solution of the matrix three-term differential-recurrence Eq. (9.12), described in Section III.C, we obtain the *exact* solution for the Laplace transform $\tilde{\mathbf{C}}_1(s)$ in terms of matrix continued fractions, namely,

$$\begin{aligned} \tilde{\mathbf{C}}_1(s) &= \tau_D[\tau_D s\mathbf{I} - \mathbf{Q}_1 - \mathbf{Q}_1^+\mathbf{S}_2(s)]^{-1} \\ &\quad \times \left\{ \mathbf{C}_1(0) + \sum_{n=2}^{\infty} \left[ \prod_{k=2}^{n} \mathbf{Q}_{k-1}^+ \mathbf{S}_k(s) (\mathbf{Q}_k^-)^{-1} \right] \mathbf{C}_n(0) \right\} \end{aligned} \tag{9.16}$$

where the matrix continued fraction $\mathbf{S}_n(s)$ is defined by Eq. (3.41). The initial condition vectors $\mathbf{C}_n(0)$ can also be calculated with the help of the matrix continued fractions $\mathbf{S}_n(0)$ (see Appendix D). On putting $s = i\omega$ in Eq. (9.16), and on using Eqs. (9.10) and (9.11) and Eq. (A61) from Appendix D, we are now able to calculate the complex dielectric susceptibility and the relaxation times. The exact matrix continued fraction solution [Eq. (9.16)] is very convenient for the purpose of computation. The greatest dimension of all the matrices involved is of the order of $10^2$, which allows us to carry out the calculations on a personal computer.

The spectra of the real and imaginary part of the complex susceptibility $\chi(\omega)$ vs. $\log_{10}(\omega\tau_D)$ and $\sigma$ are shown in Figures 69 and 70. The calculations were carried out from Eqs. (9.10) and (9.16) with $\mu^2 N_0/kT = 1$. It is clearly seen in these figures how the only diffusion mode at $\sigma = 0$ is split onto two relaxation modes for large $|\sigma|$. The first (low-frequency) mode is located at the vicinity of the average frequency of reorientation of the dipole moment vector. The characteristic frequency and the half-width of this low-frequency band are determined by $\tau^{-1}$. The second, a much weaker process, is caused by the contribution of the high-frequency intrawell relaxation modes. In spite of the large number of the high-frequency modes involved, the high-frequency relaxation process can be effectively described by one relaxation mode only with the amplitude $\Delta_W$ and a characteristic time $\tau_W$ given by (for $\sigma \gg 1$) [121]

$$\Delta_W \approx \frac{1}{4\sigma} \qquad \text{and} \qquad \tau_W \approx \frac{\tau_D}{4\sigma} \tag{9.17}$$

for positive anisotropy constant and

$$\Delta_W \approx \frac{3}{8|\sigma|} \qquad \text{and} \qquad \tau_W \approx \frac{3\tau_D}{8|\sigma|} \tag{9.18}$$

for negative anisotropy constant.

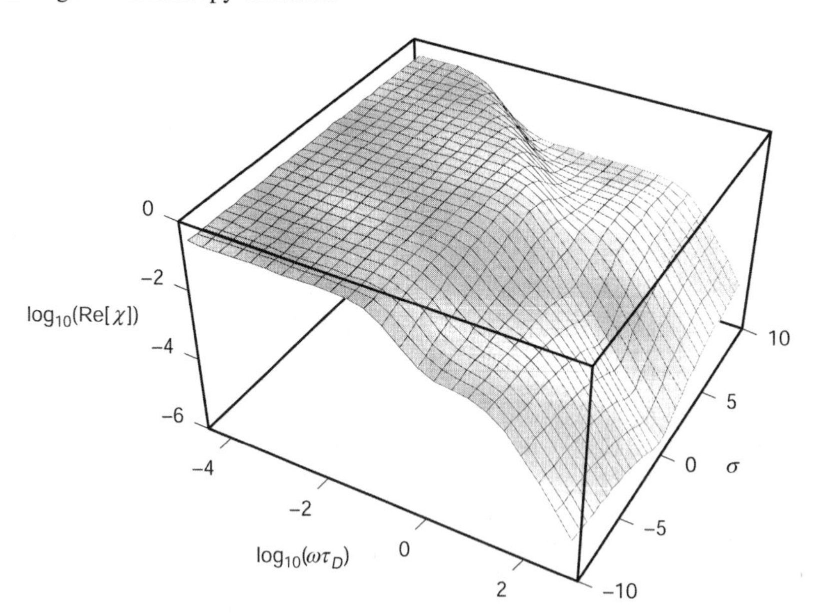


**Figure 69.** Three-dimensional plot $\log_{10}(\text{Re}\{\chi\})$ vs. $\log_{10}(\omega\tau_D)$ and $\sigma$.

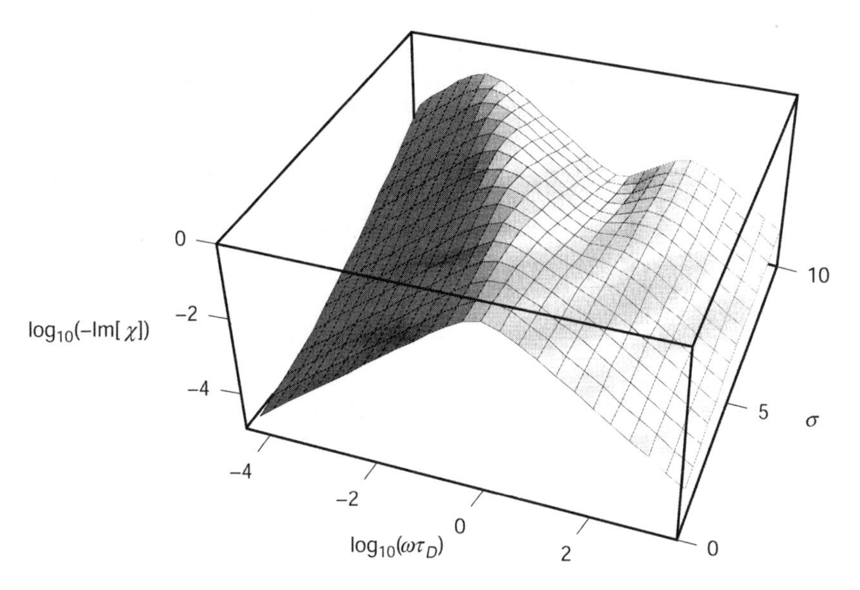


**Figure 70.** Three-dimensional plot $\log_{10}(-\text{Im}\{\chi\})$ vs. $\log_{10}(\omega\tau_D)$ and $\sigma$.

The difference between Eqs. (9.17) and (9.18) is due to the fact that the frequencies of oscillations at the minima of the cubic potentials with positive and negative anisotropy differ by a factor of two-thirds [122]. As also shown in [121], the dielectric susceptibility $\chi(\omega)$ can be evaluated from the following approximate equation:

$$\chi(\omega) \approx \frac{\mu^2 N_0}{3kT}\left[\frac{1-\Delta_W}{1+i\omega\tau} + \frac{\Delta_W}{1+i\omega\tau_W}\right] \tag{9.19}$$

where $\tau$, $\tau_W$, and $\Delta_W$ are given by Eqs. (9.11), (9.17), and (9.18). Thus, according to Eq. (9.19), which is the superposition of two relaxation modes, both low- and high-frequency relaxation processes are effectively governed by a single relaxation mode in the high-barrier approximation ($|\sigma| \gg 1$). Moreover, this two mode approximation provides satisfactory accuracy for $\chi(\omega)$ even for moderate barriers ($1 \le |\sigma| \le 3$) [121]. Equation (9.19) corresponds to the representation of the dipole autocorrelation function $C(t)$ in the time domain by a sum of two exponentials

$$C(t) \approx (1-\Delta_W)e^{-t/\tau} + \Delta_W e^{-t/\tau_W} \tag{9.20}$$

Such a simple representation in the high-barrier limit given is possible due to the fact that diffusion within potential wells is much faster than the rate transition between the wells ($\tau_W \ll \tau$). One can also use simple extrapolating formulas for the correlation time $\tau$ obtained in [121], which provide a very good approximation to the exact matrix continued fraction solutions *for all* $\sigma$, namely,

$$\tau = \tau_D \frac{(e^{\sigma} - 1)}{\sigma} \left[ \frac{\pi}{8\sqrt{2}} + \left(1 - \frac{\pi}{8\sqrt{2}}\right) 2^{-\sigma} \right]$$

for $\sigma \geq 0$ and

$$\tau = -3\tau_D \frac{(e^{-\sigma/3} - 1)}{\sigma} \left[ \frac{\pi}{4\sqrt{2}} + \left(1 - \frac{\pi}{4\sqrt{2}}\right) 2^{\sigma/3} \right]$$

for $\sigma \leq 0$.

Here, we have demonstrated how the matrix continued fraction approach to the solution of the nonlinear Langevin equations may be successfully applied to the problem of the rotational Brownian motion in a cubic potential. Other examples of the application of the approach to various problems of Brownian motion in a potential are given elsewhere [7, 8].

### B. Relaxation Processes in Superparamagnetic Particles

From a mathematical point of view, the calculation of the dielectric response of polar molecules in a mean-field potential closely resembles the problem of magnetic relaxation of single domain ferromagnetic particles [3, 6, 122, 123]. A single domain ferromagnetic particle is characterized by an internal potential having two local states of equilibrium with a potential barrier between them. If the particles are small ($\sim 10\,\text{nm}$) and, accordingly, the potential barriers are relatively low, the magnetization vector $\mathbf{M}(t)$ may cross over the barriers between one potential well and another due to thermal fluctuations [122]. The thermal instability of the magnetization results in the phenomenon of superparamagnetism [124], because each particle behaves like a paramagnetic atom having a magnetic moment $\sim 10^4$–$10^5$ BM.

The pioneering theory [125] of thermal fluctuations of the magnetization of a single domain ferromagnetic particle due to Néel was further developed by Brown [126] on the basis of the Langevin equation approach to the theory of Brownian motion. In the diffusion model, the dynamics of the magnetization vector $\mathbf{M}(t)$ of a single-domain particle is similar to the rotational Brownian motion of a macromolecule in a liquid so that it may be described by the Langevin equation for the magnetization. Brown took as the Langevin equation,

Gilbert's equation augmented by a random field term [127]

$$\frac{d}{dt}\mathbf{M}(t) = \gamma[\mathbf{M}(t) \times [\mathbf{H}(t) + \mathbf{h}(t) - \eta\dot{\mathbf{M}}(t)]] \tag{9.21}$$

where $\gamma$ is the gyromagnetic ratio, $\eta$ is the damping parameter, the magnetic field $\mathbf{H}(t)$ may consist of externally applied magnetic fields and the crystalline anisotropy field, and $\mathbf{h}(t)$ is a random Gaussian field with the white noise properties. The random field $\mathbf{h}(t)$ takes into account the thermal fluctuations of the magnetization of an individual particle. Brown derived from Eq. (9.21), the underlying Fokker–Planck equation for the distribution function $W$ of the orientations of $\mathbf{M}$ [122]

$$\frac{\partial}{\partial t}W = \frac{1}{2\tau_N}\{\beta[\alpha^{-1}\mathbf{u}\cdot(\mathrm{grad}V \times \mathrm{grad}W) + \mathrm{div}(W\mathrm{grad}V)] + \Delta W\} \tag{9.22}$$

where $\Delta$ is the Laplacian on the surface of the unit sphere; $\mathbf{u}$ is the unit vector directed along $\mathbf{M}$; $V$ is the free energy density expressed as a function of $\mathbf{M}$;

$$\tau_N = \frac{\beta\eta M_s^2(1+\alpha^{-2})}{2} \tag{9.23}$$

is the characteristic relaxation time, $\beta = \nu/kT$; $\nu$ is the volume of the particle; $M_s$ is the saturation magnetization, which is determined by the material and the temperature; and

$$\alpha = \gamma\eta M_s \tag{9.24}$$

is the dimensionless dissipation parameter. A discussion of the assumptions made in the derivation of the Fokker–Planck and Gilbert equations is given elsewhere (e.g. [122, 123]). For $\alpha = \infty$ (high damping limit), Eq. (9.22) has the same form as Eq. (2.1).

### *1. Nonlinear Response of Superparamagnetic Particles to Sudden Changes of a Strong dc Magnetic Field: The Langevin Equation Approach*

Because of the large magnitude of the magnetic dipole moment, the energy of the particle even in a moderate external magnetic field $\mathbf{H}_0$ may be of the same order as the thermal energy $kT$. The fact that the relaxation times depend on the external field (see, e.g. [122]) requires one to take into account nonlinear effects when analyzing the relaxation of magnetization due to sudden changes both in magnitude and in direction of an external dc magnetic field. If the characteristic time of the variation of the field is much shorter than that of the magnetization relaxation, one can consider these changes as instantaneous. The problem may be solved in a manner similar to that of the calculation of the after-effect function. However, so far the theory has been successfully developed only for linear response of superparamagnetic fine particles, where the changes of the particle energy due to the variations of an external magnetic field is far less than the thermal energy $kT$. The theory of nonlinear response in strong external fields has not yet been developed. For nonlinear response, the results have only been derived for a weak ac magnetic field with the help of perturbation theory (see, e.g. [128, 129]). More insight to this question has been given in recent papers [130, 131], where some problems of nonlinear response in the dynamics of the magnetization of uniaxial superparamagnetic particles were solved on applying the matrix continued fraction approach. Here, we briefly discuss the results obtained in [130, 131].

In the study of the magnetic relaxation, the quantities of interest are averages involving the spherical harmonics $Y_{l,m}$ defined by Eq. (3.17). The derivation of the stochastic equation for $Y_{l,m}$ from Gilbert's equation (9.21) and the averaging of the equation so obtained over an ensemble of particles, which all have at time $t$ the same magnetization $\mathbf{M}(t)$, was given in [132, 133], where the following equation for the sharp values $Y_{l,m}$ was obtained, namely,

$$\tau_N \frac{d}{dt} Y_{l,m} = -\frac{\beta}{2}(\mathrm{grad} V + \alpha^{-1}[\mathbf{u} \times \mathrm{grad} V]) \cdot \mathrm{grad} Y_{l,m} - \frac{l(l+1)}{2} Y_{l,m} \tag{9.25}$$

The right-hand side of Eq. (9.25) can be expressed in terms of the angular momentum operators $\mathrm{L}_Z$, $\mathrm{L}_\pm$, $\mathrm{L}^2$ [57, 60]

$$\mathrm{L}^2 = -\Delta \qquad \mathrm{L}_Z = -i\frac{\partial}{\partial\varphi} \qquad \mathrm{L}_\pm = e^{\pm i\varphi}\left(\pm\frac{\partial}{\partial\vartheta} + i\cot\vartheta\frac{\partial}{\partial\varphi}\right) \tag{9.26}$$

Thus, one obtains [132, 133]

$$\begin{aligned} \tau_N \dot{Y}_{l,m} = {} & \frac{\beta}{4}[\mathrm{L}^2(VY_{l,m}) - V\mathrm{L}^2 Y_{l,m} - Y_{l,m}\mathrm{L}^2 V] - \frac{1}{2}\mathrm{L}^2 Y_{l,m} - \frac{i\beta}{4\alpha}\sqrt{\frac{3}{2\pi}} \\ & \times \{Y_{1,1}^{-1}[(\mathrm{L}_Z V_+)(\mathrm{L}_+ Y_{l,m}) - (\mathrm{L}_+ V_+)(\mathrm{L}_Z Y_{l,m})] \\ & + Y_{1,-1}^{-1}[(\mathrm{L}_Z V_-)(\mathrm{L}_- Y_{l,m}) - (\mathrm{L}_- V_-)(\mathrm{L}_Z Y_{l,m})]\} \end{aligned} \tag{9.27}$$

where

$$\begin{aligned} & V = U - (\mathbf{M} \cdot \mathbf{H}) = V_+ + V_- \\ & V_+ = \sum_{R=1}^{\infty}\sum_{S=0}^{R} v_{R,S} Y_{R,S} \qquad V_- = \sum_{R=1}^{\infty}\sum_{S=-R}^{-1} v_{R,S} Y_{R,S} \end{aligned} \tag{9.28}$$

By making use of the theory of angular momentum, one may essentially simplify the solution of the problem under consideration because the action of the angular moment operators on $Y_{l,m}$ is given by [57]

$$\mathrm{L}_Z Y_{l,m} = mY_{l,m} \qquad \mathrm{L}_\pm Y_{l,m} = \sqrt{l(l+1) - m(m\pm 1)}Y_{l,m\pm 1} \qquad \mathrm{L}^2 Y_{l,m} = l(l+1)Y_{l,m} \tag{9.29}$$

Furthermore, the functions $Y_{1,1}^{-1}$ and $Y_{1,-1}^{-1}$ may also be considered as operators, which act on $Y_{l,m}$ in accordance with the relation [132]

$$Y_{1,\pm 1}^{-1}Y_{l,\pm m} = \sqrt{\frac{8\pi(2l+1)(l-m)!}{3(l+m)!}} \times \sum_{\substack{L=m-\varepsilon_{l,m}\\ \Delta L=2}}^{l-1} \sqrt{\frac{(2L+1)(L+m-1)!}{(L-m+1)!}}Y_{L,\pm(m-1)}, \quad (m>0) \tag{9.30}$$

where $\varepsilon_{l,m} = 1$, if the indexes $l$ and $m$ have the same order of evenness and $\varepsilon_{l,m} = 0$, otherwise. Thus, one can transform Eq. (9.27) to [132]

$$\dot{Y}_{l,m} = \sum_{l',s} d_{l',m'\pm s,l,m} Y_{l',m\pm s} \tag{9.31}$$

where

$$\begin{aligned} d_{l',m\pm s,l,m} &= \int_0^{2\pi}\int_0^{\pi} Y_{l,m} L_{\mathrm{FP}} Y^*_{l',m\pm s} \sin\vartheta d\vartheta d\varphi \\ &= \frac{l(l+1)\delta_{l,l'}\delta_{s,0}}{2\tau_N} + (-1)^m \frac{\beta}{4\tau_N}\sqrt{\frac{(2l+1)(2l'+1)}{\pi}} \\ &\times \sum_{r=s}^{\infty} v_{r,\pm s}\left\{\frac{[l'(l'+1)-r(r+1)-l(l+1)]}{2\sqrt{2r+1}}\right. \\ &\quad \times \langle l,0,l',0|r,0\rangle\langle l,m,l',-m\mp s|r,\mp s\rangle \\ &\quad + \frac{i}{\alpha}\sqrt{\frac{(2r+1)(r-s)!}{(r+s)!}} \sum_{\substack{L=s-\varepsilon_{r,s}\\ \Delta L=2}}^{r-1} \sqrt{\frac{(L+s-1)!}{(L-s+1)!}}\langle l,0,l',0|L,0\rangle \\ &\quad \times \left((m\pm s)\sqrt{(L+s)(L-s+1)}\langle l,m,l',-m\mp s|L,\mp s\rangle \mp s\right. \\ &\quad \left.\left. \times \sqrt{(l'\pm m+s)(l'\mp m-s+1)}\langle l,m,l',-m\mp s\pm 1|L,\mp s\pm 1\rangle\right)\right\} \end{aligned} \tag{9.32}$$

Here, it is assumed that $s \geq 0$ and the summation is carried out over those values of indexes, for which the Clebsch–Gordan coefficients $\langle l_1, m_1, l_2, m_2 | l, m \rangle$ do not vanish. The $Y_{l,m}$ in Eq. (9.31) are functions of the sharp values $\vartheta$ and $\varphi$, which are themselves random variables with the distribution (probability density) function $W$. Thus, we may derive the moment system for the spherical harmonics averaged over $W$, namely,

$$\frac{d}{dt}\langle Y_{l,m}\rangle(t) = \sum_{l',m'} d_{l',m',l,m}\langle Y_{l',m'}\rangle(t) \tag{9.33}$$

### 2. *Transient Nonlinear Response*

Let us suppose that both magnitude and direction of the external dc magnetic field are suddenly changed at time $t = 0$ from $\mathbf{H}_{\mathrm{I}}$ to $\mathbf{H}_{\mathrm{II}}$. We are interested in the relaxation of the system of noninteracting superparamagnetic fine particles starting from an equilibrium state I with the distribution function $W_{\mathrm{I}}(t \leq 0)$ to another equilibrium state II with the distribution function $W_{\mathrm{II}}(t \to \infty)$. The distribution functions in the states I and II are given by

$$W_{\mathrm{I}} = Z_{\mathrm{I}}^{-1} e^{-\beta[U-(\mathbf{M}\cdot\mathbf{H}_{\mathrm{I}})]} \tag{9.34}$$

and

$$W_{\mathrm{II}} = Z_{\mathrm{II}}^{-1} e^{-\beta[U-(\mathbf{M}\cdot\mathbf{H}_{\mathrm{II}})]} \tag{9.35}$$

where $U$ is the anisotropy potential, $Z_N$ ($N =$ I, II) is the partition function, $\mathbf{M}$ is the magnetization vector defined by

$$\mathbf{M} = M_S\mathbf{u} = M_S(\mathbf{i}u_X + \mathbf{j}u_Y + \mathbf{k}u_Z) \tag{9.36}$$

and $u_X, u_Y, u_Z$ are the components of $\mathbf{u}$ in a spherical coordinate system, namely,

$$u_X = \sin\vartheta\cos\varphi \qquad u_Y = \sin\vartheta\sin\varphi \qquad u_Z = \cos\vartheta \tag{9.37}$$

Just as for the problem formulated in Section IV, this problem is truly nonlinear, because the changes both in the magnitude and in the direction of a strong external dc magnetic field may be considerable. The dynamics of the magnetization is described by the normalized relaxation function:

$$f(t) = \frac{\langle M_r\rangle(t) - \langle M_r\rangle_{\mathrm{II}}}{\langle M_r\rangle_{\mathrm{I}} - \langle M_r\rangle_{\mathrm{II}}} = \frac{\langle \mathbf{r}\cdot\mathbf{u}\rangle(t) - \langle \mathbf{r}\cdot\mathbf{u}\rangle_{\mathrm{II}}}{\langle \mathbf{r}\cdot\mathbf{u}\rangle_{\mathrm{I}} - \langle \mathbf{r}\cdot\mathbf{u}\rangle_{\mathrm{II}}} \tag{9.38}$$

where $M_r$ is the projection of $\mathbf{M}$ onto the direction of a unit vector $\mathbf{r}$:

$$\mathbf{r} = \mathbf{i}v_X + \mathbf{j}v_Y + \mathbf{k}v_Z \tag{9.39}$$

and the angle brackets denote the averaging over $W_N$,

$$\langle \mathbf{r}\cdot\mathbf{u}\rangle_N = \int_0^{2\pi}\int_0^{\pi} (\mathbf{r}\cdot\mathbf{u}) W_N(\vartheta,\varphi)\sin\vartheta d\vartheta d\varphi \tag{9.40}$$

The following differential-recurrence relations for the relaxation functions $c_{l,m}(t) = \langle Y_{l,m}\rangle(t) - \langle Y_{l,m}\rangle_{\text{II}}$ can be derived from Eq. (9.33) [130]:

$$\frac{d}{dt}c_{l,m}(t) = \sum_{l'}\sum_{s} d_{l',m\pm s,l,m}c_{l',m\pm s}(t) \tag{9.41}$$

with the initial conditions

$$c_{l,m}(0) = \langle Y_{l,m}\rangle_{\text{I}} - \langle Y_{l,m}\rangle_{\text{II}} \tag{9.42}$$

The normalized nonlinear relaxation function from Eq. (9.38) can be expressed in terms of $c_{l,m}(t)$ as

$$f(t) = \frac{\sqrt{2}v_Z c_{1,0}(t) + (v_X + iv_Y)c_{1,-1}(t) - (v_X - iv_Y)c_{1,1}(t)}{\sqrt{2}v_Z c_{1,0}(0) + (v_X + iv_Y)c_{1,-1}(0) - (v_X - iv_Y)c_{1,1}(0)} \tag{9.43}$$

Equations (9.41)–(9.43) are valid for all kinds of anisotropy (uniaxial, cubic, etc.). Here, in order to simplify the mathematical calculation, the uniaxial anisotropy will be considered. The free-energy density of a single domain magnetic particle with uniaxial anisotropy in a homogeneous external magnetic field has the form [6]

$$\begin{aligned} V &= -K\cos^2\vartheta - (\mathbf{M}\cdot\mathbf{H}_N) \\ &= -H_N M_S\sqrt{\frac{2\pi}{3}}\left[(v_X^N + iv_Y^N)Y_{1,-1} + \sqrt{2}v_Z^N Y_{1,0} - (v_X^N - iv_Y^N)Y_{1,1}\right] \\ &\quad - \frac{4K}{3}\sqrt{\frac{\pi}{5}}Y_{2,0} - \frac{K}{3} \end{aligned} \tag{9.44}$$

where $K$ is the anisotropy constant, $v_X^N, v_Y^N, v_Z^N$ are the direction cosines of the vector $\mathbf{H}_N$. We introduce the notations:

$$\sigma = \beta K \qquad h_N = \frac{H_N M_S}{2K} \tag{9.45}$$

Next, one can derive from Eqs. (9.41) and (9.44) a system of coupled equations for $c_{l,m}(t)$ [130]:

$$\frac{d}{dt}c_{l,m}(t) = \sum_{l'=-2}^{2}\sum_{s=-1}^{1} d_{l+l',m+s,l,m}\, c_{l+l',m+s}(t) \qquad (t > 0) \tag{9.46}$$

where $d_{n\,\pm\,2,m\,\pm\,1,n,m} \equiv 0$. From a mathematical point of view, Eq. (9.46) is a particular case of Eq. (4.13). Thus the same method of solution can be used. We introduce the $8n$-element vector $\mathbf{C}_n(t)$ defined by Eq. (4.18). Equation (9.46) can then be transformed to a three-term matrix difference-differential equation of the form of Eq. (3.37) by replacing $\tau_D$ by $\tau_N$ [130]

$$\tau_N \frac{d}{dt}\mathbf{C}_n(t) = \mathbf{Q}_n^- \mathbf{C}_{n-1}(t) + \mathbf{Q}_n \mathbf{C}_n(t) + \mathbf{Q}_n^+ \mathbf{C}_{n+1}(t) \qquad n = 1, 2, 3, \ldots \tag{9.47}$$

where the matrix $\mathbf{Q}_n^-$ is given by

$$\mathbf{Q}_n^- = \begin{pmatrix} -\dfrac{2n+1}{2n-2}\mathbf{Z}_{2n-2}^T & \mathbf{O} \\ \mathbf{W}_{2n-1} & -\dfrac{2n}{2n-3}\mathbf{Z}_{2n-3}^T \end{pmatrix} \tag{9.48}$$

(the symbol $T$ denotes transposition) and the matrices $\mathbf{Q}_n$ and $\mathbf{Q}_n^+$, have the form of those defined by Eq. (4.22), but with different definitions for the submatrices $\mathbf{X}_l$ and $\mathbf{Y}_l$; the submatrix $\mathbf{W}_l$ is defined by Eq. (4.24) while the submatrices $\mathbf{X}_l$, $\mathbf{Y}_l$, and $\mathbf{Z}_l$ are given by

$$\mathbf{X}_l = \begin{pmatrix} x_{l,-l} & x_{l,-l}^+ & 0 & \cdots & 0 & 0 \\ x_{l,-l+1}^- & x_{l,-l+1} & x_{l,-l+1}^+ & \cdots & 0 & 0 \\ 0 & x_{l,-l+2}^- & x_{l,-l+2} & \cdots & 0 & 0 \\ \vdots & \vdots & \vdots & \ddots & \vdots & \vdots \\ 0 & 0 & 0 & \cdots & x_{l,l-1} & x_{l,l-1}^+ \\ 0 & 0 & 0 & \cdots & x_{l,l}^- & x_{l,l} \end{pmatrix}$$

$$\mathbf{Y}_l = \begin{pmatrix} y_{l,-l}^- & y_{l,-l} & y_{l,-l}^+ & \cdots & 0 & 0 & 0 \\ 0 & y_{l,-l+1}^- & y_{l,-l+1} & \cdots & 0 & 0 & 0 \\ 0 & 0 & y_{l,-l+2}^- & \cdots & 0 & 0 & 0 \\ \vdots & \vdots & \vdots & \ddots & \vdots & \vdots & \vdots \\ 0 & 0 & 0 & \cdots & y_{l,l-1} & y_{l,l-1}^+ & 0 \\ 0 & 0 & 0 & \cdots & y_{l,l}^- & y_{l,l} & y_{l,l}^+ \end{pmatrix}$$

$$\mathbf{Z}_l = \begin{pmatrix} 0 & 0 & z_{l,-l} & 0 & \cdots & 0 & 0 & 0 & 0 \\ 0 & 0 & 0 & z_{l,-l+1} & \cdots & 0 & 0 & 0 & 0 \\ \vdots & \vdots & \vdots & \vdots & \ddots & \vdots & \vdots & \vdots & \vdots \\ 0 & 0 & 0 & 0 & \cdots & z_{l,l-1} & 0 & 0 & 0 \\ 0 & 0 & 0 & 0 & \cdots & 0 & z_{l,l} & 0 & 0 \end{pmatrix}$$

The elements of the submatrices $\mathbf{W}_l, \mathbf{X}_l, \mathbf{Y}_l$, and $\mathbf{Z}_l$ are

$$w_{n,m} = d_{n-1,m,n,m} = \sigma\left[h_{\text{II}}\gamma_Z^{\text{II}}(n+1) - i\frac{m}{\alpha}\right]\sqrt{\frac{n^2-m^2}{4n^2-1}}$$

$$w_{n,m}^{+} = -(w_{n,-m}^{-})^*$$

$$= d_{n-1,m+1,n,m} = \frac{(n+1)\,\sigma h_{\text{II}}(\gamma_X^{\text{II}} - i\gamma_Y^{\text{II}})}{2}\sqrt{\frac{(n-m)(n-m-1)}{4n^2-1}}$$

$$x_{n,m} = d_{n,m,n,m} = \frac{\sigma[n(n+1)-3m^2]}{(2n-1)(2n+3)} - \frac{n(n+1)}{2} - i\frac{m\sigma h_{\text{II}}\gamma_Z^{\text{II}}}{\alpha}$$

$$x_{n,m}^{+} = -(x_{n,-m}^{-})^* = d_{n,m+1,n,m} = -i\frac{\sigma h_{\text{II}}(\gamma_X^{\text{II}} - i\gamma_Y^{\text{II}})}{2\alpha}\sqrt{(n+m+1)(n-m)}$$

$$y_{n,m} = d_{n+1,m,n,m} = -\sigma\left(h_{\text{II}}n + i\frac{m}{\alpha}\right)\sqrt{\frac{(n+1)^2-m^2}{(2n+1)(2n+3)}}$$

$$y_{n,m}^{+} = -(y_{n,-m}^{-})^* = d_{n+1,m+1,n,m} = \frac{n\sigma h_{\text{II}}(\gamma_X^{\text{II}} - i\gamma_Y^{\text{II}})}{2}\sqrt{\frac{(n+m+1)(n+m+2)}{(2n+1)(2n+3)}}$$

$$z_{n,m} = d_{n+2,m,n,m} = -\frac{\sigma n}{2n+3}\sqrt{\frac{[(n+2)^2-m^2][(n+1)^2-m^2]}{(2n+1)(2n+5)}}$$

By invoking the general method for solving tridiagonal matrix recurrence relations (Section III.C), we have an exact solution of Eq. (9.47) for the Laplace transform $\tilde{\mathbf{C}}_1(s)$, namely,

$$\tilde{\mathbf{C}}_1(s) = \tau_N[\tau_N s\mathbf{I} - \mathbf{Q}_1 - \mathbf{Q}_1^{+}\mathbf{S}_2^{\text{II}}(s)]^{-1} \times \left\{\mathbf{C}_1(0) + \sum_{n=2}^{\infty}\left(\prod_{k=2}^{n}\mathbf{Q}_{k-1}^{+}[\tau_N s\mathbf{I} - \mathbf{Q}_k - \mathbf{Q}_k^{+}\mathbf{S}_{k+1}^{\text{II}}(s)]^{-1}\right)\mathbf{C}_n(0)\right\} \tag{9.49}$$

where

$$S_n(s) = \cfrac{\mathbf{I}}{s\tau_N\mathbf{I} - \mathbf{Q}_n - \mathbf{Q}_n^+ \cfrac{\mathbf{I}}{s\tau_N\mathbf{I} - \mathbf{Q}_{n+1} - \mathbf{Q}_{n+1}^+ \cfrac{\mathbf{I}}{s\tau_N\mathbf{I} - \mathbf{Q}_{n+2} \ddots}\mathbf{Q}_{n+2}^-}\mathbf{Q}_{n+1}^-}\mathbf{Q}_n^-$$

The initial-value vectors $\mathbf{C}_n(0)$ in Eq. (9.49) are calculated in the manner described in Section IV.A. Having determined $\tilde{\mathbf{C}}_1(s)$ from Eq. (9.49), we are now able to calculate the spectrum of the relaxation function from Eq. (9.43) as well as the integral time related to this *nonlinear response*, namely,

$$\tau = \int_0^\infty f(t)dt = \frac{\sqrt{2}v_Z\tilde{c}_{1,0}(0) + (v_X + iv_Y)\tilde{c}_{1,-1}(0) - (v_X - iv_Y)\tilde{c}_{1,1}(0)}{\sqrt{2}v_Z c_{1,0}(0) + (v_X + iv_Y)c_{1,-1}(0) - (v_X - iv_Y)c_{1,1}(0)} \tag{9.50}$$

The results so obtained are valid for a system of particles having their easy axes oriented along the $Z$ axis of the laboratory system of coordinates. If the easy axes of the particles were randomly distributed in space, one should have to make an appropriate averaging in order to calculate the spectrum and the relaxation time. We have also assumed that all the particles are identical. To take polydispersity of the particles into account, it is necessary to make averaging over appropriate distribution functions (e.g., over the particle volumes).

It should be noted that for a uniaxial single-domain magnetic particle, there are several particular cases when the relaxation time $\tau$ can be calculated from an analytical equation. This is the case when the external magnetic field is parallel to the easy axis of the particle both in the states I and II. In these cases, due to the symmetry of the problem, the Fokker–Planck equation (9.22) transforms into a 1D equation [6] and the results of [32], where an exact equation for the relaxation time was found for the nonlinear response of 1D systems whose dynamics is governed by a 1D Fokker–Planck equation (see Section III.D), may be applied to the present problem as well. Thus, one can calculate the relaxation time $\tau$ from an analytical equation such as Eq. (3.90), where $\sigma_{\mathrm{I}} = \sigma_{\mathrm{II}} = \sigma$.

For the purpose of simplicity, it is sufficient to consider the case when the field $\mathbf{H}_{\mathrm{I}}$ is parallel to the easy axis of the particle and to calculate the response in the direction of the field $\mathbf{H}_{\mathrm{II}}$. Due to symmetry properties, the solution in this case is independent of the azimuthal angle, so that without loss of generality one may suppose that the direction cosines are $v_X^{\mathrm{II}} = \sin\Theta$, $v_Y^{\mathrm{II}} = 0$, $v_Z^{\mathrm{II}} = \cos\Theta$. For $\Theta \neq 0$, the results depend on the dissipation parameter $\alpha$. This dependence is due to the mutual interaction of transverse and longitudinal relaxation modes. However, this dependence qualitatively corresponds to that obtained for linear response and investigated in details in [65]. Theoretical estimates of $\alpha$ give values of the order of $\sim$0.01–0.1 [123]. Here, the calculation was made for $\alpha = 0.1$. The calculated modulus of the relaxation spectrum $|\tilde{f}|$ is represented in Figure 71 for sudden changes of the dc magnetic field as a function of the angle $\Theta$ ($h_{\text{I}} = 0, h_{\text{II}} = h$). Three peaks are visible in the spectrum of $|\tilde{f}|$. The characteristic frequency and half-width of the low-frequency peak are determined by $\tau^{-1}$. A considerably weaker relaxation peak appears at high frequencies because of the intrawell modes. Moreover, the resonance peak is visible in Figure 71 at high frequencies. This peak is due to transverse modes that have characteristic frequencies coinciding with those at which the magnetization vector precesses, namely, $\omega_0 \sim \sigma(\alpha\tau_N)^{-1}$. On decreasing $\alpha$, this high-frequency peak is narrowed and shifted to higher frequencies. At $\Theta = 0$, this peak disappears as the transverse modes do not contribute to the relaxation process. The spectrum of $|\tilde{f}|$ is also represented in Figure 72 for sudden changes of the orientation of the dc magnetic field $|\mathbf{H}_{\text{I}}| = |\mathbf{H}_{\text{II}}|$. As in Figure 71, here three peaks are visible in the spectrum of $|\tilde{f}|$. Moreover, the effect of the reduction of the low-frequency mode by the dc field at the rotation of the field vector by an angle $\Theta < \pi/2$ is clearly seen in Figure 72.

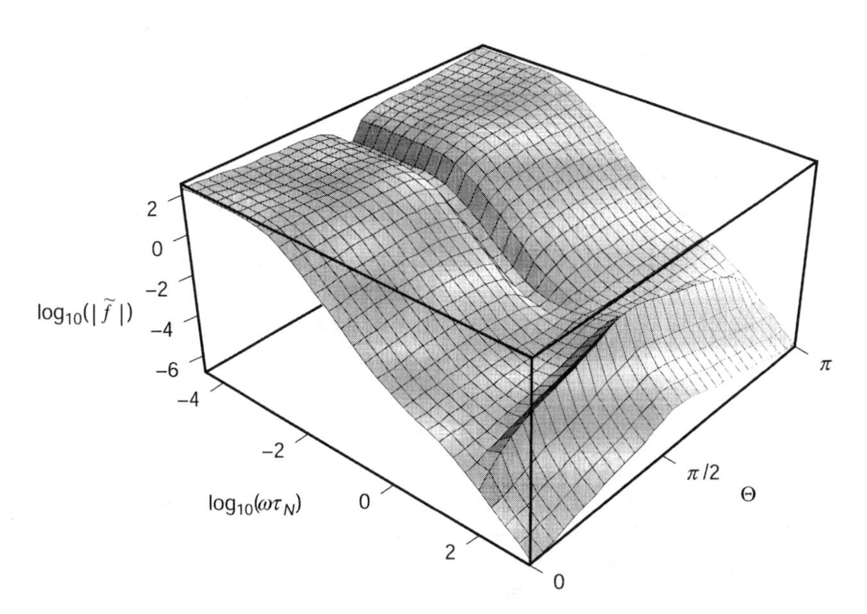


**Figure 71.** The $\log_{10}(|\tilde{f}|)$ as a function of $\log_{10}(\omega\tau_N)$ and the angle $\Theta$ for sudden switch on of dc magnetic field $h_{\text{I}} = 0$, $h_{\text{II}} = 0.1$, $\sigma = 10$, and $\alpha = 0.1$.

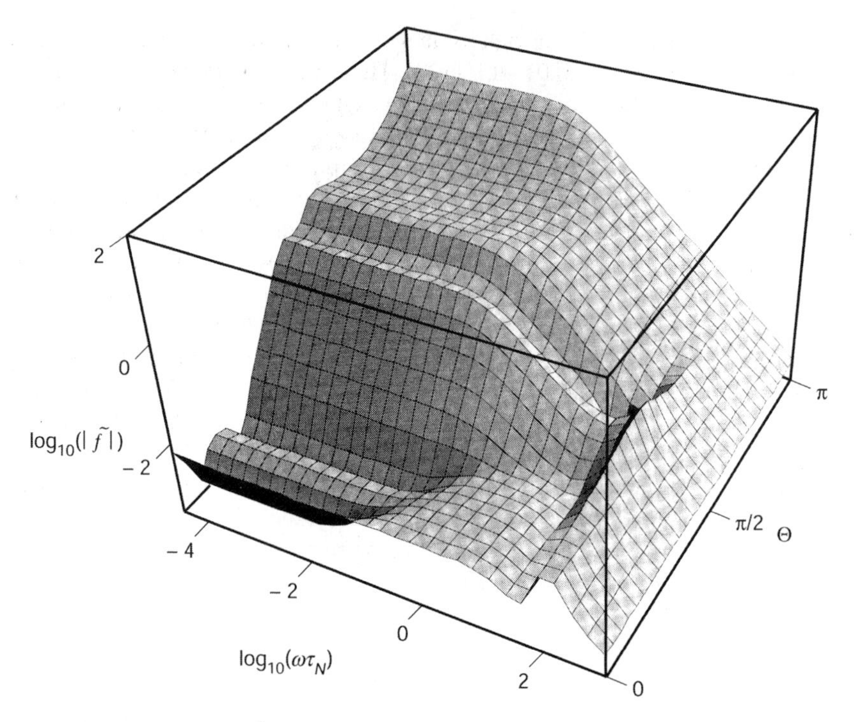


**Figure 72.** The $\log_{10}(|\tilde{f}|)$ as a function of $\log_{10}(\omega\tau_N)$ and the angle $\Theta$ for sudden change of the orientation of dc magnetic field $h = |h_I| = |h_{II}| = 0.3$ ($\sigma = 10$, and $\alpha = 0.1$).

### 3. *Nonlinear Response of Uniaxial Superparamagnetic Particles in High Superimposed ac and dc Bias Magnetic Fields*

Let us now consider a uniaxial particle subjected to superimposed strong uniform magnetic field $\mathbf{H}_0$ and a strong ac magnetic field $\mathbf{H}_1(t) = \mathbf{H}_1 \cos \omega t$ (both applied along the $Z$ axis, which coincide with the easy axis of the particle) with anisotropy energy density

$$V = -K \cos^2 \vartheta - (H_0 + H_1 \cos \omega t) M_s \cos \vartheta \tag{9.51}$$

A similar problem has been considered in Section VI.B for calculating the nonlinear response in dielectric and Kerr effect relaxation of a system of molecules acted on by strong superimposed ac and dc bias electric fields. By assuming that the magnitudes of both ac and dc fields are large enough so that the energy of the molecule in these fields can be comparable or higher than the thermal energy $kT$, one is faced to an intrinsically nonlinear problem we shall solve as follows. On expanding the distribution function $W(\vartheta, t)$ in Eq. (9.22) as a series of Legendre polynomials $P_n(\cos\vartheta)$ and using Eq. (9.51), we have the set of differential-recurrence equations for the expectation value of the Legendre polynomial of order $n, f_n(t)$ similar to Eq. (6.18) in which $\tau_D$ is replaced by $\tau_N$ and $\sigma(t)$ by $\sigma = \text{const}$, which does not depend on time, namely,

$$\tau_N \frac{d}{dt} f_n(t) + \frac{n(n+1)}{2} f_n(t) = \zeta(t)[a_n f_{n-1}(t) + b_n f_{n+1}(t)] \\ + \sigma[c_n f_{n-2}(t) + d_n f_n(t) + g_n f_{n+2}(t)] \tag{9.52}$$

where

$$f_n(t) = \langle P_n(\cos\vartheta)\rangle(t) = \int_0^{\pi} P_n(\cos\vartheta) W(\vartheta, t) \sin\vartheta d\vartheta \tag{9.53}$$

$$\zeta(t) = \xi + \xi_1 \cos\omega t \tag{9.54}$$

and

$$\xi_1 = \frac{vM_sH_1}{kT} \qquad \xi = \frac{vM_sH_0}{kT} \qquad \text{and} \qquad \sigma = \frac{vK}{kT} \tag{9.55}$$

The magnetization relaxation in the presence of the strong ac field $\mathbf{H}_1(t) = \mathbf{H}_1 \cos\omega t$, which is described by the Fokker–Planck Eq. (9.22), is a nonstationary Markovian process. Thus, in order to evaluate the stationary ac response, which is independent of the initial condition, one needs to calculate the solution of Eq. (9.52) corresponding to the stationary state. To accomplish this, one may seek (just as in Section VI.B) all the $f_n(t)$ in the form already given by Eq. (6.24) with the same Fourier amplitudes $F_k^n(\omega)$. On substituting Eq. (6.24) into Eq. (9.52), one can obtain the multiterm recurrence equations for the $F_k^n(\omega)$:

$$[2\sigma d_n - 2i\omega\tau_N k - n(n+1)]F_k^n(\omega) \\ + \xi_1 a_n[F_{k-1}^{n-1}(\omega) + F_{k+1}^{n-1}(\omega) - F_{k-1}^{n+1}(\omega) - F_{k+1}^{n+1}(\omega)] \\ + 2\xi a_n[F_k^{n-1}(\omega) - F_k^{n+1}(\omega)] + 2\sigma[c_n F_k^{n-2}(\omega) + g_n F_k^{n+2}(\omega)] = 0 \tag{9.56}$$

where $a_n$, $c_n$, and $g_n$ are given by Eqs. (6.19) and (6.20). The solution of Eq. (9.56) can be expressed in terms of a matrix continued fraction in the manner described in Section VI.B. We can introduce the column vectors $\mathbf{C}_n$ and $\mathbf{R}$ given by Eqs. (6.26) and (6.27), where the subvectors of infinite dimension $\mathbf{c}_k$ and $\mathbf{r}_1$

are given by Eq. (6.28) and the vector $\mathbf{r}_2$ is defined as

$$\mathbf{r}_2 = \begin{pmatrix} \mathbf{0} \\ 0 \\ 0 \\ 4\sigma/5 \\ 0 \\ 0 \\ \mathbf{0} \end{pmatrix} \tag{9.57}$$

Thus, the *nine-term scalar* recurrence Eq. (9.56) can be transformed into the *three-term matrix* recurrence equations (6.29) and (6.30) with different definitions for the matrices $\mathbf{Q}_n^-$, $\mathbf{Q}_n$, and $\mathbf{Q}_n^+$ as follows:

$$\mathbf{Q}_n^- = \begin{pmatrix} 2\sigma c_{2n}\mathbf{I} & \mathbf{O} \\ a_{2n-1}\mathbf{Y} & 2\sigma c_{2n-1}\mathbf{I} \end{pmatrix} \tag{9.58}$$

$$\mathbf{Q}_n = \begin{pmatrix} \mathbf{Z}_{2n} & a_{2n}\mathbf{Y} \\ -a_{2n-1}\mathbf{Y} & \mathbf{Z}_{2n-1} \end{pmatrix} \tag{9.59}$$

$$\mathbf{Q}_n^+ = \begin{pmatrix} 2\sigma g_{2n}\mathbf{I} & -a_{2n}\mathbf{Y} \\ \mathbf{O} & 2\sigma g_{2n-1}\mathbf{I} \end{pmatrix} \tag{9.60}$$

Here, $\mathbf{O}$ and $\mathbf{I}$ are the zero and identity matrices of infinite dimension, respectively, the tridiagonal submatrix $\mathbf{Y}$ is defined by Eq. (6.10) and the diagonal submatrix $\mathbf{Z}_n$ is defined as

$$\mathbf{Z}_n = \begin{pmatrix} \ddots & \vdots & \vdots & \vdots & \vdots & \vdots & ⋰ \\ \cdots & z_{n,-2} & 0 & 0 & 0 & 0 & \cdots \\ \cdots & 0 & z_{n,-1} & 0 & 0 & 0 & \cdots \\ \cdots & 0 & 0 & z_{n,0} & 0 & 0 & \cdots \\ \cdots & 0 & 0 & 0 & z_{n,1} & 0 & \cdots \\ \cdots & 0 & 0 & 0 & 0 & z_{n,2} & \cdots \\ ⋰ & \vdots & \vdots & \vdots & \vdots & \vdots & \ddots \end{pmatrix} \tag{9.61}$$

where

$$z_{n,m} = 2\sigma d_n - 2i\omega\tau_N m - n(n+1) \qquad (m = -\infty, \ldots, -1, 0, 1, \ldots, \infty)$$

Thus, we have the solution for $\mathbf{C}_1(\omega)$ in terms of matrix continued fractions given by Eq. (6.31). In order to illustrate the results of the calculation, we have plotted in Figures 73 and 74 the real and imaginary parts of the first normalized nonlinear harmonic component of the magnetization, namely,

$$\chi_1^1(\omega) = \frac{2F_1^{1^*}(\omega)}{\xi}$$

For $\xi_1 \ll 1$, the results obtained are in complete agreement with those presented in [129], where a perturbation approach to the solution of Eq. (9.56) was used.

Here, we have presented the results of the calculation of the *nonlinear* response of an assembly of *uniaxial* superparamagnetic particles. Single-domain particles with other types of anisotropy (cubic, biaxial, etc.) can be considered in a similar way. Moreover, the same approach can be applied to evaluate the *linear* response characteristics of superparamagnetic particles. For example, on using the matrix continued fraction approach, the linear magnetic susceptibility of superparamagnetic particles for various anisotropy potentials has been calculated in [64, 65, 134, 135].

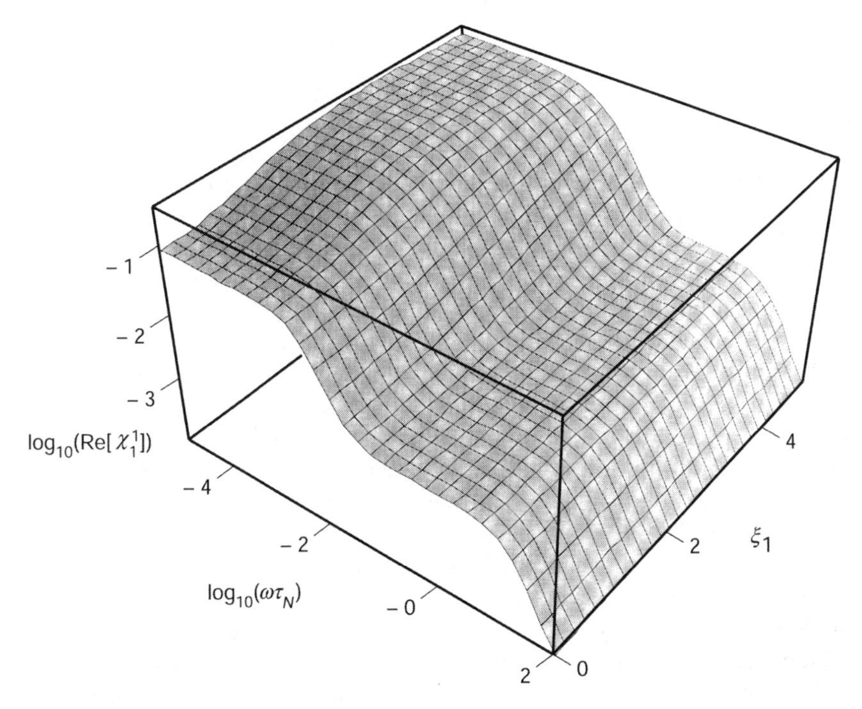


**Figure 73.** The $\log_{10}\mathrm{Re}\{\chi_1^1(\omega)\}$ as a function of $\log_{10}(\omega\tau_N)$ and $\xi(\sigma = 10, \xi = 2)$.

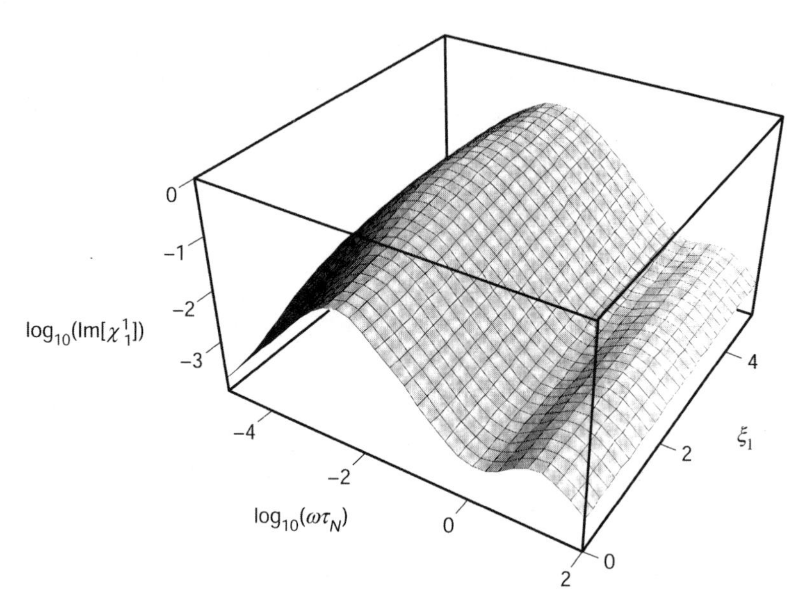


**Figure 74.** The $\log_{10}\mathrm{Im}\{\chi_1^1(\omega)\}$ as a function of $\log_{10}(\omega\tau_N)$ and $\xi$ ($\sigma = 10$, $\xi = 2$).

## X. CONCLUSIONS

The main purpose of this chapter has been to provide modern concepts in the theory of nonlinear dielectric and Kerr effect relaxation of dilute solutions of macromolecules in high electric fields based on the rotational diffusion model of Brownian particles. We have demonstrated the ability of the Langevin and/or Fokker–Planck equation approaches to solve various problems of nonlinear electrooptical responses. In order to accomplish this, we have mainly used the matrix continued fraction method that has been systematically applied by Risken [9] and later by Coffey et al. [8] to the solution of similar problems that may be described in the context of Brownian motion of a particle in an external potential. Among the various physical phenomena to which the theory of Brownian motion has been applied, one can mention [8, 9]: current–voltage characteristics of the Josephson tunneling junction, mobility of superionic conductors, line widths in NMR, quantum noise in ring laser gyroscopes, line shape of single mode semiconductor lasers, magnetic relaxation of single domain ferromagnetic particles (superparamagnetism) to quote but a few. The dynamics of all these systems may be described by an appropriate moment system whose solution may be therefore given in terms of ordinary or matrix continued fractions. As was pointed out by Risken [9], the matrix continued fraction method is especially suitable for numerical calculations and seems to be the most accurate and fastest method for treating those problems. In this chapter, we have in particular shown the efficiency of the method in the theory of linear as well as nonlinear dielectric relaxation and dynamic Kerr effect. Moreover, we have also demonstrated that this method represents a universal tool that can be used in a systematic manner for solving any of the problems involved in the orientational relaxation phenomena in high electric fields. The main results we have obtained may be summarized as follows. In Section IV, we have presented the exact analytical solution (in terms of matrix continued fractions) for the nonlinear transient response problem encountered in dielectric relaxation and dynamic Kerr effect when *both the magnitude and the direction of a strong* dc electric field may be suddenly changed at time $t = 0$ from $\mathbf{E}_{\mathrm{I}}$ to $\mathbf{E}_{\mathrm{II}}$. The theory covers all the results previously obtained for various particular transient relaxation problems such as transient responses in step-on, step-off, suddenly reversing, or suddenly rotating fields. By using the perturbation theory (Section V), we have obtained results suitable for evaluating the ac stationary solution of the dynamic birefringence and dielectric relaxation of polar and anisotropically polarizable molecules (dissolved in a nonpolar solvent) up to second order in the amplitude of a small ac electric field superimposed on a dc field of *arbitrary* strength. For the treatment of the nonlinear dielectric relaxation and dynamic Kerr response in high ac fields, where the perturbation procedure becomes inapplicable, we have shown how the matrix continued fraction approach can be applied to the calculation of the nonlinear ac stationary response of systems of rigid polar and polarizable molecules to an ac field of *arbitrary* strength (Section VI). Thus, both the nonlinear dielectric relaxation and dynamic Kerr effect response of an ensemble of *polar and anisotropically polarizable* molecules *in high ac fields* can be evaluated from an exact analytical equation in terms of matrix continued fractions. In the limit of small ac field strengths, the results are in complete agreement with those obtained by perturbation procedures. It should be noted that the theory developed allows one to calculate the nonlinear response regardless of the balance between permanent and induced dipole contributions given by the parameter $\kappa$, covering the two limiting cases, when either pure induced dipole moments ($\kappa \to \pm\infty$) or pure permanent moments ($\kappa = 0$) are taken into account. Furthermore, in Section VII we have suggested a way of generalizing the theory by including effects due to the hyperpolarizabilities of the molecules. The range of applicability of the results mentioned is restricted to the condition $\omega\tau_{\mathrm{D}} \leq 1$, as inertial effects are ignored in the model of noninertial rotational diffusion. In order to take into account the inertia of the molecules and to extend the results to higher frequencies, one should consider the Euler–Langevin equation or the underlying inertial Fokker–Planck equation in phase space. However, the theory in that context becomes much more complicated from the mathematical point of view. A possible direction of the development of the theory able to take into account inertial effects has been given in Section VIII in the particular case of the extended rotational diffusion model of molecules in a high dc electric field of arbitrary strength. Finally, we have also demonstrated in Section IX how our approach with a small modification can be used for the evaluation of the nonlinear response of analogous physical systems, whose dynamics can be described in the context of the Brownian motion of a particle in an external potential. In particular, we have presented the results of the calculation of the linear dielectric responses of molecular impurities in a cubic crystal as well as those of the nonlinear magnetic relaxation of superparamagnetic particles.

It is worth mentioning that the theory developed may be applied to the interpretation of experimental data on nonlinear transient and ac stationary responses in dielectric and Kerr effect relaxation. An example of application of the theory to the description of the transient response in high electric fields has been given in Section IV. Until now, two kinds of nonlinear ac response experiments have been carried out, namely, where either a relatively strong ac field or an ac field superimposed on a strong dc bias field were applied to the liquids (see, e.g., [15, 17, 19, 21, 42–44, 136–139]). Although the applied electric fields in those experiments were high enough ($\geq 10^6\,\mathrm{V/m}$) to observe nonlinear effects, the strengths of the fields remained still weak enough compared to the thermal energy, allowing one to use the nonlinear response equations obtained in the context of the perturbation theory, which works perfectly in these cases (see, e.g., [42–44]). Thus, the nonlinear response theory developed in the limit of weak external fields will also agree with those experimental data. However, as this theory is applicable to ac fields of arbitrary strength, it also provides a solid support for comparison with experiments on nonlinear response in high ac fields, where the perturbation approach can no longer be applied. This is the case, for example, for the optical Kerr effect encountered in the pulse-laser technique. In particular, it will be possible to carry out quantitative comparison of the theory with available molecular dynamics simulation data for high-frequency relaxation processes in strong ac fields. The use of such data is preferable for testing the theory, as in computer simulation it is much easier (than in real experiments) to achieve values of the nonlinear parameter $\xi_1 \geq 1$. For example, Evans [139] has reported the computer simulation data on the effect of a strong ac field on orientational relaxation of dipolar molecules in a picosecond timescale for $\xi_1$ going up to 12.

At this stage of our theoretical progress, we aimed for the solution of several nonlinear response problems related to the 3D Brownian motion in a potential. In particular, it should be interesting to elaborate a theory of nonlinear dielectric relaxation and dynamic Kerr effect of asymmetric top molecules in high ac and dc electric fields with account of effects of inertia and of hyperpolarizabilities. This task is a real and formidable challenge opening new directions in the electrooptical investigations of molecular mechanisms in diluted and condensed media and made nowadays possible by the wide spectral range of new experimental devices.

## Appendix A. Proof of Eq. (3.9)

Noting that the rule for changing of variables in Stratonovich differential equations is the same as in ordinary analysis [54], the equation of motion for an arbitrary differentiable function $f(\{\boldsymbol{\xi}\})$ may be obtained by cross-multiplying Eq. (3.6) by $\partial f(\{\boldsymbol{\xi}(t)\})/\partial \xi_i$, respectively, and then summing them. Thus, we obtain a stochastic equation for $f(\{\boldsymbol{\xi}(t)\})$:

$$\begin{aligned}\frac{d}{dt}f(\{\boldsymbol{\xi}(t)\}) &= h_i(\{\boldsymbol{\xi}(t)\},t)\frac{\partial}{\partial \xi_i}f(\{\boldsymbol{\xi}(t)\}) \\ &\quad + g_{ij}(\{\boldsymbol{\xi}(t)\},t)\frac{\partial}{\partial \xi_i}f(\{\boldsymbol{\xi}(t)\})\Gamma_j(t)\end{aligned} \tag{A1}$$

From a mathematical point of view, the stochastic differential equation (A1) with the δ-correlated Langevin forces $\Gamma_j(t)$ is not completely defined [8, 9]. The most satisfactory interpretation of Eq. (A1) is to regard it as the stochastic integral equations [8, 9, 54]

$$\begin{aligned}\xi_i(t+\tau) &= x_i + \int_t^{t+\tau} h_i(\{\boldsymbol{\xi}(t')\},t')dt' \\ &\quad + \int_t^{t+\tau} g_{ij}(\{\boldsymbol{\xi}(t')\},t')\Gamma_j(t')dt'\end{aligned} \tag{A2}$$

$$\begin{aligned}f(\{\boldsymbol{\xi}(t+\tau)\}) &= f(\{\mathbf{x}\}) + \int_t^{t+\tau} h_i(\{\boldsymbol{\xi}(t')\},t')\frac{\partial f(\{\boldsymbol{\xi}(t')\})}{\partial \xi_i}dt' \\ &\quad + \int_t^{t+\tau} g_{ij}(\{\boldsymbol{\xi}(t')\},t')\frac{\partial f(\{\boldsymbol{\xi}(t')\})}{\partial \xi_i}\Gamma_j(t')dt'\end{aligned} \tag{A3}$$

On supposing that the integrands in Eqs. (A2) and (A3) can be expanded in Taylor series, we obtain

$$\begin{aligned}\xi_i(t+\tau) &= x_i + \int_t^{t+\tau} h_i(\{\mathbf{x}\},t')dt' \\ &\quad + \int_t^{t+\tau} [\xi_k(t') - x_k]\frac{\partial}{\partial x_k}h_i(\{\mathbf{x}\},t')dt' \\ &\quad + \int_t^{t+\tau} g_{ij}(\{\mathbf{x}\},t')\Gamma_j(t')dt' \\ &\quad + \int_t^{t+\tau} [\xi_k(t') - x_k]\frac{\partial}{\partial x_k}g_{ij}(\{\mathbf{x}\},t')\Gamma_j(t')dt' + \cdots\end{aligned} \tag{A4}$$

$$
\begin{aligned}
f(\{\boldsymbol{\xi}_i(t+\tau)\}) = f(\{\mathbf{x}\}) &+ \int_t^{t+\tau} h_i(\{\mathbf{x}\}, t') \frac{\partial f(\{\mathbf{x}\})}{\partial x_i} dt' \\
&+ \int_t^{t+\tau} [\xi_k(t') - x_k] \frac{\partial}{\partial x_k} \left[ h_i(\{\mathbf{x}\}, t') \frac{\partial f(\{\mathbf{x}\})}{\partial x_i} \right] dt' \\
&+ \int_t^{t+\tau} g_{ij}(\{\mathbf{x}\}, t') \frac{\partial f(\{\mathbf{x}\})}{\partial x_i} \Gamma_j(t') dt' \\
&+ \int_t^{t+\tau} [\xi_k(t') - x_k] \frac{\partial}{\partial x_k} \left[ g_{ij}(\{\mathbf{x}\}, t') \frac{\partial f(\{\mathbf{x}\})}{\partial x_i} \right] \Gamma_j(t') dt' + \cdots
\end{aligned} \tag{A5}
$$

On substituting $\xi_k(t') - x_k$ from Eq. (A4) into Eq. (A5), we iterate

$$
\begin{aligned}
f(\{\boldsymbol{\xi}_i(t+\tau)\}) = f(\{\mathbf{x}\}) &+ \int_t^{t+\tau} h_i(\{\mathbf{x}\}, t') \frac{\partial f(\{\mathbf{x}\})}{\partial x_i} dt' \\
&+ \int_t^{t+\tau} \frac{\partial}{\partial x_k} \left[ h_i(\{\mathbf{x}\}, t') \frac{\partial f(\{\mathbf{x}\})}{\partial x_i} \right] \int_t^{t'} h_k(\{\mathbf{x}\}, t'') dt'' dt' \\
&+ \int_t^{t+\tau} \frac{\partial}{\partial x_k} \left[ h_i(\{\mathbf{x}\}, t') \frac{\partial f(\{\mathbf{x}\})}{\partial x_i} \right] \int_t^{t'} g_{kn}(\{\mathbf{x}\}, t'') \Gamma_n(t'') dt'' dt' \\
&+ \int_t^{t+\tau} g_{ij}(\{\mathbf{x}\}, t') \frac{\partial f(\{\mathbf{x}\})}{\partial x_i} \Gamma_j(t') dt' \\
&+ \int_t^{t+\tau} \frac{\partial}{\partial x_k} \left[ g_{ij}(\{\mathbf{x}\}, t') \frac{\partial f(\{\mathbf{x}\})}{\partial x_i} \right] \Gamma_j(t') \int_t^{t'} h_k(\{\mathbf{x}\}, t'') dt'' dt' \\
&+ \int_t^{t+\tau} \frac{\partial}{\partial x_k} \left[ g_{ij}(\{\mathbf{x}\}, t') \frac{\partial f(\{\mathbf{x}\})}{\partial x_i} \right] \Gamma_j(t') \\
&\times \int_t^{t'} g_{kn}(\{\mathbf{x}\}, t'') \Gamma_n(t'') dt'' dt' + \cdots
\end{aligned} \tag{A6}
$$

Then, averaging Eq. (A6) with account of the white noise properties (3.7) and retaining only the terms of the order of $\tau$, we have

$$
\begin{aligned}
\overline{f(\{\boldsymbol{\xi}(t+\tau)\})} = f(\{\mathbf{x}\}) &+ \int_t^{t+\tau} h_i(\{\mathbf{x}\}, t') \frac{\partial}{\partial x_i} f(\{\mathbf{x}\}) dt' \\
&+ D\delta_{jn} \int_t^{t+\tau} g_{kn}(\{\mathbf{x}\}, t') \frac{\partial}{\partial x_k} \left[ g_{ij}(\{\mathbf{x}\}, t') \frac{\partial}{\partial x_i} f(\{\mathbf{x}\}) \right] dt' + o(\tau)
\end{aligned} \tag{A7}
$$

Here, we have also used the property of the $\delta$ function, namely,

$$\int_a^b \delta(b-x)y(x)dx = \frac{y(b)}{2}$$

After obvious transformations in Eq. (A7), we obtain

$$\begin{aligned}\overline{\frac{f(\{\boldsymbol{\xi}(t+\tau)\}) - f(\{\mathbf{x}\})}{\tau}} &= h_i(\{\mathbf{x}\}, t+\tau\Theta_{iii}^{(1)})\frac{\partial}{\partial x_i}f(\{\mathbf{x}\}) \\ &+ Dg_{kj}(\{\mathbf{x}\}, t+\tau\Theta_{ijk}^{(2)}) \\ &\times \frac{\partial}{\partial x_k}\left[g_{ij}(\{\mathbf{x}\}, t+\tau\Theta_{ijk}^{(2)})\frac{\partial}{\partial x_i}f(\{\mathbf{x}\})\right] + o(1)\end{aligned} \tag{A8}$$

where $\Theta_{ijk}^{(n)}$ are constants $(0 \le \Theta_{ijk}^{(n)} \le 1)$. If we take the limit $\tau \to 0$ in Eq. (A8), we have Eq. (3.9).

## Appendix B. Integral Expression for the Relaxation Time in the Linear Case

Let us consider a stochastic system with dynamics obeying the 1D Fokker–Planck equation (3.55) with the Fokker–Planck operator $L_{\text{FP}}$ given by Eqs. (3.56)–(3.58). Then, the time dependence of the average of a dynamic variable $A(x)$ can be expressed as

$$\langle A\rangle(t) = \int_{-\infty}^{t} \Phi(t-t')F(t')dt' \tag{A9}$$

where $\Phi(t)$ is the pulse response function defined by

$$\Phi(t) = -\frac{d}{dt}C_{AB}(t) \tag{A10}$$

$C_{AB}(t)$ is the *equilibrium* (stationary) correlation function defined by Eq. (3.60). The correlation time $\tau_{AB}$ is defined as the area under the curve of the normalized correlation function

$$\tau_{AB} = \frac{1}{C_{AB}(0)}\int_0^{\infty} C_{AB}(t)dt \tag{A11}$$

We shall now show how to derive an analytic expression for $\tau_{AB}$. To accomplish this, one can use a similar method of calculation as that used in [9]. The definition of $C_{AB}(t)$ [Eq. (3.60)] implies that we can seek the solution of the 1D Fokker–Planck equation [9]

$$\begin{aligned}\frac{\partial}{\partial t}W(x,t) &= L_{\mathrm{FP}}W(x,t)\\ &= \frac{\partial}{\partial x}\left(D^{(2)}(x)\left[\frac{\partial}{\partial x}W(x,t) + W(x,t)\frac{\partial}{\partial x}U(x)\right]\right)\end{aligned} \tag{A12}$$

in the form [9]

$$W(x,t) = W_0(x) + w(x,t) \tag{A13}$$

with the initial conditions

$$w(x,0) = [B(x) - \langle B\rangle_0]W_0(x) \tag{A14}$$

and with $w(x,\infty) = 0$, $x$ being defined in the range $x_1 < x < x_2$. On taking into account that $L^0_{\mathrm{FP}}W_0 = 0$ we have the formal solution of Eq. (A12), namely,

$$w(x,t) = e^{L^0_{\mathrm{FP}}t}[B(x) - \langle B\rangle_0]W_0(x) \tag{A15}$$

and hence

$$\tau_{AB} = -\frac{\int_{x_1}^{x_2}[A(x) - \langle A\rangle_0](L^0_{\mathrm{FP}})^{-1}[B(x) - \langle B\rangle_0]W_0(x)dx}{\langle AB\rangle_0 - \langle A\rangle_0\langle B\rangle_0} \tag{A16}$$

The evaluation of

$$(L^0_{\mathrm{FP}})^{-1}[B(x) - \langle B\rangle_0]W_0(x) \tag{A17}$$

can be accomplished by taking the Laplace transform of Eq. (A12). Thus, we obtain

$$s\tilde{w}(x,s) - [B(x) - \langle B\rangle_0]W_0(x) = L^0_{\mathrm{FP}}\tilde{w}(x,s) \tag{A18}$$

where

$$\tilde{w}(x,s) = \int_0^\infty w(x,t)e^{-st}dt$$

In the limit $s \to 0$, by using the final value theorem of Laplace transformation Eq. (3.70), we have

$$\tilde{w}(x,0) = -(L_{\text{FP}}^{0})^{-1}[B(x) - \langle B\rangle_0]W_0(x) \tag{A19}$$

and Eq. (A16) becomes

$$\tau_{AB} = \frac{1}{\langle AB\rangle_0 - \langle A\rangle_0\langle B\rangle_0}\int_{x_1}^{x_2}[A(x) - \langle A\rangle_0]\tilde{w}(x,0)dx \tag{A20}$$

The quantity $\tilde{w}(x,0)$ in Eq. (A20) can be evaluated as follows. By using Eq. (A12) and taking account of Eq. (A14), one obtains

$$\begin{aligned} &-[B(x) - \langle B\rangle_0]W_0(x) \\ &\quad = \frac{d}{dx}\left(D^{(2)}(x)\left[\frac{d}{dx}\tilde{w}(x,0) + \tilde{w}(x,0)\frac{d}{dx}U(x)\right]\right) \end{aligned} \tag{A21}$$

Equation (A21) can be easily integrated to yield

$$\tilde{w}(x,0) = -W_0(x)\int_{x_1}^{x}\frac{f_B(x')dx'}{D^{(2)}(x')W_0(x')} \tag{A22}$$

where

$$f_B(x) = \int_{x_1}^{x}[B(y) - \langle B\rangle_0]W_0(y)dy \tag{A23}$$

Thus, we obtain from Eqs. (A20) and (A22)

$$\tau_{AB} = -\frac{1}{\langle AB\rangle_0 - \langle A\rangle_0\langle B\rangle_0}\int_{x_1}^{x_2}[A(x) - \langle A\rangle_0]W_0(x)\int_{x_1}^{x}\frac{f_B(x')}{D^{(2)}(x')W_0(x')}dx'dx$$

or after integrating by parts

$$\tau_{AB} = \frac{1}{\langle AB\rangle_0 - \langle A\rangle_0\langle B\rangle_0}\int_{x_1}^{x_2}\frac{f_A(x)f_B(x)dx}{D^{(2)}(x)W_0(x)} \tag{A24}$$

where

$$f_A(x) = \int_{x_1}^{x}[A(y) - \langle A\rangle_0]W_0(y)dy \tag{A25}$$

For $A = B$ and $x_1 = -\infty$, $x_2 = \infty$, Eq. (A24) reduces to Eq. (S9.14) of Risken [9].

For the problem of noninertial rotational Brownian motion of a dipolar particle in an external potential $V$, the relevant Fokker–Planck equation for the distribution function $W$ of the orientations of the particles is given by Eq. (3.76). In this case $W_0(x)$ is the Maxwell–Boltzmann distribution function

$$W_0(x) = \frac{e^{-V(x)/kT}}{Z} \tag{A26}$$

$x_1 = -1$, $x_2 = 1$, and we have already shown that the function $D^{(2)}(x)$ is given by [Eq. (3.79)]

$$D^{(2)}(x) = \frac{1 - x^2}{2\tau_D} \tag{A27}$$

Thus, on taking into account Eqs. (A26) and (A27), Eq. (A24) yields

$$\tau_{AB} = \frac{2\tau_D}{Z[\langle AB\rangle_0 - \langle A\rangle_0\langle B\rangle_0]} \int_{-1}^{1} \frac{e^{V(z)/kT} f_A(z) f_B(z) dz}{1 - z^2} \tag{A28}$$

where

$$f_A(z) = \int_{-1}^{z} [A(z') - \langle A\rangle_0] e^{-V/kT(z')} dz' \tag{A29}$$

$$f_B(z) = \int_{-1}^{z} [B(z'') - \langle B\rangle_0] e^{-V(z'')/kT} dz'' \tag{A30}$$

For the problem considered in Section V.A, $A(x)$ and $B(x)$ are given by Eqs. (5.3) and (5.5), respectively.

## Appendix C. Derivation of Eqs. (8.57) and (8.63)

In order to derive Eqs.(8.57) and (8.63), we proceed as follows. We start with the easiest expression, namely, Eq. (8.63). On noting that since $\text{sn}^4(u|m)$ is an even function of $u$, it admits the following Fourier expansion:

$$\text{sn}^4(u|m) = \tfrac{1}{2}a_0 + \sum_{n=1}^{\infty} a_n \cos\left(\frac{\pi n u}{K}\right) \tag{A31}$$

where $K = K(m)$ is the complete elliptic integral of the first kind and

$$a_n = \frac{1}{K}\int_{-K}^{K} \mathrm{sn}^4(u|m)e^{-i(\pi nu/K)}\,du \tag{A32}$$

The coefficient $a_0$ can be calculated from a table of integrals [140]:

$$a_0 = \frac{1}{K}\int_{-K}^{K} \mathrm{sn}^4(u|m)du = \frac{2}{3m}\left[\frac{2(1+m)}{m}\left(1-\frac{E}{K}\right)-1\right] \tag{A33}$$

For $n \neq 0$, the $a_n$ can be calculated as follows. Let us consider a function

$$P(z) = \mathrm{sn}^4(z|m)e^{-i(\pi nz/K)} \tag{A34}$$

in the complex plane $z = u + iv$. Then,

$$a_n = \frac{1}{K}\mathrm{Re}\left\{\int_{-K}^{K} \mathrm{sn}^4(z|m)e^{-i(\pi nz/K)}dz\right\} \tag{A35}$$

To evaluate the integral (A35), let us consider the contour ABCDA (Fig. 75). Inside this contour, the function $P(z)$ has one residue of fourth order at the point $iK'$, where $K' = K(1-m)$. Thus

$$\oint_{ABCDA} P(z)e^{-i(\pi nz/K)}dz = 2\pi i\left(\operatorname*{Res}_{z=iK'}[P(z)e^{-i(\pi nz/K)}]\right) \tag{A36}$$

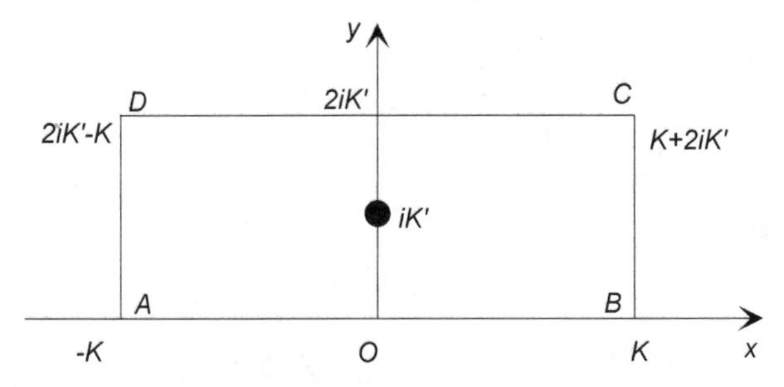


**Figure 75.** Contour of integration for the calculation of the Fourier coefficients Eqs. (A32) and (A46).

On the other hand, we have

$$\oint_{ABCDA} \mathrm{sn}^4(z|m)e^{-i(\pi nz/K)}dz = \left(\int_{-K}^{K} + \int_{2iK'+K}^{2iK'-K}\right)\mathrm{sn}^4(z|m)e^{-i(\pi nz/K)}dz$$
$$= \int_{-K}^{K} [\mathrm{sn}^4(z|m) - q^{-2n}\mathrm{sn}^4(z+2iK'|m)]e^{-i(\pi nz/K)}dz$$
$$= a_n K(1-q^{-2n}) \tag{A37}$$

Here, we have taken into account that $\mathrm{sn}(z+2iK'|m) = \mathrm{sn}(z|m)$ and the integrals along the intervals $BC$ and $DA$ are canceled. Thus

$$a_n = \frac{2\pi i}{K(1-q^{-2n})}\left\{\underset{z=iK'}{\mathrm{Res}}[\mathrm{sn}^4(z|m)e^{-i(\pi nz/K)}]\right\} \tag{A38}$$

On calculating the residue, we obtain after tedious algebra:

$$\underset{z=iK'}{\mathrm{Res}}[\mathrm{sn}^4(z|m)e^{-i(\pi nz/K)}] = \lim_{z\to iK'}\frac{d^3}{dz^3}\left[\frac{(z-iK')^4\mathrm{H}^4(z)}{6m^2\Theta^4(z)}e^{-i(\pi nz/K)}\right]$$
$$= \frac{i2\pi nq^{-n}}{3Km^2}\left[\frac{\Theta'''(iK')}{\Theta'(iK')} - 3\frac{\mathrm{H}''(iK')}{\mathrm{H}(iK')} + \frac{n^2\pi^2}{4K^2}\right] \tag{A39}$$
$$= -\frac{i2\pi nq^{-n}}{3Km^2}\left[1+m-\frac{n^2\pi^2}{4K^2}\right]$$

where $'$, $''$ and $'''$ at the theta functions designate the first-, second-, and third-order derivatives, respectively. Here we have taken into account that [39, 111]

$$\mathrm{sn}(u|m) = \frac{1}{\sqrt{m}}\frac{\mathrm{H}(u)}{\Theta(u)} \tag{A40}$$

$$\mathrm{H}(iK') = i\sqrt{\frac{2K}{\pi}\sqrt{\frac{1-m}{q}}}$$

$$\frac{\mathrm{H}'(iK')}{\mathrm{H}(iK')} = -\frac{i\pi}{2K} \qquad \frac{\mathrm{H}''(iK')}{\mathrm{H}(iK')} = 1 - \frac{E}{K} - \frac{\pi^2}{4K^2}$$

$$\Theta'(iK') = i\sqrt{\frac{2K}{\pi}\sqrt{\frac{m(1-m)}{q}}} \qquad \frac{\Theta''(iK')}{\Theta'(iK')} = -\frac{i\pi}{K}$$

and

$$\frac{\Theta'''(iK')}{\Theta'(iK')} = 3\left(1 - \frac{E}{K} - \frac{\pi^2}{4K^2}\right) - m - 1$$

Thus we obtain Eq. (8.63) from Eqs. (A38), (A33), and (A39).

A similar procedure may be used to derive Eq. (8.57). Let us consider the function

$$R(u) = \frac{\Theta(u+\alpha)\Theta(u+\beta)}{\Theta^2(u)} \tag{A41}$$

On recalling the properties of the Jacobi theta-function [39]

$$\Theta(u+2K) = \Theta(u) \tag{A42}$$

$$\Theta(u+2iK') = -q^{-1}\Theta(u)e^{-i\pi u/K} \tag{A43}$$

$$\Theta(iK') = 0 \tag{A44}$$

the Fourier series expansion of $R(u)$ [Eq. (A41)] is given by

$$R(u) = \sum_{n=-\infty}^{\infty} b_n e^{in\pi u/K} \tag{A45}$$

where

$$b_n = \frac{1}{2K}\int_{-K}^{+K} R(u)e^{-in\pi u/K}du \tag{A46}$$

In order to evaluate the Fourier coefficients (A46), we introduce again the function of the complex variable $z$:

$$R(z)e^{-in\pi z/K}$$

and integrate it over the same contour ABCDA defined above. This function has a double pole in $z = iK'$ inside this contour. Hence,

$$\int_{ABCDA} R(z)e^{-in\pi z/K}dz = 2i\pi\,\mathrm{Res}[R(z)e^{-in\pi z/K}]_{z=iK'} \tag{A47}$$

By taking into account Eq. (A43), one can transform the left-hand side of Eq. (A47) as follows:

$$\begin{aligned}\int_{ABCDA} R(z)e^{-in\pi z/K}dz &= \left[\int_{-K}^{+K} + \int_{B}^{C} + \int_{K+2iK'}^{-K+2iK'} + \int_{A}^{D}\right] R(z)e^{-in\pi z/K}dz \\ &= 2K[1 - q^{-2n}e^{-in\pi(\alpha+\beta)/K}]b_n\end{aligned} \tag{A48}$$

Here, we have used that $\int_B^C = -\int_D^A$. On the other hand, the calculation of the residue yields

$$\begin{aligned}\mathrm{Res}[R(z)e^{-in\pi z/K}]_{z=iK'} &= \lim_{z\to iK'} \frac{d}{dz}\left[\frac{(z-iK')^2\Theta(z+\alpha)\Theta(z+\beta)}{\Theta^2(z)}e^{-in\pi z/K}\right] \\ &= \frac{\pi e^{-in\pi(\alpha+\beta)/K}q^{-n}\mathrm{H}(\alpha)\mathrm{H}(\beta)}{2\sqrt{m(1-m)}K}\left[\frac{\mathrm{H}'(\alpha)}{\mathrm{H}(\alpha)} + \frac{\mathrm{H}'(\beta)}{\mathrm{H}(\beta)} - \frac{in\pi}{K}\right]\end{aligned} \tag{A49}$$

In order to obtain Eq. (A49), we have used the following formulas:

$$\begin{aligned}\Theta(u+iK') &= iq^{-1/4}e^{-i\pi u/2K}\mathrm{H}(u) \\ \Theta'(u+iK') &= iq^{-1/4}e^{-i\pi u/2K}\mathrm{H}'(u) + iq^{-1/4}e^{-i\pi u/2K}\frac{\mathrm{H}(u)}{2K} \\ \mathrm{H}'(0) &= \sqrt{\frac{2K}{\pi}\sqrt{m(1-m)}} \\ \Theta''(iK') &= q^{-1/4}\frac{\pi}{K}\sqrt{\frac{2K}{\pi}\sqrt{m(1-m)}}\end{aligned}$$

Combining Eqs. (A46)–(A49) yields

$$b_n = \frac{\pi[\mathrm{H}(\beta)\mathrm{H}'(\alpha) + \mathrm{H}(\alpha)\mathrm{H}'(\beta) - i\pi n\mathrm{H}(\alpha)\mathrm{H}(\beta)/K(m)]}{2K(m)\mathrm{H}'^2(0)\sin[\pi(\alpha+\beta+2inK(1-m))/2K(m)]} \tag{A50}$$

Equations (A50) and (A41) lead thus to Eq. (8.57).

## Appendix D. Matrices $\mathbf{Q}_n$, $\mathbf{Q}_n^+$, $\mathbf{Q}_n^-$, and their Elements in Eq. (9.12)

The matrices $\mathbf{Q}_n$, $\mathbf{Q}_n^+$, $\mathbf{Q}_n^-$ in Eq. (9.12) are given by [121]

$$\mathbf{Q}_n = \begin{pmatrix} \mathbf{A}_{4n} & \mathbf{O} & \mathbf{P}_{4n} & \mathbf{O} \\ \mathbf{O} & \mathbf{A}_{4n-1} & \mathbf{O} & \mathbf{P}_{4n-1} \\ f_{4n}\mathbf{P}_{4n}^T & \mathbf{O} & \mathbf{A}_{4n-2} & \mathbf{O} \\ \mathbf{O} & f_{4n-1}\mathbf{P}_{4n-1}^T & \mathbf{O} & \mathbf{A}_{4n-3} \end{pmatrix} \quad \text{(A51)}$$

$$\mathbf{Q}_n^+ = \begin{pmatrix} g_{4n+4}\mathbf{J}_{4n+4}^T & \mathbf{O} & f_{4n+2}\mathbf{P}_{4n+2}^T & \mathbf{O} \\ \mathbf{O} & g_{4n+3}\mathbf{J}_{4n+3}^T & \mathbf{O} & f_{4n+1}\mathbf{P}_{4n+1}^T \\ \mathbf{O} & \mathbf{O} & g_{4n+2}\mathbf{J}_{4n+2}^T & \mathbf{O} \\ \mathbf{O} & \mathbf{O} & \mathbf{O} & g_{4n+1}\mathbf{J}_{4n+1}^T \end{pmatrix} \quad \text{(A52)}$$

$$\mathbf{Q}_n^- = \begin{pmatrix} \mathbf{J}_{4n} & \mathbf{O} & \mathbf{O} & \mathbf{O} \\ \mathbf{O} & \mathbf{J}_{4n-1} & \mathbf{O} & \mathbf{O} \\ \mathbf{P}_{4n-2} & \mathbf{O} & \mathbf{J}_{4n-2} & \mathbf{O} \\ \mathbf{O} & \mathbf{P}_{4n-3} & \mathbf{O} & \mathbf{J}_{4n-3} \end{pmatrix} \quad \text{(A53)}$$

where the superscript $T$ designates the transposition and

$$f_n = -\frac{2n-11}{2n+9} \qquad g_n = -\frac{n-4}{n+1} \quad \text{(A54)}$$

The dimensions of the matrices $\mathbf{Q}_n, \mathbf{Q}_n^+, \mathbf{Q}_n^-$ are $(8n-2)\times(8n-2)$, $(8n-2)\times(8n+6)$, and $(8n-2)\times(8n-10)$, respectively. The exception is $\mathbf{Q}_1^-$, which degenerates to a column vector of dimension 6. The matrices $\mathbf{Q}_n, \mathbf{Q}_n^+, \mathbf{Q}_n^-$ consist of three-diagonal submatrices. There are two kinds of submatrices **A**, **J**, and **P** in Eqs. (A51)–(A53). The submatrices $\mathbf{A}_{4n}$, $\mathbf{A}_{4n-1}$, $\mathbf{A}_{4n-2}$, $\mathbf{A}_{4n-3}$, $\mathbf{P}_{4n-1}$, $\mathbf{P}_{4n-2}$ have the form:

$$\mathbf{X}_{4n-i} = \begin{pmatrix} x_{4n-i,-4(n-1+\delta_{i0})} & x^+_{4n-i,-4(n-1+\delta_{i0})} & 0 & \cdots & 0 \\ x^-_{4n-i,-4(n-2+\delta_{i0})} & x_{4n-i,-4(n-2+\delta_{i0})} & x^+_{4n-i,-4(n-2+\delta_{i0})} & \cdots & 0 \\ 0 & x^-_{4n-i,-4(n-3+\delta_{i0})} & x_{4n-i,-4(n-3+\delta_{i0})} & \cdots & 0 \\ \vdots & \vdots & \vdots & \ddots & \vdots \\ 0 & 0 & 0 & \cdots & x_{4n-i,4(n-1+\delta_{i0})} \end{pmatrix} \quad \text{(A55)}$$

$(i = 0, 1, 2, 3)$ and have dimension $[2(n+\delta_{0i})-1]\times[2(n+\delta_{0i})-1]$. The submatrices $\mathbf{P}_{4n}$, $\mathbf{P}_{4n-3}$, $\mathbf{J}_{4n}$, $\mathbf{J}_{4n-1}$, $\mathbf{J}_{4n-2}$, $\mathbf{J}_{4n-3}$, are defined as

$$\mathbf{X}_{4n-i} = \begin{pmatrix} x^{+}_{4n-i,-4(n-1+\delta_{i0})} & 0 & 0 & \cdots & 0 \\ x_{4n-i,-4(n-2+\delta_{i0})} & x^{+}_{4n-i,-4(n-2+\delta_{i0})} & 0 & \cdots & 0 \\ x^{-}_{4n-i,-4(n-3+\delta_{i0})} & x_{4n-i,-4(n-3+\delta_{i0})} & x^{+}_{4n-i,-4(n-3+\delta_{i0})} & \cdots & 0 \\ \vdots & \vdots & \vdots & \ddots & \vdots \\ 0 & 0 & 0 & \cdots & x^{-}_{4n-i,4(n-1+\delta_{i0})} \end{pmatrix} \tag{A56}$$

($i = 0, 1, 2, 3$) and have dimension $[2(n + \delta_{0i}) - 1] \times [2(n + \delta_{0i}]$. The submatrix elements are determined from Eq. (9.5) and are given by

$$\begin{aligned} a_{n,m} &= d_{n,m,n,m} \\ &= \sigma \frac{9(n-1)n(n+1)(n+2) - 15m^2[6n(n+1) - 5 - 7m^2]}{(2n-3)(2n-1)(2n+3)(2n+5)} - \frac{n(n+1)}{2} \end{aligned}$$

$$\begin{aligned} a^{-}_{n,m} &= a^{+}_{n,-m} = d_{n,m,n,m-4} \\ &= \frac{15\sigma\sqrt{(n+m)(n-m+4)[n^2-(m-3)^2][n^2-(m-2)^2][n^2-(m-1)^2]}}{2(2n-3)(2n-1)(2n+3)(2n+5)} \end{aligned}$$

$$p_{n,m} = d_{n,m,n-2,m} = \frac{\sigma(2n+9)(n^2-n-2-7m^2)}{(2n-5)(2n-1)(2n+3)} \times \sqrt{\frac{[n^2-m^2][(n-1)^2-m^2]}{(2n+1)(2n-3)}}$$

$$p^{-}_{n,m} = p^{+}_{n,-m} = d_{n,m,n-2,m-4} = -\frac{\sigma(2n+9)}{2(2n-5)(2n-1)(2n+3)} \times \sqrt{\frac{(n+m-5)(n+m-4)(n+m-3)(n+m)[n^2-(m-2)^2][n^2-(m-1)^2]}{(2n+1)(2n-3)}}$$

$$j_{n,m} = d_{n,m,n-4,m} = \frac{7\sigma(n+1)}{(2n-5)(2n-3)(2n-1)} \times \sqrt{\frac{[(n-3)^2-m^2][(n-2)^2-m^2][(n-1)^2-m^2][n^2-m^2]}{(2n-7)(2n+1)}}$$

$$j_{n,m}^{-} = j_{n,-m}^{+} = d_{n,m,n-4,m-4}$$
$$= \frac{\sigma(n+1)}{2(2n-5)(2n-3)(2n-1)}$$
$$\times \sqrt{\frac{(n+m-7)(n+m-6)\cdots(n+m-1)(n+m)}{(2n-7)(2n+1)}}$$

The initial condition vectors $\mathbf{C}_n(0)$ in Eq. (9.16) can also be evaluated in terms of matrix continued fractions [121]. The initial values $c_{n,m}(0)$ are given by

$$c_{n,m}(0) = \langle \cos\vartheta(0) Y_{n,m}(0)\rangle_0 = \sqrt{\frac{(n+1)^2 - m^2}{(2n+1)(2n+3)}}\langle Y_{n+1,m}\rangle_0 + \sqrt{\frac{n^2 - m^2}{(2n+1)(2n-1)}}\langle Y_{n-1,m}\rangle_0 \tag{A57}$$

According to Eq. (9.6), the equilibrium averages $\langle Y_{l,m}\rangle_0$ satisfy the recurrence relation:

$$\sum_{s=-1}^{1}\sum_{r=-2}^{2} d_{l,m,l+2r,m+4s}\langle Y_{l+2r,m+4s}\rangle_0 = 0 \tag{A58}$$

which may be written as the three-term matrix recurrence relation:

$$\mathbf{Q}_n^{-}\mathbf{R}_{n-1} + \mathbf{Q}_n\mathbf{R}_n + \mathbf{Q}_n^{+}\mathbf{R}_{n+1} = \mathbf{0}, \qquad n = 1, 2, 3, \ldots \tag{A59}$$

where

$$\mathbf{R}_n = \begin{pmatrix} \mathbf{r}_{4n} \\ \mathbf{r}_{4n-1} \\ \mathbf{r}_{4n-2} \\ \mathbf{r}_{4n-3} \end{pmatrix} \qquad \text{with} \qquad \mathbf{R}_0 = \frac{1}{\sqrt{4\pi}}$$

The column vector $\mathbf{r}_{4n-i}$ is

$$\mathbf{r}_{4n-i} = \begin{pmatrix} \langle Y_{4n-i,-4(n-1+\delta_{i0})}\rangle_0 \\ \langle Y_{4n-i,-4(n-2+\delta_{i0})}\rangle_0 \\ \vdots \\ \langle Y_{4n-i,4(n-1+\delta_{i0})}\rangle_0 \end{pmatrix} \qquad (i = 0, 1, 2, 3)$$

The solution of Eq. (A59) is

$$\mathbf{R}_n = \mathbf{S}_n(0)\mathbf{R}_{n-1} = \frac{\mathbf{S}_n(0)\mathbf{S}_{n-1}(0)\cdots\mathbf{S}_2(0)\mathbf{S}_1(0)}{\sqrt{4\pi}} \tag{A60}$$

Thus the $\mathbf{C}_n(0)$ are given by

$$\mathbf{C}_n(0) = \frac{1}{\sqrt{4\pi}}[\hat{\mathbf{K}}_n + [\mathbf{K}_n + \hat{\mathbf{K}}_{n+1}^T\mathbf{S}_{n+1}(0)]\mathbf{S}_n(0)]\mathbf{S}_{n-1}(0)\cdots\mathbf{S}_1(0)$$
$$n = 1, 2, 3, \ldots \tag{A61}$$

where

$$\mathbf{K}_n = \begin{pmatrix} \mathbf{O} & \mathbf{U}_{4n} & \mathbf{O} & \mathbf{O} \\ \mathbf{U}_{4n}^T & \mathbf{O} & \mathbf{U}_{4n-1} & \mathbf{O} \\ \mathbf{O} & \mathbf{U}_{4n-1}^T & \mathbf{O} & \mathbf{U}_{4n-2} \\ \mathbf{O} & \mathbf{O} & \mathbf{U}_{4n-2}^T & \mathbf{O} \end{pmatrix}$$

$$\hat{\mathbf{K}}_n = \begin{pmatrix} \mathbf{O} & \mathbf{O} & \mathbf{O} & \mathbf{O} \\ \mathbf{O} & \mathbf{O} & \mathbf{O} & \mathbf{O} \\ \mathbf{O} & \mathbf{O} & \mathbf{O} & \mathbf{O} \\ \mathbf{U}_{4n-3} & \mathbf{O} & \mathbf{O} & \mathbf{O} \end{pmatrix}$$

($\hat{\mathbf{K}}_1$ degenerates to a column vector of dimension 6). The matrices $\mathbf{K}_n$, $\hat{\mathbf{K}}_n$ are comprised of diagonal submatrices $\mathbf{U}_{4n-1}, \mathbf{U}_{4n-2}, \mathbf{U}_{4n-3}$, which are given by

$$\mathbf{U}_{4n-i} = \begin{pmatrix} u_{4n-i,-4(n-1)} & 0 & 0 & \cdots & 0 \\ 0 & u_{4n-i,-4(n-2)} & & \cdots & 0 \\ 0 & 0 & u_{4n-i,-4(n-3)} & \cdots & 0 \\ \vdots & \vdots & \vdots & \ddots & \vdots \\ 0 & 0 & 0 & \cdots & u_{4n-i,4(n-1)} \end{pmatrix} \tag{A62}$$

($i = 1, 2, 3$), and of the submatrix $\mathbf{U}_{4n}$, which is defined as

$$
\mathbf{U}_{4n} = \begin{pmatrix} 0 & 0 & \cdots & 0 \\ u_{4n,-4n+4} & 0 & \cdots & 0 \\ 0 & u_{4n,-4n+8} & \cdots & 0 \\ \vdots & \vdots & \ddots & \vdots \\ 0 & 0 & \cdots & u_{4n,4n-4} \\ 0 & 0 & \cdots & 0 \end{pmatrix} \tag{A63}
$$

All the elements of the submatrices (A62) and (A63) are given by

$$
u_{n,m} = \sqrt{\frac{n^2 - m^2}{4n^2 - 1}}
$$

## Acknowledgments

We thank W. T. Coffey, Yu. L. Raikher and S. V. Titov for useful discussions. A partial support of this work by the International Association for the Promotion of Co-operation with Scientists from the New Independent States of the Former Soviet Union (grant INTAS 96-0663) is gratefully acknowledged. One of us (Yu. P. K.) is very indebted to the French Ministry of Higher Education (allocation choicheur de haut niveau).

## References

1. J. R. McConnell, *Rotational Brownian Motion and Dielectric Theory*, Academic, London, 1980.
2. H. Watanabe and A. Morita, in *Advances in Chemical Physics*, Vol. 56, Wiley, New York, 1984, p. 255.
3. Yu. L. Raikher and M. I. Shliomis, in *Advances in Chemical Physics*, Vol. 87, W. T. Coffey Ed., I. Prigogine and S. A. Rice (Series Eds.), Wiley, New York, 1994, p. 595.
4. J-L. Déjardin, *Dynamic Kerr Effect*, World Scientific, Singapore, 1995.
5. J-L. Déjardin and G. Debiais, in *Advances in Chemical Physics*, Vol. 91, Wiley-Interscience, New York, 1995, p. 241.
6. L. J. Geoghegan, W. T. Coffey, and B. Mulligan, in *Advances in Chemical Physics*, Vol. 100, Wiley, New York, 1997, p. 475.
7. W. T. Coffey and Yu. P. Kalmykov, in *Advances in Liquid Crystals*, Vol. 113, A special volume of *Advances in Chemical Physics*, J. K. Vij, Ed., I. Prigogine and S. A. Rice, Series Eds., Wiley, New York, 2000, pp. 487–551.
8. W. T. Coffey, Yu. P. Kalmykov, and J. T. Waldron, *The Langevin Equation*, World Scientific, Singapore, 1996.
9. H. Risken, *The Fokker–Planck Equation*, Springer, Berlin, 1989.
10. R. E. D. McClung, *Adv. Mol. Rel. Int. Proc.* **10**, 83 (1977).
11. A. I. Burshtein and S. I. Temkin, *Spectroscopy of Molecular Rotation in Gases and Liquids*, Cambridge University Press, Cambridge, 1994.
12. H. Benoit, *Ann. Phys.* **6**, 561 (1951).
13. H. Goldstein, *Classical Mechanics*, 2nd ed., Addison-Wesley, Reading, 1980.
14. P. L. Nordio, G. Rigatti, and U. Segre, *J. Chem. Phys.* **56**, 2117 (1971).

15. H. Block and E. F. Hayes, *Trans. Faraday. Soc.* **66**, 2512 (1970).
16. R. L. Fulton, *J. Chem. Phys.* **78**, 6865 (1983); **78**, 6877 (1983).
17. M. Matsumoto, H. Watanabe, and K. Yoshioka, *J. Phys. Chem.* **74**, 2132 (1970).
18. W. M. Tolles, *J. Appl. Phys.* **46**, 991 (1975).
19. K. Tsuji and H. Watanabe, *J. Coll. Interface Sci.* **62**, 101 (1977).
20. K. Kikuchi, *J. Phys. Chem.* **88**, 6328 (1984).
21. H. Watanabe, Y. Fukuda, and T. Nakano, *J. Coll. Interface Sci.* **108**, 347 (1985).
22. W. Alexiewicz, *Mol. Phys.* **59**, 637 (1986).
23. G. B. Thurston and D. L. Bowling, *J. Coll. Interface Sci.* **30**, 34 (1969).
24. S. Ogawa and S. Oka, *J. Phys. Soc. Jpn.* **4**, 658 (1960).
25. A. Morita, *Phys. Rev. A* **34**, 1499 (1986).
26. A. Morita and H. Watanabe, *Phys. Rev. A* **35**, 2690 (1987).
27. L. D. Eskin, *Optik. Spektrosk.* **45**, 1185 (1978) [*Optic. Spectrosc.* **45**, 922 (1978)].
28. I. B. Aizenberg and L. D. Eskin, *Optik. Spektrosk.* **48**, 399 (1980) [*Optic. Spectrosc.* **48**, 222 (1980)].
29. A. Morita, *J. Phys. D* **11**, 1357 (1978).
30. J. L. Déjardin, P. Blaise, and W. T. Coffey, *Phys. Rev. E* **54**, 852 (1996).
31. W. T. Coffey, J. L. Déjardin, Yu. P. Kalmykov, and S. V. Titov, *Phys. Rev. E* **54**, 6462 (1996).
32. Yu. P. Kalmykov, J. L. Déjardin, and W. T. Coffey, *Phys. Rev. E* **55**, 2509 (1997).
33. Yu. P. Kalmykov, *Optik. Spektrosk.* **84**, 1000 (1998) [*Optic. Spectrosc.* **84**, 906 (1998)].
34. Y. H. Lee, D. Kim, S. H. Lee, and W. G. Yung, *J. Chem. Phys.* **91**, 5628 (1989).
35. Yu. L. Raikher, V. I. Stepanov, and S. V. Burylov, *Kolloid. Zh.* **52**, 887 (1990) [*Colloid J. USSR* **52**, 768 (1990)].
36. Yu. L. Raikher, V. I. Stepanov, and S. V. Burylov, *J. Coll. Interface Sci.* **144**, 308 (1991).
37. A. Morita and H. Watanabe, *J. Chem. Phys.* **77**, 1193 (1982).
38. W. Alexiewicz and B. Kasprowicz-Kielich, in *Modern Nonlinear Optics*, Part I, A Special Volume of Advances in Chemical Physics Vol. 85, M. W. Evans and S. Kielich (Eds.), Wiley, New York, 1995, p. 1.
39. *Handbook of Mathematical Functions*, M. Abramowitz and I. Stegun, Eds., Dover, New York, 1965.
40. W. T. Coffey, Yu. P. Kalmykov, and E. S. Massawe, *Advances in Chemical Physics*, Vol. 85, Part 2, I. Prigogine and S. A. Rice (Series Eds.), Wiley, New York, 1993, p. 667.
41. P. Debye, *Polar Molecules*, Chemical Catalog, Dover, New York, 1929.
42. K. De Smet, L. Hellemans, J. F. Rouleau, R. Courteau, and T. K. Bose, *Phys. Rev. E* **57**, 1384 (1998).
43. P. Kędziora, J. Jadżyn, K. De Smet, and L. Hellemans, *Chem. Phys. Lett.* **289**, 541 (1998).
44. J. Jadżyn, P. Kędziora, and L. Hellemans, *Phys. Lett.* **251**, 49 (1999).
45. W. T. Coffey and B. V. Paranjape, *Proc. R. Ir. Acad.* **78A**, 17 (1978).
46. J. L. Déjardin, *J. Chem. Phys.* **98**, 3191 (1993).
47. J. L. Déjardin, G. Debiais, and A. Ouadjou, *J. Chem. Phys.* **98**, 8149 (1993).
48. J. L. Déjardin, *J. Mag. Magn. Mater.* **122**, 187 (1993).
49. J. L. Déjardin and G. Debiais, *Phys. Rev. A* **40**, 1560 (1989).
50. J. L. Déjardin and G. Debiais, *Physica A* **164**, 182 (1990).

51. J. L. Déjardin, and Yu. P. Kalmykov, *Phys. Rev. E* **62**, 1211 (2000).
52. M. W. Evans, G. J. Evans, W. T. Coffey, and P. Grigolini, *Molecular Dynamics and the Theory of Broad Band Spectroscopy*, Wiley, New York, 1982.
53. R. L. Stratonovich, *Conditional Markov Processes and Their Application to the Theory of Optimal Control*, Elsevier, New York, 1968.
54. C. W. Gardiner, *Handbook of Stochastic Methods*, Springer, Berlin, 1985.
55. J. L. Déjardin, P. M. Déjardin, Yu. P. Kalmykov, and S. V. Titov, *Phys. Rev. E* **60**, 1475 (1999).
56. Yu. P. Kalmykov, *J. Mol. Liquids*, **69**, 117 (1996); *Khim. Fiz.* **16**(3), 130 (1997) [*Chem. Phys. Reports*, **16**, 535 (1997)].
57. R. N. Zare, *Angular Momentum. Understanding Spatial Aspects in Chemistry and Physics*, Wiley, New York, 1989.
58. A. R. Edmonds, *Angular Momentum in Quantum Mechanics*, Princeton University Press, Princeton, 1957.
59. Yu. P. Kalmykov and S. V. Titov, *Phys. Rev. Lett.* **82**, 2967 (1999).
60. L. Landau and E. Lifchitz, *Mécanique Quantique*, Mir, Moscow, 1967.
61. R. Kubo, *J. Phys. Soc. Jpn.* **12**, 570 (1957).
62. A. Szabo, *J. Chem. Phys.* **72**, 4620 (1980).
63. G. Moro and P. L. Nordio, *Mol. Phys.* **56**, 255 (1985).
64. Yu. P. Kalmykov and J. L. Déjardin, *J. Chem. Phys.* **110**, 6484 (1999).
65. Yu. P. Kalmykov and S. V. Titov, *Fiz. Tverd. Tela* (St. Petersburg) **40**, 1642 (1998) [*Phys. Solid State*, **40**, 1492 (1998)].
66. Yu. P. Kalmykov and S. V. Titov, *Zh. Eksp. Teor. Fiz.* **115**, 101 (1999) [*Sov. Phys. JETP* **88**, 58 (1999)].
67. H. S. Wall, *Continued Fractions*, Van Nostrand, Princeton, 1969.
68. L. D. Eskin, *Izv.Vysh. Uch. Zav., Ser. Math.* **9**, 131 (1977)
69. W. M. Tolles, R. A. Sanders, and G. W. Fritz, *J. Appl. Phys.* **45**, 3777 (1974).
70. J. L. Déjardin, P. M. Déjardin, and Yu. P. Kalmykov, *J. Chem. Phys.* **107**, 508 (1997).
71. R. Ullman, *J. Chem. Phys.* **56**, 1869 (1972).
72. J. L. Déjardin, *Phys. Rev. E* **54**, 2982 (1996).
73. W. T. Coffey, D. S. F. Crothers, and Yu. P. Kalmykov, *Phys. Rev. E* **55**, 4812 (1997).
74. D. A. Garanin, *Phys. Rev. E* **54**, 3250 (1996).
75. J. L. Déjardin, P. M. Déjardin, and Yu. P. Kalmykov, *J. Chem. Phys.* **106**, 5824 (1997).
76. J. L. Déjardin, P. M. Déjardin, and Yu. P. Kalmykov, *J. Chem. Phys.* **106**, 5832 (1997).
77. J. L. Déjardin and Yu. P. Kalmykov, *Phys. Rev. E* **61**, 1211 (2000).
78. J. L. Déjardin and Yu. P. Kalmykov, *J. Chem. Phys.* **112**, 2916 (2000).
79. H.-J. Breymayer, H. Risken, H. D. Vollmer, and W. Wonneberger, *Appl. Phys. B: Lasers Opt.* **28**, 335 (1982).
80. W. Schleich, C.-S. Cha, and J. D. Cresser, *Phys. Rev. A* **29**, 230 (1984).
81. L. Gammaitoni, P. Hänggi, P. Jung, and F. Marchesoni, *Rev. Mod. Phys.* **70**, 223 (1998).
82. W. T. Coffey, J. L. Déjardin, and Yu. P. Kalmykov, *Phys. Rev. E* **61**, 4599 (2000).
83. W. T. Coffey, J. L. Déjardin, and Yu. P. Kalmykov *Phys. Rev. B* **62**, 3480 (2000).
84. A. D. Buckingham and J. A. Pople, *Proc. Phys. Soc.* (*London*), **A68**, 905 (1955).
85. A. D. Buckingham and B. J. Orr, *Quart. Rev.* **21**, 195 (1967).

86. S. Kielich, *Acta Phys. Polon.* **17**, 239 (1958).
87. A. L. Andrews and A. D. Buckingham, *Mol. Phys.* **3**, 183 (1960).
88. R. J. W. Le Fèvre, *Adv. Phys. Org. Chem.* **3**, 1 (1965).
89. C. J. F. Böttcher and P. Bordewijk, *Theory of Electric Polarization*, Elsevier, Amsterdam, The Netherlands, 1978.
90. P. A. Franken and J. F. Ward, *Rev. Mod. Phys.* **35**, 23 (1963).
91. B. Kasprowicz-Kielich and S. Kielich, *Acta Phys. Polon. A* **50**, 215 (1975).
92. W. Alexiewicz, *Acta Phys. Polon.* A **72**, 753 (1987).
93. M. A. Pauley, H. W. Guan, and C. H. Wang, *J. Chem. Phys.* **104**, 6834 (1996).
94. I. Ledoux, J. Berdan, J. Zyss, A. Migus, D. Hulin, J. Etchepare, G. Grillon, and A. Antonetti, *J. Opt. Soc. Am. B* **4**, 987 (1987).
95. J. L. Déjardin, P. M. Déjardin, and Yu. P. Kalmykov, *J. Chem. Phys.* **108**, 3081 (1998).
96. C. J. Reid, *Mol. Phys.* **49**, 331 (1983).
97. H. C. Brinkman, *Physica* **22**, 29 (1956).
98. P. L. Bhatnagar, E. P. Gross, and M. Krook, *Phys. Rev.* **94**, 511 (1954).
99. Yu. P. Kalmykov, *Opt. Spektrosk.* **58**, 804 (1985) [*Opt. Spectrosc.* **58**, 493 (1985)].
100. E. P. Gross, *J. Chem. Phys.* **23**, 1415 (1955).
101. R. A. Sack, *Proc. Phys. Soc. London*, **B70**, 402 (1957).
102. R. A. Sack, *Proc. Phys. Soc. London*, **B70**, 414 (1957).
103. R. G. Gordon, *J. Chem. Phys.* **38**, 1724 (1963).
104. D. Forster, *Hydrodynamic Fluctuations, Broken Symmetry and Correlation Functions*, Benjamin, Reading, MA, 1975.
105. J. Cl. Leicknam, Y. Guissani, and S. Bratos, *Phys. Rev. A* **21**, 1005 (1980).
106. J. Cl. Leicknam and Y. Guissani, *Mol. Phys.* **42**, 1105 (1981).
107. Yu. P. Kalmykov, *Izv. Vyssh. Uchebn. Zav. Radiofiz.* **32**, 1113 (1989).
108. Yu. P. Kalmykov, *Phys. Rev. A* **45**, 7184 (1992).
109. P. M. Déjardin, *J. Chem. Phys.* **112**, 8605 (2000).
110. E. T. Whittaker and G. N. Watson, *A Course of Modern Analysis*, Cambridge University Press, Cambridge, 1927.
111. H. Hancock, *Lectures on the Theory of Elliptic Functions*, New York, Dover, 1958.
112. H. A. Kramers, *Physica* **7**, 284 (1940).
113. Y. P. Kalmykov and S. V. Titov, *J. Mol. Struct.* **479**, 123 (1999).
114. A. J. Martin, G. Meier, and A. Saupe, *Faraday Symp. Chem. Soc.* **5**, 119 (1971).
115. P. L. Nordio, G. Rigatti, and U. Segre, *Mol. Phys.* **25**, 129 (1973).
116. C. Brot and B. Lassier-Govers, *Ber. Bunsenges. Phys. Chem.* **80**, 31 (1976).
117. B. De Raedt and K. H. Michel, *Phys. Rev. B* **19**, 767 (1979).
118. G. Moro and P. L. Nordio, *Z. Phys. B* **64**, 217 (1986).
119. C. Brot and I. Darmon, *Mol. Phys.* **21**, 785 (1971).
120. R. W. Gerling and B. De Raedt, *J. Chem. Phys.* **77**, 6263 (1982).
121. J. L. Déjardin and Yu. P. Kalmykov, *J. Chem. Phys.* **111**, 3644 (1999).
122. W. F. Brown, Jr., *IEEE Trans. Mag.* **15**, 1196 (1979).
123. Yu. L. Raikher and M. I. Shliomis, *Zh. Eksp. Teor. Fiz.* **67**, 1060 (1974) [*Sov. Phys. JETP* **40**, 526 (1974)].

124. C. P. Bean and J. D. Livingston, *Suppl. J. Appl. Phys.* **30**, 1205 (1959).
125. L. Néel, *Ann. Géophys.* **5**, 99 (1949).
126. W. F. Brown, Jr., *Phys. Rev.* **130**, 1677 (1963).
127. T. L. Gilbert, *Phys. Rev.* **100**, 1243 (1956).
128. E. K. Sadykov and A. G. Isavnin, *Fiz. Tverd. Tela (St. Petersburg)*, **38**, 2104 (1996) [*Phys. Sol. State*, **38**, 1160 (1996)].
129. Yu. L. Raikher and V. I. Stepanov, *Phys. Rev. B* **55**, 15005 (1997).
130. Yu. P. Kalmykov and S. V. Titov, *Fiz. Tverd. Tela (St. Petersburg)* **42**, 893 (2000) [*Phys. Sol. State*, **42**, 918 (2000)].
131. W. T. Coffey, D. S. F. Crothers, J. L. Déjardin, Yu. P. Kalmykov, Yu. L. Raikher, V. I. Stepanov, and S. V. Titov, Proceedings of the 8th European Magnetic Materials and Applications Conference, Kiev, Ukraine, June 7–10, 2000.
132. Yu. P. Kalmykov and S. V. Titov, *Phys. Rev. Lett.* **82**, 2967 (1999).
133. Yu. P. Kalmykov and S. V. Titov, *J. Mag. Magn. Mater.* **210**, 233 (2000).
134. W. T. Coffey, D. S. F. Crothers, Yu. P. Kalmykov, and J. T. Waldron, *Phys. Rev. B* **51**, 15947 (1995).
135. Yu. P. Kalmykov, *Phys. Rev. B.* **61**, 6205 (2000).
136. P. Kędziora, J. Jadżyn, K. De Smet, and L. Hellemans, *J. Mol. Liquids* **80**, 19 (1999).
137. P. Kędziora, J. Jadżyn, K. De Smet, and L. Hellemans, *Chem. Phys. Lett.* **302**, 337 (1998).
138. T. Furukawa et K. Matsumoto, *Jpn. J. Appl. Phys.* **31**, 840 (1992).
139. M. W. Evans, *J. Chem. Phys.* **77**, 4632 (1982).
140. I. S. Gradshteyn and I. M. Ryzhik, *Table of Integrals, Series, and Products*, Academic, New York, 1980.

# CROSSOVER FORMULAS IN THE KRAMERS THEORY OF THERMALLY ACTIVATED ESCAPE RATES—APPLICATION TO SPIN SYSTEMS

W. T. COFFEY

*School of Engineering, Department of Electronic and Electrical Engineering, Trinity College, Dublin 2, Ireland*

D. A. GARANIN

*Max-Planck-Institut für Physik Komplexer Systeme, D-01187 Dresden, Germany*

D. J. McCARTHY

*School of Mathematics, Dublin Institute of Technology, Kevin St. Dublin 8, Ireland*

## CONTENTS



*Advances in Chemical Physics, Volume 117*, Edited by I. Prigogine and Stuart A. Rice.
ISBN 0-471-40542-6 

# I. INTRODUCTION

## A. Reaction Rate Theory

The origin of modern reaction rate theory which we must very briefly outline before stating the ultimate purpose of our review stems from the 1880s when Arrhenius [1] proposed, from an analysis of the experimental data, that the rate coefficient in a chemical reaction should obey the law

$$\Gamma = \nu \exp\left(-\frac{\Delta U}{kT}\right) \tag{1.1}$$

where $\Delta U$ denotes the threshold energy for activation and $\nu$ is a prefactor [1–3].

After very many developments summarized in [1–3], this equation led to the concept of chemical reactions, as an assembly of particles situated at the bottom of a potential well. Rare members of this assembly will have enough energy to escape over the potential hill due to the shuttling action of thermal agitation and never return [1] (see Fig. 1). The escape over the potential barrier represents the breaking of a chemical bond [1–3].

The Arrhenius law for the escape rate $\Gamma$ (reaction velocity in the case of chemical reactions) of particles that are initially trapped in a potential well at $A$, and that may subsequently, under the influence of thermal agitation, escape over a (high $\gg kT$) barrier of height $\Delta U$ at $C$ and never return to $A$ may be written in the context of transition state theory (TST) [1] as

$$\Gamma = \Gamma_{\mathrm{TS}} = \frac{\omega_A}{2\pi} e^{-\beta \Delta U} \tag{1.2}$$

where TS = transition state. The attempt frequency, $\omega_A$, is the angular frequency of a particle performing oscillatory motion at the bottom of a well, while $\beta = (1/kT)$ where $k$ is Boltzmann's constant and $T$ is the absolute temperature. The barrier arises from the potential function of some external force, which may be electrical, magnetic, gravitational, and so on. The formula has the form of an attempt frequency times a Boltzmann factor, which weighs the escape from the well. Reaction rate theory was firmly set in the context of non equilibrium statistical mechanics by the pioneering work of Kramers [4]. He chose (in order to take into account non equilibrium effects in the barrier crossing process which manifest themselves as a frictional dependence (i.e. a coupling to the heat bath of the prefactor in the TST formula) as a microscopic model of a chemical reaction, a classical particle moving in a one-dimensional potential (see Fig. 1). The fact that a typical particle is embedded in a heat bath is modelled by the Brownian motion, which represents (essentially through a dissipation

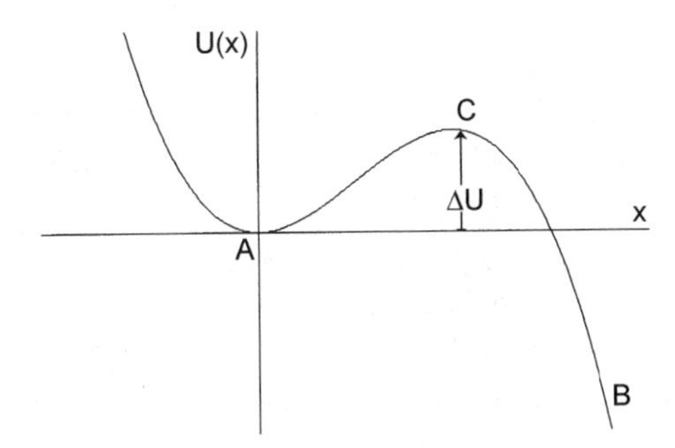


**Figure 1.** Single well potential function. The simplest example of escape over a barrier. Particles are initially trapped in the well near the point $A$ (a source of probability) by a high potential barrier at the point $C$. They very rapidly thermalize in the well. Due to thermal agitation however very few may attain enough energy to escape over the barrier into region $B$, from which they never return (a sink of probability), i.e. the barrier $C$ is assumed to be sufficiently large so that the rate of escape of particles is very small. We model a chemical reaction by introducing a co-ordinate such that $x = A$ in species $A$ and $x = B$ in species $B$. The reaction is modelled by diffusion over the boundary $C$ between the two distinct states. The *transition state* $C$ is defined [27] as an activated complex from which the system, starting at $A$, will, once it is reacted, form products specifying state $B$. $x$ is a reaction coordinate representing the distance between both fragments $A$, $B$ of a dissociated molecule, introduced by Christiansen in 1936 [1, 27].

parameter) in the single particle distribution function all the remaining degrees of freedom of the system consisting of the selected particle and heat bath which is in perpetual thermal equilibrium at temperature $T$. In Kramers' model [1, 27] the particle coordinate $x$, represents the *reaction coordinate* (i.e. the distance between two fragments of a dissociated molecule – a concept first introduced by Christiansen [1, 27] in 1936). The value of this coordinate, $x_A$, at the first minimum of the potential represents the *reactant state*, the value, $x_B$, significantly over the summit of the well at $B$ (i.e. when the particle has crossed over the summit) represents the *product state*, and the value, $x_C$, at the saddle point, represents the *transition state*. We remark that in his calculations of 1940 Kramers [4] assumed that particles are initially trapped in a well near the minimum of the potential at the point $A$. They then receive energy from the surroundings and a Maxwell-Boltzmann distribution is rapidly attained in the well (see the solution of the complete time-dependent problem in Section VIII); over a long period of time however rare particles gain energy in excess of $\Delta U$ (the barrier height). Kramers then assumes that such particles escape over the barrier $C$ (so that there is a perturbation of the Maxwell-Boltzmann distribution) and reach a minimum at $B$, which is of lower energy than $A$, and once there, never return. We list the assumptions of Kramers:

1. The particles are initially trapped in $A$ (which is a *source* of probability).
2. The barrier heights are very large compared with $kT$, (Kramers takes $k$ to be 1).
3. In the well, the number of particles with energy between $E$ and $E + dE$, is proportional to $\exp(-(E/kT))dE$, that is a Maxwell-Boltzmann distribution is attained extremely rapidly in the well (see Section VIII).
4. Quantum effects are negligible.
5. The escape of particles over the barrier is very slow (i.e. is a quasi-stationary process), so that the disturbance to the Maxwell-Boltzmann distribution of postulate 3 is almost negligible at all times.
6. Once a particle escapes over the barrier it practically never returns (i.e. $B$ is a *sink* of probability).
7. A typical particle of the reacting system may be modelled by the theory of the Brownian motion, including the inertia of the particles [5].

It is worth mentioning here that assumption 5 above relies heavily on assumption number 2. If the barrier is too low the particles escape too quickly to allow a Boltzmann distribution to be set up. If the barrier is high, on the other hand, before many particles can escape, the Boltzmann distribution is set up. As required by postulate 3 we assume, therefore, that the ratio $\Delta U/(kT)$, where $\Delta U$ is the barrier height, is at least of the order, say, 5.

This model, which yields explicit formulas for the escape rate for very low and intermediate to high dissipative coupling to the bath (so including non equilibrium effects in the TST formula), is ubiquitous in any physical system in which there is noise-activated escape from a potential well. It has recently attained new importance in connection with fields as diverse as dielectric relaxation of nematic liquid crystals, magnetic relaxation of fine ferromagnetic particles, laser physics and Josephson junctions [5, 6, 25–28].

In these new applications of the Kramers theory, situations will arise where the various formulas for escape rates characteristic of that theory will fail. An example of this is the application to relaxation of mechanical particles where the dissipation parameter lies in the crossover damping region where neither the very low damping (VLD) nor the intermediate to high damping (IHD) limits of the original Kramers theory are valid. This is of importance in the context of the Josephson junction, [1, 5], and decoherence in ring lasers, both being examples of cycle slips in phase locked loops. Another example of the failure of escape rate formulas is in the context of the rotation of the magnetic moment inside a single domain ferromagnetic particle. This is a system with two degrees of freedom namely the polar and azimuthal angles of the magnetic moment (excluding of course the bath) in contrast to the single degree of freedom of the original Kramers problem). Here for small departures of the magnetocrystalline anisotropy from axial symmetry (which in this context we shall again term the crossover region) the VLD and IHD limits of the theory (as adapted to spins [59, 70]) do not tend to the axially symmetric result for the escape rate (which is valid for all values of the dissipation parameter). Thus crossover formulas which bridge in the case of mechanical particles the VLD-IHD and in spins the axially symmetric – nonaxially symmetric escape rates must be determined. A crossover formula bridging the VLD-IHD region for particles was given [34] by Mel'nikov and Meshkov in 1986 (see Section VI) while the axially symmetric–nonaxially symmetric bridging formulas [70] for spins are given in Section VII. It is the purpose of this chapter to give a detailed account of the derivation of the various crossover formulas.

In order to accomplish this a detailed mathematical treatment of the original Kramers model and its subsequent extensions to multi-degree of freedom systems by Langer [61] (see Section IV) which is a generalization of the classic calculation of Becker and Döring in 1935 [17] (see Frenkel [7]) of the rate of condensation of a supersaturated vapour) and other investigators [1] will first be given. The range of validity of the various Kramers' asymptotic formulas for the escape rates in single and multi-dimensional systems as a function of the parameters of the system will also be discussed. Our method in the first part of the chapter will be initially to adhere as closely as possible to the original [4] paper of Kramers, adding the mathematical details where necessary. A concise and more rigorous treatment of the very high and very low damping cases first considered by him will then be presented at the end of Section II. The purpose of the review of the Kramers' work being that the later extensions of the theory, namely the crossover formulas, may be then relatively easily understood.

### B. Summary of the Chief Developments in the Theory of Brownian Motion

The development of reaction rate theory, after Arrhenius, closely follows [1] that of the development of the theory of stochastic processes, and nonequilibrium statistical mechanics, as instigated by Boltzmann in 1872 [6]. He in his attempt to demonstrate that the effect of molecular collisions, whatever the initial positions and velocities of the molecules of the gas, would be to bring about a Maxwell–Boltzmann distribution of positions and velocities, formulated his famous equation [6, 8]. This equation describes the time evolution of the density of molecules in phase space, provided that only encounters [8] between two molecules (i.e., two-body interactions) are ever of any importance.

The Boltzmann equation, which is a closed equation [6] for the *single* particle distribution function, is now the fundamental equation that allows one to describe the bath itself in a microscopic way. The particular law of binary collisions (*Stosszahlansatz*) describes the interactions between the molecules of the bath. The binary collision assumption amounts to stating that encounters with other molecules occupy only a very small part of the lifetime of a molecule. To put this statement in another way [8], it states that encounters in which *more than two* molecules take part are neglected in both number and their effect in comparison with binary encounters. Furthermore, in considering binary encounters between molecules having velocities within assigned ranges, it is assumed that both sets of molecules are distributed at random and without any correlation between velocity and position in the neighborhood of the point where the collision takes place. This is the "molecular chaos" assumption of Boltzmann [9].

The Boltzmann equation [8] is in general a nonlinear integrodifferential equation and is very difficult to solve. In certain instances, namely, where, during the course of an encounter, the velocity of a molecule may change very little (i.e. [6, 8], the molecular interactions are weak compared to the mean kinetic energy of the molecules, meaning that the deviation in velocity resulting from a collision is very small on average compared with the velocity at the start of the collision), it is possible to reduce the Boltzmann equation to a linear partial differential equation known as the Fokker–Planck equation (FPE) [6]. This reduction was first achieved without explicit reference to the Boltzmann equation by Einstein in 1905 [10, 11] who, by combining the elementary stochastic process known as the random walk [12] with the Maxwell–Boltzmann distribution, was able to explain the observed Brownian motion of particles in a colloidal suspension. By taking the jumps in the random walk of the particle as small, he obtained a PDE for the distribution of the displacements of the particle in 1D, and stipulated that such a random walk must be a natural consequence of the second law of thermodynamics. The random walk is a physical manifestation of the interaction of the macroscopic colloidal particle with its surrounding heat bath; the bath (or environment of the particle) being in perpetual thermal equilibrium at temperature $T$.

In modern terminology, the displacement process is known as the Wiener process [5], and the solution of the corresponding PDE is the probability distribution function of the Wiener process.

Einstein's approach of 1905 was complemented by the work of Langevin [13, 5]. Langevin in 1908 simply wrote down the equation of motion of the Brownian particle and assumed that the effect of the surroundings (heat bath) of the particle could be described by a systematic retardation force proportional to the velocity of the particle on which is superimposed a very rapidly fluctuating force that we now call the white noise force. Thus he was able to derive an expression for the mean-square displacement of the Brownian particle in the high friction limit that coincides with that of Einstein, who, as it turns out, implicitly assumes that the distribution of the velocities reaches equilibrium much faster than that of the displacements which is tantamount to taking the high friction limit in the averaged Langevin equation.

Subsequent important work was carried out [14, 16, 42] by Ornstein, who obtained in 1917 a formula for the mean-square displacement of a Brownian particle that is valid for all friction values. This constituted the first inclusion of inertial effects in the theory of the Brownian movement.

As far the Brownian motion in a potential is concerned, we must mention Smoluchowski, who in 1906 [6, 14, 15] considered the problem of the Brownian movement of a particle under an external force (in connexion with the study of the effect of gravity on the Brownian motion-the sedimentation problem) and Debye in 1913 [18, 19] who studied the rotational motion of dipolar molecules subjected to an external high frequency alternating field by making *the gross assumption* that in spite of their small size, they could be treated as rotating Brownian particles and so was able to give a theory of dispersion and absorption of polar molecules at microwave frequencies (which substantially agrees with experiment).

Einstein's work was reappraised [20] by Klein in 1921 [1, 20], who, by abandoning the assumption that equilibrium of the velocities had been attained, was able to construct a PDE in the complete phase space of positions and velocities which described how the number (probability) density or concentration of representative points evolves with time. This PDE in phase space which is the Boltzmann equation for a *very large number of infinitesimally small* collisions, was re-derived [4] by Kramers in 1940. His objective in rederiving the equation was to use it to calculate rigorously a prefactor (that takes account of nonequilibrium effects due to the disturbance of the Maxwell-Boltzmann distribution near the barrier) to be inserted into the TST result, from a microscopic model of the reacting system (which represents the molecules as an assembly of Brownian particles in a single well potential). Thus the Kramers formula for the escape rate from an *isolated* well as in Figure 1 would have the form:

$$\Gamma = A\frac{\omega_A}{2\pi}e^{-\beta\Delta U} \tag{1.3}$$

where the factor $A$ describes possible deviations from the transition state theory.

A related application of the escape rate method is the theory of reversal of magnetization of fine ferromagnetic particles over a potential barrier, arising from their internal magnetocrystalline anisotropy, due to thermal agitation proposed by Brown, Jr., in 1963 [21], which is an important aspect of this chapter. In addition, if the friction is allowed to become very large in the theory of magnetic relaxation, the results may be applied [36] to the dielectric relaxation (ignoring inertial effects) of nematic liquid crystals.

From a purely mathematical point of view, very substantial contributions were made by Wiener in 1923 [14, 22, 23], who proved that the displacement or Wiener process is almost everywhere continuous but nowhere differentiable and by Doob in 1942 [14, 23], who showed that as a consequence of the behavior of the Wiener process, the Langevin equation *must always be treated as an integral equation*. The interpretation of the Langevin equation as an *integral equation* lead to the development of the Stratonovitch and Itô calculi, which provide an interpretation rule for the Langevin equation [24]. A compre-

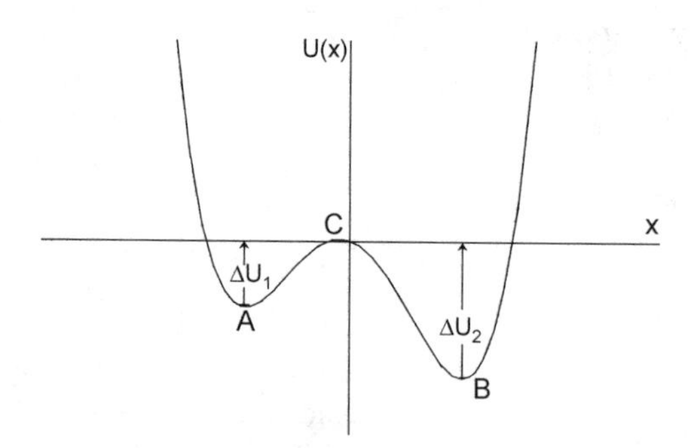


**Figure 2.** Double well potential function. A more general situation than the single well potential is the above double well potential. Particles start in the left-hand well near $A$ (a source of probability). Due to thermal agitation, they may obtain sufficient energy to escape over the barrier at $C$ into the well at $B$. From there they may remain in the well or they may return to the first well.

hensive account of these developments may be found in the standard text *The Fokker–Planck Equation* [25, 26] and in Gardiner [24]. The application of the Stratonovitch calculus (which is appropriate when white noise is taken as a limiting case of colored noise) to the Langevin equation is extensively discussed in the recent textbook *The Langevin Equation* by Coffey et al. [5] and by van Kampen [27, 28].

The reaction rate theory of Kramers is also intimately connected with the *theory of first passage times* as the escape rate over a barrier is the inverse escape time of the particle from the potential well, which is twice the mean first passage time from the bottom $A$ to the summit $C$ of the well, with, of course, a high barrier being assumed. This is extensively discussed by Gardiner [24].

Comprehensive reviews of the early work in the theory of the Brownian motion have been given by Wang and Uhlenbeck [29] and by Chandrasekhar [15]. These and other fundamental papers dealing with this subject are available in the anthology edited by Wax [14]. A further discussion of these topics is given in [5, 6, 27, 28].

### C. The Langevin and Fokker–Planck Equations

We have seen that Kramers' objective was to calculate the prefactor in the escape rate, namely,

$$\Gamma = A\frac{\omega_A}{2\pi}e^{-\beta\Delta U} \tag{1.4}$$

from a microscopic model of the chemical reaction. The fact that a microscopic model of the reacting system (viz. an assembly of Brownian particles in a potential well) is taken account of in the calculation of the prefactor $A$ means that the prefactor is closely associated both with the *stochastic differential equation* underlying the Brownian motion process, which is the Langevin equation for the evolution of the random variables (position and momentum) describing the process, and the *associated probability density diffusion equation* describing the time evolution of the density of the realizations of these random variables in phase space. This is the Fokker–Planck equation, which like the Boltzmann equation is a *closed* equation for the *single* particle or single system distribution function.

First we remark that generally a heat bath has upwards of $10^{23}$ degrees of freedom, so that it is never possible, nor indeed is it necessary to consider the trajectory of an individual representative point or molecule. Instead Langevin attempts to describe the behaviour of a Brownian particle as follows. He selects a particular Brownian particle and accounts for the effect of the molecules of the bath (environment of the particle) on that particular particle by simply regarding the particle as a single degree of freedom system and adding to its Newtonian equations of motion a systematic retarding force superimposed on which is a rapidly fluctuating random white noise. (A very detailed account of this equation is given in Coffey et al. [5]). Hence the number of degrees of freedom of the entire system consisting of heat bath (the bath or reservoir being permanently in thermal equilibrium at temperature $T$) and particle is reduced by a factor of the order of $(10^{23} - 1)$ and the Liouville equation of classical mechanics *so reduced* is *generalized* by the addition of diffusive terms accounting for the bath-particle interaction [41]. Thus we have a theory of relaxation of an assembly of such particles based on a *single* particle distribution function.

We must emphasize that both Debye [19] and Kramers [4] assume that the theory of the Brownian motion (as we have briefly outlined which accurately applies only to *macroscopic* particles the observed behaviour of which is the result of *millions* of infinitesimal molecular collisions [5]) may also be applied to molecules. One result of this assumption is that later investigators [3] (notably Grote and Hynes [33] in the reaction rate theory context) have attempted in one fashion or another to include so called memory effects in the original theory of Brownian motion.

The Langevin equation for a mechanical system with one degree of freedom subjected to a conservative force, $K = -(dU/dx)$ in the phase space of positions and momenta $(x, p)$ is

$$\dot{x} = \frac{p}{m} \tag{1.5a}$$

$$\dot{p} = -\frac{dU}{dx} - \frac{\zeta}{m}p + \lambda(t) \tag{1.5b}$$

These two equations are the usual Hamilton equations [32] supplemented by the retarding force $-(\zeta/m)p$, and the rapidly fluctuating (in comparision to $p$) white noise force $\lambda(t)$, which means, of course, that $p(t)$ must be a random variable. Furthermore, the integral of $\lambda(t)$ over an infinitesimal time $\tau$ is the Wiener process. The noise $\lambda(t)$ must satisfy

$$\overline{\lambda(t)} = 0 \tag{1.6a}$$

$$\overline{\lambda(t_1)\lambda(t_2)} = 2kT\zeta\delta(t_1 - t_2) \tag{1.6b}$$

where the statistical averages are taken over all the realizations [i.e. [5] the sample paths of $\lambda(t)$] and $\delta(t_1 - t_2)$ is the Dirac delta function.

The information concerning $\lambda(t)$ is, however, incomplete, as the random variable $\lambda(t)$ must also obey Isserlis's theorem for the mean of a number of observations (for a proof see [5]). The imposition of these conditions requires $\lambda(t)$ to be a Gaussian random variable. The technical reason for the imposition of Isserlis's theorem is that, without it, it is impossible to truncate the Taylor series expansion of the distribution function in phase space (the Kramers–Moyal expansion, [25, 26]). In like manner, if one proceeds by averaging the Langevin equation for a fixed set of positions and momenta at time $t$ over a short time interval $\tau$ so as to generate directly equations for the statistical moments as proposed by Coffey [30], it is impossible [31] to generate the set of equations for the statistical moments that are equivalent to the Fokker–Planck equation without the use of the theorem.

One can now, as is shown in Section II, write down the number of representative points in a small phase volume $dxdp$, namely

$$\rho(x, p, t)dxdp \tag{1.7}$$

where the probability density function (concentration of phase points) $\rho$ satisfies

$$\frac{\partial \rho}{\partial t} + \frac{p}{m}\frac{\partial \rho}{\partial x} - \frac{\partial}{\partial p}\left[\rho\frac{\partial U}{\partial x} + \frac{\zeta}{m}\left(\rho p + mkT\frac{\partial \rho}{\partial p}\right)\right] = 0 \tag{1.8}$$

This equation [which is the single particle Liouville equation [32] of classical mechanics supplemented by diffusion terms] is a particular example of the class of equations known as the Fokker–Planck equations and is called [25–28] the *Klein-Kramers equation*. It is fundamental in the discussion of the Kramers' escape rate problem for particles. The Klein-Kramers equation (see Section II) has the form of a *continuity* equation for the probability current density, **J**, in phase space, viz.

$$\frac{\partial \rho}{\partial t} + \nabla \cdot \mathbf{J} = 0 \tag{1.9}$$

Time independent or stationary solutions (i.e. $\partial\rho/\partial t = 0$) of this equation are then given by:

$$\nabla \cdot \mathbf{J} = 0 \tag{1.10}$$

and are important in what follows below.

In order to demonstrate this we shall now refer to the problem of escape over a single barrier, as in Figure 3. The probability density function $\rho$ in the well, which has its origin in a probability source at $A$ (see the solution of a complete time

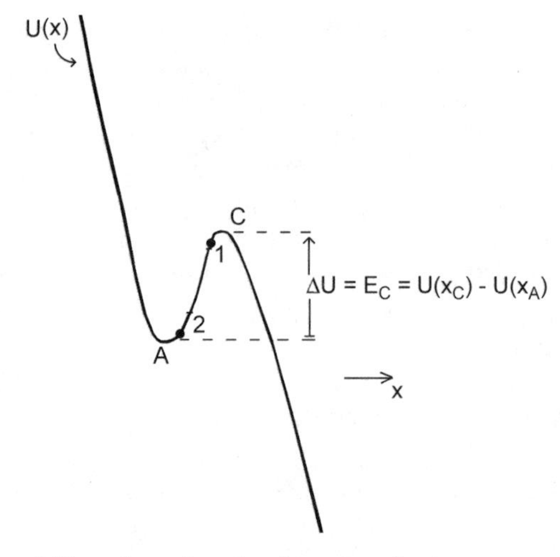


**Figure 3.** Figure 3 illustrating various damping regimes for the escape rate in the context of the potential. For intermediate to high damping the Maxwell-Boltzmann distribution due to the source of probability at $A$, which is very rapidly established and prevails in the well, persists up to a very small transition region, which is so small in terms of $x$ extension that the potential shape in this small non-equilibrium region is well approximated by the inverted parabola. If the damping is very small on the other hand as in the VLD case of Kramers, the region 1 of non-equilibrium extends much further down into the well, say to 2. Here the extent of the region 2 is such that while the Maxwell-Boltzmann distribution may be regarded as still persisting very near $A$, the parabolic approximation of the potential valid in the IHD case no longer holds. Instead the damping is now supposed so small that the conservative part of the distribution function (or the Liouville term in the Klein-Kramers equation governing the time evolution of the distribution function) vanishes just as in the absence of dissipation. Thus the *phase dependence* of the process is ignored and the escape process is then governed solely by the *slow* diffusion across the barrier at $C$ of the undamped energy trajectories of the almost periodic motion in the well with the barrier energy $E_C$, which is the critical energy needed to escape the well. The calculation of Mel'nikov and Meshkov [34] attempts to take into account in the crossover region the effect of the Liouville term (that is the dependence of the escape process on the phase which dependence is of course taken account of in the IHD region and is always ignored in the VLD region) by means of an energy-action diffusion equation. Whence they derive (by converting that equation into a Wiener-Hopf integral equation at a given action and considering the action over one period of the librational motion of those particles in the well with energy $E_C$) a formula which bridges the Kramers VLD and IHD results.

dependent problem given in Section VIII) will very quickly thermalize, that is attain the equilibrium value $\rho_{\text{eq}}$, which is the Maxwell-Boltzmann distribution

$$\rho_{\text{eq}} = \frac{e^{-\beta((p^2/2m)+U(x))}}{Z}$$

where $Z$ is the partition function and $\rho_{\text{eq}}$ satisfies the stationary Eq. (1.10). The above distribution function represents a state of *local* thermal equilibrium in the well. Near the top of the barrier however the distribution function $\rho$ deviates from the local equilibrium $\rho_{eq}$ owing to the slow draining of probability across the barrier [27, 28] due to a non zero probability current. This at first glance would suggest that one must solve the entire time dependent problem (as in Section VIII) starting from a source (delta function distribution) of probability at $A$. This is unnecessary however because the *stationary solution* will also hold in the vicinity of the barrier. Here the density, $\rho$, is exponentially small (because very few particles reach the barrier) and so also is its time derivative $\dot{\rho}$. The barrier crossing (Arrhenius) process is also exponentially slow, so that $\dot{\rho}$ has a double exponential smallness. Thus the quasi-stationary approximation $\dot{\rho} \approx 0$ is sufficient to solve the problem. The deviation from the Maxwell-Boltzmann distribution $\rho_{eq}$ obtaining in the depths of the well due to the flow across the barrier (in the TST theory the Maxwell-Boltzmann distribution holds everywhere) being accounted for by writing in the *vicinity of the barrier*

$$\rho = \rho_{eq}\zeta(x, p)$$

Moreover in the *vicinity* of the barrier $U(x)$ in $\rho_{eq}$ is to be replaced by the parabolic approximation

$$U(x) = U(x_C) - \frac{1}{2}m\omega_C^2(x - x_C)^2 \tag{1.11}$$

Here $\zeta(x, p)$ (not to be confused with the friction $\zeta$) is a crossover function, which varies rapidly in the barrier region, takes on the value unity in the depths of the well and vanishes for $x > x_C$ expressing the fact that the point $B$ in Figure 1 is a sink of probability. *This approach will fail however if the friction is so small that the deviation from the Maxwell-Boltzmann distribution is so large that x lies outside the interval, where U(x) may be approximated by an inverted parabola.* This leads us to the problem of the range of validity of escape rate formulas in the Kramers theory.

### D. Range of Validity of Escape Rate Formulas in the Kramers Theory

By supposing that $\dot{\rho} \approx 0$ (quasi-stationary) Kramers [4] discovered two asymptotic formulae for the escape rate out of a well for a system governed by the Langevin equation. The first is the intermediate-to-high damping (IHD) formula (cf. Eqs. 1.2,1.3)

$$\Gamma = \frac{\omega_A}{2\pi}\left[\sqrt{1 + \frac{\eta^2}{4\omega_C^2}} - \frac{\eta}{2\omega_C}\right]e^{-\beta\Delta U} \tag{1.12}$$

where

$$\eta = \frac{\zeta}{m}.$$

(The correction $A$ to the TST result in the prefactor of Eq. (1.12) is the positive eigenvalue (characterizing the unstable barrier crossing mode) of the Langevin equation (1.5a, 1.5b) linearized about the saddle point of the potential $U(x)$ which in the case considered by Kramers is a one dimensional maximum. For a further discussion see Section IV).

Equation (1.12) formally holds [27, 28], when the energy loss per cycle of the motion of a particle in the well with energy equal to the barrier energy $E_C = \Delta U$, is very much greater than $kT$. The energy loss per cycle of the motion of a barrier crossing particle is $\eta I(E_C)$, where $E_C$ is the energy contour through the summit or maximum of the potential and $I$ is the action [32] evaluated at $E = E_C$ (see Fig. 3). This criterion effectively follows from the Kramers VLD result, Eq. (1.14) below.

The IHD asymptotic formula is derived by supposing (see Fig. 3) that the barrier height is so high and the dissipative coupling to the bath so strong that a Maxwell-Boltzmann distribution is very rapidly established throughout almost all the well with the exception of a very small non equilibrium region near the summit where the Langevin equation may be linearized, which means that all the coefficients in the corresponding Klein-Kramers equation are linear in the positions and velocities. This is the inverted parabola approximation to the potential at the barrier alluded to above Eq. (1.11). If these simplifications can be made, then the Klein-Kramers equation, although it remains an equation in two phase variables $(x, p)$, may be integrated by introducing an *independent* variable which is a linear combination of $x$ and $p$, so that it becomes an ordinary differential equation (ODE) in a single variable. The IHD treatment was extended to several variables by Langer [61] and is treated in Section IV.

A particular case of the IHD formula is very high damping where $\Gamma$ from Eq. (1.12) becomes:

$$\Gamma = \frac{\omega_A}{2\pi}\frac{\omega_C}{\eta} e^{-\beta\Delta U} \tag{1.13}$$

in which case the quasi-stationary solution where $\dot{\rho} \approx 0$ may be obtained in integral form by quadratures and the high barrier limit of the solution which is appropriate to the escape rate may be found by the method of steepest descents. It is now unnecessary to linearize the Langevin equation, as the solution may be obtained by means of the *Smoluchowski* equation. This is an approximate equation for the distribution function in the high friction limit in configuration space $x$ only and so may be easily integrated in the stationary case. Kramers obtained this equation in heuristic fashion (Section IIE) by integrating the Klein-Kramers equation over the velocities, supposing a Maxwell-Boltzmann distribution of velocities holds approximately. A more rigorous derivation of that equation than that of Kramers' is given at the end of Section II. The Smoluchowski equation is the basis of the orginal (1905) Einstein approach to the Brownian motion, where inertial effects are neglected and which is valid for times much greater than the inertial time $\eta^{-1}$. *Thus in the IHD and very high friction regimes, when the energy loss per cycle is much greater than the thermal energy, Kramers' problem may be solved by reduction to essentially one dimensional problems.*

For small friction, such that $\eta I(E_C) \ll kT$, however the IHD formula, derived above, fails (predicting just as the TST formula, escape in the absence of coupling to the bath) because the tacit assumption that the particles approaching the barrier from the depths of the well are in thermal equilibrium is violated due to the smallness of the dissipation of energy to the bath. Thus, the spatial region of significant departure from the Maxwell-Boltzmann distribution extends far beyond the region, where the potential may be sensibly approximated by an inverted parabola.

Kramers showed how the very low damping, (VLD) case where the energy loss per cycle $\eta I(E_C) \ll kT$ may be solved by reducing the Fokker–Planck equation to a PDE in one space variable. (This is the energy or equivalently the action). Here the energy trajectories form closed loops so that they do not differ significantly from those of the undamped librational motion in a well with energy corresponding to the summit or saddle energy $\Delta U$ or $E_C$. Thus the only effect of escape is to produce a very slow spiraling of the closed energy trajectories toward the origin in phase space $(x, p)$. He solved the VLD problem by writing the FPE in angle-action variables (the angle is the phase or instantaneous state of the system along an energy trajectory) or angle-energy variables and taking a time average of the motion along a closed energy trajectory near the saddle energy trajectory. The average being along a trajectory, is, of course, equivalent to an average over the fast phase variable, thus producing a diffusion equation in the slow energy (or action) variable. The Liouville term in the Klein-Kramers equation now vanishes because the motion is almost conservative. So once again, the time derivative of $\rho$, (when $\rho$ is written as a function of the energy using the averaging procedure above), is exponentially small at the saddle point. Hence, the stationary solution in the energy variable may be used. This procedure, which is described in detail in Section II following the original approach of Kramers because of its historical interest, yields the Kramers VLD formula, namely,

$$\Gamma = \frac{\omega_A}{2\pi} \frac{\eta I(E_C)}{kT} e^{-\beta \Delta U} \tag{1.14}$$

This holds when in Eq. (1.3), $A \ll 1$ i.e.

$$\eta I(E_C) \ll kT \tag{1.15}$$

and unlike the TST result vanishes when $\eta \to 0$ as escape is impossible without coupling to the bath.

Thus, in all cases *analytical formulas for the escape rate rest on the fact that, in the damping regimes under consideration, the FPE may be reduced to an equation in a single coordinate*. The original treatment of the VLD case by Kramers given in Section II is rather lengthy and the opportunity is taken to present a more concise derivation of this result at the end of that Section.

The low-damping formula is of particular significance in that it clearly demonstrates that escape is impossible in the absence of coupling to the bath (c.f. the fluctuation dissipation theorem). Similarly, if the coupling to the bath is very large, the escape rate becomes zero. In his original paper, Kramers, made several estimates of the range of validity of both IHD and LD formulas and the region in which the transition state theory (TST) embodied in Eq. (1.2) could be

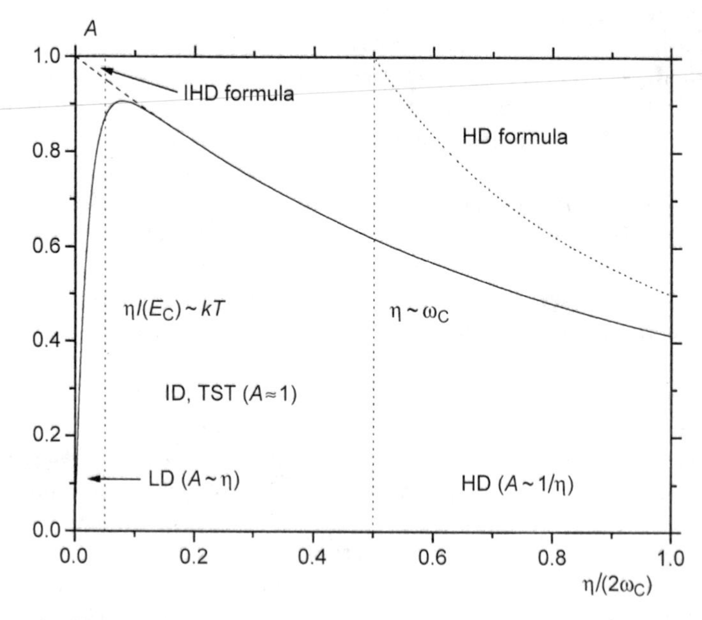


**Figure 4.** Diagram of damping regions for particles. There are three damping regions, LD, ID (TST), and HD, and two crossovers between them. The ID case is difficult to realize since the IHD correction to it is or the order $\eta/\omega_C$, and the ID case requires very small values of $\eta$ and very low temperatures in order to render both the IHD and LD corrections small.

applied with a high degree of accuracy. He was, however, unable to give a formula in the intermediate region between IHD and VLD, as there, $\eta I(E_C) \approx kT$, and there is no small perturbation parameter in the problem. A number of investigators in the period 1940–1990 have investigated this crossover problem (details in [1]). In this chapter (see Section VI), we shall summarize the work of Mel'nikov and Meshkov [34]. They took into account the contribution of the Liouville term at moderate damping by constructing an integral equation for the evolution of the energy distribution function at a given action (in this case the action in one cycle of the *librational* motion in the well at the barrier energy) which they solved using the Wiener-Hopf method [35] and obtained the formula

$$\Gamma = \exp\left(\frac{1}{2\pi}\int_{-\infty}^{\infty} \ln\left\{1 - \exp\left[-\frac{\eta I(E_C)}{kT}\left(\lambda^2 + \frac{1}{4}\right)\right]\right\}\frac{d\lambda}{\lambda^2 + (1/4)}\right)\Gamma_{\text{IHD}} \tag{1.16}$$

where $\Gamma_{\text{IHD}}$ is given by Eq. (1.12). Equation (1.16) is valid for all values of the friction $\eta$. The calculations are very long and involved and are discussed in detail in Section VI, as the details in the original paper are scanty.

### E. Application of the Kramers Theory to Rotational Brownian Motion

Subsequent to its introduction, Kramers theory was extensively employed to calculate escape rates for the rotational motion of dipoles in potentials arising from the crystalline anisotropy of the surroundings. In particular, we mention the extension of the Debye theory of dielectric relaxation to nematic liquid crystals where the Kramers theory may be used to calculate the retardation factors introduced by Maier and Saupe [36]. However, such calculations are, in essence, a particular case of the application of the theory to the magnetic relaxation of fine ferromagnetic particles over their internal potential barrier arising from the magnetocrystalline anisotropy.

This began with the work of Néel [37] who applied the TST approach to superparamagnetism (so-called because the magnetization of a single domain ferromagnetic particle behaves like an enormous Langevin paramagnet of $10^4$–$10^5$ Bohr magnetons) in order to calculate the greatest relaxation time or inverse escape rate. He considered the reversal of the direction of the magnetization vector of a small magnetic particle due to thermal agitation (see Fig. 5). This, of course, is the reorientation of a particle over the lowest saddle point corresponding to the easiest axis of magnetization on a sphere of

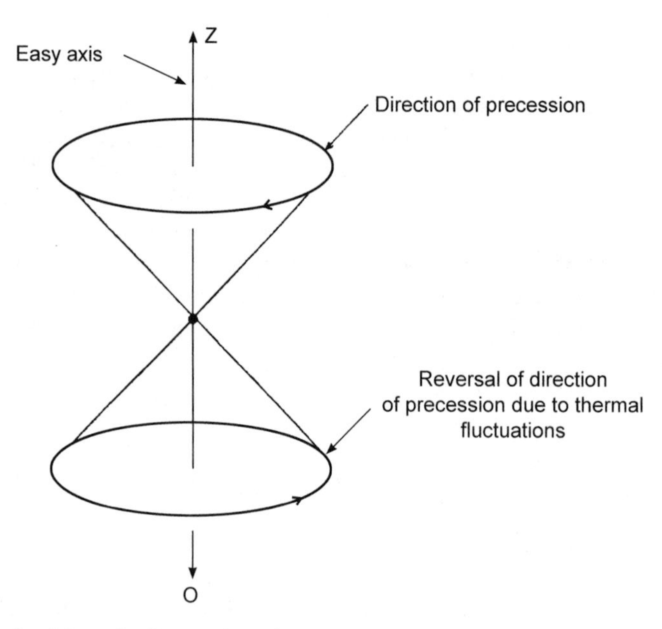


**Figure 5.** Schematic diagram for Néel relaxation [37]. Initially, the direction of magnetic precession is denoted by the upper circle. The $z$ direction is the easy axis of magnetization. Over a period of time, thermal fluctuations may cause the direction of magnetization to reverse as denoted by the lower circle. This reversal of magnetization without movement of the magnet is known as Néel relaxation.

radius $M_s$, which is the constant magnitude of the vector magnetization of the particle.

Brown, Jr., in his work [21, 39] between 1958 and 1963 criticized Néel's calculation of the greatest relaxation time on the grounds that

1. It takes no explicit account of the gyromagnetic term in the Larmor equation.
2. It relies on a discrete orientation approximation for the magnetic moments.

Thus he suggested the application of the Langevin method adapted to the motion of the magnetic moment. Hence Kramers' method (suitably adapted since the inertia of the particle plays no role, the role of inertia being sensibly played by the precessional term in the Langevin equation that arises from the Larmor precession) could be used to calculate the greatest relaxation time.

Brown's starting point is the Gilbert equation see Section III (or alternatively the Landau–Lifshitz equation) augmented by stochastic terms taken as the

Langevin equation for magnetic orientations (or spins). Brown [21] considered axial symmetry and simple uniaxial anisotropy

$$\varepsilon = \beta U = -\sigma x^2 \tag{1.17}$$

where the magnetocrystalline anisotropy depends only on the longititude $x = \cos\theta$ and obtained the Kramers escape rate (inverse of the greatest relaxation time)

$$\Gamma = \frac{2\pi a}{1+a^2}\sqrt{\frac{\sigma}{\pi}}\frac{\omega_A}{\pi}e^{-\beta\Delta U} \tag{1.18a}$$

$$= 2\pi\frac{a}{1+a^2}\sqrt{\frac{\sigma}{\pi}}\Gamma_{\text{TS}} \tag{1.18b}$$

Here $\Gamma_{\text{TS}}$ is two times greater than that of Eq. (1.2) since there are *two* wells.

The barrier height parameter is

$$\sigma = \frac{vK}{kT} = \beta\Delta U \tag{1.18c}$$

$v$ is the particle volume and

$$\omega_A = \frac{2K\gamma}{M_s} \tag{1.18d}$$

is the ferromagnetic resonance angular frequency $K$ is the anisotropy constant, $\gamma$ is the gyromagnetic ratio. See Section V for a detailed definition of the quantities involved.

The distinguishing feature of this result is that it is valid for all values of the dissipation parameter, $a$, since the FPE contains a single space variable, ($x = \cos\theta$). Here the dependence on $a$ occurs only through the free diffusion time, $\tau_N$ (see Eq. (3.17)). In order for the three cases of Kramers' problem, namely, IHD, VLD, and the crossover region [70] to appear, we have to consider nonaxially symmetric potentials, where the azimuthal angle is involved so that there is a explicit dependence on $a$ (as well as that due to $\tau_N$, e.g. Eq. (4.88)), for example,

$$\varepsilon = \beta U = -\sigma x^2 - 2\sigma h(x\cos\psi + \sqrt{1-x^2}\cos\varphi\sin\psi) \tag{1.19}$$

where

$$h = \frac{M_s \boldsymbol{H}_o}{2K}$$

which corresponds to a uniform field $\boldsymbol{H}_o$ applied in the $(x-z)$ plane at an angle, $\psi$, to the easy axis of magnetization thus the longitudinal and transverse relaxation modes are coupled. The azimuthal angle $\varphi$ is also involved of course in the *uniaxial* case, however, its only effect is to produce a steady precession of the magnetic moment, i.e., the $x$ and $\varphi$ variables are *uncoupled* as far as the longitudinal relaxation is concerned.

Note that a very interesting feature of this problem is to demonstrate how Kramers escape rates for such nonaxially symmetric potentials, which exhibit the three damping cases of Kramers' problem, because $x$ and $\varphi$ are *now coupled* may be reduced to the axially symmetric result, Eq. (1.18) that is valid for all values of the friction. The details of how this is achieved, which in the IHD case is similar to obtaining the WKBJ (Wentzel, Kramers, Brillouin, Jeffreys) [38, 40] connection formulas in quantum mechanics [38], are discussed later in this chapter (Section VII).

Another situation in which such a nonaxially symmetric potential arises is where the anisotropy is cubic in character so that the energy density has the form [62–65]

$$U(x, y, z) = K(y^2z^2 + z^2x^2 + x^2y^2) \tag{1.20}$$

Here, the question of crossover to the uniaxial case does not arise and the only crossover is that bridging the IHD and VLD results.

### F. Mean First Passage Times

We have mentioned that Kramers' theory is intimately connected with the theory of first passage times. For a domain $D$, the first passage time is approximately [1] the time for a random walker to reach the boundary of the domain for the first time from a point $x_0$ well embedded in the domain. Thus the mean first passage time (MFPT) is the average time for the random walker to reach the separatrix manifold (dividing the region where the motion is bounded from that where it is unbounded) for the first time.

Since there is a 50:50 chance that the particle, having reached the saddle point, will cross over, the MFPT is then, for the simple system of the isolated potential well of Figure 1, related to Kramers' escape rate, $\Gamma$, by

$$\Gamma = \frac{1}{2\tau_{\text{MFPT}}} \tag{1.21}$$

(assuming of course a sufficiently high barrier so that the concept of an escape rate applies).

It is our purpose in the chapter to give an extensive account of the calculation of $\Gamma$ by this method (see Sections III, V) for rotational relaxation problems. A discussion of this method in the context of the original Kramers problem is given by van Kampen [27, 28].

## II. THE KRAMERS CALCULATION OF THE REACTION RATE

### A. Plausible Derivation of the Klein–Kramers Equation

The purpose of this section is to give a detailed exposition of the original 1940 treatment of Kramers, giving all the mathematical details, where necessary. This will serve as an introduction to the application of the theory to more specialized problems in rotational Brownian motion pertaining to dielectric and magnetic relaxation. These are treated in Sections III–VIII. The original derivation of the Klein–Kramers equation by Kramers [4] is interesting from an historical point of view, and will be presented later on in this Section (B). Other derivations may be found based on the 1945 review article of Wang and Uhlenbeck [14, 29], and are given in detail in Risken [25, 26], in Coffey et al. [5], and in Coffey et al. [39], where the equation is derived starting from the definition of a stochastic process and the Markovian nature of the process is made fully apparent. Yet another novel method of derivation of the Kramers equation using functional integration will be given at the end of this Section F (5). In order to elucidate Kramers' reasoning, we shall first present a very brief derivation of the Liouville theorem of classical mechanics. Next, we shall give a plausible derivation of the Kramers equation based on Einstein's technique of 1905 [10], which consists in adding a diffusion term to the continuity equation to account for the thermal agitation due to the Brownian movement.

#### *1. The Liouville Equation*

The dynamical evolution of a conservative system is, in general, described by the Liouville equation [11, 40], which, for a system of $N$ particles, which have $3N$ degrees of freedom, with Hamiltonian

$$\mathscr{H} = \sum_{i=1}^{3N} \frac{p_i^2}{2m_i} + U(x_1, x_2, \ldots, x_{3N}) \tag{2.1}$$

is

$$\frac{D\rho}{Dt} = 0 \tag{2.2}$$

where $D/Dt$ is the total or hydrodynamical derivative defined by [40]

$$\frac{D}{Dt} \equiv \frac{\partial}{\partial t} + \mathbf{u} \cdot \text{grad} \tag{2.3}$$

and $\mathbf{u}$ is the $6N$-dimensional vector $\mathbf{x}(t), \mathbf{p}(t) = \dot{x}_1, \dot{x}_2, \ldots, \dot{x}_{3N}, \dot{p}_1, \ldots, \dot{p}_{3N}$ is the flow vector in phase space $(\mathbf{x}(t), \mathbf{p}(t))$. A very detailed account of this equation is given by Tolman [40]. The hydrodynamical derivative means the derivative evaluated at a *moving* phase point $(\mathbf{x}(t), \mathbf{p}(t))$. The statement $(D\rho/Dt = 0)$ means that there is no tendency for phase points to "crowd" into any particular region of phase space, that is, phase space behaves like an *incompressible fluid* whose representative point is $(\mathbf{x}(t), \mathbf{p}(t))$. Put in yet another way, the density in the neighborhood of any selected moving representative point is constant along the trajectory of that point. This principle is known as the *principle of conservation of density in phase*, in other words, phase points *stream*. In mathematical terms, the principle is

$$\rho(\mathbf{x}(t_0), \mathbf{p}(t_0), t_0) = \rho(\mathbf{x}(t), \mathbf{p}(t), t) \tag{2.4}$$

The second important property arising from the incompressible nature of phase space is the principle of *conservation of extension* (volume) in phase space, it is

$$dV = d\mathbf{x}(t)d\mathbf{p}(t) = d\mathbf{x}(t_0)d\mathbf{p}(t_0) \tag{2.5}$$

so that, even though the shape of the region $dV$ in phase space may alter with the course of time, it's volume does not. Here, we give a brief outline of the derivation of the Liouville equation for one degree of freedom. The position coordinate is denoted by $x$ and the momentum by $p$. These constitute a set of canonical variables satisfying Hamilton's equations, namely,

$$\dot{p} = -\frac{\partial H}{\partial x} \tag{2.6a}$$

$$\dot{x} = \frac{\partial H}{\partial p} \tag{2.6b}$$

In the absence of the creation or annihilation of particles, the rate of decrease of particles in a volume $\Delta V$ of phase space must be balanced by the flow of particles out of the volume. Hence, if $\Delta S$ is the surface bounding the volume $\Delta V$, we have

$$\frac{\partial}{\partial t} \iint\limits_{\Delta V} \rho(x, p, t) dp\, dx = -\int_{\Delta S} \rho \mathbf{u} \cdot d\mathbf{S} \tag{2.7a}$$

$$= -\iint\limits_{\Delta V} \text{div}(\rho \mathbf{u}) dp\, dx \tag{2.7b}$$

where $\mathbf{u}$ is the velocity of a particle in phase space at the point $(x, p)$. Equation (2.7b) follows from Eq. (2.7a) because of the divergence theorem. Now, since the volume element $\Delta V$ is arbitrary, Eq. (2.7b) reduces to

$$\frac{\partial \rho}{\partial t} + \text{div}(\rho \mathbf{u}) = 0 \tag{2.8}$$

which is the continuity equation of fluid mechanics. By using an elementary result in vector calculus, Eq. (2.8) becomes

$$\frac{\partial \rho}{\partial t} + \rho\, \text{div}\mathbf{u} + \mathbf{u} \cdot \text{grad}\rho = 0 \tag{2.9}$$

Now,

$$\text{div}(\rho \mathbf{u}) = \frac{\partial}{\partial x}(\rho \dot{x}) + \frac{\partial}{\partial p}(\rho \dot{p}) \tag{2.10}$$

$$= \dot{x}\frac{\partial \rho}{\partial x} + \dot{p}\frac{\partial \rho}{\partial p} + \rho\left(\frac{\partial \dot{x}}{\partial x} + \frac{\partial \dot{p}}{\partial p}\right) \tag{2.11}$$

and by using Eqs. (2.6)

$$\frac{\partial \dot{x}}{\partial x} + \frac{\partial \dot{p}}{\partial p} = \frac{\partial}{\partial x}\left(\frac{\partial H}{\partial p}\right) + \frac{\partial}{\partial p}\left(-\frac{\partial H}{\partial x}\right) \tag{2.12}$$

$$= 0 \tag{2.13}$$

from the equality of the mixed second-order partial derivatives.

Also,

$$\dot{x}\frac{\partial\rho}{\partial x}+\dot{p}\frac{\partial\rho}{\partial p}=\mathbf{u}\cdot\text{grad}\rho \tag{2.14}$$

So the equation of continuity Eq. (2.8) yields

$$\frac{\partial\rho}{\partial t}+\mathbf{u}\cdot\text{grad}\rho=0 \tag{2.15}$$

or

$$\frac{D\rho}{Dt}=0 \tag{2.16}$$

Equation (2.15) [or it's alternative form Eq. (2.16)] is known as the Liouville equation, which for $N$ particles moving in three dimensions (3D) (so that we have a $6N$-dimensional phase space or $12N$ dimensional if the rotational degrees of freedom are added) is [40]

$$\frac{\partial\rho}{\partial t}+\sum_{i=1}^{3N}\left(\frac{\partial H}{\partial p_i}\frac{\partial\rho}{\partial x_i}-\frac{\partial H}{\partial x_i}\frac{\partial\rho}{\partial p_i}\right)=0 \tag{2.17}$$

It is often written as

$$\frac{D\rho}{Dt}\equiv\frac{\partial\rho}{\partial t}+\{\rho,H\}=0 \tag{2.18}$$

where $\{\rho,H\}$ is the Poisson bracket

$$\sum_{i=1}^{3N}\left(\frac{\partial H}{\partial p_i}\frac{\partial\rho}{\partial x_i}-\frac{\partial H}{\partial x_i}\frac{\partial\rho}{\partial p_i}\right) \tag{2.19}$$

The derivation for one degree of freedom, first given in 1838 [40], is sufficient because, one must recall that Liouville's theorem is a *purely dynamical theorem*, which is *entirely equivalent* to Hamilton's equations.

### 2. *Reduction and Generalization of the Liouville Equation*

The Liouville equation is an equation with a number of variables of order $10^{23}$ and so is not tractable. In order to discuss the average dynamical behavior of a particle or system embedded in a heat bath, it is necessary [41] to modify the Liouville equation by both reducing and generalizing it: reducing it by limiting the degrees of freedom to a small but representative set with a well-defined potential and generalizing it by the addition of terms on the right-hand side of Eq. (2.17) to account for the mean interaction between this set and the remaining degrees of freedom (the background or heat bath). The first and best known such reduction and generalization of the Liouville equation is due to Boltzmann in 1872 (this is a closed integro-differential equation for the one particle distribution function) [8]. The theory of the Brownian movement instigated by Einstein, Smoluchowski, and Langevin and by Bachelier in 1900 for financial systems [5] is essentially a particular case of this reduction whereby, in a collision, the positions of the particles are unchanged and their velocities are altered by such small amounts that they can be treated as infinitesimal, so that the Boltzmann equation, supposing millions of such infinitesimal collisions because a Brownian particle is very large on a molecular scale, reduces to a linear PDE in phase space. This equation, which is a FPE, is now known as the Klein–Kramers equation for the particular case of mechanical particles. The Klein–Kramers equation is intimately connected to the Langevin equation that, for the sake of completeness, we will refer to at this juncture.

### 3. *The Langevin Equation for a System with one Degree of Freedom*

Langevin treated the Brownian motion of a free particle embedded in a heat bath at temperature $T$ by simply writing down the Newtonian equation of motion of the particle. He accounted for the interaction of the particle with the bath by adding to the Newtonian equation a systematic retarding force proportional to the velocity of the particle superimposed on which is a rapidly fluctuating force that we now call white noise. We shall slightly generalize Langevin's treatment by supposing that the particle moves in a potential $U(x)$, thus the Langevin equations are

$$\dot{p} = -\frac{dU}{dx} - \frac{\zeta p}{m} + \lambda(t) \tag{2.20a}$$

$$\dot{x} = \frac{p}{m} \tag{2.20b}$$

In Kramers' paper of 1940 [4] the force $-(\zeta/m)p + \lambda(t)$ [$X(t)$ in his notation] is termed "irregular force due to the medium". The white noise force $\lambda(t)$ has the following properties:

$$\overline{\lambda(t)} = 0 \tag{2.21a}$$

$$\overline{\lambda(t_1)\lambda(t_2)} = 2kT\zeta\delta(t_1 - t_2) \tag{2.21b}$$

The overbar means the statistical average over the realizations (that is the values it actually takes on) of $\lambda$. Since $\lambda(t)$ is a random variable, then $p(t)$ and $x(t)$ are also random variables. All averages over $p$ and $x$ are performed over a very small time interval $\tau$ with $p$ and $x$ taking sharp initial values at the starting time $t$. The statistics of $\lambda(t)$, written down above, are, however, insufficient to describe the problem fully as neither the Klein–Kramers equation nor the set of statistical moments of the system (generated by directly averaging the Langevin equation as proposed by Coffey [5, 30]) can be given [31] without supposing that $\lambda(t)$ is also Gaussian, that is, $\lambda(t)$ must obey Isserlis's theorem [5], namely,

$$\overline{\lambda_1\lambda_2\cdots\lambda_{2n}} = \overline{\lambda(t_1)\lambda(t_2)\cdots\lambda(t_{2n})} = \sum\prod_{j<k}\overline{\lambda(t_k)\lambda(t_j)} \tag{2.22a}$$

The sum is taken over all distinct products of expectation value pairs, each of which is formed by selecting $n$ pairs of subscripts from $2n$ subscripts. Also,

$$\overline{\lambda_1\lambda_2\cdots\lambda_{2n+1}} = \overline{\lambda(t_1)\lambda(t_2)\cdots\lambda(t_{2n+1})} = 0 \tag{2.22b}$$

### 4. *Effect of a Heat Bath: Intuitive Derivation of the Klein–Kramers Equation*

The intuitive derivation of the Kramers equation, which we now give, follows that of Einstein (1905) [10, 11] who included thermal agitation in the continuity equation for a particle subjected to a force $K$ by simply adding a diffusion term to the continuity equation for the number density or concentration of particles in configuration space. This enabled him to write down the Smoluchowski equation for the evolution of the number density in configuration space. We shall apply the same procedure to the Liouville equation, supposing that this equation is reduced to the Liouville equation for a single particle with the effect of the other particles i.e., the heat bath, being represented by the drift and diffusion terms that we shall add to the single particle equation so that $(D\rho/Dt)$ is no longer zero and so the phase points *diffuse*. We shall first consider the behavior of the system governed by Eqs. (2.20a) and (2.20b) without the white noise or fluctuating term (but including the damping term). Equation (2.11) becomes

$$\operatorname{div}(\rho\mathbf{u}) = \dot{x}\frac{\partial\rho}{\partial x} - \left(\frac{\zeta}{m}p + \frac{dU}{dx}\right)\frac{\partial\rho}{\partial p} + \rho\left(\frac{\partial\dot{x}}{\partial x} - \frac{\zeta}{m} - \frac{\partial}{\partial p}\frac{dU}{dx}\right) \tag{2.23}$$

and, since $\dot{x} = p/m$ and $x$ are independent variables, they play the role of generalized coordinates, and since $(dU/dx)$ is independent of $p$, Eq. (2.23) reduces to

$$\operatorname{div}(\rho \mathbf{u}) = \dot{x}\frac{\partial \rho}{\partial x} - \left(\frac{\zeta}{m}p + \frac{dU}{dx}\right)\frac{\partial \rho}{\partial p} - \frac{\zeta}{m}\rho \tag{2.24}$$

Hence, the equation of continuity Eq. (2.11) becomes

$$\frac{\partial \rho}{\partial t} + \dot{x}\frac{\partial \rho}{\partial x} - \left(\frac{\zeta}{m}p + \frac{dU}{dx}\right)\frac{\partial \rho}{\partial p} - \frac{\zeta}{m}\rho = 0 \tag{2.25}$$

To take account of the white noise we must now, following Einstein's method for the Smoluchowski equation, add on the diffusion term $D(\partial^2\rho/\partial p^2)$, where $D$ is independent of $p$, thus Eq. (2.25) becomes

$$\frac{\partial \rho}{\partial t} + \dot{x}\frac{\partial \rho}{\partial x} - \frac{dU}{dx}\frac{\partial \rho}{\partial p} - \frac{\partial}{\partial p}\left(\frac{\zeta}{m}\rho p + D\frac{\partial \rho}{\partial p}\right) = 0 \tag{2.26}$$

Now, by following Einstein we insist, that the equilibrium solution (the Maxwell–Boltzmann distribution) be a solution of Eq. (2.26). This forces

$$D = \zeta kT \tag{2.27}$$

and so Eq. (2.26) becomes

$$\frac{\partial \rho}{\partial t} + \frac{p}{m}\frac{\partial \rho}{\partial x} - \frac{dU}{dx}\frac{\partial \rho}{\partial p} - \frac{\zeta}{m}\frac{\partial}{\partial p}\left(\rho p + mkT\frac{\partial \rho}{\partial p}\right) = 0 \tag{2.28}$$

which is the Klein–Kramers equation for the evolution of the density, $\rho$, in phase space.

*Thus the effect of having a nonzero right hand side of the Liouville equation is to cause a disturbance of the streaming motion of the representative points so that they diffuse onto other energy trajectories.* The energy of a Brownian particle is no longer conserved as energy is interchanged between that particle and the heat bath.

### 5. *Calculation of the Drift and Diffusion Coefficients*

The analysis we have presented is purely a plausible derivation of the Klein–Kramers equation. A more rigorous account of the derivation of this equation, based on the modern concept of a stochastic process, may be found in Coffey et al. [5] and Risken [25, 26]. In many cases, it will be necessary to treat the multivariable form of the FPE. The multivariable form of the FPE for a process characterized by a state vector $\mathbf{y}$ is [14, 29]

$$\frac{\partial \rho}{\partial t} = -\sum_i \frac{\partial}{\partial y_i}\left[D_i^{(1)}(\mathbf{y}, t)P\right] + \frac{1}{2}\sum_{k,l} \frac{\partial^2}{\partial y_k \partial y_l}\left[D_{kl}^{(2)}(\mathbf{y}, t)P\right] \tag{2.29}$$

where the $D_i^{(1)}$ are called the drift coefficients and the $D_{kl}^{(2)}$ are called the diffusion coefficients. These coefficients are calculated from the underlying Langevin equation (postulating of Eqs. (2.22a, 2.22b) and in the case of multiplicative noise an interpretation rule [5]) for the motion of the state vector, $\mathbf{y}$. This is accomplished by writing that equation as an integral equation and, supposing that $\mathbf{y}(t)$ has a sharp value, averaging that integral equation over a time $\tau$. This time is of such short duration that (taking as an example $\mathbf{y}$ as the position and momentum of a particle) *the momentum does not significantly alter during the time $\tau$ and neither does any external conservative force*. Nevertheless, $\tau$ is supposed to be *sufficiently long that the chance that the rapidly fluctuating stochastic force $\lambda(t)$ takes on a given value at time $t + \tau$ is independent of the value that the force possessed at time $t$*. A complete account of the multivariable FPE is given in Risken [25, 26] and the calculation of the drift and diffusion coefficients are also extensively discussed in Coffey et al. [5] (*The Langevin Equation*). In writing the multivariable FPE, the summations are often dispensed with by using Einstein's notation.

Since one purpose of this chapter is to give a didactic account of the original 1940 paper of Kramers [4] and the extensions of it to subjects such as superparamagnetism, and so on, we shall now present Kramers' original derivation of the Klein–Kramers equation given by him in the context of the escape of particles over potential barriers.

## B. The Kramers' Derivation of the Klein–Kramers Equation

### 1. *The Kramers Problem*

In order to summarize Kramers' derivation of the Klein–Kramers equation we recall that Kramers [4] assumed that the effect of the interchange of energy of the particles with the heat bath could be described by the inclusion of an irregular force $X(t)$ in the single particle Newtonian equation, Isserlis' theorem being also implicitly assumed. Kramers' starting point is the equation of motion of a typical Brownian particle (in 1D) subjected to an external (conservative) field of force $K(x)$ and an irregular (Brownian) force $X(t)$ due to the heat bath. This equation is [his Eq. (1)] [4]:

$$m\ddot{x} = K(x) + X(t) \tag{2.30}$$

where $x$ is the position of the particle at time $t$
In later treatments of the problem, Eq. (2.30) is usually written

$$\dot{x} = \frac{p}{m} \tag{2.31}$$

$$\dot{p} = -\frac{dU}{dx} - \frac{\zeta}{m}p + \lambda(t), \quad K(x) = -\frac{dU}{dx} \tag{2.32}$$

where $\lambda(t)$ is a rapidly fluctuating (in comparision to $p$ and is independent of $p(t)$) white noise force with the statistical properties

$$\overline{\lambda(t)} = 0 \tag{2.33a}$$

$$\overline{\lambda(t_1)\lambda(t_2)} = 2kT\zeta\delta(t_1 - t_2) \tag{2.33b}$$

$$m\eta = \zeta,$$

and, of course, $\lambda(t)$ is a Gaussian random variable that must obey Isserlis' theorem, cited above. (The period of fluctuation of $\lambda(t)$ in a typical liquid is generally [15] of the order of the time between successive collisions of the Brownian particle and the fluid molecules.)

### 2. *Mean and Mean Square Changes in Momentum due to the Brownian Force*

The probability distribution of the random variable

$$B_\tau = \int_t^{t+\tau} X(t')dt' \tag{2.34}$$

is then considered. [Note that $B_\tau$ is the change in momentum of the Brownian particle over a time $\tau$ *due to the Brownian force X only*]. This probability distribution is assumed to be *independent* of the starting time, $t$, and *depends only on the length* of the time interval, $\tau$, i.e., we have a stationary process. If the distribution function is labelled $\varphi_\tau(B; x, p)$, and it is assumed that the moments (the overbars denote an average over the realizations of $B_\tau$)

$$\overline{B_\tau^n} \equiv \int_{-\infty}^{\infty} B^n \varphi_\tau dB \qquad \overline{B_\tau^0} = 1 \tag{2.35}$$

may be expanded as a Maclaurin series in $\tau$ then by keeping only the first-order terms in $\tau$, Einstein's original theory may be expressed as

$$\overline{B_\tau^1} = -\eta p\tau \tag{2.36a}$$

$$\overline{B_\tau^2} = \nu\tau = 2m\eta kT\zeta \tag{2.36b}$$

$$\overline{B_\tau^n} = 0 \qquad n \geq 3 \tag{2.36c}$$

Equations (2.36a)–(2.36c) given by Kramers may be clarified as follows: we have during a time $\tau$, by definition of $B_\tau$

$$\overline{B}_\tau = \overline{\int_t^{t+\tau} X(t')dt'} = \overline{-\eta\int_t^{t+\tau} p(t')dt' + \int_t^{t+\tau} \lambda(t')dt'} \tag{2.37}$$

We suppose that the change in velocity experienced in $\tau$ is very small so that (since $p(t)$ is a sharp value)

$$\overline{-\eta\int_t^{t+\tau} p(t')dt'} = -\eta p(t)\tau \tag{2.38}$$

Furthermore

$$\overline{\frac{1}{\tau}\int_t^{t+\tau} \lambda(t')dt'} = 0 \tag{2.39}$$

because

$$\overline{\lambda(t)} = 0 \tag{2.40}$$

Hence, we have for the *mean* change in momentum in the interval $\tau$ *due to the irregular force X alone*

$$\overline{B_\tau^1} = -\eta p\tau \tag{2.41a}$$

Now

$$\overline{B_\tau^2} = \overline{\left[-\eta\int_t^{t+\tau} p(t')dt' + \int_t^{t+\tau} \lambda(t')dt'\right]^2} \tag{2.41b}$$

It is apparent that, by making the same assumptions about the integrals in the first and last terms of the right-hand side of this equation, their squares are all of order $\tau^2$. Thus all the cross terms in the integral involving $\lambda$ (i.e., the second term) are zero, since $\lambda$ can only correlate with itself, and the noise, $\lambda$, is a *purely random* process with *no memory* so that the only term that survives on average is the square of the second term, if we restrict ourselves to terms linear in $\tau$.

Thus, we have for the mean square change in momentum in time $\tau$ *due to the irregular force X alone*

$$\overline{B_\tau^2} = \overline{\int_t^{t+\tau}\int_t^{t+\tau} \lambda(t')\lambda(t'')dt'dt''} \tag{2.42}$$

$$= \int_t^{t+\tau}\int_t^{t+\tau} \overline{\lambda(t')\lambda(t'')}dt'dt'' \tag{2.43}$$

$$= 2m\eta kT \int_t^{t+\tau}\int_t^{t+\tau} \delta(t'-t'')dt'dt'' \tag{2.44}$$

we note here that

$$\int_t^{t+\tau} \delta(t'-t'')dt' = 1 \qquad t < t'' < t+\tau \tag{2.45}$$

so that

$$\overline{B_\tau^2} = 2m\eta kT\tau \tag{2.46}$$

because the integral is, in effect, from $-\infty$ to $\infty$. Thus the value of the double integral is $2m\eta kT\tau$. Note that all the averages are evaluated supposing that $p(t)$ has a *sharp* or *definite* value at time $t$. Note also that $p(t+\tau)$ is a *random variable*, since the particle has suffered collisions in the time interval $\tau$, so that the initial sharp $p$ has become "blurred" during the interval $\tau$. Note also that the Langevin equation is crucial in the evaluation of the statistical moments. The odd moments for $n > 1$ all vanish because of Isserlis's theorem. The even moments for $n > 2$ vanish in like manner as Isserlis's theorem shows that all multi-time samples of the noise, on average, contribute terms of order $\tau^2$ or higher. This is explicitly illustrated in the context of the FPE by Coffey et al. [39], p414, and, in the context of the Langevin equation, by Coffey [31].

Throughout the calculation, it is assumed that the Brownian motion does not disturb the equipartition of energy. This is expressed by the equation

$$\overline{[p(t+\tau)]^2} = \overline{[p(t)]^2} \tag{2.47}$$

that is, the stochastic process $p(t)$ is *stationary*, or, the statistics (that is [29] the underlying "mechanism" or *Stosszahlansatz* which causes the fluctuations does not change in the course of time) governing $p(t)$ do not alter with the course of time.

If we now define $\mu_n$ by the equation

$$\overline{B_\tau^n} = \mu_n \tau \tag{2.48}$$

we can write, following Kramers, the diffusion equation for an ensemble of particles with density $\rho(x, p)$ in $(x, p)$ space in the following manner.

### 3. *Evolution of the Probability Density in the State Space $(x, p)$*

According to Kramers [4], the density at point $(x_1, p_1)$ at time $t + \tau$ may be thought of as being derived from the densities at a previous time $t$ along the straight line for which $x = x_2 = x_1 - (p_1/m)\tau$. If we denote by $p_2 = p_1 - K\tau$ the value that $p$ would have taken at time $t$ if no Brownian force had acted, we may write (the argument is essentially similar to that used [6] in deriving the Boltzmann equation)

$$\rho(x_1, p_1, t + \tau) = \rho\left(x_2 + \frac{p_2}{m}\tau, p_2 + K\tau, t + \tau\right) \tag{2.49}$$

$$= \int_{-\infty}^{\infty} \rho(x_2, p_2 - B, t)\varphi(B; x_2, p_2 - B)dB \tag{2.50}$$

Note the following:

1. That $p_2$ is due to the force $K$ only. The variable $p_2 - B$ takes account of the fact that an extra element of the momentum comes from the Brownian force and $\rho(x_2, p_2 - B, t)$ is the density of representative points at $x_2$ with momentum $p_2 - B$. These will arrive at the point $x_1$ with velocity $(p_1/m)$ after a time $\tau$.
2. That $\varphi(B; x_2, p_2 - B)dB$ is the fraction of points possessing a value of $B$ between $B$ and $B + dB$.
3. That $\rho(x_2, p_2 - B, t)\varphi(B; x_2, p_2 - B)dB$ is thus the fraction of points at $x_2$ possessing momentum $p_2 - B$ *and* a value of $B$ between $B$ and $B + dB$.
4. That we are assuming here that the conservative force $K$ does not sensibly vary over a large distance in the region of interest, namely, the diffusion length. We show in Section II.B.4 that this is *roughly* $(1/\eta)\sqrt{kT/m}$. This assumption allows us to accept that $p_2 = p_1 - K\tau$ is the momentum in the *absence* of Brownian forces above.

Expanding the left-hand side of Eq. (2.50) as a Taylor series in the three variables $(x_1 - x_2) = (p/m)\tau$, $(p_1 - p_2) = K\tau$, and $\tau$ and also expanding the integrand on the right-hand side as a power series in $B$, keeping only the first powers of $\tau$ and the first and second power of $B$ (details at the end of the section), we find

$$\begin{aligned}\rho(x_2, p_2, t) + \frac{\partial \rho}{\partial t}\tau + \frac{\partial \rho}{\partial p_2}K\tau + \frac{\partial \rho}{\partial x}\frac{p_2}{m}\tau \\ = \int_{-\infty}^{\infty}\left[\rho\varphi - B\frac{\partial(\rho\varphi)}{\partial p_2} + \frac{B^2}{2}\frac{\partial^2(\rho\varphi)}{\partial p_2^2} - \cdots\right]dB\end{aligned} \tag{2.51}$$

By using (2.48), and supposing that Isserlis's theorem holds so that all the moments on the right-hand side are of order $\tau^2$ for $n \geq 3$ we have on proceeding to the limit $\tau \to 0$ and writing $p_2 = p$ the Fokker–Planck equation (2.28):

$$\frac{\partial \rho}{\partial t} = -K\frac{\partial \rho}{\partial p} - \frac{p}{m}\frac{\partial \rho}{\partial x} - \frac{\partial(\mu_1\rho)}{\partial p} + \frac{1}{2}\frac{\partial^2(\mu_2\rho)}{\partial p^2} \tag{2.52}$$

This particular FPE which is now called the Klein–Kramers equation can again be written in the form of a continuity equation for the representative points in two-dimensional (2D) phase space $(x, p)$. The continuity equation is

$$\frac{\partial \rho}{\partial t} + \frac{\partial J_x}{\partial x} + \frac{\partial J_p}{\partial p} = 0 \tag{2.53}$$

where

$$J_x = \rho\frac{p}{m} \tag{2.54a}$$

$$J_p = K\rho + \mu_1 p - \frac{1}{2}\frac{\partial(\mu_2\rho)}{\partial p} + \cdots \tag{2.54b}$$

$J_x$ is the deterministic portion of the probability current density arising from the Liouville equation and $J_p$ is a drift and diffusion part arising from the stochastic forces. In fact, each of the diffusion equations we encounter will be of the form

$$\frac{\partial \rho}{\partial t} + \nabla \cdot \mathbf{J} = \mathbf{0} \tag{2.55}$$

where $\nabla$ is the differential operator in the appropriate number of dimensions and $\mathbf{J}$ is the probability current density. Kramers uses 1D operators, one in $x$ and one in $I$, the action: see [4]. Next, he assumes that the process is, practically speaking, stationary, so that $\partial\rho/\partial t$ may be taken as 0. He then finds the solution to the equation

$$\nabla \cdot \mathbf{J} = \mathbf{0} \tag{2.56}$$

One such solution is

$$\mathbf{J} = 0 \tag{2.57}$$

which corresponds to the Maxwell–Boltzmann distribution in the well. The other solution is

$$\mathbf{J} = \text{constant} \neq \mathbf{0} \tag{2.58}$$

which pertains to diffusion over the potential barrier and which after considerable manipulation, to be described below, may be used to calculate the escape rate of particles over the potential barrier. This allows Kramers to write down his Eqs. (13) and (15), which are our Eqs. (2.215) and (2.175), respectively.

The fundamental condition imposed by Kramers on the moments $\mu$ is that the Maxwell–Boltzmann distribution$(\rho_{\text{eq}})$ should be a stationary solution of the Klein–Kramers equation. That distribution is

$$\rho_{\text{eq}} = \exp\left(-\frac{(p^2/2m) + U(x)}{kT}\right) \tag{2.59}$$

where $U(x)$ is the potential function for the force field $K(x)$, that is,

$$K = -\frac{dU}{dx} \tag{2.60}$$

By using Eqs. (2.53), (2.54) and (2.59) with $\partial\rho/\partial t = 0$, we have

$$\begin{aligned}\frac{\partial}{\partial p}\Bigg\{ &-\mu_1 \exp\left(-\frac{p^2}{2mkT}\right) + \frac{1}{2}\frac{\partial}{\partial p}\left[\mu_2 \exp\left(-\frac{p^2}{2mkT}\right)\right] \\ &-\frac{1}{6}\frac{\partial^2}{\partial p^2}\left[\mu_3 \exp\left(-\frac{p^2}{2mkT}\right)\right]\Bigg\} = 0\end{aligned} \tag{2.61}$$

which on integration once with respect to $p$ and simplifying by performing one differentiation in the second and subsequent terms becomes

$$-\mu_1 - \frac{p}{2mkT}\mu_2 + \frac{1}{2}\frac{\partial\mu_2}{\partial p} - \cdots = F(x,T)\exp\left(\frac{p^2}{2mkT}\right) \tag{2.62}$$

Kramers now assumes that $\mu_1, \mu_3, \ldots$ are odd functions of $p$, while $\mu_2, \mu_4, \ldots$ are even, so that all the terms on the left-hand side are odd functions of $p$, while the right-hand side is an even function of $p$. Hence, both sides must be zero, and so we conclude that $F(x,T) = 0$. This again constitutes implicit use of Isserlis's theorem. According to Kramers, to use his own words, the simplest situation is Einstein's case. From Eqs.(2.36) and (2.46),

$$\mu_1 = -\eta p \tag{2.63a}$$

$$\mu_2 = 2m\eta kT \tag{2.63b}$$

$$\mu_3 = \mu_4 = \cdots = 0 \tag{2.63c}$$

What is meant here, however, is Einstein's treatment including the inertia of the Brownian particle as in his treatment of the Brownian motion, Einstein clearly demonstrates [6, 10] *that inertial effects are not included.* Thus again "Einstein's case..." as referred to by Kramers, means the *inertial version* of the Einstein theory first elaborated by Ornstein [16] and Uhlenbeck and Ornstein [42] (see [14].

**Justification of Eq. (2.52)**
We have by Taylor's theorem

$$\rho(x_2, p_2 - B, t)\varphi(B; x_2, p_2 - B) = \rho(x_2, p_2, t)\varphi(B; x_2, p_2) - B\frac{\partial}{\partial p_2}(\rho\varphi) + \frac{B^2}{2!}\frac{\partial^2}{\partial p_2^2}(\rho\varphi) + \cdots \tag{2.64}$$

So that, integrating both sides of Eq. (2.64) w.r.t. $B$

$$\begin{aligned}\int_{-\infty}^{\infty} \rho(x_2, p_2 - B, t)\varphi(B; x_2, p_2 - B)dB \\ = \rho(x_2, p_2, t)\int_{-\infty}^{\infty}\varphi(B; x_2, p_2)dB - \int_{-\infty}^{\infty} B\frac{\partial}{\partial p_2}(\rho\varphi)dB \\ + \frac{1}{2!}\int_{-\infty}^{\infty} B^2\frac{\partial^2}{\partial p_2^2}(\rho\varphi)dB\end{aligned} \tag{2.65}$$

[where, for brevity, we write $\rho$ for $\rho(x_2, p_2, t)$, and $\varphi$ for $\varphi(B; x_2, p_2)$]

$$= \rho(x_2, p_2, t) - \frac{\partial}{\partial p_2}\rho\int_{-\infty}^{\infty} B\varphi dB + \frac{1}{2!}\frac{\partial^2}{\partial p_2^2}\int_{-\infty}^{\infty} B^2\varphi dB \tag{2.66}$$

[since $B$ is not a function of $p_2$ and $\varphi$ is a probability density function]

$$= \rho - \frac{\partial}{\partial p_2}(\rho\mu_1) + \frac{1}{2!}\frac{\partial^2}{\partial p_2^2}(\rho\mu_2) + \cdots \tag{2.67}$$

which justifies Eq. (2.52) on replacing $p_2$ by $p$.

**Proof of Eq. (2.61)**

The stationary solution $\rho_{eq}$ satisfies the equation of continuity with $(\partial\rho/\partial t) = 0$. Thus, the probability current density obeys

$$\frac{\partial J_x}{\partial x} + \frac{\partial J_p}{\partial p} = 0 \tag{2.68}$$

Now,

$$\frac{\partial J_x}{\partial x} + \frac{\partial J_p}{\partial p} = \frac{p}{m}\rho_{eq}\beta K - K\rho_{eq}\beta\frac{p}{m} + \frac{\partial}{\partial p}(\mu_1\rho_{eq}) - \frac{1}{2}\frac{\partial^2}{\partial p^2}(\mu_2\rho_{eq}) \tag{2.69}$$

$$= \frac{\partial}{\partial p}(\mu_1\rho_{eq}) - \frac{1}{2}\frac{\partial^2}{\partial p^2}(\mu_2\rho_{eq}) \tag{2.70}$$

$$= \frac{\partial}{\partial p}\left[-\mu_1\exp\left(-\frac{(p^2/2m)+U}{kT}\right) + \frac{1}{2}\frac{\partial}{\partial p}\left\{\mu_2\exp\left(-\frac{(p^2/2m)+U}{kT}\right)\right\} + \cdots\right] = 0 \tag{2.71}$$

which is Eq. (2.61).

This completes our derivation of the Klein–Kramers equation as given originally by Kramers [4]. This derivation should be compared with that using functional integration given at the end of the Section (F.5).

### 4. *Conditions Under Which a Maxwellian Distribution in the Velocities May be Deemed to be Attained*

After having obtained the Klein–Kramers equation for the time evolution of the distribution function in the phase space $(x, p)$, Kramers proceeds to examine the conditions under which equilibrium in the velocities may be assumed to have been attained, the displacement having not yet attained its equilibrium value. The importance of such an investigation is that it allows one, for sufficiently high values of the friction parameter, to write an *approximate partial differential equation for the time evolution of the distribution function in configuration space only*. This *approximate* equation is known as the *Smoluchowski equation*. In order to explain the reasoning of Kramers on this subject, it will be useful to recall Einstein's 1905 result for the mean-square displacement of a Brownian particle, namely,

$$\langle(\Delta x)^2\rangle = \frac{2kT}{\zeta}t \qquad \zeta = m\eta \tag{2.72}$$

This equation was derived by constructing the partial differential equation for the time evolution in configuration space only and was later rederived by Langevin [13] by considering times well in excess of the inertial relaxation time $m/\zeta$, which allows him to postulate an approximate Maxwellian distribution for the velocities. Equation (2.72) has the flaw that it is not root mean square differentiable at very small times as emphasized [23, 14] by Doob (see Coffey et al. [5]). In 1930, Uhlenbeck and Ornstein [42] (Coffey et al. [5]) showed, using the Langevin equation, that the *exact* solution for the mean-square displacement of a free Brownian particle is

$$\langle(\Delta x)^2\rangle = \frac{2kT}{m\eta^2}(\eta t - 1 + e^{-\eta t}) \tag{2.73}$$

which is differentiable at short times so that Einstein's result, Eq. (2.72), is regained if $t \gg \eta^{-1}$. The mean-square displacement, which we shall write as $x^2$ for brevity, is thus governed by the two characteristic times

$$\tau_{\text{diff}} = \frac{m\eta}{2kT}x^2 \tag{2.74}$$

and the inertial time

$$\tau_{\text{in}} = \eta^{-1} \tag{2.75}$$

The ratio of these times is

$$\frac{\tau_{\text{in}}}{\tau_{\text{diff}}} = \frac{2kT}{x^2 m\eta^2} \tag{2.76}$$

Now, diffusion effects, where a Maxwellian distribution of velocities approximately holds, will predominate over inertial effects if the left-hand side of this equation is $\ll 1$, which by transposition means

$$x \ll \frac{1}{\eta}\sqrt{\frac{kT}{m}} \tag{2.77}$$

so that the quantity

$$\frac{1}{\eta}\sqrt{\frac{kT}{m}} \tag{2.78}$$

defined by Kramers is a *characteristic diffusion length* that crucially determines whether *inertial* effects or *diffusion* effects predominate.

As far as Brownian motion under the influence of a potential $U(x)$ is concerned, Kramers applies the above reasoning to this problem by supposing that the force

$$K = -\frac{dU(x)}{dx}$$

does not vary greatly over distances of the order of the diffusion length (2.78) so one would expect that starting from an arbitrary initial distribution $\rho(x, p, 0)$ a Maxwellian distribution of $p$ would be reached after time intervals $\Delta t \gg \eta^{-1}$, which allows him to postulate that

$$\rho(x, p, t) \approx \sigma(x, t) \exp\left(-\frac{p^2}{2mkT}\right) \tag{2.79}$$

The reasoning here is of crucial importance in the study of dielectric relaxation, [5, 41] where the omission of inertial effects in the Debye theory of dielectric relaxation leads to the phenomenon of "black water" or infinite integrated absorption at high frequencies.

### C. Calculation of the Reaction Rate in the IHD Limit

In order to calculate the reaction rate in what is called the IHD limit, (where the damping forces are strong enough to ensure equilibrium in the system except for a small region very close to the barrier top where the potential may be approximated to an inverted harmonic oscillator potential) we first remark that the Langevin equation may, in this limit, be linearized in the region of the maximum of the potential at $C$. This corresponds in the Klein–Kramers equation to having coefficients that are linear in momentum and displacement. Such an equation is a "linearized" Klein–Kramers equation.

### *1. Linearization of the Klein–Kramers Equation Near the Summit of the Potential Barrier*

The process is governed by diffusion in a single coordinate which, on this occasion, is a linear combination of the displacement and the momentum rather than the energy as we shall now demonstrate.

In order to proceed, we note that by using Eqs. (2.63) and (2.52) the diffusion equation becomes

$$\frac{\partial \rho}{\partial t} = -K\frac{\partial \rho}{\partial p} - \frac{p}{m}\frac{\partial \rho}{\partial x} + \eta\frac{\partial}{\partial p}\left(\rho p + mkT\frac{\partial \rho}{\partial p}\right) \tag{2.80}$$

Equation (2.80) is again the Klein–Kramers equation that we derived in Section II.B.

We now assume that the function $U$ is sufficiently well behaved to be able to expand it as a Taylor series about $x_c$ (the value of $x$ where the top of the barrier is located). We write

$$U = -\frac{1}{2}m\omega_C^2(x - x_C)^2 \tag{2.81}$$

and take the value of the potential at the top of the barrier namely $U(x_C)$ as the zero of potentials. Considering the situation as quasi-stationary i.e. very slow

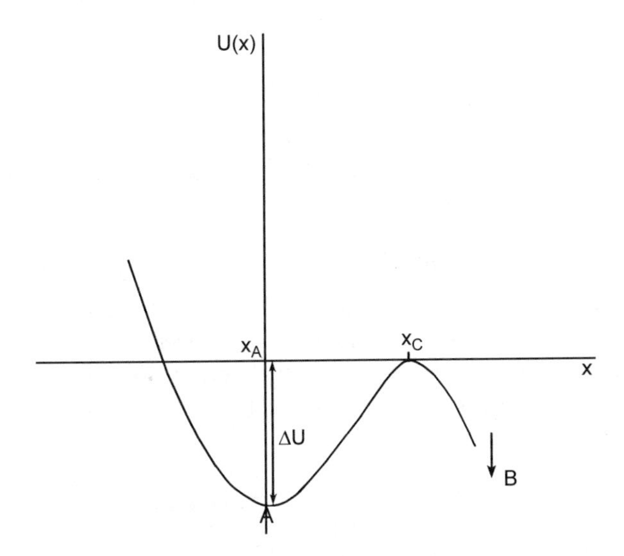


**Figure 6.** Potential energy diagram as used in the discussion of the Kramers IHD problem. The zero of potential is taken as the top of the barrier and the bottom of the well is taken at $x_A = 0$

diffusion over the barrier (i.e. $\partial\rho/\partial t \approx 0$), made possible by the condition

$$\Delta U \gg kT \qquad (\text{where } \Delta U = U(x_C) - U(x_A) = -U(x_A)) \tag{2.82}$$

we find that Eq. (2.80) reduces to the stationary equation (from now on we shall take $m$ as unity for easy comparison with the Kramers [4] calculations)

$$\omega_c^2 x' \frac{\partial \rho}{\partial p} + p \frac{\partial \rho}{\partial x'} - \eta \frac{\partial}{\partial p}\left(\rho p + kT \frac{\partial \rho}{\partial p}\right) = 0 \tag{2.83}$$

where

$$x' \equiv x - x_C \tag{2.84}$$

We now make the substitution

$$\rho \equiv \zeta \exp\left[-\left(\frac{p^2 - \omega_C^2 x'^2}{2kT}\right)\right] \tag{2.85}$$

and Eq. (2.83) becomes

$$\omega_C^2 x' \frac{\partial \zeta}{\partial p} + p \frac{\partial \zeta}{\partial x'} + \eta p \frac{\partial \zeta}{\partial p} - \eta kT \frac{\partial^2 \zeta}{\partial p^2} = 0 \tag{2.86}$$

$\zeta(x', p)$ is a crossover function which varies rapidly near the barrier and takes on the value 1 in the well and 0 over the barrier.

### 2. *Solution of the Linearized Klein–Kramers Equation*

We see immediately that $\zeta =$ constant is a solution to this equation, however, this corresponds to thermal equilibrium, and hence, to a situation of no diffusion. This would give us, of course, the Maxwell–Boltzmann distribution. We can, however, obtain another solution if we assume that $\zeta$ satisfies the condition

$$\zeta = \zeta(u) \qquad \text{where} \qquad u \equiv p - ax' \tag{2.87}$$

and $a$ is a constant to be determined. We emphasize here that our solution must satisfy the following conditions:

1. To the right of $C$, the density must go to zero in order to reflect the fact that at the beginning of the process, practically no particles have reached $B$.
2. Near point $A$ the Maxwell–Boltzmann distribution holds to a high degree of accuracy.

We take account of these two conditions by imposing the boundary conditions:

$$\zeta \to 0 \qquad \text{as} \qquad x \to \infty \qquad (\text{i.e. for } x \gg x_C) \tag{2.88a}$$

$$\zeta = 1 \qquad \text{at} \qquad x = 0 \qquad (\text{i.e. at points in the depths of the well}) \tag{2.88b}$$

On substitution of Eq. (2.87) into Eq. (2.86) we obtain

$$(ap - \omega_C^2 x' - \eta p)\zeta' + \eta kT\zeta'' = 0 \tag{2.89}$$

or

$$[(a - \eta)p - \omega_C^2 x']\zeta' + \eta kT\zeta'' = 0 \tag{2.90}$$

To solve this equation, we wish to write the equation in terms of the variables $\zeta$ and a linear combination of $x'$ and $p$ namely $u$ rather than $\zeta, x'$, and $p$. This can be achieved in a very neat way if we write

$$(a - \eta)p - \omega_C^2 x' = (a - \eta)(p - ax') = (a - \eta)u \tag{2.91}$$

which imposes on $a$ the condition (compare the eigenvalue equation of Langer, Eq. (4.20) Section IV, the condition (2.92) ensures that we have a proper ordinary differential equation in $u$):

$$\omega_C^2 = a(a - \eta) \tag{2.92}$$

or

$$a = \frac{\eta}{2} \pm \sqrt{\frac{\eta^2}{4} + \omega_C^2} \tag{2.93}$$

Equation (2.90) now takes the form of the ODE

$$\eta kT\zeta'' + (a - \eta)u\zeta' = 0 \tag{2.94}$$

which has as solution ($C$ is here a constant of integration)

$$\zeta = C \int^{u} \exp\left(-\frac{(a-\eta)u'^2}{2\eta kT}\right) du' \tag{2.95}$$

If we ignore the minus sign in Eq. (2.93), then

$$a - \eta = -\frac{\eta}{2} + \sqrt{\frac{\eta^2}{4} + \omega_C^2} = \lambda_+$$

will be positive, since

$$\sqrt{\frac{\eta^2}{4} + \omega_C^2} > \frac{\eta}{2}$$

and Eq. (2.95) will remain an error function and so will always represent a diffusion of particles over the barrier at $C$. ($(a - \eta)$ then corresponds to the unique positive eigenvalue of the Langevin equation (2.32) linearized about $C$ and characterizes the (unstable) barrier crossing mode) If we take $-\infty$ for the lower limit of the integral in Eq. (2.95), which corresponds to the condition $(x \gg x_C)$

$$\zeta \to 0 \qquad \text{as} \qquad u \to -\infty \tag{2.96}$$

[Eq. (2.96) corresponds to the situation where practically no particles are yet to be found to the right of the barrier top], in the region $x \ll x_C$ *well to the left of the barrier top* that is in the depths of the well we may extend the limits of integration in Eq. (2.95) to $-\infty, \infty$ to get

$$\zeta = C\sqrt{\frac{2\pi\eta kT}{a-\eta}}, \quad \left(\text{since } \int_{-\infty}^{\infty} e^{-\alpha x^2} dx = \sqrt{\frac{\pi}{\alpha}}\right) \tag{2.97}$$

and the density in phase space near A (the minimum of the potential) will then be the Maxwell Boltzmann distribution

$$\rho = C\sqrt{\frac{2\pi\eta kT}{a-\eta}} \exp\left(\frac{\Delta U}{kT}\right) \exp\left(-\frac{p^2 + \omega_A^2 x^2}{2kT}\right) \tag{2.98}$$

where, near the bottom of the well, $(x_A \simeq 0)$ the potential is approximated by $U(x) = -\Delta U + (1/2)\omega_A^2 x^2$.

[Note that we have made the assumption that as $x \to \infty$, $u \to -\infty$ and as $x \to -\infty$, $u \to +\infty$. Because $u \equiv p - ax'$, we can see that these conditions will hold only if $a$ is large enough to ensure that $ax'$ can go to $\infty$ quicker than $p$, and as $a$ depends on the value of $\eta$, *it is very important to note that this imposes a limitation on the range of validity of Eq. (2.107) (the escape rate) below.* We deal with the range of validity in some detail in Section II.F. Note also that some authors prefer to use $u = x' - ap$ (cf. Section VII). The form of Eq. (2.87) is retained for easier comparison with Kramers' paper.]

### 3. *Calculation of the Reaction Rate*

The number, $q$, of particles passing though the barrier top in unit time i.e. the probability current may be obtained [43] (see [1]) by integrating $\rho p$ over $p$ from $-\infty$ to $+\infty$. ($q$ in this 1D case strictly means the number of particles crossing unit area in unit time; the calculation of the current from the current density in more than one dimension is a complicated mathematical task see Section IV Eq. (4.29)) By putting $x' = 0$ (we obtain from Eqs. (2.85) and (2.87)) so that $x = x_C$ i.e. the line of flow is through the saddle point

$$q = \int_{-\infty}^{\infty} \rho p dp \tag{2.99}$$

$$= C \int_{-\infty}^{\infty} p \exp\left(-\frac{p^2}{2kT}\right) \int_{-\infty}^{p} \exp\left(-\frac{(a-\eta)p'^2}{2\eta kT}\right) dp' dp \tag{2.100}$$

$$= CkT\sqrt{\frac{2\pi\eta kT}{a}} \tag{2.101}$$

while the number of particles trapped near the minimum comes from Eq. (2.98) and is

$$n_A = C\sqrt{\frac{2\pi\eta kT}{a-\eta}} \exp\left(\frac{\Delta U}{kT}\right) \int_{-\infty}^{\infty} \int_{-\infty}^{\infty} \exp\left(-\frac{p^2 + \omega_A^2 x^2}{2kT}\right) dp dx \tag{2.102}$$

$$= C\sqrt{\frac{2\pi\eta kT}{a-\eta}} \exp\left(\frac{\Delta U}{kT}\right) \frac{2\pi kT}{\omega_A} \tag{2.103}$$

The probability of escape is, therefore,

$$\Gamma = \frac{q}{n_A} = \frac{\omega_A}{2\pi}\sqrt{\frac{a-\eta}{a}}\exp\left(-\frac{\Delta U}{kT}\right) \tag{2.104}$$

which together with Eq. (2.93) yields

$$\Gamma = \frac{\omega_A}{2\pi\omega_c}\left(\sqrt{\frac{\eta^2}{4}+\omega_C^2}-\frac{\eta}{2}\right)\exp\left(-\frac{\Delta U}{kT}\right) \tag{2.105}$$

If we now take the limit of the right-hand side of this equation, in the two cases of large and small $\eta$ (note that the terms "strong damping," "weak damping" strictly speaking refer to $\eta/2\omega_A \lessgtr 1$ however for the purposes of discussion one may suppose $\omega_A \simeq \omega_C$), we find in the first instance (strong damping) $(\eta \gg \omega_C)$:

$$\Gamma = \frac{\omega_A}{2\pi\omega_C}\left(\frac{\eta}{2}\sqrt{1+\frac{4\omega_C^2}{\eta^2}}-\frac{\eta}{2}\right)\exp\left(-\frac{\Delta U}{kT}\right) \tag{2.106a}$$

$$\approx \frac{\omega_A}{2\pi\omega_C}\left(\frac{\eta}{2}\left\{1+\frac{2\omega_C^2}{\eta^2}\right\}-\frac{\eta}{2}\right)\exp\left(-\frac{\Delta U}{kT}\right) \tag{2.106b}$$

$$= \frac{\omega_A}{2\pi\omega_C}\frac{\omega_C^2}{\eta}\exp\left(-\frac{\Delta U}{kT}\right) \tag{2.106c}$$

$$= \frac{\omega_A\omega_C}{2\pi\eta}\exp\left(-\frac{\Delta U}{kT}\right) \tag{2.106d}$$

which anticipates the result in Eq. (2.223). The result embodied in Eq. (2.106d) is, in effect, the noninertial limit where the energy dissipation in the system is so large that the inertia of the escaping particles has practically no effect. Compare the original Einstein theory of the Brownian movement with the later inertia corrected version of Uhlenbeck and Ornstein [42] (see also Coffey et al. [5]).

Many authors take the limit in the second instance weak damping ($\eta \ll 2\omega_C$) in Eq. (2.105):

$$\Gamma = \frac{\omega_A}{2\pi}\left(\sqrt{1+\frac{\eta^2}{4\omega_C^2}}-\frac{\eta}{2\omega_C}\right)\exp\left(-\frac{\Delta U}{kT}\right) \tag{2.107a}$$

$$\approx \frac{\omega_A}{2\pi}\exp\left(-\frac{\Delta U}{kT}\right) \tag{2.107b}$$

which *predicts escape in the absence of coupling to the bath and which is the TST result. However, taking the limit as* $\eta \to 0$ *is not permitted as we shall now explain.* We first remark that (cf. Section I) the IHD solution Eq. (2.107a), which we have described, relies on the assumption that the friction is large enough to ensure that the particles approaching the barrier from the depths of the well are in thermal equilibrium. If the friction coefficient becomes too small this condition is violated and the IHD solution is no longer valid *because the space interval in which the nonequilibrium behaviour prevails exceeds that where an inverted parabola approximation to the potential is valid.* Hence the need for a different treatment for very small values of the friction such that $\eta I(E_C) \ll kT$ and for crossover values of the friction where $\eta I(E_C) \approx kT$. [Put in a more simple way, in the VLD case, the coupling to the bath is so weak that the assumption of a Maxwell-Boltzmann distribution in $x$ and $p$ in a relatively large region near the top of the barrier is not valid, because the damping is so small that the motion of an escaping particle is almost that of a librating particle with energy equal to the barrier energy and without dissipation.]

**Proof of Eq. (2.101)**

$$q = C\int_{-\infty}^{\infty} p\exp\left(-\frac{p^2}{2kT}\right)\int_{-\infty}^{p}\exp\left(-\frac{(a-\eta)p'^2}{2\eta kT}\right)dp'\,dp \tag{2.108}$$

Let

$$u = \int_{-\infty}^{p}\exp\left(-\frac{(a-\eta)p'^2}{2\eta kT}\right)dp' \quad \text{and} \quad dv = p\exp\left(-\frac{p^2}{2kT}\right)dp \tag{2.109}$$

Then,

$$du = \exp\left(-\frac{(a-\eta)p^2}{2\eta kT}\right)dp \quad \text{and} \quad v = -kT\exp\left(-\frac{p^2}{2kT}\right) \tag{2.110}$$

So,

$$\begin{aligned} q = C\left[-kT\exp\left(-\frac{p^2}{2kT}\right)\int_{-\infty}^{p}\exp\left(-\frac{(a-\eta)p'^2}{2\eta kT}\right)dp'\right]_{-\infty}^{\infty} \\ + C\int_{-\infty}^{\infty} kT\exp\left(-\frac{p^2}{2kT}\right)\exp\left(-\frac{(a-\eta)p^2}{2\eta kT}\right)dp \end{aligned} \tag{2.111}$$

$$= C\int_{-\infty}^{\infty} kT\exp\left(-\frac{p^2}{2kT}\right)\exp\left(-\frac{(a-\eta)p^2}{2\eta kT}\right)dp \tag{2.112}$$

$$= C\int_{-\infty}^{\infty} kT\exp\left(-\frac{p^2}{2kT}\left\{1+\frac{a-\eta}{\eta}\right\}\right)dp \tag{2.113}$$

$$= C\int_{-\infty}^{\infty} kT\exp\left(-\frac{p^2}{2kT}\frac{a}{\eta}\right)dp \tag{2.114}$$

$$= CkT\sqrt{\frac{2\pi\eta kT}{a}} \tag{2.115}$$

as required.

Finally, we remark that the complete solution in the *vicinity* of the barrier may be easily written down from Eqs. (2.85), (2.95) and (2.97, 98). First the boundary condition that the crossover function $\zeta \to 1$ in the depths of the well so that the Maxwell-Boltzmann distribution holds there, c.f. Eq. (2.98), yields

$$C = \sqrt{\frac{(a-\eta)}{2\pi\eta kT}}\frac{\omega_A}{2\pi kT}e^{-(\Delta U/kT)} \tag{2.116}$$

because, from Eq. (2.98) for the density near $A$ and the normalization condition corresponding to one particle in the well (we could equally well use $N$)

$$1 = \int_{-\infty}^{\infty}\int_{-\infty}^{\infty} \rho dx dp = C\sqrt{\frac{(a-\eta)}{2\pi\eta kT}} e^{\Delta U/kT} \int_{-\infty}^{\infty}\int_{-\infty}^{\infty} e^{-((p^2+\omega_A^2 x^2)/2kT)} dp dx$$

$$= C\sqrt{\frac{2\pi\eta kT}{a-\eta}} e^{\Delta U/kT} \frac{2\pi kT}{\omega_A} \tag{2.117}$$

Thus with Eqs. (2.85) and (2.95) the complete solution *near the top of the barrier* ($x \simeq x_C$) is (cf. [15] Eq. (498))

$$\rho(x,p) = \frac{\omega_A}{2\pi kT}\sqrt{\frac{(a-\eta)}{2\pi\eta kT}} e^{-\Delta U/kT} \exp\left\{-\left[\frac{p^2-\omega_C^2(x-x_C^2)}{2kT}\right]\right\} \times \int_{-\infty}^{p-a(x-x_C)} e^{-(((a-\eta)u'^2)/2\eta kT)} du' \tag{2.118}$$

while (cf. [15] Eq. (499)), in the *depths of the well* from Eqs. (2.98) and (2.116)

$$\rho(x,p) = \rho_{eq} \approx \frac{\omega_A}{2\pi kT}\, e^{-((p^2+\omega_A^2 x^2)/2kT)} \tag{2.119}$$

where $Z$ the partition function is

$$Z \simeq \frac{2\pi kT}{\omega_A}$$

### D. The Case of Small Viscosity

#### 1. *The Low-Damping Regime*

In this instance, Kramers restricts the discussion to the case where the particle under consideration would perform an oscillatory motion in the well if it were not for the presence of the Brownian forces. *Small viscosity means that the Brownian forces cause only a tiny perturbation in the undamped energy during one oscillation.* This means that the Brownian forces will cause gradual changes in the distribution of the ensemble over the different energy values.

### 2. *The Klein–Kramers Equation in Energy-Phase Variables*

We now write the original Klein–Kramers equation in the variables $(x, p)$ as a diffusion equation in the energy $(E)$ and phase $(w)$. We may do this, since the energy is a *slowly varying* quantity and the phase is a *fast-varying* quantity. Thus we will be able to average the density over the *fast-phase variable* and get a diffusion equation in the *slow* (*almost conserved*) *energy variable*. We define the time average along a trajectory

$$\bar{\rho}(E, t) \equiv \frac{1}{\tau} \int_0^{\tau} \rho(E, w, t) dw \tag{2.120}$$

where in this context $\tau$ is the time required to perform one cycle of the almost periodic motion at the energy $E$. (We have assumed that the average is taken along the energy trajectory so that

$$dt = dw \tag{2.121}$$

along a trajectory). If we define the action, $I$, by the equation [32]

$$I(E) \equiv \oint_{E=\text{const.}} p dq \tag{2.122}$$

and allow the energy to vary by an amount $dE$ over a thin ring of thickness $dI$, we can account for the slow diffusion of energy. (The assumption that the damping has a negligible effect over one period means $\eta I(E_c) \ll kT$ [27, 28].) We assume that the motion of the particles in the well would always be librational, that is, have closed trajectories, in the absence of the Brownian forces (that is, if it were not for the slow diffusion of energy). The slow diffusion of the energy means in the VLD case that the trajectories of the motion are almost closed, except for a leisurely spiraling of the particles toward the minimum of the energy, due to the energy loss $dE$ per cycle. The simplest example of a librational motion arises from the energy function [32] for the harmonic oscillator:

$$E = \frac{p^2}{2m} + \frac{x^2}{(2/m\omega^2)} \tag{2.123}$$

So that the trajectories in phase space are (closed) ellipses.

Now, Eq. (2.80) with arbitrary mass $m$ is

$$\frac{\partial \rho}{\partial t} = \frac{\partial U}{\partial x}\frac{\partial \rho}{\partial p} - \frac{p}{m}\frac{\partial \rho}{\partial x} + \eta \frac{\partial}{\partial p}\left(\rho p + mkT\frac{\partial \rho}{\partial p}\right) \tag{2.124}$$

If there were no dissipation of energy, we would have by Liouville's theorem

$$\frac{\partial \rho}{\partial t} = \frac{\partial U}{\partial x}\frac{\partial \rho}{\partial p} - \frac{p}{m}\frac{\partial \rho}{\partial x} \tag{2.125}$$

(since $(\partial U/\partial x) = -\dot{p}$ and $(p/m) = \dot{x}$, give the components of the "velocity", $\mathbf{u}$, in phase space). So the remaining (diffusive) part of Eq. (2.124), namely,

$$\eta \frac{\partial}{\partial p}\left(\rho p + mkT\frac{\partial \rho}{\partial p}\right) \tag{2.126}$$

assuming that the dissipation of energy is very slow, describes the dissipation of energy. We now transform Eq. (2.124) into an equation in the energy and phase variables by using the equation

$$E = \frac{p^2}{2m} + U(x) \tag{2.127}$$

We have

$$\dot{x} = \pm\sqrt{\frac{2}{m}[E - U(x)]} \tag{2.128}$$

which, on taking the positive sign and integrating the resulting differential equation, between points $x_1$ and $x$ yields:

$$\int_{x_1}^{x} \frac{dx'}{\sqrt{(2/m)[E - U]}} = t + w \tag{2.129}$$

where $w$ is the constant of integration and now defines the phase. Now the equation

$$dw = dt \tag{2.130}$$

implies that

$$\frac{dw}{dt} = 1 \tag{2.131}$$

and since the variation in energy is very slow we have

$$\frac{dE}{dt} \approx 0 \tag{2.132}$$

that is, almost a conservative system. Also, we have the equation

$$\dot{p} = -\frac{\partial U}{\partial x} \tag{2.133}$$

Now, by the chain rule

$$\frac{\partial \rho}{\partial x} = \frac{\partial \rho}{\partial E}\frac{\partial E}{\partial x} + \frac{\partial \rho}{\partial w}\frac{\partial w}{\partial x} \tag{2.134}$$

$$\frac{\partial \rho}{\partial p} = \frac{\partial \rho}{\partial E}\frac{\partial E}{\partial p} + \frac{\partial \rho}{\partial w}\frac{\partial w}{\partial p} \tag{2.135}$$

and

$$\frac{\partial E}{\partial x} = \frac{dU}{dx} \tag{2.136}$$

$$\frac{\partial w}{\partial x} = \frac{\partial t}{\partial x} = \frac{m}{p} = \frac{1}{\dot{x}} \tag{2.137}$$

$$\frac{\partial E}{\partial p} = \frac{p}{m} \tag{2.138}$$

$$\frac{\partial w}{\partial p} = 0 \tag{2.139}$$

[by using Eq. (2.129)]
So,

$$\frac{\partial}{\partial x} = \frac{dU}{dx}\frac{\partial}{\partial E} + \frac{1}{\dot{x}}\frac{\partial}{\partial w} \tag{2.140}$$

and

$$\frac{\partial}{\partial p} = \frac{p}{m}\frac{\partial}{\partial E} \tag{2.141}$$

So, expression (2.126) becomes

$$\eta \frac{p}{m} \frac{\partial}{\partial E} \left( \rho p + pkT \frac{\partial \rho}{\partial E} \right), \tag{2.142}$$

while the Liouville equation (2.125) is simply

$$\frac{\partial \rho}{\partial t} = - \frac{\partial \rho}{\partial w} \tag{2.142a}$$

### 3. *Averaging Over the Phase Variable*

Now defining the average over one cycle of the motion as

$$\bar{\rho} = \frac{1}{\tau} \int_0^\tau \rho(E, w, t) dw \tag{2.143}$$

where $\tau$ is the periodic time, we have from Eq. (2.124)

$$\overline{\frac{\partial \rho}{\partial t}} = \overline{\frac{\partial \rho}{\partial w} + \eta \frac{p}{m} \frac{\partial}{\partial E} \left( \rho p + pkT \frac{\partial \rho}{\partial E} \right)} \tag{2.144}$$

$$= \overline{\frac{\partial \rho}{\partial w}} + \overline{\eta \frac{p}{m} \frac{\partial}{\partial E} \left( \rho p + pkT \frac{\partial \rho}{\partial E} \right)} \tag{2.145}$$

Now,

$$\overline{\frac{\partial \rho}{\partial t}} = \frac{1}{\tau} \int_0^\tau \frac{\partial}{\partial t} \rho(E, w, t) dw \tag{2.146}$$

$$= \frac{\partial}{\partial t} \frac{1}{\tau} \int_0^\tau \rho dw \tag{2.147}$$

$$= \frac{\partial (\bar{\rho})}{\partial t} \tag{2.148}$$

also

$$\overline{\frac{\partial \rho}{\partial w}} = \frac{1}{\tau}\int_0^\tau \frac{\partial \rho}{\partial w}\,dw \tag{2.149}$$

$$= \frac{1}{\tau}\int_0^\tau d\rho \tag{2.150}$$

$$= 0 \tag{2.151}$$

since the integral is taken over one complete cycle of the motion, $\rho(\tau) = \rho(0)$. Also note that Eq. (2.150) holds approximately since

$$d\rho = \frac{\partial \rho}{\partial E}dE + \frac{\partial \rho}{\partial w}dw \tag{2.152}$$

however $E$ is slowly varying, so that in one cycle $dE \approx 0$ and so

$$d\rho \approx \frac{\partial \rho}{\partial w}dw \tag{2.153}$$

Equation (2.145) then becomes

$$\frac{\partial \bar{\rho}}{\partial t} = \frac{\eta}{m}\overline{\left(p^2\frac{\partial \rho}{\partial E} + p\rho\frac{\partial p}{\partial E} + pkT\frac{\partial p}{\partial E}\frac{\partial \rho}{\partial E} + p^2kT\frac{\partial^2 \rho}{\partial E^2}\right)} \tag{2.154}$$

that, by using Eq. (2.138), simplifies to

$$\frac{\partial \bar{\rho}}{\partial t} = \eta\overline{\left(\frac{p^2}{m}\frac{\partial \rho}{\partial E} + \rho + kT\frac{\partial \rho}{\partial E} + \frac{p^2kT}{m}\frac{\partial^2 \rho}{\partial E^2}\right)} \tag{2.155}$$

Now, by using the fact that if $\alpha$ and $\beta$ are constants independent of the phase, $w$, and $P$ and $Q$ are functions of (inter alia) $w$, we have

$$\overline{\alpha P + \beta Q} = \alpha\bar{P} + \beta\bar{Q} \tag{2.156}$$

We use also the fact that

$$\overline{\frac{\partial \rho}{\partial E}} = \frac{\partial \bar{\rho}}{\partial E} \tag{2.157}$$

since

$$\overline{\frac{\partial \rho}{\partial E}} \equiv \frac{1}{\tau}\int_0^{\tau} \frac{\partial \rho}{\partial E} dw \tag{2.158}$$

$$= \frac{1}{\tau}\frac{\partial}{\partial E}\int_0^{\tau} \rho dw \tag{2.159}$$

$$= \frac{1}{\tau}\frac{\partial}{\partial E}(\bar{\rho}) \tag{2.160}$$

and finally, assuming that

$$\overline{p^2 \rho} = \overline{p^2}\bar{\rho} \tag{2.161}$$

we find that Eq. (2.155) becomes

$$\frac{\partial \bar{\rho}}{\partial t} = \eta\left(\frac{\overline{p^2}}{m}\frac{\partial \bar{\rho}}{\partial E} + \bar{\rho} + kT\frac{\partial \bar{\rho}}{\partial E} + \frac{\overline{p^2} kT}{m}\frac{\partial^2 \bar{\rho}}{\partial E^2}\right) \tag{2.162}$$

Now, Eq. (2.143) yields

$$\overline{p^2} = \frac{1}{\tau}\int_0^{\tau} p^2 dw \tag{2.163}$$

Furthermore we are taking the average along a trajectory $dw = dt$ and the periodic time in the well is $2\pi/\omega(E)$, where $\omega(E)$ is the angular frequency of oscillation, so that

$$\overline{p^2} = \frac{\omega}{2\pi}\int_0^{2\pi} p^2 dt = \frac{\omega}{2\pi}\int_0^{2\pi} pm\dot{x}dt = m\frac{\omega}{2\pi}\oint pdx = m\frac{\omega}{2\pi}I \tag{2.164}$$

Thus Eq. (2.162) becomes

$$\frac{\partial \bar{\rho}}{\partial t} = \eta\left(\bar{\rho} + kT\frac{\partial \bar{\rho}}{\partial E} + I\frac{\omega}{2\pi}\frac{\partial}{\partial E}\left(\bar{\rho} + kT\frac{\partial \bar{\rho}}{\partial E}\right)\right) \tag{2.165}$$

Now, the period of oscillation is $2\pi/\omega$, and we have

$$\omega = 2\pi\frac{dE}{dI} \tag{2.166}$$

$$\overline{p^2} = I\frac{\omega}{2\pi} \tag{2.167}$$

where $I$ is the action over one period. Thus the diffusion equation becomes

$$\frac{\partial\bar{\rho}}{\partial t} = \eta\left(1 + \frac{\omega I}{2\pi}\frac{\partial}{\partial E}\right)\left(\bar{\rho} + kT\frac{\partial\bar{\rho}}{\partial E}\right) \tag{2.168}$$

$$= \eta\left(1 + I\frac{\partial}{\partial I}\right)\left(\bar{\rho} + kT\frac{\partial\bar{\rho}}{\partial E}\right) \tag{2.169}$$

or, writing $\rho$ for $\bar{\rho}$ we obtain ($\bar{\rho} = \bar{\rho}(E, t)$)

$$\frac{\partial\rho}{\partial t} = \eta\frac{\partial}{\partial I}\left(I\rho + kTI\frac{\partial\rho}{\partial E}\right) \tag{2.170}$$

This corresponds to a diffusion in $I$ or $E$ space along the $I-$ or $E-$ coordinate; the diffusion term proper is

$$\eta\frac{\partial}{\partial I}\left(kTI\frac{\partial\rho}{\partial E}\right) = \eta\frac{\partial}{\partial I}\left(kT\frac{2\pi I}{\omega}\frac{\partial\rho}{\partial I}\right) \tag{2.171}$$

and corresponds to a diffusion coefficient $D$, given by

$$D = \frac{2\pi\eta kTI}{\omega} \tag{2.172}$$

The above paragraph constitutes an extended version of Kramers' original method of arriving at the energy diffusion equation. A more succint derivation is given in Eqs. (2.262) et seq. at the end of this chapter. Further discussion of the low damping problem will be found in van Kampen [27, 28]. A simplified discussion is also available in [2].

### 4. *Kramers' Original Calculation of the Escape Rate for Very Low Damping*

We follow, as closely as possible, the original reasoning and phraseology of Kramers. As usual, a stationary state of diffusion, that is, $(\partial\rho/\partial t) = 0$ with probability current $q$ corresponds to

$$q = -\eta\left(I\rho + kTI\frac{\partial\rho}{\partial E}\right) \tag{2.173a}$$

because the continuity equation is, in $I$ space

$$\frac{\partial \rho}{\partial t} = -\frac{\partial q}{\partial I} \tag{2.173b}$$

$$= \eta kTIe^{-\beta E}\frac{\partial}{\partial E}(\rho e^{\beta E}) \tag{2.174}$$

Integrating this equation with respect to $E$ between two points $A$ and $B$ along the $E-$ (or $I-$) coordinate, i.e. along a line of increasing energy, yields the probability current

$$q = \frac{\eta kT[\rho e^{\beta E}]_A^B}{\int_A^B (1/I)e^{\beta E}dE} \tag{2.175}$$

The probability density $\rho$ is practically constant along lines of constant energy (since a Boltzmann distribution is set up with

$$\rho = \rho_0 e^{-\beta E} \tag{2.176}$$

so $E = \text{const} \Rightarrow \rho = \text{const}$) over a range of curves that start at $A$ and that extend to energy curves that cut the $x$ axis, *not at C itself but at points D, very close to C*. This restriction is necessary if the potential function has a smooth saddle point since the frequency tends to infinity as $E$ tends to $\Delta U$. This would mean that the viscosity is no longer small in the sense used above at the beginning of Section II.D. (This restriction is unnecessary if the saddle point is not a smooth function of the space variables.) Equation (2.175) may be written

$$q = \eta kT\frac{(\rho e^{\beta E})_{\text{near}\,A} - (\rho e^{\beta E})_C}{\int_{\text{near}\,A}^C (1/I)e^{\beta E}dE} \tag{2.177}$$

We avoid integrating from point $A$ itself because at this point $E = I = 0$ and so the integral would diverge. We may take it that "near $A$" means an energy value of the order of the thermal energy $kT$, and so corresponds to points in phase space where $\rho$ is still of the same order as at $A$ itself, and where practically all the particles are trapped. The condition that particles leaving at $C$ practically never reenter the well means, according to Kramers, that $(\rho e^{\beta E})_C$ may be taken to be 0, and that the upper limit of the integral may be taken to be the barrier energy at $C$. Let us write

$$\rho_A \equiv (\rho e^{\beta E})_{\text{near}\,A} \tag{2.178}$$

Thus Eq. (2.177) becomes

$$q = \eta kT \rho_A \left\{ \int_{kT}^{\Delta U} \frac{1}{I} e^{\beta E} dE \right\}^{-1} \tag{2.179}$$

Now the main contribution to this integral for the probability current comes from energy values that differ from the barrier energy $\Delta U$ by a very small quantity (of the order of magnitude $kT$), so that we may take $I$ to have the value $I_C$ corresponding to the saddle energy trajectory through $C$. Hence, the integral in Eq. (2.179) may be written

$$\int_{kT}^{\Delta U} \frac{1}{I} e^{\beta E} dE \approx \frac{1}{I_C} e^{\beta \Delta U} \int_{kT}^{\Delta U} e^{-\beta(\Delta U - E)} dE \tag{2.180}$$

Now, let

$$\Xi \equiv \Delta U - E \tag{2.181}$$

then,

$$dE = -d\Xi \tag{2.182}$$

We now take the high barrier limit by integrating over $E$ from $-\infty$ (i.e. deep in the well) to $\Delta U$

$$\text{when} \qquad E = \Delta U, \quad \text{we have } \Xi = 0 \tag{2.183}$$

$$\text{when} \qquad E \to -\infty, \quad \text{we have } \Xi \to \infty \tag{2.184}$$

So, the above integral now becomes

$$\int_{kT}^{\Delta U} \frac{1}{I} e^{\beta E} dE \approx -\frac{1}{I_C} e^{\beta \Delta U} \int_{\infty}^{0} e^{-\beta \Xi} d\Xi \tag{2.185}$$

$$= \frac{1}{I_C} e^{\beta \Delta U} \int_{0}^{\infty} e^{-\beta \Xi} d\Xi \tag{2.186}$$

$$= \frac{kT}{I_C} e^{\beta \Delta U} \tag{2.187}$$

since $(1/\beta) = kT$. Whence, the current is

$$q \approx \eta \rho_A \, I_C e^{-\beta \Delta U}. \tag{2.188}$$

Hence, since the number of particles trapped in the well near A is:

$$n_A = \rho_A \frac{2\pi kT}{\omega_A} = \rho_A Z \tag{2.189}$$

the escape rate is

$$\Gamma = \frac{q}{n_A} = \eta \frac{I(E_C)}{kT} \frac{\omega_A}{2\pi} e^{-\beta E_C} \tag{2.190}$$

while

$$\Delta U = E_C$$

Kramers now roughly approximates the action of the almost periodic motion at the saddle point by

$$I_C = \frac{2\pi E_C}{\omega_A} \tag{2.191}$$

which is the action of a harmonic oscillator of energy equal to the barrier energy and natural frequency $f_A$ [32], so that Eq. (2.190) becomes

$$\Gamma = \eta \frac{\Delta U}{kT} e^{-\beta \Delta U} \tag{2.192}$$

which is Eq. (28) of Kramers [4]. Note the discrepancy between this equation and the low-damping limit of Eq. (2.105), that is, Eq. (2.107b), which predicts the transition state theory (TST) result Eq. (1.1) as $\eta \to 0$. We will comment further on this discrepancy later in Section II.F.

### 5. *First Passage Time Approach*

This approach is described in detail for spins in Section V (see also [45]). Here, we merely remark that Hänggi et al. [1] rederived the low-damping result above using a first passage time (FPT) approach [their Eq. (4.49)]

$$\Gamma = \eta \beta I(E_C) \frac{\omega_A}{2\pi} e^{-\beta E_C} \tag{2.193}$$

(Hänggi et al. use $\omega_0$ instead of $\omega_A$). (Note that TST theory is simple to write down, because, according to Boltzmann statistics which unlike in the treatment of Kramers is assumed to hold everywhere, the probability of a jump, that is, an event occurring is

$$P = Ke^{-\beta E} \tag{2.194}$$

where $E$ is the energy and $K$ is a constant. (In using this equation for reaction rates the constant is written as $f_A$, where $f_A$ is called the attempt frequency, so that according to TST $P = \Gamma = f_A e^{-\beta\Delta U}$). Thus in the low damping limit

$$\Gamma = \eta\beta I(E_C)\Gamma_{\text{TST}} \tag{2.195}$$

Notice that Hänggi et al. essentially used the mean-FPT method of Matkowsky et al. [44],

$$\tau = \tau(A) + 2\tau_{A\to\text{sep}} \tag{2.196}$$

(Note here that $A$ now refers to a region where a particle is trapped, e.g., a potential well and *not* the saddle energy of Kramers.) The time $\tau(A)$ is the time to go from the well to the critical energy curve (the critical energy is the energy required by a particle in order to escape the well). Thus without any extra energy a particle, on reaching the critical energy curve, may either fall back into the well region or escape, with equal probability. The parameter $\tau_{Q\to\text{sep}}$ is the time required to go from the critical energy curve to the separatrix [a curve ($g$) whose distance from the critical energy curve is infinitesimal with the difference that all particles on reaching the separatrix are on their way out of the region $A$ (see Fig. 7)]. Note that $\tau_{A\to\text{sep}}$ is considered to be negligible.

In conclusion of this section we remark that Eq. (2.190) may also be written in the form

$$\Gamma = \beta\Delta E\frac{\omega_A}{2\pi}e^{-\beta\Delta U} \tag{2.197}$$

where

$$\Delta E = \eta\oint_{E_C} pdx \ll kT \tag{2.198}$$

is the *energy loss per cycle of the motion at the saddle point* and

$$I(E_C) = \oint_{E_C} pdx \tag{2.199}$$

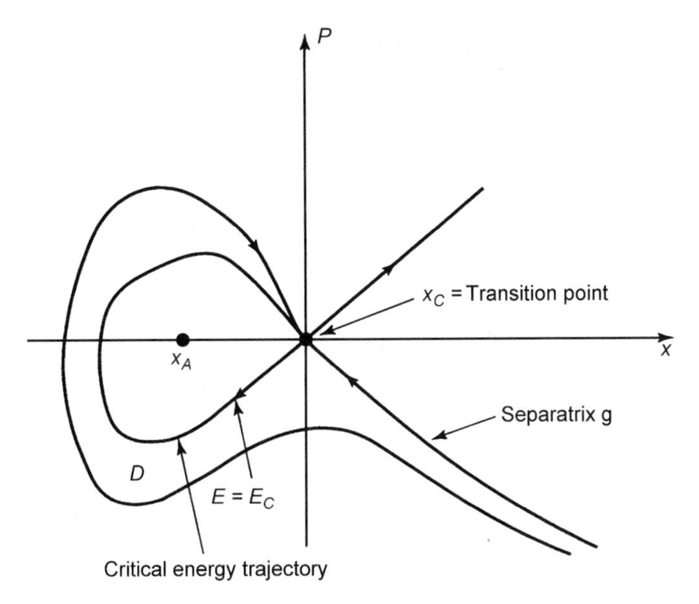


**Figure 7.** Diagram of critical energy curve and separatrix [44]. Shown here are the critical energy curve ($E = E_C$) and the separatrix ($g$) in phase space [44]. The critical energy, $E_C$, is the energy required by a particle to just escape the well. When a particle reaches this energy, it may either escape or fall back with equal probability. The separatrix separates the bounded and unbounded motions. In other words, when a particle reaches the separatrix it exits the well. The separation of these curves (greatly exaggerated in the diagram) is infinitesimally small.

is the *action of the almost periodic motion on the saddle point energy contour.* The integral in Eq. (2.199) may be evaluated explicitly by recalling that the equation defining the saddle energy, that is the equation describing the *separatrix between the bounded motion in the well and the unbounded motion after escape*, is

$$E_C = \frac{p^2}{2m} + U(x)$$

## E. The Case of Large Viscosity

### 1. The Very High Damping Regime

This is a limiting case of the IHD regime, where it is supposed that the *damping is so large that equilibrium of the velocity distribution has been sensibly attained.* In this situation, it is possible to obtain from the Klein–Kramers equation an approximate partial differential equation *for the evolution of the distribution function in configuration space only.* This approximate diffusion equation is called the Smoluchowski equation. First, we recall what we mean by *large viscosity.* By large viscosity, we mean that the *effect of the Brownian forces on the velocity of the particle is much larger than that of the external force* $K(x)$. The Smoluchowski equation may be derived, according to Kramers, by assuming that $K$ does not vary sensibly over a distance of the order of the diffusion length, that is, *roughly* $\sqrt{kT}/\eta$. We expect that, irrespective of the initial $\rho$ distribution, the distribution (with $m = 1$)

$$\rho(x, p, t) \approx \sigma(x, t) \exp\left(-\frac{p^2}{2kT}\right) \tag{2.200}$$

(i.e., a Maxwell velocity distribution) will hold after a very short time lapse ($\approx \eta^{-1}$) (cf. the arguments used by Langevin and Einstein to neglect inertial effects). The high barrier then ensures that a slow diffusion of particles over the barrier will take place, which may be expected to satisfy the Smoluchowski equation for the density $\sigma(x, t)$ in configuration space:

$$\frac{\partial \sigma}{\partial t} = -\frac{\partial}{\partial x}\left(\frac{K}{\eta}\sigma - \frac{kT}{\eta}\frac{\partial \sigma}{\partial x}\right) \tag{2.201}$$

where the diffusion coefficient is $kT/\eta$. Kramers then examines the approximate validity of Eq. (2.200). As long as no perfect temperature equilibrium is attained, Eq. (2.200) will hold only approximately, even when the external force is 0 (cf. the approximate Eq. (2.72)). He claims that in that case, while the Maxwell velocity distribution of velocities will hold *exactly* for each *particle*, it will *not* hold EXACTLY *at each value of* $x$, since *otherwise there would be no diffusion current* $\int \rho p dp$ (cf. Subsection II.F.6). Thus the *behavior is unlike that described by a single space variable FPE* such as that for axially symmetric potentials of the magnetocrystalline anisotropy, which is an *exact* equation.

### 2. *Derivation of the Smoluchowski Equation*

In order to derive Eq. (2.201) from Eq. (2.124), we first rewrite Eq. (2.124) in the form:

$$\frac{\partial \rho}{\partial t} = \eta\left(\frac{\partial}{\partial p} - \frac{1}{\eta}\frac{\partial}{\partial x}\right)\left(\rho p + kT\frac{\partial \rho}{\partial p} - \frac{K}{\eta}\rho + \frac{kT}{\eta}\frac{\partial \rho}{\partial x}\right) - \frac{\partial}{\partial x}\left(\frac{K}{\eta}\rho - \frac{kT}{\eta}\frac{\partial \rho}{\partial x}\right) \tag{2.202}$$

This can be checked most easily by rewriting Eq. (2.202) in the form of Eq. (2.124). We now integrate both sides of this equation (with respect to the momentum) along a straight line in phase space namely

$$x + \frac{p}{\eta} = x_0(\text{const}) \tag{2.203a}$$

The integration extends from $p = -\infty$ to $p = +\infty$.

Note

$$\frac{\partial}{\partial x} = \frac{\partial x_0}{\partial x}\frac{\partial}{\partial x_0} = \frac{\partial}{\partial x_0} \tag{2.203b}$$

$$\frac{\partial}{\partial p} = \frac{\partial x_0}{\partial p}\frac{\partial}{\partial x_0} = \frac{1}{\eta}\frac{\partial}{\partial x_0} \tag{2.203c}$$

and hence $(\partial/\partial p) - (1/\eta)(\partial/\partial x)$ is the zero operator along this line.

If we denote the integral of $\rho$ along this line by $\sigma(x_0)$, we obtain

$$\frac{\partial \sigma}{\partial t} = -\int_{x+\frac{p}{\eta}=x_0} \frac{\partial}{\partial x}\left(\frac{K}{\eta}\rho - \frac{kT}{\eta}\frac{\partial \rho}{\partial x}\right) dp \tag{2.204a}$$

$$\approx -\frac{\partial}{\partial x_0}\left(\frac{K(x_0)}{\eta}\sigma(x_0) - \frac{kT}{\eta}\frac{\partial \sigma(x_0)}{\partial x_0}\right) \tag{2.204b}$$

which is a diffusion equation in configuration space and is the Smoluchowski equation.

In Eq. (2.204a), note that the line integral is strictly

$$\int_{x+\frac{p}{\eta}=x_0} \frac{\partial}{\partial x}\left(\frac{K}{\eta}\rho - \frac{kT}{\eta}\frac{\partial \rho}{\partial x}\right) ds \tag{2.205}$$

where $ds$ is the element of arc length along the line in question. However, since

$$ds^2 = dx^2 + dp^2 \tag{2.206}$$

and

$$\frac{dp}{dx} = -\eta \tag{2.207}$$

or

$$dp = -\eta dx \tag{2.208}$$

and also, since we let

$$\eta \to \infty \tag{2.209}$$

we can approximate

$$ds \approx dp \tag{2.210}$$

We also use the fact that

$$\int_{x+\frac{p}{\eta}=x_0} \frac{\partial \rho}{\partial t} ds = \frac{\partial}{\partial t} \int_{x+\frac{p}{\eta}=x_0} \rho ds = \frac{\partial}{\partial t} \sigma(q_0) \tag{2.211}$$

The *approximate validity* of Eq. (2.204b) is a consequence of the *approximate validity* of Eq. (2.200) if it is also assumed that in the region of values of $p$ that dominate the integral, (that is, $|p| \lesssim \sqrt{kT}$) the variation of $x$ (which is of the order of $\sqrt{kT}/\eta$) is small compared to $x$ distances over which the force $K$ and the density in configuration space $\sigma$ undergo marked variations. These are, however, the conditions that a priori have to be imposed in order to ensure the applicability of Eq. (2.201). We remark that, since we integrate along the line $(p/\eta) = x_0 - x$ and since both $x$ and $x_0$ are of the order of the diffusion length $\sqrt{kT}/\eta$, we expect $p$ to be of the order of $\sqrt{kT}$).

**Justification of Eq. (2.204b)**

The position coordinate has the value $x \approx x_0$ along the line $x + (p/\eta) = x_0$, (if $\eta$ is large and $x$ does not sensibly vary over the range of $p$ values that contribute to the integral) so that

$$\begin{aligned} -\int_{x+\frac{p}{\eta}=x_0} \frac{\partial}{\partial x}\left(\frac{K}{\eta}\rho - \frac{kT}{\eta}\frac{\partial \rho}{\partial x}\right) dp &\approx -\frac{\partial}{\partial x_0}\frac{K(x_0)}{\eta}\int_{x+\frac{p}{\eta}=x_0} \rho dp + \frac{kT}{\eta}\frac{\partial}{\partial x_0}\frac{\partial}{\partial x_0}\int_{x+\frac{p}{\eta}=x_0} \rho dp \\ &\approx -\frac{\partial}{\partial x_0}\left(\frac{K(x_0)}{\eta}\sigma(x_0) - \frac{kT}{\eta}\frac{\partial \sigma(x_0)}{\partial x_0}\right) \end{aligned} \tag{2.212}$$

by definition of $\sigma(x_0)$.

The above paragraph describes the essentially heuristic derivation of the approximate equation for the distribution function in configuration space known as the Smoluchowski equation from the Klein–Kramers equation. The first rigorous treatment of the problem was given by Brinkman [46] who (by expanding the distribution function in appropriate sets of orthogonal functions) showed that the solution of the Klein–Kramers equation could be obtained by the method of separation of the variables. Furthermore in the small inertial limit, the results predicted by the Smoluchowski equation could be obtained. Further discussion of the problem was given by Titulaer [47] and Wilemski [48]. These discussions, which are based [8] on the Chapman–Enskog method in the kinetic theory of gases, are well summarized by van Kampen [27] (p. 216). The method of Brinkman, which proved to be of vital importance in later developments in the theory of Brownian motion, is summarized by Risken [25, 26] and Coffey et al. [5].

### 3. *Calculation of the Reaction Rate from the Smoluchowski Equation*

In this section, we shall adhere as closely as possible to the original derivation of Kramers as given in his 1940 paper since the approach taken there is reasonably didactic.

By putting $(\partial\sigma/\partial t) = 0$ in Eq. (2.204b) we see that a stationary diffusion current in configuration space obeys the law

$$q = \frac{K}{\eta}\sigma - \frac{kT}{\eta}\frac{\partial\sigma}{\partial x} = \text{const} \tag{2.213}$$

Since this expression for the current may also be written in the form

$$q = -\frac{kT}{\eta}e^{-\beta U}\frac{\partial}{\partial x}(\sigma e^{\beta U}) \tag{2.214}$$

we obtain, on integration between two points $A'$ and $B'$ on the $x$ axis:

$$q = -\frac{kT|\sigma e^{\beta U}|_{A'}^{B'}}{\int_{A'}^{B'} \eta e^{\beta U} dx} \tag{2.215}$$

By assuming a quasi-stationary state, in which practically no particles have yet arrived at $B$ (i.e., $\sigma = 0$ at $x = B$), whereas near $A$, thermal equilibrium has practically been established, application of Eq. (2.215) [letting $\sigma_A \equiv (\sigma e^{\beta U})_{\text{near}\,A}$] yields the current as

$$q = \frac{kT}{\eta}\sigma_A\left\{\int_A^B e^{\beta U} dx\right\}^{-1} \tag{2.216}$$

The number, $n_A$, of particles near $A$ may be calculated, (if we assume that the potential energy $U$ near $A$ can be represented by that of a harmonic oscillator of angular frequency $\omega_A$, in this instance for convenience we take the bottom of the well as the zero of the potential energy:

$$U \approx \tfrac{1}{2}\omega_A^2 x^2) \tag{2.217}$$

in the following manner: we have

$$n_A = \int_{-\infty}^{\infty} \sigma_A \exp\left(-\frac{\omega_A^2 x^2}{2kT}\right) dx \tag{2.218}$$

$$= \frac{\sigma_A}{\omega_A}\sqrt{2\pi kT} \tag{2.219}$$

The escape rate $\Gamma \equiv q/n_A$ denotes the chance in unit time that a particle that was originally trapped at $A$ escapes to $B$. It is given by (utilizing Eqs. (2.216) and (2.219))

$$\Gamma = \frac{\omega_A}{\eta}\sqrt{\frac{kT}{2\pi}}\left\{\int_A^B e^{\beta U} dx\right\}^{-1} \tag{2.220}$$

The main contribution to the integral comes from a small region near $C$ (since $e^{\beta U}$ falls off very quickly as we move away from the maximum of the function $U$). Thus recalling that the zero of potential is now at $x_A$ so that $\Delta U = U(x_C)$

$$\int_A^B e^{\beta U} dx \approx e^{\beta \Delta U} \int_{-\infty}^{\infty} \exp\left(-\frac{\omega_C^2 (x - x_C)^2}{2kT}\right) dx \tag{2.221}$$

$$= \frac{1}{\omega_C}\sqrt{2\pi kT} e^{\beta \Delta U} \tag{2.222}$$

and so, for the escape rate, we find

$$\Gamma \approx \frac{\omega_A \omega_C}{2\pi\eta} e^{-\beta \Delta U} \tag{2.223}$$

This equation is clearly the limiting case of Eq. (2.107b) when $\eta$ is very large. The Smoluchowski approach which we have used to derive the VHD escape rate is very useful in problems of dielectric relaxation when inertial effects are ignored because [19,41] of the complicated terms that arise from the angular momenta in the Klein–Kramers equation for rotational problems.

## F. Range of Validity of the IHD and VLD Formulas

### 1. *Behavior of the IHD Result in the Limit of Vanishing Friction*

The IHD escape rate in the limit of vanishing friction tends to the TST result, namely, Eq. (2.107b):

$$\Gamma = \frac{\omega_A}{2\pi} e^{-\beta \Delta U} \tag{2.224}$$

This limiting behavior is, however, inconsistent with the derivation of Eq. (2.107a) because, in the limit of vanishing friction, the variation of $x$ is not the same as the variation of $u$ so that the correct formula to use is Eq. (2.190), that is,

$$\Gamma = \eta \beta I(\Delta U) \frac{\omega_A}{2\pi} e^{-\beta \Delta U} \tag{2.225}$$

$$\eta \beta I(\Delta U) \ll 1 \tag{2.226}$$

### 2. *Regions of Validity of the IHD and VLD Results*

For Eq. (2.225) to hold, $\eta$ must be small compared with $\omega_A$. If $\eta = 2\omega_A$ we have aperiodic damping, and we might expect that there would be a plentiful supply of particles near the point $C$, and so the escape rate would be described by the IHD formula Eq. (2.105). Kramers [4] confesses (cf. Fig. 4) that he was unable to extend Eq. (2.225) (the VLD Result) to values of $\eta$ which were not small compared with $2\omega_A$, that is in the *crossover region* between VLD and IHD formulae.

The approximate formula Eq. (2.192) for $\Gamma$ in the VLD limit is useful for obtaining a criterion in terms of the barrier height for the ranges of friction in which the VLD and IHD Kramers formulas are valid. Equation (2.192) is (with the approximate action given by Eq. 2.191)

$$\Gamma = \Gamma_{\text{VLD}} = \eta \beta \Delta U e^{-\beta \Delta U} \tag{2.227}$$

If now we define a dimensionless friction parameter

$$\alpha = \frac{2\pi\eta}{\omega_A} \tag{2.228}$$

Eq. (2.227) becomes

$$\Gamma = \alpha \beta \Delta U \Gamma_{\text{TS}} \tag{2.229}$$

so that $\alpha\Delta U$ is approximately the energy loss per cycle. Hence the condition for the validity of Eq. (2.227) becomes

$$\alpha\beta\Delta U \ll 1 \tag{2.230}$$

while one would expect the IHD formula Eq. (2.105) to be valid if

$$\alpha\beta\Delta U \gg 1 \tag{2.231}$$

The damping region defined by

$$\alpha\beta\Delta U \approx 1$$

defining a *crossover region* where neither the VLD nor the IHD formula is valid, which is the reason for the calculation of Mel'nikov and Meshkov [34, 49] reviewed in Section VI. We shall now give a physical interpretation of the three regions identified above.

### 3. *Physical Interpretation of the Various Damping Regimes*

We may summarize the results of our calculation. In the *mechanical* Kramers problem three regimes of damping appear:

(a) Intermediate to high damping (IHD): the general picture here [27, 28] being that inside the well the distribution function is almost the Maxwell–Boltzmann distribution prevailing in the depths of the well. However, near the barrier the distribution function deviates from that equilibrium distribution due to the slow draining of particles across the barrier. The barrier region is so small that one may approximate the potential in this region by an inverted parabola.

(b) Very low damping (VLD): here the damping is so small that the assumption in (a) namely that the particles approaching the barrier region have the Maxwell–Boltzmann distribution completely breaks down. Thus the region where deviations from that distribution occur extends far beyond the region where the potential may be approximated by an inverted parabola. Thus we may now, by transforming Klein–Kramers equation into energy and phase variables [by averaging over the phase and by supposing that the motion of a particle attempting to cross the barrier is almost conservative and is the librational motion in the well of a particle with energy equal to the barrier energy,] obtain an escape rate formula. We remark that the assumption of almost conservative behaviour (meaning that the energy loss per cycle is almost negligible and is equal to the friction times the action of the undamped motion at the barrier energy) ensures that the Liouville term in the Klein–Kramers equation vanishes (unlike in IHD where there is strong coupling between the diffusive and the Liouville term). Thus only the diffusion term in the energy variable remains (the dependence on the phase having been eliminated by averaging the distribution in energy-phase variables along a closed trajectory of the energy since we assume a librational motion in the well).

(c) An intermediate (crossover) friction region where neither IHD nor VLD formulas apply: in this region neither of the above approaches may be used. In contrast to the VLD case, the Liouville term does not vanish, meaning that one cannot average out the phase dependence of the distribution function which is ultimately taken account of by constructing from the Klein–Kramers equation a diffusion equation for the distribution function with the energy and action as independent variables. This diffusion equation allows one to express the calculation of the energy distribution function at a given action, as a Fredholm integral equation which can be converted into one (or several) Wiener-Hopf equation(s) (Section VI). This procedure yields an integral formula the product of which with the IHD escape rate (cf. Eq. 1.16) provides an expression for the escape rate which is valid for all values of the damping, so allowing the complete solution of Kramers' problem. The interpolation integral derived from the Wiener-Hopf equation effectively allows for the coupling between the Liouville term and the dissipative term in the Klein–Kramers equation written in terms of energy-phase variables, which is ignored in the VLD limit.

### 4. *Numerical Verification of the Kramers Theory*

The Kramers theory may be verified numerically for large potential barrier heights by calculating the smallest nonvanishing eigenvalue of the Klein–Kramers equation [21]. This procedure is possible because of the exponential nature of the escape rate, so that, in effect, the smallest eigenvalue of the FPE is very much smaller than all the higher order eigenvalues which pertain to the fast motion in the well. Thus the Kramers escape rate is approximately given by the smallest nonvanishing eigenvalue if the barrier height is sufficiently large $> kT$. This method has been extensively used to verify the Kramers theory, in particular the application of that theory to magnetic relaxation of single domain ferromagnetic particles. We shall now present a more concise derivation of (a) the Klein–Kramers equation from the Langevin equation, (b) the Smoluchowski equation from the Klein–Kramers equation, and (c) the Kramers low-damping result.

### 5. *Alternative Derivation of the Klein–Kramers Equation*

Here, we present the derivation of the Fokker–Planck equation for a particle moving in a 1D potential, that is, the Klein–Kramers equation, from the Langevin equation. The latter is a Newtonian equation

$$\frac{dp}{dt} = -\frac{\partial U}{\partial x} - \zeta \frac{p}{m} + \lambda(t) \tag{2.232}$$

$$\frac{dx}{dt} = \frac{p}{m} \tag{2.233}$$

in which $\lambda(t)$ is a stochastic variable modelling the environment of the particle (the heat bath). The random variable $\lambda(t)$ is assumed to be white noise. A complete description of the white noise is provided by its distribution functional

$$F[\lambda(\tau)] = \frac{1}{Z_\lambda} \exp\left[-\frac{1}{2a}\int_{-\infty}^{\infty} d\tau \lambda^2(\tau)\right] \tag{2.234}$$

where $Z_\lambda$ is the partition function of the noise given by the functional integral

$$Z_\lambda = \int D\lambda \exp\left[-\frac{1}{2a}\int_{-\infty}^{\infty} d\tau \lambda^2(\tau)\right] \tag{2.235}$$

and $a = 2\zeta kT$. Here, the integration is carried out over all functions $\lambda(t)$. The average of any quantity $A$ over the realizations of the noise $\lambda$ reads

$$\langle A[\lambda] \rangle = \int D\lambda A[\lambda] F[\lambda] \tag{2.236}$$

The statistical properties of $\lambda(t)$ can be obtained as follows. With the use of the identity

$$\frac{\delta\lambda(\tau)}{\delta\lambda(t)} = \delta(\tau - t) \tag{2.237}$$

one can calculate variations of $F[\lambda(t)]$, namely,

$$\frac{\delta F[\lambda(\tau)]}{\delta\lambda(t)} = -\frac{1}{a}\lambda(t) F[\lambda(\tau)] \tag{2.238}$$

$$\frac{\delta^2 F[\lambda(\tau)]}{\delta\lambda(t)\delta\lambda(t')} = \left[\frac{1}{a^2}\lambda(t)\lambda(t') - \frac{1}{a}\delta(t-t')\right] F[\lambda(\tau)] \tag{2.239}$$

and so on. Next, we notice that the functional integral of each variation vanishes,

$$\int D\lambda \frac{\delta^n F}{\delta\lambda_1(t_1)\delta\lambda_2(t_2)\cdots\delta\lambda_n(t_n)} = 0 \tag{2.240}$$

To see this, one can discretize the time: $\lambda(t) \Rightarrow \lambda(t_n)$, then $D\lambda \Rightarrow \Pi_{n=-\infty}^{\infty} d\lambda(t_n)$. Now the integrals over $\lambda(t_n)$ can be easily evaluated, and the result vanishes since $F$ and all its derivatives vanish at $\lambda = \pm\infty$. Now, using the expressions for the variations above and the definition of the averages with respect to $\lambda$, we have

$$\langle \lambda(t) \rangle = 0 \tag{2.241a}$$

$$\langle \lambda(t)\lambda(t') \rangle = a\delta(t-t') \tag{2.241b}$$

and so on. By using higher variations of $F$, it can be proved that the averages of an odd number of $\lambda$ are zero, whereas the averages of an even number $n > 2$ of $\lambda$ can be expressed through all combinations of the pair averages above. Hence the statistics of the white noise $\lambda$ are Gaussian.

Let us now proceed to the derivation of the Klein–Kramers equation. To this end, we introduce a function

$$f = \delta(x - x(t))\delta(p - p(t)) \tag{2.242}$$

which serves to define the distribution function

$$\rho(x, p, t) = \langle f \rangle \tag{2.243}$$

The averaging is performed with respect to the realizations of $\lambda$, on which the trajectories $\{x(t), p(t)\}$ are dependent. The time derivative of $\rho$ is calculated with the help of the Langevin equation

$$\frac{\partial\rho}{\partial t} = -\left\langle \frac{\partial f}{\partial x}\frac{dx(t)}{dt} \right\rangle - \left\langle \frac{\partial f}{\partial p}\frac{dp(t)}{dt} \right\rangle = -\frac{\partial}{\partial x}\left\langle f\frac{dx(t)}{dt} \right\rangle - \frac{\partial}{\partial p}\left\langle f\frac{dp(t)}{dt} \right\rangle \tag{2.244a}$$

$$= -\frac{p}{m}\frac{\partial\rho}{\partial x} + \frac{\partial U}{\partial x}\frac{\partial\rho}{\partial p} + \frac{\zeta}{m}\frac{\partial}{\partial p}(\rho p) - \frac{\partial}{\partial p}\langle f\lambda \rangle \tag{2.244b}$$

To conclude the derivation, one has to calculate the last term:

$$\langle f\lambda\rangle = \int D\lambda f\lambda F[\lambda] = -a\int D\lambda f\frac{\delta F[\lambda]}{\delta\lambda} = a\int D\lambda\frac{\delta f}{\delta\lambda}F[\lambda] = a\left\langle\frac{\delta f}{\delta\lambda}\right\rangle \quad (2.245a)$$

$$= -a\left\langle\frac{\partial f}{\partial x}\frac{\delta x}{\delta\lambda}\right\rangle - a\left\langle\frac{\partial f}{\partial p}\frac{\delta p}{\delta\lambda}\right\rangle \quad (2.245b)$$

Now, we write a formal solution of the Langevin equation in the form

$$p(t) = p(t_0) + \int_{t_0}^{t} dt'\left[-\frac{\partial U}{\partial x} - \frac{\zeta p(t')}{m} + \lambda(t')\right] \quad (2.246)$$

hence we obtain

$$\frac{\delta p(t)}{\delta\lambda(t')} = \begin{cases} 1, & t' < t \\ 0, & t' > t \end{cases} \quad (2.247)$$

In fact, we are interested in the case $t' = t$, where the result is not defined. However, if we adopt the Stratonovich [5, 25] interpretation of stochastic processes and consider the white noise as a limiting case of colored noise, the equal-time variation is

$$\frac{\delta p(t)}{\delta\lambda(t)} = \frac{1}{2} \quad (2.248)$$

Since, according to the Langevin equation, $\lambda$ does not affect $x$ directly, one obtains

$$\frac{\delta x(t)}{\delta\lambda(t)} = 0 \quad (2.249)$$

This results in

$$\langle f\lambda\rangle = -\frac{a}{2}\frac{\partial\rho}{\partial p} = -\zeta kT\frac{\partial\rho}{\partial p} \quad (2.250)$$

so the final form of the Klein–Kramers equation becomes

$$\frac{\partial \rho}{\partial t} + \frac{p}{m}\frac{\partial \rho}{\partial x} - \frac{\partial U}{\partial x}\frac{\partial \rho}{\partial p} - \zeta \frac{\partial}{\partial p}\left(\frac{p}{m} + kT\frac{\partial}{\partial p}\right)\rho = 0 \tag{2.251}$$

It can be seen that the Maxwell–Boltzmann distribution function

$$\rho \sim \exp\left(-\frac{U + (p^2/(2m))}{kT}\right) \tag{2.252}$$

is the solution of Eq. (2.251) in equilibrium.

### 6. *Derivation of the Smoluchowski Equation*

In the limit of high damping $\zeta$, the solution of the Klein–Kramers equation

$$\frac{\partial \rho}{\partial t} + \frac{p}{m}\frac{\partial \rho}{\partial x} - \frac{\partial U}{\partial x}\frac{\partial \rho}{\partial p} - \zeta \frac{\partial}{\partial p}\left(\frac{p\rho}{m} + kT\frac{\partial \rho}{\partial p}\right) = 0 \tag{2.253}$$

approaches equilibrium as a function of $p$. Indeed, for large $\zeta$ the term in brackets should become small, which implies a nearly Maxwell–Boltzmann distribution over $p$. Thus we seek a solution in the form

$$\rho(p, x, t) = \exp\left[\frac{-p^2}{2mkT}\right] f(x, t) + \delta\rho(p, x, t) \tag{2.254}$$

where $\delta\rho(p, x, t)$ is an *essentially nonequilibrium* small correction that will be shown to be proportional to $1/\zeta$. Our aim is to derive an equation for the distribution function in purely coordinate space

$$\sigma(x, t) = \int_{-\infty}^{\infty} dp\, \rho(p, x, t) \tag{2.255}$$

Using the Klein–Kramers equation, one obtains

$$\frac{\partial \sigma}{\partial t} + \frac{1}{m}\frac{\partial}{\partial x}\int_{-\infty}^{\infty} dp\, p\rho(p, x, t) = 0 \tag{2.256}$$

It can be easily seen that the integral term in this equation is only due to the nonequilibrium correction $\delta\rho(p, x, t)$ in the distribution function. Indeed, dropping $\delta\rho(p, x, t)$ would result in the disappearance of the diffusion current that is given by the integral over $p$ in the equation above and is of order $1/\zeta$. To find $\delta\rho(p, x, t)$, we insert the expression for $\rho$ in the Klein–Kramers equation and keep only the terms of the zero order in $\zeta$

$$\exp\left(-\frac{p^2}{2mkT}\right)\left[\frac{\partial f}{\partial t}+\frac{p}{m}\left(\frac{1}{kT}\frac{\partial U}{\partial x}f+\frac{\partial f}{\partial x}\right)\right]=\zeta\frac{\partial}{\partial p}\left(\frac{p}{m}+kT\frac{\partial}{\partial p}\right)\delta\rho \qquad (2.257)$$

If we integrate this equation over $p$, we obtain $\partial f/\partial t = 0$ at this order, and thus the resulting equation

$$-\frac{m}{\zeta}\frac{\partial}{\partial p}\exp\left(-\frac{p^2}{2mkT}\right)\left(\frac{\partial U}{\partial x}f+kT\frac{\partial f}{\partial x}\right)=\frac{\partial}{\partial p}\left(p+mkT\frac{\partial}{\partial p}\right)\delta\rho \qquad (2.258)$$

can be easily integrated with the result

$$p\delta\rho=-mkT\frac{\partial}{\partial p}\delta\rho-\frac{m}{\zeta}\exp\left(-\frac{p^2}{2mkT}\right)\left(\frac{\partial U}{\partial x}f+kT\frac{\partial f}{\partial x}\right)+C \qquad (2.259)$$

Here, $C$ is an integration constant that is independent of $p$. Since all other terms of this equation vanish at $p = \pm\infty$, we conclude that $C = 0$. By using the form of $\rho$ above and neglecting terms of higher order in $1/\zeta$ one obtains

$$p\delta\rho\cong-mkT\frac{\partial}{\partial p}\delta\rho-\frac{m}{\zeta}\left(\frac{\partial U}{\partial x}\rho+kT\frac{\partial\rho}{\partial x}\right) \qquad (2.260)$$

By inserting Eq. (2.260) into the equation for $\sigma$ one obtains the Smoluchowski equation (Eq. (2.201), $\zeta = m\eta$)

$$\frac{\partial\sigma}{\partial t}-\frac{1}{\zeta}\frac{\partial}{\partial x}\left(\frac{\partial U}{\partial x}\sigma+kT\frac{\partial\sigma}{\partial x}\right)=0 \qquad (2.261)$$

### 7. *Alternative Treatment of the Case of Small Viscosity*

As we have seen, the result for the thermal activation rate $\Gamma$ obtained in the IHD case remains finite in the limit $\zeta = 0$. *This, however, cannot be true since for* $\zeta = 0$ *the coupling between the particle and the heat bath vanishes and the particle cannot therefore attain the energy needed to overcome the barrier.* Thus the approach used to derive the IHD formula breaks down in the low-damping limit. It becomes invalid since the tacit assumption that the distribution function $\rho$ is close to the equilibrium one everywhere apart from the close vicinity of the barrier (or the saddle point in the $p, x$ space) does not hold anymore. In the LD case, particles cannot equilibrate during one period of motion in the well, thus deviations from equilibrium extend into the whole range of $x$. The process that controls the escape rate $\Gamma$ for very low damping is the slow diffusion in the energy space directed from the energy minimum (the bottom of the well) to the energy $E_C$ corresponding to the top of the barrier. It is not important where exactly in the $p, x$ plane the particle reaches the energy $E_C$. Once $E_C$ is reached, the particle crosses the barrier after a much shorter time.

In his paper, Kramers could not find an approach describing both the IHD and LD cases. This was accomplished by Mel'nikov [49] and Mel'nikov and Meshkov [34] later. Instead, Kramers used physical arguments and made the *Ansatz*

$$\rho(p,x,t) \cong \rho(E,t), \qquad E = U(x) + p^2/(2m) \tag{2.262}$$

which means that the distribution function equilibrates along the lines of constant energy. This is the key assumption that helps to solve the problem. Now one can rewrite the Klein–Kramers equation in the energy space using

$$\frac{\partial \rho}{\partial p} = \frac{\partial \rho}{\partial E}\frac{\partial E}{\partial p} = \frac{\partial \rho}{\partial E}\frac{p}{m} \tag{2.263a}$$

$$\frac{\partial \rho}{\partial x} = \frac{\partial \rho}{\partial E}\frac{\partial E}{\partial x} = \frac{\partial \rho}{\partial E}\frac{\partial U}{\partial x} \tag{2.263b}$$

$$\frac{\partial^2 \rho}{\partial p^2} = \frac{\partial^2 \rho}{\partial E^2}\left(\frac{p}{m}\right)^2 + \frac{\partial \rho}{\partial E}\frac{1}{m} \tag{2.263c}$$

The conservative part of the Klein–Kramers equation now vanishes

$$\frac{p}{m}\frac{\partial \rho}{\partial x} - \frac{\partial U}{\partial x}\frac{\partial \rho}{\partial p} = 0 \tag{2.264}$$

and this equation takes on the form

$$\frac{\partial \rho}{\partial t} - \eta\left(Q + \frac{p^2}{m}\frac{\partial Q}{\partial E}\right) = 0 \tag{2.265}$$

where

$$\eta \equiv \frac{\zeta}{m}$$

and

$$Q \equiv \rho + kT\frac{\partial\rho}{\partial E} = kTe^{-\beta E}\frac{d}{dE}(\rho e^{\beta E})$$

Here, we should make an important comment. The corrections to $\rho$ that have not been taken into account and that should explicitly depend on $p$ and $x$ may contribute to the conservative part of the Klein–Kramers equation. This contribution may be comparable to the remaining dissipative terms containing the small factor $\zeta$. It is much more difficult to investigate these correction terms, and without it the whole approach is not justified.

Since the *energy diffusion is much slower* than the *conservative motion of the particle in the well*, the coefficient $p^2$ in the equation above should be averaged over the period of the motion, which leads to a closed equation for $\rho(E)$. Explicitly, we use

$$\frac{p^2}{m} \Rightarrow \left\langle \frac{p^2}{m} \right\rangle = \frac{\omega}{2\pi}\int_0^{2\pi/\omega}\frac{p^2}{m}dt = \frac{\omega}{2\pi}\oint pdx = \frac{\omega}{2\pi}I(E) \tag{2.266}$$

where $\omega = \omega(E)$ is the frequency of the almost periodic motion of the particle with energy $E$; and $I(E)$ is the action over a period of the motion that satisfies

$$\frac{\partial I}{\partial E} = \frac{2\pi}{\omega} \tag{2.267}$$

Before trying to solve the equation for $\rho(E)$, we evaluate the expression for the current of particles across a line of constant energy $E$.

$$J = \oint (J_p dx - J_x dp) \tag{2.268}$$

where the integral is taken in the clockwise direction, that is, in the direction of motion of the particle. Parametrizing this integral by time $t$ and using

$$J_x = \frac{p\rho}{m} \tag{2.269a}$$

$$J_p = -\frac{\partial U}{\partial x}\rho - \eta\left(p\rho + mkT\frac{\partial\rho}{\partial p}\right) = -\frac{\partial U}{\partial x}\rho - \eta Q \tag{2.269b}$$

together with the equation of motion one obtains

$$J = \int_0^{2\pi/\omega} dt \left( J_p \frac{dx}{dt} - J_x \frac{dp}{dt} \right) \tag{2.270a}$$

$$= -\eta Q \int_0^{2\pi/\omega} dt \frac{p^2}{m} \tag{2.270b}$$

$$= -\eta kTI(E) e^{-E/kT} \frac{d}{dE} (\rho e^{E/kT}) \tag{2.270c}$$

In equilibrium $\rho \sim e^{-E/kT}$, thus $\rho e^{E/kT} = \text{const}$ and $J = 0$. If the density of particles at higher energies is less than the equilibrium value, then $\rho e^{E/T}$ decreases with $E$ and the flow of particles is directed to the higher energies (i.e., $J > 0$).

Now one can see that the equation for $\rho$ above may be rewritten as

$$\frac{\partial \tilde{\rho}}{\partial t} + \frac{\partial J}{\partial E} = 0 \tag{2.271a}$$

where

$$\tilde{\rho} \equiv \frac{\partial I}{\partial E} \rho = \frac{2\pi\rho}{\omega(E)} \tag{2.271b}$$

where $\tilde{\rho} dE$ is the number of particles between the lines $E$ and $E + dE$.

Let us now proceed to the solution of Eqs. (2.271). As we shall see, the relevant region, as far as the thermal activation process is concerned, is that near the energy $E_C$ corresponding to the top of the barrier. In this region, $\rho$ is exponentially small at low temperatures. Since the process is exponentially slow, the derivative $\partial\rho/\partial t$ is doubly exponentially small in the relevant region and can be neglected. Then, the first integral of the Klein–Kramers equation reads

$$J = \text{const} \tag{2.272}$$

By using the definition of $J$ and integrating once more between $E_c$ and an energy $E_x$ sufficiently below the barrier, where the distribution is close to the equilibrium, one obtains

$$J \int_{E_x}^{E_C} \frac{dE}{I(E)} \exp\left(\frac{E}{kT}\right) = -\eta kT \left[\rho(E_C) \exp\left(\frac{E_C}{kT}\right) - \rho(E_x) \exp\left(\frac{E_x}{kT}\right)\right] \tag{2.273}$$

The integral on the left-hand side of this equation is determined by the narrow region $E_C - E \sim kT$, thus one can set $I(E) \cong I(E_C)$ and extend the integration to $-\infty$, which results in

$$JkT \exp \frac{E_C/kT}{I(E_C)}$$

On the right-hand side, the first term disappears, since all the particles, which have reached the top of the barrier, will cross it and never return. In the second term, one can use the equilibrium solution

$$\rho(E) = \frac{N\omega_A}{2\pi kT} \exp\left(-\frac{E}{kT}\right) \tag{2.274}$$

where $N$ is the number of particles in the well and $\omega_A$ is the frequency of small oscillations near the bottom of the well (the attempt frequency). Noticing that $J = -dN/dt$, one arrives at the equation

$$\frac{dN}{dt} = -\Gamma N \tag{2.275}$$

in which (cf. Eq. 2.190)

$$\Gamma = \frac{\omega_A}{2\pi} A \exp\left(-\frac{E_C}{kT}\right) \tag{2.276}$$

where

$$A = \frac{\eta I(E_C)}{kT}$$

Note that the quantity $\eta I(E_C)$ is the energy dissipated during one cycle of the motion with the energy $E_C$. The applicability condition of this LD result is clearly $\eta I(E_C) \ll kT$. In the opposite limit, this result becomes invalid, and the previously studied ID regime characterized by $A = 1$ is realized. The crossover between the two regimes, which occurs at $\eta I(E_C) \sim kT$, is best evaluated using Mel'nikov's [34, 49] approach to the bridging of low damping and intermediate to high damping rates in the frictional region where none of the Kramers formulas are valid (see Section VI).

## III. APPLICATION OF THE KRAMERS THEORY TO ROTATIONAL BROWNIAN MOTION

### A. Magnetic Relaxation of Single-Domain Ferromagnetic Particles

A sample of ferromagnetic material consists of several small domains in each of which the magnetization vector is pointing in a fixed direction. In the unmagnetized state, the various vectors will orient themselves more or less at random in such a way that the overall magnetization is zero (i.e., they cancel each other out). On the application of a magnetic field, however, the magnetization vectors in some of the domains align themselves with the field, and as the field is increased, more and more of these domains align themselves until saturation occurs when all the domains are aligned with the field. An increase in the field at this stage produces no further increase in the magnetization of the sample.

If a piece of magnetic material is small enough (roughly of the order of 100 Å) the sample will consist of a *single one* of these domains. Its magnetic vector will *normally be fixed in magnitude and direction*. If these samples are suspended in a fluid away from any magnetic field again, these samples will orient themselves in such a way as to cancel out the overall magnetization of the system (fluid plus suspended particles). As before, on the application of a field, more and more of these particles will orient themselves with the field until saturation occurs.

If, on the other hand, such particles are fixed in a solid, the particles *will not be able to reorient themselves*. However, Néel, around 1950, conjectured [37] using TST that, if a particle were sufficiently small that, under the influence of thermal agitation, the *magnetization vector itself could perform a type of 'magnetic' Brownian rotation*. This means, as we shall see below, that the characteristic *stability of a ferromagnet would be destroyed, and if the magnetic particles were all of this size, the magnetization of the system would revert to zero*. The main quantity of interest is thus the greatest relaxation time or inverse escape rate.

Another characteristic of magnetic particles is that of anisotropy. This means that there are preferred directions for magnetizing one of them. This anisotropy is due to several causes; among them being the crystalline structure of the material itself and the shape of the particles. If we take, for example, a particle shaped like an ellipsoid of revolution produced by rotating an ellipse about its major axis, it will be easier to magnetize this particle in the direction of this (major) axis than in any other. The theory of the Brownian motion may now be applied to this type of relaxation.

The anisotropy just mentioned means that magnetization in the direction of easiest magnetization possesses a stability characteristic of ferromagnetism, which must be overcome if it is wished to change the direction of this magnetization. This ferromagnetic stability may be accounted for mathematically by attributing to the particle a potential function per unit volume, $U$, which is the magnetocrystalline and field energy density. The simplest example of such a potential function is that of uniaxial symmetry (anisotropy), which is a very common type of magnetocrystalline anisotropy and may be written

$$U = K \sin^2\theta \tag{3.1}$$

where $K$ is a constant known as the anisotropy constant and $\theta$ is the usual polar angle of spherical polar coordinates measured from the easy axis $z$. The potential energy of a particle of volume $v$ is then

$$vU = vK \sin^2\theta \tag{3.2}$$

We also frequently use the anisotropy parameter (note that the *size* of the single domain particle is crucial in the calculation of the reversal time as the size appears, due to $Kv$, in the argument of the exponential in the TST theory so that the relaxation time may vary from *picoseconds* to *millions of years*)

$$\sigma \equiv \frac{vK}{kT} = \beta K \tag{3.2a}$$

(in fine particle magnetism $\beta$ is often written for $v/kT$ instead of $1/kT$).

In order to apply the theory of the Brownian motion to this type of magnetic relaxation (also called Néel relaxation) we start, not with Newton's second law, as in the case of the escape of particles previously considered, but with Gilbert's equation, which is the Larmor [50] equation modified by the addition of stochastic terms corresponding to the irregular force (i.e., in this case, a random magnetic field [21]) in Kramers' problem.

Gilbert's equation incorporating stochastic terms is

$$\frac{d\mathbf{M}}{dt} = \gamma\mathbf{M} \times \left(-\frac{\partial U}{\partial \mathbf{M}} - \eta\frac{d\mathbf{M}}{dt} + \boldsymbol{\Xi}\right) \tag{3.3a}$$

or ($\mathbf{n}$ is a unit vector in the direction of $\mathbf{M}$)

$$\frac{d\mathbf{n}}{dt} = \boldsymbol{\omega} \times \mathbf{n} \tag{3.3b}$$

$$\boldsymbol{\omega} = \gamma\left(\frac{\partial U}{\partial \mathbf{M}} + \eta\frac{d\mathbf{M}}{dt} - \boldsymbol{\Xi}\right) \tag{3.3c}$$

where

$\gamma$ is the gyromagnetic ratio.

$\eta$ is a damping parameter, characterizing the coupling to the heat bath.

$\boldsymbol{\Xi}$ is the white noise random magnetic field due to the surroundings.

$\mathbf{M}$ is the magnetization vector ($\mathbf{m} = \mathbf{M}v$, where $\mathbf{m}$ is the magnetic moment).

(Alternatively, we may use the Landau–Lifshitz equation [5]). Note that $-\eta(d\mathbf{M}/dt) + \boldsymbol{\Xi}$ is the irregular magnetic field due to the bath coordinates corresponding to Kramers' irregular or Brownian $X(t)$ in his mechanical problem. The random field, $\boldsymbol{\Xi}$, satisfies the equations

$$\overline{\Xi_i(t)} = 0$$

$$\overline{\Xi_i(t)\Xi_j(t')} = \frac{2\eta kT}{v}\delta_{ij}\delta(t-t')$$

where the overbar denotes taking the average over the realizations of the random (white noise) field $\Xi$.

If we take the vector product of Eq. (3.3) with $\mathbf{M}$ and use the triple vector product formula:

$$\mathbf{A} \times (\mathbf{B} \times \mathbf{C}) = \mathbf{B}(\mathbf{A}\cdot\mathbf{C}) - \mathbf{C}(\mathbf{A}\cdot\mathbf{B}) \tag{3.4}$$

we obtain an equation for $d\mathbf{M}/dt$:

$$\frac{d\mathbf{M}}{dt} = g'M_s\mathbf{M} \times (\mathbf{H} + \boldsymbol{\Xi}) + h'[\mathbf{M} \times (\mathbf{H} + \boldsymbol{\Xi})] \times \mathbf{M} \tag{3.5}$$

where $\gamma$ is modified by the factor $(1 + a^2)^{-1}$ so that

$$g' = \frac{\gamma}{(1 + a^2)M_s} \tag{3.6}$$

and

$$h' = ag' \tag{3.7}$$

measures the effect of the aligning term,

$$a = \eta\gamma M_s \tag{3.8}$$

is a dimensionless dissipation parameter

$$M_s = |\mathbf{M}| \tag{3.9}$$

$$\mathbf{H} = -\frac{\partial U}{\partial \mathbf{M}} \tag{3.10}$$

is the conservative part of the field. It is apparent from Eq. (3.5) that the leading term on the right-hand side of that equation is the Larmor equation describing the precession of **M**, however, the gyromagnetic ratio is modified by the dimensionless damping factor, $a$. The second term is an *alignment* term. Thus, in Kramers' terminology, the irregular Brownian field manifests itself in a *modification of the gyromagnetic term and in the emergence of an alignment term.*

The Gilbert equation, written in the transposed form of Eq. (3.5), has the same mathematical form [5, 39] as the earlier Landau–Lifshitz equation (if $a^2 \ll 1$ so that $g' \simeq \gamma/M_s$):

$$\frac{d\mathbf{M}}{dt} = \gamma\mathbf{M} \times (\mathbf{H} + \boldsymbol{\Xi}) + \frac{\alpha\gamma}{M_s}[\mathbf{M} \times (\mathbf{H} + \boldsymbol{\Xi})] \times \mathbf{M} \tag{3.11}$$

which has also been extensively used to treat magnetic relaxation. (We shall use both equations interchangeably throughout the text.)

Equation (3.5) will now be taken as the Langevin equation of the process. Just as in translational motion, the FPE will take on the form of a continuity equation for the current of representative points on the surface of a sphere of constant radius $M_s$. Thus the probability density of the orientations of the moments, which, in this instance, we term $W$, obeys the continuity equation

$$\frac{\partial W}{\partial t} + \nabla \cdot \mathbf{J} = 0 \tag{3.12}$$

where the components of the current density of magnetic moments, **J**, are

$$J_\theta = -\left[\left(h'\frac{\partial U}{\partial \theta} - \frac{g'}{\sin\theta}\frac{\partial U}{\partial \varphi}\right)W + k'\frac{\partial W}{\partial \theta}\right] \tag{3.13}$$

$$J_\varphi = -\left[\left(g'\frac{\partial U}{\partial \theta} + \frac{h'}{\sin\theta}\frac{\partial U}{\partial \varphi}\right)W + \frac{k'}{\sin\theta}\frac{\partial W}{\partial \varphi}\right] \tag{3.14}$$

These follow from Gilbert's equation in the absence of noise terms:

$$\frac{d\mathbf{M}}{dt} = g' M_s \mathbf{M} \times \mathbf{H} + h'[\mathbf{M} \times \mathbf{H}] \times \mathbf{M} \tag{3.15}$$

The noise term being taken account of by the addition of the diffusion term in $k'$ as in II A. After some manipulation (details in [5, 39]) we obtain the Fokker–Planck equation for the density of magnetic moment orientations on the surface of the sphere of radius $M_s$:

$$\begin{aligned}\frac{\partial W}{\partial t} = & \frac{1}{\sin\theta}\frac{\partial}{\partial\theta}\left\{\sin\theta\left[\left(h'\frac{\partial U}{\partial\theta} - \frac{g'}{\sin\theta}\frac{\partial U}{\partial\varphi}\right)W + k'\frac{\partial W}{\partial\theta}\right]\right\} \\ & + \frac{1}{\sin\theta}\frac{\partial}{\partial\varphi}\left[\left(g'\frac{\partial U}{\partial\theta} + \frac{h'}{\sin\theta}\frac{\partial U}{\partial\varphi}\right)W + \frac{k'}{\sin\theta}\frac{\partial W}{\partial\varphi}\right]\end{aligned} \tag{3.16}$$

where the diffusion coefficient is

$$k' = h'\frac{kT}{v} = \frac{1}{2\tau_N} = \frac{kT}{v}\frac{a}{1+a^2}\frac{\gamma}{M_s} \tag{3.17}$$

where $\tau_N$ is the free diffusion time.

We have indicated how Eq. (3.16) may be derived from the Langevin equation using the intuitive method of Einstein, which is to take account of the noise by adding a diffusion current to the drift current terms. A complete derivation of this equation starting from the Chapman–Kolmogorov equation for a Markov process in configuration space $(\theta, \varphi)$ is given in Coffey et al. [39]. This is essentially similar to the method presented in Kramers' paper (see Section II) with the exception that multiplicative noises are involved. The time to escape the barrier (also known as the Néel relaxation time) is the inverse of the Kramers escape rate and is also given for high barriers by

$$\tau \approx \frac{2\tau_N}{\lambda_1} \tag{3.18}$$

where $\lambda_1$, is the smallest nonvanishing eigenvalue of the FPE, Eq. (3.16).

We have assumed here that the relaxation time may be written in the form

$$\tau = 2\tau_N \sum_{i=1}^{\infty} A_i \lambda_i^{-1} \tag{3.19}$$

where

$\lambda_i$ are the eigenvalues of Eq. (3.16) labeled in ascending order.
$\lambda_1 < \lambda_2 < \cdots$.
$A_1 = k'$.
$A_i$ are constants for $i \geq 2$.

In principle, the calculation of $\lambda_1$ is usually accomplished by expressing the set of differential recurrence relations (to which the FPE may be converted by expansion in spherical harmonics) as a matrix continued fraction [5, 57], and then calculating successive convergents of the continued fraction. This procedure as applied to magnetic relaxation is exemplified by the work of Kalmykov et al. [51], where the method is applied to a cubic anisotropy potential function of the form

$$U = K(y^2z^2 + z^2x^2 + x^2y^2) \tag{3.20}$$

The matrix continued fraction method is more efficient than the straight matrix inversion of the system of recurrence relations expressed as

$$\dot{\mathbf{X}} = \mathbf{AX} \tag{3.21}$$

and truncated at a size of $\mathbf{A}$ sufficiently large to ensure convergence. This method has been used by Coffey et al. [52] to treat the problem of magnetic relaxation in a uniform field applied at an arbitrary angle to the easy axis of magnetization for the purpose of verifying the formulas for the Kramers escape rate for this problem.

## B. Dielectric Relaxation

An older application of the theory of rotational Brownian motion than that of magnetic relaxation of single domain particles is the theory of dielectric relaxation originating in the work of Debye, which began in 1913 (see [18, 19]). He considered an assembly of noninteracting rigid rotators each having a permanent dipole moment $\boldsymbol{\mu}$ where it is supposed that the inertia of the dipoles may be neglected. The dipoles are subjected both to Brownian torques arising from their heat bath and to a high-frequency alternating field (in the microwave frequency band). He then considered the mean orientation of a dipole in the presence of a field. The theory may be summarized by remarking that the angular velocity $\boldsymbol{\Omega}(t)$ of a dipole which is, of course, a random variable, satisfies the kinematic relation

$$\frac{d\boldsymbol{\mu}}{dt} = \boldsymbol{\Omega} \times \boldsymbol{\mu} \tag{3.22}$$

where $\boldsymbol{\mu}$ is the dipole moment of the molecule. In general, $\boldsymbol{\Omega}$ will be determined from the Euler equations augmented by a frictional torque proportional to the angular velocity vector and the white noise torque about each of the three principal axes of the molecule due to the Brownian motion; both torques, as usual, arising from the background or heat bath. For simplicity, if we consider a spherical molecule and we suppose that it is also subjected to an external torque arising from the dipole–dipole coupling, say, or anisotropy potential as is so in a nematic liquid crystal, then the Langevin equation will be (Debye makes the assumption that in spite of their small size the molecules may be treated as Brownian rotators)

$$I\dot{\boldsymbol{\Omega}} + \zeta\boldsymbol{\Omega} = \boldsymbol{\lambda}(t) + \boldsymbol{\mu}(t) \times \mathbf{F}(t) \tag{3.23}$$

where

$I$ is the moment of inertia of the molecule,
$I\dot{\boldsymbol{\Omega}}$ is the inertial torque,
$\zeta\boldsymbol{\Omega}$ is the damping torque due to the Brownian motion,
$\boldsymbol{\lambda}(t)$ is the white noise driving torque also due to the Brownian motion,

so that $\boldsymbol{\lambda}(t)$ has the following properties:

$$\overline{\lambda_i(t)} = 0 \tag{3.24}$$

$$\overline{\lambda_i(t)\lambda_j(t')} = 2kT\zeta\delta_{ij}\delta(t - t') \tag{3.25}$$

where

$\delta_{ij}$ is the Kronecker delta (so that torques about different axes are uncorrelated).
$i, j = 1, 2, 3$, which correspond to the Cartesian axes, $x, y, z$ fixed in the molecule.
$\delta(t - t')$ is the Dirac delta function.

The term $\boldsymbol{\mu} \times \mathbf{F}(t)$ in Eq. (3.23) is the torque due to an externally applied field $\mathbf{F}(t)$, and the overbar in Eqs. (3.24) and (3.25) denote as usual the statistical average over the realizations of $\boldsymbol{\lambda}(t)$.

Equation (3.23) includes the inertia of the molecule. The noninertial response is the response when $I \to 0$ or when $\zeta$, the friction coefficient becomes very large. In this limit, the angular velocity vector may immediately be written down from Eq. (3.23) as

$$\boldsymbol{\Omega}(t) = \frac{\boldsymbol{\lambda}(t)}{\zeta} + \frac{\boldsymbol{\mu} \times \mathbf{F}(t)}{\zeta} \tag{3.26}$$

This, combined with the kinematic relation

$$\frac{d\boldsymbol{\mu}(t)}{dt} = \boldsymbol{\Omega}(t) \times \boldsymbol{\mu}(t) \tag{3.27}$$

yields

$$\frac{d\boldsymbol{\mu}(t)}{dt} = \frac{\boldsymbol{\lambda}(t)}{\zeta} \times \boldsymbol{\mu}(t) + \frac{\boldsymbol{\mu} \times \mathbf{F}(t)}{\zeta} \times \boldsymbol{\mu}(t) \tag{3.28}$$

which, with the properties of the triple vector product, becomes

$$\frac{d\boldsymbol{\mu}(t)}{dt} = \frac{\boldsymbol{\lambda}(t)}{\zeta} \times \boldsymbol{\mu}(t) + \frac{\mu^2 \mathbf{F}(t)}{\zeta} - \frac{\boldsymbol{\mu}[\boldsymbol{\mu} \cdot \mathbf{F}(t)]}{\zeta} \tag{3.29}$$

This is the Langevin equation for the motion of $\boldsymbol{\mu}$ in the noninertial limit.

The torque Eq. (3.29) is similar to Eq. (3.5) with the exception that it lacks the gyromagnetic term characteristic of a magnet. Thus calculations for the magnetic case may be carried over into the dielectric case in the noninertial limit, provided that the Néel (free diffusion) relaxation time

$$\tau_{\mathrm{N}} = \frac{v}{2\eta kT}\left(\frac{1}{\gamma^2} + \eta^2 M_s^2\right) = \frac{vM_s}{2\gamma kT}\frac{(1+a^2)}{a} = \frac{v}{2kTh'}$$

is replaced by the Debye relaxation time $\tau_{\mathrm{D}} = \zeta/2kT$. We remark, however, that due care must be taken in transposing the results of one calculation to another, as the physics is entirely different. It is necessary to recall that the Debye problem of relaxation including inertia, in a potential exemplified, for example, by a nematic liquid crystal of the type considered by Maier and Saupe (Coffey and Kalmykov and Urban et al. [36, 83]) is the exact rotational analogue of the Kramers translational problem of Section II. Thus the noninertial limit of the rotational problem is governed by the (approximate) Smoluchowski equation and in the inertial case the three regimes of Kramers' particle problem will appear (IHD, VLD, and the frictional crossover region). The inertial case of the Debye problem, in the absence of a dipole–dipole or anisotropy potential (so that the only potential is that due to the spatially uniform but time varying electric field) has been exhaustively considered by Sack [41] and McConnell [53].

The original equation for diffusion of polar molecules due to Debye [19] which is a *true* Smoluchowski equation may be obtained from Eq. (3.16) by setting $g' = 0$

$$\frac{\zeta}{kT}\frac{\partial W}{\partial t} = \frac{1}{\sin\theta}\frac{\partial}{\partial\theta}\left(\sin\theta\frac{\partial W}{\partial\theta}\right) + \frac{1}{\sin^2\theta}\frac{\partial^2 W}{\partial\varphi^2} + \frac{1}{kT}\left[\frac{1}{\sin\theta}\frac{\partial}{\partial\theta}\left(\sin\theta W\frac{\partial U}{\partial\theta}\right) + \frac{1}{\sin^2\theta}\frac{\partial}{\partial\varphi}\left(W\frac{\partial U}{\partial\varphi}\right)\right] \tag{3.30}$$

or in the case of axial symmetry

$$\frac{\partial W}{\partial t} = \frac{1}{\sin\theta}\frac{\partial}{\partial\theta}\left[\sin\theta\left(\frac{kT}{\zeta}\frac{\partial W}{\partial\theta} + \frac{W}{\zeta}\sin\theta\frac{\partial U}{\partial\theta}\right)\right] \tag{3.31}$$

The magnetic result for the greatest relaxation time in the potential

$$U = \widetilde{K}\sin^2\theta, \quad \widetilde{K} = Kv \tag{3.32}$$

may now be directly applied to give the result for the longitudinal relaxation factor, $g$, of a nematic liquid crystal with uniaxial anisotropy. We have (here $\sigma = \widetilde{K}/kT$ so that the relaxation time does not sensitively depend on the volume as in magnetic particles)

$$\frac{1}{\lambda_1} \approx g = \frac{\tau}{2\tau_D} = \frac{\sqrt{\pi}}{4}\sigma^{-3/2}e^{\sigma} = \frac{1}{2\tau_D\Gamma}, \quad \sigma \gg 1 \tag{3.33}$$

This has been extensively discussed by Coffey and Kalmykov [36]. The Debye theory also applies [39] to magnetic relaxation of single-domain particles suspended in a fluid (ferrofluids) where it is assumed that only the *mechanical* reorientational motion of the suspended particles is of importance, that is, the "magnetic" (solid state) relaxation mechanism inside the particles or Néel relaxation is ignored. We reiterate that even though the Néel relaxation for non-axially symmetric potentials exhibits the character of inertial relaxation, the inertia itself plays no role, since the mechanism is *magnetic*; the role of inertia being mimicked by the Larmor term. In the ferrofluid problem, on the other hand, all results of the Debye theory and the inertia-corrected Debye theory of dielectric relaxation may be directly applied, since an *inertial mechanism* is the root of the relaxation process. The Debye theory should also apply accurately to ferrofluid particles since they are true (C 150 Å) Brownian particles.

### C. Application of the Kramers' Method to Axially Symmetric Potentials of the Magnetocrystalline Anisotropy

The solution for the escape rate for an arbitrary potential of the magnetocrystalline anisotropy will be given later using a method due to Langer, Section IV [61]. It is instructive, however, to give the solution for the particular case of axially symmetric potentials, as it both illustrates the application of Kramers' theory to the magnetic problem and also, the escape rate has the interesting particular property that it is valid for all values of the damping parameter, $a$, unlike the mechanical problem treated in Section II. This is a consequence of the fact that in magnetic relaxation for axial symmetry, the FPE is always effectively a one-space-variable equation. In Kramers' mechanical problem, on the other hand, the underlying equation, namely, the Klein–Kramers equation is always an equation in a 2 D state space and can only be converted to a 1 D equation in the limiting cases VLD, IHD, and VHD (where it is to be noted that the IHD case is only 1 D by virtue of the introduction of the variable $u = p - ax'$). In magnetic relaxation, the three friction regimes of Kramers' problem, namely, VLD, the crossover region, and IHD (VHD is just a limiting case of IHD) will only appear when nonaxially symmetric potentials are involved.

In the case of axial symmetry, we have

$$\frac{\partial J_\varphi}{\partial \varphi} = \frac{\partial U}{\partial \varphi} = 0 \tag{3.34}$$

and in the stationary case also $\partial W/\partial t = 0$. Hence, referring to Eq. (3.16),

$$k' \sin\theta \left[\frac{\partial W}{\partial \theta} + \frac{h'}{k'}\frac{\partial U}{\partial \theta} W\right]$$

$$= k' e^{-\beta U} \sin\theta \frac{\partial}{\partial \theta}[e^{\beta U} W] = \text{const} = k'j, \text{ say} \tag{3.35}$$

and so

$$\frac{\partial}{\partial \theta}[e^{\beta U} W] = \frac{j e^{\beta U}}{\sin\theta} \tag{3.36}$$

$$e^{\beta U} W = j \int \frac{e^{\beta U} d\theta}{\sin\theta} + C \tag{3.37}$$

where, in this instance, $C$ is a constant of integration. Suppose $W$ vanishes at the boundary ($E = E_C$) so that $W(E = E_C) = 0$, that is, particles which cross the barrier are no longer counted as is usual in FPT calculations [25, 26].

This forces $C = 0$, and so

$$j = \frac{e^{\beta U} W}{\int (e^{\beta U} d\theta / \sin\theta)} \tag{3.38}$$

or

$$W = j e^{-\beta U} \int \frac{e^{\beta U} d\theta}{\sin\theta} \tag{3.39}$$

The number of particles in the well is

$$N_0 = \int_0^{\theta_m} W \sin\theta d\theta \tag{3.40}$$

$$= j \int_0^{\theta_m} e^{-\beta U} \sin\theta \int^{\theta} \frac{e^{\beta U(\theta')} d\theta'}{\sin\theta'} d\theta \tag{3.41}$$

so that the MFPT

$$\tau = \frac{N_0}{j} \tag{3.42}$$

$$= \int_0^{\theta_m} e^{-\beta U} \sin\theta \int^{\theta} \frac{e^{\beta U(\theta')} d\theta'}{\sin\theta'} d\theta \tag{3.43}$$

This is the time to reach the top of the well provided all particles that reach the top of the well are absorbed, which is the condition that $W$ vanishes at $\theta = \theta_m$. In practice, a particle has a 50:50 chance of crossing, which means that the escape time $\tau_e$ is given by

$$\tau_e = 2\tau \tag{3.44}$$

(This implies, supposing that the integrals are taken at the high barrier limit, that $\Gamma = \tau_e^{-1} = 1/2\tau$.) These integrals may be evaluated, using the method of steepest descents [5, 54, 68] to get (on interchange of the order of integration) for the time to go from the well at $\theta = 0$, to the top of the barrier at $\theta = \theta_m$,

$$\frac{\tau(0)}{2\tau_N} = \int_0^{\theta_m} \frac{e^{\beta U(\theta)} d\theta}{\sin\theta} \int_0^{\theta} e^{-\beta U(\theta')} \sin\theta' d\theta' \tag{3.45}$$

Employing Kramers' argument [25, 26, 68] in the manner of Klein [55], the integral is now evaluated in the limit of very high potential barriers. Since almost all the particles are situated near the minimum at $\theta = 0$ then $\theta'$ is a very small angle. The inner integral in Eq. (3.45) (effectively the number of moments in the well) may then be approximated by

$$\int_0^{\theta} e^{-\beta U(\theta')} \sin\theta' d\theta' \approx \int_0^{\infty} \theta' e^{-\beta[U(0)+(U''(0)/2)\theta'^2]} d\theta' \tag{3.46}$$

The integral on the right has been extended to infinity without significant error since almost all the particles are at the origin. Because $\theta'$ is very small, the Taylor series in $U(\theta')$ can be approximated by it's first two nonvanishing terms $[U'(0) = 0]$. This is effectively a steepest descents argument [54, 56, 68]. Hence, the integral on the right-hand side in Eq. (3.46) becomes:

$$e^{-\beta U(0)} \int_0^{\infty} \theta' e^{-\beta(U''(0)/2)\theta'^2} d\theta' = \frac{e^{-\beta U(0)}}{\beta U''(0)} \tag{3.47}$$

and thus the two integrals in Eq. (3.45) effectively [26] decouple from each other.

We now evaluate the outer integral in Eq. (3.45) following the reasoning of Klein [55] for the translational double well potential problem. We have near $\theta_m$

$$U(\theta) = U(\theta_m) - \frac{|U''(\theta_m)|}{2}(\theta - \theta_m)^2 \tag{3.48}$$

and hence for the outer integral (pertaining to the barrier crossing moments)

$$\int_0^{\theta_m} \frac{e^{\beta U(\theta)} d\theta}{\sin\theta} \approx \frac{e^{\beta U(\theta_m)}}{\sin\theta_m} \int_0^{\infty} e^{\beta|U''(\theta_m)|/2(\theta-\theta_m)^2} d\theta \tag{3.49}$$

The range of integration in Eq. (3.49) may be extended to infinity since the integral has it's main contribution from values $\theta_m - \varepsilon$ to $\theta_m$ and almost no contribution from outside these values.

Now,

$$\frac{1}{\sigma\sqrt{2\pi}}\int_{0}^{\infty} e^{-(x-\mu)^2/2\sigma^2}\,dx = \frac{1}{2} \tag{3.50}$$

and with $\mu = \theta_m$ and

$$\frac{1}{\sigma^2} = \beta|U''(\theta_m)| \tag{3.51}$$

we have

$$\int_0^{\theta_m} \frac{e^{\beta U(\theta)}d\theta}{\sin\theta} \approx \frac{\sqrt{2}}{\sqrt{|\beta U''(\theta_m)|}}\frac{e^{-\beta U(\theta_m)}}{\sin\theta_m} \tag{3.52}$$

Hence, in the *high barrier limit* the MFPT $\tau(0)$ for transitions from the point domain ($\theta = 0$) is

$$\tau(0) \approx 2\tau_N \frac{1}{2}\frac{1}{\beta U''(0)}\frac{\sqrt{2\pi}}{\sqrt{|\beta U''(\theta_m)|}}\frac{e^{\beta[U(\theta_m)-U(0)]}}{\sin\theta_m} \tag{3.53}$$

and so,

$$\frac{\tau(0)}{2\tau_N} \approx \frac{1}{2}\frac{1}{\beta U''(0)}\frac{\sqrt{2\pi}}{\sqrt{|\beta U''(\theta_m)|}}\frac{e^{\beta[U(\theta_m)-U(0)]}}{\sin\theta_m} \tag{3.54}$$

and the time to go from the minimum at $\theta = \pi$ to $\theta_m$ is

$$\frac{\tau(\pi)}{2\tau_N} \approx \frac{1}{2}\frac{1}{\beta U''(\pi)}\frac{\sqrt{2\pi}}{\sqrt{|\beta U''(\theta_m)|}}\frac{e^{\beta[U(\theta_m)-U(\pi)]}}{\sin\theta_m} \tag{3.55}$$

These are the times to reach the barrier. The escape time is

$$\tau_{1,2} = \frac{1}{\Gamma_{1,2}} = 2\tau(0) \tag{3.56}$$

and similarly for $\tau(\pi)$

$$\tau_{2,1} = \frac{1}{\Gamma_{2,1}} = 2\tau(\pi) \tag{3.57}$$

and for the potential given by

$$\varepsilon = \beta U = \sigma(\sin^2\theta - 2h\cos\theta)$$

we have [59] (details in Coffey [68]):

$$2\tau_N\Gamma = \frac{2}{\sqrt{\pi}}\sigma^{3/2}(1-h^2)\{(1+h)\exp[-\sigma(1+h)^2] + (1-h)\exp[-\sigma(1-h)^2]\} \simeq \frac{1}{\lambda_1}, \qquad \sigma(1+h)^2 \gg 1, \qquad \sigma(1-h)^2 \gg 1 \tag{3.58a}$$

which in the limit of a small reduced field, $h$, $(0 \leq h < 1)$ reduces to

$$2\tau_N\Gamma = \frac{4}{\sqrt{\pi}}\sigma^{3/2}e^{-\sigma} \simeq \frac{1}{\lambda_1}, \tag{3.58b}$$

or (cf. Eq. 1.18a)

$$\Gamma = \frac{2\pi a}{1+a^2}\sqrt{\frac{\sigma}{\pi}\frac{\omega_A}{\pi}}e^{-\sigma} \tag{3.58c}$$

The reduced field, $h$, is defined as

$$h \equiv \frac{\xi}{2\sigma} = \frac{vM_sH}{2kT}$$

which pertains to a uniform field **H** applied along the polar axis, which, in turn, is the easy axis of magnetisation $\omega_A$, the attempt frequency, is the ferromagnetic resonance frequency.

Equation (3.58c) is Brown's 1963 formula [21] for the escape rate for the simplest uniaxial potential of the magnetocrystalline anisotropy namely $vK\sin^2\theta$. The steepest descents manipulations of the integrals are described by Coffey [68]. Details are not given here to avoid repetition. Note that in this reference that Eqs. (92–93) are in error by a factor of one-half, if they are regarded as the steepest descent approximations to Eq. (3.45) [Eq. (84) of Coffey] for escapes from the well at 0 and the corresponding equation for escape from the other well. This is of no consequence, however, in the numerical calculations that follow in that paper as the relaxation time for the purpose of numerical calculations was written $\tau_N/\lambda_1$ rather than the more usual $2\tau_N/\lambda_1$ so that the error cancels out.

The calculation we have just given, although presented in the manner of a first passage time using the forward Kolmogorov equation, sufficiently illustrates the application of the Kramers flux over barrier method to magnetic problems. We reiterate that this problem will be treated later using a general expression, see IV. D for the escape rate for moments on the unit sphere.

### D. The Mean First Passage Time Method of Calculating the Kramers Escape Rate

We have calculated the Kramers escape rate $\Gamma$ for axially symmetric potentials of the magnetocrystalline anisotropy or nematic potential in the case of dielectrics by calculating the time to ascend the potential hill. The escape time, $\tau_e$, is then, in the high barrier limit, twice the MFPT to the summit, $\tau_{\mathrm{MFPT}}$, and the Kramers escape rate $\Gamma = 1/\tau_e = 1/2\tau_{\mathrm{MFPT}}$, since, at the barrier, there is a 50:50 chance of the particle or rotator crossing the barrier. The calculation of MFPTs is often presented by constructing the adjoint FPE or backward Kolmogorov equation, and then the FPT, $\tau$, satisfies [26] the equation:

$$L_{\mathrm{FP}}^{+}\tau = -1 \tag{3.59}$$

with appropriate boundary conditions. Thus, if we write the Fokker–Planck (FP) operator in a state space $\{x\} = \{\theta, \varphi\} = \{p = \cos\theta, \varphi\}$, which is the space of the polar angles of the dipole vector, we have [45] (writing for convenience $g' \simeq \gamma/M_s$ so that we use the Landau–Lifshitz equation because we suppose $a^2 \ll 1$)

$$\begin{aligned} L_{\mathrm{FP}}W(p,\varphi,t) = {} & \frac{\partial}{\partial p}\left[\frac{a\gamma}{M_s}(1-p^2)H_p + \frac{\gamma}{M_s}H_\varphi\right]W + \frac{\partial}{\partial p}\left[\frac{1}{\beta}\frac{a\gamma}{M_s}(1-p^2)\frac{\partial}{\partial p}\right]W \\ & + \frac{\partial}{\partial\varphi}\left[-\frac{\gamma}{M_s}H_p + \frac{a\gamma}{M_s}(1-p^2)^{-1}H_\varphi\right]W \\ & + \frac{\partial}{\partial\varphi}\left[\frac{1}{\beta}\frac{a\gamma}{M_s}(1-p^2)^{-1}\frac{\partial}{\partial\varphi}\right]W \end{aligned} \tag{3.60}$$

$$\begin{aligned} L_{\mathrm{FP}}W = {} & \frac{\partial}{\partial p}\left[\frac{a\gamma}{M_s}(1-p^2)H_p + \frac{\gamma}{M_s}H_\varphi + 2\frac{1}{\beta}\frac{a\gamma}{M_s}p\right]W + \frac{\partial^2}{\partial p^2}\frac{1}{\beta}\frac{a\gamma}{M_s}(1-p^2)W \\ & + \frac{\partial}{\partial\varphi}\left[-\frac{\gamma}{M_s}H_p + \frac{a\gamma}{M_s}(1-p^2)^{-1}H_\varphi\right]W + \frac{\partial^2}{\partial\varphi^2}\frac{1}{\beta}\frac{a\gamma}{M_s}(1-p^2)^{-1}W \end{aligned} \tag{3.61}$$

Now in general the adjoint FPE operator (Risken's Eq. (4.97) [26]) is

$$L^{+}_{\mathrm{FP}}(\{x'\},t') = D^{(1)}_{i}(\{x'\},t')\frac{\partial}{\partial x'_i} + D^{(2)}_{ij}(\{x'\},t')\frac{\partial^2}{\partial x'_i \partial x'_j} \tag{3.62}$$

hence the Fokker–Planck adjoint operator $L^{+}_{\mathrm{FP}}$ for the present problem is given by

$$L^{+}_{\mathrm{FP}} = D^{(1)}_{p}\frac{\partial}{\partial p'} + D^{(1)}_{\varphi}\frac{\partial}{\partial \varphi'} + D^{(2)}_{pp}\frac{\partial^2}{\partial p'^2} + D^{(2)}_{\varphi\varphi}\frac{\partial^2}{\partial \varphi'^2} \tag{3.63}$$

(The application of these to magnetic relaxation is extensively discussed in Section V.) The disadvantage of using the adjoint equation to calculate $\tau_{\mathrm{MFPT}}$ is that it may be very difficult to integrate this equation and to apply the boundary conditions. An attractive alternative method of calculation of $\tau_{\mathrm{MFPT}}$, which relies on the Kramers flux over barrier method was used in Section III.C and has been given in general form by Hänggi et al. [1]. They consider a process governed by the FPE in an $N$-dimensional state space and the MFPT out of a domain $\Omega$ of the space. Next, they construct the PDE for the MFPT, $t(\mathbf{x})$, in terms of the adjoint FP operator, $L^{+}_{\mathrm{FP}}$. Hence, by direct integration over the domain $\Omega$ and, by using the properties of the adjoint operator, they arrive at their Eq. (B5). This is an equation yielding the MFPT out of $\Omega$ in terms of an integral of the zero-frequency Green function over $\Omega$. This allows the MFPT to be calculated from the Green's function of the FPE in the zero-frequency limit Section (VIII). The details of their calculation may be summarized as follows. The PDE for the MFPT, together with boundary condition, is

$$L^{+}_{\mathrm{FP}}\, t(\mathbf{x}) = -1 \qquad \mathbf{x} \in \Omega \tag{3.64}$$

$$t(\mathbf{x}) = 0 \qquad \mathbf{x} \in \partial(\Omega) \tag{3.65}$$

Green's function, $g(\mathbf{x},\mathbf{x}')$, of the FP operator on $\Omega$, is given by the boundary value problem

$$L_{\mathrm{FP}}(\mathbf{x})g(\mathbf{x},\mathbf{x}') = -k\delta^{n}(\mathbf{x}-\mathbf{x}') \qquad \mathbf{x} \in \Omega \tag{3.66}$$

$$g(\mathbf{x},\mathbf{x}') = 0 \qquad \mathbf{x} \in \partial(\Omega) \tag{3.67}$$

where the FP operator $L_{\mathrm{FP}}(\mathbf{x})$ operates on the $\mathbf{x}$ dependence of $g$ only. Green's function, as noted by Hänggi et al. [1], represents a stationary probability density of the process with an additional point source of strength $k > 0$ at the source point $\mathbf{x}'$, and perfect sinks on the boundary $\partial(\Omega)$. By defining the probability current density $\mathbf{J}(\mathbf{x}, \mathbf{x}')$ by the equation

$$\sum_i \frac{\partial J_i}{\partial x_i} = -L_{\mathrm{FP}}(\mathbf{x})g(\mathbf{x}, \mathbf{x}') \tag{3.68}$$

and by assuming the conservation of probability, we may write

$$k = \int_\Omega k\delta^n(\mathbf{x} - \mathbf{x}')d^n\mathbf{x} \tag{3.69}$$

$$= -\int_\Omega L_{\mathrm{FP}}(\mathbf{x})g(\mathbf{x}, \mathbf{x}')d^n\mathbf{x} \tag{3.70}$$

$$= \int_\Omega \sum_i \frac{\partial J_i}{\partial x_i} d^n\mathbf{x} \tag{3.71}$$

$$= \int_{\partial(\Omega)} \sum_i J_i dS_i \tag{3.72}$$

Now, by multiplying Eq. (3.66) across by $t(\mathbf{x})$, and by integrating over $\Omega$, we obtain

$$\int_\Omega t(\mathbf{x})L_{\mathrm{FP}}(\mathbf{x})g(\mathbf{x}, \mathbf{x}')d^n\mathbf{x} = -k\int_\Omega t(\mathbf{x})\delta^n(\mathbf{x} - \mathbf{x}')d^n\mathbf{x} \tag{3.73}$$

$$= -kt(\mathbf{x}') \tag{3.74}$$

If we definie the adjoint operator and use Eq. (3.64), we find

$$\int_\Omega t(\mathbf{x})L_{\mathrm{FP}}(\mathbf{x})g(\mathbf{x}, \mathbf{x}')d^n\mathbf{x} = \int_\Omega g(\mathbf{x}, \mathbf{x}')L^+_{\mathrm{FP}}(\mathbf{x})t(\mathbf{x})d^n\mathbf{x} \tag{3.75}$$

$$= -\int_\Omega g(\mathbf{x}, \mathbf{x}')d^n\mathbf{x} \tag{3.76}$$

Hence, Eqs. (3.74) and (3.76) yield

$$t(\mathbf{x}') = \frac{\int_{\Omega} g(\mathbf{x}, \mathbf{x}') d^n \mathbf{x}}{k} \tag{3.77}$$

which with Eq. (3.72) yields

$$t(\mathbf{x}') = -\frac{\int_{\Omega} g(\mathbf{x}, \mathbf{x}') d^n \mathbf{x}}{\int_{\partial(\Omega)} \sum_i J_i(\mathbf{x}, \mathbf{x}') dS_i} \tag{3.78}$$

and is Eq. (B5) of Hänggi et al. [1]. Thus they show that the MFPT may be expressed as the "population over flux", which Kramers' used to calculate the inverse of the escape rate that is, the lifetime of a particle. They note, also, that Eq. (3.78) holds whether or not $\Omega$ is a domain of attraction and whether or not the noise is weak. However, only if the noise is weak will the MFPT be essentially independent of the starting point in the domain $\Omega$.

This approach to the calculation of FPTs is particularly useful in obtaining expressions for that quantity in terms of matrix continued fractions [5, 57] (MCFs). The Laplace transform of the FPE for the transition probability in the zero-frequency limit yields $g(x, x')$ and so $\tau_{\mathrm{MFPT}}$ using Eq. (3.77). The above theorem is invaluable when $\tau_{\mathrm{MFPT}}$ is represented as a continued fraction [58], and doubly so because the MCF method is usually the only practical way in which Kolmogorov equations in more than 1D may be solved.

## E. The Integral Relaxation Time

In this section, we have shown how the axially symmetric results of Brown which in Eq. (3.58c) lead to the correction $2\pi a/(1 + a^2)\sqrt{\sigma/\pi}$ to the TST result may be derived using the method of FPTs. Another time that is used to describe the decay of a distribution of magnetization or electric polarization is the integral relaxation time [60, 68, 93, 107] (IRT), which is the area under the curve of the decay of magnetization following a sudden change in the amplitude of an applied external stimulus such as a uniform magnetic field, and so on. The integral relaxation time takes into account the contribution of all the modes of the decay of the magnetization to the relaxation process. It is virtually identical with the escape time if the configuration of the system is such that the contributions of all the other modes, save the longest lived one, are negligible. In other cases, however, the IRT may differ exponentially from the escape time, and so may not be used to estimate the escape time ($\tau_e$). The two characteristic times, $\tau_e$ and $T$ (the IRT), have been extensively discussed by Coffey [68]. Very recent work [58] shows how both the IRT and $\tau_e$ may be calculated by using MCFs from a knowledge of the transition probability or Green's function of the

FPE. The formula given here, Eq. (3.78), for example, shows how the FPT may be calculated from the Green's function.

# IV. THE EXTENSION OF KRAMERS' THEORY TO MANY DIMENSIONS IN THE INTERMEDIATE-TO-HIGH DAMPING LIMIT

## A. Introduction

We have seen that the original IHD treatment of Kramers pertained to a mechanical system of one degree of freedom specified by the coordinate $x$ with additive Hamiltonian $H = (p^2/2m) + U(x)$. Thus, the motion is separable and described by a 2D phase space with state variables $(x, p)$. The motion of the magnetic moment in a single domain particle, on the other hand, is governed by a Hamiltonian that is non-additive so that the system is nonseparable and which is simply the magnetocrystalline anisotropy energy of the particle. The relevant variables are $x = \cos\theta$, (sometimes written $p$) and the azimuthal angle, $\varphi$.

The Gilbert equation governing the process also causes *multiplicative noise* terms to appear, which complicates the calculations of the drift and diffusion coefficients in the FPE. This has been discussed at length in Coffey et al. [5, 39] (see Kalmykov and Titov [67] for the most recent treatment).

The phase-space trajectories in the Kramers problem of the underdamped motion are approximately ellipses. The corresponding trajectories in the magnetic problem are much more complicated because of the non separable form of the energy. Similar considerations hold in the extension of the Debye theory to include inertia as in this case one would usually (albeit with a separable Hamiltonian) have a six-dimensional phase space corresponding to the orientations and angular momenta of the rotator. These, and other considerations, suggest that the Kramers theory should be extended to a multidimensional phase space.

Such generalizations, having been instigated by Brinkman in 1956 [69, 71], were further developed by Landauer and Swanson in 1961 [71]. However, the most complete treatment is due to Langer 1969 [61], who considered the IHD limit.

The treatment, which we shall give of the extension of the Kramers theory to many dimensions, will loosely follow that given by Hänggi et al. [1]. As specific examples of the application of the theory, we shall apply it to the Kramers IHD limit and to the calculation of the magnetic relaxation time for an arbitrary non-axially symmetric potential of the magnetocrystalline anisotropy in that limit. This calculation was first carried out by Brown in 1979 [62], who proceeded from first principles, without reference to Langer's work, and was subsequently extensively reviewed by Geoghegan et al. [65] and applied to the important specific problem of magnetic relaxation in a uniform field applied at an angle to the easy axis of magnetization.

We remark that the first calculation of the escape rate for nonaxially symmetric potentials was, in fact, carried out by Smith and de Rozario [63] who treated relaxation in the cubic anisotropy potential essentially following the Kramers IHD calculation of II. C.

Klik and Gunther [64, 72] arrived at the same result as Brown in the IHD limit by using Langer's method. (They also seem to be the first workers to appreciate that the various Kramers' damping regimes also appear for magnetic relaxation.) We shall present their derivation in Section IV. D below.

Before proceeding we remark that a number of other interesting applications of the theory, which, as the reader will appreciate, is generally concerned with the nature of metastable states and the rates at which these states decay, have been mentioned by Langer [61] and we briefly summarize these.

Examples are

1. A supersaturated vapour which can be maintained in a metastable state for a very long time but which will eventually undergo condensation into the more stable liquid phase.
2. A ferromagnet which can persist with its magnetization pointing in a direction opposite to that of an applied magnetic field.
3. In metallurgy an almost identical problem occurs in the study of alloys whose components tend to separate on ageing or annealing.
4. The final examples quoted by Langer are the theories of superfluidity and superconductivity, where states of nonzero superflow are metastable and so may undergo spontaneous transitions to states of lower current and greater stability.

According to Langer all the phase transitions above take place by means of the nucleation and growth of some characteristic disturbance within the metastable system. Condensation of the supersaturated vapour is initiated by the formation of a sufficiently large droplet of the liquid. If this droplet is big enough it will be more likely to grow than to dissipate and will bring about condensation of the entire sample. If the nucleating disturbance appears spontaneously as a thermodynamic fluctuation it is said to be *homogeneous*. This is an intrinsic thermodynamic property of the system and is the type of disturbance described by Langer [61], which we shall review here. The other type of nucleation is *inhomogeneous nucleation* and occurs when the disturbance leading to the phase transition is caused by a foreign object, an irregularity, for example, in the walls of the container or some agent not part of the system of direct interest.

The above examples have been chosen in order to illustrate the breadth of applicability of the theory. In the present context, we remark that Langer's method, since, in effect, it can be applied to a multi-degree of freedom system, is likely to be of much use in calculating relaxation times for systems in which other types of interaction, such as exchange and dipole-dipole coupling, also appear. A useful discussion of the applicability of the theory in this context has been given recently by Braun [73]. We also emphasize that Langer's treatment of the homogeneous nucleation problem contains within it the magnetic case of the Kramers' IHD calculation. The multidimensional Kramers problem was first solved in the VHD (Smoluchowski) limit by Brinkman [69] and Landauer and Swanson [71].

We should also mention that Langer's treatment [61] is in effect also a generalization of a calculation of Becker and Döring in 1935 [17] of the rate of condensation of a supersaturated vapour. A general discussion of this problem is given in Chapter 7 of Frenkel [7] on the kinetics of phase transitions.

### B. Langer's Treatment of the IHD Limit

In order to achieve an easy comparison with previous work on the subject, we shall, as we have indicated above, loosely follow the notation of Hänggi et al. [1]. Thus we shall consider the FPE for a multidimensional process governed by a state vector $\{\eta\}$, which is [1]

$$\frac{\partial}{\partial t}P(\{\eta\},t) = \sum_{i=1}^{2N}\sum_{j=1}^{2N}\frac{\partial}{\partial \eta_i}M_{ij}\left[\frac{\partial E}{\partial \eta_j} + kT\frac{\partial}{\partial \eta_j}\right]P(\{\eta\},t) \tag{4.1}$$

Here, when the noise term in the Langevin equation is ignored, the system evolves in accordance with

$$\dot{\eta}_i = -\sum_j M_{ij}\frac{\partial E}{\partial \eta_j} \tag{4.2a}$$

where $(M_{ij})$ is called the transport matrix which, for simplicity, we shall assume to be constant.

An example of such a system is provided by Hamilton's equations for a single domain ferromagnetic particle (here, $p = \cos\theta$ and $E$ is the energy) [64, 68, 72]. In the non-stochastic limit where the dissipative coupling to the bath is entirely ignored we have

$$\dot{p} = -\frac{\gamma}{M_s}\frac{\partial E}{\partial \varphi} \qquad \left(\dot{P} = -\frac{\partial E}{\partial \varphi}\right) \tag{4.2b}$$

$$\dot{\varphi} = \frac{\gamma}{M_s}\frac{\partial E}{\partial p}, \qquad \left(\dot{\varphi} = \frac{\partial E}{\partial P}\right) \tag{4.2c}$$

where

$$P = \frac{M_s}{\gamma} p \tag{4.2d}$$

In Eq. (4.1), $E(\{\eta\})$ is a Hamiltonian (energy) function having two minima at points $A$ and $B$ separated by a saddle point $C$. Both $A$ and $B$ are surrounded by two wells. One, say that at $B$, is at a lower energy than the other. The particles have to pass over the saddle point, which acts as a barrier at $C$. We again assume that the barrier height $E_C - E_A$ is very high (at least of the order of $5\,kT$) so that the diffusion over the barrier is slow enough to ensure that a Maxwell–Boltzmann distribution is established and maintained near $A$ at all times. The high barrier also assures that the contribution to the flux over the saddle point will come mainly from a small region around $C$. The $2N$ variables $\{\eta\} = \{\eta_1, \eta_2, \ldots, \eta_{2N}\}$ are parameters, which could be the coordinates and momenta of a point in phase space or coordinates describing the orientation of the magnetization vector of a single domain ferromagnetic particle as in Eqs. (4.2).

Generally, however, the first $N$ of the $\eta_i$'s will be functions of the $N$ coordinates of position [1]

$$\eta_i = \eta(x_i) \qquad i = 1, 2, \ldots, N \tag{4.3a}$$

The second $N$ of the $\eta_i$'s will be the conjugate momenta $\pi(x)$ taken at the same points:

$$\eta_{i+N} = \pi(x_i) \qquad i = 1, 2, \ldots, N \tag{4.3b}$$

In fact, the $\eta_i$'s will often (although they need not) be the coordinates themselves; in which case (obviously):

$$\eta_i = x_i \qquad i = 1, 2, \ldots, N \tag{4.4}$$

If we define the matrices **M**, **D**, and **A** by the equations

$$\mathbf{M} = (M_{ij}) \tag{4.5}$$

so that **M** is the *transport* matrix and

$$\mathbf{D} = \tfrac{1}{2}(\mathbf{M} + \mathbf{M}^t) \tag{4.6}$$

is called the *diffusion* matrix which characterizes the thermal fluctuations due to the heat bath.

$$\mathbf{A} = \tfrac{1}{2}(\mathbf{M} - \mathbf{M}^t) \tag{4.7}$$

is a matrix which describes the motion in *the absence of the bath*, that is the inertial term in the case of mechanical particles and if **D** is not identically zero, then the dissipation of energy satisfies [1]

$$\dot{E} = -\sum_{i,j} \frac{\partial E}{\partial \eta_i} D_{ij} \frac{\partial E}{\partial \eta_j} \leq 0 \tag{4.8}$$

We consider just as in Section II a single well and suppose that at finite temperatures, a Maxwell–Boltzmann distribution is set up and the density at equilibrium is

$$\rho_{eq}(\{\eta\}) = \frac{1}{Z} e^{-\beta E(\{\eta\})} \tag{4.9}$$

where

$$Z \equiv \int_{-\infty}^{\infty} \cdots \int_{-\infty}^{\infty} e^{-\beta E} d\eta_1 \cdots d\eta_{2N} \tag{4.10}$$

is the partition function. The IHD escape rate for this multivariable problem may again be calculated by the flux over barrier method.

We make the following assumptions about $\rho(\{\eta\})$:

1. It obeys the stationary FPE (i.e., $\partial\rho/\partial t = 0$), which is (on linearization about the saddle point $\{\eta\} = \{\eta^S\}$):

$$\sum_{i,j} \frac{\partial}{\partial \eta_i} M_{ij} \left[ \sum_k e_{jk}(\eta_k - \eta_k^S) + kT \frac{\partial}{\partial \eta_j} \right] \rho(\{\eta\}) = 0 \tag{4.11}$$

where the $e_{jk}$ are the coefficients in the Taylor expansion of the energy about the saddle point truncated at the second term, namely, the quadratic (form) approximation

$$E(\{\eta\}) = E_C - \frac{1}{2} \sum_{i,j} e_{ij}(\eta_i - \eta_i^S)(\eta_j - \eta_j^S) \qquad \{\eta\} \approx \{\eta^S\} \tag{4.12}$$

and $E_C$ is the value of the energy function at the saddle point (cf. Kramers' method of Section II, in this case the saddle point is a 1D maximum). Equation (4.12) constitutes the quadratic approximation to the potential in the vicinity of the saddle point. For example, in magnetic relaxation in a uniform field with uniaxial anisotropy, the energy surface in the vicinity of the saddle point will be a hyperbolic paraboloid [65]. Equation (4.11) is the multidimensional FPE "linearized" in the region of the saddle point.

2. Due to the high barrier just as in Kramers' one-dimensional problem, a Maxwell–Boltzmann distribution is set up in the vicinity of the bottom of the well, that is, at $A$, so

$$\rho(\{\eta\}) \approx \rho_{eq}(\{\eta\}), \qquad \{\eta\} \approx \{\eta^A\} \tag{4.13}$$

3. Practically no particles have arrived at the far side of the saddle point. So that we have the sink boundary condition

$$\rho(\{\eta\}) = 0, \qquad \{\eta\} \text{ beyond } \{\eta^S\} \tag{4.14}$$

This is Kramers' condition that only rare particles of the assembly cross the barrier. Just as in the Klein–Kramers problem for one degree of freedom, we make the substitution

$$\rho(\{\eta\}) = \zeta(\{\eta\})\rho_{eq}(\{\eta\}) \tag{4.15}$$

(Again, the function $\zeta$ is known as the crossover function). Thus we obtain from Eqs. (4.9) and (4.11), as in Section II, an equation for the crossover function $\zeta(\{\eta\})$

$$\sum_{i,j} M_{ji}\left[-\sum_k e_{jk}(\eta_k - \eta_k^S) - kT\frac{\partial}{\partial\eta_j}\right]\frac{\partial}{\partial\eta_i}\zeta(\{\eta\}) = 0 \qquad \{\eta\} \approx \{\eta^S\} \tag{4.16}$$

and we postulate that $\zeta$ may be written in terms of a *single* variable $u$

$$\zeta(\{\eta\}) = \zeta(u) = \frac{1}{2\pi kT}\int_u^{\infty} \exp\left(-\frac{\beta z^2}{2}\right)dz \tag{4.17}$$

and we assume that $u$ has the form of the *linear combination*

$$u = \sum_i U_i(\eta_i - \eta_i^S) \tag{4.18}$$

This is Kramers' method of forcing the multidimensional FPE into an equation in a single variable (in his original case a linear combination of the two variables position and velocity, i.e. $u = p - ax'$ of Eq. (2.87)). In order to proceed, we have to determine the coefficients $U_i$ of the linear combination $u$ of the $\eta_j$. This is accomplished as follows: we define the matrix

$$\widetilde{\mathbf{M}} \equiv -\mathbf{M}^t \tag{4.19}$$

then we shall have the coefficients, $U_i$, of the linear combination as a solution of the *eigenvalue problem*

$$-\sum_{j,i} U_i \widetilde{M}_{ij} e_{jk} = \lambda_+ U_k \tag{4.20}$$

The eigenvalue $\lambda_+$ is the *deterministic growth rate of a small deviation from the saddle point*, and, is the positive eigenvalue of the transition (system) matrix of the noiseless Langevin equations, linearized about the saddle point. It characterizes the unstable barrier-crossing mode. Thus, in order to calculate $\lambda_+$, all that is required is a knowledge of the energy landscape and Eq. (4.20) need not, in practice, be involved (c.f. IV. C and IV. D).

Equation (4.20) is obtained essentially by substituting the linear combination $u$ i.e. Eq. (4.18), into Eq. (4.16) for the crossover function and requiring the resulting equation to be a proper ODE in the single variable $u$ [just as in Eqs. (2.89)–(2.94)] with solution given by Eq. (4.17). (The details of this are given in Appendix IV. I). Eq. (4.20) may also be written in the form

$$-\mathbf{U}\widetilde{\mathbf{M}}\mathbf{E}^S = \lambda_+ \mathbf{U} \tag{4.21a}$$

(Hänggi et al. [1] describe this equation by stating that $\mathbf{U}$ is a "left eigenvector" of the matrix $-\widetilde{\mathbf{M}}\mathbf{E}^S$. The usual eigenvalue equation of an arbitrary matrix, $\mathbf{A}$, is [74]

$$\mathbf{AX} = \lambda\mathbf{X} \tag{4.21b}$$

In the above terminology, $\mathbf{X}$ would be a "right eigenvector" of $\mathbf{A}$).

In Eq. (4.21a)

$$\mathbf{E}^S \equiv (e_{ij}) \tag{4.22}$$

is the matrix of the second derivatives of the potential evaluated at the saddle point which is used in the Taylor expansion of the energy near the saddle point. The determinant of this (Hessian) matrix is the Hessian [74] itself.

The normalization of $U_i$ is fixed so that

$$\lambda_+ = \sum_{i,j} U_i M_{ij} U_j \tag{4.23}$$

which is equivalent to

$$\sum_{i,j} U_i e_{ij}^{-1} U_j = -1 \tag{4.24}$$

[This conditions ensures that Eq. (4.17) retains the form of an error function and so may describe diffusion over a barrier c.f. Eq. (2.95) et seq. Alternatively one may say that the foregoing conditions require that the entry in the diffusion matrix in the direction of flow (that is, the unstable direction) is nonzero, that is, *we have current over the barrier, and so particles escape the well*].

Equation (4.1) is a continuity equation for the representative points just as described in Sections I and II, so that

$$\frac{\partial \rho}{\partial t} + \nabla \cdot \mathbf{J} = 0 \tag{4.25}$$

by inspection we find that the current density takes the form

$$J_i(\{\eta\}, t) = -\sum_j M_{ij} \left[ \frac{\partial E}{\partial \eta_j} + kT \frac{\partial}{\partial \eta_j} \right] \rho(\{\eta\}, t) \tag{4.26}$$

and we obtain, by using Eqs. (4.9), (4.17), and (4.18) for the *stationary* current density, that is,

$$\frac{\partial \rho}{\partial t} = 0$$

$$J_i(\{\eta\}) = \sqrt{\frac{kT}{2\pi}} \sum_j M_{ij} U_j \rho(\{\eta\}) \exp\left(-\tfrac{1}{2}\beta u^2\right) \tag{4.27}$$

We now take advantage of the condition stated above that the flux over the barrier emanates from a small region around $C$ (the summit of the potential hill). We integrate the current density *over a plane containing the saddle point but not parallel to the flow of particles* in order to obtain the total current. The plane $u = 0$ will suffice here. The total current, that is, flux of particles is

$$j = \sum_i \int_{u=0} J_i(\{\eta\}) dS_i \tag{4.28}$$

By using Eq. (4.28) with the approximation of Eq. (4.12) for the energy near the saddle point, the integration for the total flux now yields, after a very long calculation (details in Appendix IV.II),

$$j \approx \frac{\sum_{i,j} U_i M_{ij} U_j}{\sqrt{|\sum_{i,j} U_i e_{ij}^{-1} U_j|}} \frac{1}{2\pi} \frac{1}{\sqrt{|\det(\mathbf{E}^S/2\pi kT)|}} \frac{1}{Z} e^{-\beta E_C} \tag{4.29}$$

and from Eqs. (4.23) and (4.24) we obtain immediately

$$j = \frac{\lambda_+}{2\pi} |\det((2\pi kT)^{-1}\mathbf{E}^S)|^{-1/2} Z^{-1} e^{-\beta E_C} \tag{4.30}$$

Now, we assume that the energy function near $A$ may again be written in the quadratic approximation

$$E = E_A + \tfrac{1}{2} \sum_{i,j} a_{ij}(\eta_i - \eta_i^A)(\eta_j - \eta_j^A) \tag{4.31}$$

and we write

$$\mathbf{E}^A = (a_{ij}) \tag{4.32}$$

and the number of particles in the well is (details in Appendix IV.III)

$$n_A = \{\det[(2\pi kT)^{-1}\mathbf{E}^A]\}^{-1/2} Z^{-1} \tag{4.33}$$

Now, the escape rate, $\Gamma$, by the usual flux over barrier method, is defined to be

$$\Gamma \equiv \frac{j}{n_A} \tag{4.34}$$

and so from Eqs. (4.30) and (4.33) in terms of the positive eigenvalue $\lambda_+$ of the set of Langevin equations linearized about the saddle point

$$\Gamma = \frac{\lambda_+}{2\pi}\left[\frac{\det\{(2\pi kT)^{-1}\mathbf{E}^A\}}{|\det\{(2\pi kT)^{-1}\mathbf{E}^S\}|}\right]^{1/2} e^{-\beta E_C} \tag{4.35}$$

which is Langer's 1969 [61] expression in terms of the Hessians of the saddle and well energies for the escape rate for a multidimensional process in the IHD limit. The result again pertains to this limit because of our postulate that the potential in the vicinity of the saddle point may be approximated by the first two terms of its Taylor series so that, once again, that result fails for very small damping because the region of deviation from the Maxwell–Boltzmann distribution in the depths of the well extends far beyond the narrow region at the top of the barrier in which the potential may be replaced by its quadratic approximation. In passing we remark that rate theory at weak friction is generally known as "unimolecular rate theory" [1] the VLD limit of Kramers treated in Section II is an example of this. For a general discussion see Hänggi et al. [1].

### C. Kramers' Formula as a Special Case of Langer's Formula

As an example of the application of Langer's method, we shall use the method to derive the IHD result (II. C) of Kramers [4].

To recover the Kramers formula, by Langer's method, we take

$$N = 1 \tag{4.36a}$$

Thus the state variables are the position and momentum, so that

$$\eta_1 = x \tag{4.36b}$$

$$\eta_2 = p \tag{4.36c}$$

The noiseless Langevin equations are:

$$\dot{p} = -\gamma p - \frac{dU}{dx} \tag{4.37a}$$

$$\dot{x} = \frac{p}{m} \tag{4.37b}$$

$\gamma = \zeta/m$ in this Section rather than the $\eta$ of Sections I and II in order to avoid confusion with the state variables.

Now,

$$E = \frac{p^2}{2m} + U(x) \tag{4.38}$$

So,

$$\frac{\partial E}{\partial p} = \frac{p}{m} \tag{4.39a}$$

$$\frac{\partial E}{\partial x} = \frac{dU}{dx} \tag{4.39b}$$

and

$$\dot{\eta}_1 = \frac{p}{m} = \frac{\partial E}{\partial p} = \frac{\partial E}{\partial \eta_2} \tag{4.40a}$$

$$\dot{\eta}_2 = -m\gamma\frac{p}{m} - \frac{dU}{dx} = -m\gamma\frac{\partial E}{\partial p} - \frac{\partial E}{\partial x} = -\frac{\partial E}{\partial \eta_1} - m\gamma\frac{\partial E}{\partial \eta_2} \tag{4.40b}$$

Hence, we have the equation of motion in terms of the state variables $\eta_1, \eta_2$, of the general case of Langer's method above, as

$$\begin{pmatrix} \dot{\eta}_1 \\ \dot{\eta}_2 \end{pmatrix} = \begin{pmatrix} 0 & 1 \\ -1 & -m\gamma \end{pmatrix} \begin{pmatrix} \partial E/\partial \eta_1 \\ \partial E/\partial \eta_2 \end{pmatrix} \tag{4.41}$$

(Notice the extra element $-m\gamma$ denoting the contribution of the damping term) and so the transport matrix $(M_{ij})$ is (the negative of the matrix in Eq. (4.41))

$$\mathbf{M} = (M_{ij}) = \begin{pmatrix} 0 & -1 \\ 1 & m\gamma \end{pmatrix} \tag{4.42}$$

Also,

$$\widetilde{\mathbf{M}} \equiv -\mathbf{M}^t = \begin{pmatrix} 0 & -1 \\ 1 & -m\gamma \end{pmatrix} \tag{4.43}$$

If we take as in Figure 6 the origin of the coordinate system to be at the summit of the potential well, we will have

$$\eta_1^S = 0 \tag{4.44}$$

The momentum of a particle just escaping is 0 also, so

$$\eta_2^S = 0 \tag{4.45}$$

Eq. (4.12) then yields the quadratic approximation for the saddle energy, viz.

$$E = E_C - \frac{1}{2}\sum_{i,j} e_{ij}(\eta_i - \eta_i^S)(\eta_j - \eta_j^S) \tag{4.46}$$

But we have chosen the saddle energy

$$E_C = 0 \tag{4.47}$$

and

$$E = \frac{p^2}{2m} - \frac{1}{2}m\omega_C^2 x^2 \tag{4.48}$$

$$= -\frac{1}{2}m\omega_C^2(\eta_1 - \eta_1^S)^2 + \frac{1}{2}\frac{1}{m}(\eta_2 - \eta_2^S)^2 \tag{4.49}$$

Thus in Eq. (4.46)

$$e_{11} = m\omega_C^2 \tag{4.50}$$

$$e_{22} = -\frac{1}{m} \tag{4.51}$$

$$e_{12} = e_{21} = 0 \tag{4.52}$$

which are the matrix elements of the saddle Hessian matrix. We now determine $\lambda_+$.

We have the linearized equation:

$$\begin{pmatrix} \dot{\eta}_1 \\ \dot{\eta}_2 \end{pmatrix} = \begin{pmatrix} 0 & 1 \\ -1 & -m\gamma \end{pmatrix} \begin{pmatrix} \partial E^S/\partial \eta_1 \\ \partial E^S/\partial \eta_2 \end{pmatrix} \tag{4.53}$$

$$= \begin{pmatrix} 0 & 1 \\ -1 & -m\gamma \end{pmatrix} \begin{pmatrix} -m\omega_C^2 \eta_1 \\ \eta_2/m \end{pmatrix} \tag{4.54}$$

$$= \begin{pmatrix} 0 & 1 \\ -1 & -m\gamma \end{pmatrix} \begin{pmatrix} -m\omega_C^2 & 0 \\ 0 & 1/m \end{pmatrix} \begin{pmatrix} \eta_1 \\ \eta_2 \end{pmatrix} \tag{4.55}$$

Thus

$$\begin{pmatrix} \dot{\eta}_1 \\ \dot{\eta}_2 \end{pmatrix} = \begin{pmatrix} 0 & 1/m \\ m\omega_C^2 & -\gamma \end{pmatrix} \begin{pmatrix} \eta_1 \\ \eta_2 \end{pmatrix} \tag{4.56}$$

or (with **A** denoting the transition matrix)

$$\dot{\eta} = \mathbf{A}\eta \tag{4.57}$$

with secular equation

$$\det(\mathbf{A} - \lambda \mathbf{I}) = 0. \tag{4.58}$$

We thus solve the secular equation, namely

$$-\lambda(-\lambda - \gamma) - \omega_C^2 = 0 \tag{4.59}$$

to find

$$\lambda_{\pm} = -\frac{\gamma}{2} \pm \sqrt{\frac{\gamma^2}{4} + \omega_C^2}. \tag{4.60}$$

We pick the upper sign so that the solution (which is now always positive) corresponds to the *unstable barrier crossing mode* hence

$$\lambda_{+} = -\frac{\gamma}{2} + \sqrt{\frac{\gamma^2}{4} + \omega_C^2}. \tag{4.61}$$

Now the Hessian matrix of the saddle energy is [Eqs. (4.50)–(4.52)]

$$\mathbf{E}^S = \begin{pmatrix} m\omega_C^2 & 0 \\ 0 & -1/m \end{pmatrix} \tag{4.62}$$

and the Hessian matrix of the well energy is

$$\mathbf{E}^A = \begin{pmatrix} m\omega_A^2 & 0 \\ 0 & 1/m \end{pmatrix} \tag{4.63}$$

Hence

$$\det\left(\frac{\mathbf{E}^S}{2\pi kT}\right) = \frac{1}{(2\pi kT)^2}\begin{vmatrix} m\omega_C^2 & 0 \\ 0 & -1/m \end{vmatrix} \tag{4.64}$$

$$\det\left(\frac{\mathbf{E}^A}{2\pi kT}\right) = \frac{1}{(2\pi kT)^2}\begin{vmatrix} m\omega_A^2 & 0 \\ 0 & 1/m \end{vmatrix} \tag{4.65}$$

$$\det\left(\frac{\mathbf{E}^S}{2\pi kT}\right) = \frac{-\omega_C^2}{(2\pi kT)^2} \tag{4.66}$$

$$\det\left(\frac{\mathbf{E}^A}{2\pi kT}\right) = \frac{\omega_A^2}{(2\pi kT)^2} \tag{4.67}$$

and so

$$\left[\frac{\det[(2\pi kT)^{-1}\mathbf{E}^A]}{|\det[(2\pi kT)^{-1}\mathbf{E}^S]|}\right]^{1/2} = \frac{\omega_A}{\omega_C} \tag{4.68}$$

Eq. (4.35) for the escape rate is

$$\Gamma = \frac{\lambda_+}{2\pi}\left[\frac{\det[(2\pi kT)^{-1}\mathbf{E}^A]}{|\det[(2\pi kT)^{-1}\mathbf{E}^S]|}\right]^{1/2} e^{-\beta E_C} \tag{4.69}$$

which for the problem at hand is

$$= \frac{\lambda_+}{2\pi} \frac{\omega_A}{\omega_C} e^{-\beta \Delta U} \tag{4.70}$$

$$= \frac{\omega_A}{2\pi} \frac{1}{\omega_C} \left[ \left( \omega_C^2 + \frac{\gamma^2}{4} \right)^{1/2} - \frac{\gamma}{2} \right] e^{-\beta \Delta U} \tag{4.71}$$

$$= \frac{\omega_A}{2\pi} \left[ \left( 1 + \frac{\gamma^2}{4\omega_C^2} \right)^{1/2} - \frac{\gamma}{2\omega_C} \right] e^{-\beta \Delta U} \tag{4.72}$$

which is Kramers' IHD formula (Eq. 2.105).

### D. The IHD Formula for Magnetic Spins

As we have mentioned, the application of Kramers' escape rate theory to the problem of magnetic relaxation has been given in detail by Smith and de Rozario [63], Brown [62], Geoghegan et al. [65] and Klik and Gunther [64, 72] who used Langer's method just described and realized that the various Kramers damping regimes also applied to magnetic relaxation of single domain ferromagnetic particles. We show in detail how Langer's method may be used to solve this problem. Again we deal with an energy (or Hamiltonian) function, $E$ or $H$, with minima at points $A$ and $B$ separated by a barrier (saddle point) at $C$. The coordinate system we use is that of spherical polars $(\varphi, p)$ where $p = \cos\theta$, $\theta$ being the polar angle and $\varphi$ being the azimuthal angle as usual.

The noiseless Gilbert equation (Eq. (3.5) without $\Xi$), namely

$$\dot{\mathbf{n}} = -\left[ g'\mathbf{n} \times \frac{\partial H}{\partial \mathbf{n}} + h' \left( \mathbf{n} \times \frac{\partial H}{\partial \mathbf{n}} \right) \times \mathbf{n} \right] \tag{4.73}$$

has the form in these coordinates

$$\dot{p} = -h'(1 - p^2)H_p - g'H_\varphi \tag{4.74a}$$

$$\dot{\varphi} = g'H_p - h'(1 - p^2)^{-1} H_\varphi \tag{4.74b}$$

We linearize these equations about the saddle point and determine $\lambda_+$ from the transition matrix just as in IV.C.

Thus we expand the Hamiltonian $H(=E)$ as a Taylor series about the saddle point $(\varphi^S, p^S)$ where $p \equiv \cos\theta$, we obtain

$$H = E = H^S + \frac{1}{2}\left[H^S_{pp}(p - p^S)^2 + 2H^S_{p\varphi}(p - p^S)(\varphi - \varphi^S) + H^S_{\varphi\varphi}(\varphi - \varphi^S)^2\right] \tag{4.75}$$

where the subscripts denote the partial derivatives and the superscript denotes the taking of the values of the relevant (unsubscripted) functions at the saddle point. We ramark, following Klik and Gunther [64, 72] that the Hamiltonian is defined on a phase space which is a closed manifold [the space $(\varphi, p)$ is the surface of a unit sphere] and thus a local energy minimum is surrounded by two or more saddle points, depending on the symmetry of the problem. The total probability flux out of the metastable minimum equals the sum of the fluxes through all the saddle points. In an asymmetric case, e.g. when an external field is applied, some of these fluxes become exponentially small and may safely be neglected. The total flux out of the metastable minimum is, in this case, dominated by the energetically most favourable path.

Now, if the coordinates of the saddle point are $(\varphi^S, p^S)$, then

$$\frac{\partial E}{\partial p} = H^S_{pp}(p - p^S) + H^S_{p\varphi}(\varphi - \varphi^S) \tag{4.76}$$

$$\frac{\partial E}{\partial \varphi} = H^S_{p\varphi}(p - p^S) + H^S_{\varphi\varphi}(\varphi - \varphi^S) \tag{4.77}$$

Eqs. (4.73), (4.74) then yield:

$$\dot{\varphi} = g'\frac{\partial E}{\partial p} - h'\frac{\partial E}{\partial \varphi} \tag{4.78a}$$

$$\dot{p} = -h'\frac{\partial E}{\partial p} - g'\frac{\partial E}{\partial \varphi} \tag{4.78b}$$

Thus

$$\begin{pmatrix} \dot{\varphi} \\ \dot{p} \end{pmatrix} = \begin{pmatrix} -h' & g' \\ -g' & -h' \end{pmatrix}\begin{pmatrix} \partial E/\partial\varphi \\ \partial E/\partial p \end{pmatrix} \tag{4.79}$$

So the transport matrix $\mathbf{M}$ is given by:

$$\mathbf{M} = \begin{pmatrix} h' & -g' \\ g' & h' \end{pmatrix} \tag{4.80a}$$

and the matrix $\widetilde{\mathbf{M}}$ by

$$\widetilde{\mathbf{M}} = \begin{pmatrix} -h' & -g' \\ g' & -h' \end{pmatrix} \tag{4.80b}$$

Thus the linearized Eq. (4.79) has the form of the canonical equations (4.2a) and so Langer's expression, Eq. (4.35), may be used to calculate the escape rate. The equations of motion linearized at the saddle point may be written

$$\dot{\varphi} = g'\left[H_{pp}^{S}(p - p^{S}) + H_{p\varphi}^{S}(\varphi - \varphi^{S})\right] - h'\left[H_{p\varphi}^{S}(p - p^{S}) + H_{\varphi\varphi}^{S}(\varphi - \varphi^{S})\right] \tag{4.81a}$$

$$\dot{p} = -h'\left[H_{pp}^{S}(p - p^{S}) + H_{p\varphi}^{S}(\varphi - \varphi^{S})\right] - g'\left[H_{p\varphi}^{S}(p - p^{S}) + H_{\varphi\varphi}^{S}(\varphi - \varphi^{S})\right] \tag{4.81b}$$

Now, let the saddle point of interest lie on the equator $p = 0$ and make the transformation $\varphi \to \varphi - \varphi^{S}$, so that the above equations become in the notation of Klik and Gunther [64] (the superscript (1) denoting evaluation at the saddle point)

$$\dot{\varphi} = g'\left[H_{pp}^{(1)}p + H_{p\varphi}^{(1)}\varphi\right] - h'\left[H_{p\varphi}^{(1)}p + H_{\varphi\varphi}^{(1)}\varphi\right] \tag{4.82a}$$

$$\dot{p} = -h'\left[H_{pp}^{(1)}p + H_{p\varphi}^{(1)}\varphi\right] - g'\left[H_{p\varphi}^{(1)}p + H_{\varphi\varphi}^{(1)}\varphi\right] \tag{4.82b}$$

or, in matrix notation,

$$\begin{pmatrix} \dot{\varphi} \\ \dot{p} \end{pmatrix} = \begin{pmatrix} g'H_{p\varphi}^{(1)} - h'H_{\varphi\varphi}^{(1)} & g'H_{pp}^{(1)} - h'H_{p\varphi}^{(1)} \\ -h'H_{p\varphi}^{(1)} - g'H_{\varphi\varphi}^{(1)} & -h'H_{pp}^{(1)} - g'H_{p\varphi}^{(1)} \end{pmatrix} \begin{pmatrix} \varphi \\ p \end{pmatrix} \tag{4.83}$$

Eqs. (4.82a) & (4.82b) or Eq. (4.83) are the noiseless Langevin equations linearized at the saddle point given by Klik and Gunther [64] [their Eqs. (3.2)]. The secular equation of Eqs. (4.83) then leads to (just as in IV C)

$$\lambda_{\pm} = \frac{-h'\left[H_{pp}^{(1)} + H_{\varphi\varphi}^{(1)}\right] \pm \sqrt{h'^2\left[H_{pp}^{(1)} + H_{\varphi\varphi}^{(1)}\right]^2 - 4(g'^2 + h'^2)\left[H_{pp}^{(1)}H_{\varphi\varphi}^{(1)} - (H_{p\varphi}^{(1)})^2\right]}}{2} \tag{4.84}$$

The Hessian matrix of the system is

$$\begin{pmatrix} H_{\varphi\varphi} & H_{p\varphi} \\ H_{p\varphi} & H_{pp} \end{pmatrix} \tag{4.85}$$

and the Hessian itself is *negative* at the saddle point, thus, to ensure a growing disturbance at the saddle point we must again take the positive sign in Eq. (4.84). The square of the well angular frequency is (the superscript (0) denoting evaluation at the minimum)

$$\omega_A^2 = \frac{\gamma^2}{M_s^2}\left[H_{pp}^{(0)}H_{\varphi\varphi}^{(0)} - \left(H_{p\varphi}^{(0)}\right)^2\right] \tag{4.86a}$$

while the squared saddle angular frequency is

$$\omega_C^2 = -\frac{\gamma^2}{M_s^2}\left[H_{pp}^{(1)}H_{\varphi\varphi}^{(1)} - \left(H_{p\varphi}^{(1)}\right)^2\right] \tag{4.86b}$$

which, with Langer's formula, leads to the result of Klik and Gunther [64, 72]

$$\Gamma = \frac{1}{2\pi}\frac{\lambda_+\omega_A}{\omega_C}e^{-\beta E_C} \qquad (E_C \equiv \Delta U) \tag{4.87}$$

This formula underlines the power of Langer's method and shows clearly how, once the potential landscape is known, all quantities relating to the IHD escape rate may be calculated. We now choose a system of local coordinates, $(\varphi, p)$, in the vicinity of the saddle point, in which $H_{p\varphi} = 0$ at the saddle point itself. Then, writing $a \equiv (h'/g')$, we obtain a more compact expression for $\lambda_+$ namely:

$$\lambda_+ = \frac{-h'\left\{\left[H_{pp}^{(1)} + H_{\varphi\varphi}^{(1)}\right] - \sqrt{\left[H_{pp}^{(1)} - H_{\varphi\varphi}^{(1)}\right]^2 - 4a^{-2}H_{pp}^{(1)}H_{\varphi\varphi}^{(1)}}\right\}}{2} \tag{4.88}$$

Equations (4.87) and (4.88) were also derived from first principles, without recourse to Langer's work by Brown in 1979 [62] and have been reviewed by Geoghegan et al. [65]. In Brown's calculation [62] the Hamiltonian Eq. (4.75) is diagonalised so that [65]

$$H = H^S + \tfrac{1}{2}\left[c_1\varphi^2 + c_2 p^2\right] \tag{4.89}$$

$c_1$ and $c_2$ are the coefficients of the second order term of the Taylor series of the expansion of $H$ at the saddle point, $c_1^{(1)}$ and $c_2^{(1)}$ are the coefficients of the second order term in the Taylor series expansion of the energy in the well, also in Brown's notation

$$a(=g') = \frac{\gamma}{M_s(1+\alpha^2)} \tag{4.90}$$

$$b(=h') = \alpha a \tag{4.91}$$

$$\alpha(=a) = \eta\gamma M_s \tag{4.92}$$

Thus ignoring the second well of the bistable potential, Brown's result [62] reads (c.f. Eq. (5.60) of Geoghegan et al. [65], where a detailed derivation is given)

$$\Gamma = \kappa = \frac{b\left[-c_1 - c_2 + \sqrt{(c_2-c_1)^2 - 4\alpha^{-2}c_1c_2}\right]}{4\pi\sqrt{-c_1c_2}}\left[\sqrt{c_1^{(1)}c_2^{(1)}}e^{-\beta(V_0-V_1)}\right]$$

$$V_0 - V_1 \equiv \Delta U \tag{4.93}$$

We have illustrated the application of Langer's method by considering the solution of the original Kramers particle problem and the solution of the problem of magnetic relaxation of single-domain ferromagnetic particles. Langer's method provides a powerful means of calculating the escape rate in the IHD case for a multi-dimensional potential. All that is required being to evaluate the determinants of the energy function at the bottom of the well and at the saddle point. The positive eigenvalue, $\lambda_+$, (which effectively gives the correction to the TST result) characterizing the deterministic growth rate of a small deviation from the saddle point (known [1] in the chemical physics literature as the Grote-Hynes frequency [33]) is simply obtained from the secular equation of the transition (system) matrix of the noiseless equations of motion linearized about the saddle point. The eigenvalue formally corresponds to the left eigenvector $U_i$ of the matrix $\widetilde{\mathbf{M}}\mathbf{E}^S$ formed from the transport matrix and the matrix of the coefficients of the saddle energy and is formally obtained from the matrix $\widetilde{\mathbf{M}}\mathbf{E}^S$ using

$$-\mathbf{U}\widetilde{\mathbf{M}}\mathbf{E}^S = \lambda_+ \mathbf{U} \tag{4.94}$$

In practice this procedure will be quite difficult to carry out because of the problem of determining the stationary points and the second derivatives of the potential at the stationary points corresponding to the well and saddle point energies. The reader is referred to Geoghegan et al. [65] where the calculation is explicitly carried out for a uniform field at an angle to the easy axis of magnetization of a single-domain ferromagnetic particle.

Yet another example of the application of Langer's method is contained in the work of Braun [73], who successfully applied the method to calculate the greatest relaxation time when the assumption of a uniform magnetization of a single-domain ferromagnetic particle is abandoned. The extension of the theory to comprise exchange interactions (even the simplest case, two spins, gives rise to a $4 \times 4$ transition matrix), however, appears to present mathematical difficulties due to the difficulty in identifying analytically (when many variables are involved) the relevant saddle point describing reversal of the magnetization. [as emphasized by H. Kachkachi in a personal communication to the Authors].

## Appendix IV.I. Proof that U is a Left Eigenvector of the Matrix $\widetilde{\mathbf{M}}\mathbf{E}^S$ [Eq. (4.20)]

Equation (4.20) is

$$\sum_{i,j} U_i \widetilde{M}_{ij} e_j = \sum_k \lambda_k U_k$$

We start from the FPE linearized about the saddle point as

$$\sum_{i,j}(-M_{ji})\left[\sum_k e_{jk}(\eta_k - \eta_k^S) + kT\frac{\partial}{\partial \eta_j}\right]\frac{\partial \zeta}{\partial \eta_i} = 0 \tag{4.95}$$

Define

$$\widetilde{\mathbf{M}} \equiv -\mathbf{M}^t \tag{4.96}$$

Then,

$$\sum_{i,j}\widetilde{M}_{ij}\left[\sum_k e_{jk}(\eta_k - \eta_k^S) + kT\frac{\partial}{\partial \eta_j}\right]\frac{\partial \zeta}{\partial \eta_i} = 0 \tag{4.97}$$

Let,

$$\zeta \equiv \frac{1}{\sqrt{2\pi kT}} \int_u^\infty \exp\left(-\frac{\beta z^2}{2}\right) dz \tag{4.98}$$

where

$$u = \sum_i U_i(\eta_i - \eta_i^S) \tag{4.99}$$

Then,

$$\frac{\partial \zeta}{\partial \eta_i} = -\frac{\exp(-(\beta u^2/2))}{\sqrt{2\pi kT}} \frac{\partial u}{\partial \eta_i} \tag{4.100}$$

Now,

$$\frac{\partial u}{\partial \eta_i} = \sum_l U_l \delta_{li} = U_i \tag{4.101}$$

so,

$$-\sum_{i,j} \widetilde{M}_{ij} \left[ \sum_k e_{jk}(\eta_k - \eta_k^S) + kT \frac{\partial}{\partial \eta_j} \right] \frac{U_i \exp(-(\beta u^2/2))}{\sqrt{2\pi kT}} = 0 \tag{4.102}$$

$$\begin{aligned} -\frac{1}{\sqrt{2\pi kT}} \sum_{i,j} U_i \widetilde{M}_{ij} \Bigg[ \sum_k e_{jk}(\eta_k - \eta_k^S) \exp\left(-\frac{\beta u^2}{2}\right) \\ + kT\left(-\frac{\beta}{2}\frac{\partial u^2}{\partial \eta_j}\right) \exp\left(-\frac{\beta u^2}{2}\right) \Bigg] = 0 \end{aligned} \tag{4.103}$$

$$-\frac{\exp(-(\beta u^2/2))}{\sqrt{2\pi kT}} \sum_{i,j} U_i \widetilde{M}_{ij} \left[ \sum_k e_{jk}(\eta_k - \eta_k^S) - u \frac{\partial u}{\partial \eta_j} \right] = 0 \tag{4.104}$$

$$\sum_{i,j} U_i \widetilde{M}_{ij} \left[ \sum_k e_{jk}(\eta_k - \eta_k^S) - U_j \sum_k U_k(\eta_k - \eta_k^S) \right] = 0 \tag{4.105}$$

$$\sum_{i,j} U_i \widetilde{M}_{ij} \sum_k e_{jk}(\eta_k - \eta_k^S) - \sum_{i,j} U_i \widetilde{M}_{ij} U_j \sum_k U_k(\eta_k - \eta_k^S) = 0 \tag{4.106}$$

$$\sum_{k}\left[\sum_{i,j} U_i\widetilde{M}_{ij}e_{jk} - \lambda_k U_k\right](\eta_k - \eta_k^S) = 0 \tag{4.107}$$

by using the normalization of $U_i$ cf. Eq. (4.23). Now, since $\eta_k - \eta_k^S$ is an arbitrary deviation from the saddle point, we conclude

$$\sum_{i,j} U_i\widetilde{M}_{ij}e_{jk} = \lambda_k U_k \tag{4.108}$$

which is Eq. (4.20). □

## Appendix IV.II. Proof of Eq. (4.29) for the Current of Particles

We first remark that if $j$ is the current of particles over the barrier, then as justified in the next appendix [Eq. (4.164)]

$$j = \int\cdots\int \sum_{j} U_i J_i(\{\eta\})\delta(u)d\eta_1\cdots d\eta_{2N} \tag{4.109}$$

By using Eqs. (4.9) and (4.27)

$$j = \sqrt{\frac{kT}{2\pi}}\int\cdots\int\sum_{i,j} U_i M_{ij} U_j Z^{-1} e^{-\beta E} e^{-(1/2)\beta u^2}\delta(u)d\eta_1\cdots d\eta_{2N} \tag{4.110}$$

$$= \sqrt{\frac{kT}{2\pi}}\frac{\sum_{i,j} U_i M_{ij} U_j}{Z}\int\cdots\int e^{-\beta E}\delta(u)d\eta_1\cdots d\eta_{2N} \tag{4.111}$$

since

$$\exp\left(-\frac{1}{2}\beta u^2\right)\delta(u) = \delta(u) \tag{4.112}$$

$$\approx \sqrt{\frac{kT}{2\pi}}\lambda_+ Z^{-1} e^{-\beta E_C}\frac{1}{2\pi} \times \iint\cdots\int \exp\left\{-\frac{\beta}{2}\sum_{i,j} e_{ij}(\eta_i - \eta_i^S)(\eta_j - \eta_j^S) + ik\sum_{l} U_l(\eta_l - \eta_l^S)\right\} d\eta dk \tag{4.113}$$

where,

$$d\eta \equiv d\eta_1 \cdots d\eta_{2N} \tag{4.114}$$

Now, we introduce the rotation matrix

$$\mathbf{S} = (S_{ij}) \tag{4.115}$$

and define the vector $\mathbf{x} = (x_j)$ by the equation

$$\eta_j - \eta_j^S \equiv \sum_j S_{ij} x_j \tag{4.116}$$

then, we have

$$\sum_{i,j} e_{ij}(\eta_i - \eta_i^S)(\eta_j - \eta_j^S) = \sum_{l=1}^{2N} \mu_l x_l^2 \tag{4.117}$$

where,

$$\mu_1 < 0 \tag{4.118a}$$

and

$$\mu_j > 0, \qquad 2 \leq j \leq 2N \tag{4.118b}$$

(Note the difference between this and the notation of Hänggi [1].) If we write the matrices (or more correctly, the column vectors)

$$\mathbf{H} \equiv (\eta_i - \eta_i^S)_{i=1}^{2N} \tag{4.119}$$

$$\mathbf{X} \equiv (x_i)_{i=1}^{2N} \tag{4.120}$$

then, Eq. (4.116) becomes

$$\mathbf{H} = \mathbf{SX} \tag{4.121}$$

Now, the entries of $\mathbf{S}$ are constants, so

$$\frac{\partial \eta_i}{\partial x_j} = S_{ij} \tag{4.122}$$

Hence, the Jacobean of the transformation (4.116) is given by

$$J = \begin{vmatrix} \frac{\partial \eta_1}{\partial x_1} & \cdots & \frac{\partial \eta_{2N}}{\partial x_1} \\ \vdots & & \vdots \\ \frac{\partial \eta_1}{\partial x_{2N}} & \cdots & \frac{\partial \eta_{2N}}{\partial x_{2N}} \end{vmatrix} = \det(\mathbf{S}) \tag{4.123}$$

and is also a constant. So,

$$j = \sqrt{\frac{kT\lambda_+ e^{-\beta E_C}}{2\pi\, Z \quad 2\pi}} \iint \cdots \int \exp\left[-\frac{\beta}{2}\sum_j \left(\mu_j x_j^2 - \frac{2ikU_j}{\beta}\sum_l S_{jl}x_l\right)\right] J dx_1 \cdots dx_{2N} dk \tag{4.124}$$

Now,

$$\begin{aligned} &\iint \cdots \int \exp\left[-\frac{\beta}{2}\sum_j \left(\mu_j x_j - \frac{2ikU_j}{\beta}\sum_l S_{jl}x_l\right)\right] J dx_1 \cdots dx_{2N} dk \\ &= \iint \cdots \int \exp\left[-\frac{\beta}{2}\left(\mu_1 x_1^2 + \sum_{j=2}^{2N} \mu_j x_j^2 - \sum_j \sum_l \frac{2ikU_j}{\beta} S_{jl} x_l\right)\right] \\ &\quad \times J dx_1 dx_2 \cdots dx_{2N} dk \end{aligned} \tag{4.125}$$

$$= J \iint \cdots \int \exp\left[-\frac{\beta}{2}\left(\mu_1 x_1^2 + \sum_{j=2}^{2N} \mu_j x_j^2 - \sum_l \frac{2ik}{\beta}\widetilde{U}_l x_l\right)\right] dx_1 \cdots dx_{2N} dk \tag{4.126}$$

[where we define

$$\widetilde{U}_l \equiv \sum_j U_j S_{jl} \tag{4.127}$$ ]

$$\begin{aligned} &= J \int \left[\int \exp\left\{-\frac{\beta}{2}\left(\mu_1 x_1^2 - \frac{2ik}{\beta}\widetilde{U}_1 x_1^2\right)\right\} dx_1 \right. \\ &\quad \left. \times \int \cdots \int \exp\left\{-\frac{\beta}{2}\sum_{j=2}^{2N}\left(\mu_j x_j^2 - \frac{2ik}{\beta}\widetilde{U}_j x_j\right)\right\} dx_2 \cdots dx_{2N}\right] dk \end{aligned} \tag{4.128}$$

$$
\begin{aligned}
&= J\int\left[\int \exp\left\{-\frac{\beta}{2}\left(\mu_1 x_1^2 - \frac{2ik\widetilde{U}_1}{\beta}x_1\right)\right\}dx_1\right.\\
&\quad\times \int\cdots\int \exp\left\{-\frac{\beta}{2}\sum_{j=2}^{2N}\mu_j\left(x_j^2 - \frac{2ik}{\beta\mu_j}\widetilde{U}_j x_j + \frac{i^2k^2\widetilde{U}_j^2}{\beta^2\mu_j^2}\right)\right\}dx_2\cdots dx_{2N}\\
&\quad\left.\times \exp\left\{-\frac{\beta}{2}\sum_{j=2}^{2N}\frac{k^2\widetilde{U}_j^2}{\beta\mu_j}\right\}dk\right]
\end{aligned}
\tag{4.129}
$$

$$
\begin{aligned}
&= J\int\left[\int \exp\left\{-\frac{\beta}{2}\left(\mu_1 x_1^2 - \frac{2ik\widetilde{U}_1}{\beta}x_1\right)\right\}dx_1\right.\\
&\quad\left.\times \prod_{j=2}^{2N}\exp\left\{-\frac{k^2\widetilde{U}_j^2}{2\beta\mu_j}\right\}\int \exp\left\{-\frac{\beta\mu_j}{2}\left(x_j - \frac{ik\widetilde{U}_j}{\beta\mu_j}\right)^2\right\}dx_j\right]dk
\end{aligned}
\tag{4.130}
$$

$$
= J\int\left[\int \exp\left\{-\frac{\beta}{2}\left(\mu_1 x_1^2 - \frac{2ik\widetilde{U}_1}{\beta}x_1\right)\right\}dx_1\prod_{j=2}^{2N}\exp\left\{-\frac{k^2\widetilde{U}_j^2}{2\beta\mu_j}\right\}\sqrt{\frac{2\pi}{\beta\mu_j}}\right]dk
\tag{4.131}
$$

$$
= J\prod_{j=2}^{2N}\sqrt{\frac{2\pi}{\beta\mu_j}}\int \exp\left\{-\frac{\beta}{2}\left(\mu_1 x_1^2\right)\right\}\int \exp\left\{-\frac{k^2}{2\beta}\sum_{l=2}^{2N}\frac{\widetilde{U}_l^2}{\mu_l} + ik\widetilde{U}_1 x_1\right\}dk\,dx_1
\tag{4.132}
$$

$$
\begin{aligned}
&= J\prod_{j=2}^{2N}\sqrt{\frac{2\pi}{\beta\mu_j}}\int \exp\left\{-\frac{\beta}{2}\left(\mu_1 x_1^2\right)\right\}\\
&\quad\times \int \exp\left\{-\frac{1}{2\beta}\sum_{l=2}^{2N}\frac{\widetilde{U}_l^2}{\mu_l}\left(k^2 - \frac{2\beta i\widetilde{U}_1 x_1}{\left(\sum_{l=2}^{2N}\frac{\widetilde{U}_l^2}{\mu_l}\right)}k + \frac{\beta^2 i^2\widetilde{U}_1^2 x_1^2}{\left(\sum_{l=2}^{2N}\frac{\widetilde{U}_l^2}{\mu_l}\right)^2}\right)\right\}\\
&\quad\times \exp\left\{-\frac{\beta\widetilde{U}_1^2 x_1^2}{2\left(\sum_{l=2}^{2N}\frac{\widetilde{U}_l^2}{\mu_l}\right)}\right\}dk\,dx_1
\end{aligned}
\tag{4.133}
$$

$$= J\prod_{j=2}^{2N}\sqrt{\frac{2\pi}{\beta\mu_j}}\int\exp\left\{-\frac{\beta}{2}\left(\mu_1+\frac{\widetilde{U}_1^2}{\left(\sum_{l=2}^{2N}\frac{\widetilde{U}_l^2}{\mu_l}\right)}\right)x_1^2\right\}$$
$$\times\int\exp\left\{-\frac{1}{2\beta}\sum_{l=2}^{2N}\frac{\widetilde{U}_l^2}{\mu_l}\left(k-\frac{i\beta\widetilde{U}_1x_1}{\left(\sum_{l=2}^{2N}\frac{\widetilde{U}_l^2}{\mu_l}\right)}\right)^2\right\}dkdx_1 \tag{4.134}$$

$$= J\prod_{j=2}^{2N}\sqrt{\frac{2\pi}{\beta\mu_j}}\sqrt{\frac{2\pi\beta}{\sum_{l=2}^{2N}\frac{\widetilde{U}_l^2}{\mu_l}}}\sqrt{\frac{2\pi}{\beta}}\left[\mu_1+\frac{\widetilde{U}_1^2}{\sum_{l=2}^{2N}\frac{\widetilde{U}_l^2}{\mu_l}}\right]^{-1/2} \tag{4.135}$$

$$= J\prod_{j=2}^{2N}\sqrt{\frac{2\pi}{\beta\mu_j}}\sqrt{\frac{2\pi\beta}{\sum_{l=2}^{2N}\frac{\widetilde{U}_l^2}{\mu_l}}}\sqrt{\frac{2\pi}{\beta|\mu_1|}}\left[-1+\frac{\left(\frac{\widetilde{U}_1^2}{|\mu_1|}\right)}{\sum_{l=2}^{2N}\frac{\widetilde{U}_l^2}{\mu_l}}\right]^{-1/2} \tag{4.136}$$

since $\mu_1 < 0$; $\mu_l > 0$, $l \geq 2$

$$= J\prod_{j=1}^{2N}\sqrt{\frac{2\pi}{\beta|\mu_j|}}\sqrt{2\pi\beta}\left[\frac{\widetilde{U}_1^2}{|\mu_1|}-\sum_{l=2}^{2N}\frac{\widetilde{U}_l^2}{\mu_l}\right]^{-1/2} \tag{4.137}$$

$$= J\prod_{j=1}^{2N}\sqrt{\frac{2\pi}{\beta|\mu_j|}}\sqrt{2\pi\beta}\left[-\sum_{l=1}^{2N}\frac{\widetilde{U}_l^2}{\mu_l}\right]^{-1/2} \tag{4.138}$$

So,

$$j = \sqrt{\frac{kT}{2\pi}}\frac{\lambda_+}{Z}\frac{e^{-\beta E_C}}{2\pi}J\prod_{j=1}^{2N}\sqrt{\frac{2\pi}{\beta|\mu_j|}}\sqrt{2\pi\beta}\left[-\sum_{l=1}^{2N}\frac{\widetilde{U}_l^2}{\mu_l}\right]^{-1/2} \tag{4.139}$$

$$= \frac{\lambda_+}{Z}\frac{e^{-\beta E_C}}{2\pi}J\prod_{j=1}^{2N}\sqrt{\frac{2\pi}{\beta|\mu_j|}}\left[-\sum_{l=1}^{2N}\frac{\widetilde{U}_l^2}{\mu_l}\right]^{-1/2} \tag{4.140}$$

Now, let

$$\mathbf{M} = \begin{pmatrix} \mu_1 & 0 & 0 & \cdots & 0 \\ 0 & \mu_2 & 0 & \cdots & 0 \\ 0 & 0 & \mu_3 & \cdots & 0 \\ \vdots & \vdots & \vdots & \cdots & \vdots \\ 0 & 0 & 0 & \cdots & \mu_{2N} \end{pmatrix} \tag{4.141}$$

$$\mathbf{M} = \mathbf{S}^t \mathbf{E}^S \mathbf{S} \tag{4.142}$$

Hence,

$$\det \mathbf{M} = \det \mathbf{E}^S [\det(\mathbf{S})]^2 \tag{4.143}$$

$$= \det(\mathbf{E}^S) J^2 \tag{4.144}$$

but

$$\det \mathbf{M} = \prod_{j=1}^{2N} \mu_j \tag{4.145}$$

and so

$$\prod_{j=1}^{2N} \sqrt{\frac{2\pi}{\beta|\mu_j|}} = \left| \det \left[ \frac{\mathbf{E}^S}{2\pi kT} \right] \right|^{-1/2} \frac{1}{\det \mathbf{S}} \tag{4.146}$$

and

$$J \prod_{j=1}^{2N} \sqrt{\frac{2\pi}{\beta|\mu_j|}} = \left| \det \left[ \frac{\mathbf{E}^S}{2\pi kT} \right] \right|^{-1/2} \tag{4.147}$$

and

$$\mathbf{M}^{-1} = \begin{pmatrix} \frac{1}{\mu_1} & 0 & 0 & \cdots & 0 \\ 0 & \frac{1}{\mu_2} & 0 & \cdots & 0 \\ 0 & 0 & \frac{1}{\mu_3} & \cdots & 0 \\ \vdots & \vdots & \vdots & \cdots & \vdots \\ 0 & 0 & 0 & \cdots & \frac{1}{\mu_{2N}} \end{pmatrix} \tag{4.148}$$

$$= \mathbf{S}^{-1}(\mathbf{E}^S)^{-1}(\mathbf{S}^t)^{-1} \tag{4.149}$$

or

$$\mathbf{S}\mathbf{M}^{-1}\mathbf{S}^t = (\mathbf{E}^S)^{-1} = (e_{ij})^{-1} \tag{4.150}$$

Now,

$$\sum_{i,j} U_i e_{ij}^{-1} U_j = \sum_{i,j,k,l} U_i S_{ik} M_{kl}^{-1} S_{lj}^t U_j \tag{4.151}$$

$$= \sum_{k,l} \widetilde{U}_k M_{kl}^{-1} \widetilde{U}_l \tag{4.152}$$

$$= \sum_{k,l} \widetilde{U}_k \frac{1}{\mu_k} \delta_{kl} \widetilde{U}_l \tag{4.153}$$

$$= \sum_k \frac{\widetilde{U}_k}{\mu_k} \widetilde{U}_k \tag{4.154}$$

$$= \sum_k \frac{\widetilde{U}_k^2}{\mu_k} \tag{4.155}$$

Thus, (since by Eq. (4.24), $\sum_{ij} U_i e_{ij}^{-1} U_j < 0$)

$$\left[-\sum_k \frac{\widetilde{U}_k^2}{\mu_k}\right]^{-1/2} = \left|\sum_{i,j} U_i e_{ij}^{-1} U_j\right|^{-1/2} \tag{4.156}$$

and hence

$$j = \frac{\sum_{i,j} U_i M_{ij} U_j}{\left|\sum_{i,j} U_i e_{ij}^{-1} U_j\right|^{1/2}} \frac{1}{2\pi} \left|\det\left[\frac{\mathbf{E}^S}{2\pi kT}\right]\right|^{-1/2} Z^{-1} e^{-\beta E_C} \tag{4.157}$$

which is Eq. (4.29) for the current of particles. □

**Proof of Eq. (4.109), that is the Current Evaluated as an Integral over the Current Density**

Equation (4.28) for the current of particles in terms of an integral over the current density is

$$j = \sum_i \int_{u=0} J_i(\{\eta\}) dS_i \tag{4.158}$$

or

$$j = \int_{u=0} \sum_{i=1}^{2N} J_i(\{\eta\}) dS_i \tag{4.159}$$

$$= \int_{u=0} \mathbf{J} \cdot \mathbf{n} dS \tag{4.160}$$

where $\mathbf{n}$ is the unit normal to the surface $d\mathbf{S}$ (this may vary as we go around the surface $u = 0$), and $dS$ is the magnitude of the vector whose $i$th component is $dS_i$. Here, $u = 0$ is the plane

$$\sum_i U_i(\eta_i - \eta_i^S) = 0 \tag{4.161}$$

Now, from coordinate geometry, this plane

1. Is perpendicular to the vector $\mathbf{U} = U_1, U_2, \ldots, U_{2N}$.
2. Passes through the point $(\eta_1, \eta_2, \ldots, \eta_{2N})$, that is, the saddle point.

Now, on integrating any function $f(\{\eta\})$ the operation

$$\int \cdots \int f(\{\eta\}) d\eta_1 \cdots d\eta_{2N}$$

integrates it over all space. However,

$$\int \cdots \int f(\{\eta\}) \delta(u) d\eta_1 \cdots d\eta_{2N}$$

restricts this space integral to that part of space where $u = 0$, that is, to our plane. Note that $\mathbf{U}$ *need not* be a unit vector for the equation

$$\sum_i U_i(\eta_i - \eta_i^S) = 0 \tag{4.162}$$

to hold and to describe the plane, however, dividing across by $|\mathbf{U}|$ is permissible here and makes $\mathbf{U}$ a unit vector. We assume that this is done. So the total current of particles is

$$j = \int \cdots \int \mathbf{J} \cdot \mathbf{U} \delta(u) d\eta_1 \cdots d\eta_{2N} \tag{4.163}$$

$$= \int \cdots \int \sum_i U_i J_i \delta(u) d\eta_1 \cdots d\eta_{2N} \tag{4.164}$$

as required. □

## Appendix IV.III. Proof of Eq. (4.33) for the Number of Particles Trapped in the Well at $A$

$$E_A = E_1 + \frac{1}{2} \sum_{i,j} a_{ij} (\eta_i - \eta_i^A)(\eta_j - \eta_j^A) \tag{4.165}$$

Note that Hänggi et al. [1] assumes that $E_1 = 0$.

$$n_A = \int \cdots \int \rho_{eq} d\eta_1 \cdots d\eta_{2N} \tag{4.166}$$

$$= Z^{-1} \int \cdots \int e^{-\beta E^A} d\eta_1 \cdots d\eta_{2N} \tag{4.167}$$

$$= Z^{-1} \int \cdots \int e^{-\beta E_1} \exp\left( -\frac{\beta}{2} \sum_{i,j} a_{ij} (\eta_i - \eta_i^A)(\eta_j - \eta_j^A) \right) d\eta_1 \cdots d\eta_{2N} \tag{4.168}$$

As in the proof of Eq. (4.29), we let

$$\eta_i - \eta_i^A \equiv \sum_j S_{ij} x_j \tag{4.169}$$

Then,

$$\frac{1}{2} \sum_{i,j} a_{ij} (\eta_i - \eta_i^A)(\eta_j - \eta_j^A) = \frac{1}{2} \sum_l \mu_l x_l^2 \tag{4.170}$$

$$J = \det(S_{ij}) \tag{4.171}$$

$$n_A = Z^{-1}e^{-\beta E_1}\int\cdots\int \exp\left\{-\frac{\beta}{2}\sum_l \mu_l x_l^2\right\} J\,dx_1\cdots dx_{2N} \tag{4.172}$$

$$= Z^{-1}e^{-\beta E_1} J \prod_{l=1}^{2N}\sqrt{\frac{2\pi}{\beta\mu_l}} \tag{4.173}$$

Let,

$$\mathbf{M} = \begin{pmatrix} \mu_1 & 0 & 0 & \cdots & 0 \\ 0 & \mu_2 & 0 & \cdots & 0 \\ 0 & 0 & \mu_3 & \cdots & 0 \\ \vdots & \vdots & \vdots & \cdots & \vdots \\ 0 & 0 & 0 & \cdots & \mu_{2N} \end{pmatrix} \tag{4.174}$$

$$\begin{aligned} \mathbf{H} &= \left(\eta_i - \eta_i^A\right)_{i=1}^{2N}; & \mathbf{X} &= (x_i)_{i=1}^{2N}; \\ \mathbf{S} &= \left(S_{ij}\right)_{i=1 j=1}^{2N 2N}; & \mathbf{E}^A &= \left(a_{ij}\right)_{i,j=1}^{2N} \end{aligned} \tag{4.175}$$

Now, as in the proof of Eq. (4.29),

$$\mathbf{H} = \mathbf{SX} \tag{4.176}$$

$$\mathbf{S}^t\mathbf{E}^A\mathbf{S} = \mathbf{M} \tag{4.177}$$

$$[\det(\mathbf{S})]^2 \det(\mathbf{E}^A) = \det(\mathbf{M}) = \prod_{i=1}^{2N}\mu_i \tag{4.178}$$

and since

$$J = \det(\mathbf{S}) \tag{4.179}$$

we have

$$J^2 \det\left(\frac{\mathbf{E}^A}{2\pi kT}\right) = \prod_{i=1}^{2N} \frac{\beta\mu_i}{2\pi} \tag{4.180}$$

Then it follows easily that

$$J \prod_{i=1}^{2N} \sqrt{\frac{2\pi}{\beta\mu_i}} = \left[\det\left(\frac{\mathbf{E}^A}{2\pi kT}\right)\right]^{-1/2} \tag{4.181}$$

So,

$$n_A = Z^{-1} e^{-\beta E_1} J \prod_{l=1}^{2N} \sqrt{\frac{2\pi}{\beta\mu_l}} \tag{4.182}$$

$$= Z^{-1} e^{-\beta E_1} \left[\det\left(\frac{\mathbf{E}^A}{2\pi kT}\right)\right]^{-1/2} \tag{4.183}$$

$$= Z^{-1} \left[\det\left(\frac{\mathbf{E}^A}{2\pi kT}\right)\right]^{-1/2} \tag{4.184}$$

if $E_1 = 0$.

This proves Eq. (4.33). □

## V. LOW-DAMPING KRAMERS' ESCAPE RATES BY THE FIRST PASSAGE TIME METHOD

### A. Introduction

In this section, we shall summarize a method proposed by Matkowsky et al. [44] for the calculation of the low-damping Kramers escape rate, and in particular we shall give the details of the calculation of low-damping Kramers' escape rate for spins of a single domain ferromagnetic particle by this method. We shall first summarize their method of approach to the Kramers problem for mechanical particles in the low-damping limit [45].

The essence of the calculation is the fact that for a domain $D$, the first passage time is the time for a random walker to reach the boundary of the domain for the first time from a point $x_0$ well embedded in the domain. Thus the MFPT is the average time for the random walker to reach the separatrix manifold $g$ between the bounded and unbounded motions for the first time [1].

On the basis of this fact, Matkowsky et al. [44] proposed a method that they term the *uniform* asymptotic expansion of the transition rate, which may be described as follows: They first remark that the Kramers escape rate, $\Gamma$, is the reciprocal of the mean time to escape the well. This time is the sum of the mean time $\tau_1(A)$ to reach the trajectory $E = E_C$ (see Fig. 7) from the bottom of the well and the mean time to proceed from $E = E_C$ to the separatrix $g$ (see Fig. 7), and then escape the well. The latter is twice the MFPT $\tau_2(E_C)$ from $E_C$ to $g$, since trajectories that reach $E = E_C$ are equally likely to leave or to return to the well.

Hence, a formula for $\Gamma$ is, according to Matkowsky et al. [44], given by

$$\Gamma = \Gamma_{\text{unif}} = \frac{1}{\tau_1(A) + 2\tau_2(E_C)} \tag{5.1a}$$

Matkowsky et al. [44] showed that for small friction, that is, in the low-temperature limit, the time to reach the barrier energy trajectory is

$$\tau_1(A) = O\left(\frac{1}{\eta}\right) \tag{5.1b}$$

while the time to proceed from the barrier energy trajectory to the separatrix is

$$\tau_2(E_C) = O(1) \tag{5.1c}$$

so that $\tau_1(A) \gg \tau_2(E_C)$ thus $\Gamma_{\text{unif}} \approx 1/\tau_1(A) = \Gamma_1$ say.

Effectively, the latter equation was used by Hänggi et al. [1] to write down the Kramers escape rate for low-damping [their Eq. (4.48a)], since the imposition of the boundary condition that the density of particles vanishes at the top of the barrier meaning that all particles are absorbed at the barrier, is tantamount to ignoring the time to go from the critical (barrier) energy curve to the separatrix. Put more simply, the 50:50 chance of the particle returning to the well is replaced by a zero chance of returning in the low-damping case only.

If we continue the main part of this discussion, we remark that Kramers [4] (Section II.) developed [by writing the Fokker–Planck equation in angle (fast)-action (slow) variables and averaging out the fast angle (phase) variable] formula

$$\Gamma_1 = \frac{1}{\tau_1(A)} = \frac{\eta\beta I_C \omega_A}{2\pi}\exp(-\beta E_C) = \eta\beta I_C \frac{\omega_A}{2\pi} e^{-\beta\Delta U} \tag{5.2a}$$

[Here

$$\beta = \frac{1}{kT} \tag{5.2b}$$

where we recall that the system is governed by the Langevin equation

$$m\ddot{x} + \zeta\dot{x} + \frac{dU(x)}{dx} = \lambda(t) \tag{5.3a}$$

where $\lambda(t)$ is the usual white noise term and for convenience henceforth we take the mass to be 1, also

$$\omega_A = \sqrt{\frac{U''(x_A)}{m}} \tag{5.3b}$$

$$I_C = \oint_{E=E_C} \dot{x}dx \tag{5.3c}$$

$$\dot{x} = \sqrt{2[E_C - U(x)]}\,] \tag{5.3d}$$

On the other hand Matkowsky et al. [44] rederived Eq. (5.2a) using the first passage time approach. Moreover Matkowsky et al. [44] and Kramers [4] developed these formulas for point particles with energy given by

$$E = \frac{p^2}{2m} + U(x) \tag{5.4}$$

where $p$ is the momentum of the particle and $x$ is its position and again $m$ is taken equal to unity. In all the calculations, a single potential well is assumed so that a particle once having reached the separatrix never returns, which means $\rho(g) = 0$, where $\rho(g)$ is the density of particles at the separatrix $g$. However, a similar analysis also holds for orientations of the magnetization vector for magnetic particles, the role of the inertia in mechanical problems being mimicked, as stated in Section III, by the gyromagnetic term in Gilbert's equation [5]. Thus, in this section we shall calculate $\Gamma_1$ for spins by the method of Matkowsky et al. [44] giving all the details of the considerable mathematical manipulations that are involved.

In order to set this calculation in its proper context, we remark that Brown [62] in his 1979 calculation (see Sections III and IV) for nonaxially symmetric potentials of the magnetocrystalline anisotropy, only considered the IHD case when adapting the Kramers theory to spin relaxation. On the other hand Kramers [4] also showed (Section II) (by essentially treating the low-damping case as a perturbation of the zero-damping case and constructing a diffusion equation for the energy) how a simple formula Eq. (2.192) [his Eq. (28)] for the inverse relaxation time (escape rate) could be obtained in the very low-damping limit. There the condition that the energy loss per cycle of the almost periodic motion of a particle at the critical energy should be much less than the thermal energy holds. Noting this condition Klik and Gunther [64, 72] used the theory of first passage times to obtain the magnetic analogue of the Kramers low-damping formula, thus bypassing the Kramers energy controlled diffusion method entirely. Their analysis thus completed the extension of the ideas of Kramers to spin relaxation, as it was now possible to delineate a region of validity as a function of the friction of the various escape rate formulas for spins just as in the corresponding Kramers formulas for Brownian particles [68]. Subsequently, the low-damping formula has been rederived by Coffey [66] who used the energy diffusion method as slightly modified by Praestgaard and van Kampen [75] (i.e., the virial theorem [32] is invoked) and concisely by Garanin et al. [70] who (in a discussion of how nonaxially symmetric asymptotes tend to the axially symmetric asymptotes in the appropriate limits, described in detail in Section VII) again used the Kramers method. We now describe the basic PDEs underlying the problem.

### B. The Adjoint Fokker–Planck Operator and Differential Equation for the MFPT

In order to calculate the MFPT, in general (we remark that, although we shall be concerned only with the high barrier or weak noise limit here, the concept of a FPT holds [55], unlike that of an escape rate, *irrespective of the height* of any potential barrier that may be involved), it is first necessary to construct the adjoint Fokker–Planck equation (or backward Kolmogorov equation) from the Fokker–Planck (or forward Kolmogorov) equation for the distribution function of magnetic moments on the surface of the unit sphere. Risken [25, 26] gives the Fokker–Planck operator in $n$-dimensions as

$$L_{\mathrm{FP}}(\{x\},t) = -\frac{\partial}{\partial x_i} D_i^{(1)}(\{x\},t) + \frac{\partial^2}{\partial x_i \partial x_j} D_{ij}^{(2)}(\{x\},t) \tag{5.5}$$

where

$\{x\} = x_1, x_2, \ldots, x_n$ are the coordinates of a point in the state space $\{x\}$ at time $t$

$D_i^{(1)}, D_{ij}^{(2)}$ are the drift and diffusion coefficients, respectively,

and the Einstein summation convention is used.

In terms of the spherical polar coordinates $(\varphi, p)$, where $p = \cos\theta$ [64, 68, 72] the operator in Eq. (5.5) operating on $W(\varphi, p, t)$ becomes (with $a^2 \ll 1$ so that $g' \simeq (\gamma/M_s)$)

$$\begin{aligned} L_{\mathrm{FP}} W(p, \varphi, t) = & \frac{\partial}{\partial p}\left[\frac{a\gamma}{M_s}(1-p^2)H_p + \frac{\gamma}{M_s}H_\varphi\right]W + \frac{\partial}{\partial p}\left[\frac{1}{\beta}\frac{a\gamma}{M_s}(1-p^2)\frac{\partial}{\partial p}\right]W \\ & + \frac{\partial}{\partial \varphi}\left[-\frac{\gamma}{M_s}H_p + \frac{a\gamma}{M_s}(1-p^2)^{-1}H_\varphi\right]W \\ & + \frac{\partial}{\partial \varphi}\left[\frac{1}{\beta}\frac{a\gamma}{M_s}(1-p^2)^{-1}\frac{\partial}{\partial \varphi}\right]W \end{aligned} \tag{5.6}$$

(again $\beta = 1/kT$, see below for the other quantities).

The quantity $W(\varphi, p, t)$ is the concentration of particles whose magnetization vector has orientation $(\varphi, p)$.

We recall that Brown [21] derived the Fokker–Planck equation (see Section III)

$$\frac{\partial W}{\partial t} = L_{\mathrm{FP}} W$$

from Gilbert's equation including the noise terms

$$\frac{d\mathbf{M}}{dt} = \gamma\mathbf{M} \times \left[-\frac{\partial U}{\partial \mathbf{M}} - \eta\frac{d\mathbf{M}}{dt} + \boldsymbol{\Xi}(t)\right] \tag{5.7}$$

where we recall that

$U$ is the anisotropy and field energy density

$\mathbf{M}$ is the magnetization vector for a single particle whose magnitude is $M_s$

$\gamma$ is the gyromagnetic ratio

$\eta$ is a phenomenological damping constant

$$\frac{\partial U}{\partial \mathbf{M}} \equiv \frac{\partial U}{\partial M_x}\mathbf{i} + \frac{\partial U}{\partial M_y}\mathbf{j} + \frac{\partial U}{\partial M_z}\mathbf{k}$$

$-\eta\dot{\mathbf{M}}(t) + \boldsymbol{\Xi}(t)$ is the irregular magnetic field due to the bath degrees of freedom.

If we again introduce the friction $a$ by the equation

$$a = \eta\gamma M_s$$

then, an equivalent form for this equation is (if $a^2 \ll 1$ cf. Eqs. (3.5), (3.11))

$$-\frac{d\mathbf{M}}{dt} = \gamma\left[\mathbf{M} \times \left(\frac{\partial U}{\partial \mathbf{M}} - \boldsymbol{\Xi}\right)\right] + \frac{a\gamma}{M_s}\left[\mathbf{M} \times \left(\frac{\partial U}{\partial \mathbf{M}} - \boldsymbol{\Xi}\right)\right] \times \mathbf{M} \tag{5.8}$$

Hence, as in Sections III, IV, Klik and Gunther [64] wrote down the averaged (noiseless) Langevin equations (i.e., in the absence of the noise terms) for a spin in spherical polar coordinates (the results for the Gilbert equation are regained simply by replacing $\gamma$ by $\gamma/(1 + a^2)$

$$\dot{p} = -\frac{a\gamma}{M_s}(1 - p^2)H_p - \frac{\gamma}{M_s}H_\varphi \tag{5.9a}$$

$$\dot{\varphi} = \frac{\gamma}{M_s}H_p - \frac{a\gamma}{M_s}(1 - p^2)^{-1}H_\varphi, \quad a^2 \ll 1 \tag{5.9b}$$

where $H$ is the Hamiltonian $(U)$. The noiseless Langevin equations may now be used to write down the drift and diffusion coefficients in the FPE. First, we note that the diffusion term is

$$\frac{\partial}{\partial p}\left[\frac{1}{\beta}\frac{a\gamma}{M_s}(1 - p^2)\frac{\partial W}{\partial p}\right] = \frac{\partial^2}{\partial p^2}\left[\frac{1}{\beta}\frac{a\gamma}{M_s}(1 - p^2)W\right] + \frac{\partial}{\partial p}2\frac{1}{\beta}\frac{a\gamma}{M_s}pW \tag{5.10}$$

By substituting this into the equation for $L_{\mathrm{FP}}W$, and by using Eqs. (5.9a) and (5.9b) to calculate the drift and diffusion coefficients from the current density vector (the Einstein method) in the manner outlined in Section III, yields

$$\begin{aligned} L_{\mathrm{FP}}W = {} & \frac{\partial}{\partial p}\left[\frac{a\gamma}{M_s}(1 - p^2)H_p + \frac{\gamma}{M_s}H_\varphi + 2\frac{1}{\beta}\frac{a\gamma}{M_s}p\right]W + \frac{\partial^2}{\partial p^2}\frac{1}{\beta}\frac{a\gamma}{M_s}(1 - p^2)W \\ & + \frac{\partial}{\partial \varphi}\left[-\frac{\gamma}{M_s}H_p + \frac{a\gamma}{M_s}(1 - p^2)^{-1}H_\varphi\right]W + \frac{\partial^2}{\partial \varphi^2}\frac{1}{\beta}\frac{a\gamma}{M_s}(1 - p^2)^{-1}W \end{aligned} \tag{5.11}$$

which is of the form of Eq. (5.5) with drift and diffusion coefficients

$$
\begin{aligned}
D_p^{(1)} &= -\frac{a\gamma}{M_s}(1-p^2)H_p - \frac{\gamma}{M_s}H_\varphi - 2\frac{1}{\beta}\frac{a\gamma}{M_s}p \\
D_\varphi^{(1)} &= \frac{\gamma}{M_s}H_p - \frac{a\gamma}{M_s}(1-p^2)^{-1}H_\varphi \\
D_{pp}^{(2)} &= \frac{1}{\beta}\frac{a\gamma}{M_s}(1-p^2) \\
D_{\varphi\varphi}^{(2)} &= \frac{1}{\beta}\frac{a\gamma}{M_s}(1-p^2)^{-1} \\
D_{p\varphi}^{(2)} &= D_{\varphi p}^{(2)} = 0
\end{aligned}
\tag{5.12}
$$

Now, the backward Kolmogorov equation is, according to Risken [25, 26]

$$
\frac{\partial}{\partial t'}P(\{x\},t|\{x'\},t') = -L_{\text{FP}}^{+}(\{x'\},t')P(\{x\},t|\{x'\},t') \tag{5.13}
$$

$$
L_{\text{FP}}^{+}(\{x'\},t') = D_i^{(1)}(\{x'\},t')\frac{\partial}{\partial x_i'} + D_{ij}^{(2)}(\{x'\},t')\frac{\partial^2}{\partial x_i'\partial x_j'} \tag{5.14}
$$

Hence, the Fokker–Planck adjoint operator $L_{\text{FP}}^{+}$ is given by cf. Section III

$$
L_{\text{FP}}^{+} = D_p^{(1)}\frac{\partial}{\partial p'} + D_\varphi^{(1)}\frac{\partial}{\partial \varphi'} + D_{pp}^{(2)}\frac{\partial^2}{\partial p'^2} + D_{\varphi\varphi}^{(2)}\frac{\partial^2}{\partial \varphi'^2} \tag{5.15}
$$

since

$$
D_{p\varphi}^{(2)} = D_{\varphi p}^{(2)} = 0
$$

so that the adjoint operator is

$$
\begin{aligned}
L_{\text{FP}}^{+} = {} & \left[-\frac{a\gamma}{M_s}(1-p'^2)H_{p'} - \frac{\gamma}{M_s}H_{\varphi'} - 2\frac{1}{\beta}\frac{a\gamma}{M_s}p'\right]\frac{\partial}{\partial p'} \\
& + \left[\frac{\gamma}{M_s}H_{p'} - \frac{a\gamma}{M_s}(1-p'^2)^{-1}H_{\varphi'}\right]\frac{\partial}{\partial \varphi'} \\
& + \frac{1}{\beta}\frac{a\gamma}{M_s}(1-p'^2)\frac{\partial^2}{\partial p'^2} + \frac{1}{\beta}\frac{a\gamma}{M_s}(1-p'^2)^{-1}\frac{\partial^2}{\partial \varphi'^2}
\end{aligned}
\tag{5.16}
$$

The partial differential equation for the MFPT is then in terms of the source polar coordinates $(\varphi', p')$ as in Risken [25, 26] Section VIII, Eq. 8.15(a)

$$L_{\mathrm{FP}}^{+}(\{x'\})\tau(\{x'\}) = -1 \tag{5.17}$$

which must be solved subject to the boundary conditions that $\tau(\{x'\})$ must vanish at the saddle point $\{x_m\}$, which is an absorbing point or trap. Furthermore, since the poles of the sphere are reflecting boundaries situated at $p = \pm 1$, the probability current must vanish at the poles.

The backward operator, which refers to the evolution of the system starting from the source point $\{x'\}$, is written in terms of the source spherical polar coordinates $\{x'\}$ while the forward operator is written in terms of the field coordinates $\{x\}$. In general, this equation may only be solved in closed form if the magnetic moment is a function of the angle $\theta$ only, hence as in Section III the mean first passage time may be written in terms of quadratures as previously described by Coffey [68]. The resulting formula (Eqs. (3.54), (3.55)) for the MFPT is then valid for all values of the damping parameter $a$. In the general case, where the distribution function depends on both $\theta$ and $\varphi$, the MFPT is not usually used (because of the difficulties involved in integrating partial differential equations when more than one space variable is involved), rather the smallest nonvanishing eigenvalue $\lambda_1$ of the Fokker–Planck equation is calculated. The inverse of this (Section III) in the high barrier limit is then the normalized relaxation time, where the relaxation time itself is $\tau \approx 2\tau_N/\lambda_1$.

The above discussion pertains to the exact calculation of the MFPT. Here, we are only interested in the MFPT for very weak damping, (where $\tau_2$ is neglected) so that we may use the method (based on a uniform asymptotic expansion of the MFPT), which (as has been briefly described above) has been developed by Matkowsky et al. [44] for mechanical particles and which, in the context of spins, is described below.

### C. Formal Expression for the MFPT from the Uniform Asymptotic Method

Equation (5.17) now allows us, following Klik and Gunther [64, 72], to write down a formal expression for the MFPT $\tau_1$ for spins in the high barrier and very low dissipation limit for nonaxially symmetric potentials in a manner analogous to that used by Matkowsky et al. [44] to treat the Kramers very low-damping case for particles. Just as for particles we will assume a single potential well so that a spin having reached the separatrix $g$ never reverses. The extension to the actual bistable or multistable potentials characteristic of magnetic relaxation may be achieved by extending the arguments given in Section IV.D, Eqs. (4.51) et seq. of Hänggi et al. [1], see also Eqs. (5.35) et seq. of this section below.

According to Matkowsky et al. [44] in the context of particles and Klik and Gunther [64, 72] in the context of spins, the contour of critical energy, $E = E_C$, which passes through the saddle point, lies within the boundary layer near the separatrix, $g$, separating (taking the spin example) the clockwise and anticlockwise spins (see Fig. 7). Thus the uniform asymptotic expansion method relies on the introduction of a domain, $Q$, such that $E < E_C$, *so that the spin cannot reverse on the interior* of $Q$, *and* $E = E_C$ *on the boundary*, $\partial(Q)$ of the domain. It is assumed that the boundary, $\partial(Q)$, is so close to the separatrix, $g$, that the passage time, $\tau_2$, from $\partial(Q)$ to $g$ may be ignored to the first order in temperature, that is, the *low* temperature, *weak* noise limit, in comparison to the mean time, $\tau(\varphi, p)$ (i.e., $\tau_1$), to reach the boundary $\partial(Q)$ starting from a point $(\varphi, p)$ well embedded in $Q$. Thus, in Eq. (5.17), following the method of Matkowsky et al. [44] and Klik and Gunther [64, 72], we introduce an exponentially large quantity, $\tau(Q)$, independent (because the noise is assumed to be very weak corresponding to the high-barrier, low-temperature limit) of the starting point $(\varphi', p')$ in $Q$, with

$$\tau(\varphi', p') = \tau(Q) u_T(\varphi', p') \tag{5.18}$$

and $\sup_Q \{u_T(\varphi', p')\} = 1$.

Multiplying Eq. (5.17) across by $e^{-\beta H}$ and integrating over the domain $Q$ then yields

$$\iint\limits_Q e^{-\beta H} L_{\mathrm{FP}}^{+} \tau(\varphi', p') dp' d\varphi' = -\iint\limits_Q e^{-\beta H} dp' d\varphi' \tag{5.19}$$

Now using Eqs. (5.18) and (5.19), we have

$$-\iint\limits_Q e^{-\beta H} dp' d\varphi' = \tau(Q) \iint\limits_Q e^{-\beta H} L_{\mathrm{FP}}^{+} u_T(\varphi', p') dp' d\varphi' \tag{5.20}$$

since $\tau(Q)$ is constant on $Q$.

Thus

$$\tau(Q) = -\frac{\iint\limits_Q e^{-\beta H} dp' d\varphi'}{\iint\limits_Q e^{-\beta H} L_{\mathrm{FP}}^{+} u_T(\varphi', p') dp' d\varphi'} \tag{5.21}$$

Equation (5.21) is the leading term in the uniform asymptotic expansion of the first passage time of Matkowsky et al. [44] for the problem at hand. The next step in the calculation is to express the denominator of the right-hand side of Eq. (5.21), which is a surface integral over $Q$, as a line integral along the boundary of $Q$, $\partial(Q)$ using Stokes' theorem. This calculation is rather intricate, so we present it in detail in Appendix V.I and simply give the main steps in the calculation below.

## D. Expression of the Denominator of Eq. (5.21) as a Line Integral Using Stokes' Theorem

The first step in this calculation is to write out the denominator in Eq. (5.21) explicitly as follows:

$$\begin{aligned}\iint_Q e^{-\beta H} L_{\mathrm{FP}}^{+} u_T dp' d\varphi' = \iint_Q \Bigg\{ & e^{-\beta H}\left[-\frac{a\gamma}{M_s}(1-p'^2)H_{p'} - \frac{\gamma}{M_s}H_{\varphi'} - 2p'\frac{1}{\beta}\frac{a\gamma}{M_s}\right]\frac{\partial u_T}{\partial p'} \\ & + e^{-\beta H}\left[\frac{\gamma}{M_s}H_{p'} - \frac{a\gamma}{M_s}(1-p'^2)^{-1}H_{\varphi'}\right]\frac{\partial u_T}{\partial \varphi'} \\ & + e^{-\beta H}\left[\frac{1}{\beta}\frac{a\gamma}{M_s}(1-p'^2)\frac{\partial^2 u_T}{\partial p'^2} + \frac{1}{\beta}\frac{a\gamma}{M_s}(1-p'^2)^{-1}\frac{\partial^2 u_T}{\partial \varphi'^2}\right]\Bigg\} dp' d\varphi'\end{aligned} \tag{5.22}$$

Equation (5.21) may then be written with the aid of Stokes' theorem (details in Appendix V.I) as

$$\tau(Q) = -\frac{\iint_Q e^{-\beta H} dp' d\varphi'}{(1/\beta)(a\gamma/M_s) \oint_{\partial(Q)} e^{-\beta H}\left[(1-p'^2)(\partial u_T/\partial p')d\varphi' - (1-p'^2)^{-1}(\partial u_T/\partial \varphi')dp'\right]} \tag{5.23}$$

which is an expression for the first mean passage time, $\tau(Q)$, in the limit of high-barriers essentially in terms of the quantity $u_T$; $u_T$ must now be expressed in terms of known quantities. To do this, we need to construct a *boundary layer approximation* to $u_T$ near $E = E_C$ by introducing the stretching transformation:

$$\eta = \beta(E_C - E) = \beta(E_C - H) \tag{5.24}$$

The calculations for $u_T$ in the boundary layer (using the dimensionless energy variable $\eta$) are again rather involved, thus they are presented in detail in Appendices V.II and V.III.

### E. Final Result for the MFPT Using the Stretching Transformation

Since

$$u_T = 1 - e^{-\eta} \text{ (see Appendix V.III)} \tag{5.25}$$

we have

$$\frac{\partial u_T}{\partial p'} = e^{-\eta}\frac{\partial \eta}{\partial p'} = e^{-\eta}(-\beta H_{p'}) \quad \text{and} \quad \frac{\partial u_T}{\partial \varphi'} = e^{-\eta}\frac{\partial \eta}{\partial \varphi'} = e^{-\eta}(-\beta H_{p'}) \tag{5.26}$$

So the denominator in Eq. (5.23) becomes

$$\begin{aligned} &\frac{1}{\beta}\frac{a\gamma}{M_s}\oint_{\partial(Q)} e^{-\beta H}e^{-\beta E_C}e^{\beta H}[-(1-p'^2)\beta H_{p'}d\varphi' + (1-p'^2)^{-1}\beta H_{p'}dp'] \\ &= -\frac{a\gamma}{M_s}e^{-\beta E_C}\oint_{\partial(Q)}[(1-p'^2)H_{p'}d\varphi' - (1-p'^2)^{-1}H_{\varphi'}dp'] \end{aligned} \tag{5.27}$$

$$= -\frac{\gamma}{M_s}\Delta E e^{-\beta E_C} \tag{5.28}$$

where [64, 72]

$$\Delta E \equiv a\oint_{\partial(Q)}[(1-p'^2)H_{p'}d\varphi' - (1-p'^2)^{-1}H_{\varphi'}dp'] \tag{5.29}$$

and $a$ is the friction (see the Gilbert equation above).

The parameter $\Delta E$ [1] is the energy loss per cycle of the almost periodic motion at the saddle point energy, which follows by taking the time average of $\dot{M}^2$ and substituting for $p$, $\varphi$ using Eqs. (5.9a,b). Equation (5.28) represents the final simplification of the denominator.

The surface integral in the numerator may be approximately evaluated using the method of steepest descents as follows. We expand $H$ as a Taylor series about the minimum $E_0$

$$H = E_0 + \frac{(p-p_0)^2}{2!}E^0_{pp} + \frac{(\varphi-\varphi_0)^2}{2!}E^0_{\varphi\varphi} \tag{5.30}$$

where $E_0$ is the value of the energy at the minimum, $E^0_{pp}, E^0_{\varphi\varphi}$ are the partial derivatives of the energy function evaluated at the minimum, and we have

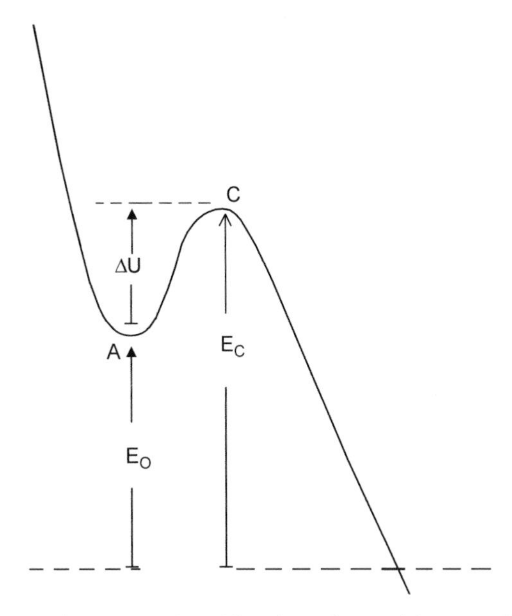


**Figure 8.** Cross-sectional (through saddle) view of potential energy diagram used in the calculations of Klik and Gunther.

chosen a system of coordinates in which the second-order mixed partial derivatives vanish.

We shall now assume that most of the particles stay in the well $E < E_C$ and that the number at large distances from the minimum is negligible. So we may extend the double integral over $Q$ to a double integral over the entire sphere without serious error. We further assume that instead of integrating over $p$ from $-1$ to $+1$ that we can integrate from $-\infty$ to $+\infty$. We also assume that we can integrate over $\varphi$ from $-\infty$ to $\infty$ instead of 0 to $2\pi$.

So the numerator of the right-hand side of Eq. (5.23) becomes

$$\int_{-\infty}^{\infty}\int_{-\infty}^{\infty} e^{-\beta E_0} e^{-(1/2)\beta(p'-p'_0)^2 E^0_{pp}} e^{-(1/2)\beta(\varphi'-\varphi'_0)^2 E^0_{\varphi\varphi}} dp' d\varphi' \approx e^{-\beta E_0} \frac{2\pi}{\beta\sqrt{E^0_{pp} E^0_{\varphi\varphi}}} \tag{5.31}$$

and the final formula for the mean first passage time from the minimum to the critical energy curve in accordance with the equation

$$\Gamma = \frac{1}{\tau_1(A)}$$

[combining Eqs. (5.28) and (5.31)] is

$$\tau(Q) = \frac{M_s}{\gamma} \frac{2\pi}{\beta \Delta E} \frac{1}{\sqrt{E_{pp}^0 E_{\varphi\varphi}^0}} e^{\beta(E_C - E_0)} \tag{5.32}$$

which on noting that the well angular frequency is given by

$$\omega_A \equiv \frac{\gamma}{M_s} \sqrt{E_{pp}^0 E_{\varphi\varphi}^0}, \tag{5.33}$$

becomes (cf. Fig. 8)

$$\tau(Q) = \frac{2\pi}{\omega_A} \frac{1}{\beta \Delta E} e^{\beta(E_C - E_0)} \qquad E_C - E_0 \equiv \Delta U \tag{5.34}$$

which is the time to reach the separatrix and differs from Eq. (4.5) of Klik and Gunther [64] by a factor of 0.5, a correction that has been checked with and agreed to by Klik [76]. The escape rate in accordance with Eqs. (5.1a–c) is then $[\tau(Q)]^{-1}$.

We remark that Eq. (5.34) has been derived under the assumption that all spins are absorbed at the boundary, that is, having reached the separatrix they *never return to their original orientation*. In practice, the potential in ferromagnetism will always have several local states of stability [5, 62]. For the purposes of illustration it is useful to consider a symmetric bistable potential that arises if a transverse field is applied to a system with simple uniaxial anisotropy. Here the Kramers escape rate is, in the low-damping limit,

$$\Gamma = \frac{\omega_A}{\pi} \beta \frac{\Delta E}{2} e^{-\beta(E_C - E_0)} \tag{5.35a}$$

$$= \beta \frac{\Delta E}{2} \Gamma_{\text{TS}} \tag{5.35b}$$

Since there are two wells. This equation takes account of crossings and recrossings of moments in a bistable potential. The escape rate from one of the wells is then

$$\Gamma_{\text{one well}} = \frac{\omega_A}{4\pi} \beta \Delta E e^{-\beta(E_C - E_0)} \tag{5.36}$$

with corresponding relaxation (escape) time

$$\tau_e = \Gamma^{-1}_{\text{one well}} = \frac{4\pi}{\omega_A}\frac{1}{\beta\Delta E}e^{\beta(E_C - E_0)} \tag{5.37}$$

The above results may be justified as follows: $\tau$, as calculated in Eq. (5.37) takes account of crossings and recrossings that occur with a probability of 0.5. The time to reach the separatrix, which is the quantity rendered by Eq. (5.34) (where it is supposed that all particles that reach the separatrix *never* return), is cf. Eq. (5.34)

$$\tau = \frac{\tau_e}{2} = \frac{2\pi}{\omega_A}\frac{1}{\beta\Delta E}e^{\beta(E_C - E_0)} \tag{5.38}$$

The Kramers escape rate as rendered by Eq. (5.35) is the *sum* of the escape rates from each of the two wells i.e.,

$$\Gamma = \left(\frac{\omega_A}{4\pi} + \frac{\omega_A}{4\pi}\right)\beta\Delta E e^{-\beta(E_C - E_0)}$$

We have illustrated how the calculation for a single well potential may be adapted to a symmetric bistable potential. The extension to an asymmetric bistable potential may be carried out in the manner described in Section IV.D of [1], which is alluded to above. The result, which is similar to that for particles, is

$$\Gamma = \beta\frac{\Delta E_1 \Delta E_2}{\Delta E_1 + \Delta E_2}\left(\Gamma^{\text{TS}}_{A\to B} + \Gamma^{\text{TS}}_{B\to A}\right) \tag{5.39a}$$

$$= \beta\frac{\Delta E_1 \Delta E_2}{\Delta E_1 + \Delta E_2}\Gamma_{\text{TS}} \tag{5.39b}$$

where

$$\Gamma^{\text{TS}}_{A\to B} = \frac{\omega_A}{2\pi}e^{-\beta(E_C - E_1)} \tag{5.40}$$

$$\Gamma^{\text{TS}}_{B\to A} = \frac{\omega_B}{2\pi}e^{-\beta(E_C - E_2)} \tag{5.41}$$

The parameter $E_i$ is the value of the energy at the minimum in well number $i$.

The parameter $\Delta E_i$ is the energy loss per cycle at the saddle point for particles in well $i$.

### F. Evaluation of the Escape Rate for a Weak Transverse Field

The evaluation of $\Gamma$ from the contour integral for $\Delta E$ the energy loss per cycle is, in general, a difficult task as one has to eliminate either $p$ or $\varphi$ by means of the equation of the separatrix $H - E_C = 0$. In general, this will involve the solution of a high-order polynomial equation particularly if there are applied fields. In many cases, the best way to proceed is to eliminate $p$, as the integral over $\varphi$ is between 0 and $2\pi$.

In the particular case (cf. Section VII, 5) of a very small uniform transverse field, that is, a small field, $h$, applied at an angle $\psi = \pi/2$ to the easy axis of magnetization, we have the potential function [70]

$$H = E = -K(p^2 + 2h\sqrt{1-p^2}\cos\varphi) \tag{5.42a}$$

$$\approx -K(p^2 + 2h\cos\varphi) \tag{5.42b}$$

Now, since the critical energy curve passes through the saddle point $C$, the critical energy is the saddle point energy. So, at the saddle point (for a transverse field) $\theta = \pi/2$ and $\varphi = 0$, and thus $p = 0$, in which case

$$E_C = -2Kh \tag{5.43}$$

Therefore the equation

$$E - E_C = 0 \tag{5.44}$$

implies that

$$2Kh - K(p^2 + 2h\cos\varphi) = 0 \tag{5.45}$$

which, on solving for $p$, yields

$$p = \pm\sqrt{2h(1 - \cos\varphi)} \tag{5.46}$$

$$= \pm 2\sqrt{h}\sin\frac{\varphi}{2} \tag{5.47}$$

and

$$dp = \pm\sqrt{h}\cos\frac{\varphi}{2}d\varphi \tag{5.48}$$

Now, using Eq. (5.42), we see that

$$\frac{\partial E}{\partial p} = -2Kp \qquad \frac{\partial E}{\partial \varphi} = 2Kh \sin \varphi \tag{5.49}$$

and so, substituting these results into Eq. (5.28), suppressing the primes, and remembering that the major contribution to the integral comes from the vicinity of the saddle point where $p \approx 0$, the energy loss per cycle of the motion may be seen to be approximately

$$\Delta E \approx a \oint_{E=E_C} \left[ \frac{\partial E}{\partial p} d\varphi - \frac{\partial E}{\partial \varphi} dp \right] \tag{5.50}$$

$$= a \int_0^{2\pi} \left[ 4K\sqrt{h} \sin \frac{\varphi}{2} - 2Kh^{3/2} \sin \varphi \cos \frac{\varphi}{2} \right] d\varphi \tag{5.51}$$

$$\approx 4aK\sqrt{h} \int_0^{2\pi} \sin \frac{\varphi}{2} d\varphi \tag{5.52}$$

(since $h$ is small), thus

$$\beta \Delta E = 16 a \sigma \sqrt{h} \tag{5.53}$$

In Garanin's notation this yields the correction to the TST result as [70] cf. Eq. (7.126)

$$A = \left( \frac{\beta \Delta E}{2} \right) = 8 a \sigma \sqrt{h}.$$

An interesting consequence of this result is that as $h \to 0$, the Brown axially symmetric result (Eq. (1.18a)) for the $K \sin^2 \theta$ potential which is valid for all values of $a$ is *not* regained as $A$ of Eq. (5.53) tends to zero rather than $2\pi a \sqrt{\sigma / \pi}$. *This is an example of the breakdown of the nonaxially symmetric asymptotes for small departures from axial symmetry*, which is discussed at length in Section VII.

### G. Conclusion

In this section, in view of the extreme importance of accurate theoretical expressions for the superparamagnetic relaxation time for the purpose of interpretation of experimental results on individual small ferromagnetic particles

(Wernsdorfer et al. [77, 78]) we have given in detail the calculation of the low-damping Kramers' escape rate using the uniform asymptotic expansion of the transition rate proposed by Matkowsky et al. [44] and adapted to spins by Klik and Gunther [64]. The analysis in this section provides the details of the complicated mathematical manipulations that are required in order to establish the low-damping formula for spins and that have not been given hitherto.

The results of the present detailed analysis verify the calculations of Klik and Gunther [64] and are in agreement with derivations of the low-damping formula using the entirely different energy diffusion method of Kramers as adapted to spins that have been independently carried out by Coffey [66] using the method of Praestgaard and van Kampen [75] and by Garanin et al. [70] using a method involving a change of variables to energy and azimuthal angle. It should be noted that the LD calculation is peculiar to magnetic relaxation essentially because of the *non separable* form of the Hamiltonian. In the case of dielectric relaxation where the Hamiltonian is *additive* as in the original Kramers problem the appropriate formulation is the extension of Kramers' theory for mechanical particles to inertial rotators. Thus the three cases of the Kramers theory will again appear with the VHD limit being governed by the Smoluchowski equation for diffusion in the space of Euler angles. The diffusion equation developed by Debye (see Section III) is a particular case of this equation. Thus, the only formula from the theory of magnetic relaxation in nonaxially symmetric potentials that is common to dielectric relaxation is the IHD formula (Eq. 4.87) developed in Section IV with the gyromagnetic term set equal to 0. This will then yield a formula for the Kramers' escape rate for a rotator in a nonaxially symmetric nematic potential, say, if the inertia of the rotator is neglected. The principal practical use of such a formula would be to estimate [83] the retardation ($g$) factors [36] for dielectric relaxation of nematic liquid crystals in non-axially symmetric potentials.

It is apparent from the brief survey of the calculations of Matkowsky et al. [44] of the Kramers escape rate for particles and our detailed account of the calculations of the corresponding quantity for spins, that the MFPT method provides a powerful method of attack on the problem of the Kramers low-damping escape rate. The integration of the adjoint FPE or the calculation of the Green's function of the FPE in the zero-frequency limit, which will also yield the MFPT, as described in Section III, is, however, beset with the usual mathematical difficulties associated with PDEs in two space variables or higher. Thus, if one wishes to calculate MFPTs for arbitary energy functions, one must usually express the solution of the relevant PDE as a recurrence (Section VIII) relation as suggested by Coffey [58]. We shall, in Section VI, briefly outline the problems that arise in the Kramers theory in regions where no expansion in a small parameter is possible. The best known such transition region is the region between VLD and IHD where the energy loss per cycle is of the order of the thermal energy.

## Appendix V.I. Derivation of Eq. (5.23) from Eq. (5.21)

We expand the terms on the right-hand side of Eq. (5.22) as follows:

$$-e^{-\beta H}\frac{a\gamma}{M_s}(1-p'^2)H_{p'}\frac{\partial u_T}{\partial p'} = \frac{a\gamma}{M_s}kT(1-p'^2)\frac{\partial e^{-H/kT}}{\partial p'}\frac{\partial u_T}{\partial p'}\cdots \tag{A}$$

$$-\frac{\gamma}{M_s}H_{\varphi'}e^{-\beta H}\frac{\partial u_T}{\partial p'} = \frac{1}{\beta}\frac{\gamma}{M_s}(-\beta H_{\varphi'}e^{-\beta H})\frac{\partial u_T}{\partial p'} = \frac{1}{\beta}\frac{\gamma}{M_s}\frac{\partial e^{-\beta H}}{\partial \varphi'}\frac{\partial u_T}{\partial p'}\cdots \tag{B}$$

$$-2p'\frac{1}{\beta}\frac{a\gamma}{M_s}e^{-\beta H}\frac{\partial u_T}{\partial p'} = +\frac{1}{\beta}\frac{a\gamma}{M_s}\left\{\frac{\partial}{\partial p'}(1-p'^2)\right\}e^{-\beta H}\frac{\partial u_T}{\partial p'}\cdots \tag{C}$$

$$\frac{\gamma}{M_s}H_{p'}e^{-\beta H}\frac{\partial u_T}{\partial \varphi'} = -\frac{1}{\beta}\frac{\gamma}{M_s}(-\beta H_{p'}e^{-\beta H})\frac{\partial u_T}{\partial \varphi'} = -\frac{1}{\beta}\frac{\gamma}{M_s}\frac{\partial e^{-\beta H}}{\partial p'}\frac{\partial u_T}{\partial \varphi'}\cdots \tag{D}$$

$$-e^{-\beta H}\frac{a\gamma}{M_s}(1-p'^2)^{-1}H_{\varphi'}\frac{\partial u_T}{\partial \varphi'} = \frac{1}{\beta}\frac{a\gamma}{M_s}(1-p'^2)^{-1}\frac{\partial e^{-\beta H}}{\partial \varphi'}\frac{\partial u_T}{\partial \varphi'}\cdots \tag{E}$$

Referring to the first term in the final square bracket of the above double integral on the right-hand side of Eq. (5.22) as $\ldots F$, and to the second term as $\ldots G$

Collecting together $A + C + F$ yields $\quad \displaystyle\frac{1}{\beta}\frac{a\gamma}{M_s}\frac{\partial}{\partial p'}\left[(1-p'^2)e^{-\beta H}\frac{\partial u_T}{\partial p'}\right]$

Collecting together $E + G$ yields $\quad \displaystyle\frac{1}{\beta}\frac{a\gamma}{M_s}\frac{\partial}{\partial \varphi'}\left[(1-p'^2)^{-1}e^{-\beta H}\frac{\partial u_T}{\partial \varphi'}\right]$

Collecting together $B + D$ yields $\quad \displaystyle\frac{1}{\beta}\frac{\gamma}{M_s}\left[\frac{\partial e^{-\beta H}}{\partial \varphi'}\frac{\partial u_T}{\partial p'} - \frac{\partial e^{-\beta H}}{\partial p'}\frac{\partial u_T}{\partial \varphi'}\right]$

So,

$$\iint\limits_Q e^{-\beta H} L_{\mathrm{FP}}^{+} u_T dp' d\varphi' = \frac{1}{\beta}\frac{\alpha\gamma}{M_s}\iint\limits_Q \left\{ \frac{\partial}{\partial p'}\left[(1-p'^2)e^{-\beta H}\frac{\partial u_T}{\partial p'}\right] + \frac{\partial}{\partial \varphi'}\left[(1-p'^2)^{-1}e^{-\beta H}\frac{\partial u_T}{\partial \varphi'}\right]\right\} dp' d\varphi' + \frac{1}{\beta}\frac{\gamma}{M_s}\iint\limits_Q \left[\frac{\partial e^{-\beta H}}{\partial \varphi'}\frac{\partial u_T}{\partial p'} - \frac{\partial e^{-\beta H}}{\partial p'}\frac{\partial u_T}{\partial \varphi'}\right] dp' d\varphi' \tag{5.54}$$

Now, on applying Stokes' theorem in the plane [79, 80] (also known as Green's theorem in the plane) to the second integral on the right-hand side we find

$$\frac{1}{\beta}\frac{\gamma}{M_s}\iint\limits_Q \left[\frac{\partial e^{-H/kT}}{\partial \varphi'}\frac{\partial u_T}{\partial p'} - \frac{\partial e^{-H/kT}}{\partial p'}\frac{\partial u_T}{\partial \varphi'}\right] dp' d\varphi' = \frac{1}{\beta}\frac{\gamma}{M_s}\oint_{\partial(Q)} e^{-\beta H}\left[\frac{\partial u_T}{\partial \varphi'}d\varphi' + \frac{\partial u_T}{\partial p'}dp'\right] \tag{5.55}$$

$$= \frac{1}{\beta}\frac{\gamma}{M_s}\oint_{\partial(Q)} e^{-\beta H} du_T \tag{5.56}$$

and, since on $\partial(Q)$ we have $H = E_C = \text{const}$, we find

$$\frac{1}{\beta}\frac{\gamma}{M_s}\iint\limits_Q \left[\frac{\partial e^{-\beta H}}{\partial \varphi'}\frac{\partial u_T}{\partial p'} - \frac{\partial e^{-\beta H}}{\partial p'}\frac{\partial u_T}{\partial \varphi'}\right] dp' d\varphi' = \frac{1}{\beta}\frac{\gamma}{M_s} e^{-\beta E_c}\oint_{\partial(Q)} du_T \tag{5.57}$$

$$= 0 \tag{5.58}$$

Again, applying the same theorem to the first integral on the right-hand side above we find

$$\frac{1}{\beta}\frac{a\gamma}{M_s}\iint\limits_Q \left\{ \frac{\partial}{\partial p'}\left[(1-p'^2)e^{-\beta H}\frac{\partial u_T}{\partial p'}\right] + \frac{\partial}{\partial \varphi'}\left[(1-p'^2)^{-1}e^{-\beta H}\frac{\partial u_T}{\partial \varphi'}\right]\right\} dp' d\varphi' = \frac{1}{\beta}\frac{a\gamma}{M_s}\oint_{\partial(Q)} \left[(1-p'^2)e^{-\beta H}\frac{\partial u_T}{\partial p'}d\varphi' - (1-p'^2)^{-1}e^{-\beta H}\frac{\partial u_T}{\partial \varphi'}dp'\right] \tag{5.59}$$

and the desired result [Eq. (5.23)] follows.

### Appendix V.II. Justification for the Use of the Boundary Layer Approximation for $u_T$

The Langevin equations in the absence of noise arise from Gilbert's equation, ignoring the noise terms, we have $a^2 \ll 1$ so that we are effectively using the Landau-Lifshitz equation [5] cf. Eqs. (4.73), (4.74) [64]:

$$\begin{aligned}\dot{p} &= -\frac{a\gamma}{M_s}H_p(1-p^2) - \frac{\gamma}{M_s}H_\varphi \\ \dot{\varphi} &= \frac{\gamma}{M_s}H_p - \frac{a\gamma}{(1-p^2)M_s}H_\varphi\end{aligned} \tag{5.60}$$

Now, $u_T$ [44] satisfies approximately the equation:

$$L_{\text{FP}}^{+}u_T = 0 \tag{5.61}$$

which yields [cf. Eq. (5.16)]:

$$\begin{aligned}&\left[\frac{a\gamma}{M_s}(1-p^2)H_p + \frac{\gamma}{M_s}H_\varphi + 2\frac{a\gamma}{\beta M_s}p\right]\frac{\partial u_T}{\partial p} + \left[-\frac{\gamma}{M_s}H_p + \frac{a\gamma H_\varphi}{M_s(1-p^2)}\right]\frac{\partial u_T}{\partial \varphi} \\ &\quad - \frac{a\gamma}{\beta M_s}(1-p^2)\frac{\partial^2 u_T}{\partial p^2} - \frac{a\gamma}{\beta M_s}\frac{1}{1-p^2}\frac{\partial^2 u_T}{\partial \varphi^2} = 0\end{aligned} \tag{5.62}$$

which in the low-temperature limit further reduces to

$$\left[-\frac{a\gamma}{M_s}(1-p^2)H_p - \frac{\gamma}{M_s}H_\varphi\right]\frac{\partial u_T}{\partial p} + \left[\frac{\gamma}{M_s}H_p - \frac{a\gamma}{M_s}\frac{H_\varphi}{1-p^2}\right]\frac{\partial u_T}{\partial \varphi} = 0 \tag{5.63}$$

and which, by using Eq. (5.60), yields:

$$\frac{\partial u_T}{\partial p}\dot{p} + \frac{\partial u_T}{\partial \varphi}\dot{\varphi} = 0 \tag{5.64}$$

Hence, we have proved that

$$\frac{du_T}{dt} = 0 \tag{5.65}$$

where $(\varphi(t), p(t))$ is any trajectory of Eqs. (5.60), so $u_T$ is constant on any such trajectory. Now, all such trajectories converge to the point $A$ of stable equilibrium, and since $u_T$ is continuous for $E < E_C$

$$u_T(\varphi, p) = u_T(A) = \text{const} \qquad \text{for} \quad E < E_C \tag{5.66}$$

We normalize this constant to 1. Now, $u_T$ must also satisfy the boundary condition

$$u_T = 0 \qquad \text{for } E = E_C \text{ [cf. the boundary condition after Eq. (5.17)]} \tag{5.67}$$

So $u_T \equiv 1$ cannot be a valid solution near $E = E_C$.

This justifies the construction of a boundary layer approximation to $u_T$ near $E = E_C$ as carried out in Appendix V.III.

## Appendix V.III. Derivation and Solution of the Boundary Layer Equation

The partial derivatives of $\eta$ are

$$\eta_p = -\beta H_p \qquad \eta_{pp} = -\beta H_{pp} \qquad \eta_\varphi = -\beta H_\varphi \qquad \text{and} \qquad \eta_{\varphi\varphi} = -\beta H_{\varphi\varphi} \tag{5.68}$$

Also,

$$\frac{\partial}{\partial p} = \frac{\partial \eta}{\partial p}\frac{\partial}{\partial \eta} = -\beta H_p \frac{\partial}{\partial \eta} \tag{5.69}$$

$$\frac{\partial}{\partial \varphi} = \frac{\partial \eta}{\partial \varphi}\frac{\partial}{\partial \eta} = -\beta H_\varphi \frac{\partial}{\partial \eta} \tag{5.70}$$

$$\frac{\partial^2}{\partial p^2} = \frac{\partial}{\partial p}\left(\frac{\partial \eta}{\partial p}\frac{\partial}{\partial \eta}\right) = \frac{\partial^2 \eta}{\partial p^2}\frac{\partial}{\partial \eta} + \left(\frac{\partial \eta}{\partial p}\right)^2 \frac{\partial^2}{\partial \eta^2} = -\beta H_{pp}\frac{\partial}{\partial \eta} + \beta^2 H_p^2 \frac{\partial^2}{\partial \eta^2} \tag{5.71}$$

$$\frac{\partial^2}{\partial \varphi^2} = \frac{\partial}{\partial \varphi}\left(\frac{\partial \eta}{\partial \varphi}\frac{\partial}{\partial \eta}\right) = \frac{\partial^2 \eta}{\partial \varphi^2}\frac{\partial}{\partial \eta} + \left(\frac{\partial \eta}{\partial \varphi}\right)^2 \frac{\partial^2}{\partial \eta^2} = -\beta H_{\varphi\varphi}\frac{\partial}{\partial \eta} + \beta^2 H_\varphi^2 \frac{\partial^2}{\partial \eta^2} \tag{5.72}$$

So that Eq. (5.62) for $u_T$ becomes

$$\left[\frac{a\gamma}{M_s}(1-p^2)\beta H_p^2 + 2\frac{a\gamma}{M_s}pH_p + \frac{a\gamma}{M_s}\frac{\beta H_\varphi^2}{(1-p^2)} - \frac{a\gamma}{M_s}(1-p^2)H_{pp} - \frac{a\gamma}{M_s}\frac{H_{\varphi\varphi}}{1-p^2}\right]\frac{\partial u_T}{\partial \eta} + \left[\frac{a\gamma}{M_s}(1-p^2)\beta H_p^2 + \frac{a\gamma}{M_s}\frac{\beta H_\varphi^2}{(1-p^2)}\right]\frac{\partial^2 u_T}{\partial \eta^2} = 0 \tag{5.73}$$

By dividing across by $\beta$ and by simplifying and ignoring terms of order $\beta^{-1}$, we have

$$\left[\frac{a\gamma}{M_s}(1-p^2)H_p^2 + \frac{a\gamma}{M_s}\frac{H_\varphi^2}{1-p^2}\right]\left(\frac{\partial u_T}{\partial \eta} + \frac{\partial^2 u_T}{\partial \eta^2}\right) = 0 \tag{5.74}$$

yielding the so-called boundary layer equation for $u_T$

$$\frac{\partial^2 u_T}{\partial \eta^2} + \frac{\partial u_T}{\partial \eta} = 0 \tag{5.75}$$

which must be solved subject to the boundary conditions:

$$\begin{aligned} u_T &= 0 \qquad \text{when} \qquad \eta = 0 \\ u_T &\to 1 \qquad \text{as} \qquad \eta \to \infty \end{aligned} \tag{5.76}$$

which has the solution

$$u_T = 1 - e^{-\eta} \tag{5.77}$$

## VI. THE CROSSOVER BETWEEN THE VERY LOW DAMPING AND THE INTERMEDIATE-TO-HIGH DAMPING REGIMES IN THE KRAMERS THEORY

### A. The Crossover in the Kramers Theory for Particles

We have seen throughout this treatment of the Kramers problem that Kramers was able to obtain asymptotic formulas for the escape rate for VLD and IHD. However, such methods fail when the energy loss per cycle for a typical particle is of the order of the thermal energy $kT$. This statement is used to define the crossover friction region between the VLD (where the coupling between the Liouville and dissipative terms in the FPE is ignored) and IHD regimes. The definition follows by recalling that in the Kramers approach for very low damping, the unperturbed motion of a particle with energy equal to the saddle energy in a well is librational, that is, the phase space trajectories are closed and we have the energy loss per cycle, $\eta I(E_C) \ll kT$, while in the IHD region we have $\eta I(E_C) \gg kT$. Thus, the crossover region is defined to be the region (Section II.F) such that

$$\eta I(E_C) \approx kT \tag{6.1}$$

or utilizing the approximate Eq. (2.229)

$$\eta \frac{E_C}{f_A} \approx kT \tag{6.2}$$

that is,

$$\alpha\beta E_C \approx 1 \tag{6.3}$$

Several attempts to extend the Kramers theory to the crossover range of friction have been made since 1940, particularly in the decade 1980–1990, because of the particular application to cycle slips in phase-locked loops and Josephson junctions. The various investigations have been briefly summarized in the comprehensive review of Hänggi et al. [1] (see their p. 286). The best known of these is that [34] due to Mel'nikov and Meshkov, who show that, provided the energy loss per cycle can be computed to a fair degree of accuracy, then an integral formula (which takes account of the Liouvile terms) may be used to connect the IHD and the VLD solutions (see Fig. 4).

They proceed by converting the FPE in the stationary case into an integral equation for the energy, which they solve by the Wiener–Hopf method [35, 81]. (Their notation is adhered to as far as possible for the purpose of easy comparison with the original work)

### B. Expression for the Prefactor of the Escape Rate as a Departure from the TST Expression

Equation (2.105) for the escape rate in the IHD regime is

$$\Gamma = \frac{\omega_A}{2\pi}\left[\left(1+\frac{\eta^2}{4\omega_C^2}\right)^{1/2} - \frac{\eta}{2\omega_C}\right] e^{-\beta\Delta U} \tag{6.4}$$

which, as we have seen in Sections I and II, holds when ($U_1 = \Delta U$ in their notation since the zero of potential is the top of the barrier as in Figure 9)

$$\eta \gg \frac{\omega_C kT}{U_1} \sim \frac{kT}{S_1} \tag{6.5}$$

since $U_1 \sim \omega_C S_1$ and they suppose that $\omega_A$, $\omega_C$ have comparable orders of magnitude.

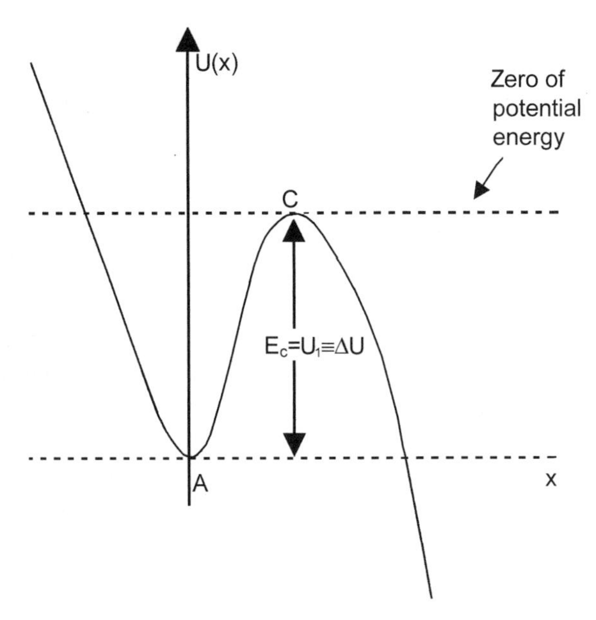


**Figure 9.** Potential energy diagram as used in Section VI.

Equation (2.193) for the escape rate in the VLD regime is

$$\Gamma = \frac{\omega_A}{2\pi}\frac{\eta I(E_C)}{kT}e^{-\beta\Delta U} \tag{6.6}$$

which holds when

$$\eta \ll \frac{\omega_C kT}{U_1} \sim \frac{kT}{S_1} \tag{6.7}$$

Now, of course $E_C = U_1$ and $I(E_C) = S_1$, and, writing

$$\delta_1 \equiv \eta S_1 \tag{6.8a}$$

and

$$\Delta \equiv \frac{\delta_1}{kT} \tag{6.8b}$$

we may write Eq. (6.6), the VLD escape rate in the TST form, as

$$\Gamma = \Delta\frac{\omega_A}{2\pi}e^{-\beta\Delta U} \tag{6.9}$$

We now seek an expression for $\Gamma$, which holds for values of $\eta$ between the two damping regimes, $\eta \ll (kT/S_1)$ and $\eta \gg (kT/S_1)$. This expression will, obviously have to tend to the VLD and IHD forms in the limits of small and large $\Delta$, respectively.

We note that both the IHD [Eq. (6.4)] and the VLD [Eq. (6.9)] escape rates are of the form

$$\Gamma = A' \frac{\omega_A}{2\pi} e^{-\beta \Delta U} = A' \Gamma_{\mathrm{TS}} \tag{6.10}$$

where $A'$ is some prefactor expressing the departure from the TST behavior. If we assume, following Mel'nikov and Meshkov, note that in our notation cf. Eq. (1.3), $A'$ corresponds to $A$ of that equation, that

$$A' = A(\Delta)\left[\left(1 + \frac{\eta^2}{4\omega_C^2}\right)^{1/2} - \frac{\eta}{2\omega_C}\right] \tag{6.11}$$

where $A(\Delta)$ in the notation of this Section is a function that interpolates between the VLD and IHD solutions. we see that, as $\eta \to 0$

$$\left(1 + \frac{\eta^2}{4\omega_C^2}\right)^{1/2} - \frac{\eta}{2\omega_C} \to 1 \tag{6.12}$$

which is the TST result and which is invalid as $\eta \to 0$ (Section II.F) and so we need (cf. Eq. (6.9))

$$A(\Delta) \to \Delta \quad \text{as} \quad \eta \to 0, \text{ or equivalently } \Delta \ll 1 \tag{6.13a}$$

$$A(\Delta) \to 1 \quad \text{as} \quad \eta \to \infty, \text{ or equivalently } \Delta \gg 1 \tag{6.13b}$$

so that the IHD result is regained. Thus, if we know the functional form of $A(\Delta)$ in the crossover region, the "naïve product of it with the IHD prefactor will give the prefactor $A'$ within the whole damping range" [34].

### C. The Fokker–Planck Equation in Energy-Action Variables

Mel'nikov and Meshkov now proceed by writing the FPE as a PDE in energy-action variables. Their calculation proceeds as follows.

We first assume that particles escaping over the barrier have total energy $\varepsilon$ less than or of the order of the thermal energy $kT$. In the high barrier $(U_1 \gg kT)$ limit, and with the assumption that the *energy loss per cycle*, $\delta_1$, due to friction is *very much less* than $U_1$, the barrier energy, the motion is again almost conservative. This is the *Ansatz* underlying the Mel'nikov and Meshkov treatment. The various approaches to the crossover problem have been criticized by Hänggi et al. [1]. We now use the equations (justified in Appendix VI.I).

$$\left.\frac{\partial}{\partial p}\right|_x = \frac{p_+(\varepsilon, x)}{m} \left.\frac{\partial}{\partial \varepsilon}\right|_x \quad \text{if } p > 0 \tag{6.14a}$$

$$= \frac{p_-(\varepsilon, x)}{m} \left.\frac{\partial}{\partial \varepsilon}\right|_x \quad \text{if } p < 0 \tag{6.14b}$$

$$\left.\frac{\partial}{\partial x}\right|_p = \left.\frac{\partial}{\partial x}\right|_\varepsilon + \frac{dU}{dx}\left.\frac{\partial}{\partial \varepsilon}\right|_x \tag{6.14c}$$

to write the Fokker–Planck equation in the limit $|U(x)| \gg |\varepsilon|$ as:

$$\frac{\partial f^R}{\partial x} = \sqrt{-2mU(x)}\left\{\eta \frac{\partial}{\partial \varepsilon}\left[f^R + kT\frac{\partial f^R}{\partial \varepsilon}\right]\right\} \tag{6.15a}$$

$$\frac{\partial f^L}{\partial x} = -\sqrt{-2mU(x)}\left\{\eta \frac{\partial}{\partial \varepsilon}\left[f^L + kT\frac{\partial f^L}{\partial \varepsilon}\right]\right\} \tag{6.15b}$$

(Eqs. (6.15) ultimately allow us to write the FPE as a diffusion equation in the energy and the action) where

$$f^R(\varepsilon, x) = F(p_+(\varepsilon, x), x) \tag{6.16a}$$

$$f^L(\varepsilon, x) = F(p_-(\varepsilon, x), x) \tag{6.16b}$$

$$p_\pm = \pm\{2m[\varepsilon - U(x)]\}^{1/2} \tag{6.17}$$

$f^R$ gives the density of particles moving in the positive $x$ direction and
$f^L$ gives the density of particles moving in the negative $x$ direction.

We also note that ($x_1$ being the left hand point $U(x_1) \simeq 0$ in their terminology)

$$f^L = f^R \qquad \text{if } x \approx x_1 \tag{6.18a}$$

$$f^L = f^R \qquad \text{if } x \approx 0 \text{ and } \varepsilon < 0 \tag{6.18b}$$

$$f^L = 0 \qquad \text{if } x \approx 0 \text{ and } \varepsilon > 0 \tag{6.18c}$$

Equation (6.18b) is a consequence of the fact that all *particles with negative energy are reflected from the barrier*, and (6.18c) of the fact that *particles with positive energy escape over the barrier.*

If we now define the action, $s$, by the equation

$$s = \oint p dx \tag{6.19a}$$

with of course

$$\varepsilon = \frac{p^2}{2m} + U(x) \tag{6.19b}$$

[Mel'nikov and Meshkov [34] use the definition

$$\left(\frac{ds}{dx}\right)^2 = -2mU(x) \tag{6.19c)]}$$

we find, using the simple chain rule, that the FPE takes on the form of a diffusion equation in the energy with the *action*, $s$, and the *energy*, $\varepsilon$, as the independent variables

$$\frac{\partial f}{\partial s} = \eta \frac{\partial}{\partial \varepsilon}\left[f + kT\frac{\partial f}{\partial \varepsilon}\right] \tag{6.20}$$

In what follows, we shall also need the action $S(\varepsilon)$ in one oscillation, namely,

$$S(\varepsilon) \equiv \oint \{2m[\varepsilon - U(x)]\}^{1/2} dx \tag{6.21}$$

Now, if $\varepsilon$ is small, it is justified in Appendix VI.II that

$$S(0) - S(\varepsilon) = \frac{2\pi|\varepsilon|}{\omega_A(\varepsilon)} \approx \frac{\varepsilon}{2\omega_C}\ln\frac{U_1}{m\omega_C^2} = \left|\frac{\varepsilon}{\omega_C}\right|\ln\left|\frac{U_1}{|\varepsilon|}\right| \qquad \text{if } |\varepsilon| \ll U_1 \tag{6.22}$$

which tends to 0 in the limit of small, $\varepsilon$ by L'Hôpital's rule. So we may take

$$S(\varepsilon) \approx S(0) \equiv S_1, \quad \text{say} \tag{6.23}$$

Thus

$$S_1 = \oint \{-2mU(x)\}^{1/2} dx$$
$$= 2\int_{x_1}^{0} \{-2mU(x)\}^{1/2} dx \tag{6.24}$$

We note that $S$ is the quantity labeled as $I$ and $S_1$ is the quantity $I(E_C)$ in Kramers paper [4] and in Sections I and II. Also note that, by approximating the action in order of magnitude by

$$S_1 \sim \frac{U_1}{\omega_C} \tag{6.25}$$

we have

$$\frac{kT}{S_1} \sim \frac{\omega_C kT}{U_1} \tag{6.26}$$

We also remark that, if we define the function

$$f(\varepsilon) \equiv f^R(\varepsilon, 0) \qquad \text{for } \varepsilon > 0 \tag{6.27a}$$

$$\equiv f^R(\varepsilon, x(\varepsilon)) \quad \text{for } \varepsilon < 0 \tag{6.27b}$$

where $x(\varepsilon)$ is a root of the equation

$$U(x) = \varepsilon \qquad x_m < x < 0 \tag{6.27c}$$

then $f(\varepsilon)$ gives the rate of escape for $\varepsilon > 0$, and the rate of reflection from the barrier for $\varepsilon < 0$. These reflected particles make up the population of left-going particles $f^L$. They are subsequently reflected from the other side of the potential well at which point the function $f^L$ becomes the function $f^R$. When these particles reach the barrier again, they reproduce the initial distribution $f(\varepsilon)$.

It now remains to solve Eq. (6.20), which is best accomplished by writing it as an integral equation. This equation is later converted to a Wiener–Hopf equation by use of the Green's function method, which is accomplished by first finding the Green's function for the energy diffusion equation (6.20).

### D. Green's Function of the Energy Diffusion Equation

We first note that if $g(\varepsilon - \varepsilon', s - s')$ is the Green's function of Eq. (6.20) so that the initial condition is

$$g(\varepsilon - \varepsilon', 0) = \delta(\varepsilon - \varepsilon') \tag{6.28}$$

then the solution of the diffusion equation Eq. (6.20) for $g$ gives (details in Appendix VI.III)

$$g(\varepsilon - \varepsilon', s) = \frac{1}{\sqrt{4\pi kT\eta s}} \exp\left[-\frac{(\varepsilon - \varepsilon' + \eta s)^2}{4kT\eta s}\right] \tag{6.29}$$

We now introduce the energy loss in one period, by writing

$$\eta S_1 = \delta_1 (\equiv \eta\ I(E_C)) \tag{6.30}$$

It follows by the principle of superposition from Eq. (6.29), written in the form

$$g(\varepsilon - \varepsilon') \equiv g(\varepsilon - \varepsilon', S_1) = \frac{1}{\sqrt{4\pi kT\delta_1}} \exp\left[-\frac{(\varepsilon - \varepsilon' + \delta_1)^2}{4kT\delta_1}\right] \tag{6.31}$$

together with the boundary condition, that deep in the well the distribution is Maxwell–Boltzmann, (cf. our discussion in Sections I and II)

$$f_0(\varepsilon) = \frac{\omega_A}{2\pi kT} \exp\left[-\frac{\varepsilon + U_1}{kT}\right] \qquad \varepsilon \ll -kT \tag{6.32}$$

so that

$$f(\varepsilon) = \int_{-\infty}^{0} g(\varepsilon - \varepsilon') f(\varepsilon') d\varepsilon' \tag{6.33}$$

which is the desired integral equation for the energy distribution function $f(\varepsilon)$ and that has the form of a Wiener–Hopf equation [35, 81]. [Note that, in view of the fact that the exponential factor in $g(\varepsilon - \varepsilon')$ decays so quickly, we suffer no great error in replacing the lower limit, which should strictly be $-U_1$, by $-\infty$. This fact is important as otherwise the problem could not be posed as the solution to a Wiener–Hopf equation.]

If we now *normalize the distribution to one particle in the well*, as in Section II, Eq. (2.117) the decay rate, by the flux over barrier method is

$$\Gamma = \tau^{-1} = J = \int_0^\infty f(\varepsilon)d\varepsilon \tag{6.34}$$

This expression follows from Eq. (2.99) (for the number of particles crossing the barrier top in unit time), the fact that $d\varepsilon = (p/m)dp$, and the fact that in order for a particle to escape, its momentum must be positive. In Appendix VI.IV we justify that the variance of the energy is

$$\langle(\varepsilon - \langle\varepsilon\rangle)^2\rangle = 2kT\eta s \tag{6.35a}$$

where the braces denote an average over the distribution of Eq. (6.31). This formula is similar to Einstein's formula for the mean-square displacement of a Brownian particle. We also show there that the mean energy loss is

$$\varepsilon' - \langle\varepsilon\rangle = \eta s \tag{6.35b}$$

which is a form of the fluctuation-dissipation theorem.

## E. Solution of the Integral Equation for the Energy Distribution Function by the Wiener–Hopf Method

To solve Eq. (6.33) we define, in the manner of the Wiener–Hopf method [35, 81], the one-sided Fourier transforms (the transform variable is $\lambda$ and the transform of $f(\varepsilon)e^{\varepsilon/2kT}$ is taken),

$$\varphi^+(\lambda) \equiv \frac{2\pi}{\omega_A}\exp\left(\frac{U_1}{kT}\right)\int_0^\infty f(\varepsilon)\exp\left[\frac{(i\lambda + (1/2))\varepsilon}{kT}\right]d\varepsilon \tag{6.36a}$$

$$\varphi^-(\lambda) \equiv \frac{2\pi}{\omega_A}\exp\left(\frac{U_1}{kT}\right)\int_{-\infty}^0 f(\varepsilon)\exp\left[\frac{(i\lambda + (1/2))\varepsilon}{kT}\right]d\varepsilon \tag{6.36b}$$

The transforms play a crucial role in the calculations since (on account of the fact that in the crossover region we may assume that the IHD prefactor is of order 1, i.e., the TST value) the correction to the prefactor $A'$ in Eq. (6.11) is by transposition of Eq. (6.10) ($\Gamma = \tau^{-1}$)

$$A = \frac{2\pi\tau^{-1}}{\omega_A}\exp\left(\frac{U_1}{kT}\right) \tag{6.37}$$

and by Eq. (6.34)

$$\tau^{-1} = \int_0^\infty f(\varepsilon)d\varepsilon \tag{6.38}$$

Thus Eq. (6.36a) yields on evaluation at $\lambda = i/2$ so that the argument of the exponential vanishes, an expression for $A$ in terms of the Fourier transform $\varphi^+$, namely

$$A = \varphi^+\left(\frac{i}{2}\right) \tag{6.39}$$

The first step in determining $\varphi^+(\lambda)$, is to find the Fourier transform, $\varphi_0^-(\lambda)$, for the function $f_0(\varepsilon)$ corresponding to the boundary condition (6.32) namely deep in the well the probability distribution must be the Mawell-Boltzmann distribution. We have

$$\varphi_0^-(\lambda) = \frac{2\pi}{\omega_A}\exp\left(\frac{U_1}{kT}\right)\int_{-\infty}^0 \frac{\omega_A}{2\pi kT}\exp\left[-\frac{\varepsilon + U_1}{kT}\right]\exp\left[\frac{(i\lambda + (1/2))\varepsilon}{kT}\right]d\varepsilon \tag{6.40}$$

$$= \frac{1}{kT}\int_{-\infty}^0 \exp\left(i\left[\lambda + \frac{i}{2}\right]\frac{\varepsilon}{kT}\right)d\varepsilon \tag{6.41}$$

By using the substitution $u = \varepsilon/kT$, we have

$$\varphi_0^-(\lambda) = \int_{-\infty}^0 \exp\left(i\left[\lambda + \frac{1}{2}\right]u\right)du \tag{6.42}$$

Letting $\lambda = \lambda_1 + i\lambda_2$, we find the Fourier transform of $f_0(\varepsilon)$:

$$\varphi_0^-(\lambda) = \frac{1}{i(\lambda + (i/2))}e^{i\lambda_1 u}e^{-(\lambda_2+(1/2))u}\Big|_{u=-\infty}^0 \tag{6.43}$$

$$= -\frac{i}{\lambda + (i/2)} \qquad \text{if } \mathrm{Im}(\lambda) < -\frac{1}{2} \tag{6.44}$$

This is the value of $\varphi^-(\lambda)$ corresponding to the boundary value of $f(\varepsilon)$ namely $f_0(\varepsilon)$ given in Eq. (6.32).

*Now the function* $f(\varepsilon)$, *dealt with, holds deep in the well where* $\varepsilon \ll -kT$, *so we require the contribution to* $\varphi_0^-$ *coming from the interval* $(\alpha, 0)$, *where* $\alpha$ *is finite but* $\ll -kT$, *to be negligible*. This will be so if

$$\left|\lambda + \frac{i}{2}\right| \ll 1 \tag{6.45}$$

We may justify this as follows, because by taking $\alpha = -nkT$, say, where $\alpha$ is any positive number, we find (with $u = \varepsilon/kT$)

$$\frac{1}{kT}\int_{-nkT}^{0} e^{i[\lambda+(i/2)](\varepsilon/kT)}d\varepsilon = \int_{-n}^{0} e^{i[\lambda+(i/2)]u}du \tag{6.46}$$

$$= \frac{1}{i(\lambda + (i/2))}\left[1 - e^{-ni(\lambda+(i/2))}\right] \tag{6.47}$$

$$= -n\frac{1 - e^{-ni(\lambda+(i/2))}}{-ni(\lambda + (i/2))} \tag{6.48}$$

$$= n \qquad \text{as} \left|i\left(\lambda + \frac{i}{2}\right)\right| \to 0 \tag{6.49}$$

In Eq. (6.48), we used the fact that

$$\lim_{t\to 0}((e^t - 1)/t) = 1$$

That is,

$$\frac{1}{kT}\int_{-nkT}^{0} e^{i(\lambda+(i/2))(\varepsilon/kT)}d\varepsilon$$

is obviously finite as $|i(\lambda + (i/2))| \to 0$ whereas $|-[i/(\lambda + (i/2))]|$ becomes very large as

$$\left|i\left(\lambda + \frac{i}{2}\right)\right| \to 0,$$

and thus

$$\frac{1}{kT}\int_{-nkT}^{0} e^{i(\lambda+(i/2))(\varepsilon/kT)}d\varepsilon = n$$

which is obviously finite, is *negligible compared with*

$$\varphi_0^- = \left| -\frac{i}{\lambda + (1/2)} \right|$$

So, we need the two conditions

$$\mathrm{Im}(\lambda) < -\frac{1}{2} \tag{6.50a}$$

$$\left| \lambda + \frac{i}{2} \right| \ll 1 \tag{6.50b}$$

in order to justify Eq. (6.44), and so $\varphi_0^-(\lambda)$ becomes

$$\varphi_0^-(\lambda) = -\frac{i}{\lambda + (i/2)} \tag{6.51}$$

under these restrictions. Equation (6.51) is central to the application of the Wiener–Hopf method to our problem [see the paragraph following Eq. (6.73) below].

We shall now proceed to determine $\varphi^+(\lambda)$, which, when combined with Eq. (6.51), will allow us to determine $\varphi^-(\lambda)$ by the Wiener–Hopf method and thus the quantity $A$ as we shall see below. We recall that the probability distribution function of the energy is given by the Wiener–Hopf equation because by the principle of superposition

$$f(\varepsilon) = \int_{-\infty}^{0} g(\varepsilon - \varepsilon')f(\varepsilon')d\varepsilon' \tag{6.52}$$

On using Eq. (6.31) for the Green's function $g$, we then have

$$f(\varepsilon) = \int_{-\infty}^{0} (4\pi\delta_1 kT)^{-1/2} \exp\left[-\frac{(\varepsilon - \varepsilon' + \delta_1)^2}{4kT\delta_1}\right] f(\varepsilon')d\varepsilon' \tag{6.53}$$

so that the desired Fourier transform, $\varphi^+(\lambda)$, by Eq. (6.36a) is

$$\begin{aligned}\varphi^+(\lambda) = {} & \frac{2\pi}{\omega_A\sqrt{4\pi\delta_1 kT}} e^{\beta U_1} \int_0^{\infty}\int_{-\infty}^{0} \exp\left[-\frac{(\varepsilon - \varepsilon' + \delta_1)^2}{4kT\delta_1}\right] f(\varepsilon') \\ & \times \exp\left[\frac{(i\lambda + (1/2))\varepsilon}{kT}\right] d\varepsilon d\varepsilon'\end{aligned} \tag{6.54}$$

$$= \frac{2\pi}{\omega_A \sqrt{4\pi\delta_1 kT}} e^{\beta U_1} \int_{-\infty}^{0} f(\varepsilon') \exp\left[-\frac{(\delta_1 - \varepsilon')^2}{4kT\delta_1}\right]$$

$$\times \int_0^\infty \exp\left[-\frac{\varepsilon^2 + \{2(\delta_1 - \varepsilon') - 4\delta_1(i\lambda + (1/2))\}\varepsilon}{4kT\delta_1}\right] d\varepsilon\, d\varepsilon' \quad (6.55)$$

The innermost integral may be evaluated as follows. We have

$$\int_0^\infty \exp\left[-\frac{\varepsilon^2 + \{2(\delta_1 - \varepsilon') - 4\delta_1(i\lambda + (1/2))\}\varepsilon}{4kT\delta_1}\right] d\varepsilon$$

$$= \exp\left[\frac{(-2i\lambda\delta_1 - \varepsilon')^2}{4kT\delta_1}\right] \int_0^\infty \exp\left[-\frac{(\varepsilon - 2i\lambda\delta_1 - \varepsilon')^2}{4kT\delta_1}\right] d\varepsilon \quad (6.56)$$

$$= \exp\left[\frac{(2i\lambda\delta_1 + \varepsilon')^2}{4kT\delta_1}\right] \sqrt{4kT\delta_1}\,\frac{\sqrt{\pi}}{2} \quad (6.57)$$

$$= \sqrt{\pi kT\delta_1} \exp\left[\frac{(2i\lambda\delta_1 + \varepsilon')^2}{4kT\delta_1}\right] \quad (6.58)$$

Hence, the desired Fourier transform is rendered by the single integral

$$\varphi^+(\lambda) = \frac{2\pi}{\omega_A \sqrt{4\pi\delta_1 kT}} e^{\beta U_1} \int_{-\infty}^{0} f(\varepsilon') \exp\left[-\frac{(\delta_1 - \varepsilon')^2}{4kT\delta_1}\right] \sqrt{\pi kT\delta_1}$$
$$\times \exp\left[\frac{(2i\lambda\delta_1 + \varepsilon')^2}{4kT\delta_1}\right] d\varepsilon' \quad (6.59)$$

$$= \frac{\pi}{\omega_A} e^{\beta U_1} \exp\left[-\frac{\delta_1(1 + 4\lambda^2)}{4kT}\right] \int_{-\infty}^{0} f(\varepsilon') \exp\left[\frac{(4i\lambda + 2)\varepsilon'}{4kT}\right] d\varepsilon' \quad (6.60)$$

and defining the function (the *two* sided Fourier transform of Green's function at $\varepsilon' = 0$)

$$\tilde{g}(\lambda) \equiv \exp\left[-\frac{\delta_1(\lambda^2 + (1/4))}{kT}\right] = \int_{-\infty}^{\infty} g(\varepsilon) e^{(i\lambda + (1/2))(\varepsilon/kT)} d\varepsilon \quad (6.61)$$

it follows that in terms of $\varphi^{-}(\lambda)$ and $\tilde{g}(\lambda)$

$$\varphi^{+}(\lambda) = \tfrac{1}{2}\tilde{g}(\lambda)\varphi^{-}(\lambda) \tag{6.62}$$

Now, just as Eq. (6.54)

$$\begin{aligned} \varphi^{-}(\lambda) = {} & \frac{2\pi}{\omega_A\sqrt{4\pi\delta_1 kT}} e^{\beta U_1} \int_{-\infty}^{0}\int_{-\infty}^{0} \exp\left[-\frac{(\varepsilon - \varepsilon' + \delta_1)^2}{4kT\delta_1}\right] f(\varepsilon') \\ & \times \exp\left[\frac{(i\lambda + (1/2))\varepsilon}{kT}\right] d\varepsilon\, d\varepsilon' \end{aligned} \tag{6.63}$$

and by almost identical mathematical manipulation, we find that in terms of $\varphi^{+}(\lambda)$ and $\tilde{g}(\lambda)$

$$\varphi^{-}(\lambda) = \tfrac{1}{2}\tilde{g}(\lambda)\varphi^{-}(\lambda) \tag{6.64}$$

and so, by combining Eqs. (6.62) and (6.64) we arrive at an equation that is ubiquitous in the Wiener–Hopf method namely

$$\varphi^{+}(\lambda) + \varphi^{-}(\lambda) = \tilde{g}(\lambda)\varphi^{-}(\lambda) \tag{6.65}$$

or (cf. Eq. (8.5.51) of Morse and Feshbach [81] or Section 11.17 of [35])

$$\varphi^{+}(\lambda) + [1 - \tilde{g}(\lambda)]\varphi^{-}(\lambda) = 0 \tag{6.66}$$

which, on letting $G(\lambda) \equiv 1 - \tilde{g}(\lambda)$, becomes, again as is usual in the Wiener–Hopf method [35, 81]

$$\varphi^{+}(\lambda) + G(\lambda)\varphi^{-}(\lambda) = 0 \tag{6.67}$$

The Fourier transforms $\varphi^{+}(\lambda)$ and $\varphi^{-}(\lambda)$ may be determined in terms of $G(\lambda)$ as follows. We first remark that Eq. (6.67) may be written in the form, again following the Wiener–Hopf method

$$\ln[-\varphi^{+}(\lambda)] = \ln\varphi^{-}(\lambda) + \ln G(\lambda) \tag{6.68}$$

If we define (once again following the Wiener–Hopf method) using Cauchy's integral formula [80],

$$\ln G^{+}(\lambda) \equiv \frac{1}{2\pi i}\lim_{\theta\to 0}\int_{-\infty}^{\infty} \frac{\ln G(\lambda')}{\lambda' - \lambda - i\theta} d\lambda' \tag{6.69}$$

$$\ln G^{-}(\lambda) \equiv -\frac{1}{2\pi i}\lim_{\theta\to 0}\int_{-\infty}^{\infty}\frac{\ln G(\lambda')}{\lambda'-\lambda+i\theta}d\lambda' \tag{6.70}$$

then,

$$\ln G(\lambda) = \ln G^{+}(\lambda) + \ln G^{-}(\lambda) \tag{6.71}$$

and

$$\ln[-\varphi^{+}(\lambda)] - \ln G^{+}(\lambda) = \ln \varphi^{-}(\lambda) + \ln G^{-}(\lambda) \tag{6.72}$$

The function $G^{+}(\lambda)$ has no zeros in the half-plane $\mathrm{Im}\,\lambda > 0$, since $\ln G^{+}(\lambda)$ is analytic there. The function $G^{+}(\lambda)$ has no zeros in the half-plane $\mathrm{Im}\,\lambda < 0$, since $\ln G^{-}(\lambda)$ is analytic there. Both $G^{+}(\lambda)$ and $G^{-}(\lambda) \to 1$ as $\lambda \to \infty$, and

$$G(\lambda) = G^{+}(\lambda)G^{-}(\lambda) \tag{6.73}$$

As the functions on both sides of Eq. (6.72) are analytic in different half-planes of $\lambda$, both sides equal a function that is analytic in the entire plane and which we may choose (cf. [35, 81]) to coincide with $\varphi_0^{-}(\lambda)$ in Eq. (6.51) for

$$|\lambda + (i/2)| \ll 1$$

We then automatically arrive by the Wiener–Hopf method at the following solution to Eq. (6.66) of which details are given in Appendix VI.V:

$$\varphi^{+}(\lambda) = \frac{iG^{+}(\lambda)G^{-}(-i/2)}{\lambda + (i/2)} \tag{6.74a}$$

$$\varphi^{-}(\lambda) = -\frac{iG^{-}(-i/2)}{G^{-}(\lambda)(\lambda + (i/2))} \tag{6.74b}$$

Now, the prefactor

$$A = \varphi^{+}\left(\frac{i}{2}\right) \tag{6.75}$$

$$= \frac{iG^{-}(-i/2)G^{+}(i/2)}{(i/2)+(i/2)} \tag{6.76}$$

$$= \left|G^{+}\left(\frac{i}{2}\right)\right|^{2} \tag{6.77}$$

since (the overbar denoting complex conjugation)

$$\overline{\ln G^{-}\left(-\frac{i}{2}\right)} = \frac{1}{2\pi i}\int_{-\infty}^{\infty} \frac{\overline{\ln G(\lambda')}}{\lambda' - (i/2) - i0} d\lambda' \tag{6.78}$$

$$= \frac{1}{2\pi i}\int_{-\infty}^{\infty} \frac{\ln G(\lambda')}{\lambda' - (i/2) - i0} d\lambda' \tag{6.79}$$

$$= \ln G^{+}\left(\frac{i}{2}\right) \tag{6.80}$$

by Cauchy's integral formula (Eq. (6.79) is true since the integration is along the line $\operatorname{Im}\lambda' = 0$) and we find that

$$\overline{G^{-}\left(-\frac{i}{2}\right)} = G^{+}\left(\frac{i}{2}\right) \tag{6.81}$$

From this, Eq. (6.77) for $A$ follows easily.

## F. Integral Formula for the Prefactor $A$ ($\Delta$)

We have seen that the prefactor $A$ in the crossover region is given by the equation

$$A = \left|G^{+}\left(\frac{i}{2}\right)\right|^2 \tag{6.82}$$

which may be given in a more explicit integral form as follows. First we note that, by Cauchy's integral formula

$$\ln G^{+}\left(\frac{i}{2}\right) = \frac{1}{2\pi i}\lim_{\theta\to 0}\int_{-\infty}^{\infty} \frac{\ln G(\lambda')d\lambda'}{\lambda' - (i/2) - i\theta} \tag{6.83}$$

$$= -\frac{i}{2\pi}\int_{-\infty}^{\infty} \frac{\ln[1 - \exp\{-(\delta_1/kT)(\lambda'^2 + (1/4))\}]}{\lambda' - (i/2)} d\lambda' \tag{6.84}$$

$$= \frac{1}{2\pi}\int_{-\infty}^{\infty} \frac{\ln[1 - \exp\{-(\delta_1/kT)(\lambda'^2 + (1/4))\}](-i\lambda' + (1/2))}{\lambda'^2 + (1/4)} d\lambda' \tag{6.85}$$

and so

$$G^{+}\left(\frac{i}{2}\right)=\exp\left[\frac{1}{2\pi}\int_{-\infty}^{\infty}\frac{\ln[1-\exp\{-(\delta_1/kT)(\lambda'^2+(1/4))\}](-i\lambda'+(1/2))}{\lambda'^2+(1/4)}d\lambda'\right] \tag{6.86}$$

Now, using the fact that

$$|e^z| = e^{\mathrm{Re}z} \tag{6.87}$$

for any complex number $z$, we see that Eq. (6.82) may be written

$$\left|G^{+}\left(\frac{i}{2}\right)\right|^2 = \exp\left[\frac{2}{2\pi}\int_{-\infty}^{\infty}\frac{\ln[1-\exp\{-(\delta_1/kT)(\lambda'^2+(1/4))\}](1/2)}{\lambda'^2+(1/4)}d\lambda'\right] \tag{6.88}$$

and so, by writing $\Delta \equiv \delta_1/kT$, we see that the prefactor $A$ becomes the integral

$$A(\Delta) = \exp\left[\frac{1}{2\pi}\int_{-\infty}^{\infty}\frac{\ln[1-\exp\{-\Delta(\lambda'^2+(1/4))\}]}{\lambda'^2+(1/4)}d\lambda'\right] \tag{6.89}$$

Now, using the substitution

$$\lambda' = \frac{1}{2}\tan x \tag{6.90}$$

we see that $A(\Delta)$ may also be written

$$A(\Delta) = \exp\left[\frac{2}{\pi}\int_0^{\pi/2}\ln\left[1-\exp\left\{-\frac{\Delta}{4}\sec^2 x\right\}\right]dx\right] \tag{6.91}$$

Now, if we write

$$\begin{aligned} I &= \int_{-\infty}^{\infty}\frac{\ln[1-\exp\{-\Delta(\lambda'^2+(1/4))\}]}{\lambda'^2+(1/4)}d\lambda' \\ &= 2\int_0^{\infty}\frac{\ln[1-\exp\{-\Delta(\lambda'^2+(1/4))\}]}{\lambda'^2+(1/4)}d\lambda' \end{aligned} \tag{6.92}$$

and by using the expansion

$$\ln(1-t) = -\sum_{n=1}^{\infty}\frac{t^n}{n} \qquad \text{for} \qquad |t|<1 \tag{6.93}$$

we see

$$\ln\left[1 - \exp\left\{-\Delta\left(\lambda'^2 + \frac{1}{4}\right)\right\}\right] = -\sum_{n=1}^{\infty} n^{-1}\left\{\exp\left[-\Delta\left(\lambda'^2 + \frac{1}{4}\right)\right]\right\}^n \quad (6.94)$$

$$= -\sum_{n=1}^{\infty} n^{-1} \exp\left[-n\Delta\left(\lambda'^2 + \frac{1}{4}\right)\right] \quad (6.95)$$

Thus,

$$I = -2\int_0^{\infty} \sum_{n=1}^{\infty} n^{-1} \exp\left[-n\Delta\left(\lambda'^2 + \frac{1}{4}\right)\right] \frac{d\lambda'}{\lambda'^2 + (1/4)} \quad (6.96)$$

$$= -2\sum_{n=1}^{\infty} n^{-1} \int_0^{\infty} \exp\left[-n\Delta\left(\lambda'^2 + \frac{1}{4}\right)\right] \frac{d\lambda'}{\lambda'^2 + (1/4)} \quad (6.97)$$

$$= -2\sum_{n=1}^{\infty} n^{-1} \exp\left(-\frac{n\Delta}{4}\right) \int_0^{\infty} \frac{\exp(-n\Delta\lambda'^2)d\lambda'}{\lambda'^2 + (1/4)} \quad (6.98)$$

Next, by using the fact that

$$\int_0^{\infty} \frac{e^{-at^2}dt}{t^2 + x^2} = \frac{\pi}{2x} e^{ax^2} \operatorname{erfc}\sqrt{a}x \qquad \text{for} \qquad a > 0 \text{ and } x > 0 \quad (6.99)$$

we find

$$I = -2\pi \sum_{n=1}^{\infty} n^{-1} \operatorname{erfc} \frac{\sqrt{n\Delta}}{2} \quad (6.100)$$

yielding $A$ in the form of the exponential series

$$A(\Delta) = \exp\left[-\sum_{n=1}^{\infty} n^{-1} \operatorname{erfc} \frac{\sqrt{n\Delta}}{2}\right] \quad (6.101)$$

We may also write Eq. (6.91) in the form

$$A(\Delta) = \Delta \exp\left[\sqrt{\frac{\Delta}{\pi}} \sum_{n=0}^{\infty} \frac{\zeta((1/2) - n)}{n!(2n+1)} \left(-\frac{\Delta}{4}\right)^n\right] \quad (6.102)$$

where

$$\zeta(s) \equiv \frac{1}{s-1} + \frac{1}{2} + \sum_{k=1}^{\infty} \frac{B_{2k}}{2k} \frac{\Gamma(s+2k-1)}{(2k-1)!\Gamma(s-1)} \tag{6.103}$$

is the Riemann Zeta function [82] and $B_n$ are the Bernoulli numbers [82], Eq. (23.2.3). The series in Eq. (6.103) converges inside the circle $|\Delta| = 8\pi$ (see Appendix VI.VI).

The series representation allows us to turn our attention to the behavior of the function $A(\Delta)$ for large and small values of $\Delta$. For large values of $\Delta$, we use the form obtained in Eq. (6.103). We first note that the series in Eq. (6.101) converges for large values of $\Delta$ (details in Appendix VI.VII). Also note that (details in Appendix VI.VIII)

$$\operatorname{erfc}\frac{\sqrt{\Delta}}{2} \approx \frac{2}{\sqrt{\pi\Delta}} \exp\left(-\frac{\Delta}{4}\right) \tag{6.104}$$

So,

$$A(\Delta) = \exp\left(-\sum_{n=1}^{\infty} \frac{1}{n} \operatorname{erfc}\frac{\sqrt{n\Delta}}{2}\right) \tag{6.105}$$

$$\approx 1 - \sum_{n=1}^{\infty} \frac{1}{n} \operatorname{erfc}\frac{\sqrt{n\Delta}}{2} \tag{6.106}$$

since if $\Delta \gg 1$, then $\operatorname{erfc}(\sqrt{n\Delta}/2) \ll 1$ if $n > 0$, and so

$$A(\Delta) \approx 1 - \operatorname{erfc}\frac{\sqrt{\Delta}}{2} \tag{6.107}$$

$$\approx 1 - \frac{2}{\sqrt{\pi\Delta}} \exp\left(-\frac{\Delta}{4}\right) \tag{6.108a}$$

$$\approx 1 \tag{6.108b}$$

if $\Delta$ is very large. Thus the IHD result is regained by Eq. (6.11).

Now, for $\Delta \ll 1$ we have, by using Eq. (6.102) and taking only the term $n = 0$ in the series,

$$A(\Delta) \approx \Delta\left[1 + \sqrt{\frac{\Delta}{\pi}}\zeta\left(\frac{1}{2}\right)\right] \tag{6.109}$$

$$\approx \Delta[1 - 0.82\sqrt{\Delta}] \tag{6.110a}$$

$$\approx \Delta \tag{6.110b}$$

if $\Delta$ is very small. Thus, the VLD result is regained by Eq. (6.11). Since $A(\Delta)$ approaches the appropriate limit for large and small values of $\Delta$, the expression for the escape rate is now given by taking the product of $A(\Delta)$ with the right-hand side of Eq. (6.4). This gives a final result for the escape rate, $\Gamma$, from a single potential well as

$$\Gamma = \frac{\omega_A}{2\pi}\left[\left(1 + \frac{\eta^2}{4\omega_C^2}\right)^{1/2} - \frac{\eta}{2\omega_C}\right]A(\Delta)e^{-\beta\Delta U} \tag{6.111}$$

### G. Bridging Formula for a Double Well Potential

We now consider the case of a double well potential cf. Figure 2, such that, when particles escape over the barrier, they enter another well of finite depth. There is now a finite nonnegligible probability for the particle to return to the initial well in the underdamped case. In this case, *the particle in the second well loses its energy so slowly that, after several oscillations, the white noise force may give it sufficient energy to send it back over the barrier.*

To handle this additional problem, we introduce the distributions $f_1(\varepsilon)$ and $f_2(\varepsilon)$ of particles moving toward the barrier from the first and second wells, respectively. Green's functions for the Fokker–Planck equations for the two wells are

$$g_j(\varepsilon - \varepsilon') = \frac{1}{\sqrt{4\pi\delta_j kT}}\exp\left[-\frac{(\varepsilon - \varepsilon' + \delta_j)^2}{4\delta_j kT}\right] \tag{6.112}$$

$$j = 1, 2 \qquad \delta_j \equiv \gamma S_j$$

where $S_j$ is the action per oscillation of the particle with $\varepsilon = 0$ in well number $j$.

$$S_1 = 2\int_{x_1}^{0} [-2mU(x)]^{1/2} dx \tag{6.113a}$$

$$S_2 = 2\int_{0}^{x_2} [-2mU(x)]^{1/2} dx \tag{6.113b}$$

There are two distinct contributions to each $f_j(\varepsilon)$, one from the particles reflected from the barrier one period earlier with distribution $f_j(\varepsilon)\theta(-\varepsilon)$, the other from particles that have passed over the barrier one period earlier with distribution $f_k(\varepsilon)\theta(\varepsilon); k \neq j$. The function $\theta(x)$ is the usual step function. The full system is then

$$f_1(\varepsilon) = \int_{-\infty}^{\infty} g_1(\varepsilon - \varepsilon')[f_1(\varepsilon')\theta(-\varepsilon') + f_2(\varepsilon')\theta(\varepsilon')]d\varepsilon' \tag{6.114a}$$

$$f_2(\varepsilon) = \int_{-\infty}^{\infty} g_2(\varepsilon - \varepsilon')[f_2(\varepsilon')\theta(-\varepsilon') + f_1(\varepsilon')\theta(\varepsilon')]d\varepsilon' \tag{6.114b}$$

The boundary condition

$$f_1(\varepsilon) = \frac{\omega_A}{2\pi kT}\exp\left[-\frac{\varepsilon + U_1}{kT}\right] \qquad \varepsilon \ll -kT \tag{6.115}$$

holds for $f_1$, but since there were initially no particles in the second well, $f_2(\varepsilon)$ does not have a Boltzmann distribution deep in well number 2.

If we follow the definitions of Eqs. (6.36), we introduce the one-sided Fourier transforms $\varphi_1^{\pm}(\lambda)$ and $\varphi_2^{\pm}(\lambda)$ of $f_1(\varepsilon)$ and $f_2(\varepsilon)$, respectively. We show in Appendix VI. IX that (again in the manner of the Wiener–Hopf method)

$$\varphi_1^{+}(\lambda) + \varphi_1^{-}(\lambda) = [1 - G_1^{-}(\lambda)][\varphi_1^{-}(\lambda) + \varphi_2^{+}(\lambda)] \tag{6.116a}$$

$$\varphi_2^{+}(\lambda) + \varphi_2^{-}(\lambda) = [1 - G_2^{-}(\lambda)][\varphi_2^{-}(\lambda) + \varphi_1^{+}(\lambda)] \tag{6.116b}$$

where

$$G_j(\lambda) \equiv 1 - \exp\left[-\Delta_j\left(\lambda^2 + \frac{1}{4}\right)\right] \qquad \Delta_j \equiv \frac{\delta_j}{kT} \tag{6.117}$$

Now deep in the first well, $f_1(\varepsilon)$ may be written

$$f_1(\varepsilon) = \frac{\omega_A}{2\pi kT} \exp\left[-\frac{\varepsilon + U_1}{kT}\right] \tag{6.118}$$

and so

$$\varphi_1^-(\lambda) = \frac{2\pi}{\omega_A} e^{\beta U_1} \int_{-\infty}^{0} \frac{\omega_A}{2\pi kT} \exp\left[-\frac{\varepsilon + U_1}{kT}\right] \exp\left(\frac{i\lambda + (1/2)}{kT}\varepsilon\right) d\varepsilon \tag{6.119}$$

$$= -\frac{i}{\lambda + (i/2)} \qquad \text{if } \operatorname{Im}\lambda < -\tfrac{1}{2} \tag{6.120}$$

$\varphi_1^+(\lambda), \varphi_2^+(\lambda)$, and $\varphi_2^-(\lambda)$ are analytic, the first two in the upper half-plane and the third in the lower.

The system (6.116) may be written, again following the Wiener–Hopf method (details in Appendix VI. X):

$$\varphi^+(\lambda) + \frac{G_1(\lambda)G_2(\lambda)}{G_{12}(\lambda)} \varphi^-(\lambda) = 0 \tag{6.121a}$$

$$\psi^+(\lambda) + \psi^-(\lambda) = 0 \tag{6.121b}$$

where

$$\varphi^{\pm}(\lambda) \equiv \varphi_1^{\pm}(\lambda) - \varphi_2^{\pm}(\lambda) \tag{6.122a}$$

$$\psi^{\pm}(\lambda) \equiv G_1(\lambda)\varphi_1^{\pm}(\lambda) + G_2(\lambda)\varphi_2^{\pm}(\lambda) \tag{6.122b}$$

$$G_{12}(\lambda) \equiv 1 - \exp\left[-(\Delta_1 + \Delta_2)\left(\lambda^2 + \frac{1}{4}\right)\right] \tag{6.122c}$$

The flux over the barrier is

$$\tau^{-1} = \int_0^{\infty} [f_1(\varepsilon) - f_2(\varepsilon)] d\varepsilon \tag{6.123a}$$

$$= \frac{\omega_A}{2\pi} e^{-\beta U_1} \varphi_1^+\left(\frac{i}{2}\right) - \frac{\omega_A}{2\pi} e^{-\beta U_1} \varphi_2^+\left(\frac{i}{2}\right) \tag{6.123b}$$

$$= \frac{\omega_A}{2\pi} e^{-\beta U_1} \varphi^+\left(\frac{i}{2}\right) \tag{6.123c}$$

and so

$$A = A(\Delta_1, \Delta_2) = \frac{2\pi}{\omega_A} e^{\beta U_1} \tau^{-1} = \varphi^+\left(\frac{i}{2}\right) \tag{6.124}$$

Equation (6.122a) yields

$$\ln[-\varphi^+(\lambda)] = \ln G_1(\lambda) + \ln G_2(\lambda) - \ln G_{12}(\lambda) + \ln \varphi^-(\lambda) \tag{6.125}$$

Now, with the notation of Eqs. (6.59) and (6.60) let

$$\ln G_1^+(\lambda) \equiv \frac{1}{2\pi i} \int_{-\infty}^{\infty} \frac{\ln G_1(\lambda')}{\lambda' - \lambda - i0} d\lambda' \tag{6.126a}$$

$$\ln G_1^-(\lambda) \equiv -\frac{1}{2\pi i} \int_{-\infty}^{\infty} \frac{\ln G_1(\lambda')}{\lambda' - \lambda + i0} d\lambda' \tag{6.126b}$$

and similarly for $\ln G_2^+(\lambda), \ln G_2^-(\lambda)$, $\ln G_{12}^+(\lambda)$, and $\ln G_{12}^-(\lambda)$. The functions $\ln G_1^+(\lambda), \ln G_2^+(\lambda)$, and $\ln G_{12}^+(\lambda)$ are analytic in the upper half-plane. The functions $\ln G_1^-(\lambda), \ln G_2^-(\lambda)$, and $\ln G_{12}^-(\lambda)$ are analytic in the lower half-plane, and

$$\begin{aligned} &\ln[-\varphi^+(\lambda)] - \ln G_1^+(\lambda) - \ln G_2^+(\lambda) + \ln G_{12}^+(\lambda) \\ &\quad = \ln G_1^-(\lambda) + \ln G_2^-(\lambda) - \ln G_{12}^-(\lambda) + \ln \varphi^-(\lambda) \end{aligned} \tag{6.127}$$

Each side is analytic in separate half-planes and so must equal an entire function, say $f(\lambda) = \ln h(\lambda)$. This implies

$$\ln\left[\frac{-\varphi^+(\lambda)G_{12}^+(\lambda)}{G_1^+(\lambda)G_2^+(\lambda)}\right] = \ln(h(\lambda)) = \ln\left(\frac{\varphi^-(\lambda)G_1^-(\lambda)G_2^-(\lambda)}{G_{12}^-(\lambda)}\right) \tag{6.128}$$

and so

$$\varphi^-(\lambda) = \frac{h(\lambda)G_{12}^-(\lambda)}{G_1^+(\lambda)G_2^+(\lambda)} \tag{6.129}$$

When $|\lambda + (i/2)| \ll 1$ and $\operatorname{Im}\lambda < -\frac{1}{2}$, we want

$$\varphi^-(\lambda) = -\frac{i}{\lambda + (i/2)} \tag{6.130}$$

and so

$$h(\lambda) = -\frac{iG_1^-(\lambda)G_2^-(\lambda)}{G_{12}^-(\lambda)(\lambda + (i/2))} \tag{6.131}$$

This yields

$$\varphi^+(\lambda) = i\frac{G_1^-(\lambda)G_2^-(\lambda)G_1^+(\lambda)G_2^+(\lambda)}{G_{12}^-(\lambda)G_{12}^+(\lambda)(\lambda + (i/2))} \tag{6.132a}$$

$$\varphi^-(\lambda) = -i\frac{G_1^-(\lambda)G_2^-(\lambda)G_{12}^-(\lambda)}{G_{12}^-(\lambda)G_2^+(\lambda)(\lambda + (i/2))G_1^+(\lambda)} \tag{6.132b}$$

Now let $A(\Delta_1), A(\Delta_2)$, and $A(\Delta_1 + \Delta_2)$ be defined as in Eq. (6.89). Now Eq. (6.125) yields

$$A(\Delta_1, \Delta_2) = \varphi^+\left(\frac{i}{2}\right) \tag{6.133a}$$

$$= \frac{|G_1^+(i/2)|^2 |G_2^+(i/2)|^2}{|G_{12}^+(i/2)|^2} \tag{6.133b}$$

Now,

$$A(\Delta_1) = \left| G_1^+\left(\frac{i}{2}\right)\right|^2 \tag{6.134a}$$

$$A(\Delta_2) = \left| G_2^+\left(\frac{i}{2}\right)\right|^2 \tag{6.134b}$$

$$A(\Delta_{12}) = \left| G_{12}^{+}\left(\frac{i}{2}\right) \right|^2 \tag{6.134c}$$

Hence,

$$A(\Delta_1, \Delta_2) = \frac{A(\Delta_1)A(\Delta_2)}{A(\Delta_1 + \Delta_2)} \tag{6.135}$$

## H. Decay Rate of the Metastable State in the Whole Damping Range

The final expression for the decay rate of the metastable state of Section VI.G is given by

$$\tau_1^{-1} = \frac{\omega_A(1)}{2\pi} \left[ \left(1 + \frac{\eta^2}{4\omega_C^2}\right)^{1/2} - \frac{\eta}{2\omega_C} \right] \frac{A(\Delta_1)A(\Delta_2)}{A(\Delta_1 + \Delta_2)} e^{-\beta \Delta U} \tag{6.136}$$

where $\omega_A(1)$ is the attempt frequency in well number 1. This formula holds for arbitrary damping $\eta$. Furthermore, if $S_2 \gg S_1$ then Eq. (6.136) reduces to Eq. (6.6) since for large values of $S_2$ we see that $A(\Delta_2) \approx 1$.

If we substitute $\omega_A(1)$ (the attempt frequency in well 1) by $\omega_A(2)$ (the attempt frequency in well 2) and $U_1$ by $U_2$ in Eq. (6.136) we get the escape rate $\tau_2^{-1}$ from well 2 into well 1.

The lifetimes $\tau_1$ and $\tau_2$, derived by solving Kramers' problem, can be used to solve the phenomenological rate equations for the well populations

$$\frac{dN_1}{dt} = -\frac{N_1}{\tau_1} + \frac{N_2}{\tau_2} \tag{6.137a}$$

$$\frac{dN_2}{dt} = \frac{N_1}{\tau_1} - \frac{N_2}{\tau_2} \tag{6.137b}$$

adding (6.137a) and (6.137b) shows that $N_1 + N_2 = \text{constant}$. The solution is given by

$$N_1(t) = \frac{N_1(0)[1 + (\omega_A(1)/\omega_A(2))\exp((U_2 - U_1)/kT - (t/\tau))] + N_2(0)[1 - \exp(-t/\tau)]}{1 + (\omega_A(1)/\omega_A(2))\exp((U_2 - U_1)/kT)} \tag{6.138}$$

where

$$\tau^{-1} = \tau_1^{-1} + \tau_2^{-1}$$
$$= \left[\left(1 + \frac{\eta^2}{4\omega_C^2}\right)^{1/2} - \frac{\eta}{2\omega_C}\right] \frac{A(\Delta_1)A(\Delta_2)}{A(\Delta_1 + \Delta_2)} \left[\frac{\omega_A(1)}{2\pi} e^{-\beta\Delta U_1} + \frac{\omega_A(2)}{2\pi} e^{-\beta\Delta U_2}\right] \tag{6.139}$$

The proof of Eq. (6.138) is given in Appendix VI.XI.

The above bridging expression yields results that are correct to within approximately $\leq 20\%$ of the numerically precise answer obtained by computing the smallest nonvanishing eigenvalue of the FPE in the crossover region [1, 100]. The above considerations should also apply to dielectric relaxation of particles in bistable potentials, including inertial effects, as that problem is the exact rotational analogue of the Kramers mechanical problem.

## I. Application to Magnetic Relaxation

The IHD and VLD results for magnetic relaxation, which we have obtained in Sections IV and V, although they have a completely different physical origin to those of the mechanical problem, as they arise from the gyromagnetic term in the Langevin equation, have, however, exactly the same mathematical form as those of the Kramers problem for particles. Hence, as suggested in Garanin et al. [70], the results of Mel'nikov and Meshkov [34, 49] should be applicable without change to the magnetic problem (this conjecture has recently been proved by Dejardin et al. [84]), provided one is not concerned with small departures from axial symmetry. Then, the formalism developed for the magnetic problem in Sections IV and V is no longer valid, and a second type of crossover problem for the Kramers theory emerges namely, how does one bridge the gap between the axially symmetric solution for $\Gamma$ of Brown, given in Section III (see also Eq. 7.32a) which is valid for all values of the friction, and the nonaxially symmetric solution. There the three cases of Kramers' problem (viz., the IHD, see Eq. (7.4) VLD, Section V see also Eqs. (7.126) and (6.140), and ID or crossover region) arise.

It should be noted that in the magnetic problem, the calculation of the parameters $\delta_1$ and $\delta_2$ must be carried out by accurate computation of the line integral giving $\Delta E$. Taking the simple case of a transverse field as an example, cf. Section VII, we have

$$A = \frac{a}{2}\oint_{E=E_C} \left[(1 - x^2)\frac{\partial E}{\partial x} d\varphi - \frac{1}{1 - x^2}\frac{\partial E}{\partial \varphi} dx\right] \tag{6.140}$$

where

$$A = \frac{\delta}{2kT}$$

and $\delta$ is the energy loss per cycle for a magnetic particle and $a$, as before, is the dimensionless friction parameter.

The interpolation problem posed by the crossover between axially symmetric and nonaxially symmetric solutions is the subject of Section VII.

### Appendix VI.I. Proof of Eqs. (6.14) for the Transformation of the Right-Hand Side FPE Operator to an Energy Diffusion Operator

Let

$$V = V(p, x, \varepsilon) \tag{6.141}$$

be an arbitrary function of the state variables and the energy.

Now, we have defined in the text

$$p_{+} \equiv \{2m[\varepsilon - U(x)]\}^{1/2} \tag{6.142}$$

$$p_{-} \equiv -\{2m[\varepsilon - U(x)]\}^{1/2} \tag{6.143}$$

Also, we have that

$$\varepsilon = \varepsilon(p, x) \tag{6.144}$$

or precisely

$$\varepsilon = \frac{p^2}{2m} + U(x) \tag{6.145}$$

which yields

$$\frac{\partial \varepsilon}{\partial p} = \frac{p}{m} \qquad \text{and} \qquad \frac{\partial \varepsilon}{\partial x} = \frac{dU}{dx} \tag{6.146}$$

From the chain rule

$$\left.\frac{\partial V}{\partial p}\right|_x = \left.\frac{\partial V}{\partial \varepsilon}\right|_x \frac{\partial \varepsilon}{\partial p} = \frac{p}{m}\left.\frac{\partial V}{\partial \varepsilon}\right|_x \tag{6.147a}$$

$$\left.\frac{\partial V}{\partial x}\right|_p = \left.\frac{\partial V}{\partial \varepsilon}\right|_p \frac{\partial \varepsilon}{\partial x} + \left.\frac{\partial V}{\partial x}\right|_\varepsilon \tag{6.147b}$$

and, on using Eqs. (6.146), (6.14) follow easily.

## Appendix VI.II. Expression for the Action in One Oscillation

$$S(0) - S(\varepsilon) = \oint [\{-2mU(x)\}^{1/2} - \{2m(\varepsilon - U(x))\}^{1/2}]dx \qquad \varepsilon < 0 \tag{6.148}$$

$$= (2m)^{1/2} \oint [\{-U(x)\}^{1/2} - \{\varepsilon - U(x)\}^{1/2}]dx \tag{6.149}$$

$$= (2m)^{1/2} \oint \left[ \{-U(x)\}^{1/2} - \{-U(x)\}^{1/2} \left\{1 - \frac{\varepsilon}{U(x)}\right\}^{1/2} \right] dx \tag{6.150}$$

since $|U(x)| > |\varepsilon|$, we may now expand the binomial taking only the first-order term in the binomial expansion. Hence,

$$S(0) - S(\varepsilon) \approx (2m)^{1/2} \oint \left[ \{-U(x)\}^{1/2} \left\{1 - 1 + \frac{\varepsilon}{2U(x)}\right\} \right] dx \tag{6.151}$$

$$= (2m)^{1/2} \oint \left[ \{-U(x)\}^{1/2} \frac{\varepsilon}{2U(x)} \right] dx \tag{6.152}$$

$$= \frac{\sqrt{2m}}{2} \oint \left[ \frac{\varepsilon}{\sqrt{-U(x)}} \right] dx \tag{6.153}$$

We assume that the potential energy near the barrier may be taken to approximate well to that of an inverted harmonic oscillator, that is, $U = -\frac{1}{2}m\omega^2x^2$ $(\omega = \omega_C)$, and so

$$S(0) - S(\varepsilon) = \left[\frac{m}{2}\right]^{1/2} \oint \frac{dx}{2\sqrt{m}\omega x} \tag{6.154}$$

$$= \frac{\varepsilon}{2\omega} \oint \frac{dx}{x} \tag{6.155}$$

$$= \frac{\varepsilon}{2\omega} 2 \ln x \Big|_{\alpha}^{\alpha+\sqrt{U_1/m\omega^2}} \tag{6.156}$$

$$\approx \frac{\varepsilon}{\omega} \ln \sqrt{\frac{U_1}{m\omega^2}} \tag{6.157}$$

$$= \frac{\varepsilon}{2\omega} \ln \frac{U_1}{m\omega^2} \tag{6.158}$$

so that $S(\varepsilon) \approx S(0)$ for small $\varepsilon$.

## Appendix VI.III. Proof of Eq. (6.29) for the Green's Function of the Energy Diffusion Equation

Equation (6.20) for the energy diffusion may be written

$$\eta kT \frac{\partial^2 f}{\partial \varepsilon^2} + \eta \frac{\partial f}{\partial \varepsilon} - \frac{\partial f}{\partial s} = 0 \tag{6.159}$$

or on dividing across by $\eta kT$

$$\frac{\partial^2 f}{\partial \varepsilon^2} + \frac{1}{kT}\frac{\partial f}{\partial \varepsilon} - \frac{1}{\eta kT}\frac{\partial f}{\partial s} = 0 \tag{6.160}$$

We seek a solution to the diffusion equation with point sources

$$\frac{\partial^2 g}{\partial \varepsilon^2} + \frac{1}{kT}\frac{\partial g}{\partial \varepsilon} - \frac{1}{\eta kT}\frac{\partial g}{\partial s} = -4\pi\delta(\varepsilon - \varepsilon')\delta(s - s') \tag{6.161}$$

with the initial condition

$$g(\varepsilon - \varepsilon', 0) = \delta(\varepsilon - \varepsilon') \tag{6.162}$$

which is best achieved using Fourier transforms [35].
Let,

$$R = \varepsilon - \varepsilon' \tag{6.163a}$$

and

$$\tau = s - s' \tag{6.163b}$$

Define the function (not to be confused with the escape rate) $\Gamma(p, \tau)$, which is the Fourier transform over the $R$ or energy variable, so that the Green's function $g$ is given by the inverse Fourier transform

$$g(R, \tau) = \frac{1}{2\pi} \int_{-\infty}^{\infty} e^{ipR} \Gamma(p, \tau) dp \tag{6.164}$$

hence by the properties of Fourier transforms [35]

$$\frac{\partial^2 g}{\partial \varepsilon^2} = -\frac{1}{2\pi} \int_{-\infty}^{\infty} p^2 e^{ipR} \Gamma(p, \tau) dp \tag{6.165}$$

$$\frac{1}{kT} \frac{\partial g}{\partial \varepsilon} = \frac{i}{2\pi kT} \int_{-\infty}^{\infty} p e^{ipR} \Gamma(p, \tau) dp \tag{6.166}$$

$$\frac{1}{\eta kT} \frac{\partial g}{\partial s} = \frac{1}{2\pi \eta kT} \int_{-\infty}^{\infty} e^{ipR} \frac{\partial \Gamma}{\partial s} dp \tag{6.167}$$

Now, on substituting Eqs. (6.165)–(6.167) into the left-hand side of Eq. (6.161), we obtain the left-hand side of Eq. (6.161) in terms of $\Gamma$

$$\frac{\partial^2 g}{\partial \varepsilon^2} + \frac{1}{kT} \frac{\partial g}{\partial \varepsilon} - \frac{1}{\eta kT} \frac{\partial g}{\partial s} = \frac{1}{2\pi} \int_{-\infty}^{\infty} e^{ipR} \left[ -p^2 \Gamma + \frac{ip}{kT} \Gamma - \frac{1}{\eta kT} \frac{\partial \Gamma}{\partial s} \right] dp \tag{6.168}$$

and since the Dirac delta function has the representation

$$\delta(R) = \frac{1}{2\pi} \int_{-\infty}^{\infty} e^{ipR} dp \tag{6.169}$$

we see that the Fourier transform $\Gamma$ satisfies the first-order differential equation

$$\frac{1}{\eta kT}\frac{\partial\Gamma}{\partial s} + \left(p^2 - \frac{ip}{kT}\right)\Gamma = 4\pi\delta(\tau) \tag{6.170}$$

Now, taking the Fourier transform of Eq. (6.164)

$$\Gamma(p,\tau) = \int_{-\infty}^{\infty} e^{-ipR} g(R,\tau)dR \tag{6.171}$$

we have, setting $\tau = 0$, which corresponds to setting $s = s'$

$$\Gamma(p,0) = \int_{-\infty}^{\infty} e^{-ipR} g(R,0)dR \tag{6.172}$$

$$= \int_{-\infty}^{\infty} e^{-ipR} \delta(R)dR \tag{6.173}$$

$$= 1 \tag{6.174}$$

Now,

$$\frac{\partial\Gamma}{\partial s} + \eta kT\left(p^2 - \frac{ip}{kT}\right)\Gamma = 4\pi\eta kT\delta(\tau) \tag{6.175}$$

yields, utilizing the integrating factor,

$$\frac{\partial}{\partial s}\left[\Gamma \exp\left(\eta kT\left\{p^2 - \frac{ip}{kT}\right\}s\right)\right] = 4\pi\eta kT\delta(\tau)\exp\left(\eta kT\left\{p^2 - \frac{ip}{kT}\right\}s\right) \tag{6.176}$$

On integration, and using the properties of the Dirac delta function, we obtain

$$\Gamma \exp\left(\eta kT\left\{p^2 - \frac{ip}{kT}\right\}s\right) = 4\pi\eta kT \exp\left(\eta kT\left\{p^2 - \frac{ip}{kT}\right\}s'\right) + C \tag{6.177}$$

where $C$ is a constant of integration. This equation may be simplified to

$$\Gamma = 4\pi\eta kT \exp\left(-\eta kT\left\{p^2 - \frac{ip}{kT}\right\}\tau\right) + C\exp\left(-\eta kT\left\{p^2 - \frac{ip}{kT}\right\}s\right) \tag{6.178}$$

On letting $s = 0$ and by using Eq. (6.174), we obtain

$$C = (1 - 4\pi\eta kT)\exp\left\{-\eta kT\left(p^2 - \frac{ip}{kT}\right)s'\right\} \tag{6.179}$$

hence, in the $p$–domain

$$\Gamma = \exp\left\{-\eta kT\left(p^2 - \frac{ip}{kT}\right)\tau\right\} \tag{6.180}$$

Now, on substituting for $\Gamma$ in Eq. (6.164) we find that in the $R$–domain

$$g = \frac{1}{2\pi}\int_{-\infty}^{\infty} e^{ipR}\exp\left\{-\eta kT\left(p^2 - \frac{ip}{kT}\right)\tau\right\}dp \tag{6.181}$$

or

$$g(R,\tau) = \frac{1}{2\pi}\int_{-\infty}^{\infty}\exp\left\{-\eta kT\left(p^2 - i\frac{\eta\tau + R}{\eta kT\tau}p\right)\tau\right\}dp \tag{6.182}$$

$$= \frac{1}{2\pi}\exp\left(-\frac{(\eta\tau + R)^2}{4\eta kT\tau}\right) \times \int_{-\infty}^{\infty}\exp\left\{-\eta kT\left(p^2 - i\frac{\eta\tau + R}{\eta kT\tau}p + \left[-i\frac{\eta\tau + R}{2\eta kT\tau}\right]^2\right)\tau\right\}dp \tag{6.183}$$

$$= \frac{1}{2\pi}\exp\left(-\frac{(\eta\tau + R)^2}{4\eta kT\tau}\right)\sqrt{\frac{\pi}{\eta kT\tau}} \tag{6.184}$$

and so the Green's function is

$$g(R,\tau) = \frac{1}{\sqrt{4\pi\eta kT\tau}}\exp\left(-\frac{[\varepsilon - \varepsilon' + \eta(s - s')]^2}{4\eta kT(s - s')}\right) \tag{6.185}$$

and

$$g(R, s) = \frac{1}{\sqrt{4\pi\eta kTs}} \exp\left( -\frac{[\varepsilon - \varepsilon' + \eta s]^2}{4\eta kTs} \right) \tag{6.186}$$

which is Eq. (6.29)

## Appendix VI.IV. To Show that the Variance of the Energy $\langle(\epsilon - \langle\epsilon\rangle)^2\rangle = 2kT\eta s$

Now,

$$\langle\varepsilon\rangle = \int_0^\infty \frac{\varepsilon}{\sqrt{4\pi kT\eta s}} \exp\left[ -\frac{(\varepsilon - \varepsilon' + \eta s)^2}{4\pi kT\eta s} \right] d\varepsilon \tag{6.187}$$

and on setting

$$u \equiv \frac{\varepsilon - \varepsilon' + \eta s}{\sqrt{4kT\eta s}} \tag{6.188}$$

we find

$$\langle\varepsilon\rangle = \frac{\sqrt{4kT\eta s}}{\sqrt{4\pi kT\eta s}} \int_{-\infty}^{\infty} \left[ u\sqrt{4kT\eta s} + \varepsilon' - \eta s \right] \exp(-u^2) du \tag{6.189}$$

$$= \sqrt{\frac{4kT\eta s}{\pi}} \int_{-\infty}^{\infty} u \exp(-u^2) du + \frac{(\varepsilon' - \eta s)}{\sqrt{\pi}} \int_{-\infty}^{\infty} \exp(-u^2) du \tag{6.190}$$

The integrand in the first integral is odd, and so the integral is zero, while the second integral is simply $\sqrt{\pi}$, and so

$$\langle\varepsilon\rangle = \varepsilon' - \eta s \tag{6.191}$$

Thus,

$$\varepsilon - \langle\varepsilon\rangle = \varepsilon - \varepsilon' + \eta s \tag{6.192}$$

Now,

$$\langle(\varepsilon - \langle\varepsilon\rangle)^2\rangle = \int_{-\infty}^{\infty} \frac{(\varepsilon - \varepsilon' + \eta s)^2}{\sqrt{4kT\eta s}} \exp\left[ -\frac{(\varepsilon - \varepsilon' + \eta s)^2}{\sqrt{4kT\eta s}} \right] d\varepsilon \tag{6.193}$$

using the substitution

$$t = \frac{\varepsilon - \varepsilon' + \eta s}{\sqrt{4kT\eta s}} \tag{6.194}$$

Eq. (6.193) becomes

$$\langle(\varepsilon - \langle\varepsilon\rangle)^2\rangle = \frac{4kT\eta s}{\sqrt{\pi}} \int_{-\infty}^{\infty} t^2 \exp(-t^2)dt \tag{6.195}$$

and on integration by parts, the desired result follows.

Also note that it is easily shown that the mean energy loss is connected to $s$ by the relation

$$2kT(\varepsilon' - \langle\varepsilon\rangle) = 2kT\eta s \tag{6.196}$$

so that the results we have proved are a form of the fluctuation–dissipation theorem.

## Appendix VI.V. Proof of Eqs. (6.74) for the Fourier Transforms $\varphi^+$ and $\varphi^-$ in the Wiener–Hopf Method

From Eq. (6.72) and the following paragraph we have

$$\ln[-\varphi^+(\lambda)] - \ln G^+(\lambda) = \ln \varphi^-(\lambda) + \ln G^-(\lambda) = f(\lambda) \tag{6.197}$$

where $f(\lambda)$ is an entire function. Let,

$$f(\lambda) = \ln h(\lambda) \tag{6.198}$$

so

$$-\varphi^+(\lambda) = h(\lambda)G^+(\lambda) \tag{6.199}$$

and

$$\varphi^-(\lambda) = \frac{h(\lambda)}{G^-(\lambda)} \tag{6.200}$$

$$= -\frac{i}{\lambda + (i/2)} \tag{6.201}$$

for

$$\left|\lambda + \frac{i}{2}\right| \ll 1 \qquad \text{and} \qquad \text{Im}\,\lambda < -\frac{1}{2}$$

Thus,

$$h(\lambda) = -\frac{iG^{-}(\lambda)}{\lambda + (i/2)} \tag{6.202}$$

for

$$\left|\lambda + \frac{i}{2}\right| \ll 1 \qquad \text{and} \qquad \text{Im}\,\lambda < -\frac{1}{2}$$

Pick

$$h(\lambda) = -\frac{iG^{-}(-(i/2))}{\lambda + (i/2)} \tag{6.203}$$

and Eq. (6.74) follow easily.

## Appendix VI.VI. Proof of the Fact that the Radius of Convergence of the Series in Eq. (6.102) for A (Δ) is 8π

Let

$$S(\Delta) \equiv \sum_{n=0}^{\infty} \frac{\zeta((1/2) - n)}{n!(2n+1)} \left(-\frac{\Delta}{4}\right)^{n} \equiv \sum_{n=0}^{\infty} a_n, \text{ say} \tag{6.204}$$

This series converges if

$$\lim_{n\to\infty} \left|\frac{a_{n+1}}{a_n}\right| < 1 \tag{6.205}$$

So, we need

$$\lim_{n\to\infty} \left|\frac{\zeta(-(1/2) - n)}{\zeta((1/2) - n)} \frac{n!}{(n+1)!4} \frac{(2n+1)}{(2n+3)} \Delta\right| < 1 \tag{6.206}$$

or

$$\lim_{n\to\infty} \left|\frac{\zeta(-(1/2) - n)}{\zeta((1/2) - n)} \frac{1}{4(n+1)} \Delta\right| < 1 \tag{6.207}$$

Define the function

$$\xi(s) \equiv \frac{1}{2}s(s-1)\pi^{-(s/2)}\Gamma\left(\frac{s}{2}\right)\zeta(s) \tag{6.208}$$

which has the property

$$\xi(s) = \xi(1-s) \tag{6.209}$$

so that

$$\pi^{(1/2)-s}\Gamma\left(\frac{s}{2}\right)\zeta(s) = \Gamma\left(\frac{1-s}{2}\right)\zeta(1-s) \tag{6.210}$$

Equation (6.210) implies that

$$\pi^{-n}\Gamma\left(\frac{n}{2}+\frac{1}{4}\right)\zeta\left(n+\frac{1}{2}\right) = \Gamma\left(\frac{1}{4}-\frac{n}{2}\right)\zeta\left(\frac{1}{2}-n\right) \tag{6.211}$$

$$\pi^{-n-1}\Gamma\left(\frac{n}{2}+\frac{3}{4}\right)\zeta\left(n+\frac{3}{2}\right) = \Gamma\left(-\frac{1}{4}-\frac{n}{2}\right)\zeta\left(-\frac{1}{2}-n\right) \tag{6.212}$$

and so we obtain that

$$\zeta\left(\frac{1}{2}-n\right) = \pi^{-n}\frac{\Gamma((n/2)+(1/4))}{\Gamma((1/4)-(n/2))}\zeta\left(n+\frac{1}{2}\right) \tag{6.213}$$

$$\zeta\left(-\frac{1}{2}-n\right) = \pi^{-n-1}\frac{\Gamma((n/2)+(3/4))}{\Gamma(-(1/4)-(n/2))}\zeta\left(n+\frac{3}{2}\right) \tag{6.214}$$

Hence, Eq. (6.207) becomes

$$\lim_{n\to\infty}\frac{1}{4\pi(n+1)}\frac{\Gamma((n/2)+(3/4))}{\Gamma(-(1/4)-(n/2))}\frac{\Gamma((1/4)-(n/2))}{\Gamma((n/2)+(1/4))}\frac{\zeta(n+(3/2))}{\zeta(n+(1/2))}|\Delta| < 1 \tag{6.215}$$

or

$$\begin{aligned}\lim_{n\to\infty}\frac{1}{4\pi(n+1)}&\frac{((n/2)-(1/4))\Gamma((n/2)-(1/4))}{\Gamma(-(1/4)-(n/2))}\\ &\times\frac{\Gamma((1/4)-(n/2))}{\Gamma((n/2)+(1/4))}\frac{\zeta(n+(3/2))}{\zeta(n+(1/2))}|\Delta| < 1\end{aligned} \tag{6.216}$$

and thus

$$\lim_{n\to\infty} \frac{n}{8\pi(n+1)} \frac{\Gamma(n/2)}{\Gamma(-(n/2))} \frac{\Gamma(-(n/2))}{\Gamma(n/2)} \frac{\zeta(n)}{\zeta(n)} |\Delta| < 1 \tag{6.217}$$

From this, it follows that for the series to converge, we need

$$|\Delta| < 8\pi \tag{6.218}$$

## Appendix VI.VII. Proof that the Series $\sum_{n=1}^{\infty}$ (1/n) erfc ($\sqrt{n\Delta}/2$) in the Series Expression (6.104) for A(Δ) Converges

It may be shown that [82] [Eq. (7.1.13)]

$$\int_x^\infty e^{-t^2} dt \leq \frac{e^{-x^2}}{x + \sqrt{x^2 + (4/\pi)}} \tag{6.219}$$

thus

$$\frac{2}{\sqrt{\pi}} \int_{(\sqrt{n\Delta}/2)}^\infty \exp(e^{-t^2}) dt \leq \frac{(2/\sqrt{\pi}) e^{-(n\Delta/4)}}{(\sqrt{n\Delta}/2) + \sqrt{(n\Delta/4) + (4/\pi)}} \tag{6.220}$$

Now,

$$\sqrt{\frac{n\Delta}{4} + \frac{4}{\pi}} > \sqrt{\frac{n\Delta}{4}} \tag{6.221}$$

and so

$$\frac{\sqrt{n\Delta}}{2} + \sqrt{\frac{n\Delta}{4} + \frac{4}{\pi}} > \sqrt{n\Delta} \geq \sqrt{n} \qquad \text{if } \Delta \gg 1 \tag{6.222}$$

Hence, Eq. (6.220) yields

$$\operatorname{erfc} \frac{\sqrt{n\Delta}}{2} < \frac{(2/\sqrt{\pi}) e^{-(n\Delta/4)}}{\sqrt{n}} < \frac{1}{\sqrt{n}} \tag{6.223}$$

if

$$\frac{2}{\sqrt{\pi}} e^{-(n\Delta/4)} < 1 \tag{6.224}$$

or

$$n\Delta > 4\ln\frac{\sqrt{\pi}}{2} \tag{6.225}$$

and

$$\frac{1}{n}\operatorname{erfc}\frac{\sqrt{n\Delta}}{2} < n^{-(3/2)} \tag{6.226}$$

hence,

$$\sum_{n=1}^{\infty}\frac{1}{n}\operatorname{erfc}\frac{\sqrt{n\Delta}}{2}$$

converges.

## Appendix VI.VIII. Proof that erfc $(\sqrt{\Delta}/2) \approx (2/\sqrt{\pi\Delta})\exp(-(\Delta/4))$ for $\Delta \gg 1$

It may be shown that, for $x > 0$ [82] [Eq. (7.1.13)]

$$\frac{1}{x+\sqrt{x^2+2}} < e^{x^2}\int_x^{\infty} e^{-t^2}dt \leq \frac{1}{x+\sqrt{x^2+(4/\pi)}} \tag{6.227}$$

By letting $x = (\sqrt{\Delta}/2)$, we obtain

$$\frac{e^{-(\Delta/4)}}{(\sqrt{\Delta}/2)+\sqrt{(\Delta/4)+2}} < \int_{(\sqrt{\Delta}/2)}^{\infty}\exp(-t^2)dt \leq \frac{e^{-(\Delta/4)}}{(\sqrt{\Delta}/2)+\sqrt{(\Delta/4)+(4\pi)}} \tag{6.228}$$

or

$$\frac{(2/\sqrt{\Delta})e^{-(\Delta/4)}}{1+\sqrt{1+(8/\Delta)}} < \int_{(\sqrt{\Delta}/2)}^{\infty}\exp(-t^2)dt \leq \frac{(2/\sqrt{\Delta})e^{-(\Delta/4)}}{1+\sqrt{1+(16/\Delta\pi)}} \tag{6.229}$$

By letting $\Delta \to \infty$, we see that

$$\int_{(\sqrt{\Delta}/2)}^{\infty}\exp(-t^2)dt \approx \frac{1}{\sqrt{\Delta}}e^{-(\Delta/4)} \tag{6.230}$$

and hence

$$\mathrm{erfc}\,\frac{\sqrt{\Delta}}{2} \equiv \frac{2}{\sqrt{\pi}}\int_{(\sqrt{\Delta}/2)}^{\infty} \exp(-t^2)dt \approx \frac{2}{\sqrt{\pi\Delta}}e^{-(\Delta/4)} \tag{6.231}$$

## Appendix VI.IX. Proof of Eqs. (6.116) Connecting the Fourier Transforms $\varphi^+(\lambda)$ and $\varphi^-(\lambda)$ in the Wiener–Hopf Method Applied to the Double Well Potential

$$\varphi_1^+(\lambda) = \frac{2\pi}{\omega_A}e^{\beta U_1}\int_0^{\infty} f_1(\varepsilon)\exp\left[\left(\frac{i\lambda + (1/2)}{kT}\right)\varepsilon\right]d\varepsilon \tag{6.232}$$

$$= \frac{2\pi}{\omega_A}e^{\beta U_1}\int_0^{\infty}\exp\left[\left(\frac{i\lambda + (1/2)}{kT}\right)\varepsilon\right] \times \left[\int_{-\infty}^{0} g_1(\varepsilon - \varepsilon')f_1(\varepsilon')d\varepsilon' + \int_0^{\infty} g_1(\varepsilon - \varepsilon')f_2(\varepsilon')d\varepsilon'\right]d\varepsilon \tag{6.233}$$

$$= \frac{(2\pi/\omega_A)e^{\beta U_1}}{\sqrt{4\pi\delta_1 kT}}\left\{\int_0^{\infty}\int_{-\infty}^{0}\exp\left[\left(\frac{i\lambda + (1/2)}{kT}\right)\varepsilon\right] \times \exp\left[-\frac{(\varepsilon - \varepsilon' + \delta_1)^2}{4\delta_1 kT}\right]f_1(\varepsilon')d\varepsilon' d\varepsilon + \int_0^{\infty}\int_0^{\infty}\exp\left[\left(\frac{i\lambda + (1/2)}{kT}\right)\varepsilon\right] \times \exp\left[-\frac{(\varepsilon - \varepsilon' + \delta_1)^2}{4\delta_1 kT}\right]f_2(\varepsilon')d\varepsilon' d\varepsilon\right\} \tag{6.234}$$

Now the first double integral in Eq. (6.234) may be written

$$\int_{-\infty}^{0} f_1(\varepsilon')\exp\left[-\frac{(\varepsilon' - \delta_1)^2}{4\delta_1 kT}\right]\int_0^{\infty}\exp\left[-\frac{\varepsilon^2}{4\delta_1 kT}\right] \times \exp\left[\frac{2(\varepsilon' - \delta_1) + 4\delta_i(i\lambda + (1/2))}{4\delta_1 kT}\varepsilon\right]d\varepsilon d\varepsilon' \tag{6.235}$$

$$= \int_{-\infty}^{0} f_1(\varepsilon') \exp\left[-\frac{(\varepsilon' - \delta_1)^2}{4\delta_1 kT}\right] \exp\left[\frac{(\varepsilon' + 2i\lambda\delta_1)^2}{4\delta_1 kT}\right] \times \int_{0}^{\infty} \exp\left[-\frac{(\varepsilon - \varepsilon' - 2i\lambda\delta_1)^2}{4\delta_1 kT}\right] d\varepsilon d\varepsilon' \quad (6.236)$$

$$= \frac{\sqrt{4\pi\delta_1 kT}}{2} \int_{-\infty}^{0} f_1(\varepsilon') \exp\left[-\frac{(\varepsilon' - \delta_1)^2}{4\delta_1 kT}\right] \exp\left[\frac{(\varepsilon' + 2i\lambda\delta_1)^2}{4\delta_1 kT}\right] d\varepsilon' \quad (6.237)$$

$$= \frac{\sqrt{4\pi\delta_1 kT}}{2} \exp\left[-\Delta_1\left(\lambda^2 + \frac{1}{4}\right)\right] \int_{-\infty}^{0} f_1(\varepsilon') \exp\left[-\frac{i\lambda + (1/2)}{kT}\varepsilon'\right] d\varepsilon' \quad (6.238)$$

$$= \frac{\sqrt{4\pi\delta_1 kT}}{2} \exp\left[-\Delta_1\left(\lambda^2 + \frac{1}{4}\right)\right] \frac{\omega_A}{2\pi} e^{-\beta U_1} \varphi_1^-(\lambda) \quad (6.239)$$

By a similar set of calculations, the second double integral in Eq. (6.234) may be shown to be

$$\frac{\sqrt{4\pi\delta_1 kT}}{2} \exp\left[-\Delta_1\left(\lambda^2 + \frac{1}{4}\right)\right] \frac{\omega_A}{2\pi} e^{-\beta U_1} \varphi_2^+(\lambda) \quad (6.240)$$

Hence,

$$\varphi_1^+(\lambda) = \frac{1}{2} \exp\left[-\Delta_1\left(\lambda^2 + \frac{1}{4}\right)\right] [\varphi_1^-(\lambda) + \varphi_2^+(\lambda)] \quad (6.241)$$

Now

$$\varphi_1^-(\lambda) = \frac{2\pi}{\omega_A} e^{\beta U_1} \int_{-\infty}^{0} f_1(\varepsilon) \exp\left[\left(\frac{i\lambda + (1/2)}{kT}\right)\varepsilon\right] d\varepsilon \quad (6.242)$$

$$= \frac{2\pi}{\omega_A} e^{\beta U_1} \int_{-\infty}^{0} \exp\left[\left(\frac{i\lambda + (1/2)}{kT}\right)\varepsilon\right] \times \left[\int_{-\infty}^{0} g_1(\varepsilon - \varepsilon') f_1(\varepsilon') d\varepsilon' + \int_{0}^{\infty} g_1(\varepsilon - \varepsilon') f_2(\varepsilon') d\varepsilon'\right] d\varepsilon \quad (6.243)$$

$$= \frac{(2\pi/\omega_A)e^{\beta U_1}}{\sqrt{4\pi\delta_1 kT}} \left\{ \int_{-\infty}^{0} \int_{-\infty}^{0} \exp\left[\left(\frac{i\lambda + (1/2)}{kT}\right)\varepsilon\right] \right.$$

$$\times \exp\left[-\frac{(\varepsilon - \varepsilon' + \delta_1)^2}{4\delta_1 kT}\right] f_1(\varepsilon')d\varepsilon' d\varepsilon$$

$$+ \int_{-\infty}^{0} \int_{0}^{\infty} \exp\left[\left(\frac{i\lambda + (1/2)}{kT}\right)\varepsilon\right]$$

$$\left. \times \exp\left[-\frac{(\varepsilon - \varepsilon' + \delta_1)^2}{4\delta_1 kT}\right] f_2(\varepsilon')d\varepsilon' d\varepsilon \right\} \tag{6.244}$$

$$\varphi_1^-(\lambda) = \frac{1}{2}\exp\left[-\Delta_1\left(\lambda^2 + \frac{1}{4}\right)\right][\varphi_1^-(\lambda) + \varphi_2^+(\lambda)] \tag{6.245}$$

and so

$$\varphi_1^+(\lambda) + \varphi_1^-(\lambda) = \exp\left[-\Delta_1\left(\lambda^2 + \frac{1}{4}\right)\right][\varphi_1^-(\lambda) + \varphi_2^+(\lambda)] \tag{6.246}$$

$$= [1 - G_1(\lambda)][\varphi_1^{-1}(\lambda) + \varphi_2^+(\lambda)] \tag{6.247}$$

By similar calculations,

$$\varphi_2^+(\lambda) + \varphi_2^-(\lambda) = [1 - G_2(\lambda)][\varphi_2^-(\lambda) + \varphi_1^+(\lambda)] \tag{6.248}$$

## Appendix VI.X. Proof of Eqs. (6.121) for the Function G(λ) Occurring in the Wiener–Hopf Method Applied to the Double Well Potential

$$G_{12}(\lambda) = 1 - \exp\left[-(\Delta_1 + \Delta_2)\left(\lambda^2 + \frac{1}{4}\right)\right] \tag{6.249}$$

$$= 1 - \exp\left[-\Delta_1\left(\lambda^2 + \frac{1}{4}\right)\right]\exp\left[-\Delta_2\left(\lambda^2 + \frac{1}{4}\right)\right] \tag{6.250}$$

$$= 1 - [G_1 - 1][G_2 - 1] \tag{6.251}$$

$$= G_1 + G_2 - G_1 G_2 \tag{6.252}$$

Now, from Eq. (6.116), we find

$$\varphi_1^+ = \varphi_2^+ - G_1(\varphi_1^- + \varphi_2^+) \tag{6.253a}$$

$$\varphi_2^+ = \varphi_1^+ - G_1(\varphi_2^- + \varphi_1^+) \tag{6.253b}$$

which may be written

$$\varphi^+ = -G_1(\varphi_1^- + \varphi_2^+) \tag{6.254}$$

$$-\varphi^+ = -G_1(\varphi_2^- + \varphi_1^+) \tag{6.255}$$

and which yield

$$(G_1 + G_2)\varphi^+ = -G_1G_2(\varphi_1^- + \varphi_2^+ - \varphi_2^- - \varphi_1^+) \tag{6.256}$$

$$= G_1G_2\varphi^+ - G_1G_2\varphi^- \tag{6.257}$$

This may be written as

$$G_{12}\varphi^+ + G_1G_2\varphi^- = 0 \tag{6.258}$$

and Eq. (6.121a) follows easily.
Now,

$$\psi^+ + \psi^- = G_1\varphi_1^+ + G_2\varphi_2^+ + G_1\varphi_1^- + G_2\varphi_2^- \tag{6.259}$$

on substituting Eq. (6.253a) into the right-hand side of Eq. (6.253b)

$$\varphi_2^+ = \varphi_2^+ - G_1(\varphi_2^- + \varphi_1^+) - G_2(\varphi_2^- + \varphi_1^+) \tag{6.260}$$

Eq. (6.121b) follows.

## Appendix VI.XI. Proof of Eq. (6.138) for the Population in the Double Well Potential

From Eqs. (6.137a) and (6.137b), we see that

$$N_1(t) + N_2(t) = N_1(0) + N_2(0) \tag{6.261}$$

$$N_2(t) = N_1(0) + N_2(0) - N_1(t) \tag{6.262}$$

and so

$$\frac{dN_1}{dt} = -\frac{N_1}{\tau_1} + \frac{N_1(0) + N_2(0)}{\tau_2} - \frac{N_2}{\tau_2} \tag{6.263}$$

$$= -\frac{1}{\tau}N_1 + \frac{1}{\tau_2}[N_1(0) + N_2(0)] \tag{6.264}$$

This equation may be written

$$\frac{dN_1}{(1/\tau)N_1 - (1/\tau_2)[N_1(0) + N_2(0)]} = -dt \tag{6.265}$$

and by integrating we obtain

$$\tau \ln\left[\frac{N_1}{\tau} - \frac{N_1(0) + N_2(0)}{\tau_2}\right] = -t + C \tag{6.266}$$

When $t = 0$, we have $N_1 = N_1(0)$, so

$$C = \tau \ln\left[\frac{N_1(0)}{\tau} - \frac{N_1(0) + N_2(0)}{\tau_2}\right] \tag{6.267}$$

$$= \tau \ln\left[\frac{N_1(0)}{\tau_1} - \frac{N_2(0)}{\tau_2}\right] \tag{6.268}$$

Thus,

$$\ln\left[\frac{N_1}{\tau} - \frac{N_1(0) + N_2(0)}{\tau_2}\right] = \frac{t}{\tau} + \ln\left[\frac{N_1(0)}{\tau_1} - \frac{N_2(0)}{\tau_2}\right] \tag{6.269}$$

or

$$\frac{N_1}{\tau} - \frac{N_1(0) + N_2(0)}{\tau_2} = \left[\frac{N_1(0)}{\tau_1} - \frac{N_2(0)}{\tau_2}\right] \exp\left[-\frac{t}{\tau}\right] \tag{6.270}$$

Hence,

$$N_1 = \tau\left[\left\{\frac{N_1(0)}{\tau_1} - \frac{N_2(0)}{\tau_2}\right\} \exp\left[-\frac{t}{\tau}\right] + \frac{N_1(0) + N_2(0)}{\tau_2}\right] \tag{6.271}$$

$$= \tau\left[N_1(0)\left\{\frac{1}{\tau_2}+\frac{1}{\tau_1}\exp\left(-\frac{t}{\tau}\right)\right\}+\frac{N_2(0)}{\tau_2}\left\{1-\exp\left(-\frac{t}{\tau}\right)\right\}\right] \quad (6.272)$$

Now, by using the expression for $\tau_1^{-1}$ in Eq. (6.136) and the corresponding ones for $\tau_2^{-1}$ and $\tau^{-1}$, we obtain Eq. (6.139).

# VII. INTERPOLATION FORMULAS BETWEEN KRAMERS' ESCAPE RATES FOR AXIALLY SYMMETRIC AND NONAXIALLY SYMMETRIC POTENTIALS FOR SINGLE DOMAIN FERROMAGNETIC PARTICLES

## A. Introductory Concepts

### 1. *Introduction*

It is apparent from the theory of first passage times that exact integral representations (see the axially symmetric results of Section III and [68]) of the reversal time of the magnetic moment of a single domain ferromagnetic particle or calculations of the inverse of the smallest nonvanishing eigenvalue, $\lambda_1$, of the FPE for the density of magnetic moment orientations by their very nature allow continuity between the reversal times for spins in axially symmetric and nonaxially symmetric potentials of the magnetocrystalline anisotropy.

Since, in principle, it is impossible to calculate $\lambda_1$ exactly in closed form, and since the exact integral representations of the reversal time are *purely formal* [68], with the exception of axial symmetry, where the distribution function is independent of the longitude, asymptotic solutions of the relaxation time problem have usually proceeded by calculation of the Kramers escape rate. This method, initiated by Brown [21], has the advantage that it yields closed form expressions for the reversal time that constitute with certain exceptions which are considered here an accurate approximation to the true solution in the limit of high potential barriers, thus yielding formulas that may readily be compared with experiment [52, 68, 85, 86].

We have mentioned earlier that the exact solution for the escape rate based on the calculation of $\lambda_1$ ensures continuity between axially symmetric and nonaxially symmetric solutions. Another important concept here is that of the blocking temperature, $T_B$. This is the temperature at which the magnetization of a single domain ferromagnetic particle may be caused to disappear and which is generally determined from measurements of the reversal time. Our intuitive expectation is that the imposition of a uniform magnetic field will lower $T_B$ as it lowers the potential barrier. Very recently an attempt was made by Kachkachi et al. [86] to calculate $T_B$ from the nonaxially symmetric formulas for $\Gamma$ for uniaxial anisotropy with a uniform field applied at an arbitrary angle to the anisotropy axis. This has the effect in the IHD limit of producing a *rise* in $T_B$ for small values of the applied field, and then a *decrease*, which initially appears to concur with experiment. However, this behavior is not manifested by the exact solution based on $\lambda_1$, which indicates a *monotonic decrease* in $T_B$ for all fields. Thus, the behavior predicted by the nonaxially symmetric Kramers escape rate namely the asymptotic $\lambda_1$ [viz., a (counterintuitive) rise in $T_B$ for small fields, rather than the expected decrease], must arise as a result of a failure of that solution to tend to the axially symmetric result (as one would expect, by inspection of the FPE) for low fields and high friction. In other words, the nonaxially symmetric asymptotic solution *diverges* as axial symmetry is approached.

Hence, suitable interpolation formulas must be found that can bridge the gap between the axially symmetric potential and nonaxially symmetric potential asymptotes, so that an asymptotic solution should reproduce the behavior of the exact solution. This is yet another special aspect of Kramers' problem for spins in so far as not only does one have to derive an interpolation formula between the various regimes of friction (in the manner of Mel'nikov and Meshkov [34] for the mechanical problem) but one also must derive interpolation formulas between the axially symmetric and nonaxially symmetric potential asymptotes for each of the three friction regimes. An ultimate goal of the theory, therefore, would be to achieve for spins a single formula uniting the transition between axial and nonaxial symmetry and the two friction regimes, VLD and IHD. We remark that this has been effectively achieved for mechanical particles (where the friction is the only variable and symmetry considerations are of course irrelevant) by Mel'nikov and Meshkov [34] [their Eq. (6.1); see Section VI].

The purpose of this section is to review the progress that has been made in obtaining interpolation formulas between the axially symmetric and nonaxially symmetric potential asymptotes for various values of the friction. The IHD limit will mainly be treated as the existing solution [70] is for the transverse field only. The present treatment shows how the existing solution may be generalized to an arbitrary field angle. The problems attendant on the other cases, namely, the crossover region between VLD and IHD, will also be alluded to. The treatment we shall give is much more general than that given in Section III, as one of the consequences of it will be the derivation of an exact integral expression for the Kramers escape rate for spins, which is valid for an arbitrary potential of the magnetocrystalline anisotropy.

### 2. *General Formalism*

The particular nonaxially symmetric system, which we shall use to illustrate the theory, is an assembly of single domain ferromagnetic particles having simple uniaxial anisotropy with a uniform field, $\mathbf{H}_o$, applied at an arbitrary angle to the anisotropy axis which, without loss of generality, we may assume to be applied in the $xz$ plane. Thus the model Hamiltonian is

$$H = -\widetilde{K}n_z^2 - \mu_0 \mathbf{n} \cdot \mathbf{H}_o, \qquad |\mathbf{n}| = 1, \qquad \mathbf{n} = \frac{\mathbf{M}}{M_s} \tag{7.1}$$

$\mu_0 = vM_s$ is the magnetic moment.
$\widetilde{K} = v\,K$ is the uniaxial anisotropy energy constant of a particle for ease of comparison with the earlier paper of Garanin et al. [70], and we shall retain the notation of that paper.
$v =$ the particle volume.
$M_s =$ the magnitude of the magnetization of the particle.
$K =$ the anisotropy constant.
$\beta = (kT)^{-1}$

In order to preserve, once again, a close relationship with the original paper [70]. on this subject we shall, in this instance, proceed from the stochastic Landau–Lifschitz equation rather than the Gilbert equation. The stochastic Landau–Lifshitz equation for the motion of the magnetic moment vector of an individual particle is (note the noise field is omitted from the alignment term as in Ref. [70], this does not however affect the final result.)

$$\dot{\mathbf{n}} = \gamma[\mathbf{n} \times (\mathbf{H}_{eff} + \boldsymbol{\Xi})] - \gamma a[\mathbf{n} \times (\mathbf{n} \times \mathbf{H}_{eff})] \tag{7.2}$$

where, just as in the Gilbert equation

$$\gamma = \text{the gyromagnetic ratio} \qquad \mathbf{H}_{eff} = -\frac{1}{\mu_0}\frac{\partial H}{\partial \mathbf{n}} \qquad a = \eta\gamma M_s \tag{7.3}$$

$\eta =$ a damping constant chosen so that $a$ is dimensionless.
$\boldsymbol{\Xi} =$ the white noise random field arising from the thermal motion of the surroundings with correlation functions

$$\langle \Xi_\alpha(t)\Xi_\beta(t')\rangle = \frac{2akT}{\gamma\mu_0}\delta_{\alpha\beta}\delta(t - t') \qquad \alpha, \beta = 1, 2, 3 \tag{7.4}$$

about each of the three Cartesian axes, associated with the magnetic moment, specified by the numbers $\alpha, \beta$.

$\delta_{\alpha\beta} =$ Kronecker's delta.
$\delta(t - t')$ is the Dirac delta function.

The Fokker–Planck equation for the distribution function $f(\mathbf{N}, t) = \langle \delta(\mathbf{N} - \mathbf{n}(t)) \rangle$ of the orientations $\mathbf{n}(t)$, of the magnetic moments, on the sphere $|\mathbf{N}| = 1$, where the average is taken over the realizations of $\boldsymbol{\Xi}$, is

$$\frac{\partial f}{\partial t} + \frac{\partial}{\partial \mathbf{N}} \left\{ \gamma [\mathbf{N} \times \mathbf{H}_{eff}] - \gamma a [\mathbf{N} \times (\mathbf{N} \times \mathbf{H}_{eff})] + \frac{\gamma a k T}{\mu_0} \left[ \mathbf{N} \times \left( \mathbf{N} \times \frac{\partial}{\partial \mathbf{N}} \right) \right] \right\} f = 0 \tag{7.5}$$

which is simply Eq. (7.7a) below written in vector form. The equilibrium solution has the form

$$f_{\rm eq}(\mathbf{N}) = Z^{-1} \exp(-\beta H) \tag{7.6}$$

where $\beta = (1/kT)$ and $Z$ is the partition function.
If we represent as usual the orientation of the magnetic moment by the polar angles $(\theta, \varphi)$, Eq. (7.5) takes on the familiar form of the continuity equation:

$$\frac{\partial f}{\partial t} + \frac{\partial j_x}{\partial x} + \frac{\partial j_\varphi}{\partial \varphi} = 0 \tag{7.7a}$$

where

$$x \equiv \cos \theta \tag{7.7b}$$

$$j_x = \frac{\gamma}{\mu_0} \left[ -\frac{\partial H}{\partial \varphi} f - a(1 - x^2) \left( \frac{\partial H}{\partial x} + kT \frac{\partial}{\partial x} \right) f \right] \tag{7.8a}$$

$$j_\varphi = \frac{\gamma}{\mu_0} \left[ \frac{\partial H}{\partial x} f - \frac{a}{1 - x^2} \left( \frac{\partial H}{\partial \varphi} + kT \frac{\partial}{\partial \varphi} \right) f \right] \tag{7.8b}$$

Here, $j_x$ and $j_\varphi$ are the components of the probability current in the $(x, \varphi)$ space. The components of the probability current on the sphere are given by

$$J_x = \frac{j_x}{\sqrt{1 - x^2}} \tag{7.9a}$$

$$J_\varphi = j_\varphi \sqrt{1 - x^2} \tag{7.9b}$$

($x = p$ in the calculation of Sections IV and V).

We have seen (Section III) that Gilbert's equation, where the damping is directly incorporated into the effective field, may also be used to derive the Fokker–Planck equation [21]. Gilbert's equation, incorporating stochastic terms, is

$$\frac{d\mathbf{M}}{dt} = \gamma\mathbf{M} \times \left( -\frac{\partial H}{\partial \mathbf{M}} - \eta \frac{d\mathbf{M}}{dt} + \boldsymbol{\Xi} \right) \tag{7.10}$$

where the magnetization vector

$$\mathbf{M} = M_s \mathbf{n} \tag{7.11}$$

We reiterate that this equation also leads to an equation of the same mathematical form as Eq. (7.7). For ease of comparison with our recent paper [70], we shall use the Landau–Lifshitz equation.

To solve the FPE for low temperatures, $kT \ll \Delta U$, where $\Delta U$ is the barrier height (i.e., the difference in the values of $H$ at the summit and at the minimum) we represent $f$ in the form

$$f(\mathbf{N}, t) = f_{eq}(\mathbf{N}) g(\mathbf{N}, t) \tag{7.12}$$

Since we are concerned with a quasistationary process we may neglect the exponentially small $\dot{g}$ to obtain

$$\begin{aligned} \frac{\partial H}{\partial \varphi}\frac{\partial g}{\partial x} - \frac{\partial H}{\partial x}\frac{\partial g}{\partial \varphi} + a\Bigg[ \left( -\frac{\partial H}{\partial x} + kT\frac{\partial}{\partial x} \right)(1 - x^2)\frac{\partial g}{\partial x} \\ + \frac{1}{1 - x^2}\left( -\frac{\partial H}{\partial \varphi} + kT\frac{\partial}{\partial \varphi} \right)\frac{\partial g}{\partial \varphi} \Bigg] = 0 \end{aligned} \tag{7.13}$$

If we now consider just as in Section II (in accordance with our quasistationary assumption) $g$ as a crossover function that takes on the values $g_1$ and $g_2$ in the wells of a bistable potential [created, e.g., by Eq.(7.1)] and changes in a narrow region about the top of the barrier, then $g$ may be represented as

$$g(x, \varphi) = g_1 + (g_2 - g_1)\zeta(x, \varphi) \tag{7.14}$$

where $\zeta$ assumes the values 0 and 1 in the first and second wells, respectively. The resulting equation for $\zeta$ is effectively Eq. (4.11) adapted to spins. We can now use Eqs. (7.13) and (7.14) to obtain a general expression for the escape rate for spins in integral form. The advantage of this is that it allows one to determine crossover formulas that bridge the gap between axially symmetric and nonaxially symmetric asymptotes in the region of small departures from axial symmetry. The number of particles in the wells are normalized to satisfy $N_1 + N_2 = 1$. These particles are given by

$$N_1 = g_1 N_{1,\text{eq}} \tag{7.15a}$$

$$N_2 = g_2 N_{2,\text{eq}} \tag{7.15b}$$

where $N_{i,eq} = Z_i/Z$ are the equilibrium values and $Z_1$ and $Z_2$ are the partition functions in each of the wells.

The change in the number of particles in the first well, $N_1$, arises from the flow of particles from the first to the second well through the line $x = \text{const}$ (say the equator $x = 0$):

$$\dot{N} = -\int_0^{2\pi} j_x \, d\varphi \tag{7.16}$$

By using

$$f_{\text{eq}} \frac{\partial H}{\partial \varphi} = -kT \frac{\partial f_{\text{eq}}}{\partial \varphi} \tag{7.17}$$

we can integrate the first term of $j_x$ by parts (details in Appendix VII.II) to obtain

$$\dot{N}_1 = \frac{\gamma kT}{\mu_0} \int_0^{2\pi} f_{\text{eq}} \left[ a(1 - x^2) \frac{\partial g}{\partial x} + \frac{\partial g}{\partial \varphi} \right] d\varphi \tag{7.18}$$

Finally, by using Eqs. (7.14) and (7.15) (again, see Appendix VII.II), we have the kinetic equations

$$\dot{N}_1 = -\dot{N}_2 = \Gamma(N_2 N_{1,\text{eq}} - N_1 N_{2,\text{eq}}) \tag{7.19}$$

where the escape rate $\Gamma$ is given by

$$\Gamma = \left[ \frac{1}{Z_1} + \frac{1}{Z_2} \right] \frac{\gamma kT}{\mu_0} \int_0^{2\pi} e^{-\beta H} \left[ a(1 - x^2) \frac{\partial \zeta}{\partial x} + \frac{\partial \zeta}{\partial \varphi} \right] d\varphi \tag{7.20}$$

which should be compared with the corresponding formulas for particles, namely, Eqs. (2.8)–(2.11) of Mel'nikov and Meshkov [34] which we give in *their notation* below.

The equation for mechanical particles, corresponding to Eq. (7.16) is

$$\dot{N} = J = \int_{-\infty}^{\infty} \frac{p}{m} F(p, x, t) dp \tag{7.21a}$$

or in terms of the crossover function $\zeta$ as

$$\int_{-\infty}^{\infty} \frac{p}{m} e^{-\beta H} \zeta \, dp \tag{7.21b}$$

In Eq. (7.21a) $F(p, x, t)$ is the density of representative points having momentum $p$ and position $x$ at time $t$. The escape rate is given by

$$\Gamma = \tau^{-1} = \frac{J}{N} \tag{7.22}$$

where $N$ is the number of particles trapped in the well in the vicinity of the point $A$. Now in the quasistationary approximation, $(\partial \rho / \partial t) \approx 0$, the Klein–Kramers equation for $F = F(p, x)$, where $F(p, x)$ is given by

$$F(p, x, t) = F(p, x) e^{-\frac{t}{\tau}}$$

is

$$\frac{p}{m}\frac{\partial F}{\partial x} - \frac{\partial}{\partial p}\left[F\frac{dU}{dx} + \gamma\left(pF + mkT\frac{\partial F}{\partial p}\right)\right] = 0 \tag{7.23}$$

We then use in order to calculate $\Gamma$ taking the IHD case of Section II as a definite example

1. The inverted oscillator approximation for the potential energy at $c$ (the zero of potential energy)

$$U(x) = -\frac{1}{2} m \omega_C^2 x^2 \tag{7.24}$$

2. The boundary condition that the particles, once over the barrier, never return, that is the particles are absorbed at the barrier top. We state this last condition mathematically as

$$F(p, x, t) \to 0 \qquad \text{as} \qquad x \to \infty \tag{7.25}$$

which is equivalent to

$$F(p,x) \to 0 \qquad \text{as} \qquad x \to \infty \tag{7.26}$$

and finally

3. The fact that deep in the well, the distribution approximates very closely to the Maxwell–Boltzmann distribution. We obtain, by using the same calculation as in Section II, Section II.C.2, Kramers' solution in the form cf. Eq. (2.118):

$$F(p,x) = \text{const}\exp\left(-\frac{p^2}{2mkT} + \frac{m\omega_C^2 x^2}{2kT}\right)\int_{x-\frac{\lambda p}{m\omega^2}}^{\infty} \exp\left(-\frac{m\omega_C^2\xi^2}{2\gamma\lambda_+ kT}\right)d\xi \tag{7.27}$$

where ($\gamma = (\zeta/m)$ is the friction parameter)

$$\lambda_+ = \left(\omega_C^2 + \frac{1}{4}\gamma^2\right)^{\frac{1}{2}} - \frac{1}{2}\gamma \tag{7.28}$$

Now, again referring to spins, Eq.(7.20) is an exact formula for the escape rate regardless of the damping. The calculation of the axially symmetric result of Section III for the particular potential $Kv\sin^2\theta$ is given in Appendix VII.IV.

### 3. *Divergence of the Nonaxially Symmetric IHD Formulas for Small Departures from Axial Symmetry*

In order to illustrate the problem of finding suitable crossover formulas between the axially symmetric and nonaxially symmetric potential escape rates we shall, as mentioned above, consider the particular problem of a uniform field, $\mathbf{H}_0$, applied at an arbitrary angle $\psi$ to the anisotropy axis, with Hamiltonian $H$, so that

$$\varepsilon = \beta H = -\sigma[x^2 + 2h(x\cos\psi + \sqrt{1-x^2}\cos\varphi\sin\psi)] \tag{7.29}$$

where we have introduced the dimensionless variables, $\varepsilon$, the barrier height parameter

$$\sigma = \beta\widetilde{K} = \frac{Kv}{kT} \tag{7.30}$$

and the reduced field parameter

$$h = \frac{\mu_0 H_0}{2\widetilde{K}} = \frac{M_s H_0}{2K} = \frac{\xi}{2\sigma} \tag{7.31}$$

This Hamiltonian, normalized by the thermal energy, supposes that the field, $\mathbf{H}_0$, is applied in the $xz$ plane. If the reduced field $h \to 0$, the escape rate is of the form described by Brown [21] for uniaxial anisotropy, namely (see either Appendix VII.IV, where the expression that follows is directly derived from Eq. (7.20), or Section III, where the expression (namely 3.58b) is derived from Brown's general formula 3.56, 3.57 for axially symmetric potentials)

$$\Gamma = 2\pi a\sqrt{\frac{\sigma}{\pi}}\frac{\omega_1}{\pi}e^{-\sigma} = 2\pi a\sqrt{\frac{\sigma}{\pi}}\Gamma_{\mathrm{TS}} \tag{7.32a}$$

because we have two wells,
where

$$\omega_1^2 = \frac{2K\gamma}{M_s} \equiv \omega_A \tag{7.32b}$$

is the ferromagnetic resonant angular frequency (well angular frequency). If the Gilbert equation, favored by Brown [21], is used, $\gamma$ is to be replaced by $\gamma/(1+a^2)$.

Equation (7.32a) was originally derived by Brown [21], in the manner described in Section III, and is valid for all values of the damping, since the underlying FPE is effectively one dimensional. We say effectively one dimensional, in this case, because the $\varphi$ variation only appears as a steady precession so that, as far as the longitudinal motion is concerned, the only relevant variable is $\theta$. Unlike the nonaxially symmetric case there is *no dynamical coupling between the transverse and longitudinal motions*. In general, it is convenient to write the escape rate as

$$\Gamma = A_1\frac{\omega_1}{2\pi}e^{-\beta\Delta U_1} + A_2\frac{\omega_2}{2\pi}e^{-\beta\Delta U_2} \tag{7.33}$$

(where $\Delta U_1$ and $\Delta U_2$ are the respective barrier heights) for a bistable potential with wells numbered 1 and 2. The variables $A_1$ and $A_2$ are constants that describe the deviation from the Arrhenius (transition state theory) result. An illustration of this is provided by Brown's IHD formula [62, 65], which we have derived (Eq. 4.93) using Langer's general formalism in Section IV, namely, (we retain the notation of [65])

$$\Gamma = \tau^{-1} \approx \frac{\Omega_0}{2\pi\omega_0}[\omega_1\exp\{-\beta\Delta U_1\} + \omega_2\exp\{-\beta\Delta U_2\}] \tag{7.34}$$

which is valid provided

$$
\begin{aligned}
a\beta\Delta U_1 > 1 \qquad & \text{with} \qquad \beta\Delta U_1 \gg 1 \\
& \text{and} \\
a\beta\Delta U_2 > 1 \qquad & \text{with} \qquad \beta\Delta U_2 \gg 1
\end{aligned} \tag{7.35}
$$

We note that VLD means $a\beta\Delta U_i \ll 1$ with $\beta\Delta U_i \gg 1$, $i = 1, 2$.
The crossover region where neither formula applies is

$$a\beta\Delta U_i \approx 1 \qquad \text{with} \qquad \beta\Delta U_i \gg 1 \qquad i = 1, 2 \tag{7.35c}$$

In Eq. (7.34), the squares of the well angular frequencies are

$$\omega_i^2 = \frac{\gamma^2}{M_s^2} c_1^{(i)} c_2^{(i)} \quad i = 1, 2 \tag{7.36}$$

the square of the saddle angular frequency is

$$\omega_0^2 = -\frac{\gamma^2}{M_s^2} c_1 c_2 \tag{7.37}$$

and the damped saddle angular frequency is (ignoring terms $O(a^2)$ in $h'$ or $b$)

$$\Omega_0 = \frac{a\gamma}{2M_s} \left[ -c_1 - c_2 + \sqrt{(c_2 - c_1)^2 - 4a^{-2} c_1 c_2} \right] \tag{7.38}$$

$$c_j \equiv c_j^{(0)} \qquad j = 1, 2 \tag{7.39}$$

and $c_j^{(i)} (i = 0, 1, 2;\; j = 1, 2)$ are the coefficients of the quadratic terms in the Taylor expansion of the potential energy, ($= H$ the Hamiltonian), about the saddle point ($i = 0$) and about the minima ($i = 1, 2$). The details of the derivation of this general result, which are rather lengthy, are given in Geoghegan et al. [65]. This derivation employs Brown's [62] original method, which proceeded from first principles without appeal to Langer's work [61] used by us in Section IV. Having determined a general formula for $\Gamma$, Eq. (7.34), considerable mathematical manipulation is needed to specialize it to the potential of Eq. (7.1), that is, a uniform field at an angle to the easy axis of a particle having simple uniaxial anisotropy. Thus from the calculations presented in [65], one may show that the $c$'s are determined by the roots of the quartic equation [65]

$$x^4 + 2(h\cos\psi)x^3 - (1 - h^2)x^2 - 2(h\cos\psi)x - h^2\cos^2\psi = 0 \tag{7.40}$$

Hence, one may show (details in Appendix VII.I) that the saddle angular frequency, $\omega_0$, diverges like $h^{-1/2}$ for small $h$ so that the escape rate becomes *infinite* as $h \to 0$. *Hence, the axially symmetric or uniaxial limit is not approached for very small h.* The net effect of the divergence of the saddle frequency, if it is used to calculate the blocking temperature, produces a spurious rise in $T_B$ as the field is initially increased from 0. The divergence of the IHD formula may be simply addressed by generalizing a method used to treat the problem for a purely transverse field due to Garanin et al. [70] to a field at an arbitrary angle, described below.

In what follows, it is convenient to write Eqs. (7.34)–(7.39) in the notation of Geoghegan et al. [65]. Thus, Eq. (7.34), namely, the high-energy barrier approximation to the smallest nonvanishing eigenvalue of the Fokker–Planck equation, becomes [65].

$$\Gamma = p_1 = h' \frac{-c_1 - c_2 + \sqrt{(c_2 - c_1)^2 - 4a^{-2}c_1c_2}}{4\pi\sqrt{-c_1c_2}} \times \left[\sqrt{c_1^{(1)}c_2^{(1)}}e^{-\beta\Delta U_1} + \sqrt{c_1^{(2)}c_2^{(2)}}e^{-\beta\Delta U_2}\right] \tag{7.41}$$

where

$$h' = \frac{a\gamma}{M_s} \tag{7.42}$$

since we are using the Landau–Lifshitz equation $(1 + a^2) \simeq 1$.

So,

$$p_1 = A\left[\frac{\omega_1}{2\pi}e^{-\beta\Delta U_1} + \frac{\omega_2}{2\pi}e^{-\beta\Delta U_2}\right] \tag{7.43}$$

where

$$A = \frac{\Omega_0}{\omega_0} \quad \left(\equiv \frac{\Omega_C}{\omega_C}\right) \tag{7.44}$$

represents the deviation from the transition state theory result. The derivation of the transition state theory result from our general expression for the escape rate will be presented later (Eq. 7.72).

We shall now study the particular cases $\psi = (\pi/2)$ and $\psi = (\pi/4)$. The general case (arbitrary $\psi$) is studied in Appendix VII.I.

1.

$$\psi = \frac{\pi}{2}$$

From Geoghegan et al. [65] [Eq. (5.60) or Eq. (7.41)] with $h = (\mu_0 H_0)/(2\widetilde{K}) < 1$, so that the bistable structure of the energy is maintained, i.e., $h < h_c (= 1)$

$$c_1 = 2\widetilde{K}h \tag{7.45a}$$

$$c_2 = -2\widetilde{K}(1 - h) \tag{7.45b}$$

So, the IHD result, for spins for a transverse field, is

$$A = \frac{a}{2}\left[\frac{-2\widetilde{K}h + 2\widetilde{K}(1-h)}{2\widetilde{K}\sqrt{h(1-h)}} + \frac{2\widetilde{K}\sqrt{1 + 4a^{-2}h(1-h)}}{2\widetilde{K}\sqrt{h(1-h)}}\right] \tag{7.46}$$

with limiting forms

$$A = \begin{cases} 1, \quad a^2 \ll h(1-h) \quad \text{(TST)} \\ \sqrt{1-h}(1+a^2)/a, \quad 1-h \ll 1, a^2 \\ a/\sqrt{h}, \quad h \ll 1, a^2 \quad \text{(HD)} \end{cases} \tag{7.46a}$$

The last line of Eq. (7.46a) which is of greatest interest is found as follows: we have

$$A \approx \frac{a}{2}\left[\frac{1-2h}{\sqrt{h(1-h)}} + \frac{1}{\sqrt{h(1-h)}}\right] \tag{7.47}$$

$$= a\frac{1-h}{\sqrt{h(1-h)}} \tag{7.48}$$

$$= a\frac{\sqrt{1-h}}{\sqrt{h}} \tag{7.49}$$

$$\to \frac{\text{const}}{\sqrt{h}} \quad \text{as} \quad h \to 0 \tag{7.50}$$

The constant being of the order $a$ if $a^2 \ll 1$. The first line of Eq. (7.46a) is the TST result, an upper bound for which is $a^2 = h$

2.

$$\psi = \frac{\pi}{4}$$

Again from Geoghegan et al. [65] (Here the critical or nucleation field $h_c = (1/2)$)

$$c_1 = \widetilde{K}\left[1 + (h - \sqrt{h^2+2})\sqrt{\frac{1 - h^2 - h\sqrt{h^2+2}}{2}}\right] \tag{7.51}$$

$$\approx \widetilde{K}\sqrt{2h} \qquad \text{for small h} \tag{7.52}$$

$$c_2 = -2\widetilde{K}\sqrt{h^2+2}\sqrt{\frac{1 - h^2 - h\sqrt{h^2+2}}{2}} \tag{7.53}$$

$$\approx -2\widetilde{K}(1 - \sqrt{2}h) \qquad \text{for small } h \tag{7.54}$$

So,

$$\sqrt{-c_1 c_2} = \sqrt{2\sqrt{2}h\widetilde{K}^2(1 - h\sqrt{2})} \tag{7.55}$$

$$\approx 2^{3/4}\widetilde{K}\sqrt{h} \qquad \text{for small } h \tag{7.56}$$

and

$$\Omega_0 = \frac{h'}{2}\left[-\sqrt{2}\widetilde{K}h + 2\widetilde{K} - 2\sqrt{2}\widetilde{K}h + \sqrt{4\widetilde{K}^2\left(1 - \frac{h}{\sqrt{2}}\right) + 16a^{-2}2\sqrt{2}\widetilde{K}h(1 - \sqrt{2}h)}\right] \tag{7.57}$$

Now, if $a$ is large (again in the IHD case)

$$\Omega_0 \approx \frac{h'}{2}\left[2\widetilde{K}\left(1 - \frac{3\sqrt{2}}{2}h\right) + 2\widetilde{K}\left(1 - \frac{h}{\sqrt{2}}\right)\right] \tag{7.58}$$

$$\approx 2\widetilde{K}h'(1-\sqrt{2}h) \tag{7.59}$$

So,

$$\frac{\Omega_0}{\omega_0} \approx \frac{2\widetilde{K}a}{2^{3/4}\widetilde{K}\sqrt{h}} \tag{7.60}$$

$$\approx \frac{2^{1/4}a}{\sqrt{h}} \tag{7.61}$$

Thus, again

$$A \to \frac{\text{const}}{\sqrt{h}} \qquad \text{as} \qquad h \to 0 \tag{7.62}$$

We show that this is true for arbitrary $\psi$ in Appendix VII.I. *Thus the nonaxially symmetric asymptotes for finite h do not approach the uniaxial asymptote in the limit of h tending to 0 as they must do*. The behavior exhibited by the transition state theory result also shows that this result is incapable of bridging the gap between axially symmetric and nonaxially symmetric solutions, as we shall now demonstrate. Similar considerations apply to the VLD limit of the nonaxially symmetric solution. (For example, we showed in Section V that

$$A = \frac{\beta\Delta E}{2} = 8a\sigma\sqrt{h} \tag{7.63}$$

for $\psi = (\pi/2)$, which in the limit of small $h$, cannot tend to Eq. (7.32a) as it vanishes in the limit of vanishing $h$.)

In the *transition state theory*, we ignore $a$. In this case, our general escape rate Eq. (7.20) yields

$$\Gamma = \left[\frac{1}{Z_1} + \frac{1}{Z_2}\right] \frac{\gamma kT}{\mu_0} \int_0^{2\pi} e^{-\beta H} \frac{\partial\zeta}{\partial\varphi} d\varphi \tag{7.64}$$

In order to integrate by parts, we let

$$u = e^{-\beta H} \qquad \text{and} \qquad dv = \frac{\partial\zeta}{\partial\varphi} d\varphi \tag{7.65}$$

then,

$$du = \frac{\partial}{\partial\varphi} e^{-\beta H} d\varphi \qquad \text{and} \qquad v = \zeta \tag{7.66}$$

Now $H$ is periodic in $\varphi$, while the crossover function satisfies $\zeta(x, 2\pi) = \zeta(x, 0)$ and so is also periodic in $\varphi$. Hence, the term $uv|_0^{2\pi} = 0$, and so

$$\Gamma = -\left[\frac{1}{Z_1} + \frac{1}{Z_2}\right] \frac{\gamma kT}{\mu_0} \int_0^{2\pi} \zeta \frac{\partial}{\partial\varphi} e^{-\beta H} d\varphi \tag{7.67}$$

Along the line $x = 0$ (i.e., the equator); $\zeta = 1$ where the particles go from the first well to the second (i.e., from the saddle point to the maximum of $H$), and $\zeta = 0$ otherwise.

Now the saddle point occurs when $\varphi = 0$ and $x = 0$, and the maximum at $\varphi = \pi$ and $x = 0$, hence Eq. (7.67) becomes (see Appendix VII.III)

$$\Gamma = -\left[\frac{1}{Z_1} + \frac{1}{Z_2}\right] \frac{\gamma kT}{\mu_0} e^{-\beta H}\Big|_0^{\pi} \tag{7.68}$$

Now for the transverse field model with small $h$ (again see Appendix VII.III) the partition functions are

$$Z_1 = Z_2 \approx \frac{\pi e^{\sigma}}{\sigma} \tag{7.69}$$

So,

$$\frac{1}{Z_1} + \frac{1}{Z_2} \approx \frac{2\sigma}{\pi} e^{-\sigma} = \frac{2vK}{\pi kT} e^{-\sigma} \tag{7.70}$$

in addition,

$$\mu_0 = vM_s \tag{7.71}$$

Hence,

$$\Gamma = -\frac{2vK}{\pi kT} e^{-\sigma} \frac{\gamma kT}{vM_s} \left[\exp\left(-\beta U_{\max}\right) - \exp\left(-\beta U_{\text{sad}}\right)\right] \tag{7.72}$$

where $U_{\max}$ is the value of $H$ at the maximum and, similarly, $U_{\text{sad}}$ is the value of $H$ at the saddle point.

Now, by using the following results from Appendix VII.III

$$\beta U_{\max} = \xi \tag{7.73a}$$

$$\beta U_{\text{sad}} = -\xi \tag{7.73b}$$

$$\beta U_{\min} = -\sigma \tag{7.73c}$$

we obtain

$$\beta U_{\text{sad}} - \beta U_{\min} = \sigma - \xi \tag{7.74}$$

so

$$-\sigma = -\beta\Delta U - \xi \tag{7.75}$$

$$= -\beta\Delta U + \beta U_{\text{sad}} \tag{7.76}$$

Also, by using the fact that

$$\omega_1 = \omega_A = 2\frac{K\gamma}{M_s} \tag{7.77}$$

we obtain

$$\Gamma = -\frac{\omega_A}{\pi} e^{-\beta\Delta U} \exp\left(\beta U_{\text{sad}}\right)\left[\exp\left(-\beta U_{\max}\right) - \exp\left(-\beta U_{\text{sad}}\right)\right] \tag{7.78}$$

$$= -\frac{\omega_A}{\pi} e^{-\beta\Delta U}\left[1 - \exp\left\{\beta U_{\max} - \beta U_{\text{sad}}\right\}\right] \tag{7.79}$$

and by using Eqs. (7.73a) and (7.73b), which pertain to the transverse model, this yields

$$\Gamma = -\frac{\omega_A}{\pi} e^{-\beta\Delta U}[1 - \exp\{2\xi\}] \tag{7.80}$$

Thus,

$$A = 1 - \exp\left[-\frac{U_{\max} - U_{sad}}{kT}\right] \tag{7.81}$$

which, again, if the field is transverse, yields

$$A = 1 - e^{-2\xi} \tag{7.82}$$

which becomes zero in the limit of vanishing $h$.

Thus the transition state theory predicts that the prefactor $A$ *should vanish in the limit of small fields and not tend to Eq. (7.32a) as it must do*.

Note that we have neglected to include the dissipative component of the current density. Now it is *impossible to have particles crossing the barrier without dissipation*, (cf the fluctuation-dissipation theorem) and, in addition, this neglected component of the current density provides a crossover to the uniaxial result of Eq. (7.32a). Furthermore, a more detailed analysis shows that this component is important in the range $\xi \lesssim 1$. If we take account of these arguments, we come to the conclusion that Eq. (7.81) has no applicability region.

Before proceeding, we remark that the problem of determining interpolation formulas between the axially symmetric result Eq. (7.32a), (which, of course, is valid for all values of the damping parameter) and the nonaxially symmetric asymptotes is compounded by the fact that in the *nonaxially symmetric case, the three regimes of damping, identified by Kramers, naturally appear* (since the problem is governed by a FPE in two space variables). Thus, for each of the three damping regimes, VLD, ID [34], and IHD, it is necessary to have *separate* interpolation formulas. This is relatively easily accomplished in the IHD case for the problem at hand for an arbitrary angle. As far as the VLD is concerned, the solution has been partially determined for a transverse field only, Garanin et al. [70]. The Progress to date is described in Section IIB. We shall now describe how the IHD case may be solved.

#### 4. *Crossover Formulas Bridging the Axially Symmetric and Nonaxially Symmetric Solutions for High Damping*

To remove the discrepancy between the two solutions, we write the escape rate formula, noting that for very small $h$, both wells are almost the same depth, as

$$\Gamma = A\frac{\omega_1}{\pi}e^{-\beta\Delta U} = A\Gamma_{\text{TS}} \tag{7.83}$$

Here, in general, $A$ represents some function of $h$ that tends to $2\pi a\sqrt{\sigma/\pi}$ as $h$ tends to 0 (thus giving the axially symmetric result) and tends to constant$/\sqrt{h}$ for $h$ finite but small enough so that $\zeta$ and $\partial\zeta/\partial x$ do not differ significantly from the uniaxial form as detailed in Appendix VII.IV, that is,

$$\frac{\partial\zeta}{\partial x} = 2\pi a\sqrt{\frac{\sigma}{\pi}} \tag{7.84}$$

To derive this crossover formula, we consider Eqs. (7.20) and (7.84) and by using the fact that for axial symmetry we have $\partial\zeta/\partial\varphi = 0$, we have, in Eq. (7.33),

$$A_1 = A_2 = \pi a\sqrt{\frac{\sigma}{\pi}} \tag{7.85}$$

We now consider two cases $\psi = (\pi/2)$ [70] and arbitrary $\psi$.

**Case (a) $\psi = \pi/2$**

The potential function now becomes

$$\varepsilon = \beta H = -\sigma x^2 - \xi\sqrt{1-x^2}\cos\varphi \tag{7.86}$$

so

$$\Gamma = \left(\frac{1}{Z_1} + \frac{1}{Z_2}\right)\frac{\gamma}{\beta\mu_0}\int_0^{2\pi} e^{\sigma x^2 + \xi\sqrt{1-x^2}\cos\varphi} a(1-x^2)\sqrt{\frac{\sigma}{\pi}} e^{-\sigma x^2}\, d\varphi \tag{7.87}$$

which, using the formula for the partition functions (see Appendix VII.III)

$$Z_1 = Z_2 = \frac{\pi}{\sigma\sqrt{1-h^2}} e^{\sigma(1+h^2)} \tag{7.88}$$

yields

$$\Gamma = \frac{2\sigma\sqrt{1-h^2}}{\pi} e^{-\sigma(1+h^2)} \frac{\gamma a}{\beta\mu_0}\sqrt{\frac{\sigma}{\pi}}\int_0^{2\pi} e^{\xi\cos\varphi}\, d\varphi \tag{7.89}$$

Using [82]

$$\int_0^{2\pi} e^{\xi\cos\varphi}\, d\varphi = 2\pi I_0(\xi) \tag{7.90}$$

where $I_0$ is the modified Bessel function of the first kind of zero order [82], and defining ($\beta = (kT)^{-1}$ in this case)

$$\omega_1 \equiv \frac{2\sigma\gamma}{\beta\mu_0}\sqrt{1-h^2} \equiv \omega_A \tag{7.91}$$

(the ferromagnetic resonance angular frequency), yields

$$\Gamma = 2\pi a\sqrt{\frac{\sigma}{\pi}} I_0(\xi)\frac{\omega_1}{\pi} e^{-\sigma(1+h^2)} \tag{7.92}$$

$$= 2\pi a\sqrt{\frac{\sigma}{\pi}} e^{-2\sigma h} I_0(\xi)\frac{\omega_1}{\pi} e^{-\sigma(1-2h+h^2)} \tag{7.93}$$

$$= 2\pi a\sqrt{\frac{\sigma}{\pi}} e^{-\xi} I_0(\xi)\frac{\omega_1}{\pi} e^{-\beta\Delta U} \tag{7.94}$$

where $\Delta U$ is the barrier height. This gives

$$A = 2\pi a \sqrt{\frac{\sigma}{\pi}} e^{-\xi} I_0(\xi) \tag{7.95}$$

Now, for relatively large $\xi \gg 1$ (or $h \gg (1/\sigma)$) we have [82]

$$e^{-\xi} I_0(\xi) \approx \frac{1}{\sqrt{2\pi\xi}} = \frac{1}{\sqrt{4\pi\sigma h}} \tag{7.96}$$

so (cf. Eq. (7.50))

$$A \approx 2\pi a \sqrt{\frac{\sigma}{\pi}} \frac{1}{\sqrt{4\pi\sigma h}} = \frac{a}{\sqrt{h}} \tag{7.97}$$

Thus at relatively high fields $h \gtrsim a^2 (a \ll 1)$ a crossover to the HD formula, the third line of Eq. (7.46) is achieved. Equation (7.46) applied if $2\sigma h \gg 1$ which means that Eq. (7.94) applies if $2\sigma a^2 \gg 1$ i.e., if $\alpha = a\sqrt{\sigma} \gg 1$ in order of magnitude. On the other hand for small $\xi$ [82]

$$e^{-\xi} I_0(\xi) \approx 1 \tag{7.98}$$

yields

$$A \approx 2\pi a \sqrt{\frac{\sigma}{\pi}} \tag{7.99}$$

that is, the uniaxial result. Thus, Eq. (7.95) is the required crossover function for $\psi = (\pi/2)$. This result involving the modified Bessel functions resembles the corresponding interpolation formulas when the WKJB method is used in quantum mechanics, where the interpolation function is an Airy function [38].

**Case (b) Arbitrary $\psi$, Small $h$**

We have (details in Appendix VII.III)

$$Z_1 = Z_2 = \frac{2\pi e^{\sigma + \xi\cos\psi}}{2\sigma + \xi\cos\psi} \tag{7.100}$$

so that Eq. (7.20) becomes

$$\Gamma = \frac{2\sigma + \xi\cos\psi}{\pi} e^{-\sigma-\xi\cos\psi} \frac{\gamma a}{\beta\mu_0}\sqrt{\frac{\sigma}{\pi}} \times \int_0^{2\pi} (1-x^2)e^{-\sigma x^2} e^{\sigma x^2 + \xi(x\cos\psi + \sqrt{1-x^2}\cos\varphi\sin\psi)} d\varphi \tag{7.101}$$

Now, in the region of the saddle point we have $x = O(h)$, and by defining the ferromagnetic resonance frequency

$$\omega_1 \equiv (2\sigma + \xi\cos\psi)\frac{\gamma}{\beta\mu_0} \equiv \omega_A \tag{7.102}$$

and by noting that the barrier height is

$$\beta\Delta U \equiv \beta(U_{\text{sad}} - U_{\text{min}}) = \sigma + \xi\cos\psi - \xi\sin\psi \tag{7.103}$$

where the subscripts denote the points at which $U$ is evaluated, we have

$$\Gamma = 2\pi a\sqrt{\frac{\sigma}{\pi}}\frac{\omega_1}{\pi} e^{-\beta\Delta U} e^{-\xi\sin\psi} I_0(\xi\sin\psi) \tag{7.104}$$

so that

$$A = 2\pi a\sqrt{\frac{\sigma}{\pi}} e^{-\xi\sin\psi} I_0(\xi\sin\psi) \tag{7.105}$$

For relatively large $\xi$, we have

$$A \approx \frac{a}{\sqrt{h\sin\psi}} \tag{7.106}$$

that is, $A$ behaves as

$$\frac{\text{const}}{\sqrt{h}}$$

while for small $\xi$

$$A \approx 2\pi a\sqrt{\frac{\sigma}{\pi}} \tag{7.107}$$

again the uniaxial result. Thus $A$ is the desired crossover function.

And so the problem of determining suitable interpolation formulas in the HD limit is completely resolved, and the *different temperature dependences of the uniaxial and nonaxially symmetric HD asymptotes can now be understood.*

### 5. *Need for Interpolation Formulas in the VLD Limit*

We reiterate that the VLD limit has been examined by Klik and Gunther [64, 72], by Coffey [66], by McCarthy and Coffey [45], and by Garanin et al. [70] and it has been shown that the nonaxially symmetric VLD formula fails in the region of small departures from axial symmetry (see Section V). The VLD formula for spins has essentially the same form as that for particles. For the purposes of explanation, we remark that by simply generalizing the results of Section V to a bistable potential we have as in Section V.

$$\Gamma = \beta \frac{\Delta E_1 \Delta E_2}{\Delta E_1 + \Delta E_2} \left( \frac{\omega_1}{2\pi} e^{-\beta \Delta U_1} + \frac{\omega_2}{2\pi} e^{-\beta \Delta U_2} \right) \tag{7.108}$$

where

$\Delta E_1$ is the energy loss per cycle of a particle in well number 1.
$\Delta E_2$ is the energy loss per cycle of a particle in well number 2.
$\beta \Delta U_1$ is the reduced barrier height in well number 1.
$\beta \Delta U_2$ is the reduced barrier height in well number 2.

For the particular case of two identical wells that are relevant to the present problem i.e. a transverse applied field, we have

$$\Gamma = \frac{\beta \Delta E_1}{2} \frac{\omega_1}{\pi} e^{-\beta \Delta U_1} \tag{7.109a}$$

$$= \frac{\beta \Delta E_1}{2} \Gamma_{TS} \tag{7.109b}$$

since there are two wells.
Thus,

$$A = \frac{\beta \Delta E_1}{2} \tag{7.110}$$

with [45, 70] and Section V, Eq. (5.29)

$$\Delta E_1 = a \oint_{E=E_C} \left[ (1 - x^2) \frac{\partial H}{\partial x} d\varphi - \frac{\partial H}{\partial \varphi} \frac{dx}{1 - x^2} \right] \tag{7.111}$$

where $E = E_C$ is the critical energy curve [70]. The critical energy is that energy required by a particle to barely escape the well.

The line integral may be evaluated on the critical energy contour $H = E = E_C$ by expressing $x$ in terms of $\varphi$. This is, in general, a difficult task to accomplish analytically as it usually involves the solution of a quartic equation. The sole exception is the case of a small transverse field when the behavior for small fields may be evaluated analytically, as described in Section V. Clearly, that result does not tend to the axially symmetric one in the limit of $h$ tending to 0 just as in the IHD case.

### B. Interpolation Formulas in the VLD Limit

#### 1. *Justification of the Energy Diffusion Method*

In Section V we derived the VLD limit by the method of first passage times. For the purpose of our treatment here and to indicate how the Kramers method may be applied to the calculation of the VLD formula for spins, we shall briefly indicate how the problem may be attacked by the energy diffusion method. The advantage of using this method is that in certain cases (the potential of the uniaxial anisotropy with transverse field is one) the method lends itself to the determination of interpolation formulas.

#### 2. *The Energy Diffusion Method*

The application of the energy controlled diffusion method to spins may be briefly described as follows. We change the variables in the FPE to energy (slow) and azimuthal angle $\varphi$ (fast) variables as described for particles in Section II, there the relevant variables were the energy and the phase, playing the role of slow and fast variables respectively. The transformation to energy and azimuthal angle variables is accomplished using the formulas using the formulas given in our Appendix VII.VII.

This procedure leads, to a diffusion equation for the energy and Kramers' method may be applied as in Section II. We summarize the calculations as in Garanin et al. [70]. Thus, if the damping constant $a$ becomes smaller than some characteristic value, (so as to render $\Delta E < kT$), the IHD formula becomes inapplicable because the tacit assumption that the particles or magnetic spins approaching the barrier from the well are in thermal equilibrium is violated. For these values of $a$, they cannot thermalize during one period of the motion. (see our extensive discussion in Section II.F)

*In the LD limit, on the other hand, the diffusion of energy becomes very slow and thus the distribution function reaches equilibrium along the constant energy lines*. This implies that $\zeta$ [in Eq. (7.14)] may be approximated as

$$\zeta(x, \varphi) \approx \zeta(\varepsilon) \tag{7.112a}$$

with

$$\varepsilon = \varepsilon(x, \varphi) = \beta H \tag{7.112b}$$

This ensures that the first two terms of Eq. (7.13) cancel. The rest may be averaged out over the phase variable, that is, over the constant energy line, to yield (see Appendix VII.V)

$$\frac{d^2\zeta}{d\varepsilon^2} = \left[1 - \frac{A(\varepsilon)}{B(\varepsilon)}\right]\frac{d\zeta}{d\varepsilon} \tag{7.113}$$

where

$$A(\varepsilon) \equiv \left\langle \frac{\partial}{\partial x}(1 - x^2)\frac{\partial \varepsilon}{\partial x} + \frac{1}{1 - x^2}\frac{\partial^2 \varepsilon}{\partial \varphi^2} \right\rangle \tag{7.114a}$$

$$B(\varepsilon) \equiv \left\langle (1 - x^2)\left(\frac{\partial \varepsilon}{\partial x}\right)^2 + \frac{1}{1 - x^2}\left(\frac{\partial \varepsilon}{\partial \varphi}\right)^2 \right\rangle \tag{7.114b}$$

The averaging is defined by the contour integral

$$\langle \cdots \rangle = \frac{\oint \cdots (d\mathbf{l} \cdot \mathbf{n}_{\|})}{\oint (d\mathbf{l} \cdot \mathbf{n}_{\|})} \tag{7.115}$$

where $dl_{\|} \equiv d\mathbf{l} \cdot \mathbf{n}_{\|}$ is an infinitesimal length along the constant energy line. In terms of the polar and azimuthal angles $(\theta, \varphi)$ the line element, $d\mathbf{l}$, is given by

$$d\mathbf{l} = d\theta \mathbf{n}_\theta + \sin\theta d\varphi \mathbf{n}_\varphi \tag{7.116}$$

where $\mathbf{n}_\theta$ and $\mathbf{n}_\varphi$ are unit vectors in the directions of increasing $\theta$ and $\varphi$, respectively. Now, by using the fact that if $\mathbf{n}_x$ is a unit vector in the direction of *increasing* $x$, then

$$\mathbf{n}_x = -\mathbf{n}_\theta \tag{7.117}$$

and also that

$$dx = -\sin\theta\, d\theta \tag{7.118}$$

we have

$$d\mathbf{l} = \frac{\mathbf{n}_x dx}{\sqrt{1-x^2}} + \mathbf{n}_\varphi \sqrt{1-x^2} d\varphi \tag{7.119}$$

Furthermore, $\mathbf{n}_\parallel$ is one of two orthogonal unit vectors:

$$\mathbf{n}_\varepsilon = \frac{\nabla \varepsilon}{|\nabla \varepsilon|} = \frac{\sqrt{1-x^2}(\partial\varepsilon/\partial x)\mathbf{n}_x + (1/\sqrt{1-x^2})(\partial\varepsilon/\partial\varphi)\mathbf{n}_\varphi}{\sqrt{(1-x^2)(\partial\varepsilon/\partial x)^2 + (1/1-x^2)(\partial\varepsilon/\partial\varphi)^2}} \tag{7.120}$$

$$\mathbf{n}_\parallel = \frac{\sqrt{1-x^2}(\partial\varepsilon/\partial x)\mathbf{n}_\varphi - (1/\sqrt{1-x^2})(\partial\varepsilon/\partial\varphi)\mathbf{n}_x}{\sqrt{(1-x^2)(\partial\varepsilon/\partial x)^2 + (1/1-x^2)(\partial\varepsilon/\partial\varphi)^2}} \tag{7.121}$$

Since $\nabla\varepsilon$ is a vector perpendicular to the surface $\varepsilon = \text{const}$, we see that $\mathbf{n}_\varepsilon$ at any point is perpendicular to the constant energy surface at that point, and hence $\mathbf{n}_\parallel$ is tangential to it at that point. Now, we have

$$d\mathbf{l} \cdot \mathbf{n}_\parallel = \frac{(1-x^2)(\partial\varepsilon/\partial x)d\varphi - (1/1-x^2)(\partial\varepsilon/\partial\varphi)dx}{\sqrt{(1-x^2)(\partial\varepsilon/\partial x)^2 + (1/1-x^2)(\partial\varepsilon/\partial\varphi)^2}} \tag{7.122}$$

Normally, in Eq. (7.113) the term $(A(\varepsilon)/B(\varepsilon))$ is of the order of $(kT/E_{\text{char}})$ (where $E_{\text{char}}$ is a characteristic energy) and may be neglected at low temperatures. Then, the first integration of Eq. (7.113) yields

$$\frac{\partial\zeta}{\partial\varepsilon} = \tfrac{1}{2}\exp(\varepsilon - \varepsilon_C) \tag{7.123}$$

where $\varepsilon_C$ is the reduced saddle point energy, and the integration constant comes from the fact that $\zeta$ changes from 0 to $\frac{1}{2}$ as $\varepsilon$ changes from $-\infty$ in the first well to $\varepsilon_C$ at the saddle point and then back to $-\infty$ in the second well. Note the factor of $\frac{1}{2}$ in Eq. (7.123) which comes from the situation of having two symmetric wells. In the case $kT \ll \Delta U_1 \ll \Delta U_2$, that is the heavily biased case, the current of particles from the second well to the first may be ignored, the function $\zeta$ attains its limiting value of 1 at the separatrix, and the coefficient in Eq. (7.123) is equal to 1.

The time derivative of the number of particles in the first well may be written as the line integral along the part of the separatrix $\varepsilon = \varepsilon_C$, which surrounds the first well, i.e.

$$\dot{N}_1 = -\oint (d\mathbf{l} \cdot \mathbf{n}_\parallel)(\mathbf{J} \cdot \mathbf{n}_\varepsilon) \tag{7.124}$$

where the current on the sphere is defined by Eq. (7.9). By using Eq. (7.112a, b), we may simplify the scalar product $\mathbf{J} \cdot \mathbf{n}_\varepsilon$ to

$$-\frac{\gamma akT}{\mu_0} f_{\mathrm{eq}}(g_2 - g_1)\frac{\partial\zeta}{\partial\varepsilon}\sqrt{(1-x^2)\left(\frac{\partial\varepsilon}{\partial x}\right)^2 + \frac{1}{1-x^2}\left(\frac{\partial\varepsilon}{\partial\varphi}\right)^2} \tag{7.125}$$

which cancels out the denominator in Eq. (7.122). Now, with the use of Eqs. (7.15) and (7.123), we come to the kinetic equation (7.19), where the escape rate $\Gamma$ is given by Eq. (7.83) with

$$A = \frac{a}{2}\oint_{\varepsilon=\varepsilon_C}\left[(1-x^2)\frac{\partial\varepsilon}{\partial x}d\varphi - \frac{1}{1-x^2}\frac{\partial\varepsilon}{\partial\varphi}dx\right] \tag{7.126}$$

The result is in the form $A = (\delta/2kT)$ where $\delta$ is the energy dissipated in one period of the motion at the saddle point energy in the LD case. This coincides exactly with the corresponding result for particles.

The application of this formula to the particular case of a transverse field with uniaxial anisotropy leads to

$$A = 8a\sigma\sqrt{h} \tag{7.127}$$

so that, as we have seen, the *asymptotes fail at small departures from axial symmetry.* This situation may, however, be ameliorated if we treat the energy diffusion equation more accurately, that is, we do not neglect $A(\varepsilon)$ and $B(\varepsilon)$ in our calculations.

### 3. Uniaxial/LD Crossovers

By referring back to Eq. (7.114), we see that if $h$ is small,

$$\varepsilon \approx -\sigma x^2 \tag{7.128}$$

so

$$\frac{\partial\varepsilon}{\partial x} \approx -2\sigma x \tag{7.129a}$$

$$\frac{\partial^2\varepsilon}{\partial x^2} \approx -2\sigma \tag{7.129b}$$

$$\frac{\partial\varepsilon}{\partial\varphi} \approx 0 \tag{7.129c}$$

$$\frac{\partial^2 \varepsilon}{\partial \varphi^2} \approx 0 \tag{7.129d}$$

hence, the functions $A(\varepsilon)$ and $B(\varepsilon)$ of Eq. (7.114) are approximately

$$A(\varepsilon) \approx \left\langle \frac{\partial^2 \varepsilon}{\partial x^2} \right\rangle \approx -2\sigma \tag{7.130a}$$

$$B(\varepsilon) \approx \left\langle \left( \frac{\partial \varepsilon}{\partial x} \right)^2 \right\rangle \approx -4\sigma\varepsilon \tag{7.130b}$$

With these approximations, we find [Eq. (7.113) and Appendix VII.VI]

$$\frac{\partial^2 \zeta}{\partial \varepsilon^2} = \left[ 1 - \frac{1}{2\varepsilon} \right] \frac{\partial \zeta}{\partial \varepsilon} \tag{7.131}$$

which, on integration, becomes

$$\frac{\partial \zeta}{\partial \varepsilon} = \frac{C}{\sqrt{-\varepsilon}} \exp(\varepsilon - \varepsilon_C) \tag{7.132}$$

where the constant of integration, $C$, is given by

$$C \int_{-\infty}^{\varepsilon_C} \frac{d\varepsilon}{\sqrt{-\varepsilon}} \exp(\varepsilon - \varepsilon_C) = \frac{1}{2} \tag{7.133}$$

where the $\frac{1}{2}$ is a consequence of the two equivalent wells
and

$$\varepsilon_C = -\xi \tag{7.134}$$

[cf. Eq. (7.122)]

[Note here that the difference between this case and the standard LD case is that for $\xi \lesssim 1$ the term $(1/2\varepsilon)$ in Eq. (7.131) cannot be neglected, since $\varepsilon \approx \varepsilon_C \approx \xi$].

We now find

$$A \approx aC \lim_{\varepsilon \to \varepsilon_C = -\xi} \frac{\oint (\partial \varepsilon / \partial x) d\varphi}{\sqrt{-\varepsilon}} \tag{7.135}$$

which takes on the form

$$A = 2\pi a \sqrt{\frac{\sigma}{\pi}} f(\xi) Q \tag{7.136}$$

where

$$Q = \lim_{\varepsilon \to -\xi} \frac{1}{2\pi} \int_0^{2\pi} \sqrt{\frac{-\varepsilon - \xi \cos\varphi}{-\varepsilon}} d\varphi \tag{7.137}$$

and

$$f(\xi) = \sqrt{\pi} \left( \int_0^\infty \frac{e^{-t} dt}{\sqrt{\xi + t}} \right)^{-1} = e^{-\xi} \left( \operatorname{erfc} \sqrt{\xi} \right)^{-1} \tag{7.138}$$

If $\xi = 0$ we find that (details in Appendix VII.VIII)

$$Q = 1 \tag{7.139a}$$

If $\xi \neq 0$, then

$$Q = 2^{(3/2)} \pi^{-1} \approx 0.900 \tag{7.139b}$$

The asymptotic expressions for $f(\xi)$ are (see Appendix VII.IX):

$$f(\xi) \approx 1 + 2\sqrt{\frac{\xi}{\pi}} \qquad \xi \ll 1 \tag{7.140a}$$

$$f(\xi) \approx \sqrt{\pi \xi} \qquad \xi \gg 1 \tag{7.140b}$$

We can see that for $\xi = 0$, the uniaxial limit $A = 2\pi a \sqrt{\sigma/\pi}$ is recovered. In the region $\xi \approx 1$, the function $f(\xi)$ describes the crossover to the standard LD result $A = 8a\sigma\sqrt{h}$. The form of $Q$ above shows that our treatment has failed to describe another crossover at smaller fields, resulting in the jump in the value of $Q$ between Eqs. (7.140a) and (7.140b). We see later that this crossover takes place for $\xi \approx \sqrt{\alpha}$ where $\alpha \equiv a\sqrt{\sigma}$.

Thus we can see that the parameters governing the uniaxial crossover are $\xi$ and $\alpha$. For $\sigma \gg 1$, in the relevant region ($\xi \approx 1$), we have $h \ll 1$. Thus in the equation satisfied by $\zeta$ we may neglect the terms $h \cos\varphi$, $a(\partial\zeta/\partial\varphi)$, and $\partial^2\zeta/\partial\varphi^2$, since the derivatives with respect to $\varphi$ are much smaller than those with respect to $x$. By letting $x' \equiv x\sqrt{\sigma}$, and, as mentioned above, $\alpha \equiv a\sqrt{\sigma}$, the resulting equation is

$$\alpha\left(\frac{\partial^2\zeta}{\partial x'^2} + 2x'\frac{\partial\zeta}{\partial x'}\right) + 2x'\frac{\partial\zeta}{\partial\varphi} + \xi\sin\varphi\frac{\partial\zeta}{\partial x'} = 0 \tag{7.141}$$

and the boundary conditions are

$$\zeta = 0 \qquad \text{when} \qquad x' \to -\infty \tag{7.142a}$$

$$\zeta = 1 \qquad \text{when} \qquad x' \to \infty \tag{7.142b}$$

If $\alpha \gg 1$, then, Eq. (7.141) becomes

$$\frac{\partial^2\zeta}{\partial x'^2} + 2x'\frac{\partial\zeta}{\partial x'} = 0 \tag{7.143}$$

which, on multiplying across by $e^{x^2}$ may be written

$$\frac{\partial}{\partial x'}\left[\frac{\partial\zeta}{\partial x'}\exp\left(x'^2\right)\right] = 0 \tag{7.144}$$

On integrating, we obtain

$$\frac{\partial\zeta}{\partial x'} = C\exp(-x'^2) \tag{7.145}$$

and, integrating again yields

$$\zeta = C\int_{-\infty}^{x'}\exp(-u^2)du \tag{7.146}$$

the lower limit is taken at $-\infty$, because of the boundary condition (7.142a). Condition (7.142b) forces

$$C = \sqrt{1/\pi}. \tag{7.147}$$

So the solution for $\zeta$ is *approximately* the uniaxial solution.

The "phase diagram" in Figure 10 describes the various crossover situations for the escape rate when small transverse fields ($h \ll 1$) are applied to magnetic particles with uniaxial anisotropy.

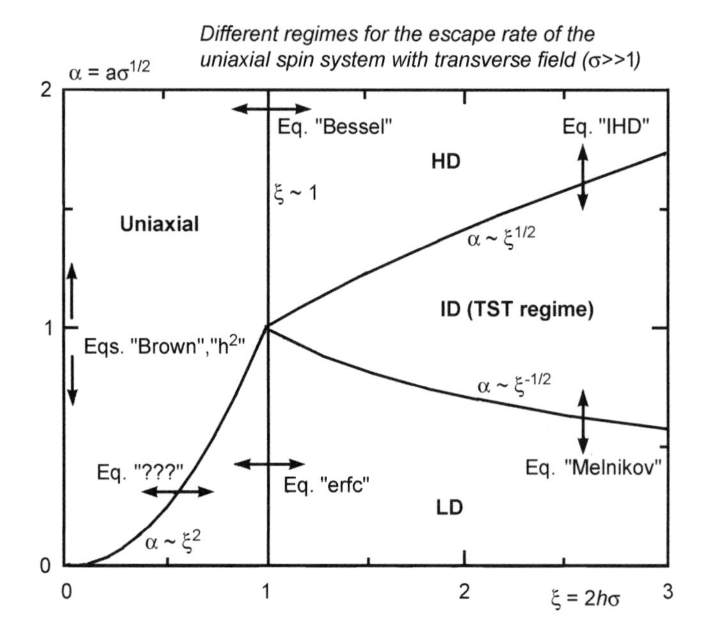


**Figure 10.** Crossover regions for spins [70]. The diagram shows the various crossover regions discussed in the text. (1) The HD/uniaxial (when $\alpha \gg 1$, $\xi \approx 1$) [see Eq. (7.147)] (2) The ID/LD (when $\alpha \approx \xi^{1/2}$) [see Eqs. (6.10) and (6.11)]. (3) The ID/HD (when $\alpha \approx \sqrt{\xi}$) [see Eq. (7.43)]. (4) The LD/uniaxial (when $\xi \approx 1$) [see Eqs. (7.135)–(7.139)]. (5) The other LD/uniaxial case ($\alpha \approx \xi^2$) has not yet been discovered.

Thus, in HD limit the main effect of the transverse field is contained in the factor $e^{\xi \cos \varphi}$, which leads to Eq. (7.95), namely,

$$A = 2\pi a \sqrt{\frac{\sigma}{\pi}} e^{-\xi} I_0(\xi) \tag{7.148}$$

The ID/HD crossover is described by Eq. (7.46) cf. Eq. (7.46a) and occurs at $a \approx \sqrt{h}$ or $\alpha \approx \sqrt{\xi}$, at which value TST approximately applies.

The ID/LD crossover occurs if in the VLD Eq. (7.126) $A \approx 1$ (TST value) which amounts to $\alpha \approx \frac{1}{\sqrt{\xi}}$. This crossover for mechanical particles is described by the Mel'nikov-Meshkov formulae Eq. (6.11).

The HD/uniaxial crossover occurs when $\alpha \gg 1$ and $\xi \approx 1$, and is described by Eqs. (7.148).

For $\alpha \ll 1$, there are two LD/uniaxial crossovers. One appears around $\xi = 1$, and is described by Eqs. (7.136)–(7.140), the other at $\alpha = \xi/2$

Equations (7.95) and (7.136) as evidenced in Fig. 11 compare very favourably with numerical calculation of the smallest non vanishing eigenvalue of the FPE Eq. (7.5).

### 4. *Perturbations about the Uniaxial Case*

To the second order in the field parameter $\xi$, the solution of Eq. (7.141) may be written as

$$\zeta(x', \varphi) \approx \zeta_0(x') + \xi\zeta_1(x', \varphi) + \xi^2\zeta_2(x', \varphi) \tag{7.149}$$

where

$$\zeta_1 = A(x') \sin\varphi + B(x') \cos\varphi \tag{7.150a}$$

and

$$\zeta_2 = C(x') + \cdots \tag{7.150b}$$

Define the functions $Y$ and $Z$ by the equations

$$Y \equiv A + iB \tag{7.151a}$$

$$C' \equiv \mathrm{Re}Z \tag{7.151b}$$

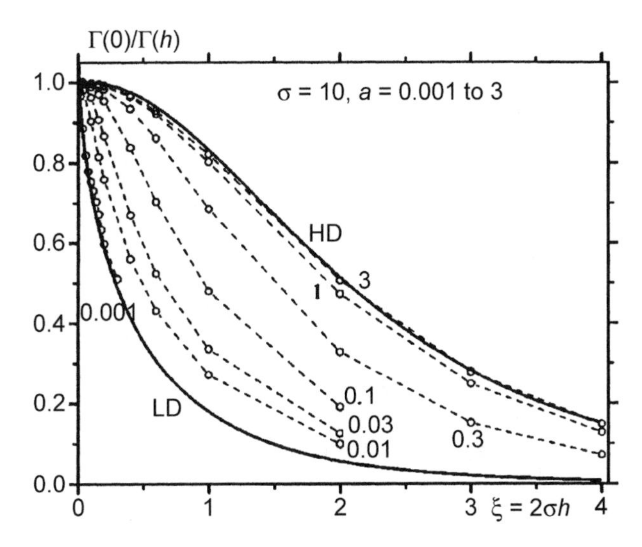


**Figure 11.** Transverse-field dependence of the inverse relaxation rate for $\sigma = 10$ and different values of the damping constant $a$. Solid lines represent Eq. (7.83) namely $\Gamma = \frac{AW_1}{\pi} e^{-\beta\Delta U} = A\Gamma_{TS}$ with $A$ given by Eqs. (7.95) and (7.135) with $Q = 2^{3/2}/\pi \approx 0.900$.

These satisfy the differential equations (see Appendix VII.X)

$$\alpha(Y'' + 2x'Y') + 2ix'Y = -\zeta_0' = -\pi^{-1/2}\exp(-x'^2) \tag{7.152a}$$

$$\alpha(Z' + 2x'Z) = -\tfrac{1}{2}Y' \tag{7.152b}$$

Now define the Fourier transforms $Y_k, Z_k$ of $Y$, and $Z$, and the function $\widetilde{Y}_k$ by the equations:

$$Y(x') \equiv \frac{1}{2\pi}\int_{-\infty}^{\infty} Y_k \exp(-ikx')\exp\left(-\frac{k^2}{4}\right)dk \tag{7.153a}$$

$$Z(x') \equiv \frac{1}{2\pi}\int_{-\infty}^{\infty} Z_k \exp(-ikx')\exp\left(-\frac{k^2}{4}\right)dk \tag{7.153b}$$

$$\widetilde{Y}_k \equiv (1 - \alpha k)Y_k \tag{7.153c}$$

Equations (7.152) transform as

$$\frac{d\widetilde{Y}_k}{dk} - \frac{1}{2}\frac{k}{1-\alpha k}\widetilde{Y}_k = -\frac{1}{2} \tag{7.154a}$$

$$2\alpha\frac{dZ_k}{dk} = -\frac{1}{2}\frac{k}{1-\alpha k}\widetilde{Y}_k = \frac{d\widetilde{Y}_k}{dk} - \frac{1}{2} \tag{7.154b}$$

If we define the function

$$\psi(u) \equiv \ln|1-u| + u \tag{7.155}$$

then the solutions to Eqs. (7.154) are given by

$$\widetilde{Y}_k = \frac{1}{2}\exp\left(-\frac{\psi(\alpha k)}{2\alpha^2}\right)\int_k^{\frac{1}{\alpha}}\exp\left(\frac{\psi(\alpha k')}{2\alpha^2}\right)dk' \tag{7.156a}$$

$$2\alpha Z_k = -\widetilde{Y}_k + \widetilde{Y}_0 - \frac{k}{2} \tag{7.156b}$$

We give details of the justification of Eqs. (7.154) and their solutions in Appendix VII.XI. The constant of integration in Eq. (7.156a) has been chosen to remove the pole at $k = 1/\alpha$, which would result in an oscillatory type of behavior of $Y(x')$. That in Eq. (7.156b) has been determined from the condition

$$C(\infty) - C(-\infty) \approx \int_{-\infty}^{\infty} Z(x')dx' = Z_0 = 0 \tag{7.157}$$

Now, expanding Eq. (7.20) up to $\xi^2$ and integrating over $\varphi$, one obtains (details in Appendix VII.XII),

$$\frac{\Gamma(h)}{\Gamma(0)} = 1 + \frac{\xi^2}{4} F(\alpha) = 1 + h^2\sigma^2 F(\alpha) \tag{7.158}$$

where (see Appendix VII.XIII)

$$F(\alpha) = 1 + \frac{2\sqrt{\pi}}{\alpha}(A + \alpha B' + 2\alpha C')|_{x'=0} \tag{7.159a}$$

$$= 1 + \frac{2\sqrt{\pi}}{\alpha}\frac{1}{2\pi}\int_{-\infty}^{\infty}(\widetilde{Y}_k + 2\alpha Z_k)\exp\left(-\frac{k^2}{4}\right)dk \tag{7.159b}$$

By using Eqs. (7.156), we arrive at the final result as in Appendix VII.XIV

$$F(\alpha) = 1 + \frac{1}{\alpha^2}\int_0^1 \exp\left[\frac{\psi(u)}{2\alpha^2}\right] du \tag{7.160a}$$

$$= 1 + 2(2\alpha^2 e)^{1/2\alpha^2}\gamma\left(1 + \frac{1}{2\alpha^2}, \frac{1}{2\alpha^2}\right) \tag{7.160b}$$

where

$$\gamma(a, z) \equiv \int_0^z t^{a-1}e^{-t}dt \tag{7.160c}$$

is an incomplete gamma function. The limiting forms of the above results are (see Appendix VII.XV).

$$F \approx 1 + \frac{1}{\alpha^2} - \frac{1}{(2\alpha)^2} + \cdots \qquad \alpha \gg 1 \tag{7.161}$$

$$F \approx \frac{\sqrt{\pi}}{\alpha} - \frac{1}{3} + \frac{\alpha\sqrt{\pi}}{6} + \cdots \qquad \alpha \ll 1 \tag{7.162}$$

The last formula shows that for $\alpha \ll 1$, the escape rate given by Eq. (7.158) essentially deviates from its uniaxial value if $\alpha \approx \xi^2$. This defines a crossover that was mentioned above. The function describing the crossover has not yet been worked out.

### C. Summary of Section VII

In this section, we have outlined the problems that occur when the Kramers escape rates for nonaxially symmetric potentials of the magnetocrystalline anisotropy are used in the region of small departures from axial symmetry and we have shown how interpolation formulas may be obtained in the HD limit by referring to the particular example of relaxation of a particle possessing simple uniaxial anisotropy in a magnetic field. The VLD case that is likely to be of importance in a discussion on macroscopic quantum tunneling of the magnetization has been solved for the particular case of a transverse field for a wide range of parameters $\xi$. However, the general solution of this problem for an arbitrary directed field has yet to be accomplished. In addition, the application to spins of the Mel'nikov and Meshkov formalism (Section VI) for the escape rate as a function of the friction should be tested by comparison with the results of numerical calculations of $\lambda_1$ as a function of the friction. This has very recently been accomplished by Déjardin et al. [84] and the results compare reasonably with the numerical value of $\lambda_1$ in the whole range of friction avoiding of course regions of small departure from axial symmetry.

## Appendix VII.I.

## DIVERGENCE OF ESCAPE RATES

Equation (6.113) of Geoghegan et al. [65] gives

$$x^4 + 2ux^3 - (1 - h^2)x^2 - 2ux - u^2 = 0 \tag{7.163}$$

where

$$u = h\cos\psi$$

which can be factorized as

$$\left[x^2 + (u - E)x + \frac{1}{2}(y_1 - F)\right]\left[x^2 + (u + E)x + \frac{1}{2}(y_1 + F)\right] = 0 \tag{7.164}$$

where

$$E = \sqrt{y_1 + u^2 + 1 - h^2} \tag{7.165}$$

$$F = \sqrt{y_1^2 + 4u^2} \tag{7.166}$$

and $y_1$ is a root of the cubic equation

$$y^3 + (1 - h^2)y^2 - 4u^2h^2 \sin^2\psi = 0 \tag{7.167}$$

The roots of Eq. (7.164) are:

$$x_{\pm 1} = -\frac{u + E}{2} \pm \frac{\sqrt{(u + E)^2 - 2(y_1 + F)}}{2} \tag{7.168a}$$

$$x'_{\pm 1} = \frac{E - u}{2} \pm \frac{\sqrt{(E - u)^2 + 2(F - y_1)}}{2} \tag{7.168b}$$

We seek the value of $x$ at the saddle point, it can be checked [52] that this is in fact $x_1$.

To solve Eq. (7.167): compare the coefficients with the standard form of a cubic equation

$$y^3 + a_1 y^2 + a_2 y + a_3 = 0$$

we have

$$a_1 = 1 - h^2 \tag{7.169a}$$

$$a_2 = 0 \tag{7.169b}$$

$$a_3 = -4h^4 \sin^2\psi \cos^2\psi \tag{7.169c}$$

If we define [87]

$$p = a_2 - \frac{a_1^2}{3} \tag{7.170a}$$

$$q = \frac{2a_1^3}{27} - \frac{a_1 a_2}{3} + a_3 \tag{7.170b}$$

$$P = \sqrt[3]{-\frac{q}{2} + \sqrt{\frac{p^3}{27} + \frac{q^2}{4}}} \tag{7.170c}$$

$$Q = \sqrt[3]{-\frac{q}{2} - \sqrt{\frac{p^3}{27} + \frac{q^2}{4}}} \tag{7.170d}$$

then, the required root of (7.167) is [87]

$$y_1 = P + Q - \frac{a_1}{3}. \tag{7.171}$$

This yields, for small values of $h$

$$y_1 \approx -1 + h^2 \tag{7.172}$$

So,

$$E \approx u \tag{7.173}$$

$$F \approx 1 - h^2(1 - 2\cos^2\psi) \tag{7.174}$$

$$E + u \approx 2u \tag{7.175}$$

$$E - u \approx 0 \tag{7.176}$$

$$F + y_1 \approx 2h^2 \cos^2\psi \tag{7.177}$$

$$F - y_1 \approx 2(1 - h^2 \sin^2\psi) \tag{7.178}$$

Thus, the values of $x$ satisfying (7.164) are

$$x_1 \approx -h\cos\psi \tag{7.179a}$$

$$x_{-1} \approx -h\cos\psi \tag{7.179b}$$

$$x'_1 \approx 1 - \frac{1}{2}h^2\sin^2\psi \tag{7.179c}$$

$$x'_{-1} \approx -1 + \frac{1}{2}h^2\sin^2\psi \tag{7.179d}$$

It can be checked [Geoghegan et al. [65]] that $x'_1$ and $x'_{-1}$ are the values of $x$ at the minima, and that $x_1(\approx x_{-1})$ as already stated is the value of $x$ at the saddle point. Now the important factor in determining the rate of divergence of the eigenvalue asymptote is the factor $\sqrt{-c_1c_2}$ in the denominator, effectively the saddle angular frequency. We now see that for small values of $h$

$$c_1 \approx 2\widetilde{K}h\sin\psi\sqrt{1 - h^2\cos^2\psi} \tag{7.180}$$

$$\approx 2\widetilde{K}h\sin\psi \tag{7.181}$$

$$c_2 \approx -2\widetilde{K}(1 - h\sin\psi) \tag{7.182}$$

So,

$$\sqrt{-c_1c_2} \approx \sqrt{4\widetilde{K}^2 h\sin\psi(1 - h\sin\psi)} \tag{7.183}$$

$$\approx 2\widetilde{K}\sqrt{h}\sqrt{\sin\psi} \tag{7.184}$$

In particular, for $\psi = \pi/4$ we have

$$\sqrt{-c_1c_2} \approx 2^{\frac{3}{4}}\widetilde{K}\sqrt{h} \tag{7.185}$$

and for $\psi = \pi/2$

$$\sqrt{-c_1c_2} \approx 2\widetilde{K}\sqrt{h} \tag{7.186}$$

So the escape rate *diverges as* $h^{-(1/2)}$ *as* $h \to 0$ *for any angle.*

## Appendix VII.II. Derivation of Eq. (7.20) from Eq. (7.16)

Equation (7.16) for the rate of change of the number of particles in the first well is implies

$$\dot{N}_1 = \frac{\gamma}{\mu_0} \int_0^{2\pi} \left[ \frac{\partial H}{\partial \varphi} f + a(1 - x^2) \left( \frac{\partial H}{\partial x} + kT \frac{\partial}{\partial x} \right) f \right] d\varphi \tag{7.187}$$

By using Eqs. (7.6) and (7.12), we have

$$\int_0^{2\pi} \frac{\partial H}{\partial \varphi} f d\varphi = \int_0^{2\pi} g f_{\text{eq}} \frac{\partial H}{\partial \varphi} d\varphi \tag{7.188}$$

$$= -kT \int_0^{2\pi} g \frac{\partial f_{\text{eq}}}{\partial \varphi} d\varphi \tag{7.189}$$

$$= -kT \left[ g f_{\text{eq}} \Big|_0^{2\pi} - \int_0^{2\pi} f_{\text{eq}} \frac{\partial g}{\partial \varphi} d\varphi \right] \tag{7.190}$$

$$= kT \int_0^{2\pi} f_{\text{eq}} \frac{\partial g}{\partial \varphi} d\varphi \tag{7.191}$$

since $f_{\text{eq}}$ is periodic in $\varphi$. Also,

$$\frac{\partial f_{\text{eq}}}{\partial x} = -\frac{1}{Z} \beta \frac{\partial H}{\partial x} e^{-\beta H} \tag{7.192}$$

$$= -f_{\text{eq}} \beta \frac{\partial H}{\partial x} \tag{7.193}$$

So,

$$f \frac{\partial H}{\partial x} + kT \frac{\partial f}{\partial x} = -kTg \frac{\partial f_{\text{eq}}}{\partial x} + kT \frac{\partial}{\partial x} (f_{\text{eq}} g) \tag{7.194}$$

$$= kT f_{\text{eq}} \frac{\partial g}{\partial x} \tag{7.195}$$

Hence,

$$\dot{N}_1 = \frac{\gamma T}{\mu_0} \int_0^{2\pi} f_{\mathrm{eq}} \left[ \frac{\partial g}{\partial \varphi} + a(1 - x^2) \frac{\partial g}{\partial x} \right] d\varphi \tag{7.196}$$

which is Eq. (7.18), and yields

$$\dot{N}_1 = \frac{\gamma kT}{\mu_0} \frac{1}{Z} \int_0^{2\pi} e^{-\beta H} \left[ \frac{\partial g}{\partial \varphi} + a(1 - x^2) \frac{\partial g}{\partial x} \right] d\varphi \tag{7.197}$$

Now Eq. (7.14) yields

$$\frac{\partial g}{\partial x} = (g_2 - g_1) \frac{\partial \zeta}{\partial x} \tag{7.198}$$

$$\frac{\partial g}{\partial \varphi} = (g_2 - g_1) \frac{\partial \zeta}{\partial \varphi} \tag{7.199}$$

So, in terms of the crossover function $\zeta$

$$\dot{N}_1 = \frac{\gamma kT}{\mu_0} \frac{1}{Z} (g_2 - g_1) \int_0^{2\pi} e^{-\beta H} \left[ a(1 - x^2) \frac{\partial \zeta}{\partial x} + \frac{\partial \zeta}{\partial \varphi} \right] d\varphi \tag{7.200}$$

Now, since the number of particles in the two wells is normalized to 1, we have

$$N_1 + N_2 = 1 \tag{7.201}$$

This equation also holds in equilibrium, so

$$N_{1,\mathrm{eq}} + N_{2,\mathrm{eq}} = 1 \tag{7.202}$$

and by using Eqs. (7.15)

$$g_1 N_{1,\mathrm{eq}} + g_2 N_{2,\mathrm{eq}} = 1 \tag{7.203}$$

Now,

$$N_2 N_{1,\mathrm{eq}} - N_1 N_{2,\mathrm{eq}} = (g_2 - g_1) N_{1,\mathrm{eq}} N_{2,\mathrm{eq}} \tag{7.204}$$

so

$$g_2 - g_1 = \frac{1}{N_{1,\mathrm{eq}} N_{2,\mathrm{eq}}} (N_2 N_{1,\mathrm{eq}} - N_1 N_{2,\mathrm{eq}}) \tag{7.205}$$

$$= \frac{N_{1,\mathrm{eq}} + N_{2,\mathrm{eq}}}{N_{1,\mathrm{eq}} N_{2,\mathrm{eq}}} (N_2 N_{1,\mathrm{eq}} - N_1 N_{2,\mathrm{eq}}) \tag{7.206}$$

$$= \left( \frac{1}{N_{1,\mathrm{eq}}} + \frac{1}{N_{2,\mathrm{eq}}} \right) (N_2 N_{1,\mathrm{eq}} - N_1 N_{2,\mathrm{eq}}) \tag{7.207}$$

$$= Z \left( \frac{1}{Z_1} + \frac{1}{Z_2} \right) (N_2 N_{1,\mathrm{eq}} - N_1 N_{2,\mathrm{eq}}) \tag{7.208}$$

Substituting this result into Eq. (7.200) and comparing with Eq. (7.19) yields the escape rate, Eq. (7.20).

## Appendix VII.III. Evaluation of Partition Functions by the Method of Steepest Descents

We calculate the partition function

$$Z = \int_0^{2\pi} \int_0^{\theta_s} e^{-\beta H(\theta,\varphi)} \sin\theta d\theta d\varphi \tag{7.209}$$

where $\theta_s$ is the value of the polar angle at the saddle point. We assume that practically all the particles stay at or near the minimum $(\theta_1)$, that is, the high barrier situation, and we write

$$\varepsilon = \beta H = \varepsilon(\theta_1, \varphi_1) + \frac{(\theta - \theta_1)^2}{2} \frac{\partial^2 \varepsilon}{\partial \theta^2} \bigg|_{(\theta_1,\varphi_1)} \tag{7.210}$$

$$\sin\theta \approx \theta \tag{7.211}$$

($\varepsilon$ is independent of $\varphi$ in the cases with which we deal)

### Case (1) Arbitrary $\psi$

In what follows, the subscripts denote partial derivatives and $x \equiv \cos\theta$.

$$\varepsilon = \beta H = -\sigma x^2 - 2\sigma h (x \cos\psi + \sqrt{1 - x^2} \cos\varphi \sin\psi) \tag{7.212}$$

$$\varepsilon_x = -2\sigma x - 2\sigma h \cos\psi + 2\sigma h \frac{x}{\sqrt{1 - x^2}} \cos\varphi \sin\psi \tag{7.213}$$

$$\varepsilon_{\varphi} = 2\sigma h\sqrt{1-x^2}\sin\varphi\sin\psi \tag{7.214}$$

$$\varepsilon_{xx} = -2\sigma + 2\sigma h\frac{1}{(1-x^2)^{(3/2)}}\cos\varphi\sin\psi \tag{7.215}$$

$$\varepsilon_{\varphi\varphi} = 2\sigma h\sqrt{1-x^2}\cos\varphi\sin\psi \tag{7.216}$$

$$\varepsilon_{x\varphi} = -2\sigma h\frac{x}{\sqrt{1-x^2}}\sin\varphi\sin\psi \tag{7.217}$$

For a maximum, minimum or saddle point $\varepsilon_x = 0$ and $\varepsilon_{\varphi} = 0$.
Now,

$$\varepsilon_{\varphi} = 0 \Rightarrow \varphi = 0 \text{ or } \pi, \text{ or } x = \pm 1 \tag{7.218}$$

since the field is applied in the $xz$ plane we take as the solution $\varphi = 0$. Then $\varepsilon_x$ becomes

$$\varepsilon_x = -2\sigma x - 2\sigma h\cos\psi + 2\sigma h\frac{x}{\sqrt{1-x^2}}\sin\psi \tag{7.219}$$

$$= 0$$

if

$$\xi\frac{x}{\sqrt{1-x^2}}\sin\psi = 2\sigma x + \xi\cos\psi \tag{7.220}$$

Squaring and manipulating both sides leads to the equation

$$x^4 + 2hx^3\cos\psi - (1-h^2)x^2 - 2hx\cos\psi - h^2\cos^2\psi = 0 \tag{7.221}$$

which, for small $h$, reduces to

$$x^4 - x^2 = 0 \tag{7.222}$$

with solutions

$$x = 0 \text{ and } x = \pm 1 \tag{7.223}$$

Now, by checking the sign of the Hessian $\Delta \equiv \varepsilon_{xx}\varepsilon_{\varphi\varphi} - \varepsilon_{x\varphi}^2$, we find

$$x = 0\left(\text{or } \theta = \frac{\pi}{2}\right), \varphi = 0 \text{ gives a saddle point}$$

$$x = 0\left(\text{or } \theta = \frac{\pi}{2}\right), \quad \varphi = \pi \text{ gives a maximum}$$

$$x = \pm 1(\text{or } \theta = 0 \text{ or } \pi), \quad \varphi = 0 \text{ are minima}$$

Hence,

$$\varepsilon_{\text{max}} = \xi \tag{7.224}$$

$$\varepsilon_{\text{sad}} = -\xi \tag{7.225}$$

$$\varepsilon_{\text{min}} = -\sigma \tag{7.226}$$

Now, to approximate $\varepsilon$ by the Taylor series about the minimum $\theta = 0$, $\varphi = 0$, we note

$$\varepsilon(0,0) = -\sigma - \xi \cos\psi \tag{7.227}$$

$$\varepsilon_{\theta\theta}(0,0) = 2\sigma + \xi\cos\psi \tag{7.228}$$

$$\varepsilon_{\varphi\varphi}(0,0) = \varepsilon_{x\varphi}(0,0) = 0 \tag{7.229}$$

So,

$$\varepsilon(\theta,\varphi) \approx -\sigma - \xi\cos\psi + \tfrac{1}{2}(2\sigma + \xi\cos\psi)\theta^2 \tag{7.230}$$

and

$$Z = e^{\sigma+\xi\cos\psi} \int_0^{2\pi} \int_0^{(\pi/2)} e^{-(1/2)(2\sigma+\xi\cos\psi)\theta^2} \sin\theta \, d\theta d\varphi \tag{7.231}$$

Now we assume that almost all of the particles stay in the vicinity of the minimum. So the major part of the contribution to the integral comes from the vicinity of $\theta = 0$. Thus, we can approximate $\sin\theta \approx \theta$. We can also extend the upper limit to $+\infty$ without serious error, since the integrand goes to 0 so rapidly away from the minimum.

So,

$$Z = 2\pi e^{\sigma+\xi\cos\psi} \int_0^\infty e^{-(1/2)(2\sigma+\xi\cos\psi)}\theta \, d\theta \tag{7.232}$$

$$= \frac{2\pi e^{\sigma+\xi\cos\psi}}{2\sigma + \xi\cos\psi} \tag{7.233}$$

**Case (2) $\psi = \pi/2$**

We see, on substitution of $\psi$, in the above formula that

$$Z = \frac{\pi e^{\sigma}}{\sigma} \tag{7.234}$$

Moreover to any order in $h$ one may easily show that in this instance

$$Z = \frac{\pi}{\sigma\sqrt{1-h^2}} e^{\sigma(1+h^2)}$$

### Appendix VII.IV. Calculation of $(\partial\zeta/\partial x)$ and $\Gamma$ for Axial Symmetry

For axial symmetry

$$\frac{\partial H}{\partial \varphi} = \frac{\partial f}{\partial \varphi} = 0 \tag{7.235}$$

and also

$$\frac{\partial g}{\partial \varphi} = 0 \tag{7.236}$$

so that for the potential of Eq. (7.1) with $h = 0$ (simple uniaxial anisotropy) Eq. (7.13) becomes

$$a\left[2kT\sigma x + kT\frac{\partial}{\partial x}\right](1 - x^2)(g_2 - g_1)\frac{\partial \zeta}{\partial x} = 0 \tag{7.237}$$

By dividing across by $kTa(g_2 - g_1)$ and by letting $R(x) = (1 - x^2)(\partial\zeta/\partial x)$, we find

$$\frac{dR}{dx} + 2\sigma x R = 0 \tag{7.238}$$

which has as solution

$$R = (1 - x^2)\frac{\partial\zeta}{\partial x} = C_1 \exp(-\sigma x^2) \tag{7.239}$$

near the saddle point $1 - x^2 \approx 1$, so

$$\frac{\partial\zeta}{\partial x} \approx C_1 \exp(-\sigma x^2) \tag{7.240}$$

Hence,

$$\zeta = C_1 \int_{-\infty}^{x} \exp(-\sigma t^2) dt \tag{7.241}$$

(since $\zeta \to 0$ when $x \to -\infty$, i.e., deep in the first well) now as $x \to \infty$ (i.e., deep in the second well) $\zeta \to 1$. This yields

$$C_1 = \sqrt{\frac{\sigma}{\pi}} \tag{7.242}$$

so that

$$\frac{\partial\zeta}{\partial x} = \sqrt{\frac{\sigma}{\pi}} \exp(-\sigma x^2) \tag{7.243}$$

and using the fact that $Z_1 = Z_2 = (\pi e^{\sigma}/\sigma)$, this yields the escape rate as

$$\Gamma = 2\pi a \sqrt{\frac{\sigma}{\pi}} \frac{\omega_1}{\pi} e^{-\sigma} \tag{7.244}$$

that is Brown's original result [21], cf. Eq. (3.58c) of Section III

## Appendix VII.V. Proof of Eq. (7.113) for the Crossover Function $\zeta$

Equation (7.13) for the quasi-stationary function $g$ is

$$\frac{\partial H}{\partial \varphi}\frac{\partial g}{\partial x} - \frac{\partial H}{\partial x}\frac{\partial g}{\partial \varphi} + a\left[\left(-\frac{\partial H}{\partial x} + kT\frac{\partial}{\partial x}\right)(1-x^2)\frac{\partial g}{\partial x} + \frac{1}{1-x^2}\left(-\frac{\partial H}{\partial \varphi} + kT\frac{\partial}{\partial \varphi}\right)\frac{\partial g}{\partial \varphi}\right] = 0 \tag{7.245}$$

Now, by using the fact that the Hamiltonian $H = kT\varepsilon$ and the chain rule, we obtain

$$kT\frac{\partial \varepsilon}{\partial \varphi}\frac{\partial g}{\partial \varepsilon}\frac{\partial \varepsilon}{\partial x} - kT\frac{\partial \varepsilon}{\partial x}\frac{\partial g}{\partial \varepsilon}\frac{\partial \varepsilon}{\partial \varphi} + a\left[\left(-kT\frac{\partial \varepsilon}{\partial x} + kT\frac{\partial}{\partial x}\right)(1-x^2)\frac{\partial g}{\partial \varepsilon}\frac{\partial \varepsilon}{\partial x} + \frac{1}{1-x^2}\left(-kT\frac{\partial \varepsilon}{\partial \varphi} + kT\frac{\partial}{\partial \varphi}\right)\frac{\partial g}{\partial \varepsilon}\frac{\partial \varepsilon}{\partial \varphi}\right] = 0 \tag{7.246}$$

which yields

$$\left[\left(-\frac{\partial \varepsilon}{\partial x} + \frac{\partial}{\partial x}\right)(1-x^2)\frac{\partial g}{\partial \varepsilon}\frac{\partial \varepsilon}{\partial x} + \frac{1}{1-x^2}\left(-\frac{\partial \varepsilon}{\partial \varphi} + \frac{\partial}{\partial \varphi}\right)\frac{\partial g}{\partial \varepsilon}\frac{\partial \varepsilon}{\partial \varphi}\right] = 0 \tag{7.247}$$

Now, since by definition of $\zeta$ [Eq. (7.14)]

$$g = g_1 + (g_2 - g_1)\zeta \tag{7.248}$$

it follows that

$$\frac{\partial g}{\partial \varepsilon} = (g_2 - g_1)\frac{\partial \zeta}{\partial \varepsilon} \tag{7.249}$$

and hence on substitution into Eq. (7.247) and dividing across by $g_2 - g_1$, we obtain:

$$-\left[(1-x^2)\left(\frac{\partial \varepsilon}{\partial x}\right)^2 + \frac{1}{1-x^2}\left(\frac{\partial \varepsilon}{\partial \varphi}\right)^2\right]\frac{\partial \zeta}{\partial \varepsilon} + \frac{\partial}{\partial x}(1-x^2)\frac{\partial \zeta}{\partial \varepsilon}\frac{\partial \varepsilon}{\partial x} + \frac{1}{1-x^2}\frac{\partial}{\partial \varphi}\frac{\partial \zeta}{\partial \varepsilon}\frac{\partial \varepsilon}{\partial \varphi} = 0 \tag{7.250}$$

Again applying the chain rule to the last two terms on the left-hand side of

Eq. (7.250) yields

$$-\left[(1-x^2)\left(\frac{\partial\varepsilon}{\partial x}\right)^2+\frac{1}{1-x^2}\left(\frac{\partial\varepsilon}{\partial\varphi}\right)^2\right]\frac{\partial\zeta}{\partial\varepsilon}+\left(\frac{\partial}{\partial x}(1-x^2)\frac{\partial\varepsilon}{\partial x}\right)\frac{\partial\zeta}{\partial\varepsilon}$$
$$+(1-x^2)\left(\frac{\partial\varepsilon}{\partial x}\right)^2\frac{\partial^2\zeta}{\partial\varepsilon^2}+\frac{1}{1-x^2}\frac{\partial^2\varepsilon}{\partial\varphi^2}\frac{\partial\zeta}{\partial\varepsilon}+\frac{1}{1-x^2}\left(\frac{\partial\varepsilon}{\partial\varphi}\right)^2\frac{\partial^2\zeta}{\partial\varepsilon^2}=0 \tag{7.251}$$

which simplifies to

$$-\left[(1-x^2)\left(\frac{\partial\varepsilon}{\partial x}\right)^2+\frac{1}{1-x^2}\left(\frac{\partial\varepsilon}{\partial\varphi}\right)^2\right]\frac{\partial\zeta}{\partial\varepsilon}$$
$$+\left(\frac{\partial}{\partial x}(1-x^2)\frac{\partial\varepsilon}{\partial x}+\frac{1}{1-x^2}\frac{\partial^2\varepsilon}{\partial\varphi^2}\right)\frac{\partial\zeta}{\partial\varepsilon} \tag{7.252}$$
$$+\left[(1-x^2)\left(\frac{\partial\varepsilon}{\partial x}\right)^2+\frac{1}{1-x^2}\left(\frac{\partial\varepsilon}{\partial\varphi}\right)^2\right]\frac{\partial^2\zeta}{\partial\varepsilon^2}=0$$

We may now average this last equation along a line of constant energy, since $\zeta=\zeta(\varepsilon)$, we find that $\zeta$, $(\partial\zeta/\partial\varepsilon)$, and $(\partial^2\zeta/\partial\varepsilon^2)$, are all constants, and hence the energy diffusion equation Eq. (7.113) with $A(\varepsilon)$ and $B(\varepsilon)$ given as in Eq. (7.114) follows easily.

## Appendix VII.VI. Justification of Eqs. (7.131)–(7.133)

Let

$$R=\frac{\partial\zeta}{\partial\varepsilon} \tag{7.253}$$

Then

$$\frac{\partial R}{\partial\varepsilon}=\left[1-\frac{1}{2\varepsilon}\right]R \tag{7.254}$$

or

$$\frac{dR}{R}=\left[1-\frac{1}{2\varepsilon}\right]d\varepsilon \tag{7.255}$$

and hence

$$\ln R=\varepsilon-\tfrac{1}{2}\ln|\varepsilon|+A \tag{7.256}$$

and

$$\ln R + \ln \sqrt{\varepsilon} = \ln C_1 + \varepsilon \tag{7.257}$$

where

$$A \equiv \ln C_1 \tag{7.258}$$

Hence,

$$\ln \frac{R\sqrt{\varepsilon}}{C_1} = \varepsilon \tag{7.259}$$

and it follows that

$$R = \frac{C_1 e^{\varepsilon}}{\sqrt{|\varepsilon|}} \tag{7.260}$$

Now defining the constant $C$ by the equation

$$C_1 \equiv C e^{-\varepsilon_C} \tag{7.261}$$

we obtain on noting that $\varepsilon < 0$

$$R = \frac{d\zeta}{d\varepsilon} = C \frac{e^{\varepsilon - \varepsilon_C}}{\sqrt{-\varepsilon}} \tag{7.262}$$

which is Eq. (7.132). Now

$$\int_{-\infty}^{\varepsilon_C} \frac{d\zeta}{d\varepsilon} d\varepsilon = \zeta|_{-\infty}^{\varepsilon_C} \tag{7.263}$$

also

$$\int_{-\infty}^{\infty} \frac{d\zeta}{d\varepsilon} d\varepsilon = \int_{-\infty}^{\varepsilon_C} \frac{d\zeta}{d\varepsilon} d\varepsilon + \int_{\varepsilon_C}^{\infty} \frac{d\zeta}{d\varepsilon} d\varepsilon \tag{7.264}$$

If the wells are symmetric, then

$$\int_{-\infty}^{\varepsilon_C} \frac{d\zeta}{d\varepsilon} d\varepsilon = \int_{\varepsilon_C}^{\infty} \frac{d\zeta}{d\varepsilon} d\varepsilon \tag{7.265}$$

and so

$$\int_{-\infty}^{\varepsilon_C} \frac{d\zeta}{d\varepsilon} d\varepsilon = \frac{1}{2}\int_{-\infty}^{\infty} \frac{d\zeta}{d\varepsilon} d\varepsilon = \frac{1}{2}\zeta|_{-\infty}^{\infty} \tag{7.266}$$

Now,

$$\zeta(-\infty) = 0 \tag{7.267a}$$

$$\zeta(\infty) = 1 \tag{7.267b}$$

and so

$$\int_{-\infty}^{\varepsilon_C} \frac{d\zeta}{d\varepsilon} d\varepsilon = \frac{1}{2} \tag{7.268}$$

Hence,

$$C\int_{-\infty}^{\varepsilon_C} \frac{e^{\varepsilon-\varepsilon_C}}{\sqrt{-\varepsilon}} d\varepsilon = \frac{1}{2} \tag{7.269}$$

which proves Eq. (7.133)

## Appendix VII.VII. Justification of Eqs. (7.135)–(7.138)

From Eq. (7.20), we know that

$$\Gamma = \left[\frac{1}{Z_1} + \frac{1}{Z_2}\right] \frac{\gamma kT}{\mu_0} \int_0^{2\pi} a(1-x^2)\frac{\partial \zeta}{\partial x} e^{-\beta H} d\varphi \tag{7.270}$$

$$= \left[\frac{1}{Z_1} + \frac{1}{Z_2}\right] \frac{\gamma kT}{\mu_0} \int_0^{2\pi} a(1-x^2)\frac{\partial \zeta}{\partial \varepsilon}\frac{\partial \varepsilon}{\partial x}|_{\varepsilon=\varepsilon_C} e^{-\varepsilon} d\varphi \tag{7.271}$$

which, using Eq. (7.132) becomes:

$$\Gamma = \left[\frac{1}{Z_1} + \frac{1}{Z_2}\right] \frac{\gamma kT}{\mu_0} \int_0^{2\pi} a(1-x^2)C\frac{e^{\varepsilon-\varepsilon_C}}{\sqrt{-\varepsilon}}\frac{\partial \varepsilon}{\partial x}|_{\varepsilon=\varepsilon_C} e^{-\varepsilon} d\varphi \tag{7.272}$$

$$= \left[\frac{1}{Z_1} + \frac{1}{Z_2}\right] \frac{a\gamma kT}{\mu_0} \int_0^{2\pi} (1-x^2)C\frac{e^{-\varepsilon_C}}{\sqrt{-\varepsilon}}\frac{\partial \varepsilon}{\partial x}|_{\varepsilon=\varepsilon_C} d\varphi \tag{7.273}$$

now, using the fact that near the saddle point $x \approx 0$

$$\Gamma \approx \left[\frac{1}{Z_1} + \frac{1}{Z_2}\right] \frac{a\gamma kT}{\mu_0} C \frac{e^{\xi}}{\sqrt{-\varepsilon}} |_{\varepsilon=\varepsilon_C} \int_0^{2\pi} \frac{\partial \varepsilon}{\partial x} |_{\varepsilon=\varepsilon_C} d\varphi \tag{7.274}$$

Since the partition functions are given by

$$Z_1 = Z_2 = \frac{\pi}{\sigma\sqrt{1-h^2}} e^{\sigma(1+h^2)} \tag{7.275}$$

we have

$$\Gamma = \frac{2\sigma\sqrt{1-h^2}}{\pi} e^{-\sigma(1+h^2)} \frac{a\gamma kT}{\mu_0} C \frac{e^{2\sigma h}}{\sqrt{-\varepsilon}} |_{\varepsilon=\varepsilon_C} \int_0^{2\pi} \frac{\partial \varepsilon}{\partial x} |_{\varepsilon=\varepsilon_C} d\varphi \tag{7.276}$$

$$= \frac{2\sigma\gamma kT\sqrt{1-h^2}}{\mu_0 \pi} e^{-\sigma(1+h^2-2\sigma h)} \frac{aC}{\sqrt{-\varepsilon}} |_{\varepsilon=\varepsilon_C} \int_0^{2\pi} \frac{\partial \varepsilon}{\partial x} |_{\varepsilon=\varepsilon_C} d\varphi \tag{7.277}$$

Since

$$\sigma = \frac{Kv}{kT} \tag{7.278}$$

$$\mu_0 = vM_s \tag{7.279}$$

and the attempt frequency that is the ferromagnetic resonance frequency is

$$\omega_A = \frac{2K\gamma}{M_s}\sqrt{1-h^2} = \omega_1 \tag{7.280}$$

we have

$$\Gamma = \frac{2vK\gamma\sqrt{1-h^2}}{vM_s\pi} e^{-\sigma(1-h)^2} \lim_{\varepsilon \to \varepsilon_C = -\xi} \frac{aC}{\sqrt{-\varepsilon}} \int_0^{2\pi} \frac{\partial \varepsilon}{\partial x} d\varphi \tag{7.281}$$

$$= \frac{\omega_A}{\pi} e^{-\beta \Delta U} \lim_{\varepsilon \to \varepsilon_C = -\xi} \frac{aC}{\sqrt{-\varepsilon}} \int_0^{2\pi} \frac{\partial \varepsilon}{\partial x} d\varphi \tag{7.282}$$

and so, writing

$$\Gamma = \frac{\omega_A}{\pi} A e^{-\beta \Delta U} \tag{7.283}$$

we see that

$$A = \lim_{\varepsilon \to \varepsilon_C = -\xi} \frac{aC}{\sqrt{-\varepsilon}} \int_0^{2\pi} \frac{\partial \varepsilon}{\partial x} d\varphi \tag{7.284}$$

which is Eq. (7.135) for the corrected VLD prefactor.

Now, we determine $C$ as follows, we have

$$C = \frac{1}{2} \left[ \int_{-\infty}^{\varepsilon_C} \frac{e^{\varepsilon - \varepsilon_C}}{\sqrt{-\varepsilon}} d\varepsilon \right]^{-1} \tag{7.285}$$

By letting

$$t = \varepsilon_C - \varepsilon \tag{7.286}$$

we find

$$dt = -d\varepsilon \tag{7.287}$$

$$-\varepsilon = t - \varepsilon_C = t + \xi \tag{7.288}$$

when

$$\varepsilon \to -\infty, \qquad \text{we see} \qquad t \to +\infty \tag{7.289}$$

when

$$\varepsilon = \varepsilon_C, \qquad \text{we see} \qquad t = 0 \tag{7.290}$$

and so

$$C = \frac{1}{2} \left[ \int_0^{\infty} \frac{e^{-t} dt}{\sqrt{\xi + t}} \right]^{-1} \tag{7.291}$$

$$= \frac{1}{2} e^{-\xi} \left[ \int_0^{\infty} \frac{e^{-(\xi + t)} dt}{\sqrt{\xi + t}} \right]^{-1} \tag{7.292}$$

Now, by letting

$$u = \xi + t \tag{7.293}$$

we find

$$C = \frac{1}{2} e^{-\xi} \left[ \int_{\xi}^{\infty} \frac{e^{-u} du}{\sqrt{u}} \right]^{-1} \tag{7.294}$$

and by letting

$$u = v^2 \tag{7.295}$$

we obtain

$$C = \frac{1}{2} e^{-\xi} \left[ \int_{\sqrt{\xi}}^{\infty} e^{-v^2} dv \right]^{-1} \tag{7.296}$$

$$= \frac{1}{2} e^{-\xi} \left[ \sqrt{\pi}\, \mathrm{erfc}(\sqrt{\xi}) \right]^{-1} \tag{7.297}$$

## Appendix VII.VIII. Justification of Eq. (7.139)

Now, for $\xi \neq 0$

$$\lim_{\varepsilon \to -\xi} \int_0^{2\pi} \sqrt{1 + \frac{\xi}{\varepsilon} \cos \varphi d\varphi} = \int_0^{2\pi} \sqrt{1 - \cos \varphi d\varphi} \tag{7.298}$$

$$= \int_0^{2\pi} \sqrt{2} \sin \frac{\varphi}{2} d\varphi \tag{7.299}$$

$$= -2\sqrt{2} \cos \frac{\varphi}{2} \bigg|_0^{2\pi} \tag{7.300}$$

$$= 2^{5/2} \tag{7.301}$$

and so

$$Q = \frac{1}{2\pi} 2^{5/2} = \frac{2^{3/2}}{\pi} \approx 0.900 \tag{7.302a}$$

If $\xi = 0$, it can easily be seen that

$$Q = 1 \tag{7.302b}$$

## Appendix VII.IX. Justification of Eq. (7.140)

The asymptotic expansion of $f(\xi)$

1. **$\xi \ll 1$**

   Let,

$$t = \sqrt{\xi} \tag{7.303}$$

and let

$$g(t) = f(t^2) = f(\xi) = e^{-\xi}\left[\operatorname{erfc}\sqrt{\xi}\right]^{-1} = e^{-t^2}[\operatorname{erfc}(t)]^{-1} \tag{7.304}$$

Now, expanding $g(t)$ as a Maclaurin series as far as the first-order term, we have

$$g(t) \approx g(0) + tg'(0) \tag{7.305}$$

$$= 1 + tg'(0) \tag{7.306}$$

Now,

$$g'(t) = -2te^{-t^2}[\operatorname{erfc}(t)]^{-1} + \frac{2}{\sqrt{\pi}}e^{-t^2}[\operatorname{erfc}(t)]^{-2} \tag{7.307}$$

and so

$$g'(0) = \frac{2}{\sqrt{\pi}} \tag{7.308}$$

Hence, for small $\xi$,

$$f(\xi) \approx 1 + 2\sqrt{\frac{\xi}{\pi}} \tag{7.309}$$

2. **If $\xi \gg 1$**

   By using asymptotic expansion formula (7.1.23) of Abramowitz and Stegun [82], we find

$$\sqrt{\pi}\sqrt{\xi}e^{\xi}\operatorname{erfc}(\sqrt{\xi}) \approx 1 \tag{7.310}$$

and hence

$$e^{-\xi}\left[\operatorname{erfc}(\sqrt{\xi})\right]^{-1} \approx \sqrt{\pi\xi} \tag{7.311}$$

yielding, in the limit of large $\xi$,

$$f(\xi) \approx \sqrt{\pi\xi} \tag{7.312}$$

which is Eq. (7.140b).

## Appendix VII.X. Proof of Eq. (7.152)

$$\zeta_1(x', \varphi) \equiv \frac{\partial\zeta}{\partial\xi}(x', \varphi, 0)$$

is written as a Fourier series. We keep the first-order terms in $\varphi$ only.

$$\zeta_2(x', \varphi) \equiv \frac{\partial^2\zeta}{\partial\xi^2}(x', \varphi, 0)$$

is also written as a Fourier series, however, we keep only the zero-order term as higher order terms will not be needed. We now set up the differential equations, that is, Eqs. (7.152).

$$\frac{\partial\zeta}{\partial x'} = \zeta_0'(x') + \xi[A'(x')\sin\varphi + B'(x')\cos\varphi] + \xi^2 C'(x') \tag{7.313}$$

$$\frac{\partial^2\zeta}{\partial x'^2} = \zeta_0''(x') + \xi[A''(x')\sin\varphi + B''(x')\cos\varphi] + \xi^2 C''(x') \tag{7.314}$$

$$\frac{\partial\zeta}{\partial\varphi} = \xi[A(x')\cos\varphi - B(x')\sin\varphi] \tag{7.315}$$

Equation (7.141) implies

$$\begin{aligned}
&\alpha[\zeta_0''(x') + \xi[A''(x')\sin\varphi + B''(x')\cos\varphi] + \xi^2 C''(x') \\
&\quad + 2x'\{\zeta_0'(x') + \xi[A'(x')\sin\varphi + B'(x')\cos\varphi] + \xi^2 C'(x')\}] \\
&\quad + 2x'\xi[A(x')\cos\varphi - B(x')\sin\varphi] \\
&\quad + \xi\sin\varphi\{\zeta_0'(x') + \xi[A'(x')\sin\varphi + B'(x')\cos\varphi] + \xi^2 C'(x')\} = 0
\end{aligned} \tag{7.316}$$

By equating the coefficients of the various powers of $\xi$ to zero we have

$$\alpha[\zeta_0'' + 2x'\zeta_0'] = 0 \tag{7.317}$$

$$\alpha\{A''(x')\sin\varphi + B''(x')\cos\varphi + 2x'[A'(x')\sin\varphi + B'(x')\cos\varphi]\} \\ + 2x'[A(x')\cos\varphi - B(x')\sin\varphi] + \zeta_0'\sin\varphi = 0 \tag{7.318}$$

$$\alpha[C'' + 2x'C'] + \sin\varphi[A'(x')\sin\varphi + B'(x')\cos\varphi] = 0 \tag{7.319}$$

Differentiating the left-hand side of Eq. (7.318) partially with respect to $\varphi$ gives

$$\alpha\{A''(x')\cos\varphi - B''(x')\sin\varphi + 2x'[A'(x')\cos\varphi - B'(x')\sin\varphi]\} \\ - 2x'[A(x')\sin\varphi + B(x')\cos\varphi] + \zeta_0'\cos\varphi = 0 \tag{7.320}$$

We now add $-i$ times Eq. (7.320) to Eq. (7.318) to get

$$\alpha\{A''(\sin\varphi - i\cos\varphi) + iB''(\sin\varphi - i\cos\varphi) \\ + 2x'[A'(\sin\varphi - i\cos\varphi) + iB'(\sin\varphi - i\cos\varphi)] \\ + 2x'[A(\sin\varphi - i\cos\varphi) + iB(\sin\varphi - i\cos\varphi)] \\ + \zeta_0'(\sin\varphi - \cos\varphi) = 0 \tag{7.321}$$

Now, by dividing across by $\sin\varphi - i\cos\varphi$ and writing $Y = A + iB$, we obtain

$$\alpha(Y'' + 2x'Y') + 2ix'Y = -\zeta_0' \tag{7.322}$$

Now, in the uniaxial case

$$\frac{\partial\zeta}{\partial x} = \sqrt{\frac{\sigma}{\pi}}e^{-\sigma x^2} \tag{7.323}$$

so

$$\sqrt{\sigma}\frac{\partial\zeta}{\partial x'} = \sqrt{\frac{\sigma}{\pi}}e^{-x'^2} \tag{7.324}$$

$$\frac{\partial\zeta}{\partial x'} = \sqrt{\frac{1}{\pi}}e^{-x'^2} \tag{7.325}$$

and it follows then that

$$\alpha(Y'' + 2x'Y') + 2ix'Y = -\pi^{-(1/2)}e^{-x'^2} \tag{7.326}$$

Now the coefficient of $\xi^2$ in Eq. (7.316) is

$$\alpha(C'' + 2x'C') + A' \sin^2\varphi + B' \sin\varphi \cos\varphi = 0 \tag{7.327}$$

On integrating Eq. (7.327) across with respect to $\varphi$ from 0 to $2\pi$ we obtain

$$2\pi\alpha(C'' + 2x'C') + A'\pi = 0 \tag{7.328}$$

or

$$\alpha(C'' + 2x'C') + \tfrac{1}{2}A' = 0 \tag{7.329}$$

which is the real part of the equation

$$\alpha(Z' + 2x'Z) = -\tfrac{1}{2}Y' \tag{7.330}$$

which is Eq. (7.152)

## Appendix VII.XI. Justification of Eqs. (7.154)–(7.156)

$$Y(x') = \frac{1}{2\pi}\int_{-\infty}^{\infty} e^{-ikx'} e^{-k^2/4} Y_k dk \tag{7.331}$$

$$Y'(x') = -\frac{1}{2\pi}\int_{-\infty}^{\infty} ike^{-ikx'} e^{-k^2/4} Y_k dk \tag{7.332}$$

$$Y''(x') = -\frac{1}{2\pi}\int_{-\infty}^{\infty} k^2 e^{-ikx'} e^{-k^2/4} Y_k dk \tag{7.333}$$

Now,

$$\alpha 2x'Y' = -\frac{\alpha}{2\pi}\int_{-\infty}^{\infty} 2k(ix')e^{-ikx'} e^{-k^2/4} Y_k dk \tag{7.334}$$

$$= \frac{\alpha}{2\pi}\int_{-\infty}^{\infty} 2k\left(\frac{d}{dk}e^{-ikx'}\right) e^{-k^2/4} Y_k dk \tag{7.335}$$

Now in order to integrate by parts, we let

$$u = 2ke^{-k^2/4}Y_k \qquad \text{and} \qquad dv = \frac{d}{dk}e^{-ikx'}dk \tag{7.336}$$

then,

$$du = \left[2e^{-k^2/4}Y_k + 2k\left(-\frac{k}{2}\right)e^{-k^2/4}Y_k + 2ke^{-k^2/4}\frac{dY_k}{dk}\right]dk \qquad \text{and} \qquad v = e^{-ikx'} \tag{7.337}$$

and so

$$\alpha 2x'Y' = \frac{\alpha}{2\pi}\left[2ke^{-k^2/4}Y_k e^{-ikx'}|_{-\infty}^{\infty} - \int_{-\infty}^{\infty} e^{-ikx'}du\right] \tag{7.338}$$

$$= -\frac{\alpha}{2\pi}\int_{-\infty}^{\infty} e^{-ikx'}du \tag{7.339}$$

$$\begin{aligned} &= -\frac{2\alpha}{2\pi}\int_{-\infty}^{\infty} e^{-ikx'}e^{-k^2/4}Y_k dk + \frac{\alpha}{2\pi}\int_{-\infty}^{\infty} k^2 e^{-ikx'}e^{-k^2/4}Y_k dk \\ &\quad -\frac{2\alpha}{2\pi}\int_{-\infty}^{\infty} ke^{-ikx'}e^{-k^2/4}\frac{dY_k}{dk}dk \end{aligned} \tag{7.340}$$

So,

$$\alpha(Y'' + 2x'Y') = -\frac{\alpha}{\pi}\left[\left[\int_{-\infty}^{\infty} ike^{-ikx'}e^{-k^2/4}Y_k dk + \int_{-\infty}^{\infty} ke^{-ikx'}e^{-k^2/4}\frac{dY_k}{dk}dk\right]\right. \tag{7.341}$$

$$2ix'Y = \frac{1}{2\pi}\int_{-\infty}^{\infty} 2ix'e^{-ikx'}e^{-k^2/4}Y_k dk \tag{7.342}$$

$$= \frac{1}{\pi}\int_{-\infty}^{\infty}\left(\frac{d}{dk}e^{-ikx'}\right)e^{-k^2/4}Y_k dk \tag{7.343}$$

Let,

$$u = e^{-k^2/4}Y_k \qquad \text{and} \qquad dv = \frac{d}{dk}e^{-ikx'}dk \tag{7.344}$$

then

$$du = \left[-\frac{k}{2}e^{-k^2/4}Y_k + e^{-k^2/4}\frac{dY_k}{dk}\right]dk \qquad \text{and} \qquad v = e^{-ikx'} \tag{7.345}$$

So,

$$2ix'Y = -\frac{1}{\pi}\left[e^{-k^2/4}e^{-ikx'}Y_k|_{-\infty}^{\infty} - \int_{-\infty}^{\infty} e^{-ikx'}\left(-\frac{k}{2}e^{-k^2/4}Y_k + e^{-k^2/4}\frac{dY_k}{dk}\right)dk\right] \tag{7.346}$$

$$= \frac{1}{\pi}\int_{-\infty}^{\infty} e^{-ikx'}\left(-\frac{k}{2}e^{-k^2/4}Y_k + e^{-k^2/4}\frac{dY_k}{dk}\right)dk \tag{7.347}$$

$$= -\frac{1}{2\pi}\int_{-\infty}^{\infty} ke^{-ikx'}e^{-k^2/4}Y_k dk + \frac{1}{\pi}\int_{-\infty}^{\infty} e^{-ikx'}e^{-k^2/4}\frac{dY_k}{dk}dk \tag{7.348}$$

Hence,

$$\begin{aligned}\alpha(Y'' + 2x'Y') + 2ix'Y &= \frac{1}{\pi}\int_{-\infty}^{\infty} e^{-ikx'}e^{-(k^2/4)}\frac{dY_k}{dk}(1 - \alpha k)dk \\ &\quad - \frac{1}{\pi}\int_{-\infty}^{\infty}\left(\alpha + \frac{k}{2}\right)e^{-ikx'}e^{-k^2/4}Y_k dk\end{aligned} \tag{7349}$$

$$= \frac{1}{\pi}\int_{-\infty}^{\infty} e^{-ikx'}e^{-k^2/4}\frac{d\widetilde{Y}_k}{dk}dk - \frac{1}{\pi}\int_{-\infty}^{\infty} e^{-ikx'}e^{-k^2/4}Y_k dk \tag{7.350}$$

Now, we define the function $u_k$ by the equation

$$\frac{e^{-x^2}}{\sqrt{\pi}} = \frac{1}{2\pi}\int_{-\infty}^{\infty} e^{-ikx'}e^{-k^2/4}u_k dk \tag{7.351}$$

and so

$$e^{-k^2/4}u_k = \int_{-\infty}^{\infty}\frac{e^{-x'^2}}{\sqrt{\pi}}e^{-ikx'}dx' \tag{7.352}$$

$$= \frac{1}{\sqrt{\pi}}\int_{-\infty}^{\infty} e^{-x'^2}e^{-ikx'}e^{-(1/2ik)^2}e^{-k^2/4}dx' \tag{7.353}$$

$$= \frac{e^{-k^2/4}}{\sqrt{\pi}}\int_{-\infty}^{\infty} e^{(x'+ik/2)^2}dx' \tag{7.354}$$

$$= \frac{e^{-k^2/4}}{\sqrt{\pi}}\sqrt{\pi} \tag{7.355}$$

Hence,

$$u_k \equiv 1 \tag{7.356}$$

and

$$\frac{e^{-x^2}}{\sqrt{\pi}} = \frac{1}{2\pi}\int_{-\infty}^{\infty} e^{-ikx'}e^{-k^2/4}dk \tag{7.357}$$

So,

$$\alpha(Y'' + 2x'Y') + 2ix'Y = -\frac{e^{-x^2}}{\sqrt{\pi}} \tag{7.358}$$

implies

$$\frac{1}{\pi}\left[\int_{-\infty}^{\infty} e^{-ikx'}e^{-k^2/4}\left\{\frac{d\widetilde{Y}_k}{dk} - \frac{1}{2}kY_k + \frac{1}{2}\right\}dk\right] = 0 \tag{7.359}$$

Hence,

$$\frac{d\widetilde{Y}_k}{dk} - \frac{1}{2}\frac{k}{1-\alpha k}\widetilde{Y}_k + \frac{1}{2} = 0 \tag{7.360}$$

or

$$\frac{d\widetilde{Y}_k}{dk} - \frac{1}{2}\frac{k}{1-\alpha k}\widetilde{Y}_k = -\frac{1}{2} \tag{7.361}$$

Now,

$$Z(x') = \frac{1}{2\pi}\int_{-\infty}^{\infty} e^{-ikx'}e^{-k^2/4}Z_k dk \tag{7.362}$$

$$Z'(x') = \frac{1}{2\pi}\int_{-\infty}^{\infty} ike^{-ikx'}e^{-k^2/4}Z_k dk \tag{7.363}$$

$$2x'Z = \frac{2}{2\pi}\int_{-\infty}^{\infty} x'e^{-ikx'}e^{-k^2/4}Z_k dk \tag{7.364}$$

$$= -\frac{i}{\pi}\int_{-\infty}^{\infty} ix' e^{-ikx'} e^{-k^2/4} Z_k dk \tag{7.365}$$

$$= \frac{i}{\pi}\int_{-\infty}^{\infty} \left(\frac{d}{dk} e^{-ikx'}\right) e^{-k^2/4} Z_k dk \tag{7.366}$$

By letting

$$u = e^{-k^2/4} Z_k \qquad \text{and} \qquad dv = \frac{d}{dk} e^{-ikx'} dk \tag{7.367}$$

we have

$$du = \left(-\frac{k}{2} e^{-k^2/4} Z_k + e^{-k^2/4} \frac{dZ_k}{dk}\right) dk \qquad \text{and} \qquad v = e^{-ikx'} \tag{7.368}$$

So,

$$2x'Z = -\frac{i}{\pi}\left[\int_{-\infty}^{\infty} \left(-\frac{k}{2}\right) e^{-ikx'} e^{-k^2/4} Z_k dk + \int_{-\infty}^{\infty} e^{-ikx'} e^{-k^2/4} \frac{dZ_k}{dk} dk\right] \tag{7.369}$$

$$= \frac{i}{2\pi}\int_{-\infty}^{\infty} k e^{-ikx'} e^{-k^2/4} Z_k dk - \frac{i}{\pi}\int_{-\infty}^{\infty} e^{-ikx'} e^{-k^2/4} \frac{dZ_k}{dk} dk \tag{7.370}$$

$$Z' + 2x'Z = -\frac{i}{\pi}\int_{-\infty}^{\infty} e^{-ikx'} e^{-k^2/4} \frac{dZ_k}{dk} dk \tag{7.371}$$

and by using Eq. (7.322)

$$\alpha(Z' + 2x'Z) + \frac{1}{2}Y' = 0 \tag{7.372}$$

implies

$$-\frac{i}{\pi}\int_{-\infty}^{\infty} e^{-ikx'} e^{-k^2/4} \left(\alpha\frac{dZ_k}{dk} + \frac{1}{4}kY_k\right) dk = 0 \tag{7.373}$$

or

$$\alpha\frac{dZ_k}{dk} + \frac{1}{4}kY_k = 0 \tag{7.374}$$

and so by using the definition of $\widetilde{Y}_k$ and Eq. (7.361) we obtain

$$2\alpha\frac{dZ_k}{dk} = -\frac{1}{2}\frac{k}{1-\alpha k}\widetilde{Y}_k = -\frac{d\widetilde{Y}_k}{dk} - \frac{1}{2} \tag{7.375}$$

It follows easily that

$$\frac{d}{dk}(2\alpha Z_k + \widetilde{Y}_k) = -\frac{1}{2} \tag{7.376}$$

direct integration yields

$$2\alpha Z_k + \widetilde{Y}_k = -\frac{1}{2}k + A \tag{7.377}$$

where $A$ is a constant of integration.
Now if $k = 0$ we have $Y_k = Y_0$, and $Z_k = Z_0 = 0$, and so

$$2\alpha Z_k = -\widetilde{Y}_k - \frac{1}{2}k + Y_0 \tag{7.378}$$

which is Eq. (7.156b)

## Appendix VII.XII. Justification of Eq. (7.158)

Near the saddle point we have that $x \approx 0$, hence $x' \equiv x\sqrt{\sigma} = 0$, and so

$$\varepsilon \approx -\xi \cos\varphi \tag{7.379}$$

Hence,

$$e^{-\varepsilon} \approx 1 - \xi\cos\varphi + \frac{1}{2}\xi^2\cos^2\varphi \tag{7.380}$$

and so

$$e^{-\varepsilon}\left[a(1-x^2)\frac{\partial\zeta}{\partial x'} + \frac{\partial\zeta}{\partial\varphi}\right] \approx \left(1 + \xi\cos\varphi + \frac{\xi^2}{2}\cos^2\varphi\right) \\ \times \left[a\left\{\frac{\partial\zeta_0}{\partial x'} + \xi(A'\sin\varphi + B'\cos\varphi) + \xi^2 C'\right\} + \xi(A\cos\varphi - B\sin\varphi)\right] \tag{7.381}$$

$$\approx \frac{a}{\sqrt{\pi}} + \left[\frac{a}{\sqrt{\pi}}\cos\varphi + aA'\sin\varphi + aB'\cos\varphi + A\cos\varphi - B\sin\varphi\right]\xi$$
$$+ \left[aC' + \frac{a}{2\sqrt{\pi}}\cos^2\varphi + \cos\varphi(aA'\sin\varphi + aB'\cos\varphi + A\cos\varphi - B\sin\varphi)\right]\xi^2 \tag{7.382}$$

where the functions $A, B, C$ and their derivatives are evaluated near $x' = 0$, and so, the integration of this function with respect to $\varphi$ may now be readily performed, yielding

$$\int_0^{2\pi} e^{-\varepsilon}\left[a(1-x^2)\frac{\partial\zeta}{\partial x'} + \frac{\partial\zeta}{\partial\varphi}\right]d\varphi = 2a\sqrt{\pi} + \xi^2\pi\left[A + aB' + 2aC' + \frac{a}{2\sqrt{\pi}}\right]_{x'=0} \tag{7.383}$$

Now,

$$\frac{1}{Z_1} + \frac{1}{Z_2} = \approx \frac{\sigma\sqrt{1-h^2}}{\pi}e^{-\sigma(1-h^2)} \tag{7.384}$$

so using Eq. (7.20) for $\Gamma$

$$\Gamma(h) \approx \frac{2\sigma\sqrt{1-h^2}}{\pi}\frac{\gamma kT}{\mu_0}e^{-\sigma(1+h^2)}\left\{2a\sqrt{\pi} + \xi^2\pi\left[A + aB' + 2aC' + \frac{a}{2\sqrt{\pi}}\right]_{x'=0}\right\} \tag{7.385}$$

Now, when $h = 0$ we have $\xi = 2\sigma h = 0$, and so

$$\Gamma(0) \approx \frac{2\sigma}{\pi}\frac{\gamma kT}{\mu_0}e^{-\sigma}2a\sqrt{\pi} \tag{7.386}$$

and hence

$$\frac{\Gamma(h)}{\Gamma(0)} = \sqrt{1-h^2}e^{-\sigma}\left\{1 + \frac{\xi^2}{4}\left[1 + \frac{2\sqrt{\pi}}{a}(A + aB' + 2aC')\right]_{x'=0}\right\} \tag{7.387}$$

$$\approx 1 + \frac{\xi^2}{4}\left[1 + \frac{2\sqrt{\pi}}{a}(A + aB' + 2aC')\right]_{x'=0} \tag{7.388}$$

if $h$ is small, which is Eq. (7.158)

## Appendix VII.XIII. Justification of Eq. (7.159b)

$$2\alpha C'|_{x'=0} = \mathrm{Re}2\alpha Z|_{x'=0} \tag{7.389}$$

$$= \mathrm{Re}\frac{1}{2\pi}\int_{-\infty}^{\infty} e^{-ikx'}e^{-k^2/4}2\alpha Z_k dk|_{x'=0} \tag{7.390}$$

$$= \mathrm{Re}\frac{1}{2\pi}\int_{-\infty}^{\infty} e^{-k^2/4}2\alpha Z_k dk \tag{7.391}$$

Now,

$$\frac{1}{2\pi}\int_{-\infty}^{\infty} e^{-ikx'}e^{-k^2/4}\widetilde{Y}_k dk = \frac{1}{2\pi}\int_{-\infty}^{\infty} e^{-ikx'}e^{-k^2/4}(Y_k - \alpha k Y_k)dk \tag{7.392}$$

Also

$$\frac{1}{2\pi}\int_{-\infty}^{\infty} e^{-ikx'}e^{-k^2/4}\alpha k Y_k dk = \frac{i}{2\pi}\int_{-\infty}^{\infty} e^{-ikx'}e^{-k^2/4}(-ik)\alpha Y_k dk \tag{7.393}$$

$$= i\alpha\frac{d}{dx'}\frac{1}{2\pi}\int_{-\infty}^{\infty} e^{-ikx'}e^{-k^2/4}Y_k dk \tag{7.394}$$

$$= i\alpha\frac{d}{dx'}Y(x') \tag{7.395}$$

and hence

$$\frac{1}{2\pi}\int_{-\infty}^{\infty} e^{-ikx'}e^{-k^2/4}\widetilde{Y}_k dk = Y(x') - i\alpha Y'(x') \tag{7.396}$$

$$= A + iB - i\alpha(A' + iB') \tag{7.397}$$

$$= A + \alpha B' + i(B - \alpha A') \tag{7.398}$$

So,

$$A + \alpha B'|_{x'=0} = \mathrm{Re}\frac{1}{2\pi}\int_{-\infty}^{\infty} e^{-k^2/4}\widetilde{Y}_k dk \tag{7.399}$$

And finally

$$F(\alpha) = 1 + \frac{2\pi}{\sqrt{\alpha}}(A + \alpha B' + 2\alpha C')|_{x'=0} \tag{7.400}$$

$$= 1 + \frac{2\pi}{\sqrt{\alpha}}\mathrm{Re}\frac{1}{2\pi}\int_{-\infty}^{\infty} e^{-ikx'}e^{-k^2/4}(\widetilde{Y}_k + 2\alpha Z_k)dk \tag{7.401}$$

which is Eq. (7.159b).

## Appendix VII.XIV. Justification of Eqs. (7.160)

Equation (7.156b) may be written in the form

$$2\alpha Z_k + \widetilde{Y}_k = \widetilde{Y}_0 - \frac{k}{2} \tag{7.402}$$

while Eq. (7.156a) yields, with $k = 0$

$$\widetilde{Y}_0 = \frac{1}{2}\int_0^{\frac{1}{\alpha}} \exp\left[\frac{\ln|1 - \alpha k'| + \alpha k'}{2\alpha^2}\right]dk' \tag{7.403}$$

By using the substitution

$$u = \alpha k' \tag{7.404}$$

we obtain

$$\widetilde{Y}_0 = \frac{1}{2\alpha}\int_0^1 \exp\left[\frac{\ln|1 - u| + u}{2\alpha^2}\right]du \tag{7.405}$$

Now Eq. (7.159b) may be manipulated as follows:

$$F(\alpha) = 1 + \frac{2\sqrt{\pi}}{\alpha}\mathrm{Re}\int_{-\infty}^{\infty}\frac{1}{2\pi}e^{-k^2/4}\left(\widetilde{Y}_0 - \frac{k}{2}\right)dk \tag{7.406}$$

$$= 1 + \frac{2\sqrt{\pi}}{\alpha}\left[\widetilde{Y}_0\int_{-\infty}^{\infty}\frac{1}{2\pi}e^{-k^2/4}dk - \int_{-\infty}^{\infty}\frac{1}{2\pi}\frac{k}{2}e^{-k^2/4}dk\right] \tag{7.407}$$

$$= 1 + \frac{2\sqrt{\pi}}{\alpha}\frac{1}{2\pi}\widetilde{Y}_0\sqrt{\pi} - \frac{2\sqrt{\pi}}{\alpha}\frac{1}{2\pi}\int_{-\infty}^{\infty}\frac{k}{2}e^{-k^2/4}dk \tag{7.408}$$

Now the integrand in the last term is odd, hence the last term is zero, and it follows, by using Eq. (7.405), that

$$F(\alpha) = 1 + \frac{2}{\alpha}\widetilde{Y}_0 \tag{7.409}$$

$$= 1 + \frac{1}{\alpha^2}\int_0^1 \exp\left[\frac{\ln|1-u| + u}{2\alpha^2}\right] du \tag{7.410}$$

Hence,

$$F(\alpha) - 1 = \frac{1}{\alpha^2}\int_0^1 \exp\left[\frac{\ln|1-u|}{2\alpha^2} + \frac{u}{2\alpha^2}\right] du \tag{7.411}$$

$$= \frac{1}{\alpha^2}\int_0^1 \exp\left(\frac{u}{2\alpha^2}\right) \exp(\ln|1-u|^{1/(2\alpha^2)})du \tag{7.412}$$

$$= \frac{1}{\alpha^2}\int_0^1 (1-u)^{1/(2\alpha^2)} \exp\left(\frac{u}{2\alpha^2}\right) du \tag{7.413}$$

Now, we define the incomplete gamma function [82]

$$\gamma(a, z) \equiv \int_0^z e^{-t} t^{a-1} dt \tag{7.414}$$

and so

$$\gamma\left(1 + \frac{1}{2\alpha^2}, \frac{1}{2\alpha^2}\right) = \int_0^{1/(2\alpha^2)} e^{-t} t^{1/(2\alpha^2)} dt \tag{7.415}$$

Now, if in Eq. (7.413), we use the substitution

$$t = 1 - u \tag{7.416}$$

we obtain

$$\alpha^2[F(\alpha) - 1] = \int_0^1 t^{1/(2\alpha^2)} \exp\left(\frac{1-t}{2\alpha^2}\right) dt \tag{7.417}$$

$$= \exp\left(\frac{1}{2\alpha^2}\right) \int_0^1 t^{1/(2\alpha^2)} \exp\left(-\frac{t}{2\alpha^2}\right) dt \tag{7.418}$$

and now by using the substitution

$$v = \frac{t}{2\alpha^2} \tag{7.419}$$

we obtain

$$\alpha^2[F(\alpha) - 1] = 2\alpha^2(2\alpha^2 e)^{1/(2\alpha^2)} \int_0^{1/(2\alpha^2)} v^{1/(2\alpha^2)} \exp(-v) dv \tag{7.420}$$

$$= 2\alpha^2(2\alpha^2 e)^{1/(2\alpha^2)} \gamma\left(1 + \frac{1}{2\alpha^2}, \frac{1}{2\alpha^2}\right) \tag{7.421}$$

and Eq. (7.160b) follows.

### Appendix VII.XV. Justification of Eqs. (7.161) and (7.162)

$$F(\alpha) = 1 + \frac{1}{\alpha^2} \int_0^1 \exp\left[\frac{\ln(1-u) + u}{2\alpha^2}\right] du \tag{7.422}$$

Now if $\alpha$ is large

$$\int_0^1 \exp\left[\frac{\ln(1-u) + u}{2\alpha^2}\right] du \approx \int_0^1 \left[1 + \frac{\ln(1-u) + u}{2\alpha^2}\right] du \tag{7.423}$$

$$\approx u|_0^1 + \frac{1}{2\alpha^2} \int_0^1 \ln(1-u) du + \frac{u^2}{4\alpha^2}\Big|_0^1 \tag{7.424}$$

The integral on the right-hand side may be integrated by parts, and using the fact that

$$\lim_{x \to 0} x \ln x = 0 \tag{7.425}$$

we obtain a value of $-1$ for the integral itself. Hence, for large $\alpha$

$$F(\alpha) \approx 1 + \frac{1}{\alpha^2}\left[1 - \frac{1}{4\alpha^2}\right] \tag{7.426}$$

and Eq. (7.161) follows easily.

Now, to derive the asymptotic expression Eq. (7.162) for $F(\alpha)$ in the limit of small $\alpha$, we refer to Eq. (7.160a).

$$F(\alpha) = 1 + \frac{1}{\alpha^2}\int_0^1 \exp\left[\frac{\ln(1-u)+u}{2\alpha^2}\right]du \tag{7.427}$$

now for $u \in [0, 1)$

$$\ln(1-u) = -u - \frac{u^2}{2} - \frac{u^3}{3} - \cdots \tag{7.428}$$

and so

$$\ln(1-u) + u = -\left[\frac{u^2}{2} + \frac{u^3}{3} + \cdots\right] \tag{7.429}$$

and on writing

$$\chi = \frac{1}{2\alpha^2} \tag{7.430}$$

we obtain

$$\exp\left[\frac{\ln(1-u)+u}{2\alpha^2}\right] \approx \exp\left[-\chi\left(\frac{u^2}{2} + \frac{u^3}{3} + \frac{u^4}{4} + \frac{u^5}{5}\right)\right] \tag{7.431}$$

$$= \exp\left(-\frac{\chi u^2}{2}\right)\exp\left(-\frac{\chi u^3}{3}\right)\exp\left(-\frac{\chi u^4}{4}\right)\exp\left(-\frac{\chi u^5}{5}\right) \tag{7.432}$$

and, keeping only terms up to $u^6$ in the second and subsequent exponentials, we find

$$\exp\left[\frac{\ln(1-u)+u}{2\alpha^2}\right] \approx \exp\left(-\frac{\chi u^2}{2}\right)\left[1 - \frac{\chi u^3}{3} + \frac{\chi^2 u^6}{18}\right]\left[1 - \frac{\chi u^4}{4}\right]\left[1 - \frac{\chi u^5}{5}\right] \tag{7.433}$$

$$\approx \exp\left(-\frac{\chi u^2}{2}\right)\left[1 - \frac{\chi u^3}{3} - \frac{\chi u^4}{4} - \frac{\chi u^5}{5} + \frac{\chi^2 u^6}{18}\right] \tag{7.434}$$

Now,

$$\begin{aligned}\int_0^1 \exp\left[\frac{\ln(1-u)+u}{2\alpha^2}\right] du \approx& \int_0^\infty \exp\left(-\frac{\chi u^2}{2}\right) du - \frac{\chi}{3}\int_0^\infty u^3 \exp\left(-\frac{\chi u^2}{2}\right) du \\ &- \frac{\chi}{4}\int_0^\infty u^4 \exp\left(-\frac{\chi u^2}{2}\right) du - \frac{\chi}{5}\int_0^\infty u^5 \exp\left(-\frac{\chi u^2}{2}\right) du \\ &- \frac{\chi}{6}\int_0^\infty u^6 \exp\left(-\frac{\chi u^2}{2}\right) du + \frac{\chi^2}{18}\int_0^\infty u^6 \exp\left(-\frac{\chi u^2}{2}\right) du\end{aligned} \tag{7.435}$$

The range of integration has been extended to $\infty$ since the integrands converge rapidly to zero beyond $u = 1$.

Now,

$$\int_0^\infty \exp\left(-\frac{\chi u^2}{2}\right) du = \sqrt{\frac{2}{\chi}}\frac{\sqrt{\pi}}{2} = 2\alpha\frac{\sqrt{\pi}}{2} = \alpha\sqrt{\pi} \tag{7.436}$$

To evaluate the other integrals, we integrate by parts as follows [we evaluate the second integral on the right-hand side of Eq. (7.435), the other integrals are evaluated similarly]:
Let,

$$I_2 = \int_0^\infty u^3 \exp\left(-\frac{\chi u^2}{2}\right) du \tag{7.437}$$

Let,

$$v = u^2 \qquad \text{and} \qquad dw = u \exp\left(-\frac{\chi u^2}{2}\right) du \tag{7.438}$$

then

$$dv = 2u du \qquad \text{and} \qquad w = -\frac{1}{\chi}\exp\left(-\frac{\chi u^2}{2}\right) \tag{7.439}$$

and so

$$I_2 = -\frac{u^2}{\chi}\exp\left(-\frac{\chi u^2}{2}\right)\Bigg|_0^\infty + \frac{2}{\chi}\int_0^\infty u \exp\left(-\frac{\chi u^2}{2}\right) du \tag{7.440}$$

The first term on the right-hand side is zero and the second term may be readily integrated to yield

$$I_2 = -\frac{2}{\chi^2} \exp\left(-\frac{\chi u^2}{2}\right)\Big|_0^\infty \tag{7.441}$$

$$= \frac{2}{\chi^2} \tag{7.442}$$

$$= 8\alpha^4 \tag{7.443}$$

and hence

$$-\frac{\chi}{3} I_3 = -\frac{4}{3}\alpha^2 \tag{7.444}$$

similarly, applying this technique to the other terms, repeatedly if necessary, yields

$$-\frac{\chi}{4}\int_0^\infty u^4 \exp\left(-\frac{\chi u^2}{2}\right) du = -\frac{3}{2}\alpha^3\sqrt{\pi} \tag{7.445}$$

$$-\frac{\chi}{5}\int_0^\infty u^5 \exp\left(-\frac{\chi u^2}{2}\right) du = -\frac{32}{5}\alpha^2 \tag{7.446}$$

$$-\frac{\chi}{6}\int_0^\infty u^6 \exp\left(-\frac{\chi u^2}{2}\right) du = 20\alpha^6\sqrt{\pi} \tag{7.447}$$

$$\frac{\chi^2}{18}\int_0^\infty u^6 \exp\left(-\frac{\chi u^2}{2}\right) du = \frac{10}{6}\alpha^3\sqrt{\pi} \tag{7.448}$$

and so

$$\begin{aligned} F(\alpha) &= 1 + \frac{1}{\alpha^2}\int_0^1 \exp\left[\frac{\ln(1-u)+u}{2\alpha^2}\right] du \\ &\approx 1 + \frac{\sqrt{\pi}}{\alpha} - \frac{4}{3} - \frac{3}{2}\alpha\sqrt{\pi} + \frac{10}{6}\alpha\sqrt{\pi} + \cdots \end{aligned} \tag{7.449}$$

$$= \frac{\sqrt{\pi}}{\alpha} - \frac{1}{3} + \frac{1}{6}\alpha\sqrt{\pi} + \cdots \tag{7.450}$$

which is Eq. (7.162).

# VIII. CONNECTION OF THE KRAMERS THEORY WITH THE DYNAMIC RESPONSE ILLUSTRATED BY THE CALCULATION OF ESCAPE TIMES FOR RIGID BROWNIAN ROTATORS IN A BISTABLE POTENTIAL FROM THE TIME EVOLUTION OF THE GREEN FUNCTION

## A. Introduction

We have seen throughout this chapter that the dynamics of a rotator subjected to thermal agitation in a potential having two stable stationary points separated by a potential barrier is of central importance in the theory of dielectric and magnetic relaxation, and that in particular (Sections II, V) one of the parameters used to characterize the dynamics of the system is the escape time, $T_e$, or the time for the rotator to reverse its direction of rotation, that is, to traverse the potential barrier between the two stable stationary states. This time has usually been calculated by finding the inverse of the smallest nonvanishing eigenvalue of the associated Fokker–Planck equation for the probability density of the orientations of the rotator on the unit sphere. Furthermore, for barrier heights significantly greater than the mean thermal energy $kT$ of the rotator we have seen that asymptotic expressions for the escape time $T_e$ may be obtained from the Kramers theory of escape of particles over potential barriers (Sections III, IV, VII). These expressions have lately been applied with considerable success in the analysis of experimental observations of dielectric and magnetic relaxation [83, 85].

We have also seen in Sections III and V that yet another approach to the calculation of the escape time is to determine, from the stationary solution of the Fokker–Planck equation (that is the solution describing the fact that a Maxwell–Boltzmann distribution has been established in each of the two wells with rare transitions between the wells due to barrier crossing) the time for a rotator to reach the summit of the barrier from one of the stable stationary points assuming that the summit is an absorbing (i.e., the rotators are no longer counted if they have crossed the barrier) barrier. This time is the mean first passage time say $T_m$ to the barrier summit. The escape time from a particular well in the high barrier limit is then twice the mean first passage time as the rotator has a 50:50 chance, having reached the summit, to either fall right back into the well or to fall down into the next well.

The advantage of formulating the escape time problem in this way is that for systems characterized by a single state variable say the colatitude $\theta$, the escape time for an arbitrary bistable potential $U(\theta)$ may be obtained in integral form from the FPE by quadratures. The disadvantage of such a method on the other hand is that *it may not be extended to systems described by more than one state variable as the FPE cannot then be integrated by quadratures*. Furthermore, it is not entirely obvious how the escape time problem has its origin in the time-dependent solution of the FPE that describes the evolution of the system from an arbitrary nonequilibrium initial state.

In view of these difficulties, an alternative method of calculation of the escape time based on the time-dependent solution that may be easily extended to multidegree-of-freedom systems was proposed by Coffey [58] by expanding the solution of the FPE in an appropriate set of elementary functions in the space variable. Such an approach to set it in a historical context is in effect an extension of the Floquet–Hill method of expansion in Fourier series originally used to solve Hill's periodic differential equation arising in the motion of the lunar perigee, without the introduction of special functions in this case the Mathieu functions [88] (such an approach is a particular case of the Sturm–Liouville problem [89]). The transitions between the wells have their origin in an additional point source (delta function distribution) in one of the wells so as to give rise to a nonzero probability current. The Floquet method as applied to the FPE leads to a set of differential recurrence relations in the time variable describing the time evolution of the system from an initial very cold state (delta function distribution of orientations in a well). This is the Green function (transition probability) of the FPE, whence the escape time may be calculated from (effectively the after effect solution of) the FPE by taking the zero-frequency limit of the Laplace transform of the set of differential recurrence relations. In the present context, the results also constitute an analytic proof of the Kramers theory for the particular case of a bistable potential.

We remark that as far as single degree of freedom systems are concerned, our method often [5] allows one to express the zero-frequency limit of the solution of the set of differential recurrence relations as a series of special functions especially if the recurrence relations are three-term ones. On the other hand, for multidegree of freedom systems the solutions may always be obtained [5] as a sum of products of matrix continued fractions. The representation of the calculation of the escape time $T_e$ as a set of recurrence relations also arises naturally from the Langevin equation underlying the problem by averaging that equation over its realizations at a time $t + t_1$ given that at time $t$, the state variables had sharp values. Here $t_1$ is a time which, in accordance with the basic premises (Sections I, II) underlying the theory of the Brownian motion, is so small that both the external torque and the state variables may not alter, yet $t_1$ is so large that the rotator will have undergone very many collisions so that the integrals of the collision torque evaluated at times $t$ and $t + t_1$ are independent of each other (the Wiener process) (see Section II B).

Since the hierarchy of recurrence relations arises naturally by averaging the Langevin equation over its set of realizations, it follows that the escape time may be directly calculated from the Langevin equation without reference to the FPE. The averaging of the Langevin equation given sharp values at time $t$ in effect amounts to finding the Fourier series (here a Fourier–Laplace series) expansion of the transition probability (Green's function of the FPE).

In this section, we shall illustrate the application of the Floquet–Hill method just described, by considering the simplest 3D problem (see Section III), which is ubiquitous in the theory of dielectric [5] and magnetic relaxation, namely, a rotator in the bistable potential

$$Kv \sin^2\theta = Kv(1 - z^2) \tag{8.1}$$

We shall regard this potential in a physical sense as that of a single domain ferromagnetic particle characterized by internal anisotropy $K$. Here $\theta$ and $\varphi$ are the usual polar angles specifying the orientation of the magnetization vector $\mathbf{M}$. We desire, in the case of Eq. (8.1), just as in Section IIIC, the time taken for a rotator situated at a latitude near a minimum of the potential to reach the summit of the potential at $\theta = \pi/2$. The escape time or time to cross the summit is, in the high barrier limit, twice this time. The minimum point in this problem refers to $0_+$ and $\pi_-$ in order to ensure that they remain interior points of the sphere. A more general case of axially symmetric potentials than Eq. (8.1) is Eq. (8.1) with an applied field $\mathbf{H}_0$ parallel to the $z$ axis so that

$$vU = Kv \sin^2\theta - vM_s H_0 \cos\theta \tag{8.2}$$

where $vM_s$ is the dipole moment.

### B. Integral Expression for the Escape Time in Terms of the Green Function

We have seen Section III D that the calculation of escape times from the stationary Fokker–Planck equation is usually achieved by constructing the adjoint FPE or backward Kolmogorov equation, and then the mean first passage time $T_{\text{MFPT}} = T_m$ satisfies the equation:

$$L_{\text{FP}}^{+} T_m = -1 \tag{8.3}$$

with appropriate boundary conditions—where $L_{\text{FP}}^{+}$ is the adjoint Fokker–Planck operator and is written in terms of the source set of state variables (source coordinates).

The disadvantage of using the adjoint equation to calculate the escape time $T_e$ as twice the MFPT is that it may be very difficult to integrate this equation and to apply the boundary conditions. An attractive alternative method of calculation of $T_m$ from the stationary solution of the FPE, which relies on the Kramers flux over barrier method has been given by Hänggi et al. [1] (see section III D), which corroborates the results of the present section (see Section III D).

The method of solution for $T_e$ to be described in this section does not rely on either of the two methods based on the stationary solution that we have just mentioned. These two methods postulate zero escape probability density at the summit of the wells and calculate the escape time as twice the mean first passage to the summit assuming that stationary conditions prevail, namely, a Maxwell–Boltzmann distribution in the wells and occasional transitions between the wells arising from an additional point source in each of the two wells so as to give rise to a steady probability current. The method of solution proposed here commences by postulating an initial delta function distribution of orientations in one of the wells, so that the entire time evolution of the system may be studied, and evaluates the escape time by integrating the zero-frequency limit of the Laplace transform of the resulting time-dependent distribution (Green's function) over a well.

The method may be briefly described as follows. The Green function of the Fokker–Planck equation for the time evolution of the probability density due to a delta function source at $\mathbf{x}'$ in a state space $\mathbf{X}$ may be written as

$$\frac{\partial}{\partial t} g(\mathbf{x}, t|\mathbf{x}', 0) = L_{\mathrm{FP}} g \tag{8.4}$$

The time-dependent part (after effect solution for delta function initial conditions) of the solution of Eq. (8.4), that is, disregarding the equilibrium Maxwell–Boltzmann solution to which $g$ ultimately tends (if that solution exists), is denoted by $g_t$ and satisfies

$$\frac{\partial}{\partial t} g_t(\mathbf{x}, t|\mathbf{x}', 0) = L_{\mathrm{FP}} g_t \tag{8.5}$$

The Laplace transform of Eq. (8.5) is, with

$$\mathcal{L}\{g_t\} = \tilde{g}_s(\mathbf{x}, s|\mathbf{x}', t = 0) \tag{8.6}$$

$$s\mathcal{L}\{g_t\} - \delta(\mathbf{x} - \mathbf{x}') = L_{\mathrm{FP}} \mathscr{L}\{g_t\} \tag{8.7}$$

We may now take the zero-frequency limit of this after effect solution for delta function initial conditions to get, using the final value theorem [90] of Laplace transformation (see below).

$$-\delta(\mathbf{x} - \mathbf{x}') = L_{\mathrm{FP}} \tilde{g}_0(\mathbf{x}, 0|\mathbf{x}', t = 0) \tag{8.8}$$

Our method of calculating escape times is based on the fact that Eq. (8.8), which is the zero-frequency limit of the time-dependent Green function resulting from a delta function situated at $\mathbf{x}'$ in a domain $\Omega$ of $\mathbf{X}$ is formally the same [58] as that of Risken [25, 26] (his Eqs. (8.8) and (8.9)) for the first passage time distribution $g(\mathbf{x}, \mathbf{x}')$ if a *unit rate* of probability is injected into the system at $\mathbf{x}'$. The boundary conditions for $\tilde{g}_0$ and $\tilde{g}(\mathbf{x}, \mathbf{x}')$ are, however, different as in the present solution, which regards the escape time as the relaxation time, and it is unnecessary to assume zero probability density at the boundaries. This assumption is required if Risken's method [25, 26] is applied.

The time, $T_s$, to reach the boundaries of a domain $\Omega$ of $\mathbf{X}$ specified by $\mathbf{x}_1$ and $\mathbf{x}_2$ may now be calculated (analogously to the MFPT) by simply integrating $\tilde{g}_0$ over $\mathbf{x}_1$ and $\mathbf{x}_2$ to get

$$T_s = \int_{\mathbf{x}_1}^{\mathbf{x}_2} \tilde{g}_0(\mathbf{x}, 0|\mathbf{x}', 0)d\mathbf{x} \tag{8.9}$$

In the high barrier limit, referring to escape from a bistable potential, the escape time $T_e$ is then

$$T_e = 2T_s \tag{8.10}$$

as the particle has an equal chance of either returning to the bottom of its source well or falling down into the next well.

The procedure we have just outlined for the calculation of the escape time (for an exact comparison, see Section VII.G below), is entirely analogous to the calculation of the characteristic time of the probability evolution (the integral of the configuration space probability density function with respect to the position coordinate for a particle undergoing translational diffusion in a potential; a concept originally used by Malakhov and Pankratov [104]. The introduction of the characteristic time of the probability evolution enabled them to obtain exact solutions of the Kramers one-dimensional translational escape rate problem for piecewise parabolic potentials, when formulated as the solution of a Sturm-Liouville equation. Their Sturm-Liouville method should be compared with that of Brinkman [25, 26, 46, 106] involving the recurrence relations for the decay functions favored here and which arises naturally by averaging [5], the Langevin equation over its realizations.

### C. Fokker–Planck Equation for the Green Function as the Zero Frequency Limit of the Time-Dependent Fokker–Planck Equation with Delta Function Initial Distribution of Orientations

The Fokker–Planck equation for the probability density $W(\theta, \varphi, t)$ of the orientations of a rigid rotator on the unit sphere for an axially symmetric potential

$vU(\theta)$ is, as we have already seen in Section III (cf. Eq. 3.16),

$$2\tau\frac{\partial W}{\partial t} = \frac{1}{\sin\theta}\frac{\partial}{\partial\theta}\left[\sin\theta\left(\frac{\partial W}{\partial\theta} + \frac{v}{kT}\frac{\partial U}{\partial\theta}W\right)\right] + \frac{1}{\sin\theta}\frac{\partial}{\partial\varphi}\left\{\frac{v}{kT}\frac{1}{a}\frac{\partial U}{\partial\varphi}W + \frac{1}{\sin\theta}\frac{\partial W}{\partial\varphi}\right\} \tag{8.11}$$

In Eq. (8.11), $\tau$ is the characteristic free diffusion relaxation time. Thus, in magnetic relaxation $\tau$ is the Néel diffusion relaxation time, while in dielectric relaxation $\tau$ is the Debye relaxation time [19], the last term in $1/a$ is the gyromagnetic term that arises only in magnetic relaxation, and $a$ is the usual dimensionless damping constant. Here, we shall be concerned for the most part with axially symmetric potentials of the crystalline anisotropy and with the longitudinal relaxation so that Eq. (8.11) becomes, disregarding the dependence of $W$ and $U$ on $\varphi$,

$$2\tau\frac{\partial W}{\partial t} = \frac{1}{\sin\theta}\frac{\partial}{\partial\theta}\left[\sin\theta\left(\frac{\partial W}{\partial\theta} + \frac{v}{kT}\frac{\partial U}{\partial\theta}W\right)\right] \tag{8.12}$$

Equation (8.12) may now be used to determine the Laplace transform of the Fokker–Planck equation in the zero-frequency limit above as we shall describe below, hence the escape time or greatest relaxation time may be calculated from Eqs. (8.9) and (8.10).

We first remark that Eq. (8.12) (cf. Eq. (2.55) of Section II) admits of two types of solutions (1) time-independent solutions with $\dot{W} = 0$, which describe the steady state (equilibrium) that is attained a very long time after the application of an initial stimulus to the system and (2) the time-dependent solution governing the relaxation of the system to the steady-state.

The steady-state or thermal equilibrium state is characterized by a Maxwell–Boltzmann distribution of orientations within the wells and occasional transitions (reversals of directions of rotation) of the rotators over the potential barrier generated by $U(\theta)$. The nature of the solutions may be understood by recalling that Eq. (8.11) is a continuity equation, namely,

$$\frac{\partial W}{\partial t} + \mathrm{div}\mathbf{J} = 0 \tag{8.13}$$

for the number density, $W$, of representative points [i.e., the tips of the rotators with coordinates $(1, \theta, \varphi)$ on the unit sphere where $\mathbf{J}$ is the probability current density of such representative points]. The time-independent solutions are then characterized by (see Section II)

$$\dot{W} = -\text{div}\mathbf{J} = 0 \tag{8.14}$$

so that

$$\mathbf{J} = \mathbf{0} \tag{8.15}$$

and

$$\mathbf{J} = \text{constant} \tag{8.16}$$

are possible time-independent solutions. The solution

$$\mathbf{J} = \mathbf{0} \tag{8.17}$$

corresponding to a value of $W$, which we shall label $W_{\text{eq}}$, yields the Maxwell–Boltzmann (thermal) distribution for the steady-state distribution of the rotators in the wells of the potential $U(\theta)$. The solution $\mathbf{J}$ = constant on the other hand describes the reversal of the rotators, that is, barrier crossing, over the potential barriers, the distribution of the rotators in the wells having attained its thermal equilibrium value $W_{\text{eq}}$.

The time-independent solution $W_{\text{eq}}$ with zero probability current obeys

$$\frac{1}{\sin\theta}\frac{\partial}{\partial\theta}\left[\sin\theta\left(\frac{\partial W_{\text{eq}}}{\partial\theta} + \beta\frac{\partial U}{\partial\theta}W_{\text{eq}}\right)\right] = 0 \tag{8.18}$$

$$\beta = \frac{v}{kT}$$

and is the Maxwell-Boltzmann distribution:

$$\frac{e^{-\beta U(\theta)}\sin\theta d\theta d\varphi}{2\pi\int_0^\pi e^{-\beta U(\theta)}\sin\theta d\theta} \tag{8.19}$$

The time-dependent solution $W_t$ (into which the second time-independent solution with nonzero current density describing barrier crossing is absorbed) obeys

$$2\tau\frac{\partial W_t}{\partial t} = \frac{1}{\sin\theta}\frac{\partial}{\partial\theta}\left[\sin\theta\left(\frac{\partial W_t}{\partial\theta} + \beta\frac{\partial U}{\partial\theta}W_t\right)\right] \tag{8.20}$$

Here, $t$ refers to the time elapsed since the application of an initial external stimulus at $t = 0$. The complete solution is then

$$W = W_t + W_{\text{eq}} \tag{8.21}$$

The quantity of interest to us is then

$$W - W_{\text{eq}} = W_t = g_t \tag{8.22}$$

where $g_t$ denotes the time-dependent distribution consequent on a delta function (an extreme nonequilibrium state) initial distribution of orientations that is the Green's function or the transition probability

$$W_t(\theta, t|\theta_0, 0)$$

Let us now take the Laplace transform of Eq. (8.20) with $W_t = g_t$. We have with,

$$\tilde{g}_s = \tilde{g}_s(\theta, s) = \mathcal{L}\{g_t(\theta, t)\} = \int_0^\infty g_t e^{-st} dt \tag{8.23}$$

$$2\tau[s\tilde{g}_s - g_0(\theta, 0)] = \frac{1}{\sin\theta}\frac{\partial}{\partial\theta}\left[\sin\theta\left(\frac{\partial\tilde{g}_s}{\partial\theta} + \beta\frac{\partial U}{\partial\theta}\tilde{g}_s\right)\right] \tag{8.24}$$

where

$$g_0 = \frac{\delta(\theta - \theta_0)}{2\pi\sin\theta} \tag{8.25}$$

Equation (8.25) follows because $g_0$ is a probability density function, hence

$$\int_0^{2\pi} d\varphi \int_0^{\pi} \frac{A\delta(\theta - \theta_0)\sin\theta\, d\theta}{2\pi\sin\theta} = 1 \tag{8.26}$$

$$A = \frac{1}{2\pi}$$

We now note the final value theorem of Laplace transformation [90], namely,

$$\lim_{s\to 0} s\tilde{f}(s) = \lim_{t\to\infty} f(t) = f(\infty) \tag{8.27}$$

$$f \in \mathcal{L}^2$$

and take the zero-frequency limit of Eq. (8.24) to get

$$\frac{\delta(\theta - \theta_0)}{2\pi \sin\theta} = \frac{1}{\sin\theta}\frac{\partial}{\partial\theta}\left[\sin\theta\left(\frac{\partial\tilde{g}_0(\theta,0)}{\partial\theta} + \beta\frac{\partial U}{\partial\theta}\tilde{g}_0(\theta,0)\right)\right] \tag{8.28}$$

Equation (8.28) is precisely Eq. (8.8), which governs the behavior of the distribution in the zero-frequency limit resulting from a delta function source at $\theta_0 = 0_+$, specialized to the rotational problem at hand. The methods that have been previously applied [5] to obtain the solution of the time-dependent equation may then be directly applied to Eq. (8.28), which describes the zero-frequency behavior of the system by taking the zero-frequency limit of the Laplace transform of the time-dependent solution, which, in turn, will yield the Kramers result in the high barrier limit. Thus, the results of his escape rate theory emerge (as is essential) as a limiting case of the dynamic response. We proceed as outlined below.

## D. Recurrence Relations in the Zero-Frequency Limit

We shall assume that the entire distribution function in the interval $(0, \pi)$ is of the form

$$W(\theta, t) = W_t + W_{\text{eq}} = \sum_{n=0}^{\infty} a_n(t)P_n(\cos\theta) \tag{8.29}$$

where the $P_n$ are the Legendre polynomials [91].

We also note that

$$\overline{P_n(\cos\theta)}(t) = \frac{\int_0^{2\pi}\int_0^{\pi} P_n(\cos\theta)W\sin\theta d\theta d\varphi}{\int_0^{2\pi}\int_0^{\pi} W\sin\vartheta d\theta d\varphi} = \frac{1}{2n+1}\frac{a_n(t)}{a_0} \tag{8.30}$$

Furthermore, since $W$ is a probability density function we again must have

$$\int_0^{2\pi}\int_0^{\pi} W\sin\vartheta d\theta d\varphi = 1 \tag{8.31}$$

so that

$$a_0 = \frac{1}{4\pi} \tag{8.32}$$

The time-dependent distribution function is then

$$W_t = \sum_{n=0}^{\infty} [a_n(t) - a_n(\infty)] P_n(\cos\theta) \tag{8.33}$$

Now we shall suppose, in order to consider a definite problem, that $U$ is given by Eq. (8.2), thus we find [5] that the decay (after effect) functions $f_n(t)$ defined by

$$f_n(t) = \frac{a_n(t) - a_n(\infty)}{(2n+1)a_0} = \overline{P_n(\cos\theta)}(t) - \overline{P_n(\cos\theta)}(\infty) \tag{8.34}$$

obey the set of differential recurrence relations

$$\begin{aligned} \frac{2\tau}{n(n+1)}\dot{f}_n(t) + \left[1 - \frac{2\sigma}{(2n-1)(2n+3)}\right] f_n(t) \\ = \frac{\xi}{2n+1}[f_{n-1}(t) - f_{n+1}(t)] + \frac{2\sigma(n-1)}{(2n+1)(2n-1)} f_{n-2}(t) \\ - \frac{2\sigma(n+2)}{(2n+1)(2n+3)} f_{n+2}(t) \end{aligned} \tag{8.35}$$

where

$$\sigma = Kv/kT \qquad \xi = vM_sH_0/kT$$

are the barrier height and external field parameters, respectively . The Green function $g_t$ of the time-dependent distribution may then be determined by solving the set of recurrence relations Eq. (8.35), postulating a sharp set of initial values for the $f_n(t)$.

In order to proceed, we remark that the Laplace transform of Eq. (8.35) is

$$\begin{aligned} \frac{2\tau}{n(n+1)}[s\tilde{f}_n(s) - f_n(0)] + \left[1 - \frac{2\sigma}{(2n-1)(2n+3)}\right]\tilde{f}_n(s) \\ = \frac{\xi}{2n+1}[\tilde{f}_{n-1}(s) - \tilde{f}_{n+1}(s)] \\ + \frac{2\sigma(n-1)}{(2n+1)(2n-1)}\tilde{f}_{n-2}(s) - \frac{2\sigma(n+2)}{(2n+1)(2n+3)}\tilde{f}_{n+2}(s) \end{aligned} \tag{8.36}$$

which becomes in the zero-frequency limit since $f_n(\infty) = 0$ by dint of Eq. (8.34), noting the final value theorem above,

$$
\begin{aligned}
-\frac{2\tau}{n(n+1)} f_n(0) + \left[1 - \frac{2\sigma}{(2n-1)(2n+3)}\right]\tilde{f}_n(0) \\
= \frac{\xi}{2n+1}[\tilde{f}_{n-1}(0) - \tilde{f}_{n+1}(0)] + \frac{2\sigma(n-1)}{(2n+1)(2n-1)}\tilde{f}_{n-2}(0) \\
- \frac{2\sigma(n+2)}{(2n+1)(2n+3)}\tilde{f}_{n+2}(0)
\end{aligned}
\tag{8.37}
$$

In order to determine the initial values $f_n(0)$, we write

$$\frac{\delta(\theta - \theta_0)}{2\pi \sin\theta} = \sum_{n=0}^{\infty} [a_n(0) - a_n(\infty)] P_n(\cos\theta) \tag{8.38}$$

so that by orthogonality

$$\frac{P_n(\cos\theta_0)}{2\pi} = \frac{2}{2n+1}[a_n(0) - a_n(\infty)] \tag{8.39}$$

hence by Eqs. (8.32) and (8.34)

$$\operatorname*{Lim}_{t\to\infty} f_n(t) = f_n(0) = P_n(\cos\theta_0) \tag{8.40}$$

and we note that if $\theta_0$ is in the vicinity of the minimum at $\theta = 0$ so that $\theta_0 = 0_+$. The $0_+$ ensures that $\theta_0$ remains an interior point of the domain $(0, \pi)$, which is the case of greatest interest, we have

$$f_n(0) = 1 \tag{8.41}$$

The problem of calculating the Green function $\tilde{g}_0(\theta, 0)$ in Eq. (8.28) is thus reduced to the solution of Eq. (8.37) for $\tilde{f}_n(0)$ with the initial condition, Eq. (8.41). The time $T_s$, to reach the summit $\theta_m$ from a point in the domain $(0, \theta_m)$, may then be calculated simply by integrating $\tilde{g}_o(\theta, 0)$ between 0 and $\theta_m$ in accordance with Eq. (8.9), where $\tilde{g}_0$ is rendered by the Laplace series:

$$\tilde{g}_0(\theta, 0) = \sum_{n=1}^{\infty} \frac{(2n+1)}{4\pi} \tilde{f}_n(0) P_n(\cos\theta) \tag{8.42}$$

The summation commencing at $n = 1$ since $\tilde{f}_0(0)$ vanishes again by dint of Eq. (8.34), because

$$\mathcal{L}\{f_0(t)\} = \mathcal{L}\{a_0(t) - a_0(\infty)\} = 0 \tag{8.43}$$

since $a_0$ is a constant.

### E. Series Expression for the Time to Reach the Summit

Having solved the recurrence relation Eq. (8.37) of the $\tilde{f}_n(0)$ (which in general must be carried out using matrix continued fractions as detailed in Coffey et al. [5] and Kalmykov and Titov [67] with the exception of the special case $\xi = 0$, which will be given as a simple example of the present method) $T_s$ may be determined in series form by integration of Eq. (8.42), as follows.

From Eq. (8.9) we have for the time to reach the summit from the minimum at $0_+$

$$T_s = \int_0^{2\pi} \int_0^{\theta_m} \tilde{g}_0(\theta, 0) \sin\theta d\theta d\phi \tag{8.44}$$

Thus, $T_s$ is given by ($z = \cos\theta$)

$$T_s = \int_{z_m}^{1} \sum_{n=1}^{\infty} \frac{(2n+1)}{2} \tilde{f}_n(0) P_n(z) dz \tag{8.45}$$

Equation (8.45) may be radically simplified in the particular case where the maximum is at $\theta_m = \pi/2$, which corresponds to the symmetric bistable potential of Eq. (8.1), that is, we set the external field $H_0 = 0$ in Eq. (8.2). We then have, utilizing the result [91]

$$\begin{aligned} \int_0^1 P_n(z)dz &= 1, \qquad n = 0 \\ &= 0, \qquad n = 2, 4, 6, \ldots \\ &= \frac{(-1)^{n-1}}{2^n((n+1)/2)!((n-1)/2)!} \end{aligned} \tag{8.46}$$

for odd $n$ so that

$$\int_0^1 P_{2n+1}(z)dz = \frac{(-1)^n (2n)!}{2^{2n+1} n!(n+1)!}$$

Thus $T_s$ for transitions from zero to $\pi/2$ is given by the series

$$T_s = \sum_{n=0}^{\infty} \frac{(-1)^n (2n)!((4n+3)/2)\tilde{f}_{2n+1}(0)}{2^{2n+1}(n+1)!n!} \tag{8.47}$$

Equation (8.47), to which only the odd $\tilde{f}_n$ contribute, will now be used in conjunction with the solution of the recurrence relation of Eq. (8.37) to obtain $T_e$ in series form for the potential of Eq. (8.1).

### F. Explicit Expression for the Escape Time for Simple Uniaxial Anisotropy

The solution of Eq. (8.37) with $\xi = 0$ for arbitrary initial conditions has been given in Coffey et al. [5, 54] and is

$$\begin{aligned}\frac{\tilde{f}_1(s)}{f_1(0)} &= \frac{\tau}{\tau s + 1 - \frac{2}{5}\sigma + \frac{2}{3}\sigma \widetilde{S}_3(s)} \\ &\times \left[1 + \frac{4}{3}\sum_{n=1}^{\infty}(-1)^n \frac{f_{2n+1}(0)}{f_1(0)} \frac{(n+\frac{3}{4})\Gamma(n+\frac{1}{2})}{\Gamma(n+2)\Gamma(\frac{1}{2})} \prod_{k=1}^{n} \widetilde{S}_{2k+1}(s)\right]\end{aligned} \tag{8.48}$$

where $\widetilde{S}_n(s)$ is the continued fraction $(n > 1)$($\Gamma$ is the gamma function)

$$\widetilde{S}_n(s) = \frac{\frac{2\sigma(n-1)}{4n^2-1}}{\frac{2\tau s}{n(n+1)} + 1 - \frac{2\sigma}{(2n-1)(2n+3)} + \frac{2\sigma(n+2)}{(2n+1)(2n+3)} \widetilde{S}_{n+2}(s)} \tag{8.49}$$

and which, when evaluated at $s = 0$, has the explicit form

$$\widetilde{S}_n(0) = \frac{2(n-1)\sigma}{(4n^2-1)} \frac{M((n+1)/2, n+\frac{3}{2}, \sigma)}{M((n-1)/2, n-\frac{1}{2}, \sigma)}, \tag{8.50}$$

where

$$M(a, b, z) = 1 + \frac{a}{b}z + \frac{a(a+1)}{b(b+1)}\frac{z^2}{2!} + \frac{a(a+1)(a+2)}{b(b+1)(b+2)}\frac{z^3}{3!} \cdots \tag{8.51}$$

is Kummer's function [82].

Equation (8.50) now permits us to give an explicit form for $T_e$ just as was accomplished in Coffey et al. [5, 54, 56] for the correlation time. We first note that successive $f_n(s)$, $n > 1$, are determined, from

$$\tilde{f}_n(s) = \widetilde{S}_n(s)\tilde{f}_{n-2}(s) + q_n(s)$$

where

$$q_n = \frac{\frac{2\tau f_n(0)}{n(n+1)} - \frac{2\sigma(n+2)}{(2n-1)(2n+3)} q_{n+2}}{\frac{2\tau s}{n(n+1)} + 1 - \frac{2\sigma}{(2n-1)(2n+3)} + \frac{2\sigma(n+2)}{(2n+1)(2n+3)} \widetilde{S}_{n+2}(s)} \tag{8.52}$$

In particular, for $n = 1$ we have

$$\tilde{f}_1(s) = \frac{1}{G(\sigma, s)} \left[ \tau f_1(0) - \frac{2}{5} \sigma q_3 \right]$$

where

$$G(\sigma, s) = s\tau + 1 - \frac{2}{5}\sigma + \frac{2}{5}\sigma \widetilde{S}_3(s) \tag{8.53}$$

We desire the zero-frequency limit if the various $\tilde{f}_n(s)$. In the case $n = 1$, we have with $s = 0$, utilizing Eq. (8.50) and proceeding as in Coffey et al. [5]

$$\tilde{f}_1(0) = \tau \sum_{n=0}^{\infty} f_{2n+1}(0) \frac{(-\sigma)^n \left(n + \frac{3}{4}\right) \Gamma\left(n + \frac{1}{2}\right) M\left(n + 1, 2n + \frac{5}{2}, \sigma\right)}{(n+1)\Gamma\left(2n + \frac{5}{2}\right)} \tag{8.54}$$

which is an exact expression for $\tilde{f}_1(0)$ for *arbitrary* initial conditions. The zero frequency Fourier component $\tilde{f}_1(0)$, which corresponds to Green's function with $\theta_0 = 0_+$, is obtained by setting $f_{2n+1}(0) = 1$ as required by Eq. (8.41).

In like manner, we have with the aid of our recurrence relation [Eq. (8.37) with $\xi = 0$]

$$\tilde{f}_3(0) = \frac{5}{2\sigma} f_1(0) - \frac{5}{2\sigma} \left(1 - \frac{2\sigma}{5}\right) \tilde{f}_1(0). \tag{8.55}$$

The successive $\tilde{f}_{2n+1}(0)$ required in Eq. (8.47) may be explicitly evaluated from Eq. (8.52) so that we ultimately have an exact closed-form expression in terms of Kummer's functions for $T_s$. The resulting expression may be checked by considering the limiting values of Eq. (8.47) for very high and very low barriers and comparing them with the known results of these two limiting cases.

We first consider the behavior of $\tilde{f}_1(0)$ for very high barriers. In order to accomplish this, we note that in the limit of large $z$ [82]

$$M(a, b, z) \simeq \frac{\Gamma(b)}{\Gamma(a)} e^z z^{a-b} [1 + O(|z|^{-1})] \qquad \mathrm{Re}(z) > 0 \tag{8.56}$$

Thus, in the high barrier limit Eq. (8.54) becomes

$$\tilde{f}_1(0) \cong \tau\sigma^{-3/2}e^{\sigma}\sum_{n=0}^{\infty}(-1)^n \frac{f_{2n+1}(0)\left(n+\frac{3}{4}\right)\Gamma\left(n+\frac{1}{2}\right)}{(n+1)!} \qquad \sigma \gg 1 \qquad (8.57)$$

[Note a question of uniform convergence [56] would in general arise here as $n$ and $\sigma$ simultaneously become large. This problem has been satisfactorily addressed by Coffey et al. [5, 54], who justify the steps involved in replacing the confluent hypergeometric function $M(n+1, 2n+5/2, \sigma)$ by its asymptotic value and summing the resulting series by representing the $M$ function in integral form so removing the $n$ dependence]. The asymptotic equation (8.57), with the initial condition $f_{2n+1}(0) = 1$ appropriate to Green's function, then becomes

$$\tilde{f}_1(0) \cong \tau\frac{\sqrt{\pi}}{2}\sigma^{-3/2}e^{\sigma} \qquad \sigma \gg 1 \qquad (8.58)$$

because [5, 54]

$$\sum_{n=0}^{\infty}\frac{(-1)^n}{(n+1)!}\left(n+\frac{3}{4}\right)\Gamma\left(n+\frac{1}{2}\right) = \frac{\sqrt{\pi}}{2} \qquad (8.59)$$

Furthermore, Eq. (8.55) implies that the behavior of $\tilde{f}_3(0)$ and all higher order $\tilde{f}_{2n+1}(0)$ in the higher barrier limit is given by Eq. (8.58) so that, in the high barrier limit, $T_s$ asymptotically becomes

$$T_s \cong \tau\frac{\sqrt{\pi}}{2}\sigma^{-3/2}e^{\sigma}\sum_{n=0}^{\infty}\frac{(-1)^n((4n+3)/2)(2n)!}{2^{2n+1}(n+1)!n!} \qquad (8.60)$$

The series in Eq. (8.60) may be summed [5, 54] by noting that

$$\frac{(2n)!}{2^{2n}(n+1)!n!} = \frac{1}{\sqrt{\pi}}\frac{\Gamma\left(n+\frac{1}{2}\right)}{(n+1)!} \qquad (8.61)$$

so that

$$\sum_{n=0}^{\infty}\frac{(-1)^n(2n)!\left(n+\frac{3}{4}\right)}{2^{2n+1}(n+1)!n!} = \sum_{n=0}^{\infty}\frac{(-1)^n\Gamma\left(n+\frac{1}{2}\right)\left(n+\frac{3}{4}\right)}{2\sqrt{\pi}(n+1)!}$$

which is, by Eq. (8.59) above,

$$\frac{1}{2\sqrt{\pi}}\frac{\sqrt{\pi}}{2} = \frac{1}{4} \tag{8.62}$$

Hence,

$$T_s \cong \frac{\tau\sqrt{\pi}}{4}\sigma^{-3/2}e^{\sigma} \qquad \sigma \gg 1 \tag{8.63}$$

Since the rotator has a 50:50 chance of either crossing the equator or falling all the way back down into the well at $\theta = 0_+$ when it reaches $\theta_m = \pi/2$, then the escape time or greatest relaxation time, $T_e$, is (here $\Gamma$ denotes the Kramers escape rate)

$$T_e \cong \tau\frac{\sqrt{\pi}}{2}\sigma^{-3/2}e^{\sigma} = \frac{1}{\Gamma} \tag{8.64}$$

Equation (8.64) which is the time to leave the domain 0, $\pi/2$ and, which results from the solution of the time dependent FPE given an initial delta function distribution of orientations in the first well, is in complete agreement with the result predicted by the Kramers theory of escape of particles over potential barriers (Section III Eq. 3.58b). Thus Eqs. (8.54–8.60) constitute an analytical method of verifying the Kramers theory, which emerges from the exact solution for $T_e$ as the asymptotic value of a series, rather than the customary numerical method that determines the smallest nonvanishing eigenvalue of the Fokker–Planck equation.

As well as reducing to the Kramers asymptote in the high barrier limit, namely, Eq. (8.64), Eq. (8.47) must also yield the result of Klein [55] (Eq. 82 of Coffey [68]), namely, for the time to reach the equator

$$T_{\text{MFPT}} = T_m = 2\tau \ln 2 \tag{8.65}$$

as the barrier height $\sigma$ approaches zero. In order to regain this result from our approach to the problem we note that Eq. (8.47) for the time to reach the summit becomes in the zero barrier limit [by setting $\sigma = 0$ in Eq. (8.37)]

$$T_s = 2\tau\sum_{n=0}^{\infty}\frac{(-1)^n(2n)!\left(n+\frac{3}{4}\right)}{2^{2n+1}(2n+1)(n+1)n!(n+1)!} \tag{8.66}$$

This series is best summed (the summation is reasonably involved so we give the full details) by writing it as a hypergeometric function of the form ${}_pF_q$.

Thus, we first note that [82]

$$(2n)! = \Gamma(1+2n) = (2\pi)^{-1/2} 2^{2n+1/2} \Gamma\left(n+\frac{1}{2}\right)\Gamma(n+1)$$
$$= (2\pi)^{1/2} 2^{n+1/2} \sqrt{\pi}\left(\frac{1}{2}\right)_n n!$$

so that Eq. (8.66) becomes

$$T_s = 2\tau \sum_{n=0}^{\infty} \frac{(4n+3)}{4} \frac{(-1)^n \left(\frac{1}{2}\right)^n}{(n+1)!(2n+1)(2n+2)}$$

and using partial fractions we then have

$$\begin{aligned} T_s &= \frac{2\tau}{4}\left[\sum_{n=0}^{\infty} \frac{\left(\frac{1}{2}\right)_n (-1)^n}{(n+1)!(2n+1)} + \sum_{n=0}^{\infty} \frac{\left(\frac{1}{2}\right)_n (-1)^n}{(n+1)!(2n+2)}\right] \\ &= \frac{2\tau}{8} \sum_{n=0}^{\infty} \frac{\left(\frac{1}{2}\right)_n (-1)^n}{(n+1)!}\left(\frac{1}{n+\frac{1}{2}} + \frac{1}{n+1}\right) \\ &= \frac{\tau}{4}\left[\sum_{n=0}^{\infty} \frac{\left(\frac{1}{2}\right)_n (-1)^n (1)_n \left(\frac{1}{2}\right)_n}{(2)_n n! \frac{1}{2}\left(\frac{3}{2}\right)_n} + \sum_{n=0}^{\infty} \frac{\left(\frac{1}{2}\right)_n (-1)^n (1)_n (1)_n}{(2)_n n! (2)_n}\right] \end{aligned} \tag{8.67}$$

$$T_s = \frac{\tau}{4}\left[2\,{}_3F_2\left(1, \frac{1}{2}, \frac{1}{2}; 2, \frac{3}{2}; -1\right) + {}_3F_2\left(1, \frac{1}{2}, 1; 2, 2; -1\right)\right] \tag{8.68}$$

However, from Prudnikov et al. [92] (p. 573, Vol. 3, Eq. 258),

$${}_3F_2\left(1, \frac{1}{2}, 1; 2, 2; 1\right) = -4\left[1 - \sqrt{2} + \ln\left(\frac{1+\sqrt{2}}{2}\right)\right] \tag{8.69}$$

Furthermore, again from Prudnikov et al. [92] (Vol. 3, p. 571, Eq. 209),

$${}_3F_2\left(1, \frac{1}{2}, \frac{1}{2}; \frac{3}{2}, 2; 1\right) = \lim_{z \to -1}\left\{\frac{2}{z}\left[\sqrt{1-z} + \sqrt{z}\sin^{-1}\sqrt{z} - 1\right]\right\} \tag{8.70}$$

$$= \ln(\sqrt{2} - 1)$$

Thus

$$T_s = \tau \ln 2 \tag{8.71}$$

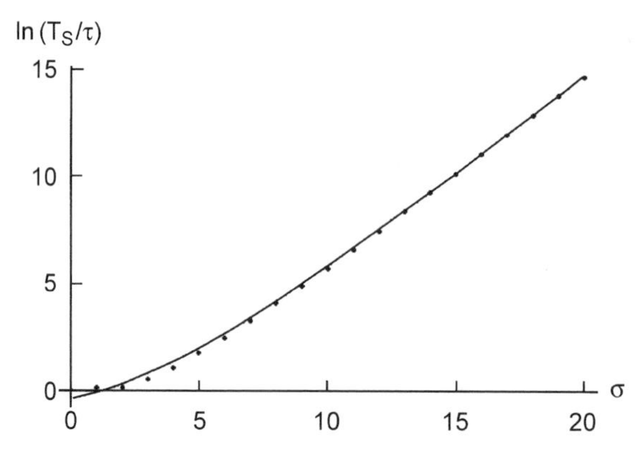


**Figure 12.** $\ln(T_s/\tau)$ as a function of $\sigma$. Solid curve corresponds to Eq. (8.47) and dots corresponds to Eq. (8.63).

Hence,

$$T_e = 2T_s = 2\tau \ln 2 \tag{8.72}$$

Equation (8.71) is in full agreement with the result of Klein [55] derived [68], Eq. (82), from the adjoint FPE [Eq. (5.16)].

### G. Discussion of Results and Conclusion

The calculation we have just illustrated is a particularly simple case of the theory as only one barrier height is involved since we have a symmetric bistable potential. Furthermore, $\tilde{f}_{2n+1}(0)$ is available in closed form for any barrier height. If the external field parameter $\xi \neq 0$, as in Eq. (8.2) so that we have an *asymmetric* bistable potential, the solution is more complicated because (a) $\tilde{f}_n(0)$ must be determined from a matrix continued fraction so that it is difficult to find the exact closed form solution for $\tilde{f}_n(0)$; (b) in the calculation of $T_e$, two barrier heights are involved, namely, $[U(\theta_m) - U(0)]$ and $[U(\theta_m) - U(\pi)]$ so that Eqs. (8.44) and (8.45) now comprise two integrals so that

$$T_e = \int_0^{\theta_m} \tilde{G}_0 \sin\theta d\theta + \int_{\pi}^{\theta_m} \tilde{G}_0 \sin\theta d\theta \tag{8.73}$$

Apart from (a) and (b), no new principles are involved and our method may be easily extended to nonaxially symmetric potentials $U(\theta, \varphi)$ and may be checked by reference to known asymptotes for such potentials, e.g. Eq. (4.87).

An advantage of the present formulation of the calculation of the greatest relaxation time as a sum over the zero-frequency decay modes is that it may be easily related to another time that is used to describe the decay of a distribution of magnetization or electric polarization. This is the integral relaxation time [93, 107], which is (Section III.E) the area under the curve of the decay function of the electric polarization or the magnetization following a sudden change in the amplitude of an applied external stimulus such as a uniform field, and so on. In the case of linear response to the applied stimulus, the IRT is the correlation time of the slowest decay mode.

We remark in passing that once the Laplace transform of the time-dependent Green function is known as a continued fraction or in general as a matrix continued fraction the response to an arbitrary nonequilibrium initial distribution may always be calculated by integrating the Green function over that initial distribution.

As far as the IRT and the escape time are concerned, it is important to note that $T_e$ is governed by a *sum* of the decay modes $\tilde{f}_n(0)$ for delta function initial conditions that are all connected to each other in nonlinear fashion by the underlying recurrence relation. The IRT on the other hand is determined by the equation (taking for simplicity the IRT in linear response that is the correlation time)

$$T_c = \tau \frac{\tilde{f}_1(0)}{f_1(0)} \tag{8.74}$$

Here, $f_1(t)$ is the decay function of the lowest decay mode following a small change in the amplitude of an applied stimulus, with equilibrium conditions having been established prior to the change. Thus, $T_c$ unlike $T_e$ is ostensibly governed by the zero-frequency Laplace transform of the *single* mode $\tilde{f}_1(0)$ only, however $\tilde{f}_1(0)$ is again connected to all other modes by the underlying recurrence relation. We remark that the correlation time is virtually identical to the escape time, $T_e$, if the configuration of the system is such that the contributions of all the other modes, save the longest lived one, are negligible. In other cases, however, the IRT may differ exponentially from the escape time, a fact that has been extensively discussed by Coffey [68]. The present section emphasizes how both $T_c$ and $T_e$ may be calculated in terms of the decay functions of the spherical harmonics using either scalar or matrix continued fractions from a knowledge of the transition probability or Green's function of the FPE.

Another advantage of formulating the calculation of the greatest relaxation time by studying the approach to equilibrium of the system is that the connection with other models of approach to equilibrium based on random walks such as the Ehrenfest urn model [6, 94] now becomes apparent. The connection to such models of the relaxation process is also likely to be important in computer simulations of the approach to equilibrium.

We remark that the continued fraction approach used here to determine $T_s$ should be contrasted with that of Kalmykov [95] who has used a similar method in order to determine the smallest nonvanishing eigenvalue of the FPE and with that of Garanin [96] who has formulated the problem of calculation of the relaxation time in terms of the relaxation time of the well populations. A clear discussion of the multidimensional relaxation problem is given by Braun [101]. Secondly, we remark that an approach to the calculation of the escape time which is essentially equivalent to ours has been given by Malakhov and Pankratov [102–104] in their exact solution of the Kramers one-dimensional translational escape rate problem for piecewise parabolic potentials, which they obtained by posing the solution as a Sturm-Liouville problem. This procedure enabled them to find an expression for the escape rate in terms of the parabolic cylinder functions [82]. Their definition of the relaxation time may be summarized as follows [105]. They start from the configuration space density $W(x, t)$ at time $t$, resulting from an initial distribution of particles situated near the point $x = x_0$, where $x$ is the particle coordinate, they then select some boundary point $x = L$ and assume the evolution probability to be of the form

$$P(t) = \int_{-\infty}^{L} W(x, t)dx, \tag{8.75}$$

whence it is evident that W corresponds to our $g$. They then define [104, 105] the characteristic time of the probability evolution to be

$$T = \frac{\int_0^\infty [P(t) - P(\infty)]dt}{P(0) - P(\infty)} \tag{8.76}$$

or in terms of the Laplace transform as

$$T = \lim_{s \to 0} \left\{ \frac{s\hat{P}(s) - P(\infty)}{s[P(0) - P(\infty)]} \right\} \tag{8.77}$$

where

$$\hat{P}(s) = \int_0^\infty P(t)e^{-st}dt \tag{8.78}$$

and $P(\infty)$ denotes the final equilibrium value of $P(t)$. It is apparent from these equations that by choosing $x = L$ as the summit of the potential that their definition of the time to reach the summit coincides with our Eq. (8.9) as specialized to one state variable. The particular problem we have treated is also of interest from another point of view, namely the so called stochastic resonance phenomenon [97] (which is intimately connected with the Kramers escape rate) as it is an example of a magnetic system which exhibits such behavior and has been treated in this context by Raikher and Stepanov [98]. See also the review by Coffey and Kalmykov of the dielectric properties of nematic liquid crystals [36]. Yet another closely related problem of profound interest in this context is stochastic resonance in non axially symmetric potentials of the magnetocrystalline anisotropy as then, see for example Eq. (4.84) of Section IV, both longitudinal and transverse modes of the relaxation process will contribute to the phenomenon producing pronounced nonlinear effects in the dynamical response.

In conclusion, we remark that the work of this section clearly demonstrates how the Kramers results for bistable potentials emerge as a limit of the dynamic response.

## IX. GENERAL CONCLUSIONS

In this review, we have attempted to give a didactic account i.e., we give the details of all the mathematical manipulations which are required of the Kramers theory of escape of particles over potential barriers and its modern extensions. Amongst the most important of these are:

(i) The extension of the theory to many state variables in the IHD limit by Langer [61] which constitutes a statistical theory of the decay of metastable states and which may be applied to a very wide variety of problems where a phase transition is involved.

(ii) The extension by Mel'nikov and Meshkov [34, 49] of the Kramers treatment to all values of the dissipation parameter, which appears (for the cases in which it has been numerically compared with the rate calculated by numerical solution of the FPE) to provide a reasonable description of the reaction rate. [Déjardin et al. [84], and Risken et al. [100].

(iii) The application of the theory to rotational Brownian motion initiated by Einstein in 1905 (see Chapter 1 of Coffey et al. *The Langevin Equation* [5] and Fürth [10]) and Debye [19] for mechanical rotators with a separable Hamiltonian and generalized to include inertial effects by Gross and Sack [5, 41, 53].

(iv) Finally the extension of the theory to single-domain ferromagnetic particles by W.F. Brown [21] where the Hamiltonian is, of course, no longer separable, where the inertia plays no role but where the role of inertia is mimicked in some part by the gyromagnetic term. Stochastic Resonance phenomena in these systems and associated nonlinear effects particularly in non axially symmetric potentials are worthy of future investigation.

The review has focused to some extent on problems of rotational relaxation and in that context reflects the interests of the authors. A detailed account of recent applications of reaction rate theory in chemical physics may be found in the *Proceedings of the 20th. Solvay Conference on Chemistry* Vol. 101 of this series [97]. The reader is also referred to Hänggi et al. [1].

## Acknowledgments

The authors wish to acknowledge the assistance of Forbairt (Basic Research Grant SC/97/701), and the Dublin Institute of Technology. They also wish to thank D. S. F. Crothers, P. M. Déjardin, H. Kachkachi, Yu. P. Kalmykov and S. V. Titov for helpful conversations, and D. Gallagher for help in preparing this manuscript.

## References

1. P. Hänggi, P. Talkner, and M. Borkovec, *Rev. Mod. Phys.* **62**, 251 (1990).
2. G. D. Billing and K. V. Mikkelsen, *Molecular Dynamics and Chemical Kinetics, Vol. 1*. Wiley-Interscience, New York, 1996.
3. R. A. Marcus, *Adv. Chem. Phys.* **101**, 391 (1997).
4. H. A. Kramers, *Physica* (Utrecht) **7**, 284 (1940).
5. W. T. Coffey, Yu. P. Kalmykov, and J. T. Waldron, *The Langevin Equation*, World Scientific, Singapore, 1996, Reprinted 1998.
6. W. T. Coffey, *Adv. Chem. Phys.* **63**, 65 (1985).
7. J. Frenkel, *The Kinetic Theory of Liquids*, Oxford University Press, London, 1946. Reprinted Dover, New York, 1955.
8. S. Chapman and T. G. Cowling, *The Mathematical Theory of Non-Uniform Gases*, Cambridge University Press, London, 1939, Third ed., 1970.
9. J. H. Jeans, *The Dynamical Theory of Gases*, Cambridge University Press, London, 4th. ed. 1923. Reprinted Dover, New York, 1954.
10. A. Einstein, in *Investigations on the Theory of the Brownian Movement*, R. H. Fürth, Ed., Methuen, London, 1926, Reprinted Dover, New York, 1954.
11. G. Joos, *Theoretical Physics*, Blackie, London, 1934.
12. G. H. Weiss, *Aspects and Applications of the Random Walk*, Elsevier, Amsterdam, 1994.
13. P. Langevin, *C. R. Acad. Sci.* **146**, 530 (1908).
14. N. Wax, *Selected Papers on Noise and Stochastic Processes*, Dover, New York, 1954.
15. S. Chandrasekhar, *Rev. Mod. Phys.* **15**, 1 (1943). Reprinted in [14].
16. L. S. Ornstein, *Versl, Acad. Amst.* **26**, 1005 (1917) (=1919 *Proc. Acad. Amst.* **21**, 96).
17. R. Becker and W. Döring, *Ann. Phys.* (Leipzig) **24**, 719 (1935).
18. P. Debye, *The Collected Papers of Peter J. W. Debye*. Interscience, New York, 1954.
19. P. Debye, *Polar Molecules*. Chem. Catalog 1929, New York, Reprinted: Dover, New York, 1954.
20. O. Klein, *Ark. Math., Astron. Fysik.* **16**, 5 (1921).
21. W. F. Brown, Jr., *Phys.Rev.* **130**, 1677 (1963).
22. N. Wiener and R. E. A. C. Paley, *Fourier Transforms in the Complex Domain*, Amer. Math. Soc. Coll. Pub. XIX.

23. J. L. Doob, *Ann. Math.* **43**, 351 (1942). Reprinted in [14].
24. C. W. Gardiner, *Handbook of Stochastic Methods*, Springer, Berlin, 1985.
25. H. Risken, *The Fokker–Planck Equation*, Springer, Berlin, 1984.
26. H. Risken, *The Fokker–Planck Equation*, Springer, 2nd ed., 1989.
27. N. G. van Kampen, *Stochastic Processes in Physics and Chemistry*, North-Holland, Amsterdam, 1981.
28. N. G. van Kampen, *Stochastic Processes in Physics and Chemistry*, 2nd ed., North-Holland, Amsterdam, 1992.
29. M. C. Wang and G. E. Uhlenbeck, *Rev. Mod. Phys.* **17**, 323 (1945). Also reprinted in [14].
30. W. T. Coffey, *J. Chem. Phys.* **93**, 724 (1990).
31. W. T. Coffey, *J. Chem. Phys.* **99**, 3014 (1993).
32. H. Goldstein, *Classical Mechanics*, 2nd ed., Addison Wesley, Reading, Ma., 1980.
33. R. F. Grote and G. T. Hynes, *J. Chem. Phys.* **73**, 2715 (1980).
34. V. I. Mel'nikov and S. V. Meshkov, *J. Chem. Phys.* **85**, 1018 (1986).
35. E. C. Titchmarsh, *Introduction to the Theory of Fourier Integrals*, Oxford University Press, London, 1937.
36. W. T. Coffey and Yu. P. Kalmykov, *Adv. Chem. Phys.* **113**, 487 (2000).
37. L. Néel, *Ann Géophys.* **5**, 99 (1949).
38. E. Fermi, *Notes on Quantum Mechanics* Phoenix Press, Chicago, 1965.
39. W. T. Coffey, P. J. Cregg, and Yu. P. Kalmykov, *Adv. Chem. Phys.* **83**, 263 (1993).
40. R. C. Tolman, *The Principles of Statistical Mechanics*, Oxford University Press, London, 1938.
41. R. A. Sack, *Proc. Phys. Soc. London* **B70**, 402, 414 (1957).
42. G. E. Uhlenbeck and L. S. Ornstein, *Phys. Rev.* **36**, 823 (1930). Also reprinted in Wax [14].
43. L. Farkas, *Z. Phys. Chem.* (Leipzig) **125**, 236 (1927).
44. B. J. Matkowsky, Z. Schuss, and C. Tier, *J. Stat. Phys.* **35**, 443 (1984).
45. D. J. McCarthy and W. T. Coffey, *J. Phys. Cond. Matter* **11**, 10586 (1999).
46. H. C. Brinkman, *Physica* (Utrecht) **22**, 29 (1956).
47. U. M. Titulaer, *Physica A* (Utrecht) **91**, 321 (1978).
48. G. Wilemski, *J. Stat. Phys.* **14**, 153 (1976).
49. V. I. Mel'nikov, *Physica A* (Utrecht) **130**, 606 (1985).
50. A. Abragam, *The Principles of Nuclear Magnetism*, Oxford University Press, London, 1961.
51. Yu. P. Kalmykov, S. V. Titov, and W. T. Coffey, *Phys. Rev. B* **58**, 3267 (1998).
52. W. T. Coffey, D. S. F. Crothers, J. L. Dormann, L. J. Geoghegan, and E. C. Kennedy, *Phys. Rev. B* **58**, 3249 (1998).
53. J. R. McConnell, *Rotational Brownian Motion and Dielectric Theory*, Academic, New York, 1980.
54. W. T. Coffey, D. S. F. Crothers, Yu. P. Kalmykov, E. S. Massawe, and J. T. Waldron, *Phys. Rev. E*, **49**, 1869 (1994).
55. G. Klein, *Proc. R. Soc. London, Ser. A* **211**, 431 (1952).
56. W. T. Coffey, D. S. F. Crothers, and J. T. Waldron, *Physica A* (Utrecht) **203**, 600 (1994).
57. W. T. Coffey, Yu. P. Kalmykov, and J. T. Waldron, *Physica A* (Utrecht) **208**, 462 (1994).
58. W. T. Coffey, *J. Chem. Phys.* **111**, 8350 (1999).

59. W. T. Coffey, D. S. F. Crothers, J. L. Dormann, L. J. Geoghegan, E. C. Kennedy, and W. Wernsdorfer, *J. Phys. Cond. Matter* **10**, 9093 (1998).
60. W. T. Coffey, J. L. Déjardin, Yu. P. Kalmykov, and S. V. Titov, *Phys. Rev. E* **54**, 6462 (1996).
61. J. S. Langer, *Ann. Phys.* **54**, 258 (N.Y.) (1969).
62. W. F. Brown, Jr., *IEEE Trans. Magn.* **MAG-15**, 196 (1979).
63. D. A. Smith and F. A. de Rozario, *J. Magn. Magn. Mater.* **3**, 219 (1976).
64. I. Klik and L. Gunther, *J. Stat. Phys.* **60**, 473 (1990).
65. L. J. Geoghegan, W. T. Coffey, and B. Mulligan, *Adv. Chem. Phys.* **100**, 475 (1997).
66. W. T. Coffey, *J. Mol. Struct.* **479**, 261 (1999).
67. Yu. P. Kalmykov and S. V. Titov, *Phys. Rev. Lett.* **82**, 2967 (1999).
68. W. T. Coffey, *Adv. Chem. Phys.* **103**, 259 (1998).
69. H. C. Brinkman, *Physica* (Utrecht) **22**, 149 (1956).
70. D. A. Garanin, E. C. Kennedy, D. S. F. Crothers, and W. T. Coffey, *Phys. Rev. E.* **60**, 6499 (1999).
71. R. Landauer and J. A. Swanson, *Phys. Rev.* **121**, 1668 (1961).
72. I. Klik and L. Gunther, *J. Appl. Phys.* **67**, 4505 (1990).
73. H. B. Braun, *J. Appl. Phys.* **76** 6310 (1994).
74. H. W. Turnbull and A. C. Aitken, *An Introduction to the Theory of Canonical Matrices*, Blackie, London, 1932.
75. E. Praestgaard and N. G. van Kampen, *Mol. Phys.* **43**, 33 (1981).
76. I. Klik, Personal communication (1997).
77. W. Wernsdorfer, E. Bonet Orozco, K. Hasselbach, A. Benoit, B. Barbara, N. Demoncy, A. Loiseau, D. Boivin, H. Pascard, and D. Mailly, *Phys. Rev. Lett.* **78**, 1791 (1997).
78. W. Wernsdorfer, E. Bonet Orozco, K. Hasselbach, A. Benoit, D. Mailly, O. Kubo, and B. Barbara, *Phys. Rev. Lett.* **79**, 4014 (1997).
79. G. B. Arfken and H. L. Weber, *Mathematical Methods for Physicists*, 4th ed., Academic, New York, 1995.
80. M. R. Spiegel, *Advanced Calculus, Schaum's Outline Series*, McGraw-Hill, New York, 1974.
81. P. M. Morse and H. Feshbach, *Methods of Theoretical Physics*, Vol 1, McGraw-Hill, New York, 1953.
82. M. Abramowitz and I. Stegun, *Handbook of Mathematical Function*, Dover, New York, 1972.
83. S. Urban, D. Busing, A. Würflinger, and B. Gestblom, *Liquid Crystals* **25**, 253 (1998).
84. P. M. Déjardin, D. S. F. Crothers, W. T. Coffey, and D. J. McCarthy, *Phys. Rev. E*, in the press (2001).
85. M. Respaud, M. Goiran, J. M. Broto, F. Lionti, L. Thomas, B. Barbara, T. Ould Ely, C. Amiens, and B. Chaudret, *Europhys. Lett.* **47**, 122 (1998).
86. H. Kachachi, W. T. Coffey, D. S. F. Crothers, A. Ezzir, E. C. Kennedy, M. Noguès, and E. Tronc, *J. Phys. Cond. Matter* **12**, 3077 (2000).
87. W. S. Burnside and A. W. Panton, *The Theory of Equations*, 4th ed., Longmans-Green, London, 1899.
88. E. T. Whittaker and G. N. Watson, *Modern Analysis*, 4th ed., Cambridge University Press, London, 1927.
89. E. L. Ince, *Ordinary Differential Equations*, Longmans-Green, London, 1927.
90. J. A. Aseltine, *Transform Methods in Linear System Analysis*, McGraw-Hill, New York, 1958.

91. T. M. MacRobert, *Spherical Harmonics*, 3rd ed., revised by I. N. Sneddon, Pergamon, London, 1967.
92. A. D. Prudnikov, Yu. A. Brychkov, and O. I. Marichev, *Integrals and Series. Special Functions*, Vol. 3, Gordon and Breach, New York, 1990.
93. D. A. Garanin, *Phys. Rev. E* **54**, 3250 (1996).
94. M. Kac, *Am. Math. Month* **43**, 369 (1947), reprinted in [14].
95. Yu. P. Kalmykov, *Phys. Rev. E* **61**, 6320 (2000).
96. D. A. Garanin, *Europhys. Lett.* **48**, 486 (1999).
97. L. Gammaitoni, P. Hänggi, P. Jung, and F. Marchesoni, *Rev. Mod. Phys.* **70**, 223 (1998).
98. Yu. L. Raikher and V. Stepanov, *J. Phys. Cond. Matter* **6**, 4137 (1994).
99. P. Gaspard, I. Burghardt, eds., *Adv. Chem. Phys.* 101 (1997).
100. H. Risken, K. Vogel, and H. D. Vollmer, *IBM J. Res. Dev.* **32**, 112 (1988).
101. H. B. Braun, *Phys. Rev. B* **50**, 16501 (1994).
102. A. L. Pankratov, *Phys. Lett. A* **255**, 17 (1999)
103. A. N. Malakhov and A. L. Pankratov, *Physica C* (Utrecht) **269**, 46 (1996).
104. A. N. Malakhov and A. L. Pankratov, *Physica A* (Utrecht) **229**, 109 (1996).
105. N. V. Agondov and A. N. Malakhov, *Radiophys. Quantum Electron*, **36**, 97 (1993).
106. C. Blomberg, *Physica A* (Utrecht) **86**, 49 (1977).
107. D. A. Garanin, V. V. Ischenko, and L. V. Panina, *Teor Mat. Fiz.* **82**, 242 (1990) [*Theor. Math. Phys.* **82**, 169 (1990)].

# AUTHOR INDEX

Numbers in parentheses are reference numbers and indicate that the author's work is referred to although his name is not mentioned in the text. Numbers in *italic* show the pages on which the complete references are listed.

# SUBJECT INDEX